T0259041

ELECTRONICS

BASIC, ANALOG, AND DIGITAL

with PSpice®

ELECTRONICS
BASIC, ANALOG, AND DIGITAL
with PSpice®

NASSIR H. SABAH

American University of Beirut
Lebanon

 CRC Press
Taylor & Francis Group
Boca Raton   London   New York

CRC Press is an imprint of the
Taylor & Francis Group, an **informa** business

CRC Press
Taylor & Francis Group
6000 Broken Sound Parkway NW, Suite 300
Boca Raton, FL 33487-2742

© 2010 by Taylor and Francis Group, LLC
CRC Press is an imprint of Taylor & Francis Group, an Informa business

No claim to original U.S. Government works

Printed in the United States of America on acid-free paper
10 9 8 7 6 5 4 3 2 1

International Standard Book Number: 978-1-4200-8707-9 (Hardback)

Library of Congress Cataloging-in-Publication Data

Sabah, Nassir H.
   Electronics : basic, analog, and digital with PSpice / author, Nassir H. Sabah.
      p. cm.
   "A CRC title."
   Includes bibliographical references and index.
   ISBN 978-1-4200-8707-9 (hardcover : alk. paper)
   1. Electronics. I. Title.

TK7816.S23 2010
621.381--dc22
                                                                 2009036999

Visit the Taylor & Francis Web site at
http://www.taylorandfrancis.com

and the CRC Press Web site at
http://www.crcpress.com

To Gharid

In loving appreciation of her continued support and forbearance,

and in fond remembrance of

E.E. Ward, my first teacher of electronics.

Everything should be made as simple as possible, but not simpler.

—**Albert Einstein**

Brief Contents

Preface... xxi
Introduction xxv
Convention for Symbols... xxxiii

Chapter 1   Basic Diode Circuits 1

Chapter 2   Basic Principles of Semiconductors... 55

Chapter 3   *pn* Junction and Semiconductor Diodes 105

Chapter 4   Semiconductor Fabrication.. 149

Chapter 5   Field-Effect Transistors... 179

Chapter 6   Bipolar Junction Transistor.. 241

Chapter 7   Two-Port Circuits, Amplifiers, and Feedback 295

Chapter 8   Single-Stage Transistor Amplifiers.. 335

Chapter 9   Multistage and Feedback Amplifiers... 405

Chapter 10   Differential and Operational Amplifiers 467

Chapter 11   Power Amplifiers and Switches 531

Chapter 12   Basic Elements of Digital Circuits 587

Chapter 13   Digital Logic Circuit Families 647

Appendix A: Reference Material 703

Appendix B: Basic PSpice Models 705

References and Bibliography 709

Index.. 711

Contents

Preface xxi
Introduction xxv
Convention for Symbols xxxiii

1. Basic Diode Circuits 1
Learning Objectives 1
1.1　Ideal and Practical Diodes 2
　　　Ideal Diode 2
　　　Ideal Si *pn* Junction Diode 2
　　　Practical Diodes 3
　　　Incremental Diode Resistance 4
1.2　Basic Analysis of Diode Circuits 5
　　　Piecewise Linear Approximation 5
　　　Bias Point 6
　　　Small-Signal Model 6
1.3　Rectifier Circuits 10
　　　Half-Wave Rectifier 10
　　　Full-Wave Rectifier 11
　　　Smoothing of Output 14
　　　Capacitor-Input Filter 14
　　　Approximate Analysis of Capacitor-Input Filter 16
1.4　Zener Voltage Regulator 24
　　　Voltage–Current Characteristic 24
　　　Analysis of Zener Regulator 25
　　　Load Regulation and Line Regulation 28
1.5　Diode Voltage Limiters 32
　　　Surge Protection 36
1.6　Diode-Capacitor Circuits 36
　　　dc Restorer 36
　　　Voltage Multiplication 41
Summary of Main Concepts and Results 42
Learning Outcomes 43
Supplementary Examples and Topics on Website 43
Problems and Exercises 44

2. Basic Principles of Semiconductors 55
Learning Objectives 55
2.1　Excerpts from Wave Mechanics 56
　　　Some Basic Ideas 56
　　　Schrödinger's Equation 58
　　　Wave Packets 59
　　　Free Atoms 64
　　　Energy Band Structure 64
2.2　Electric Conduction in Semiconductors 66

Electrons in a Periodic Potential 66
Effective Mass 68
Hole Conduction 69
Density of Electron States 71
Fermi–Dirac Distribution 73
2.3 Intrinsic and Extrinsic Semiconductors 74
Crystal Structure 74
Intrinsic Semiconductor 75
Extrinsic Semiconductors 77
2.4 Electrochemical Potential 79
Expression for Chemical Potential 79
Expression for Electrochemical Potential 81
Generalized Ohm's Law 82
State of Equilibrium 83
Fermi Level and Electrochemical Potential 84
2.5 Carrier Concentrations in Semiconductors 85
Carrier Generation and Recombination 87
Intrinsic Semiconductor 88
Position of Fermi Level 89
n-Type Semiconductor 89
p-Type Semiconductor 90
2.6 Carrier Mobility 94
Variation with Dopant Concentration 94
Variation with Temperature 95
2.7 Carrier Recombination 95
Minority Carrier Lifetime 97
Summary of Main Concepts and Results 98
Learning Outcomes 99
Supplementary Examples and Topics on Website 100
Problems and Exercises 101

3. *pn* Junction and Semiconductor Diodes 105
Learning Objectives 105
3.1 The *pn* Junction at Equilibrium 106
Junction Potential 106
Depletion Approximation 107
3.2 The Biased *pn* Junction 112
The *pn* Junction as a Rectifier 112
Width of Depletion Region 113
Charge Distributions and Currents under Bias 113
Current–Voltage Relation 117
Charge–Current Relation 118
pn Junction Capacitances 118
Temperature Effects 120
Junction Breakdown 121
3.3 Semiconductor Photoelectric Devices 124
Photoconductive Cell 124
Photodiode 125

Photocell 126
3.4   Light-Emitting Diodes 128
3.5   Tunnel Diode 131
3.6   Contacts between Dissimilar Materials 134
Metal–Metal Contacts 135
Metal–Semiconductor Contacts 136
Schottky Diode 139
3.7   Heterojunction 139
Summary of Main Concepts and Results 143
Learning Outcomes 144
Supplementary Examples and Topics on Website 144
Problems and Exercises 145

4. Semiconductor Fabrication 149
Learning Objectives 149
4.1   Preparation of Silicon Wafer 150
High-Purity Silicon 150
Crystal Growth 150
Wafer Production 151
Oxidation 151
4.2   Patterning Processes 152
Preparation of Masks 152
Lithography 153
Etching 156
4.3   Deposition Processes 157
Ion Implantation 157
Diffusion 157
Chemical Vapor Deposition 158
Metallization 158
Clean Room Environment 159
4.4   Packaging 160
Package Types 160
4.5   Fabrication of Simple Devices 161
Resistors 161
pn Junction Diodes 162
Capacitors 163
4.6   CMOS Fabrication 163
Transistor Isolation 163
Transistor Formation 164
Silicon-on-Insulator Technology 168
4.7   Fabrication of Bipolar Junction Transistors 169
Basic BJT Structure 169
Modified BJT Structures 170
BiCMOS 171
4.8   Miscellaneous Topics 172
SiGe Technology 172
Crystal Defects 173
Si–SiO$_2$ Interface 174

Summary of Main Concepts and Results 175
Learning Outcomes 176
Supplementary Examples and Topics on Website 177
Problems and Exercises 177

5. Field-Effect Transistors 179
Learning Objectives 179
5.1   Amplifiers 180
       Hypothetical Amplifying Device 181
5.2   Basic Operation of the MOSFET 183
       Structure 183
       Operation of Enhancement-Type MOSFETs 184
       Current–Voltage Relations 187
       p-Channel MOS Transistor 190
       Small-Signal Operation 191
5.3   Secondary Effects in MOSFETs 196
       Channel-Length Modulation 196
       Transconductance 199
       Overdrive Voltage 201
       Temperature Effects 201
       Breakdown 201
       Body Effect 202
       Capacitances 205
       Unity-Gain Bandwidth 207
       Short-Channel Effects 209
             Carrier Velocity Saturation and Hot Carriers 210
             Reduced Output Resistance and Threshold 211
             Other Effects 211
       Improved Performance 212
5.4   Depletion-Type MOSFETs 214
       Diode Connection 217
5.5   Complementary MOSFETs 218
       CMOS Amplifier 219
       CMOS Transmission Gate 222
       CMOS Inverter 225
5.6   Junction Field-Effect Transistor 225
       Structure 225
       Operation 226
       Current–Voltage Relation 228
5.7   Metal–Semiconductor Field-Effect Transistor 232
       Structure 233
       Operation 233
       High-Mobility Devices 234
Summary of Main Concepts and Results 235
Learning Outcomes 235
Supplementary Examples and Topics on CD 236
Problems and Exercises 236

6. Bipolar Junction Transistor **241**
    Learning Objectives 241
    6.1   Basic Operation of the BJT 242
           Common-Base dc Current Gains 245
           Typical Structure 246
           Common-Emitter Configuration 247
           Small-Signal Current Gains 250
           Small-Signal Equivalent Circuits 254
    6.2   Secondary Effects in BJTs 256
           Base-Width Modulation 256
           Hybrid-π Equivalent Circuit 260
           Variation of i_C with v_{be} 265
           h-Parameter Equivalent Circuit 265
           Temperature Effects 265
           Breakdown 266
           Punchthrough 268
           BJT Capacitances 268
           Unity-Gain Bandwidth 271
    6.3   BJT Large-Signal Models 275
           Ebers–Moll Model 275
           Saturation Mode 278
           The BJT as a Switch 281
           Diode Connection 282
           Regenerative Pair 282
           Augmented Models 284
    6.4   Heterojunction Bipolar Transistor 284
    6.5   Noise in Semiconductors 286
    6.6   Comparison of BJTs and FETs 287
    Summary of Main Concepts 288
    Learning Outcomes 288
    Supplementary Examples and Topics on Website 289
    Problems and Exercises 289

7. Two-Port Circuits, Amplifiers, and Feedback **295**
    Learning Objectives 295
    7.1   Two-Port Circuits 296
           Interpretation of Parameters 296
           Equivalent Circuits 298
    7.2   Ideal Amplifiers 298
    7.3   Negative Feedback 300
           Series–Shunt Feedback 301
           Signal-Flow Diagrams 303
           Series–Series Feedback 305
           Shunt–Series Feedback 307
           Shunt–Shunt Feedback 308
    7.4   Ideal Operational Amplifier 310
           Noninverting Configuration 311

Unity-Gain amplifier 313
Inverting Configuration 314
Integrator 317
Differentiator 318
Extraneous Signals 321
Gain-Bandwidth Product 322
Summary of Main Concepts and Results 325
Learning Outcomes 326
Supplementary Topics and Examples on Website 326
Problems and Exercises 326

8. Single-Stage Transistor Amplifiers 335
Learning Objectives 335
8.1   Transistor Biasing 336
Biasing of Discrete Transistors 336
Current Mirror 341
BJT Current Mirror 341
MOSFET Current Mirror 344
8.2   Basic Amplifier Configurations 346
General Considerations 346
Common-Emitter Amplifier 348
Common-Source Amplifier 350
Common-Drain Amplifier 350
Common-Collector Amplifier 353
Common-Source Amplifier with Source Resistor 356
Common-Emitter Amplifier with Emitter Resistor 358
Common-Gate Amplifier 361
Common-Base Amplifier 363
8.3   High-Frequency Response 368
Miller's Theorem 368
Poles and Zeros of Transfer Function 369
Common Emitter/Source Amplifier 370
Common Collector/Drain Amplifier 373
Common-Emitter/Source Amplifier with Feedback Resistor 375
Common-Base/Gate Amplifier 376
8.4   Composite Transistor Connections 379
Darlington Pair 379
Common Collector-Common Emitter Cascade 381
Cascode Amplifier 384
MOSFET Cascode 384
BJT Cascode 386
BiCMOS Cascode 389
8.5   Cascode Current Sources and Mirrors 391
Summary of Main Concepts and Results 392
Learning Outcomes 393
Supplementary Examples and Topics on Website 393
Problems and Exercises 394

9. Multistage and Feedback Amplifiers **405**
  Learning Objectives 405
  9.1 Cascaded Amplifiers 406
       dc Level Shifting 409
  9.2 *RC*-Coupled Amplifiers 411
       Common-Source Amplifier 411
       Common-Emitter Amplifier 413
  9.3 Feedback Amplifiers 416
       Series–Shunt Feedback 417
       Series–Series Feedback 424
       Shunt–Series Feedback 428
       Shunt–Shunt Feedback 433
  9.4 Closed-Loop Stability 436
       Stability from Bode Magnitude Plots 438
       Frequency Compensation 441
       Feedback Oscillators 445
       Wien-Bridge Oscillator 445
       High-*Q* Oscillator 446
  9.5 Tuned Amplifiers 448
  9.6 *LC* Oscillators 451
  9.7 Crystal Oscillators 456
  Summary of Main Concepts and Results 457
  Learning Outcomes 458
  Supplementary Examples and Topics on Website 458
  Problems and Exercises 459

10. Differential and Operational Amplifiers **467**
  Learning Objectives 467
  10.1 Differential Pair 468
        Basic Operation 468
        Transfer Characteristic 470
        Small-Signal Differential Operation 471
        High-Frequency Response 472
        Small-Signal Common-Mode Response 472
        High-Frequency Response 474
        Input Bias Currents 476
        Input Offset Voltage 476
        Current-Mirror Load 478
        MOSFET Differential Pair 483
  10.2 Two-Stage CMOS Op Amp 486
        Input-Offset Voltage 486
        Voltage Gain and Output Swing 487
        Common-Mode Response 488
        Frequency Response 488
        Slew Rate 491
  10.3 Folded Cascode CMOS Op Amp 492
        Common-Mode Input Range 492

Output Voltage Swing..494
Voltage Gain 494
Frequency Response 495
Slew Rate 495
10.4 CMOS Current and Voltage Biasing 498
Self-Biasing 498
10.5 BJT Op Amps..501
Input Resistance and Bias Current 502
Input Offset Voltage 503
10.6 Some Basic Practical Op-Amp Circuits...507
Inverting and Noninverting Op-amp Circuits......................................507
Integrator 509
Difference Amplifier 511
Instrumentation Amplifier 514
10.7 Switched-Capacitor Circuits 516
10.8 Digital-Analog Conversion 519
Summary of Main Concepts and Results...522
Learning Outcomes..523
Supplementary Examples and Topics on Website....................................523
Problems and Exercises..524

11. **Power Amplifiers and Switches**..**531**
Learning Objectives 531
11.1 General Considerations 532
Safe Operating Limits 532
Thermal Resistance 533
Thermal Stability 535
Nonlinear Distortion..536
Power-Conversion Efficiency 537
11.2 Class A Operation 539
Transformer Coupling 542
Emitter Follower 543
11.3 Class B Operation...545
11.4 Class AB Operation...546
11.5 Class C Operation...549
11.6 Power Operational Amplifiers 550
11.7 Power Switching...552
Class D Amplifier...552
Switched Regulated Supplies 553
dc-to-ac Converters 556
11.8 Power Diodes...560
11.9 Power Transistors...563
Bipolar Junction Transistors 563
MOSFETs...567
Insulated Gate Bipolar Transistors (IGBTs)..569
11.10 Power Latches...572
Thyristor 572
Summary of Main Concepts and Results...577

Learning Outcomes.. 578
Supplementary Examples and Topics and on Website............................. 578
Problems and Exercises.. 579

12. Basic Elements of Digital Circuits.. 587
    Learning Objectives 587
    12.1   Digital Signals and Processing 588
            Digital Signals.. 588
            Boolean Algebra 590
    12.2   Logic Gates.. 592
            Gate Types... 592
            CMOS Gate Examples 595
            Gate Performance 596
    12.3   Flip-Flops... 602
            Basic Latch... 602
            SR Latch... 604
            JK Flip-Flop 606
            D Flip-Flop 608
    12.4   Digital System Memories 611
            Classification of Semiconductor Memories 612
            Organization of RAM 613
    12.5   Read/Write Memory.. 614
            Static Memory Cell.. 614
            Dynamic Memory Cell 616
            Sense Amplifier and Precharge Circuit............................. 617
            Row Decoder 620
            Column Decoder 622
    12.6   Read Only Memory 623
            Mask ROM 623
            Programmable ROM.. 625
            Erasable Programmable ROM.. 626
            Flash ROM.. 627
    12.7   Ferroelectric RAM 631
    12.8   Metallic Interconnect.. 632
            Capacitance 632
            Resistance 632
            Distributed Models 633
    Summary of Main Concepts and Results.. 636
    Learning Outcomes.. 637
    Supplementary Examples and Topics on Website................................... 637
    Problems and Exercises.. 637

13. Digital Logic Circuit Families 647
    Learning Objectives 647
    13.1   CMOS... 648
            CMOS Inverter... 648
            Static Behavior 648
            Noise Margins.. 651

Propagation Delay.. 653
Power Dissipation 655
CMOS Gates.. 657
    NAND and NOR Gates 657
    CMOS Gate Design.. 659
    Effects of Sizing and Scaling 661
Low-Power CMOS 663
Summary 663
13.2 Pseudo NMOS 664
Static Operation 665
Dynamic Operation... 667
13.3 Pass-Transistor Logic 668
13.4 Dynamic Logic.. 674
Basic Configuration... 674
Limitations of Dynamic Logic.. 675
Domino Logic 676
Pipelined Single-Phase Clock Architecture 677
13.5 BiCMOS Logic 678
Basic Operation.. 678
Propagation Delay... 680
BiCMOS Gates 680
13.6 Transistor–Transistor Logic 682
Basic TTL Inverter... 682
Advanced Low-Power Schottky TTL.................................... 685
13.7 Emitter-Coupled Logic 686
Basic Circuit 686
ECL 100k 687
ECL Gates.. 691
Summary of Main Concepts and Results............................. 693
Learning Outcomes.. 694
Supplementary Examples and Topics on Website................ 694
Problems and Exercises... 694

Appendix A: Reference Material.. 703

Appendix B: Basic PSpice Models 705

References and Bibliography 709

Index... 711

Preface

This book differs significantly from other current introductory textbooks on electronics in its coverage and approach. Most of the current textbooks pay only scant attention to basic electronics and the underlying theory of semiconductors. This, I believe, is a serious shortcoming, because some knowledge of the fundamental physical concepts involved is needed for two main reasons: first, in order to gain more than a superficial understanding of the behavior of semiconductor devices, and, second, to appreciate the nature of continuing improvements to the performance of these devices and therefore be able to follow the ongoing and projected advances in microelectronics. Given that the theory of electric conduction in semiconductors is well above the level of an introductory textbook, the challenge is to present the essentials of this theory in a way that gives at least a good qualitative understanding of the fundamental concepts involved. Fundamentals and concepts are strongly emphasized throughout this book, as in the author's book *Electric Circuits and Signals*, CRC press, Boca Raton, Florida, 2008.

Pedagogy

The following are specific features of the book's pedagogy:

1. Diffusion and drift in semiconductors and contacts between dissimilar materials are discussed in terms of electrochemical potential, which is shown to be a relatively simple but fundamental and unifying generalization of electric potential in the presence of concentration gradients. Most textbooks on electronics ignore electrochemical potential, and, if they go this far, consider instead Fermi and quasi-Fermi levels whose nature and role in semiconductors are not clearly addressed. Consequently, students find it difficult to understand why the Fermi level should be the same throughout a system at equilibrium, or the meaning of statements such as "Current is maintained by a gradient of the quasi-Fermi level," or "A voltmeter actually measures the difference between quasi-Fermi levels." An added benefit of introducing the concept of electrochemical potential is that electrophysiology of membranes of living cells is almost invariably discussed in terms of electrochemical potential. The concept of electrochemical potential provides an instructive link between semiconductor and ionic systems at a time when electrical engineering students are being increasingly exposed to biological systems.

2. An important objective in presenting semiconductor basics and operation of semiconductor devices is to "demystify" concepts and relations that are usually stated as facts, without at least some "plausible" explanation as to "the reason why." It is shown how many of the basic relations can be quite simply derived from fundamental considerations. Where this represents a significant departure from the mainstream of the discussion, it is left to Special Examples and Topics on the website of the book at:
www.crcpress.com/product/isbn/9781420087079

3. The discussion of transistors starts with the MOSFET; then the JFET and its derivative, the MESFET; and lastly the BJT. The discussion of BJT and MOSFET circuits is integrated as much as possible so as to highlight the common underlying

principles, particularly as concerns negative feedback. GaAs is briefly considered and its main characteristics, advantages, and limitations are highlighted.

4. State-of-the-art technologies, such as silicon-on-insulator and SiGe technologies, are presented.

5. The discussion of power electronics includes a brief treatment of power switching and *npnp* devices.

6. The Introduction presents a brief history of electronics and its impact.

7. Practical, real-life material is included wherever appropriate, mainly in the form of "Application Windows" that apply theory and provide additional motivation for students.

8. PSpice® simulations based on schematic capture and using OrCAD® 16.0 Demo Software are presented in some detail, including the simulation procedure, and are integrated within the discussion (OrCAD, PSpice, SPECTRA for OrCAD, and Cadence are registered trademarks of Cadence Design Systems, Inc., San Jose, California). The PSpice simulation examples are included on the website of the book. A free download of OrCAD Demo Software is announced at the Cadence website:
https://www.cadence.com/products/orcad/pages/downloads.aspx.

9. Answers for all problems are included on the website of the book.

It is assumed that the reader has good familiarity with circuit analysis and basic physics.

Organization

Chapter 1 covers basic diode circuits, the diode being introduced simply as a two-terminal device having a specified current–voltage characteristic. This chapter serves as a "soft," application-oriented, and motivating introduction to electronics, instead of the rather abstract and unfamiliar principles of semiconductor physics. After having aroused the reader's interest and curiosity as to how a semiconductor diode operates, Chapter 2 presents the basic background on the physics of semiconductors and the principles associated with electric conduction in semiconductors. Building on this background, Chapter 3 explains the theory behind the *pn* junction and the operation of the *pn* junction diode and several of its derivatives. To satisfy the reader's curiosity as to how *pn* junction diodes are fabricated, Chapter 4 discusses semiconductor fabrication in general.

Chapter 5 discusses transistor operation for field-effect transistors, namely, MOSFETs, the JFET, and the MESFET, and Chapter 6 discusses the same for the bipolar junction transistor. The discussion in Chapters 5 and 6 is oriented toward both analog and digital applications. Before considering transistor circuits, Chapter 7 introduces some basic notions on two-port circuits, amplifiers, and negative feedback, illustrating these concepts with ideal operational amplifiers.

Chapters 8 through 10 are dedicated to conventional analog electronics, starting with single-stage amplifiers (Chapter 8), then multistage and feedback amplifiers (Chapter 9), and, finally, differential and operational amplifiers (Chapter 10). Chapter 11 is concerned with power amplification and switching, including class D amplifiers and power latches. Chapters 12 and 13 deal with digital electronics; Chapter 12 covers the basic elements of digital circuits, whereas Chapter 13 discusses the most important logic families at present.

Every chapter begins with a brief overview and is followed by learning objectives. These are divided into two groups, inspired by Bloom's taxonomy on cognitive learning: the "to be familiar with" group and the "to understand" group that the student must pay particular attention to. Learning outcomes are stated at the end of every chapter together with a summary of the main concepts and results.

Good use is made of the website. Almost half as much material as in the book is included on the website in the form of supplementary examples (having the section number prefixed by SE) and supplementary topics (prefixed by ST). A given topic is included on the website, rather than in the text, if it is not basic enough, in the sense that it may be somewhat advanced, or it may not introduce additional fundamental concepts, or it might be outside the mainstream of the discussion.

Solutions Manual and Classroom Presentations

A solutions manual for all problems and exercises is available with its own companion CD that includes classroom presentations. These are Microsoft Word files for each chapter that present, in the form of colored bulleted text and figures, the main ideas discussed in the chapters, as well as the examples illustrating these ideas. The files are intended to be projected in the classroom by instructors and used as a basis for explaining the chapter material. They can be modified by instructors as they may deem appropriate for their own purposes.

MATLAB® is a registered trademark of The MathWorks, Inc. For product information, please contact:

The MathWorks, Inc.
3 Apple Hill Drive
Natick, MA 01760-2098 USA
Tel: 508 647 7000
Fax: 508-647-7001
E-mail: info@mathworks.com
Web: www.mathworks.com

Acknowledgments

I am as indebted as ever to my students for their interactions and for asking the sorts of questions that would have never crossed my mind, and which would call for better and clearer explanations of some aspects of the course material. I am very grateful to my colleagues in the ECE department, Professor Ayman Kayssi and Dr. Mariette Awad, for commenting on the manuscript, and to Professor Kayssi for allowing me to include some of his particularly instructive problems. I would also like to express my sincere appreciation of the efforts of the staff of CRC Press and their associates in producing and promoting this book, particularly Nora Konopka, Publisher, Engineering and Environmental Sciences, for her considerate support, Glenon Butler, Project Editor, and Arunkumar Aranganathan, Project Manager, SPi Technologies, India, for their adept professionalism. Special thanks to Shayna Murry for her artfully creative cover design.

Introduction

Brief History of Electronics

Electronics is the branch of science and engineering concerned with devices whose operation depends on phenomena associated with the flow of electric charges in nonmetals. These devices, referred to as **electronic devices**, comprise diodes, transistors, and a number of other devices, which, at present, are mainly made from semiconductors. Electronics is considered to have started in 1904 with the vacuum-tube diode, although the German physicist Karl Ferdinand Braun (1850–1918) discovered in 1874 the rectifying properties of a point contact between a metal wire and a galena crystal, a lead-sulfide semiconductor. This effect was used over 30 years later in the form of the "cat's whisker" crystal radio detector and formed the background for the experiments that led to the point-contact transistor in 1948. Braun also developed the cathode-ray tube in 1897, consisting of an unheated (cold) cathode, a high-voltage anode, and a phosphor screen. The electron beam was formed by passing electrons through a hole in an aluminum plate and was deflected by a current-carrying coil.

Following the discovery in 1893 by the American inventor Thomas Alva Edison (1857–1931) that a heated filament in vacuum emits electrons to a metal nearby (the Edison effect), the English electrical engineer and physicist John Ambrose Fleming (1849–1945) constructed a **vacuum-tube diode** (the **kenotron**) in 1904 having two main terminals: an anode and a cathode that was connected to one end of the heater, with the other end of the heater being brought out to a third terminal. Shortly thereafter, in 1906, the American inventor Lee De Forest (1873–1961) created the **vacuum-tube triode** (the **audion**) by adding a third electrode (the grid) to the vacuum-tube diode to control the anode current by applying a voltage between the grid and the cathode. The performance of the vacuum-tube triode was later improved in the late 1920s and early 1930s by adding a second grid (the **tetrode**) and a third grid (the **pentode**).

Until the advent of the transistor in the 1950s, vacuum-tube devices were used in electronic amplifiers and circuits in radio, television, wireless communications, radar, control systems, and many other applications. Nevertheless, because of their relative fragility and the short life of their heating filaments, vacuum-tube devices were not considered robust or reliable enough for many industrial and military applications, in addition to their consuming too much power and generating considerable heat. Other types of amplifiers were developed for special purposes, such as magnetic amplifiers, fluidic amplifiers, and rotating-machine amplifiers (the **amplidyne**). The earliest electronic digital computer, the Electronic Numerical Integrator and Calculator (**ENIAC I**), inaugurated in 1946, had about 19,000 vacuum tubes. Because of the burnout of filaments of these tubes, trouble-free operations did not, at best, exceed five days even after "high-reliability" vacuum tubes were used.

Because of the disadvantages of vacuum tubes, there were many attempts to replace them by solid-state devices. In 1925, the Austro-Hungarian physicist Julius Edgar Lilienfeld (1881–1963) filed a patent for a field-effect amplifier similar to the MESFET (Section 5.7, Chapter 5), which he followed in 1928 by a device similar to the MOSFET (Section 5.2, Chapter 5). The German physicist/electrical engineer and inventor Oskar Heil (1908–1994) also described in a patent application in 1934 how current through a semiconductor can be controlled by a transverse electric field. But none of these devices was practical because of the detrimental effects of surface states. The selenium rectifier was introduced in 1933 and consisted of an aluminum or steel electrode, plated with a thin layer of bismuth or nickel, on which was deposited a relatively thick layer of selenium, which formed the other electrode. The rectifier could handle larger currents than the vacuum-tube diode but was not a very efficient rectifier.

An intense, comprehensive, and concerted effort was launched in 1945 at Bell Telephone Laboratories to develop a solid-state alternative to the mechanical relays in telephone switching equipment. The team of about 12 members was headed by William Shockley (1910–1989) and included Walter Brattain (1902–1987) and John Bardeen (1908–1991), all physicists, as well as chemists, engineers, and other physicists. After much experimental work and theoretical interpretations, they succeeded in building what became known as a **point contact transistor** in 1947, consisting of two closely spaced gold foils that made point contacts with a germanium crystal. The term transistor was coined as a combination of **trans**resistance and var**istor**, a nonlinear resistor element. One contact (the emitter) was biased positively with respect to the crystal (the base), while the other contact (the collector) was biased negatively with respect to the base. In an early version, a voltage amplification of 2 and a power amplification of 330 were obtained. Bardeen, Brattain, and Shockley received the Nobel Prize in physics in 1956, "for their researches on semiconductors and their discovery of the transistor effect."

The point contact transistor was commercially produced but suffered from several disadvantages. Its voltage and power gains were low, it could handle only small currents, it was difficult to manufacture, and it was prone to instability under certain operating conditions. Early in 1948, Shockley conceived of an improved transistor, the bipolar junction transistor (BJT; Section 6.1, Chapter 6), based on the theory of pn junctions. It took another two years, however, before the device could be made, and BJTs were first commercially produced by Texas Instruments in 1954. The semiconductor used in early BJTs was germanium, but after methods for large-scale purification of silicon were developed, silicon BJTs prevailed in the early 1960s because of their superior performance. The BJT was the transistor of choice for both discrete and integrated circuits for almost 30 years.

Alongside the development of the point contact and bipolar junction transistors, work on field-effect transistors continued in an attempt to mimic the behavior of the vacuum-tube amplifier in the solid state. Following the contributions of Lilienfeld and Heil, Shockley described in his patent application for the BJT in 1948 a field-effect device having an insulated gate over a pn junction. In 1952, Shockley described a junction field-effect transistor (JFET; Section 5.6, Chapter 5). In 1958, M.M. Atalla and coworkers at Bell Telephone Laboratories reported on how silicon dioxide, thermally grown on a silicon surface, can passivate the silicon surface, thereby overcoming some of the problems associated with this surface. In 1960 he, along with D. Kahng, described what is essentially a p-channel MOSFET that they built using this technique. Silicon-based, discrete MOSFETs were first produced commercially in 1964 by Fairchild Semiconductor Corporation and Radio Corporation of America (RCA). However, discrete MOSFETs were not widely used because BJTs had a superior performance.

Meanwhile, integrated circuits were developed and became increasingly important for meeting the demands of a fast-growing digital computer industry. It was soon recognized that MOSFETs had decisive advantages over BJTs in integrated circuits, as they: (1) were easier to manufacture; (2) could be packed more densely on a chip without seriously affecting performance; and (3) dissipated less power, particularly when combined with complementary MOSFETs in what became known as CMOS technology. By 1980, the MOSFET eclipsed the BJT, mainly because of its adoption in digital integrated circuits.

The first "integrated circuit" that combined circuit elements and electronic devices in a single package was introduced in 1926 by a German company, Loewe Radio A.G., and consisted of three vacuum-tube triodes, two capacitors, and four resistors in a single glass envelope. The motivation for this combination was a German tax on radio receivers based on the number of glass envelopes in the receiver. An early attempt at integration was made by the British engineer and automation pioneer John Sargrove (1906–1974), who experimented in 1936 and 1937 with creating a low-cost radio receiver by molding resistors, capacitors, inductors, and their interconnections in bakelite, so that all that remained was to insert the vacuum tubes in their sockets and connect the loudspeaker, thereby reducing the amount of labor involved and lowering the cost of production. His larger goal was to develop what he called "electronic circuit making equipment." Although his work succeeded technically, it failed commercially. In 1949, the German mechanical engineer–physicist Werner Jacobi (1904–1985) filed a patent for a two-stage "semiconductor amplifier" consisting of five transistors on a common substrate. This, however, remained largely unknown and was not commercially exploited. In 1952, a British electrical engineer at the Royal Radar Establishment, Geoffrey Dummer (1905–2002), proposed to have "electronic equipment in a solid block with no connecting wires," with the block consisting of layers of insulating and conducting materials (forming passive electrical components) combined with electronic devices. In 1958, Jack S. Kilby (1923–2005) of Texas Instruments constructed a flip-flop consisting of two BJTs on a single chip of germanium, with gold wire connections between the transistors. He shared the 2000 Nobel Prize in physics, "for his part in the invention of the integrated circuit," with Herbert Kroemer and Zores I. Alferov "for developing semiconductor heterostructures used in high-speed- and opto-electronics." Kilby's integrated circuit was not monolithic since the gold wire interconnections were not formed on the chip itself. American physicist Robert N. Noyce (1927–1990) of Fairchild Semiconductor demonstrated in 1959 the first monolithic integrated circuit consisting of silicon BJTs and resistors on a single silicon chip that were interconnected with aluminum conductors produced by photolithography (Section 4.2, Chapter 4).

In 1968, Noyce and chemist–physicist Gordon E. Moore left Fairchild Semiconductor to found Intel Corporation, which is the largest semiconductor company in the world today. Based on his experience with chip design and manufacturing, Moore had enunciated in 1965 what became known as "Moore's law," according to which, the number of transistors on a chip roughly doubles every two years or so, based on three factors: (1) the increase in chip size, (2) the shrinking of transistor dimensions, and (3) improved layout on the chip so as to reduce "dead space" as much as possible, as by merging devices and circuit elements in the chip and increasing the number of layers for interconnections on the chip. Remarkably, the development of semiconductor chips has conformed to Moore's law over the years, and even exceeded it at times.

To complement the preceding brief history of electronic devices and integrated circuits, Table 1 chronicles some important developments relevant to electronics and its applications.

Table 1 Dates for Some Important Developments Relevant to Electronic Applications

1822 English mathematician, philosopher, inventor, and mechanical engineer Charles Babbage (1791–1871) expounded the principles of the difference engine, a mechanical computer designed to compute values of polynomial functions, and whose basic architecture is very similar to that of modern digital computers.

1847   English mathematician and philosopher George Boole (1815–1864) promulgated Boolean algebra.

1849   Italian inventor Antonio Meucci (1808–1889) demonstrated a "talking telegraph," the first phone system in which voice is sent over wires.

1873   Scottish mathematician and theoretical physicist James Clerk Maxwell (1831–1879) published the four partial differential equations on electromagnetism, known since as Maxwell's equations.

1876   Scottish-born inventor Alexander Graham Bell (1847–1922) developed his practical telephone. He founded the Bell Telephone Company in 1877.

1882   American physicist and inventor Amos Emerson Dolbear (1837–1910) demonstrated wireless communication through earth via buried metal rods connected to telephones. The transmission range was about half a mile.

1895   Italian inventor Guglielmo Marconi (1874–1937) communicated via radio signals near Bologna, Italy, using a spark transmitter that could only send telegraph-type signals. He was able to increase the transmission range to about 1.5 km by increasing the length of the transmitting and receiving antennas and by orienting them vertically, with one end touching ground. That same year, Alexander S. Popov claimed to have sent radio signals over 600 yards. Nikola Tesla is claimed to have demonstrated radio transmission in 1893. In 1885, Sir William Pierce had demonstrated wireless transmission by magnetic induction. Marconi shared the 1909 Nobel Prize in physics with Karl Ferdinand Braun, "in recognition of their contributions to the development of wireless telegraphy."

1898   Danish engineer Valdemar Poulsen (1869–1942) built the first working magnetic recorder, which he called the Telegraphone, having magnetic wire wrapped around a drum.

1900   Canadian inventor Reginald Aubrey Fessenden (1866–1932) transmitted speech by radio over a distance of about a mile using a high-frequency spark transmitter. In 1906, he transmitted music by radio using a continuous-wave alternator–transmitter.

1907   Russian scientist and inventor Boris Lvovich Rosing (1869–1933) demonstrated a primitive TV system featuring a mechanical scanning disc and an early form of a cathode-ray tube.

1918   American electrical engineer and inventor Edwin Howard Armstrong (1890–1954) patented the superheterodyne receiver.

1921   The U.S. Department of Commerce licensed the first commercial broadcasting station, Westinghouse's KDKA station, in Pittsburgh. The first broadcast was the returns in the Harding vs. Cox presidential race.

1923   Russian-born, American inventor and engineer Vladimir Kozmich Zworykin (1889–1982) filed a patent for the "iconoscope," a rudimentary TV camera in which charge is stored on a large number of picture elements in proportion to light from the image. The elements are then scanned by an electron beam to reproduce the charge pattern. In 1929, Zworykin patented the "kinescope," the precursor to the modern CRT television tube.

1925 Scottish engineer and inventor John Logie Baird (1888–1946) demonstrated the world's first working television system using a mechanical scanning disc patented in 1884 by the German inventor Paul Gottlieb Nipkow (1860–1940).

1927 American electrical engineer Harold Stephen Black (1898–1983) used negative feedback to improve the performance of an electronic amplifier.

1928 American inventor Philo Taylor Farnsworth (1906–1971) publicly demonstrated an all-electronic television system using a working electronic camera tube.

1933 American electrical engineer and inventor Edwin Howard Armstrong (1890–1954) patented wideband frequency modulation (FM) radio.

1937 British engineer Alec Harvey Reeves (1902–1971) patented pulse-code modulation.

1939 RCA demonstrated the first black-and-white television set.

1940 English physicist Henry Albert Howard Boot (1917–1983) developed a high-power cavity magnetron for use in centimetric radar, which allowed the detection of much smaller objects and the use of much smaller antennas than the earlier lower frequency radars. A basic multi-cavity resonant magnetron had been developed in 1935 by German electronic specialist Hans Erich Hollmann (1899–1960).

1941 NBC launched commercial television broadcasting in the United States.

1941 RCA announced an electronic color TV system that was potentially compatible with black-and-white television.

1943 Austrian inventor Paul Eisler (1907–1995) patented the printed circuit board, first used in the proximity fuse intended to bring down German V1 rockets toward the end of World War II.

1946 The Electronic Numerical Integrator and Computer (ENIAC I), the first general-purpose, Turing-complete, electronic digital computer, unveiled at the University of Pennsylvania, conceived and designed by the American physicist John William Mauchly (1907–1980) and the American electrical engineer John Presper Eckert (1919–1995). Programs were hard-wired by means of switches and cables. The ENIAC I was intended to calculate artillery firing tables for the U.S. Army's Ballistic Research Laboratory.

1948 The Electronic Delay Storage Automatic Calculator (EDSAC), the first practical, stored program, electronic digital computer, was unveiled at the University of Cambridge Mathematical Laboratory, England.

1949 American chemist Gordon Kidd Teal (1907–2003) of Bell Laboratories developed, with help from mechanical engineer John Little and technician Ernest Buehler, a method of applying the Czochralski process to produce extremely pure germanium single crystals for making greatly improved transistors. He later extended this method to silicon. After moving to Texas Instruments in 1952, he demonstrated in 1954 a silicon BJT.

1950 Color TV was publicly demonstrated by CBS.

1951 The Lyons Electronic Office I (LEO I), the first commercial digital computer, was introduced in Britain by J. Lyons and Co. and modeled after the EDSAC. The first commercial digital computer in the United States was the UNIVersal Automatic

Computer I (UNIVAC I), made by Remington Rand and delivered to the U.S. Census Bureau in 1952.

1952   Andrew Kay of Non Linear Systems, San Diego, California introduced the first commercial digital voltmeter.

1954   The Regency TR-1, the first commercial transistor radio, was announced by the Regency Division of Industrial Development Engineering Associates (I.D.E.A.) of Indianapolis, Indiana. It was patented by the Dutch physicist Heinz De Koster, an employee of the company.

1957   Japanese physicist Leo Esaki of Tokyo Tsushin Kogyo (now known as Sony) invented the tunnel diode, also known as the Esaki diode (Section 3.5, Chapter 3). Esaki shared the 1973 Nobel Prize in physics with Ivar Giaever and Brian David Josephson, "for their discoveries regarding tunneling phenomena in solids."

1960   American physicist Theodore Harold Maiman (1927–2007) of Hughes Research Laboratories demonstrated the first working laser, based on a synthetic ruby crystal.

1961   Fairchild Semiconductor introduced resistor-transistor logic (RTL) ICs, the earliest class of digital logic ICs.

1962   Nick Holonyak Jr. of General Electric Company developed the first practical visible-spectrum (red) LED.

1963   The first CMOS circuit was described by Frank Wanlass of Fairchild Semiconductor.

1964   Sony Corporation introduced the first all-transistor, desktop electronic calculator using a neon-tube display. In 1970, a "cigarette pack"–sized calculator was introduced by Bowmar (United States), featuring an LED display and rechargeable batteries.

1965   The μA709, the first commercially successful IC operational amplifier, was developed by Robert J. Widlar of Fairchild Semiconductor. The μA702, also developed by Widlar at Fairchild Semiconductor about a year earlier, was not a commercial success because of many limitations.

1968   A one-transistor DRAM cell was patented by Robert Dennard of IBM.

1969   Intel introduced a 64-bit high-speed static random access memory (RAM), the 3101, using Shottky BJTs. Later, in 1969, Intel introduced the 1101, a 256-bit static RAM, the first memory chip based on MOSFETs.

1969   Charge-coupled devices were invented by George Smith and Willard Boyle of Bell Laboratories.

1970   Intel introduced the first 1K dynamic RAM, the 1103, using three PMOS transistors per memory cell.

1970   Intel introduced the 2048-bit mask programmable MOS, the 1301, and the 1701 chip, a 256-byte erasable read-only memory (EROM).

1970   Hoffmann-LaRoche of Switzerland applied for a patent on the twisted nematic (TN) field effect in liquid crystals. Shortly thereafter, in 1971, James Fergason, of Westinghouse Research Laboratories, Pittsburgh, applied for an identical patent in the United States. LCDs based on the TN effect soon superseded earlier LCDs because of their lower operating voltages and power consumption.

1970 Robert Maurer, Donald Keck, and Peter Schultz of Corning Glass Works introduced a low-loss optical fiber. The first fiber optic, semiflexible gastroscope, was patented in 1956 by Basil Hirschowitz, C. Wilbur Peters, and Lawrence E. Curtiss of the University of Michigan.

1971 Intel introduced the first microprocessor, the 4004, based on 4-bit PMOS technology.

1971 Dov Frohman of Intel invented the Erasable PROM (EPROM) using a floating gate.

1973 The first public telephone call was placed on a portable cellular phone by Martin Cooper, general manager of Motorola's Communications Systems Division and inventor of the first modern portable handset.

1977 Micro Instrumentation and Telemetry Systems (MITS), New Mexico, commercialized the Altair 8800, the first low-cost, personal computer using a single-chip microprocessor, Intel's 8080. It shipped in kit form.

1981 IBM launched its mass-produced, open-architecture personal computer based on Intel's 8088 microprocessor.

1982 GRiD Compass, the first laptop using the clamshell design used today by almost all laptops, was introduced by Grid Systems Corporation of Fremont, California. The computer was designed in 1979 by the British industrial engineer William Moggridge, an employee of the company, and featured an Intel 8086 processor, a 320×200-pixel (CGA) electroluminescent display, a 340 kB magnetic bubble memory, and a 1200 bit/s modem. Hard drives, floppy drives, and other devices could be connected via the 488 I/O GPIB (General Purpose Instrumentation Bus).

1983 The PF-3000, the first PDA, was released by CASIO, for holding memos, addresses, and telephone numbers.

1984 Flash memory was developed by Fujio Masuoka of Toshiba. Flash memory was commercialized by Intel in 1988.

1987 Ching Tang and Steven Van Slyke of Eastman Kodak Company developed the first organic light-emitting diode.

1994 Apple Computer Inc. introduced the QuickTake 100 camera, the first digital camera for the consumer-level market that worked with a home computer via a serial cable.

2006 Intel's first announcement of the Tukwila server chip, billed as the word's first 2 billion transistor chip.

Impact of Electronics

The transistor, billed as the greatest invention of the twentieth century, was a big improvement over vacuum tubes. Initially, transistors suffered failures because of surface effects. But once these problems were overcome, transistors had a decisive advantage over vacuum tubes as they had a long life, greater robustness, much smaller size, lower cost of production, and much less power consumption. Not only did transistors quickly and almost totally replace vacuum tubes, but they made possible a whole new range of industrial, military, commercial, and consumer applications. For example, portable transistor radios of small size and low battery consumption became available, transforming the lives of many people all over the world.

Electronics and its applications did not really explode, however, until integrated circuits came along and brought not only continuing miniaturization but dramatic cost reduction as well. General-purpose operational amplifiers, for example, of a near-million voltage gain, now cost less than a few cents (of a U.S. dollar) each. The cost of digital memories of huge capacities is less than a cent per megabit. Moreover, the impact of integrated circuits was vastly enhanced by their wedding to digital system techniques. This brought: (1) the economy and ease of design and manufacturing of modules and repeating circuit units, (2) the versatility of software control, and (3) the advantages of digital mass data storage, and digital processing such as data compression, error detection, and correction. It became possible to control the operation of all kinds of equipment by embedded microprocessors and microcontrollers. The wedding of integrated circuits and digital circuit techniques heralded the Third Industrial Revolution, that of information technology based on computers and communications—the First Industrial Revolution being that of mechanization using mainly steam power, and the Second Industrial Revolution being mainly that of electrification in factories, workplaces, and homes. Data processing hardware comprising computers and communications equipment is today the largest electronics application sector in terms of monetary value. Consumer electronics is highly visible in our everyday lives in the form of cellular phones, smartphones, media players, personal digital assistants (PDAs), TVs and radio receivers, recorders, digital cameras, etc. Even at home, electronic control is used in many appliances such as refrigerators, washing machines, and vacuum cleaners. Electronics is a vital constituent of all types of military hardware, biomedical devices and equipment, and process control in chemical plants and manufacturing plants of all kinds. Integrated circuits have opened new application areas in aviation, cars, robots, instrumentation, electric motor and power system control, and equipment for many fields of research. Novel applications are being envisioned all the time that would have been unimaginable in the 1950s or would have been deemed as science fiction. Moreover, microelectronic fabrication techniques have made possible micro-electro-mechanical systems (MEMS) and have paved the way to nanotechnology, one of the major technologies of the twenty-first century. In short, electronics, in its myriad forms, is becoming ever more intricately woven into the fabric of our lives.

Convention for Symbols

Current and Voltage Symbols

- Capital letter with capital subscript denotes a dc, or average, quantity. Example: I_C.
- Capital letter with m subscript denotes the peak value of an alternating quantity. Example: $I_m \sin \omega t$.
- Lowercase letter with capital subscript denotes a total instantaneous quantity. Example: i_D.
- Lowercase letter with lowercase subscript denotes a small signal of zero average value. Example: i_d.
- Double subscript in a voltage symbol denotes a voltage drop from the node or terminal designated by the first subscript to the node or terminal designated by the second subscript. Example: v_{ce}.

Gate Capacitance Symbols

- Asterisk denotes value of a circuit parameter per unit area. Example: C^*.
- Prime denotes value of a circuit parameter per unit length. Example: C'.

1

Basic Diode Circuits

Semiconductor diodes are by far the most important examples of passive, two-terminal, nonlinear devices, as evidenced by their wide use in a variety of important applications in electric and electronic circuits. Some of the more basic applications of diodes are considered in this chapter, treating the diode as a circuit element having a specified current–voltage characteristic. This approach is essentially an extension of circuit analysis to circuits having nonlinear elements. The basic principles underlying the operation of semiconductor devices are covered in Chapter 2 and are applied to various types of semiconductor diodes in Chapter 3.

Just as linear passive elements are idealized in order to highlight their salient characteristics and the way they affect circuit responses, diodes are idealized for the same reasons. However, this idealization can take several forms, depending on the diode circuit under consideration. It is essential, therefore, to appreciate the nature and implications of the approximations inherent in these idealizations. This is true of electronic circuits in general so that gaining insight into the behavior of these circuits is particularly important. Except in relatively few cases, "exact" solutions are usually tedious and unwarranted, considering the usual tolerances on electronic devices. One should therefore strive to acquire a "feel" for electronic circuits in order to make "judicious" approximations. For more accurate results, simulations are usually resorted to. In this and the following chapters on electronic circuits, approximations and shortcuts are emphasized. More exact solutions are mentioned mainly to appreciate the fundamental principles involved and the nature of the approximations made.

Learning Objectives

❖ To be familiar with:

- The terminology associated with semiconductor diodes and their circuits.

❖ To understand:

- The implications of the characteristics of ideal diodes and *pn* junction diodes.
- The principles of operation of rectifier circuits and the methods used to smooth the output.
- The application of zener diodes to voltage regulation.
- The basic operation of voltage limiters, dc restorers, and voltage multipliers.

1.1 Ideal and Practical Diodes

Ideal Diode

Definition: *An ideal diode is a two-terminal device that behaves as a short circuit for current flow in one direction (the **forward** direction) and as an open circuit for current flow in the opposite direction (the **reverse** direction).*

The ideal diode behaves like an ideal switch that is controlled by the direction of current flow, much like a one-way, or no-return, valve that allows fluid to flow in one direction but not in the opposite direction.

The symbol used in this book for an ideal diode switch has an unfilled arrowhead (Figure 1.1.1a), the forward direction being that of the arrow, with the assigned positive directions of diode current and voltage as indicated. The *i–v* characteristic has the right-angled shape depicted in Figure 1.1.1b, and represents an ideal **unilateral** element, that is, an element that conducts in one direction only. The diode terminals are denoted as **anode** and **cathode**, the direction of forward current through the diode being from anode to cathode. This terminology dates back to one of the earliest forms of diodes, the vacuum-tube diode (see Introduction).

Although the ideal diode switch may seem quite unrealistic, it is shown in the next section that it can serve as a good approximation under certain conditions.

Ideal Si *pn* Junction Diode

The most common diode is the silicon (Si) *pn* junction diode. This can take one of two forms: a discrete diode that is fabricated as a two-terminal device, or a diode-connected transistor in an integrated circuit (IC), as explained in Section 6.3, Chapter 6.

An ideal, Si *pn* junction diode has the *i–v* relation given by (Equation 3.2.3, Chapter 3):

$$i_D = I_S(e^{v_D/\eta V_T} - 1) \tag{1.1.1}$$

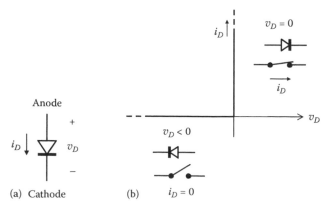

FIGURE 1.1.1
Ideal diode switch: (a) symbol and assigned positive directions; (b) *i–v* characteristic.

where I_S is a small **saturation current** and $V_T = kT$ is the **thermal voltage**, k being Boltzmann's constant in eV/K (8.62×10^{-5}). At $T = 300$K, $V_T = 25.852$ mV. It is generally assumed that $\eta = 2$ for discrete diodes and $\eta = 1$ for IC diodes. Note that $V_T = k_J T/|q|$, where $|q|$ is the magnitude of the electronic charge and k_J is Boltzmann's constant in J/K.

Exercise 1.1.1

(a) At what voltage will the reverse current of an ideal Si diode reach 99% of its saturation value, assuming $\eta = 2$ and $V_T = 26$ mV? (b) At what voltage will the forward current be $100I_S$?

   Answers: (a) -0.24 V so that the exponential can be neglected for $v_D < -0.24$ V and i_D considered to be very nearly equal to $-I_S$; (b) $+0.24$ V; under these conditions $I_D \gg I_S$, or the exponential is much greater than unity so that $i_D \cong I_S e^{v_D/\eta V_T}$. ∎

   If $v_D > 0$, the diode is **forward-biased**. According to Equation 1.1.1, i_D increases slowly at first and then very rapidly as v_D increases beyond about 0.5–0.6 V (Figure 1.1.2). The voltage below which the forward current is negligible is the **cut-in voltage** V_γ.

   If $v_D < 0$, the diode is **reverse-biased**. i_D rapidly approaches $-I_S$, as shown in Figure 1.1.2 for a much more expanded current scale in the negative direction of current. The symbol for a diode other than the ideal diode switch of Figure 1.1.1 has a filled arrowhead (Figure 1.1.2).

Practical Diodes

The i–v characteristics of practical pn junction diodes differ from Equation 1.1.1 in the following respects depending on the diode type, structure, and method of fabrication:

1. η, the **emission coefficient**, is in the range of 1–2 and varies with the diode current. Theoretically, as discussed in Sections ST3.6 and ST3.7, Chapter 3, $\eta = 2$ at low values and at high values of current, and $\eta = 1$ at intermediate values.

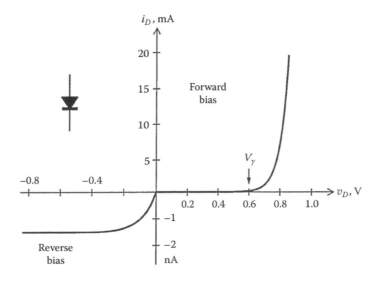

FIGURE 1.1.2
i–v characteristic of ideal Si pn junction diode. Inset is the symbol for a diode other than an ideal diode switch.

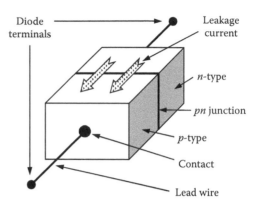

Diode terminals — Leakage current

n-type

pn junction

p-type

Contact

Lead wire

FIGURE 1.1.3
Diagrammatic illustration of a *pn* junction diode.

2. A *pn* junction diode essentially consists of a *p*-type semiconductor and an *n*-type semiconductor in the same crystal. The boundary between the two types of semiconductor is the *pn* junction, as diagrammatically illustrated in Figure 1.1.3. v_D in Equation 1.1.1 is, strictly speaking, the voltage drop across the *pn* junction only. When the diode is forward biased, the total voltage drop across the diode includes an additional voltage drop due to the flow of current through the **bulk resistance** of the diode. As its name implies, the bulk resistance comprises all the diode resistive elements, including the resistances of the *p*- and *n*-regions on either side of the metallurgical junction, the resistance of the contacts to these regions, and the resistance of the lead wires that connect to the external diode terminals. At high forward currents, the voltage drop due to the bulk resistance is appreciable, so that the voltage across the diode is larger than v_D, for a given i, and varies more linearly with i.

3. The reverse current I_R exceeds I_S because of a surface leakage current I_{leak} around the *pn* junction, as indicated in Figure 1.1.3 by the block arrows. The magnitude of I_{leak} increases with the reverse voltage and its temperature dependence is different from that of I_S (Section 3.2, Chapter 3). The total reverse current I_R is therefore:

$$I_R = I_S + I_{\text{leak}} \tag{1.1.2}$$

I_{leak} can be accounted for by a large resistance across the diode. In the forward direction, I_{leak} is insignificant compared to the normal diode forward current.

4. When the reverse voltage exceeds a certain value, the *pn* junction breaks down and the current increases rapidly, as illustrated in Figure 1.4.1.

Incremental Diode Resistance

Because Equation 1.1.1 is nonlinear, an **incremental diode resistance** r_d is defined as the reciprocal of the slope of diode *i–v* characteristic at any given point. From Equation 1.1.1,

$$r_d = \frac{1}{di_D/dv_D} = \frac{\eta V_T}{I_S e^{v_D/\eta V_T}} = \frac{\eta V_T}{i_D + I_S} \cong \frac{\eta V_T}{i_D} \tag{1.1.3}$$

Exercise 1.1.2

(a) What is r_d of an ideal, discrete Si diode at a forward current of 10 mA if $I_S = 10$ nA and $V_T = 26$ mV? (b) What is the corresponding v_D? (c) What is v_D for $i_D = 20$ mA?
   Answers: (a) 5.2 Ω; (b) 0.718 V; (c) 0.754 V. ∎

1.2 Basic Analysis of Diode Circuits

Piecewise Linear Approximation

In practical applications, the presence of the diode can be accounted for in different ways depending on the diode circuit under consideration. In many cases, when the diode is conducting an appreciable forward current, the forward characteristic in Figure 1.1.2 can be approximated by two straight lines L_1 and L_2, defined as:

$$L_1: i_D = 0, \quad v_D \leq V_{D0}$$
$$L_2: i_D = \frac{1}{r_D}(v_D - V_{D0}), \quad v_D > V_{D0} \tag{1.2.1}$$

where the intercept of L_2 on the voltage axis is the diode **offset voltage** V_{D0}, and $1/r_D$ is the slope of L_2 (Figure 1.2.1a). The diode can then be represented *when conducting in the forward direction* by a linear circuit consisting of a battery of voltage V_{D0} in series with a resistance r_D (Figure 1.2.1b). An ideal diode switch is added to make the reverse current zero.

Exercise 1.2.1

Using an appropriate computer program, plot the forward characteristics of two ideal, discrete Si diodes, (a) one having $I_S = 1$ μA, (b) the other having $I_S = 1$ nA, and estimate V_γ, V_{D0}, and r_D. Assume $V_T = 26$ mV.
   Answers: (a) $V_\gamma \cong 0.18$ V, $V_{D0} \cong 0.42$ V; (b) $V_\gamma \cong 0.5$ V, $V_{D0} \cong 0.78$ V. r_D is approximately 5.5 Ω in both cases. Note that the larger I_S, the smaller are V_γ and V_{D0}. ∎

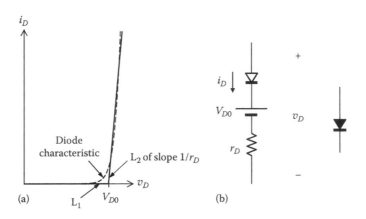

FIGURE 1.2.1
Piecewise linear approximation of diode characteristic: (a) approximation lines; (b) circuit representation.

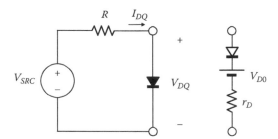

FIGURE 1.2.2
Determination of bias point.

Bias Point

Consider a diode connected in series with a dc voltage source V_{SRC} and a resistance R (Figure 1.2.2). From Kirchhoff's voltage law (KVL):

$$V_{DQ} + RI_{DQ} = V_{SRC} \qquad (1.2.2)$$

where V_{DQ} and I_{DQ} are the dc values of the diode voltage and current, respectively. They define what is referred to in electronic circuits as the **bias point** or the **quiescent point**.

Given V_{SRC}, R, and the diode nonlinear characteristic, the bias point can be determined in one of two general ways:

1. If the diode characteristic is known analytically, in the form of Equation 1.1.1, for example, I_{DQ} can be determined by successive iterations. The iterations are started by determining I_{DQ} from Equation 1.2.2 using a reasonable initial value of V_{DQ}, say V_{D0}. From this value of I_{DQ}, a new value is found for V_{DQ} from Equation 1.1.1, which is then substituted in Equation 1.2.2, and so on. The procedure rapidly converges after two or three iterations.

2. Alternatively, the straight-line approximation of Equation 1.2.1 can be solved with Equation 1.2.2 to determine V_{DQ} and I_{DQ}. Both methods are illustrated in Example 1.2.1.

Graphical representations can be very helpful in providing insight into the behavior of nonlinear systems and in visualizing the interrelations of the variables involved. In the case of the diode circuit of Figure 1.2.2, for example, Equation 1.2.2 can be represented by a straight line having a voltage intercept V_{SRC} and a slope $-1/R$ (Figure 1.2.3a). Such a line, when drawn on the graph of the diode characteristic, is a **load line**. The point of intersection Q of this line with the diode characteristic is the graphical solution for the bias point because it satisfies both Equation 1.2.2 and the diode characteristic.

Small-Signal Model

If a small ac signal, say $V_m \sin\omega t$, is applied in series with the dc source in Figure 1.2.2, then this signal is superimposed on V_{SRC}, the instantaneous source voltage V_{SRC} being $V_{SRC} + V_m \sin\omega t$. The load line oscillates horizontally about its zero-signal position, the voltage intercept varying between $V_{SRC} - V_m$ and $V_{SRC} + V_m$, as illustrated in exaggerated form in Figure 1.2.3b. The diode voltage will vary by a small amount v_d around its value

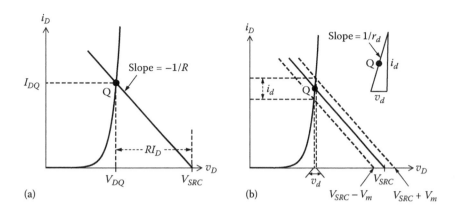

FIGURE 1.2.3
(a) Load-line construction. (b) Small-signal excursions about quiescent point.

of V_{DQ}, the corresponding small-signal variation in diode current being i_d. For small enough variations, the operating point can be considered to move along the tangent to the curve at Q. v_d and i_d will then be related by the diode incremental resistance $v_d = r_d i_d$, as illustrated by the inset in Figure 1.2.3b.

Concept: *For small signals, the diode appears to the small-signal source as a resistance equal to the incremental resistance r_d (Equation 1.1.3).*

In other words, the **small-signal model** of the diode is simply r_d.

Example 1.2.1 Bias Point Calculation

It is required to determine the bias point in the circuit of Figure 1.2.2 when $V_{SRC} = 10$ V and $R = 2$ kΩ, using two diode models: (a) Equation 1.1.1 with $I_S = 1.5$ nA, $\eta = 2$, and (b) the linear approximation to this characteristic with $r_D = 2.2$ Ω and $V_{D0} = 0.77$ V.

SOLUTION

(a) The iterative solution, which is a near-exact solution according to Equation 1.1.1, is based on Equation 1.2.2 expressed as:

$$I_{DQ} = \frac{1}{2}(10 - V_{DQ})\,\text{mA} \qquad (1.2.3)$$

and on Equation 1.1.1 expressed as:

$$V_{DQ} = \eta V_T \ln\left(\frac{I_{DQ} + I_S}{I_S}\right) \cong \eta V_T \ln\left(\frac{I_{DQ}}{I_S}\right) \qquad (1.2.4)$$

when $I_{DQ} \gg I_S$. The value of V_T used is 0.025852 V to obtain good agreement with the simulation in Example 1.2.2. The iteration proceeds as follows:

	V_{DQ}, V	I_{DQ}, mA
First iteration	0.7 (arbitrary)	4.65 (from Equation 1.2.3)
Second iteration	0.7728 (from Equation 1.2.4)	4.614 (from Equation 1.2.3)
Third and subsequent iterations	0.7724 (from Equation 1.2.4)	4.614 (from Equation 1.2.3)

The iterations can be readily performed using the copy and paste commands of a spreadsheet program.

(b) According to the linear approximation, $V_D = RI_D + V_{D0}$. Substituting in Equation 1.2.2 gives $I_{DQ} = \dfrac{(10 - V_{D0})\text{V}}{(2 + r_D)\text{k}\Omega} = 4.610\,\text{mA}$, and $V_{DQ} = 0.77 + 0.0022 \times I_{DQ} = 0.7801$ V. Clearly, as $r_D = 2.2\ \Omega \ll 2$ kΩ, neglecting r_D altogether does not significantly affect I_{DQ}. If V_{D0} is also neglected, $I_{DQ} = 5$ mA, which is about 8.5% too high. This percentage error decreases for larger values of V_{SRC}.

Example 1.2.1 illustrates a useful concept based on the straight line approximation:

Concept: *In a given diode circuit, r_D can be neglected if it is in series with a much larger resistance, and V_{D0} can be neglected if it is in series with a much larger voltage. When r_D and V_{D0} are neglected, the diode can be considered an ideal diode switch.*

Simulation Example 1.2.2 Diode Bias Point Using PSpice

It is required to simulate the bias point calculation of Example 1.2.1 and illustrate in the process how ideal diode switches can be simulated in PSpice.

SIMULATION

Appendix SA on the website explains the basic mechanics of PSpice simulation and provides a convenient reference for some general features of PSpice.

(a) The first simulation is based on Equation 1.1.1 using the values of I_S and η as in Example 1.2.1. Enter the schematic shown in Figure 1.2.4. For the diode, select Dbreak from the BREAKOUT library under PSpice. Click on the diode symbol with the left mouse button to select the diode, right-click on the selected diode and choose Edit PSpice Model. The following line appears in the editor window:

```
.model Dbreak D Is = 1e-14 Cjo = .1pF Rs = .1
```

This indicates that the diode $i\text{--}v$ relation satisfies Equation 1.1.1 with $I_S = 10^{-14}$ A. The default value of η is 1, which can be changed to 2 by entering $n = 2$. In addition, the diode has a junction capacitance at zero bias of 0.1 pF and a bulk resistance of 0.1 Ω. Choose a new name for the diode instead of Dbreak, such as DS1, change I_S to 1.5e−9, and R_S to 0. After making the changes, choose File/Save and then File/Exit. The diode name changes to DS1 on the schematic.

Select Bias Point and General Settings in the Simulation Profile and run the simulation. Press the V and I buttons to display the voltages and currents in the circuit. These values

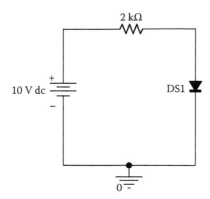

FIGURE 1.2.4
Figure for Simulation Example 1.2.2.

are $V_{DQ} = 772.8$ mV and $I_{DQ} = 4.614$ mA, in agreement with the results of Example 1.2.1, the default temperature being 27°C, or 300K.

(b) To simulate the piecewise linear approximation with the diode conducting in the forward direction, and the ideal diode switch acting as a short circuit, we need to only use a 0.77 V battery in series with a 2.2 Ω resistor. We will, however, add an ideal diode switch, as in Figure 1.2.2, and simulate it using: (1) a diode model having a very small η (Exercise 1.2.3), or (2) a voltage controlled switch. To do this we will open two additional pages in Capture. In the file hierarchical block, click on Design Resources, then all the way to SCHEMATIC1. Click the right mouse button on SCHEMATIC1 and select New Page from the menu that pops up. Enter a different page name, if desired, and click the OK button. Select the new page, copy/paste the schematic of Figure 1.2.4 to the new page, and add the 0.77 V battery and the 2.2 Ω resistor (Figure 1.2.5). In the PSpice Model Editor, change the diode name to DS2 and *n* to 1e−6. Run the simulation and press the V and I buttons. The indicated values are $V_{DQ} = 780.1$ mV and $I_{DQ} = 4.61$ mA, in agreement with Example 1.2.1. Note that the voltage across the diode is zero.

To model the ideal diode switch as a voltage-controlled switch, copy/paste the schematic of Figure 1.2.5 and replace the diode by the switch SBreak from the BREAK-OUT library (Figure 1.2.6). In the PSpice Model Editor change the name to SD1, Ron to 1u, and Von to 0.01m. With these values, the switch will close when the voltage across it, in the forward direction, is 0.01 mV and will open when the voltage across it drops to zero, the on resistance being 1 μΩ. Run the simulation and verify the values of voltage and current.

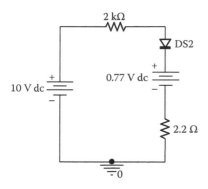

FIGURE 1.2.5
Figure for Simulation Example 1.2.2.

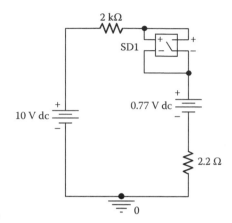

FIGURE 1.2.6
Figure for Simulation Example 1.2.2.

Exercise 1.2.2

PSpice includes models for some commercially available *pn* junction diodes, such as 1N4002, 1N4148, and 1N914. These are listed in the EVAL library prefixed by D. Look up I_S, η, and R_S listed in the PSpice models for these diodes.

    Answers: D1N4002: $I_S = 14.11$ nA, $\eta = 1.984$, $R_S = 0.03389$ Ω; D1N4148: $I_S = 2.682$ nA, $\eta = 1.836$, $R_S = 0.5664$ Ω; D1N914: $I_S = 168.1 \times 10^{-21}$ A, $\eta = 1$, $R_S = 0.1$ Ω.   ■

Exercise 1.2.3

An ideal diode switch can be approximated in PSpice by using small values for I_S and η. If $I_S = 1$ pA and $\eta = 10^{-6}$, for example, what is i_D for $v_D = 1$ μV?

    Answer: 5.1×10^4 A.   ■

1.3 Rectifier Circuits

Half-Wave Rectifier

Definition: *A rectifier circuit converts an alternating, or bidirectional, current or voltage to a unidirectional one.*

    A common application of a rectifier circuit is to provide a dc supply for electronic circuits from the ac mains supply. Being unidirectional devices, diodes are evidently suitable for this purpose. We will start with the simplest possible circuit, that of Figure 1.2.2, with the source being an ac voltage $V_m\sin\omega t$, instead of dc, and R denoted as a load R_L (Figure 1.3.1a).

    Assuming a linear approximation for the diode in the forward direction, the diode starts to conduct on the positive half-cycle, when $V_m\sin\omega t = V_{D0}$. While the diode conducts, $i_L = (V_m\sin\omega t - V_{D0})/(R_L + r_D)$. The diode stops conducting when $V_m\sin\omega t$ again equals V_{D0}, remains off until the next positive half-cycle, and so on (Figure 1.3.1b). It follows that during a cycle,

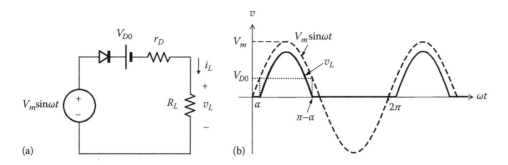

FIGURE 1.3.1
(a) Half-wave rectifier circuit. (b) Input and load voltage waveforms.

$$v_L = R_L i_L = \frac{R_L}{R_L + r_D}(V_m \sin\omega t - V_{D0}), \quad 2m\pi + \alpha \leq \omega t \leq (2m+1)\pi - \alpha,$$
$$= 0, \quad \text{otherwise} \tag{1.3.1}$$

where $\alpha = \sin^{-1}(V_{D0}/V_m)$ and $m = 0, 1, 2, \ldots$

The dc, or average, value of v_L is given by:

$$V_{DC} = \frac{1}{2\pi}\int_\alpha^{\pi-\alpha} v_L d(\omega t) = \frac{R_L[2V_m \cos\alpha + (2\alpha - \pi)V_{D0}]}{2\pi(R_L + r_D)} \tag{1.3.2}$$

When $V_{D0} \ll V_m$ and $r_D \ll R_L$, then V_{D0} and r_D can be neglected, and v_L will consist of complete, sinusoidal, positive half-cycles. Such a waveform has a dc value, $V_{DC} = V_m/\pi$, a fundamental of frequency $\omega/2\pi$, and even harmonics. If the diode connection is reversed, conduction occurs during negative half-cycles, which reverses the polarity of V_{DC}. Because conduction is limited to half-cycles of one polarity, the circuit is a **half-wave rectifier**.

The half-wave rectifier has two undesirable features: (1) It does not utilize supply cycles of both polarities, which reduces the dc output voltage and increases the ripple in the output. (2) The ac supply is usually derived from the secondary winding of a step-down transformer, in which case the dc current passes through the transformer secondary and produces a dc flux in the core. If, as is most likely at power frequencies, the core is ferromagnetic, the superposed ac flux may take the core into saturation (Sabah 2008, p. 216), which causes a relatively large magnetizing current and reduces the output voltage. The full-wave rectifier overcomes these disadvantages.

Full-Wave Rectifier

Concept: *In a full-wave rectifier, diodes are connected in such a way that current flows through the load in the same direction during both half-cycles of the supply.*

There are two basic full-wave rectifier circuits: (1) center-tapped transformer (CTT), and (2) bridge. In the former circuit, two identical secondary windings are used, the dot markings being as in Figure 1.3.2a, where $v_1 = v_2 = V_m \sin\omega t$. Assume that the polarities of

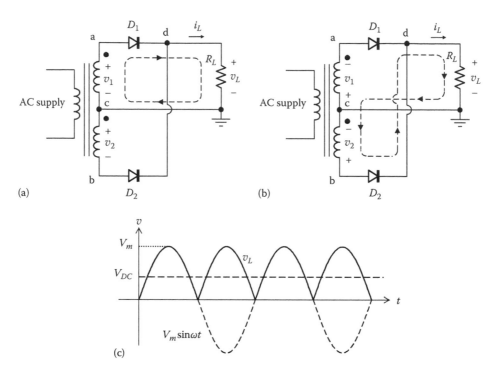

FIGURE 1.3.2
Full-wave CTT rectifier: (a) current flow during one half-cycle; (b) current flow during the other half-cycle;
(c) input and load voltage waveforms.

v_1 and v_2 during positive half-cycles are as indicated, so that node a is positive with respect
to node c, which is positive in turn with respect to node b. Diode D_1 is forward biased
and current flows through the transformer secondary, D_1, and R_L as shown. While D_1
conducts, nodes a and d are effectively connected together, assuming the diode is an
ideal switch. A reverse voltage $v_{D2} = v_1 + v_2$ appears across D_2, which therefore does not
conduct.

During negative half-cycles, the polarities of v_1 and v_2 are reversed (Figure 1.3.2b). D_2
now conducts, while D_1 does not. i_L flows in the same direction in R_L during the two half-
cycles. Both half-cycles of the supply contribute to the output, producing a full-wave
rectified waveform $|V_m\sin\omega t|$, neglecting V_{D0} and r_D (Figure 1.3.2c). The average value of
$|V_m\sin\omega t|$ is $2V_m/\pi$ and its fundamental frequency is twice that of the supply.

The current that flows in each half of the secondary is a half-wave rectified waveform of
amplitude $I_m = V_m/R_L$ and rms value $I_m/2$. Moreover, the currents in each half of the
secondary winding flow in opposite directions relative to the dot markings, so there is
no net dc polarization of the core over a complete cycle of the supply.

An important parameter of a rectifier circuit is the **peak inverse voltage** (PIV) to which
the diodes are subjected. This is because a practical diode has a maximum reverse voltage
rating, beyond which breakdown occurs and the reverse current rapidly increases. It was
mentioned earlier that when a diode does not conduct, the reverse voltage is $v_1 + v_2 =
2V_m|\sin\omega t| > 0$, neglecting V_{D0}. Hence, the PIV is $2V_m$. If V_{D0} is taken into account, the
peak load voltage is $V_m - V_{D0}$, whereas the PIV is this voltage plus V_m, which is $2V_m - V_{D0}$.

The full-wave bridge rectifier is illustrated in Figure 1.3.3. Assume that during positive
half-cycles the polarity of the secondary voltage $V_m\sin\omega t$ is as indicated in Figure 1.3.3a.

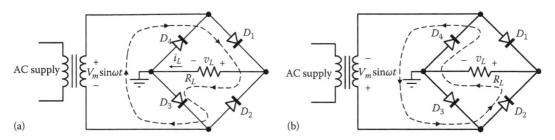

FIGURE 1.3.3
Full-wave bridge rectifier: (a) current flow during one half-cycle; (b) current flow during the other half-cycle.

Diodes D_1 and D_3 are forward biased and i_L flows as shown. While D_1 and D_3 conduct, v_L appears as a reverse voltage across D_2 and D_4, which will therefore be nonconducting. During negative half-cycles, the polarity of $V_m\sin\omega t$ is reversed (Figure 1.3.3b). Diodes D_2 and D_4 conduct, whereas D_1 and D_3 do not. If V_{D0} is taken into account, the peak load voltage is $V_m - 2V_{D0}$, because two diodes conduct in series, whereas the PIV is $v_L + V_{d0}$, which is $V_m - V_{D0}$. Neglecting V_{D0} and r_D, the transformer secondary current is sinusoidal of rms value $I_m/\sqrt{2}$.

If V_{D0} and r_D are taken into account, then for a CTT rectifier,

$$V_{DC} = \frac{1}{\pi} \int_{\alpha}^{\pi-\alpha} v_L d(\omega t) = \frac{R_L[2V_m\cos\alpha + (2\alpha - \pi)V_{D0}]}{\pi(R_L + r_D)} \tag{1.3.3}$$

where v_L is given by Equation 1.3.1. V_{DC} is now twice that given by Equation 1.3.2. Equation 1.3.3 applies to a bridge rectifier, with V_{D0} and r_D multiplied by 2 and $\alpha = \sin^{-1}(2V_{D0}/V_m)$. If $r_D \ll R_L$ and α is small, Equation 1.3.3 reduces to $(2/\pi)V_m - V_{D0}$ for a CTT circuit and to $(2/\pi)V_m - 2V_{D0}$ for a bridge circuit. The two types of full-wave rectifier having resistive loads are compared in Table 1.3.1. The CTT circuit is more suitable for low voltages because for a given V_m the load voltage is reduced by V_{D0} instead of $2V_{D0}$. For higher voltages, the bridge circuit is preferred because the PIV is $V_m - V_{D0}$ instead of $2V_m - V_{D0}$. The CTT circuit requires a transformer of higher VA rating than a bridge circuit. The core loss is the same in both cases, assuming the same number of turns in the primary winding, because the core loss depends on the flux, which in turn depends on the primary voltage and the number of turns of the primary winding. The copper loss in the transformer secondary depends on the rms value of the current of the secondary windings. The VA rating of the transformer is considered as the product of the rms voltage and the rms current. If the peak load voltage and current are V_m and I_m, respectively, then

TABLE 1.3.1

Comparison of CTT and Bridge Circuits Having Resistive Loads

Parameter	CTT	Bridge
Peak load voltage	$V_m - V_{D0}$	$V_m - 2V_{D0}$
V_{DC} (for small α)	$\cong (2/\pi)V_m - V_{D0}$	$\cong (2/\pi)V_m - 2V_{D0}$
Peak inverse voltage	$2V_m - V_{D0}$	$V_m - V_{D0}$
Relative VA rating of transformer secondary	1	0.7

the total secondary rating in the CTT case is $2 \times (V_m/\sqrt{2}) \times (I_m/2) = V_m I_m/\sqrt{2}$, whereas in the bridge case it is $(V_m/\sqrt{2}) \times (I_m/\sqrt{2}) = V_m I_m/2$, which is 0.7 times that of the CTT. In other words, a transformer of smaller rating, and hence size, is required for the bridge rectifier, because the transformer secondary in the CTT case is underutilized as current flows in each half-secondary for one half-cycle only.

Bridge rectifiers are available in a single package that includes four matched diodes and has two terminals for the ac input and two terminals for the rectified output.

Smoothing of Output

The full-wave rectified waveform of Figure 1.3.2c is far from being a smooth dc voltage, the peak-to-peak ripple in the output being V_m. There are two approaches to smoothing the output: (1) using some form of low-pass (LP) filtering, or (2) using a polyphase supply, usually having a large number of phases.

The simplest types of filters are the capacitor-input filter and the inductor-input filter. In a **capacitor-input filter**, a large capacitor is connected in parallel with the load so that the combined impedance of the capacitor and load is small. The rectifier thus applies a pulsating *current* to the capacitor–load combination. The dc component of this pulsating current passes through the resistor and establishes the dc load voltage across it, whereas the ac component develops only a small voltage across the small impedance of the capacitor and load.

In an **inductor-input filter**, a large inductor is connected in series with the load so that the combined impedance of the inductor and load is large. The rectifier thus applies a pulsating *voltage* to the inductor–load combination. The dc component of this pulsating voltage appears across the resistor and establishes the dc load current through it, whereas the ac component develops only a small current through the large impedance of the inductor and load.

Where load currents or load voltages are high, filtering inductors or capacitors become impractical. The filter is dispensed with altogether and a polyphase supply is used to provide a sufficiently smooth dc output without the need for additional filtering. In electroplating and aluminum industries, supplies of 48 phases or more are used to give a sufficiently smooth, high-current dc supply. Smoothing using polyphase supplies is discussed in Section ST1.2.

Capacitor-Input Filter

A full-wave rectifier circuit with a capacitor-input filter is shown in Figure 1.3.4a. To get better insight into circuit operation, we will analyze the circuit based on some simplifying assumptions before presenting an approximate analysis that is used in practical design.

Assume ideal diode switches, an ideal transformer, and an ideal supply of zero source impedance. C will charge to V_m, the peak of the supply voltage, during the first quarter cycle. If the output is open-circuited ($R_L \to \infty$), C cannot discharge through the diodes in the reverse direction. The capacitor voltage will therefore remain at V_m. If R_L is finite, the capacitor discharges through R_L when the diodes are nonconducting. The diodes stop conducting when the instantaneous transformer secondary voltage falls below the capacitor voltage.

In the steady state, C charges during part of the cycle and discharges through the load during the rest of the cycle (Figure 1.3.4b). When a diode conducts, v_L follows the supply voltage, so that:

$$v_L = V_m \cos \omega t, \quad -\theta_2 \leq \omega t \leq \theta_1 \tag{1.3.4}$$

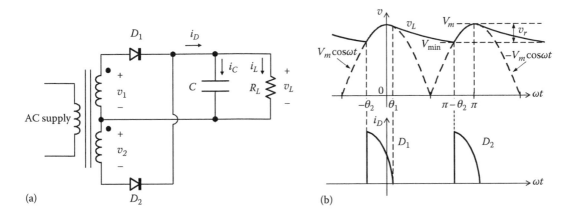

FIGURE 1.3.4
(a) Full-wave rectifier with capacitor-input filter. (b) Voltage and current waveforms.

During conduction, $i_D = i_L + i_C$, where $i_L = v_L/R_L$ and $i_C = C dv_L/dt$. This gives:

$$i_D = \frac{V_m}{R_L}\cos\omega t - \omega C V_m \sin\omega t, \quad -\theta_2 \le \omega t \le \theta_1 \tag{1.3.5}$$

When the diode stops conducting at $\omega t = \theta_1$, $i_D = 0$. Equation 1.3.5 gives:

$$\tan\theta_1 = \frac{1}{\omega C R_L} \tag{1.3.6}$$

During the interval $\theta_1 \le \omega t \le \pi - \theta_2$, both diodes are nonconducting. C discharges exponentially through R_L, with a time constant $R_L C$, from an initial value $V_m\cos\theta_1$ at $\omega t = \theta_1$:

$$v_L = V_m\cos\theta_1 e^{-(t-\theta_1/\omega)/CR_L} = V_m\cos\theta_1 e^{-(\omega t-\theta_1)/\omega CR_L}, \quad \theta_1 \le \omega t \le \pi - \theta_2 \tag{1.3.7}$$

Conduction resumes at $\omega t = \pi - \theta_2$ when the decaying exponential intersects the rising negative half-cycle $v_L = -V_m\cos\omega t$. Substituting $v_L = -V_m\cos\theta_2$ in Equation 1.3.7,

$$\cos\theta_2 = \cos\theta_1 e^{-[\pi-(\theta_1+\theta_2)]/\omega CR_L} \tag{1.3.8}$$

where $(\theta_1 + \theta_2)$ is the total conduction angle. Equation 1.3.8 is a transcendental equation that can be solved numerically for θ_2, once θ_1 has been determined from Equation 1.3.6. Knowing θ_1 and θ_2, the dc, or average value of v_L is:

$$V_L = \frac{V_m}{\pi}\left[\int_{-\theta_2}^{\theta_1}\cos\omega t d(\omega t) + \int_{\theta_1}^{\pi-\theta_2}\cos\theta_1 e^{-(\omega t-\theta_1)/\omega CR_L} d(\omega t)\right] \tag{1.3.9}$$

Concept: *Under idealized conditions, the behavior of a capacitor-input filter depends entirely on the product $\omega C R_L$, which determines θ_1 and θ_2.*

However, the analysis of a practical capacitor-input filter involves some rather awkward computations, so an approximate analysis is usually resorted to, as presented below.

Exercise 1.3.1

Derive Equation 1.3.6 by considering that v_L and dv_L/dt are continuous at $\omega t = \theta_1$. How do you explain this continuity? ∎

Exercise 1.3.2

Given a full-wave rectifier having a peak secondary voltage of 50 V, 50 Hz, $C = 5$ μF, and $R_L = 10$ kΩ. (a) Determine $\omega C R_L$, θ_1, θ_2, and V_{min}; over what fraction of the half-cycle does each diode conduct? (b) Repeat (a) if R_L is increased so that $\omega C R_L = 50$.
   Answers: (a) 15.71, $\theta_1 = 3.64°$, $\theta_2 = 31.8°$, $V_{min} = 42.5$ V, 19.7%; (b) $\theta_1 = 1.15°$, $\theta_2 = 19.0°$, $V_{min} = 47.3$ V, 11.2%. ∎

Approximate Analysis of Capacitor-Input Filter

The approximate analysis is based on the fact that in a well-designed power supply, the peak-to-peak ripple is small, typically not exceeding 5% of the dc voltage. The conduction interval is small, the diode current flows in narrow pulses (Figure 1.3.4b), and the time constant $R_L C$ is large compared to the period of the supply, $\omega C R_L$ being typically more than 50. This makes the conduction angle less than about 10% of half a period of the supply (Exercise 1.3.2). Consequently, a number of simplifying assumptions can be made, leading to the following steps:

1. Because conduction occurs during a short interval near the peak of the supply waveform, the peak input voltage is reduced from V_m to V_m' by a diode voltage drop V_D that depends on the rectifier circuit and the peak diode current. For a CTT circuit and a moderate load, V_D is typically 1–1.5 V. The supply voltage can then be considered $V_m' \cos\omega t$ instead of $V_m \cos\omega t$ (Figure 1.3.5). In other words, where ωt is small, in the vicinity of the peak of the sinusoid, $V_m \cos\omega t - V_D \cong V_m \cos\omega t - V_D \cos\omega t = (V_m - V_D)\cos\omega t = V_m' \cos\omega t$.

2. Because $\omega R_L C \gg 1$, $\theta_1 \cong 0$ (Equation 1.3.6) so that C begins to discharge at the peak value V_m'. Moreover, the capacitor voltage decreases almost linearly and at a relatively slow rate from V_m' to V_{min} (Figure 1.3.5). In other words, the initial part of the exponential discharge is approximated by a straight line when the time constant is large. The capacitor discharge current, $-Cdv_L/dt$, is then nearly constant and equals I_{DC}, the dc load current. Because the period of conduction $(\theta_1 + \theta_2)$ is small compared to π, the discharge period can be considered to be nearly $T/2$, or $1/2f$, where T is the supply period and f is the supply frequency.

3. During discharge, the charge lost by C is $I_{DC}/2f$ and the drop in capacitor voltage, from V_m' to V_{min}, is $I_{DC}/2fC$, which is the peak-to-peak ripple v_r (Figure 1.3.5). Hence,

$$v_r = \frac{I_{DC}}{2fC} \tag{1.3.10}$$

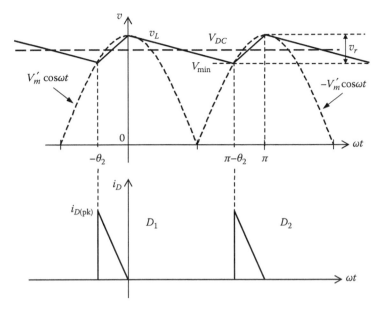

FIGURE 1.3.5
Voltage and current waveforms of full-wave rectifier with capacitor-input filter when the filter time constant is large.

4. When the conduction period is small, the variation of v_L during conduction can also be considered linear. v_L will be almost triangular in shape (Figure 1.3.5), so that,

$$V_{DC} = V'_m - \frac{v_r}{2} = V'_m - \frac{I_{DC}}{4fC} \tag{1.3.11}$$

The following should be noted about what has been discussed so far:

a. Given V'_m, C, and f, the three unknowns V_{DC}, I_{DC}, and v_r are determined by Equations 1.3.10 and 1.3.11 and the dc relation for the load, $V_{DC} = R_L I_{DC}$.

b. Strictly speaking, if I_{DC} is constant during capacitor discharge, then I_{DC} flowing through R_L implies a constant $V_L = V_{DC} = R_L I_{DC}$. In fact, v_L during capacitor discharge is part of an exponential, and $i_{L1} = -C\dfrac{dv_L}{dt}\Big|_{\omega t=\theta_1} > i_{L2} = -C\dfrac{dv_L}{dt}\Big|_{\omega t=\pi-\theta_2}$.

c. Normally, in a well-smoothed supply, v_r is small compared to V_{DC} or V'_m. Thus, if v_r is 5% of V'_m, then $v_r/2$ in Equation 1.3.11 is only 2.5% of V'_m. To a first approximation, $V_{DC} \cong V'_m$.

d. Even if all resistances in series with the load are neglected, then as long as Equation 1.3.11 applies, the capacitor-input filter has an effective dc source resistance of $1/4fC$. However, V'_m changes with the diode current and hence I_{DC}. Thus, when $I_{DC}=0$, the capacitor charges to the peak supply voltage V_m, not V'_m, and remains charged, neglecting the reverse diode current.

e. The average diode current over a supply period is $I_{DC}/2$. This is because the average capacitor current over a supply period is zero so that each diode supplies half the dc load current.

5. To calculate the conduction angle θ_2, we note that $V_{\min} = V'_m\cos\theta_2$. Hence,

$$v_r = V'_m - V_{\min} = V'_m(1 - \cos\theta_2) \quad \text{or} \quad 1 - \cos\theta_2 = \frac{v_r}{V'_m} \tag{1.3.12}$$

We next use the approximation $\cos\theta_2 \cong 1 - \theta_2^2/2$ for small θ_2, or $(1 - \cos\theta_2) \cong \theta_2^2/2$. Equation 1.3.12 then gives:

$$\theta_2 = \sqrt{\frac{2v_r}{V'_m}} \tag{1.3.13}$$

The conduction time T_{cd} is θ_2/ω so that:

$$T_{cd} = \frac{1}{2\pi f}\sqrt{\frac{2v_r}{V'_m}} \tag{1.3.14}$$

6. An important quantity in a capacitor-input filter is the peak diode current $i_{D(pk)}$. During conduction, $i_D = Cdv_L/dt + v_L/R_L$. When the ripple is small, $v_L/R_L \cong I_{DC}$. Moreover, $dv_L/dt = -V'_m\omega\sin\omega t$. It is seen from Figure 1.3.5 that maximum dv_L/dt occurs at the beginning of conduction, when $\omega t = -\theta_2 = -\omega T_{cd}$. As this is small, we can consider $-\sin\omega t = \sin\theta_2 \cong \omega T_{cd}$. It follows that $i_{D(pk)} = \omega^2 CT_{cd}V'_m + I_{DC}$. From Equation 1.3.14, $\omega T_{cd} = \theta_2 = \sqrt{2v_r/V'_m}$, and from Equation 1.3.10, $\omega C = \pi I_{DC}/v_r$. Substituting in the expression for i_D,

$$i_{D(pk)} = I_{DC}\left(1 + 2\pi\sqrt{\frac{V'_m}{2v_r}}\right) \cong 2\pi I_{DC}\sqrt{\frac{V'_m}{2v_r}} \tag{1.3.15}$$

The diode current waveform is almost a right triangle of height $i_{D(pk)}$ at the instant the diode begins to conduct. Problem P1.2.22 gives an alternative derivation of Equation 1.3.15.

Exercise 1.3.3

Let the *average diode current over the conduction interval* be $i_{D(av)cd}$. Then the charge delivered by the diode during conduction, $i_{D(av)cd}T_{cd}$, equals the charge supplied to the load, $I_{DC}T_{cd}$, plus that supplied to the capacitor. The latter is equal to that lost during nonconduction, which can be considered as $I_{DC}((T/2) - T_{cd})$. Using Equation 1.3.14, deduce that $i_{D(av)cd} = (I_{DC}/2)(T/T_{cd}) = I_{DC}\pi\sqrt{V'_m/2v_r} \cong i_{D(pk)}/2$. Since the average diode current over a whole period is $I_{DC}/2$, the average during conduction must be $(I_{DC}/2)(T/T_{cd})$ irrespective of the waveform of the diode current. ∎

Exercise 1.3.4

Derive Equations 1.3.8, 1.3.11, 1.3.14, 1.3.15, and $i_{D(av)cd}$ (Exercise 1.3.3) for a half-wave rectifier. How do these values compare with those for a full-wave rectifier and why?

Answers: $\cos\theta_2 = \cos\theta_1 e^{-[2\pi-(\theta_1+\theta_2)]/\omega CR_L}$; $V_{DC} = V'_m - I_{DC}/2fC$; $T_{cd} = 1/2\pi f\sqrt{2v_r/V'_m}$; $i_{D(pk)} = I_{DC}\left(1 + 2\pi\sqrt{2V'_m/v_r}\right)$; $i_{D(av)cd} = I_{DC}\pi\sqrt{2V'_m/v_r}$. ∎

The basic characteristics of a capacitor-input filter can be deduced from the preceding relations and summarized as follows:

Summary

1. *For a given V'_m and a given load, increasing C: (a) reduces the ripple, (b) increases V_{DC}, and (c) improves the regulation, that is, the variation of V_{DC} with I_{DC}, but at the cost of reducing the conduction interval and increasing the diode peak current*
2. *For a given V'_m and a given C, decreasing R_L increases the ripple and decreases V_{DC}.*

Simulation Example 1.3.1 Capacitor-Input Filter

SIMULATION

The circuit is shown in Figure 1.3.6. Instead of a CTT, two voltage sources $15\sin100\pi t$ V are used because the PSpice model of a CTT is available in the full version of OrCAD but not in the Demo version. 1N4002 diodes from the EVAL library are used rather than ideal diodes for a more realistic simulation.

In the first simulation, $C = 2000$ μF, which gives $\omega C R_L = 20\pi$. The resulting plots are shown in Figure 1.3.7. Only positive voltages are shown in these plots by selecting, in the Schematics page, Plot/Axis Settings/Y-Axis, then selecting User Defined under Data Range, and entering 0 as the lower limit and 20 as the upper limit. The scale for diode currents is expanded by selecting Trace/Add Traces and entering under Trace Expression 2*I(D1),2*I(D2).

The following values are obtained from the plots of Figure 1.3.7: $V_{DC} = 13.81$ V, $v_r = 0.6$ V, diode voltage drop at peak input $= 0.93$ V, so that $V'_m = 14.17$ V. A design based on $V'_m = 14$ V gives, by solving Equations 1.3.10 and 1.3.11, with $I_{DC} = V_{DC}/100$: $V_{DC} = 13.7$ V and $v_r = 0.69$ V, in good agreement with the simulation. Equation 1.3.15 gives for the calculated peak current: $i_{D(pk)} \cong 0.137 \times 2\pi\sqrt{14/(2 \times 0.69)} = 2.7$ A, compared to a simulated value of 1.93 A. The larger value of the calculated peak current is due to neglecting the diode resistance, and represents erring on the safe side. Equation 1.3.14 gives for the calculated conduction time $T_{cd} = (1/2\pi \times 50)\sqrt{2 \times 0.69/14} = 1$ ms, compared to a simulated value of about 1.6 ms.

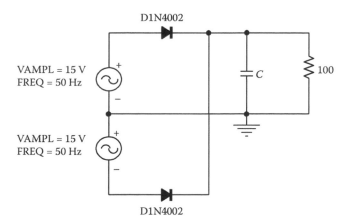

VAMPL = 15 V
FREQ = 50 Hz

VAMPL = 15 V
FREQ = 50 Hz

D1N4002

100

C

D1N4002

FIGURE 1.3.6
Figure for Simulation Example 1.3.1.

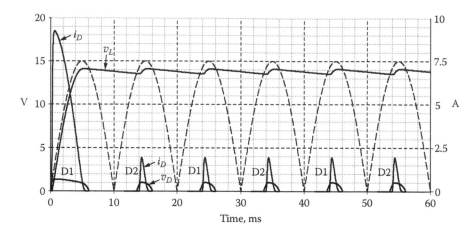

FIGURE 1.3.7
Figure for Simulation Example 1.3.1.

The peak diode current upon switching is 9.2 A. The calculated value is $Cd(15\sin100\pi t)/dt = 9.4$ A at $t = 0$. This is not a worst-case condition. The worst-case condition occurs when the circuit is switched at the peak of the supply voltage (Problem P1.2.24).

For the second simulation, $C = 100$ μF, which makes $\omega CR_L = \pi$ and invalidates the approximate analysis. The resulting plots are shown in Figure 1.3.8. The simulation gives $\theta_1 = 23.4°$ and $|\theta_2| = 57.6°$, compared to calculated values of 18.2° and 56.1°, respectively.

Design Example 1.3.2 Regulated Power Supply

It is required to design a regulated power supply that delivers 15 V dc at 0.5 A. The supply is $220 \pm 5\%$ V, 50 Hz. Assume a CTT rectifier.

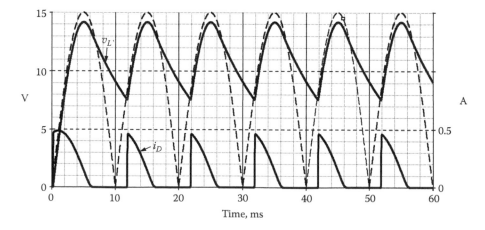

FIGURE 1.3.8
Figure for Simulation Example 1.3.1.

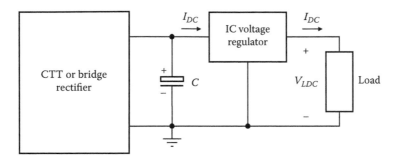

FIGURE 1.3.9
Figure for Design Example 1.3.2.

SOLUTION

In practice, the filter capacitor is almost invariably of the electrolytic type. These capacitors have a large capacitance in a small volume, but are polarized, that is, the voltage across them should be of one polarity only, as is the output voltage of a rectifier. The filter capacitor is commonly followed, in low-power dc supplies, by an IC regulator, which provides an output voltage that is virtually ripple-free and constant, despite variations in load current and ac supply voltage within specified ranges. Figure 1.3.9 illustrates the basic circuit. Given a suitable regulator, what concerns us here is the design of a rectifier circuit.

An important characteristic of the regulator is the minimum input voltage $V_{I(min)}$ to the regulator that will give V_{LDC} at the regulator output. For $V_{LDC} = 15$ V, $V_{I(min)}$ is typically 17.6 V. When the input voltage to the regulator is within specifications, the input and output currents of the regulator are practically the same. Low dropout voltage regulators are available that need a voltage of less than 0.5 V across the regulator. However, the input and output currents of the regulator are not nearly equal, and a capacitor is generally required at the regulator output for frequency compensation (Section 9.4, Chapter 9).

When the regulator is supplied from a capacitor-input filter, $V_{I(min)} = V_{DC(min)} - v_r/2$, where $V_{DC(min)}$ is the minimum dc output of the filter. To determine $V_{DC(min)}$, we have to assume an appropriate value for v_r, which can be checked later. Let v_r be 5% of $V_{DC(min)}$. It follows that $V_{DC(min)}(1 - 0.025) = 17.6$, which gives $V_{DC(min)} \cong 18$ V. Hence, the rectifier/capacitor-input filter is required to give 18 V at a dc current of 0.5 A, with 0.9 V peak-to-peak ripple when the ac supply is at its minimum, that is, $220\% - 5\% = 209$ V rms.

The next step is to determine the filter capacitor. From Equation 1.3.10, $C = I_{DC}/2fv_r = 0.5/(100 \times 0.9) \equiv 5555$ μF. We have to choose a standard value of the electrolytic capacitor that is larger than 5555 μF. A common value is $6800 \pm 20\%$ μF, which gives a minimum value of C of 5400 μF. Although this is about 2.8% less than 5555 μF, it is usually acceptable in view of the approximations made. With this value of C, $v_r = I_{DC}/2fC = 0.5/(100 \times 5400 \times 10^{-6}) = 0.93$ V. We should at this stage check the validity of the approximate analysis. Thus, $\omega CR_L = 2\pi \times 50 \times 5400 \times 10^{-6} \times 18/0.5 = 61$. As this is greater than 50, the approximate analysis is justified. $V'_m = 18 + 0.93/2 \cong 18.5$ V. For the CTT rectifier, the peak half-secondary open-circuit voltage $V_m = V'_m + 1 = 19.5$ V, where the 1 V allows for the voltage drop in the diode and in the transformer at full load. This gives a half-secondary open-circuit rms voltage of $19.5/\sqrt{2} = 13.8$ V. If the primary voltage is 209 V rms, then the primary-to-half-secondary turns ratio is $209/13.8 = 15.14$. We should take this ratio as 15, that is, the largest integer that is smaller than 15.14 V in order to have a larger secondary open-circuit voltage. It is seen that with this turns ratio, $V_m = 209\sqrt{2}/15 = 19.7$ V and $V'_m = 18.7$ V. Hence, $V_{DC(min)} = 18.7 - 0.93/2 = 18.23$ V, which is slightly larger than the minimum voltage required, and is on the safe side.

Let us now check the design at the other extreme, when the ac voltage is $220 + 5\%$ $V = 231$ V. This gives $V_m = 231\sqrt{2}/15 = 21.8$ V and $V'_m = 21.8 - 1 = 20.8$ V. With $C = 6800 + 20\% = 8160$ μF, $v_r = 0.5/(100 \times 8160 \times 10^{-6}) = 0.61$ V. The maximum dc voltage is $V_{DC(max)} = 20.8 - 0.31 \cong 20.5$ V. We are now ready to specify the components of the filter.

Electrolytic capacitor: C is nominally 6800 μF. The voltage rating of the capacitor should be the lowest standard value that exceeds $V_m = 21.8$ V at $220 + 5\%$ V, to allow for the possibility of disconnection of the load. From manufacturer's specifications, this value is 25 V. The specifications also quote a maximum rms ripple current through the capacitor. Since v_r is nearly triangular, its rms value is $v_r/2\sqrt{3} = 0.93/2\sqrt{3} = 0.27$ V (Sabah 2008, p. 350). At this ripple voltage and $C = 5400$ μF, the ripple current is $0.27\omega C = 0.27 \times 100\pi \times 5400 \times 10^{-6} = 0.46$ A, which is less than the maximum allowed, because the ripple is small. A small radio frequency (rf) polystyrene or ceramic capacitor of say 10 nF may be connected in parallel with the electrolytic capacitor to protect against voltage surges in the power supply (Section 1.5).

Transformer: The primary-to-half-secondary turns ratio is 15. To determine the VA of the transformer, we could calculate the rms value of the diode current, which is a rather awkward calculation. A rule of thumb is usually used instead, according to which the transformer rating is 1.7 times the dc power for a CTT rectifier and 1.4 times for a bridge rectifier. The maximum dc power is $20.5 \times 0.5 = 10.25$ W. Hence, the transformer rating should be at least $1.7 \times 10.25 = 17.5$ VA. Small transformers are usually specified in terms of the secondary current. The secondary voltage is $220 \times 2/15 = 29.3$ V rms, so that the secondary current is $(17.5 \text{ VA})/(29.3 \text{ V}) \cong 0.6$ A rms.

Rectifier diode: The diode is normally chosen on the basis of the average diode current for a capacitive or resistive/inductive load, and the peak repetitive reverse voltage. In the present example, the average diode current is 0.25 A per diode. The largest peak repetitive inverse voltage is $21.8 + 20.8 = 41.8$ V. A small, general-purpose rectifier diode would be appropriate in this case. The 1N4001 general-purpose rectifier diode, for example, has an average rectified forward current of 1 A for a resistive/inductive load, and a peak repetitive reverse voltage of 50 V. A derating of 20% is applied to the average rectified forward current of 1 A if the load is capacitive in order to allow for current peaking.

An important consideration in capacitor-input filters is the current surge at power-on. Initially, the capacitor is discharged and behaves essentially as a short circuit, so the switching transient is limited by the series resistance in the circuit, ignoring the small leakage inductance of the transformer. Under worst conditions, the supply is applied at its peak value, so the diode current at that instant is $i_{D(surge)} = \dfrac{V_m}{R_T + r_{D(bulk)}}$, where $r_{D(bulk)}$ is the bulk resistance of the diode and R_T is the resistance in the transformer secondary circuit, neglecting the leakage reactance of the transformer and the small impedance reflected from the primary side. The surge capability of a diode is quoted in manufacturer's data sheets as I_{FSM}, defined as the **nonrepetitive peak surge current** for half a cycle of the supply at rated load conditions. The 1N4001 has an I_{FSM} of 30 A. With maximum $V_m = 21.8$ V, which is a relatively low voltage, $i_{D(surge)}$ would be small. Even if the total resistance $R_T + r_{D(bulk)}$ is 1 Ω, $i_{D(surge)} = 21.8$ A, which is less than I_{FSM} of the 1N4001. In higher voltage circuits, it may be necessary to add a resistance between the cathodes of the diodes and the capacitor in order to limit the surge current.

In contrast to capacitor-input filters, the current in inductor-input filters flows continuously throughout the cycle, and there is no current surge at power-on. Inductor-input filters are discussed in Section ST1.1. As smoothing in both capacitor-input and inductor-input filters depends on reactance, then for a given filter and load, increasing the frequency of the supply makes the output smoother (Application Window 1.3.1).

Exercise 1.3.5

Consider the rectifier circuit of Example 1.3.2. Determine: (a) the effective source resistance over the range of I_{DC} for which Equation 1.3.11 is valid; (b) the largest value of $I_{D(pk)}$.
   Answers: (a) 0.5 Ω; (b) 13.5 A.   ■

Exercise 1.3.6

Repeat Example 1.3.2 assuming a bridge rectifier.   ■

Application Window 1.3.1 X-Ray Tube

An X-ray tube is diagrammatically illustrated in Figure 1.3.10. A filament heats the cathode to a temperature of about 2000°C. Electrons acquire sufficient energy at this temperature to overcome the electric potential barrier at the surface, which results in **thermionic emission**. Once electrons escape from the cathode, they are attracted to the anode by the high dc voltage applied between anode and cathode. The electrons hit the anode with sufficient energy to generate X-rays. The anode is usually made of tungsten or is a tungsten insert in copper. The anode current could be as high as a few hundred mA, and the anode-to-cathode voltage could be well over 100 kV. For best results, this dc voltage should be as smooth as possible so as to produce an X-ray beam of uniformly high energy. At these high voltages, the use of a large smoothing capacitor at the supply frequency is impractical. For high-quality X-ray production, a relatively small filtering capacitor is used with a high frequency supply of a few tens of kHz (Equation 1.3.10). In less expensive equipment, a sufficiently smooth output can be derived from a three-phase supply, as discussed in Section ST1.2.

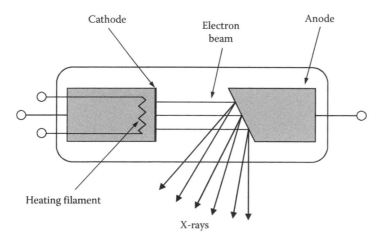

FIGURE 1.3.10
Figure for Application Window 1.3.1.

1.4 Zener Voltage Regulator

Voltage–Current Characteristic

In addition to being smooth, it is often required in practice that a dc supply be regulated as well, which means that the load voltage should remain constant despite variations of load current and supply voltage between specified limits. Available IC voltage regulators can provide fixed or adjustable, positive or negative, output voltages (Example 1.3.2). The operation of IC voltage regulators is explained in Example SE11.9, Chapter 11.

A basic voltage-regulating device is the zener diode, which can also be used to provide a reference voltage both in discrete-element circuits and in ICs. A zener diode is a *pn* junction diode that normally operates in the breakdown region, at a closely controlled breakdown voltage (Figure 1.4.1). In the forward direction, it behaves like a normal *pn* junction diode.

It is more convenient when working with zener diodes to draw the *i–v* characteristic in the first quadrant, with the reverse voltage v_Z and the reverse current i_Z considered positive quantities, as shown in Figure 1.4.2 together with the symbol for the zener diode. The *i–v* characteristic is almost a straight line beyond the **knee current** I_{ZK}, specified in the diode data sheet together with the voltage V_{ZK} at the current I_{ZK}. A voltage V_{ZT} across the diode is quoted at a specified test current I_{ZT}. V_{ZT} is considered to be the nominal regulation voltage of the zener diode and could be in the range of a few volts to a few hundred volts. The sign and magnitude of the temperature coefficient of V_{ZT} depend on the value of V_Z. The temperature coefficient of V_{ZT} is negative for V_{ZT} less than about 5 V, and is positive for V_{ZT} greater than about 7 V (Section 3.2, Chapter 3).

The maximum current of the diode $I_{Z(\max)}$ depends on its power rating. $I_{Z(\max)}$ for a 0.5 W, 12 V zener diode, for example, may be about 30 mA, taking into account the increase of V_Z with current and with the higher temperature associated with the larger current. Over the operating region, between I_{ZK} and $I_{Z(\max)}$, the variation of i_Z with v_Z is almost linear. The corresponding resistance r_Z can be considered constant for a given diode and is normally in the range of a few ohms to a few tens of ohms. r_Z is also the **incremental** or **dynamic** resistance of the diode.

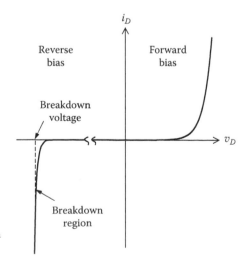

FIGURE 1.4.1
i–v characteristic of a *pn* juction diode showing breakdown under reverse bias.

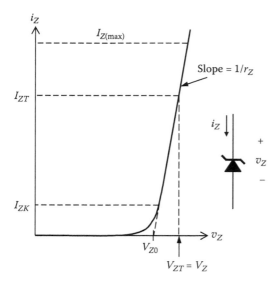

FIGURE 1.4.2
i–v characteristic of zener diode. Inset shows zener diode symbol and assigned positive directions.

As in the case of a forward-biased diode, the *i–v* characteristic in the operating region can be approximated by a straight line having a voltage intercept V_{Z0} and slope $1/r_Z$ (Figure 1.4.2), which leads to an equivalent circuit consisting of a battery of voltage V_{Z0} in series with a resistance r_Z. In particular cases, r_Z can be neglected.

Analysis of Zener Regulator

A typical circuit of a zener diode regulator is shown in Figure 1.4.3. v_{SRC} could be the filtered output of a rectifier circuit and can be regarded as the sum of a dc voltage V_{DC} and a ripple voltage v_r. When the zener diode is replaced by V_{Z0} in series with r_Z, the circuit becomes linear, and superposition applies. Thus, we can consider the complete response to be the sum of: (1) the response due to the dc sources V_{DC} and V_{Z0} acting alone, with $v_r = 0$, and (2) the response due to v_r acting alone, with $V_{DC} = V_{Z0} = 0$. In the latter case, the circuit reduces to a source v_r connected to a resistive voltage divider consisting of R_s in series with

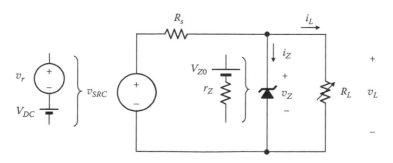

FIGURE 1.4.3
Typical zener diode regulator.

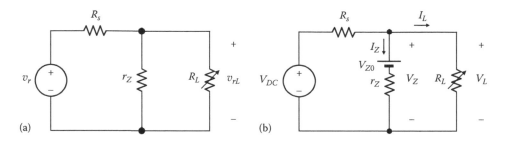

FIGURE 1.4.4
(a) ac source of Figure 1.4.3 applied alone; (b) dc sources applied alone.

the parallel combination of r_Z and R_L (Figure 1.4.4a). It follows that v_{rL}, the ripple in the output is given by:

$$v_{rL} = \frac{r_Z \| R_L}{R_s + r_Z \| R_L} v_r \cong \frac{r_Z}{R_s} v_r \tag{1.4.1}$$

because r_Z is normally small compared to either R_L or R_s. Typically, the ripple is reduced by a factor of at least 10.

Consider next the dc sources acting alone, that is, with $v_r = 0$ and with the diode operating in the linear region (Figure 1.4.4b). The diode equation is:

$$V_Z = V_{Z0} + r_Z I_Z \tag{1.4.2}$$

From KVL,

$$V_{DC} = V_Z + R_s(I_L + I_Z) \tag{1.4.3}$$

Moreover,

$$V_Z = V_L = R_L I_L \tag{1.4.4}$$

Equations 1.4.2 through 1.4.4 are the governing equations for the circuit. In a typical design of a voltage-regulated supply, V_Z, maximum I_L, and maximum v_{rL} are specified. V_{DC}, v_r, and R_s are to be determined and the zener diode selected. In most cases, the rectifier and regulator are considered together, so as to optimize the overall design.

For purpose of analysis, we can consider the source voltage v_{SRC}, V_{Z0}, r_Z, R_s, and R_L or i_L to be known, the unknown quantities being v_L and i_Z. If the load branch in Figure 1.4.3 is replaced by a current source i_L, in accordance with the substitution theorem (Sabah 2008, p. 121), and superposition is applied to the three sources, it follows that:

$$v_L = \frac{r_Z}{R_s + r_Z} v_{SRC} + \frac{R_s}{R_s + r_Z} V_{Z0} - (r_Z \| R_s) i_L \tag{1.4.5}$$

As $r_Z \to 0$, $v_L \to V_{Z0}$ independently of i_L and v_{SRC}, as expected for a perfect regulator.

Exercise 1.4.1
Derive Equation 1.4.5 without replacing the load branch by a current source i_L. ∎

Exercise 1.4.2
Considering v_{SRC} and V_{Z0} to be the only sources in the circuit, as in Figure 1.4.4b, apply superposition to show that: ∎

$$v_L = \frac{r_Z||R_L}{R_s + r_Z||R_L}v_{SRC} + \frac{R_s||R_L}{r_z + R_s||R_L}V_{Z0} \qquad (1.4.6)$$

Exercise 1.4.3
Since the circuit is linear, changes in v_{SRC} and v_L are the same as small signal v_r and v_{rL}, respectively. By considering changes in v_{SRC} and v_L only in Equation 1.4.6, show that the relation between Δv_{SRC} and Δv_L is given by Equation 1.4.1. ∎

If r_Z is small, we can consider $V_Z \cong V_{Z0}$, and Equation 1.4.3 can be written as:

$$I_L + I_Z \cong \frac{V_{DC} - V_{Z0}}{R_s} \qquad (1.4.7)$$

The sum of I_L and I_Z is nearly constant. When I_L is maximum, I_Z is minimum, and conversely. Hence, the following concept applies:

Concept: *In a zener regulator, the range of I_L defines two important design limits: at maximum I_L, I_Z should not fall below some $I_{Z(min)}$, whereas at minimum I_L, I_Z should not increase to the point where the power rating of the diode is exceeded.*

In practice, $I_{z(min)} > I_{ZK}$ so as to have some safety margin that ensures operation well into the breakdown region. Although in normal operation I_L may not fall below a certain limit, the load may be accidentally disconnected, so I_L becomes zero. The power rating of the diode should not be exceeded under these conditions so as not to damage the diode because of overheating.

The interrelations between the various quantities involved can be illustrated graphically by means of a simple modification of the load-line construction. On the graph of the diode $i–v$ characteristic, a line of slope $-1/R_s$ is drawn passing through the point $P(V_{DC}, -I_L)$ as in Figure 1.4.5. The equation of this line is $I_L + I_Z = (V_{DC} - V_Z)/R_s$, which is the same as Equation 1.4.3. The intersection of this line with the diode characteristic gives the bias point Q. Segment QT is $(I_L + I_Z)$ and segment PT is $(V_{DC} - V_Z)$. If I_L is zero, P is on the voltage axis, and $I_{Z(max)}$ flows through the diode. As I_L increases, the load line moves vertically downward and I_Z decreases. As I_Z varies between $I_{Z(min)}$ and $I_{Z(max)}$, the maximum excursion in V_Z, and hence in the load voltage, is ΔV_Z. The reduction in ripple at the output can also be derived graphically (Section ST1.3).

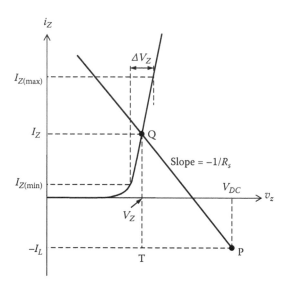

FIGURE 1.4.5
Graphical construction for zener diode regulator with
load current.

In summary, the main performance parameters of the zener regulator are:

Summary

1. *The ripple in the output is reduced by a factor of nearly r_Z/R_s.*
2. *$R_{s(max)}$ is determined by $I_{Z(min)}$, with minimum V_{DC} and maximum I_L.*
3. *For a given R_s, the power rating of the zener diode is determined by $V_{DC(max)}$ and $I_{L(min)}$, which is usually zero.*

Load Regulation and Line Regulation
The change in output voltage as the current changes from no load to full load, at constant supply voltage, is the **load regulation**. Taking differentials in Equation 1.4.5, with v_{src} constant, the load regulation is:

$$\Delta v_L = -(r_Z \| R_s)\Delta i_L \tag{1.4.8}$$

where
  $(r_Z \| R_s)$ is the effective small-signal source resistance (Thevenin resistance) seen by the load
  Δi_L is the change in current

The change in v_L with the supply voltage v_{src}, at constant load current, is the **line regulation**. Taking differentials in Equation 1.4.5, with i_L constant, the line regulation is:

$$\Delta v_L = \frac{r_Z}{R_s + r_Z}\Delta v_{src} \tag{1.4.9}$$

Exercise 1.4.4

Eliminate V_Z between Equations 1.4.2 and 1.4.3 to obtain an expression for I_Z:

$$I_Z = \frac{V_{DC} - V_{Z0} - R_s I_L}{R_s + r_Z} \tag{1.4.10}$$

Assume that all quantities on the RHS of Equation 1.4.10 can vary between maximum and minimum limits. Express $I_{Z(max)}$ and $I_{Z(min)}$ in terms of these limiting values and interpret them using the graphical construction of Figure 1.4.5.

Answers: $I_{Z(max)} = \dfrac{V_{DC(max)} - V_{Z0(min)} - R_{s(min)} I_{L(min)}}{r_{Z(min)} + R_{s(min)}}$, $I_{Z(min)} = \dfrac{V_{DC(min)} - V_{Z0(max)} - R_{s(max)} I_{L(max)}}{r_{Z(max)} + R_{s(max)}}$ ∎

Exercise 1.4.5

Using the results of Exercise 1.4.4, deduce how: (a) $R_{s(max)}$ varies with $V_{DC(min)}$, $V_{Z0(max)}$, $r_{Z(max)}$, $I_{Z(min)}$, and $I_{L(max)}$; (b) $R_{s(min)}$ varies with $V_{DC(max)}$, $V_{Z0(min)}$, $r_{Z(min)}$, $I_{Z(max)}$, and $I_{L(min)}$. Verify these deductions using the graphical construction of Figure 1.4.5. ∎

Design Example 1.4.1 Zener Voltage Regulator

Given a full-wave rectified dc supply of $18 \pm 10\%$ V. A zener regulator is required to supply a load of 0–20 mA at a nominally constant voltage of 12 V.

SOLUTION

The first step is to estimate the maximum power dissipation of the diode in order to select a suitable diode. Let us assume $I_{Z(min)} = 1$ mA, well above the knee of the $i\text{--}v$ characteristic. According to the previous discussion, I_Z is minimum when $V_{DC} = 18 \times 0.9 = 16.2$ V and $I_L = 20$ mA. Substituting in Equation 1.4.3, assuming a nominal zener voltage of 12 V: $16.2 = 12 + R_s(20 + 1)$, which gives $R_s = (4.2\ \text{V})/(21\ \text{mA}) = 200\ \Omega$. Note that this is $R_{s(max)}$ (Exercise 1.4.5), and that it is advantageous to use $R_{s(max)}$, because this will limit $I_{z(max)}$ when V_{DC} is maximum and I_L is minimum, which results in the smallest power rating of the zener diode.

I_Z is maximum when $V_{DC} = 18 \times 1.1 = 19.8$ V and $I_L = 0$. Equation 1.4.3 gives: $19.8 = 12 + 0.2 I_{Z(max)}$, or $I_{Z(max)} = 39$ mA. The minimum power rating of the diode is $12 \times 0.039 \cong 0.47$ W.

From manufacturer's data sheets, a 0.5 W zener diode is chosen that has $V_z = 12$ V at 10.5 mA. r_Z for this diode is 11.5 Ω and $I_{ZK} = 0.25$ mA. The equation of the $i\text{--}v$ characteristic over the operating range is: $\dfrac{(V_Z - 12)\ \text{V}}{(I_Z - 10.5)\ \text{mA}} = 0.0115$, or,

$$V_Z = 0.0115 I_Z + 11.88\ \text{V} \tag{1.4.11}$$

At $I_Z = 1$ mA, $V_Z = 11.89$ V. We can now determine the actual $R_{s(max)}$. Substituting $I_Z = 1$ mA and $V_Z = 11.89$ V in Equation 1.4.3: $16.2 = 11.89 + R_{s(max)} \times 21$. This gives $R_{s(max)} = 0.205\ \text{k}\Omega \equiv 205\ \Omega$. A smaller standard value is chosen, say 200 Ω with a tolerance of $\pm 2\%$.

We can recheck the figures, taking resistance tolerances into account. At $I_L = 20$ mA, I_Z is minimum at the higher value of resistance, that is, $200 \times 1.02 = 204\ \Omega$. Equation 1.4.3 gives for this value of resistance and the lower supply voltage: $16.2 = V_Z + 0.204(20 + I_Z)$. When this is solved with Equation 1.4.9, the result is: $V_{Z(min)} = 11.89$ V, $I_{Z(min)} = 1.11$ mA, which is well above the knee of the $i\text{--}v$ characteristic. At $I_L = 0$ mA, I_Z is maximum at the lower value of resistance, that is, $200 \times 0.98 = 196\ \Omega$. Equation 1.4.3 gives for this value of resistance and the higher supply voltage: $19.8 = V_Z + 0.196 I_Z$. When this is solved with Equation 1.4.9, the result is: $V_{Z(\max)} = 12.32$ V, $I_{Z(\max)} = 38.2$ mA. The maximum power dissipation in the diode is $V_{Z(\max)} \times I_{Z(\max)} = 0.47$ W, neglecting the temperature variation of V_Z.

The maximum variation of the voltage across the load, due to variations in both the supply voltage and the load current, has been reduced to $V_{Z(\max)} - V_{Z(\min)} = 0.43$ V, or 3.6% of the nominal value of 12 V. If the peak-to-peak ripple in the supply is 1 V, the peak-to-peak ripple in the current is approximately $1/(11.5 + 196) \cong 4.8$ mA, and the peak-to-peak ripple across the load is $4.8 \times 11.5 \cong 55$ mV, or $(55 \times 10^{-3}/12) \times 100 = 0.46\%$ of the nominal load voltage of 12 V. The waveform of the ripple is very nearly triangular. If the peak-to-peak ripple at the output is v_{rL}, the rms value of the ac component is $v_{rL}/2\sqrt{3}$ (Sabah 2008, p. 350), and the ac ripple power dissipated in the diode is $\dfrac{v_{rL(\mathrm{rms})}^2}{r_D} = \dfrac{(55 \text{ mV})^2}{12 \times 11.5 \ \Omega} = 0.02$ mW, which is quite negligible. The rms value of the current in R_s when $I_L = 0$ is: $\sqrt{(38.2)^2 + (4.8/2\sqrt{3})^2} \cong 38.2$ mA, and the maximum power dissipated in this resistor is $196 \times (38.2)^2 \cong 0.3$ W.

Simulation Example 1.4.2 Zener Diode Characteristics Using PSpice

It is required to obtain, using PSpice, the $i{-}v$ forward and reverse characteristics of a zener diode at $-50°$C, $27°$C, and $100°$C.

Simulation

The zener diode used is the 1N750, 4.7 V diode, available as D1N750 from the EVAL library. The schematic is entered as in Figure 1.2.4, with the zener diode oriented in the same direction. Click on the diode symbol with the left mouse button, and choose Edit/PSpice model from the Orcad Capture menu to view the diode parameters.

In the Simulation Settings, choose dc Sweep, check the Primary Sweep box, select Voltage source, and enter V1 for name. Under Sweep type, select Linear and enter -50 V for Start value, 15 V for End value, and 0.01 for Increment. Check the Secondary Sweep box, select Temperature, select Value list under Sweep type, and enter -50, 27, 100.

Run the simulation and use a current marker for the diode current or select I(D1) under Add Trace. The diode current is displayed vs. the source voltage. To display the diode current vs. diode voltage, select Plot/Axis Settings, press the Axis Variable button under X Axis and select the diode voltage V1(D1). A plot appears displaying i_D vs. v_D from -5 to 1 V for the three temperatures. To display the forward characteristics, change the Data Range to user defined 0.4–0.9 V for the x-axis and 0–10 mA for the y-axis. The plot appears as in Figure 1.4.6.

To display the reverse characteristic, change the Data Range to user defined -4.0 to -4.8 V for the x-axis and 0 to -50 mA for the y-axis. The plot appears as in Figure 1.4.7. To determine which characteristic is for which temperature, run a separate simulation for one of the extreme temperatures, $-50°$C or $100°$C. At constant current, the voltage magnitude has a negative temperature coefficient in both the forward and reverse directions.

It is evident that if the diode voltage is to remain constant, as when the zener diode is used as a voltage reference, the diode current must be kept constant. This can be achieved by supplying the diode at nearly constant current, or from a nearly constant voltage through a resistor, and isolating the zener diode from load variations (Section 7.4, Chapter 7).

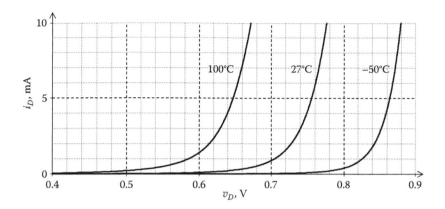

FIGURE 1.4.6
Figure for Simulation Example 1.4.2.

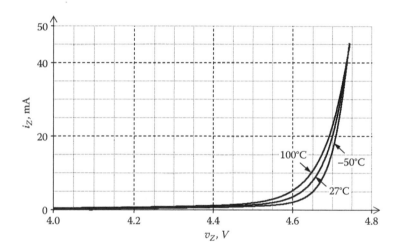

FIGURE 1.4.7
Figure for Simulation Example 1.4.2.

TABLE 1.4.1

Diodes for Voltage Regulation

Range of Regulated Voltage (V)	Type of Regulating Diode
2.5–200	Zener
1.5–2	LED
0.6–0.8	Si *pn*
0.2–0.4	Si Schottky

For values of regulated voltage smaller than those for which zener diodes are available, *forward-biased* diodes of various types, discussed in Chapter 3, can be used, as indicated in Table 1.4.1. Series combinations of these diodes can be used for intermediate values of regulated voltage or to achieve desired temperature coefficients.

1.5 Diode Voltage Limiters

Definition: *A voltage-limiting circuit prevents voltage excursions beyond a certain level, which could be positive or negative.*

Diode-resistor circuits can be used for this purpose, a basic circuit being shown in Figure 1.5.1a, where V_{LIM}^{+} is the nominal desired limiting voltage. The diode can be represented under forward bias by a voltage V_{D0} in series with a resistance r_D and can be considered an open circuit under reverse bias.

In analyzing diode-limiting circuits, it is usually more convenient to consider the range of output voltage over which the diode does not conduct, and then derive the corresponding range of input voltage. In Figure 1.5.1a, as long as the open-circuit voltage $v_O < V_{\mathrm{LIM}}^{+} + V_{D0}$, the diode does not conduct, no current flows in R_s, and $v_{SRC} = v_O$. If $v_O > V_{\mathrm{LIM}}^{+} + V_{D0}$, the diode is forward biased. Assuming that the load current is zero, the current i that flows in the circuit is:

$$i = \frac{v_{SRC} - V_{D0} - V_{\mathrm{LIM}}^{+}}{R_s + r_D}, \quad v_{SRC} > V_{\mathrm{LIM}}^{+} + V_{D0} \tag{1.5.1}$$

and,

$$v_O = V_{\mathrm{LIM}}^{+} + V_{D0} + r_D i \tag{1.5.2}$$

Substituting for i from Equation 1.5.1,

$$v_O = \frac{R_s}{R_s + r_D}\left(V_{\mathrm{LIM}}^{+} + V_{D0}\right) + \frac{r_D}{R_s + r_D} v_{SRC}, \quad v_{SRC} > V_{\mathrm{LIM}}^{+} + V_{D0} \tag{1.5.3}$$

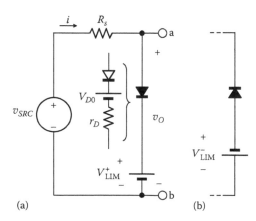

(a)   (b)

FIGURE 1.5.1
Diode voltage limiter: (a) limiting at positive voltage levels; (b) limiting at negative voltage levels.

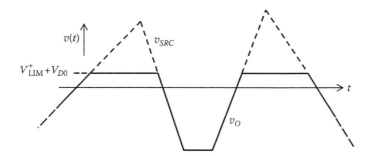

FIGURE 1.5.2
Diode limiting assuming piecewise-linear approximation of diode characteristic.

Equation 1.5.3 can, of course, be derived by superposition: v_{SRC} is applied, with $V^+_{LIM} + V_{D0} = 0$, then V^+_{LIM} and V_{D0} are applied with $v_{SRC} = 0$. If, as is normally the case, $r_D \ll R_s$, then $v_O = V^+_{LIM} + V_{D0}$. The output v_O follows v_{SRC} as long as $v_{SRC} < (V^+_{LIM} + V_{D0})$ but cannot exceed a positive level of $V^+_{Lim} + V_{D0}$ (Figure 1.5.2). Because voltages are clipped at a certain level, voltage limiting circuits are also known as **clipping** circuits. The smooth variation of current with voltage in practical diodes results in gradual clipping rather than abrupt clipping (Figure 1.5.4). Moreover, if r_D is not negligible compared to R_s, the clipped level varies somewhat with v_{SRC}.

When a load R_L is connected to the output of the voltage-limiting circuit, the transfer characteristic can be readily obtained by assuming values of output voltage and working backward to determine the currents and the corresponding input voltages. If the output voltage in the limiting range is required for a specific input voltage, Thevenin's equivalent circuit (TEC) can be derived as seen at the load terminals (Exercise 1.5.1).

Exercise 1.5.1

Show that while the diode is conducting in the circuit of Figure 1.5.1a, the circuit between terminals ab is equivalent to a series combination of an input source $v_{SRC} \, r_D/(R_s + r_D)$, a battery $(V^+_{LIM} + V_{D0})R_s/(R_s + r_D)$, and a resistance $R_s \| r_D$. ∎

If the polarities of the battery and diode are reversed (Figure 1.5.1b), limiting occurs at a negative level $-(V^+_{LIM} + V_{D0})$. Limiting at both negative and positive levels is achieved by combining the two subcircuits of Figure 1.5.1a and b.

The diode and battery of Figure 1.5.1a can be replaced by a zener diode, whereby $(V^+_{LIM} + V_{D0})$ and r_D are replaced by V_{Z0} and r_Z, respectively. Limiting at a positive level and a negative level can be obtained by connecting two zener diodes back-to-back, that is, in series opposition, so that when one diode is in its breakdown region, the other diode is forward biased (Example 1.5.1). The limiting voltages are $+(V_{Z01} + V_{D01})$ and $-(V_{Z02} + V_{D02})$, and the resistances are $r_Z + r_D$ for the respective diodes. If the positive and negative limiting levels are to be equal in magnitude, either the two zener diodes must be closely matched, or a diode bridge is used with a single zener diode (Problem P1.3.3).

Simulation Example 1.5.1 Voltage Limiting Using Zener Diodes

It is required to simulate voltage limiting using two back-to-back zener diodes.

SIMULATION

The schematic is shown in Figure 1.5.3. The diodes are 1N750, of the type whose $i-v$ characteristic was investigated in Example 1.4.2. A VSIN source of 15 V amplitude and 50 Hz frequency is applied. A Time Domain (Transient) analysis is performed, the resulting plots of the source voltage, output voltage across the diodes, and current in the circuit being as shown in Figure 1.5.4. The output voltage begins to deviate significantly from the input voltage at about 4.6 V. The limiting voltage varies from about 5.3 V to about 5.43 V at the peak of the sinusoid, when the current is about 9.57 mA.

Exercise 1.5.2

A clipping circuit consists of two back-to-back zener diodes, as in Figure 1.5.3 with $R_s = 1\,\text{k}\Omega$ and a load resistance of 4 kΩ. The diodes have $V_{D0} = 0.7$ V and $V_{Z0} = 9.1$ V, the diode resistance being negligible. If $v_{SRC} = 15\sin\omega t$ V, determine the output voltage.
Answer: $v_O = 9.8$ V, $35.2° \le \omega t \le 144.8°$, $v_O = -9.8$ V, $215.2° \le \omega t \le 324.8°$, and $v_O = 12\sin\omega t$ V for the rest of the period. ∎

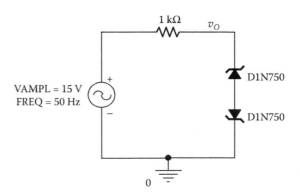

FIGURE 1.5.3
Figure for Simulation Example 1.5.1.

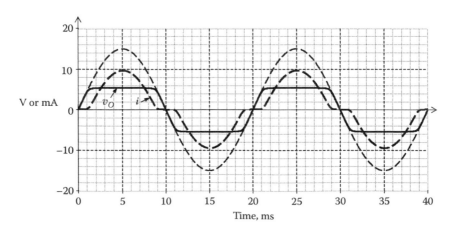

FIGURE 1.5.4
Figure for Simulation Example 1.5.1.

Design Example 1.5.2 Overload Protection of Meter

A moving coil meter has a basic movement of 100 Ω resistance and 1 mA full-scale deflection (FSD). The meter is to read full-scale when 10 V dc is applied. It is required to design a diode circuit that protects the meter against excessive overloads.

SOLUTION

A simple diode protection circuit is shown in Figure 1.5.5. It can be seen from Figure 1.1.2 that the $i-v$ characteristic of a *pn* junction diode, as given by Equation 1.1.1, shows a rapid increase in current at $V_D \cong V_{D0}$ V. We can choose R_1 so that at FSD the voltage across D_1 is 0.7 V and the diode current I_D should not exceed 10 μA, or 1% of FSD, so as not to introduce appreciable error. Thus, $(R_1 + 0.1) \times 1 = 0.7$, which gives $R_1 = 0.6$ kΩ. From Equation 1.1.1, if $V_T = 26$ mV and $\eta = 2$, then a diode of $I_S = 15$ pA will have $I_D \cong 10$ μA at $V_D = 0.7$ V. The reverse current I_S of D_2 is negligible in comparison. Neglecting the 10 μA through R_2 at FSD, compared to 1 mA, R_2 must be such that $R_2 \times 1 + 0.7 = 10$. This gives $R_2 = 9.3$ kΩ.

If an input voltage of 100 V is accidentally applied, V_D and I_D can be determined from TEC at the diode terminals. $V_{Th} = 100 \times (0.6 + 0.1)/(0.6 + 0.1 + 0.93) = 7$ V and $R_{Th} = 0.7 || 9.3 = 0.65$ kΩ. To determine I_D, the equation $V_D = 7 - 0.65I$ has to be solved with the diode equation $V_D \cong 0.052 \times \ln (I_D/(15 \times 10^{-9}))$, where I_D is in mA and V_D is in volts. After a few iterations, starting with $V_D = 0.8$ V, we find $V_D = 1.05$ V, $I_D = 9.15$ mA. At this value of V_D, the current through the meter is $1.05/0.7 = 1.5$ mA, Thus, an input voltage, of either polarity, whose magnitude is ten times FSD overloads the meter by 50%. The meter is unlikely to be damaged by this overload.

It is possible to prevent a reverse current through the meter by connecting diode D_2 in series with R_2 and adjusting the value of R_2 accordingly (Problem P1.3.5). But D_2 must then withstand the maximum negative applied voltage. With the two diodes connected in inverse parallel, the PIV of one diode is the maximum forward voltage drop of the other.

Application Window 1.5.1 Protection of Transistor Switch

A transistor is sometimes used to switch an inductive load, such as the coil of an electromagnetic relay (Figure 1.5.6a). When the transistor switch is turned on, a connection is made, with a very small voltage drop between the relay coil and ground. When the transistor switch opens, which can occur very rapidly, a large voltage, $v_L = Ldi_L/dt$, appears across the coil. The polarity of v_L, as shown, is such as to keep i_L flowing through the circuit. The resulting voltage across the transistor is $v_L + V_{CC}$ and could be high enough to damage the transistor.

A diode connected as in Figure 1.5.6b, and referred to as a **freewheeling diode**, protects the transistor. When the switch opens, the current i_L is diverted through the diode, thus limiting v_L to the small forward voltage drop of the diode. i_L decays with a

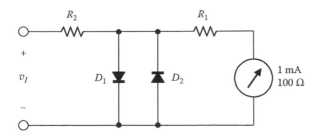

FIGURE 1.5.5
Figure for Design Example 1.5.2.

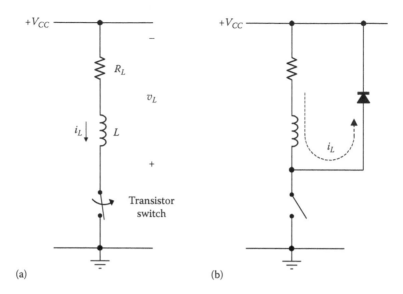

FIGURE 1.5.6
Figure for Application Window 1.5.1. Diode protection of transistor switch. (a) switch is about to open; (b) switch open.

time constant $L/(R_L + r_D)$, which introduces some delay in the opening of the relay. During normal operation, the diode is reverse biased and has no significant effect on the circuit.   ∎

Voltage-limiting circuits are used for protection against overvoltage and for wave shaping, as in the clipping of a triangular waveform at positive and negative levels to produce a trapezoidal waveform.

Surge Protection

Voltage surges of large amplitude and short duration often occur in power systems. Because of their rapid rise time, these surges can be transmitted via the interwinding capacitance of mains transformers of equipment, and can damage any electrolytic capacitors and IC voltage regulators that may be present (Example 1.3.2). Fortunately, these components can be adequately protected by **surge protectors**, or **transient voltage suppression** (TVS) diodes, as they are also called. In one form of surge protectors, special zener diodes are connected in parallel with the devices to be protected. The zener diodes are designed to turn on very rapidly and to withstand a fairly large amount of power for a very short time.

1.6 Diode-Capacitor Circuits

dc Restorer

Consider the circuit shown in Figure 1.6.1a in which an ideal diode switch is connected in series with a capacitor, v_{SRC} being the rectangular waveform of Figure 1.6.1b. If C is replaced by a resistor, the circuit is a voltage limiter that clips the negative excursions to

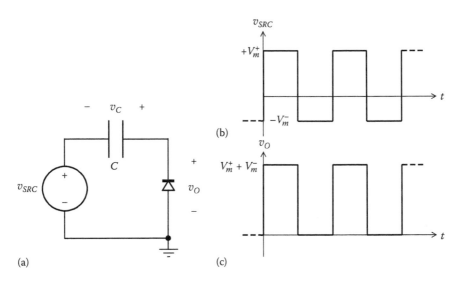

FIGURE 1.6.1
(a) Diode clamp; (b) applied waveform; (c) clamped waveform.

zero without affecting the positive parts of the waveform. With C connected and initially uncharged, then when v_{SRC} is positive, no current flows through C, assuming an open-circuit at the output. When v_{SRC} is negative, C charges to V_m^- and remains charged to this voltage because it cannot discharge through the ideal diode in the reverse direction. From KVL, $v_O = v_{SRC} + v_C = v_{SRC} + V_m^-$. So when v_{SRC} becomes V_m^+, $v_O = V_m^+ + V_m^-$. During the next negative half-cycle, the voltage across the diode, and hence I, is zero. Thus:

$$v_O = 0, \quad v_{SRC} < 0 \tag{1.6.1}$$

$$v_O = V_m^+ + V_m^-, \quad v_{SRC} > 0 \tag{1.6.2}$$

The input voltage is shifted in the positive direction so that the most negative input is at zero level (Figure 1.6.1c). The circuit is a **diode clamp**, a **clamping circuit**, or a **dc restorer**.

In ac-coupled amplifiers (Section 9.2, Chapter 9), series-connected capacitors block the dc component of voltage so as to avoid problems associated with drift in dc amplifiers. A dc restorer provides a dc component of output voltage. In **pulse-width modulation**, the pulse width of successive pulses in a pulse train is varied, or modulated, in accordance with the information to be transmitted. Varying the pulse width, while keeping the amplitude and period fixed, varies the average, or dc, level in direct proportion to the pulse width. The pulse-width-modulated signal is amplified and transmitted with the dc component removed. To recover this component, and hence the modulating information, the pulse-width-modulated signal is dc restored.

Exercise 1.6.1

Suppose that in Figure 1.6.1, $C = 10$ nF, $V_m^+ = 2$ V, $V_m^- = 1$ V, and the period of the waveform is 10 ms. Assuming an open-circuited output, and a reverse diode current of 1 nA, by what percentage will v_C drop during the time when the diode is nonconducting? (Hint: consider the charge leaked when the diode is reverse biased.)
   Answer: 0.05%.   ∎

When practical diodes are used in the circuit of Figure 1.6.1a and v_O displayed, say on an oscilloscope, it is found that the negative level is clamped at slightly below zero. The following simulation example explains why.

Simulation Example 1.6.1 Clamping Circuit

It is required to simulate the behavior of a clamping circuit, with and without load.

SIMULATION

The simulated circuit (Figure 1.6.1a) uses a D1N914 diode from the EVAL library, $C = 1$ nF, with v_{SRC} from a VPULSE source being a square wave of 2 ms period having $V_m^- = V_m^+ = 2$ V.

The output at no load is shown in Figure 1.6.2a. On the first negative half-cycle, with the capacitor initially uncharged, the diode conducts heavily and rapidly, charging the capacitor to a voltage of 2 V less the diode voltage drop. After the initial rapid charge, the current and the diode voltage drop decrease in magnitude, while the capacitor voltage increases slowly, because the current is still a charging current. When the input changes to $+2$ V, the capacitor voltage does not change instantaneously, and the output voltage increases by 4 V to $+4$ V less the diode voltage. On the next negative half-cycle, the diode conducts and the capacitor receives an additional small charge. Figure 1.6.2b, shows v_O one second after the source is first applied. The most negative excursion of v_O is approximately -60 mV in the steady state.

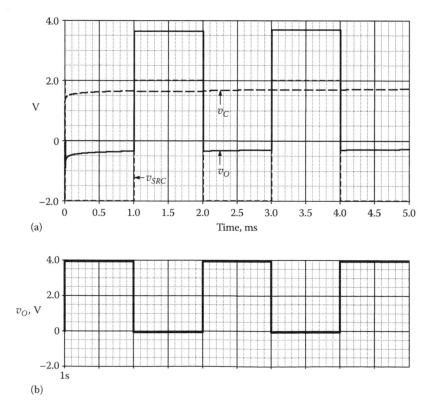

FIGURE 1.6.2
Figure for Simulation Example 1.6.1. (a) Beginning at $t = 0$; (b) beginning at $t = 1$s.

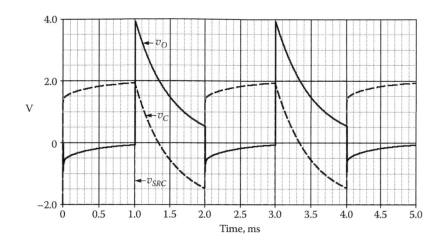

FIGURE 1.6.3
Figure for Simulation Example 1.6.1.

In practice, the reason why v_O is not zero during negative half-cycles of v_{SRC} is because of some leakage of the charge on C during positive half-cycles of v_{SRC} due to: (1) the reverse leakage current of the diode, (2) finite insulation resistance of C, and (3) finite resistance of the oscilloscope probe used to display v_O. In the steady state, the capacitor voltage is such that the charge lost during the positive half-cycles of v_{SRC} is equal to the charge gained during the negative input half-cycles (Example 1.6.2). This means that there will be a steady-state forward voltage drop across the diode during negative half-cycles so that the most negative excursion of v_O is below the zero level. It must not be supposed that the final voltage across C is $V_m^- - V_{D0}$. If this were the case, then the voltage across the diode during the negative phase is V_{D0} and the diode will still have a forward current that continues to charge the capacitor. In the absence of all leakage currents, the capacitor is fully charged only when the diode current is zero, which means that the diode voltage is zero, since the diode characteristic passes through the origin.

The output waveform when $R = 500$ kΩ is shown in Figure 1.6.3. On the first negative half-cycle, the diode conducts heavily, and the capacitor is rapidly charged to a voltage $(2 - v_D)$, as in the case of no load. The capacitor receives additional charge through the resistor during the negative half-cycle, and v_O decreases in magnitude to about -0.07 V at 1 ms. At the start of the positive half-cycle, v_{SRC} increases by 4 V and the output voltage increases by this amount to about 3.935 V because the voltage across the capacitor does not change instantaneously. During the positive half-cycle, the diode is reverse biased and the capacitor discharges through the load resistor with a time constant $RC = 0.5$ ms, v_O reaching about 0.54 V at 2 ms. When v_{SRC} changes to -2 V, a large forward voltage is impressed across the diode, which rapidly charges the capacitor to about 1.4 V. The capacitor is then charged slowly, v_O reaching about -0.07 V at 3 ms. The cycle repeats thereafter.

The following example illustrates an important concept that was applied in the case of the capacitor-input filter and referred to in Example 1.6.1.

Concept: *When a capacitor is alternately charged and discharged, then in the steady state, the charge gained in the charging phase is equal to the charge lost in the discharging phase.*

Example 1.6.2 Capacitor Charging and Discharging with Different Time Constants

v_{SRC} in Figure 1.6.4 is a periodic, rectangular waveform that is at a positive level V_m^+ for a duration T_1, and at a negative level $-V_m^-$ for a duration T_2. It is required to determine, in the steady state, v_O and the average voltage V_C across C, assuming that C is very large.

SOLUTION

Because of large C, the capacitor voltage is nearly constant at a value V_C. Since v_{SRC} is periodic, then in the steady state, V_C is such that the charge lost by C during the positive part of the cycle must be equal to the charge gained during the negative part of the cycle. Otherwise, there will be a net gain or loss of charge, and V_C would not have reached a steady state. The charge lost during the positive phase of the cycle is $(V_m^+ + V_C)T_1/R_1$. Similarly, the charge gained during the negative phase of the cycle is $(V_m^- - V_C)T_2/R_2$. Equating these charges gives:

$$V_C = \frac{V_m^- R_1 T_2 - V_m^+ R_2 T_1}{R_1 T_2 + R_2 T_1} \tag{1.6.3}$$

Thus, $V_C < 0$ if $V_m^+ R_2 T_1 > V_m^- R_1 T_2$, and $V_C > 0$ if $V_m^+ R_2 T_1 < V_m^- R_1 T_2$. If $R_1 \rightarrow \infty$, $V_C = V_m^-$, as argued above in connection with Figure 1.6.1a. Moreover, if $R_1 = R_2$, $V_C = (V_m^- T_2 - V_m^+ T_1)/(T_2 + T_1)$, which is the negative of the dc, or average, value of v_{SRC}. That is, V_C is equal and opposite to the dc component of the input, so that the dc component of v_O is zero. In other words, the dc component of v_{SRC} is blocked by the capacitor, as in a linear circuit. However, when $R_1 \neq R_2$, the charging and discharging capacitor currents are unequal because of the diodes; rectification occurs, and the output voltage has a dc component equal to the algebraic sum of the dc components of v_{SRC} and V_C.
$v_O = v_{SRC} + V_C$, so that during the positive phase,

$$v_O^+ = V_m^+ + V_C = \frac{R_1 T_2 \left(V_m^+ + V_m^- \right)}{R_1 T_2 + R_2 T_1} \tag{1.6.4}$$

During the negative phase,

$$v_O^- = -V_m^- + V_C = -\frac{R_2 T_1 \left(V_m^+ + V_m^- \right)}{R_1 T_2 + R_2 T_1} \tag{1.6.5}$$

$v_O^+ - v_O^- = V_m^+ + V_m^-$, as it should be.

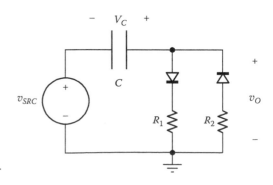

FIGURE 1.6.4
Figure for Example 1.6.2.

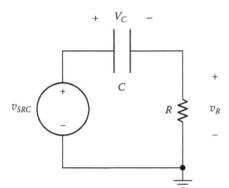

FIGURE 1.6.5
Figure for Exercise 1.6.3.

The other extreme case, where the time constants are small compared to the duration of the positive and negative phases is considered in Problem P1.4.5.

Exercise 1.6.2

Assume that in Figure 1.6.4, $V_m^+ = 10$ V, $V_m^- = 5$ V, $T_1 = T_2$, $R_1 = 3$ kΩ in series with a 4 V dc source, and $R_2 = 1$ kΩ in series with a 6 V dc source, the negative terminals of both dc sources being connected to ground. Determine V_C, justifying any assumptions made.
   Answer: $V_C = 6.75$ V. ∎

Exercise 1.6.3

Let v_{SRC} in Figure 1.6.5 be a periodic waveform. KVL gives:

$$v_{SRC} = v_C + v_R \tag{1.6.6}$$

where the voltages are instantaneous values. (a) By taking the averages of each term in Equation 1.6.6, deduce that $V_{SRC} = V_C + V_R$, where the voltages are dc values. (b) Deduce from equality of the charge gained by C during one part of the cycle to the charge lost during the remainder of the cycle, that $V_R = 0$, which means that C blocks the dc component of v_{SRC}. ∎

Voltage Multiplication

Consider the circuit of Figure 1.6.6, and let the positive half-cycle be of the polarity indicated. Assuming that the capacitors are initially uncharged, C_1 charges through D_1

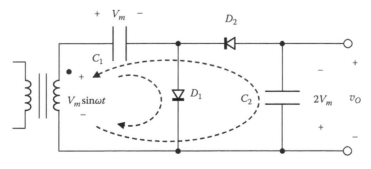

FIGURE 1.6.6
Basic voltage doubler.

during the first positive half-cycle to a voltage V_m, as indicated. During the following negative half-cycle, C_1 partly discharges as C_2 is charged, so that C_2 charges to less than $2V_m$. However, C_1 charges again during the next positive half-cycle, and C_2 receives additional charge during the next negative half-cycle. Eventually, in the steady state, and neglecting the load current and the reverse currents of the diodes, C_1 is charged to a steady voltage V_m and C_2 is charged to a steady voltage $2V_m$, as indicated. The voltage across D_1 is zero at the peak of positive half-cycles and is in the reverse direction the rest of the time. Similarly, the voltage across D_2 is zero at the peak of negative half-cycles and is in the reverse direction the rest of the time.

The circuit can be extended to provide multiplication by any integer n. In Figure 1.6.7, two more modules C_3D_3 and C_4D_4 are added, so that the voltages across C_3 and C_4 are $2V_m$, as indicated, neglecting the load current and the reverse currents of diodes, and assuming a steady state when all capacitor voltages have reached their final values. The $2V_m$ across C_3, the $2V_m$ across C_2, and the V_m across C_1 make the voltage across D_3 equal to zero at the peak of the positive half-cycles, with D_3 reverse-biased the rest of the time. Similarly, the $2V_m$ across C_4, the $2V_m$ across C_3, the $2V_m$ across C_2, and the V_m across C_1, make the voltage across D_4 equal to zero at the peak of the negative half-cycles, with D_4 reverse-biased the rest of the time.

Summary of Main Concepts and Results

- An ideal diode is a passive, unidirectional element that behaves like a short circuit in the forward direction and an open circuit in the reverse direction.
- In a given diode circuit, r_D can be neglected if it is in series with a much larger resistance, and V_{D0} can be neglected if it is in series with a much larger voltage.
- A rectifier circuit converts a bidirectional input to a unidirectional one. Full-wave rectifier circuits are based on a CTT secondary or on a diode bridge. The CTT circuit is better suited for lower voltages.

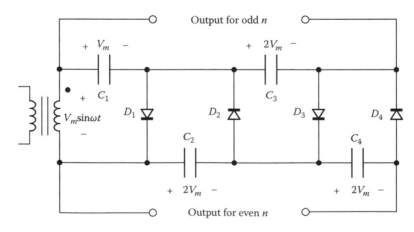

FIGURE 1.6.7
Modular voltage multiplier.

- The rectified output is smoothed using inductors or capacitors as low-pass filter elements, or by using a polyphase supply of appropriate number of phases.
- Zener diodes can be used to provide a fixed dc voltage for regulation or for reference purposes. In a regulated power supply, the zener diode maintains a nominally constant load voltage and reduces the ripple across the load.
- Diode-resistor circuits can be used to limit voltage excursions to within specified limits.
- Diode-capacitor circuits can be used for dc restoration and for voltage multiplication.
- When a capacitor is alternately charged and discharged, then in the steady state, the charge gained during charging is equal to the charge lost during discharge.

Learning Outcomes

- Analyze and design basic diode rectifiers, voltage regulators, limiters, and dc restorers.

Supplementary Examples and Topics on Website

SE1.1 Capacitor-Input Filter with Constant-Current Load. Analyzes this simpler case in which the load is a current source rather than a fixed resistor.

SE1.2 Diode Transmission Gate. Analyzes a diode transmission gate consisting of a diode bridge that allows transmission of an input signal to a load when a control signal is of one polarity, and prevents transmission when the polarity of the control signal is reversed.

SE1.3 Diode Logic Gates. The basic Boolean OR and AND operations are discussed using diode gates.

SE1.4 Diode-*LC* Circuit. Investigates the effect of including a diode in an *LC* circuit.

ST1.1 Inductor Filters. Several types of inductor filters are analyzed, and their main features highlighted, including the simple inductor-input filter, L-section filter, and π-section filter.

ST1.2 Peak-to-Peak Ripple in Polyphase Systems. Derives an expression for this ripple, as a percentage of the peak value, and in terms of the number of peaks per supply cycle.

ST1.3 Graphical Analysis of Zener Voltage Regulator. Illustrates how the change in V_Z is graphically related to the change in V_{DC}.

ST1.4 Diode Demodulator. Discusses the use of what is essentially a simple half-wave rectifier circuit for demodulating an amplitude-modulated signal.

ST1.5 Diode Mixer. As a nonlinear element, a diode can be used in a mixer, which is a circuit that generates sum and difference frequencies from two signals of different frequencies.

Problems and Exercises

P1.1 Diode Characteristics and Bias Point Calculations

In the following problems, assume $V_T = 0.026$ V and $\eta = 2$, unless otherwise specified.

P1.1.1   The terminals of a diode are unmarked. How would you use a simple ohmmeter to determine which diode terminal is the anode? How would you use the ohmmeter to determine if a diode is open-circuited or short-circuited?

P1.1.2   A diode has $i_D = 1$ A at $v_D = 0.7$ V and $i_D = 2$ A at $v_D = 0.73$ V. Determine I_S and η. What is r_d in both cases?

P1.1.3   Consider Equation 1.1.1 with $\eta = 1$. Determine v_D for which: (a) $e^{v_D/\eta V_T} = 10$; (b) $e^{v_D/\eta V_T} = 100$; (c) i_D is 80% of I_S; (d) i_D is 95% of I_S. How do these values change for $\eta = 2$?

P1.1.4   Consider Equation 1.1.1 with $\eta = 1$. If $i_D = 1$ mA, determine I_S for a v_D of: (a) 0.3 V; (b) 0.5 V; (c) 0.7 V.

P1.1.5   Assuming that in an ideal Si diode $e^{v_D/\eta V_T} \gg 1$, (a) what is the change in v_D that multiplies i_D by a factor k? Deduce the change in v_D that: (b) doubles i_D assuming $\eta = 2$; (c) multiplies i_D by a factor of 10 assuming: (i) $\eta = 2$, and (ii) $\eta = 1$. Note that these values of v_D are independent of i_D as long as $e^{v_D/\eta V_T} \gg 1$. If Δv_D is the change in v_D that causes a 10-fold change in i_D, the diode is referred to as a Δv_D/decade diode. What is η for a 0.1 V/decade diode?

P1.1.6   Two *pn* junction diodes are connected in inverse parallel to protect the differential inputs of an op amp. If the diodes have $I_S = 0.1$ nA and the differential input does not exceed ± 1 mV in normal operation, what is the minimum effective incremental resistance of the two diodes?

P1.1.7   Two identical ideal Si diodes having $I_S = 10$ nA are connected in series opposition across a 10 V dc supply. Make a reasonable assumption, which you check later, to determine the current in the circuit and the voltage across each diode.

P1.1.8   The diodes of Problem P1.1.7 are both connected in the forward direction in series with an ideal current source of 1 mA. Determine the voltage across the diodes, assuming $\eta = 1$. Repeat with a resistor connected in parallel with the two diodes having a value of: (a) 10 kΩ; (b) 1 kΩ. How do you interpret the results? Simulate with PSpice.

P1.1.9   Determine the voltage across each diode in Problem P1.1.8 if the current source is 5 mA and the 1 kΩ resistor is connected across one of the diodes. Simulate with PSpice.

DP1.1.10 A voltage reference of 1.4 V is to be derived from *pn* junction diodes supplied from a current source. If the diodes available have $\eta = 1.98$ and $I_S = 14.11$ nA (1N4002), what should be the source current? What will be the small-signal output resistance of the voltage reference source? What is the significance of this output resistance? If the current source varies by $\pm 10\%$ what will be the variation in the reference voltage? If the available current source has twice the required value, what would be an appropriate diode arrangement? Simulate with PSpice.

DP1.1.11 Suppose that a voltage reference of 1.3 V is required, and the designer opts to use the same current source and diodes as in Problem DP1.1.10 and connect a resistor R across one of the diodes; determine R. What will be the small-signal output resistance in this case? Can you suggest how the output resistance can be reduced to a low value?

P1.1.12 Five identical *pn* junction diodes are connected in the forward direction in series with a 3 kΩ resistor across a 9 V supply. If each diode has a 0.7 V drop at 1 mA and can be approximated as a 0.1 V/decade diode (see Problem P1.1.5), determine the voltage across the diode string.

P1.1.13 The diodes in Exercise 1.2.1 are connected in the forward direction in series with a 1 kΩ resistance across a 10 V supply. Assuming the linear approximation for the diodes, determine the current and the voltage across each diode. Compare the results with those obtained from Equation 1.1.1. What is the power dissipated in each diode?

P1.1.14 Determine the diode currents in Figure P1.1.14.

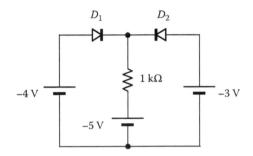

FIGURE P1.1.14

P1.1.15 Determine the current in each diode and v_O in Figure P1.1.15.

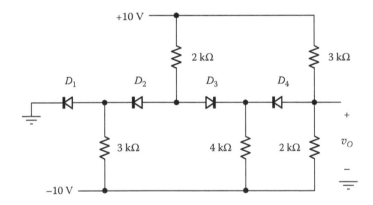

FIGURE P1.1.15

P1.1.16 In the circuit of Figure P1.1.16, the diode can be represented in the forward direction by $V_{D0} = 0.7$ V and $r_D = 10$ Ω. Determine i_D and v_D if $v_{SRC1} = A\cos100\,\pi t$ V and $V_{SRC2} = 5$ V. For what amplitude of the sinusoidal signal will i_D just fall to zero during the cycle? If the amplitude of the sinusoidal signal is twice this value, determine i_D and v_D.

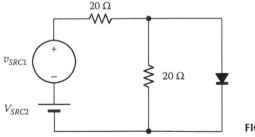

FIGURE P1.1.16

P1.1.17 Consider the OR gate of Figure P1.1.17. Assume that v_A is a sinusoidal signal of amplitude A, v_B is a triangular signal of the same amplitude, and v_C is a rectangular signal of amplitude $A/2$, all signals having zero average. The three signals have the same frequency and phase, that is, their positive (or negative) cycles coincide. Sketch the output of the OR gate. Can you generalize the output of the OR gate to inputs of different amplitudes?

P1.1.18 Repeat Problem P1.1.17 for the AND gate of Figure P1.1.18.

P1.1.19 Assume that in the OR gate of Problem P1.1.17, $R = 10$ kΩ, V_A is derived from a +10 V source in series with 0.5 kΩ, V_B is derived from a +15 V source in series with 1 kΩ, and $+V_C$ is derived from a 20 V source in series with 2 kΩ. Determine V_Y.

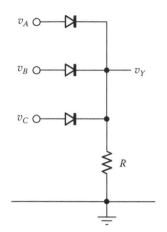

FIGURE P1.1.17

P1.1.20 Repeat Problem P1.1.19 for the AND gate of Figure P1.1.18, with $V_{CC} = 30$ V.

P1.2 Rectifiers and Regulated Power Supplies

P1.2.1 In a high-voltage unregulated power supply with a capacitor-input filter, a bleeder resistor is sometimes added across the capacitor. Why?

DP1.2.2 Figure DP1.2.2 shows a rectifier circuit that is commonly used to provide positive and negative voltage supplies of equal magnitude. Develop this circuit starting with two CTT circuits, one that provides a positive load voltage, the other a negative load voltage of equal magnitude and with respect to the same common ground. Combine the two CTT transformers into one, and redraw the circuit so that the four rectifier diodes are connected in a bridge.

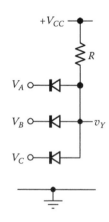

FIGURE P1.1.18

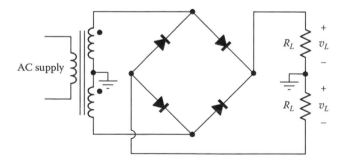

FIGURE DP1.2.2

P1.2.3 What happens in a CTT or a bridge rectifier if one of the diodes becomes: (a) open-circuited, or (b) short-circuited?

P1.2.4 Assume that in Example 1.3.2 the waveform of the diode current is a right triangle having an amplitude of $i_{D(pk)}$ and a duration equal to that of the conduction period. Determine the rms value of current in: (a) each transformer secondary, and (b) the transformer primary. If a fuse is to be connected in series with the transformer primary to protect against any faults that might develop in the rectifier circuit, what should be the rating of the fuse? It is recommended that such a fuse be of the "slow blow" type. Why?

P1.2.5   In a CTT rectifier with a resistive load (Figure 1.3.2), the diodes have $V_{D0} = 0.7$ V and $r_D = 10$ Ω. Assume a source resistance of 2 Ω for each transformer secondary and a load resistance of 1 kΩ. If the transformer open-circuit voltage is 20 V rms, determine: (a) the peak output voltage; (b) the fraction of the cycle during which a diode conducts; (c) the dc output voltage; and (d) the PIV. Simulate with PSpice.

P1.2.6   Repeat Problem P1.2.5 assuming the same diodes connected as a bridge rectifier and supplied with the same secondary voltage. Simulate with PSpice.

DP1.2.7   A bridge rectifier is to be used for charging 12 V car batteries. Design a suitable circuit consisting of a transformer, a bridge rectifier, and a resistor R in series with the battery, subject to the following assumptions: (1) the maximum charging voltage is to be 13.5 V at the battery terminals when the charging current is zero; (2) the maximum peak charging current is to be limited to 10 A when the battery voltage is 10.5 V; (3) the diode voltage drop may be assumed to be 1 V when the diode is conducting, with the battery voltage less than 12 V, and 0.5 V when the battery voltage is 12 V; (4) the diode resistance, source resistance, and battery resistance are negligible. Specify the transformer secondary voltage, R, the PIV of the diodes, and the largest average diode current. When the battery voltage is 12 V, what are the peak diode current, the average diode current, and the conduction period as a fraction of the supply cycle?

DP1.2.8   Consider a CTT rectifier with a 1 kΩ load supplied from a 220 V, 50 Hz supply through a transformer having a primary-to-half-secondary turns ratio of 1. Assume a diode voltage drop of 0.7 V, zero diode resistance, and zero source resistance. Determine the peak load voltage. Determine the minimum C that should be used if the minimum load voltage is not to fall below its maximum by (1) 10 V; (2) 1 V. For each case, assuming that the approximate analysis applies, determine: (a) the dc output voltage; (b) the average diode current; (c) the fraction of the cycle during which a diode conducts. Simulate with PSpice.

P1.2.9   In a bridge rectifier with capacitive input filter, $v_{SRC} = 12\cos100\,\pi t$ V, the diodes drop 0.8 V when conducting, $R_L = 1$ kΩ, and $C = 1000$ μF. Determine the peak-to-peak ripple in the output voltage, neglecting all resistances in the transformer and diodes. Simulate with PSpice.

DP1.2.10   The bridge rectifier of Problem P1.2.9 with a capacitive input filter supplies an IC voltage regulator that maintains its output at 12 V across a 100 Ω load as long as its input does not drop below 14 V. What is the minimum value of C that should be used? Assume that when the regulator is operating properly, its input current is the same as its output current.

DP1.2.11   In a bridge rectifier with a capacitive input filter, the dc load is 5 V, 50 mA, and the ripple across the load has a frequency of 100 Hz and a peak-to-peak amplitude of 200 mV. Determine C, the average diode current, and the peak diode current. If the maximum voltage drop during conduction is 1 V, and the primary supply voltage is 220 V, what should be the transformer turns ratio? Simulate with PSpice.

P1.2.12   A load is to be supplied at 1 kV, 100 mA dc. Only a 1 μF capacitor can be conveniently used for filtering purposes. If the peak-to-peak ripple across the load should not exceed 1 V, what frequency of a single phase supply should be used, assuming a full-wave rectifier with negligible voltage drops? If a polyphase supply is to be used with the 1 μF capacitor at 50 Hz, what should be the effective number of phases (refer to Section ST1.2)?

P1.2.13   A bridge rectifier with capacitive input filter is to be used to supply a dc load at 12 V, 1 A, from a 50 Hz supply. If the peak diode current is not to exceed 40 times the average diode current over the whole cycle, what is the peak-to-peak ripple in the output voltage and the maximum size of capacitor that can be used? Assume to begin with that $V'_m = V_{DC}$ and determine v_r. Using this value of v_r determine a new value of V'_m. Repeat the iterations until you get consistent values of V'_m and v_r.

P1.2.14   A zener diode is specified to have $V_Z = 15$ V at $I_{ZT} = 8.5$ mA, $r_Z = 16$ Ω, and a power rating of 0.5 W. The diode is used as a regulator with $R_s = 300$ Ω, the minimum zener diode current being 2.5 mA. (a) Determine V_Z at $I_{Z(min)}$; (b) if the load current is 20 mA, what is the lowest supply voltage for which the diode current is $I_{Z(min)}$? (c) At no load, what is the highest supply voltage at which the power rating of the zener diode is not exceeded? (d) For a 20 V dc input, what is the maximum load current for which the diode current is $I_{Z(min)}$?

DP1.2.15   A zener regulator is required to provide a dc supply of 10 V nominal voltage at a current of 0 to 0–25 mA, the maximum dc voltage available being 12 V. A zener diode is chosen for this purpose that has $V_Z = 10$ V at $I_{ZT} = 10$ mA, $r_Z = 8$ Ω, $I_{ZK} = 1$ mA, and a power rating of 1 W. The maximum peak-to-peak ripple in the output is to be 50 mV. Design a suitable regulator, specifying the series resistance, the minimum dc input voltage and the maximum peak-to-peak ripple. Consider $I_{Z(min)} = 2$ mA.

DP1.2.16   Repeat Problem DP1.2.15 for the same load current but using two 1N750 diodes in series. Assume the 1N750 is rated at 500 mW, has $V_Z = 4.7$ V at $I_Z = 20$ mA, $r_Z = 2.5$ Ω, and $I_{Z(min)} = 5$ mA. Simulate with PSpice.

P1.2.17   A bridge rectifier with capacitor input filter and zener regulator has the following nominal values: secondary voltage 50 V rms, $V_{Z0} = 30$ V, $r_Z = 5$ Ω, $C = 10{,}000$ μF, $R_L = 500$ Ω, and $R_s = 75$ Ω. The diode voltage drop when conducting may be assumed to be 1.5 V and the source resistance can be neglected. Determine the worst-case ripple voltage across the load, assuming the following tolerances: transformer voltage ±10%; V_{Z0}, r_Z, and R_s ±5%; C ±20%.

P1.2.18   A zener diode is packaged so that it can dissipate to its surroundings 10 mW/°C temperature rise. If the temperature of the diode is not to increase by more than 15°C what is the maximum zener current at a voltage of 10 V?

P1.2.19   The voltage across a zener diode is 10 V when the current through the diode is 10 mA, the diode resistance being 10 Ω. The diode is used in a regulator circuit having a supply voltage of 20 ± 5 V, a series resistance of 200 Ω, and a load of 0–20 mA. (a) Determine the change in v_L as the supply voltage changes between 15 and 25 V at a constant i_L of: (i) zero, (ii) 20 mA. Show that the change in v_L is the same in both cases and is given by the line regulation. (b) Determine the change in v_L as the load current changes between 0 and 20 mA at a constant supply voltage of: (i) 15 V, (ii) 25 V. Show that the change in v_L is the same in both cases and is given by the load regulation. (c) Determine the worst-case change in v_L as the supply voltage changes between 15 and 25 V, and the load current changes between 0 and 20 mA. Show that ΔV_L is the sum of what was determined in (a) and (b). How do you justify this result?

P1.2.20   Given a zener diode regulator using a zener diode of $V_{Z0} = 9$ V and $I_{Z(min)} = 2$ mA. The input varies between $V_{I(max)}$ and $V_{I(min)}$, where $V_{I(max)} - V_{I(min)} = 5$ V. A variation of the load current between 0 and 20 mA, at constant input voltage, changes the load voltage by 0.2 V. A change in the input voltage between $V_{I(max)}$ and $V_{I(min)}$ changes the load voltage by 0.1 V, at a constant load current of 20 mA. Determine: (a) R_s; (b) r_Z; (c) V_Z at $I_Z = 10$ mA; (d) $V_{I(min)}$; and (e) minimum power rating of zener diode.

P1.2.21   A 24-phase rectifier is to supply a resistive load at 50 A, 20 V dc as in Figure P1.2.21, where each phase is derived from a transformer secondary, having one terminal connected to a diode, the other to ground. What is the peak load voltage? If the diode voltage drop is 1.5 V, what is the open-circuit phase voltage? What is the transformer secondary rating per phase (rms voltage×rms current)?

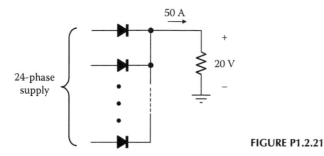

FIGURE P1.2.21

P1.2.22 Express Equation 1.3.5 as: $i_D = (V'_m/R_L)\sqrt{(\omega CR_L)^2+1} \times \cos[\omega t + \tan^{-1}(\omega CR_L)]$, $-\theta_2 \le \omega t \le \theta_1$. Show that if $\omega CR_L \gg 1$, $i_D \cong -V'_m\omega C\sin\omega t$. The maximum value over the given range occurs at $\omega t = -\theta_2$. Deduce from the value of θ_2 derived from Equation 1.3.13, and using Equation 1.3.10, that $i_{D(pk)} \cong 2\pi I_{DC}\sqrt{V'_m/2v_r}$ (Equation 1.3.15).

P1.2.23 A **ripple factor** r is defined as $r = v_{r(rms)}/V_{DC}$. Show that for a full-wave rectifier with capacitive input filter for which the approximate analysis is valid, $r = 1/4fCR_L\sqrt{3}$. Although the ripple factor is not commonly used with simple capacitor-input filters, it is useful for specifying the performance of inductor-input and more elaborate types of filters.

P1.2.24 Repeat Simulation Example 1.3.1 considering the worst-case switching transient. To do so, use VSIN sources with a phase angle of 90° and insert a normally open switch between the diodes and the filter with a closure delay of 1 μs.

P1.3 Clipping Circuits

P1.3.1 $v_I = -6$ V in the circuit of Figure P1.3.1. Determine v_O and all the currents in the circuit.

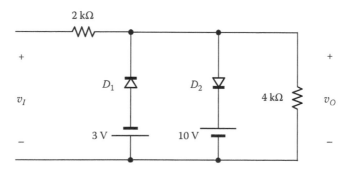

FIGURE P1.3.1

P1.3.2 In the circuit of Figure 1.5.1a, a variable V^+_{LIM} can be obtained by replacing the supply V^+_{LIM} by a supply V_{DC} having a potentiometer across it and connecting the cathode terminal of the diode to the slider of the potentiometer. If the resistance between the slider of a potentiometer and common reference is αR_P, $0 \le \alpha \le 1$, where R_P is the total resistance of the potentiometer, modify Equation 1.5.1 to take into account the presence of the potentiometer, assuming the dc source is of negligible resistance. What is the maximum resistance between the slider of the potentiometer and common reference?

P1.3.3 Positive and negative limiting levels of equal magnitude are to be obtained by connecting a zener diode to two terminals of a diode bridge, with the other two bridge terminals connected between R_s and common reference in Figure 1.4.3. Draw the circuit showing the appropriate polarities of the bridge diodes. Can a bridge rectifier package be used for this purpose? If each diode in the bridge has a temperature coefficient of -2 mV/°C, what should be the temperature coefficient of the zener diode so that the limiting voltages are independent of temperature? If the zener diode selected has a temperature coefficient of $+6$ mV/°C, how would you make the limiting voltages independent of temperature?

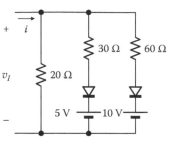

FIGURE P1.3.6

P1.3.4   Determine the transfer characteristic v_O vs. v_I in Problem P1.3.3 if $R_s = 400\ \Omega$, the bridge diodes are represented by $V_{D0} = 0.7$ V, $r_D = 10\ \Omega$, and the zener diode by $V_{Z0} = 9$ V, $r_Z = 20\ \Omega$.

P1.3.5   If one of the diodes in Example 1.5.2 is connected in series with R_2 to prevent a current of negative polarity through the meter, determine the new value of R_2. What would be the overload on the meter in case 100 V is applied?

P1.3.6   Sketch i in Figure P1.3.6 as v_I varies from -20 V to $+20$ V. Simulate with PSpice.

P1.3.7   Sketch i in Figure P1.3.6 as a function of time if v_I is a periodic triangular waveform defined over one period as: $v_I = 20t$, $0 \le t \le 1$ s, $v_I = -20(t-2)$, $1 \le t \le 3$ s, $v_I = 20(t-4)$, $3 \le t \le 4$ s. Simulate with PSpice.

P1.3.8   Determine the input characteristic i vs. v_I of the circuit of Problem P1.3.6 if each diode is represented by $V_{D0} = 0.7$ V and $r_D = 10\ \Omega$.

P1.3.9   Repeat Problem P1.3.6 if the diode in series with 30 Ω is reversed.

P1.3.10 Repeat Problem P1.3.9 if the diode in series with 60 Ω is also reversed.

P1.3.11 Repeat Problem P1.3.6 if the polarity of the 10 V battery is reversed.

P1.3.12 Suppose that in the circuit of Figure 1.5.1a, $v_{SRC} = 10\cos\omega t$ V, $R_s = 100\ \Omega$, and $V_{LIM}^+ = 5$ V. Determine the transfer characteristic and v_O, assuming an ideal diode switch.

P1.3.13 Suppose that branches (a) and (b) in Figure 1.5.1 are paralleled together, with $R_s = 200\ \Omega$, $V_{LIM}^+ = 5$ V, $V_{LIM}^+ = -5$ V, and the diodes represented by $V_{D0} = 0.7$ V and $r_D = 10\ \Omega$. Determine the transfer characteristic. Simulate with PSpice.

P1.3.14 Consider a circuit consisting of the two branches of Figure 1.5.1a and b connected in parallel with a load resistance of 4 kΩ, with $R_s = 1$ kΩ, $V_{LIM}^+ = -4$ V, $V_{LIM}^- = -2$ V and v_{SRC} a triangular waveform of 4 s period and amplitude ± 10 V. Determine the load voltage and the source current, assuming the diodes are ideal.

P1.3.15 Determine the transfer characteristic (v_O vs. v_I) for the circuit of Figure P1.3.15.

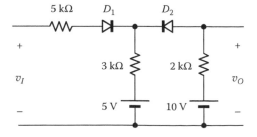

FIGURE P1.3.15

P1.3.16 Repeat Problem P1.3.15 with D_1 reversed.
P1.3.17 Repeat Problem P1.3.15 with D_2 reversed.
P1.3.18 Repeat Problem P1.3.15 with D_1 and D_2 reversed.

P1.4 Clamping Circuits

P1.4.1 A source voltage is defined as follows: $v_{SRC} = 0$, $t < 0$; $v_{SRC} = 10$ V, $0 < t < 1$ ms; $v_{SRC} = -5$ V, $1 < t < 2$ ms, and the waveform repeats thereafter. The waveform is applied to the circuit of Figure 1.6.1a with $C = 1$ μF and a load resistance of 100 kΩ. Determine the output waveform in the steady state, assuming an ideal diode switch. Simulate with PSpice.

P1.4.2 Repeat Problem P1.4.1 with the diode reversed.

P1.4.3 Repeat Problem P1.4.1 with a battery of 3 V connected in series with the diode, with the negative pole of the battery connected to ground. Simulate with PSpice.

P1.4.4 Generalize Exercise 1.6.2 to the case where the positive and negative phases of the waveform have amplitudes denoted by V_m^+ and V_m^-, respectively.

P1.4.5 Consider Example 1.6.2 but with $CR_1 \ll T_1$ and $CR_2 \ll T_2$. (a) Argue that under these conditions v_C very nearly reaches the steady values V_m^+ and V_m^- of v_{SRC} so that v_C is given by: $v_C = -V_m^+ + V_{pp}e^{-t/\tau_1}, 0 \le t \le T_1$, and $v_C = V_m^- - V_{pp}e^{-t'/\tau_2}, 0 \le t' \le T_2$, where $V_{pp} = V_m^+ + V_m^-$. (b) By integrating v_C with respect to time, show that the average of v_C is $V_C \cong -\dfrac{V_m^+T_1 - V_m^-T_2}{T_1 + T_2} + \dfrac{\tau_1 - \tau_2}{T_1 + T_2}V_{pp} \cong -\dfrac{V_m^+T_1 - V_m^-T_2}{T_1 + T_2}$. (c) Deduce that $v_O = V_{pp}e^{-t/\tau_1}, 0 \le t \le T_1$, and $v_O = V_{pp}e^{-t'/\tau_2}, 0 \le t' \le T_2$, and show by integrating with respect to time that, in terms of average values, $V_{SRC} + V_C = V_O$. (d) Deduce that the currents in the two branches are $i_1 = \dfrac{V_{pp}}{R_1}e^{-t/\tau_1}, 0 \le t \le T_1$ and $i_2 = \dfrac{V_{pp}}{R_2}e^{-t'/\tau_2}, 0 \le t' \le T_2$ (Figure P1.4.5). By integrating the currents with respect to time, show that the charge lost in the positive phase is equal to that gained in the negative phase.

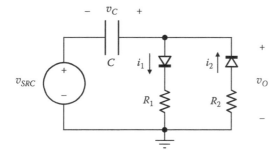

FIGURE P1.4.5

P1.4.6 Based on the results of Problem P1.4.5, determine the average values of v_C and v_O in Figure P1.4.6, assuming $V_m^+ = 12$ V for 2 ms, and $V_m^- = 8$ V for 3 ms.

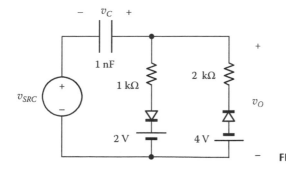

FIGURE P1.4.6

P1.4.7 The periodic waveform whose first period is illustrated in Figure P1.4.7 is applied to the circuit shown. The diodes are identical and can be considered to have a forward voltage of 1 V and a breakdown voltage of 5 V for all forward or reverse currents, respectively. The capacitor is so large that its voltage can be considered to remain nearly constant in the steady state. Determine the steady-state values of: (a) the average voltage V_C across the capacitor; (b) the average output voltage V_O.

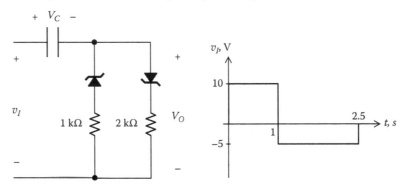

FIGURE P1.4.7

P1.4.8 In Figure P1.4.8, the zener diode has $V_{Z0} = 5$ V, with negligible r_Z. In the forward direction, $V_{D0} = 1$ V with negligible r_D. If v_{SRC} is the square voltage shown, determine the average, or dc, voltage v_O, assuming a very large capacitor.

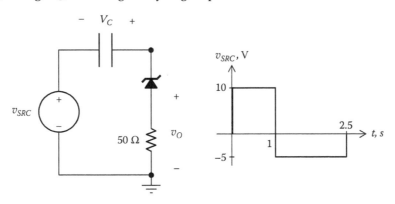

FIGURE P1.4.8

P1.4.9 Determine V_{C1} and V_{C2} in the voltage-doubler circuit of Figure P1.4.9.

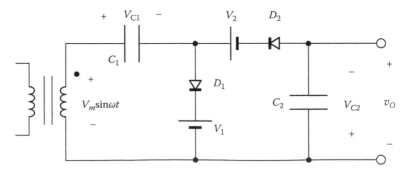

FIGURE P1.4.9

P1.5 Miscellaneous

DP1.5.1 A diode-resistor network is to have the following transfer characteristic: $v_O = 1$ V, $v_{SRC} \leq 0$; $v_O = 0.5 v_{SRC} + 1$ V, $2 \leq v_{SRC} \leq 3$ V; and $v_O = 4$ V, $v_{SRC} \geq 3$ V. Design a suitable diode-resistor-battery circuit, assuming ideal diodes and open-circuit output.

P1.5.2 The inductor in Figure P1.5.2 is initially uncharged and $v_{SRC} = V_m \sin \omega t$. Analyze the circuit beginning $t = 0$ and show that the inductor current is "pumped" every cycle by $2V_m/\omega L$.

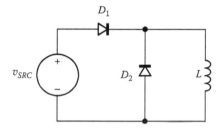

FIGURE P1.5.2

P1.5.3 Refer to Equation ST1.1.3 (Section ST1.1) for the simple inductor-input filter, neglect harmonics having $n > 2$, and assume that $(2\omega L/R_L)^2 \gg 1$. Show that the ripple factor (Problem P1.2.23) is $R_L/3\sqrt{2}\omega L$. A full-wave rectifier is to supply 10 A at 50 V with a ripple that should not exceed 0.2 V rms. Determine a suitable value for L, assuming $f = 50$ Hz.

P1.5.4 Consider an L-section filter (Figure ST1.1.2, Section ST1.1) subject to the assumptions made in calculating I_{LC}. Show that the ripple factor (Problem P1.2.23) is $1/6\sqrt{2}\omega^2 LC$. Note that the ripple is inversely proportional to ω^2 and that for a given ripple factor, more flexibility is provided in the choice of the filter components L and C. Choose filter parameters for a ripple factor of 0.1%, considering the dc supply of Problem P1.5.3.

P1.5.5 Consider the π-section filter (Figure ST1.1.3, Section ST1.1). Subject to the approximation made in the analysis of the capacitor-input and L-section filters, and using Equation ST1.1.6, show that the ripple factor (Problem P1.2.23) is $\sqrt{2}/8\omega^3 R_L CC_1 L_1$. Choose filter parameters for a ripple factor of 0.01%, considering the dc supply of P15.3. If a simple capacitor-input filter were to be used for the same ripple factor, what size of capacitor would be required? What would be the peak diode current?

P1.5.6 Verify Equation 1.4.6 for incremental changes using the graphical construction of Section ST1.3.

2

Basic Principles of Semiconductors

A rigorous discussion of electric conduction in semiconductors is well beyond the scope of this book, as it relies heavily on wave mechanics. However, some phenomena of fundamental importance in semiconductors, such as hole conduction and tunneling, cannot be explained by classical mechanics. One of the aims of this chapter is to present in a simplified manner some of the basic ideas of wave mechanics so as to give at least a plausible explanation of phenomena that cannot be explained by classical mechanics. This chapter also lays the foundation for understanding some of the important concepts related to electric conduction in semiconductors.

A critical difference between metals and semiconductors is the much lower conductivity of semiconductors. As a consequence, concentration gradients can exist in a semiconductor, leading to current flow by diffusion, in addition to the familiar current flow by drift under the influence of an applied electric field. In the presence of gradients of concentration and of electric potential, the fundamental concept of electrochemical potential provides a unifying and meaningful framework for understanding the behavior of semiconductors.

Another critical difference between metals and semiconductors, which underlies the practical importance of semiconductors, is the wide latitude that semiconductors provide in selecting the predominant type of current carrier, whether negatively charged electrons or positively charged holes, as well as the carrier concentration. It is important, therefore, to consider aspects such as carrier concentrations, mobility, and lifetime.

Learning Objectives

❖ To be familiar with:

• Some of the basic ideas of wave mechanics.

❖ To understand:

• The basis for the energy band structure of semiconductors.
• The basis for tunneling.
• The concept of effective mass and the basis for hole conduction.
• The concept of electrochemical potential and its implications for semiconductors at equilibrium and under bias.
• How carrier concentrations in a semiconductor can be determined.
• The factors affecting carrier mobility and lifetime.

2.1 Excerpts from Wave Mechanics

Some Basic Ideas

Quantum theory, or quantum mechanics, provides the most rigorous and powerful tool available for investigating physical phenomena. A particular formulation of quantum theory, **wave mechanics**, is customarily used to analyze electric conduction in crystals. It is well beyond the scope of this chapter to go beyond a brief and sketchy introduction to concepts needed for gaining some understanding of the electrical behavior of semi-conductors.

Concept: *Radiant energy, like matter, is quantized.*

In other words, energy in the form of radiation is not emitted or absorbed continuously but in discrete packets. Each packet is a **quantum** having energy E_q given by:

$$E_q = h\nu \tag{2.1.1}$$

where
  h is Planck's constant (6.626×10^{-34} Js)
  ν is the frequency of the radiation

In the case of ultraviolet light of $\nu = 3 \times 10^{15}$ Hz, for example, $E_q \cong 2 \times 10^{-18}$ J, which is exceedingly small. A quantum of electromagnetic radiation is a **photon**.

According to the Special Theory of Relativity, energy and mass are equivalent in accordance with Einstein's famous relation $E = mc^2$, where c is the speed of light in vacuum. Hence, a photon of energy $E_q = h\nu$ has an equivalent mass $h\nu/c^2$. If traveling at the speed of light in vacuum its momentum is:

$$|\boldsymbol{p}| = mc = \frac{h\nu}{c} = \frac{h}{\lambda} \tag{2.1.2}$$

where
  λ is the wavelength of the electromagnetic radiation
  $|\boldsymbol{p}|$ denotes the magnitude of the momentum vector $\boldsymbol{p}$

The fact that light behaves as a wave is well established and dominated interpretations of optical phenomena until the theory of light as a stream of particle-like photons was advanced by Einstein early in the twentieth century to explain photoelectric emission.

The converse assertion, that is, that of the behavior of particles as waves, is **De Broglie's hypothesis**, according to which a particle of total energy E and momentum $\boldsymbol{p}$ has *associated with it a hypothetical* De Broglie wave of frequency ν and wavelength λ such that:

$$\frac{E}{\nu} = \frac{1}{(1/\lambda)} |\boldsymbol{p}| = h \tag{2.1.3}$$

This gives $|\boldsymbol{p}| = h/\lambda$, as for a photon (Equation 2.1.2).

That electrons behave as particles is well known. A beam of electrons can be deflected by electric or magnetic fields just like a stream of particles. That electrons behave as waves is also well established. Electrons passing through metal crystals produce a diffraction pattern in the same way as light is diffracted by appropriate optical structures. This dual particle/wave behavior of particles and of radiation is central to wave mechanics:

Concept: *Particles and radiant energy may behave as particles or as waves, depending upon the nature of the system under consideration, including the means of observation of the phenomenon in question.*

The inclusion of the means of observation as an integral part of the system has far-reaching implications. In classical mechanics it is quite in order, for example, to specify, independently, both the position of a particle and its momentum at a particular instant of time, and to devise methods for measuring these two quantities simultaneously. However, if one attempts to do so with a subatomic particle such as an electron, a fundamental difficulty arises. Suppose we direct a beam of light at the electron in order to determine its position; the beam of light, consisting of photons, interacts with the electron. Although we may be able to determine the position of the electron precisely by this means, the interaction with photons changes the electron's momentum, making it impossible to account precisely for this interaction, even if it were only with a single photon. Hence, the electron's position and momentum cannot be determined simultaneously. The same sort of interaction occurs with any other method we may devise. In other words, *the act of observation alters the state of what is being observed*.

Interactions with the means of observation are insignificant at the macroscopic level but become critical at the subatomic level. For according to quantum mechanics, it is meaningless to specify both the position and the momentum of a particle at any given instant if they cannot be measured simultaneously. Such ambiguities are embodied in **Heisenberg's uncertainty principle**, according to which the product of the uncertainties in position and momentum is at least of the order of Planck's constant. So if we know either of these two quantities precisely, that is, with zero uncertainty, we cannot say anything about the other quantity (infinite uncertainty). Heisenberg's uncertainty principle applies not only to position and momentum but to any two quantities, such as energy and time, whose product has the dimensions of position $\times$ momentum (ml^2/t). Hence, it would be more appropriate to speak of the probability of the particle being at a certain location, or the probability of its having a certain momentum, and so on.

Summary: *Wave mechanics is the more general and rigorous form of mechanics. It reduces to classical mechanics when considering large-scale phenomena or situations where the dimensions of the system are very large compared to the De Broglie wavelength, so that the variation of potential energy over distances comparable to this wavelength is quite negligible. If this is not the case, classical mechanics breaks down, because it does not take into account the effect of the means of observation on what is being observed.*

The De Broglie wavelength is calculated in Section ST2.1 under macroscopic and microscopic conditions to illustrate the validity of classical mechanics.

Exercise 2.1.1

Calculate the De Broglie wavelength of an electron of 1 eV energy, where an eV is the change in energy of an electron in moving through a voltage difference of 1 V and is $|q| \times 1$ J. How does the wavelength vary with the energy?

Answer: 1.23×10^{-9} m; inversely as the square root of the energy. ∎

Exercise 2.1.2

Verify that energy × time has the same dimensions as momentum × distance. These dimensions are those of a quantity known as **action**. If the uncertainty in the instant at which a missile is launched is 1 ms, what is the uncertainty in the initial energy?

Answer: Approximately 6.63×10^{-31} J, assuming $\Delta E \times \Delta t = h$. ∎

Schrödinger's Equation

In wave mechanics, the De Broglie wave is represented by a wave function $\psi(r, t)$ that must satisfy **Schrödinger's equation**, where r denotes the coordinates of a point in three dimensions. Once this equation is solved for $\psi(r, t)$, then the probability of finding the particle at time t in a volume $d\Lambda$ centered at r is given by $|\psi(r, t)|^2 \, d\Lambda = \psi(r, t)\psi^*(r, t)d\Lambda$, where $\psi^*(r, t)$ is the complex conjugate of $\psi(r, t)$. That $\psi(r, t)$ is in general complex serves to emphasize the hypothetical nature of the De Broglie waves. Schrödinger's equation is to wave mechanics what Newton's equation of motion is to classical mechanics.

In fact, there are two Schrödinger's equations, the first is time dependent, and the second is time independent. If the potential energy in the system does not vary with time, then the solution to Schrödinger's time-dependent equation in one dimension is of the form:

$$\psi(x, t) = \psi(x)e^{-j\omega t} \tag{2.1.4}$$

where $\omega = 2\pi\nu = 2\pi E/h$ (Equation 2.1.3), E is the total energy, that is, the sum of potential and kinetic energies, and $\psi(x)$ satisfies the second Schrödinger's equation in one dimension:

$$-\frac{h^2}{8\pi^2 m}\frac{d^2\psi(x)}{dx^2} + U(x)\psi(x) = E\psi(x) \tag{2.1.5}$$

with $U(x)$ representing the potential energy as a function of position.

Consider a free electron, that is, an electron of total energy E in a region where the potential energy $U(x)$ is constant everywhere, so that the electron is not subjected to any electrical force. Since the zero of potential energy is arbitrary, $U(x)$ can be set to zero. Equation 2.1.5 reduces to $-\dfrac{h^2}{8\pi^2 m}\dfrac{d^2\psi(x)}{dx^2} = E\psi(x)$. A solution to this equation is of the form:

$$\psi(x) = Ae^{j|k|x} \tag{2.1.6}$$

where A and k are constants with respect to x; k is a vector, which means that in the one-dimensional case being considered, it can have positive or negative values. Boldfacing also serves to distinguish it from Boltzmann's constant k. Substituting Equation 2.1.6 in Schrödinger's equation, with $U(x) = 0$,

$$E = \frac{h^2}{8\pi^2 m}|k|^2 \tag{2.1.7}$$

Since the potential energy is zero everywhere, E is purely kinetic energy:

$$E = \frac{1}{2}m|u|^2 = \frac{1}{2m}|p|^2 \tag{2.1.8}$$

It follows from Equations 2.1.7 and 2.1.8 that:

$$p = \frac{h}{2\pi}k \tag{2.1.9}$$

From Equations 2.1.3 and 2.1.9,

$$|k| = \frac{2\pi}{\lambda} \tag{2.1.10}$$

k is the **wave vector** and is a vector quantity because it is related to the momentum vector by Equation 2.1.9. Its magnitude is also given by Equation 2.1.10 in accordance with De Broglie's hypothesis. $1/\lambda$ is the wave number, or the number of wavelengths per meter. $|k|$ is thus the number of radians per meter along the length of the wave.

Exercise 2.1.3

If the kinetic energy of the particle is E, show from Equations 2.1.7 and 2.1.10 that $\lambda = h/\sqrt{2mE}$, in accordance with De Broglie's hypothesis (Equation 2.1.3). ∎

Exercise 2.1.4

Calculate $|k|$ for a free electron of energy 1 eV.
   Answer: 5.1×10^9 rad/m. ∎

Wave Packets

The motion of an electron in wave mechanics is described in terms of a **wave packet** that is formed by the superposition of a large number of wave functions $\psi(r, t)$ (Equation 2.1.4 in three dimensions) having different energies. Because Schrödinger's equation is linear in the form being considered, the superposition of wave functions of different amplitudes, frequencies, and phases that satisfy Schrödinger's equation would still satisfy the equation. The superposed wave functions are chosen so that their sum is nonzero where the electron is known to be but is zero elsewhere, resulting in a wave packet, as diagrammatically illustrated in Figure 2.1.1. Both the spread in energy of the superposed wave functions forming the wave packet, and the spread in size of the wave packet over several De Broglie wavelengths, are a reflection of Heisenberg's uncertainty principle. As t increases, the individual waves reinforce or cancel at some new values of x. The wave packet therefore moves with time at the **group velocity** $u_g = d\omega/dk$, which, for a free electron, is in fact equal to the particle velocity u of the electron, as shown in Section ST2.2.

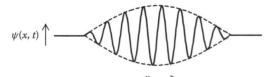

$\psi(x, t)$ ↑

$x \longrightarrow$

FIGURE 2.1.1
Representation of a wave packet.

Summary: *An electron can be considered as a particle whose behavior is described by a wave equation–Schrödinger's equation–whose solution gives the probability of finding the electron in a certain region at a given time. This solution has the form of a wave packet that has a nonzero amplitude over the region where the electron is known to be, and has zero amplitude elsewhere.*

Example 2.1.1 Electron in Infinite Potential Well

It is required to solve Schrodinger's equation for an electron in a one-dimensional "box" of length L, the electric potential energy being zero inside the box and infinite outside it.

SOLUTION

Let the box extend from $x = 0$ to $x = L$. $U(x) \to +\infty$ presents an insurmountable energy barrier to the electron, so the probability of finding the electron outside the box is zero. Hence ψ $(x) = 0$, for $x \leq 0$ and $x \geq L$. Inside the box, Equation 2.1.5 reduces to:

$$-\frac{h^2}{8\pi^2 m}\frac{d^2\psi(x)}{dx^2} = E\psi(x) \qquad (2.1.11)$$

$\psi(x)$ must be zero at $x = 0$ and $x = L$ so as to be continuous with $\psi(x)$ outside the box. Otherwise, a discontinuity at $x = 0$ or $x = L$ would lead to different probabilities of finding the electron at the same point, which is not acceptable. The solution of Equation 2.1.11 subject to the boundary conditions that $\psi(x) = 0$ at $x = 0$ and $x = L$ is:

$$\psi(x) = A\sin n\frac{\pi}{L}x \qquad (2.1.12)$$

where
  A is a constant
  n is an integer

A period of this function of x is the wavelength λ and is equivalent to an angle of 2π radians. This means that the boundary conditions are satisfied if L is an integral multiple of half-wavelengths, that is, $L = n\lambda/2$, where $n = 1, 2, 3, \ldots$ (Figure 2.1.2a). Since $|k| = 2\pi/\lambda$, Equation 2.1.12 can be written as:

$$\psi(x) = A\sin |k|x, \quad |k| = n\frac{\pi}{L} \qquad (2.1.13)$$

Substituting Equation 2.1.13 in Equation 2.1.11,

$$E = \frac{n^2 h^2}{8mL^2} = \frac{h^2}{8\pi^2 m}|k|^2 \qquad (2.1.14)$$

The value of A is obtained from the normalization condition that $\int_0^L \psi(x)\psi^*(x)dx = 1$, since the electron is known to be in the box. This gives $\int_0^L A^2 \sin^2 kx\, dx = 1$, or $A = \sqrt{2/L}$.

$|\psi(x)|^2$ is plotted in Figure 2.1.2b for $n = 1, 2$, and 3. According to Equation 2.1.14, energy levels are quantized: they can assume only discrete values corresponding to integer values of n, as illustrated in Figure 2.1.3a.

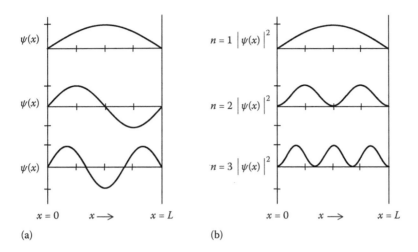

FIGURE 2.1.2
Plots of $\psi(x)$ (a) and $|\psi(x)|^2$ (b) for stationary states of a particle in an infinite potential well.

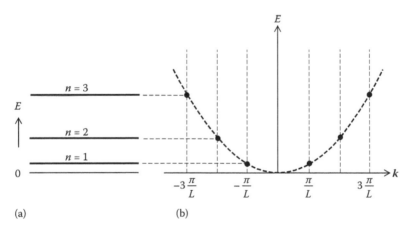

FIGURE 2.1.3
Energy levels of the stationary states of Figure 2.1.2: (a) discrete energy levels; (b) allowed values of k and E.

Equation 2.1.14 is of the same form as Equation 2.1.7 with the difference that the values of $|k|$ in Equation 2.1.14 are restricted to integer multiples of π/L, in accordance with Equation 2.1.13, whereas in Equation 2.1.7 the values of $|k|$ are unrestricted. In Figure 2.1.3b the dashed curve is a parabolic plot of E vs. k according to Equation 2.1.7. Equation 2.1.14 gives a series of points on this parabola corresponding to the allowed values of k. For the lowest energy state ($n = 1$), $E_1 = \dfrac{h^2}{8mL^2}$ J $= \dfrac{h^2}{8mL^2|q|}$ eV $= \dfrac{3.76 \times 10^{-19}}{L^2}$ eV, where L is in meters.

For macroscopic objects, where L is at least of the order of microns, $E_1 \cong 0$, in agreement with classical mechanics, according to which the lowest energy is for a particle at rest. But for L of the order of atomic dimensions, say $L = 3$ Å $= 3 \times 10^{-10}$ m, $E_1 = 4.18$ eV, which is not negligible. Moreover, for macroscopic objects, the separation of energies for different values of n is extremely small, so that energy values become essentially continuous, as in classical mechanics.

Note that $n=0$ gives $E=0$ and $\psi(x)=0$ (Equation 2.1.12). But $\psi(x)=0$ inside the box means zero probability of finding the particle inside the box, which leads to a contradiction, since the particle is assumed to be in the box.

Because energy for each n is precisely defined, time is indeterminate, according to Heisenberg's uncertainty principle. Each of these states is a **stationary state**, having $|\psi|^2$ independent of time. $|\psi|^2\Delta x$ for each stationary state gives the probability of finding the electron in a given interval Δx. For $n=1$, for example, the electron is most likely to be found around the middle of the box. Since time is indeterminate, the motion of the electron inside the box cannot be followed for a stationary state. Doing so would inevitably disturb the electron and remove it from the stationary state concerned. To follow the motion of the electron when not in a stationary state, a wave packet, of size smaller than L, should be built up from the superposition of wave functions for a number of stationary states.

Example 2.1.2 Finite Energy Barrier and Tunneling

An electron of total energy E traveling in the positive x-direction in a region of zero potential energy encounters at $x=0$ a potential barrier of infinite width in the positive x-direction, and of height U, where $U>E$ (Figure 2.1.4). It is required to derive the solution of Schrödinger's equation for the two regions $x \leq 0$ and $x \geq 0$.

SOLUTION

(a) Region $x \leq 0$. Equation 2.1.5, with $U(x)=0$, can be written as:

$$\frac{d^2\psi_1(x)}{dx^2} + k_1^2\psi_1(x) = 0 \qquad (2.1.15)$$

where $k_1^2 = 8\pi^2 mE/h^2$. The solution of Equation 2.1.15 is of the form:

$$\psi_1(x) = A\cos k_1 x + B\sin k_1 x \qquad (2.1.16)$$

as can be readily verified by substitution, where A and B are constants.

(b) Region $x \geq 0$. According to classical mechanics, a particle of total energy E cannot penetrate a potential energy barrier of height $U>E$, because the kinetic energy of the particle in this region, $E-U$, will be negative. In wave mechanics, on the other hand, Schrödinger's equation has a valid solution $\psi(x)$ in this region, which means that there is a

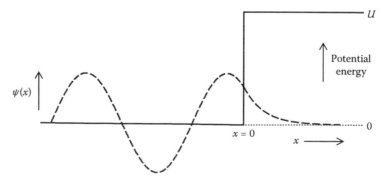

FIGURE 2.1.4
Figure for Example 2.1.2.

finite probability, as determined by $|\psi(x)|^2$, of finding the electron beyond the barrier. However, this probability is expected to decrease with x and with increasing $U - E$.

For $x \geq 0$, Equation 2.1.15 can be written as:

$$\frac{d^2\psi_2(x)}{dx^2} - k_2^2\psi_2(x) = 0 \tag{2.1.17}$$

where $k_2^2 = \frac{8\pi^2 m}{h^2}(U - E)$. The solution of Equation 2.1.17 is of the form:

$$\psi_2(x) = Ce^{-k_2x} + De^{k_2x} \tag{2.1.18}$$

as can be verified by substitution, where C and D are constants. Moreover, $D=0$, or else the probability of finding the electron inside of the barrier increases with x, which is impossible.

At $x=0$, $\psi(x)$ must be continuous, so that $\psi_1(0) = \psi_2(0)$, or else, as noted for the infinite potential well, there will be two different probabilities for finding the electron at $x=0$. Applying $\psi_1(0) = \psi_2(0)$ to Equations 2.1.16 and 2.1.18, with $D=0$, gives $A=C$.

Equation 2.1.5 can be integrated from $x=0^-$ on one side of the barrier to $x=0^+$ on the other side to the barrier to give:

$$\int_{0^-}^{0^+} \frac{d^2\psi}{dx^2} = \int_{0^-}^{0^+} \frac{8\pi^2 m}{h}(U - E)\psi(x)dx$$

or,

$$\left.\frac{d\psi}{dx}\right|_{x=0^+} - \left.\frac{d\psi}{dx}\right|_{x=0^-} = \int_{0^-}^{0^+} \frac{8\pi^2 m}{h}(U - E)\psi(x)dx \tag{2.1.19}$$

The integral on the RHS vanishes as long as $(U - E)$ is finite. This means that $d\psi/dx$ is continuous at $x=0$. It may be noted that in Example 1.1.2 $d\psi/dx$ is not continuous at $x=0$ and $x=L$ because the potential barrier is of infinite height.

Applying $\left.\frac{d\psi_1}{dx}\right|_{x=0} = \left.\frac{d\psi_2}{dx}\right|_{x=0}$ to Equations 2.1.16 and 2.1.18, with $D=0$, gives $B = -\frac{k_2}{k_1}A$.
The complete solution is therefore given by:

$$\psi_1(x) = A\left(\cos k_1 x - \frac{k_2}{k_1}\sin k_1 x\right), \quad x \leq 0 \tag{2.1.20}$$

$$\psi_2(x) = Ae^{-k_2x}, \quad x \geq 0 \tag{2.1.21}$$

These functions are plotted in Figure 2.1.4. It is seen that the larger k_2 the more rapid is the decrease of $\psi_2(x)$ with x and the smaller is the probability of finding the particle where $x > 0$.

When the electron is known to be in a given region $x \leq 0$ at a given time, a wave packet is set up over this region. The normalization condition discussed for the preceding Example gives the value of A. As t increases, the wave packet moves in the positive x-direction until it encounters the barrier. The probability is unity that the wave packet will be reflected at the barrier of infinite width, corresponding to the electron "bouncing" back in the negative x-direction. However, if the barrier width is small, and $U - E$ in the barrier region is not too large, there is a finite probability that the electron will penetrate the barrier and move to the other side. This phenomenon, known as **tunneling,** cannot be explained by classical mechanics but is explainable by the probabilistic interpretation of $|\psi(\mathbf{r}, t)|^2$, where $\psi(\mathbf{r}, t)$ is the wave function solution of Schrodinger's equation.

Exercise 2.1.5

An electron is in an infinite potential well 10 Å wide. (a) Determine the energy of the state for which $n = 3$; (b) if the electron drops to the lowest energy state, or ground state ($n = 1$), what is the frequency of the radiation emitted?

Answers: (a) 3.39 eV; (b) 7.28×10^{14} Hz. ∎

Exercise 2.1.6

Given an electron of 10 eV energy and a barrier of 50 eV. Determine k_1 and k_2.

Answers: $k_1 = 16.2 \times 10^9$ rad/m; $k_2 = 32.4 \times 10^9$ rad/m. ∎

Free Atoms

A free atom is an isolated atom that is not subjected to any external influences, as in a dilute gas. Schrödinger's equation can be solved exactly for a free atom only in very simple cases, such as that of a nucleus and a single electron. Solutions in this case are valid only for *integer* values of three parameters n, l, and m_l referred to as **quantum numbers** (Section ST2.3), which arise naturally from the three-dimensional spherical coordinates, just like the quantized value of n in the one-dimensional case of the infinite potential well. When considering atoms having more than one electron, quantization of energy still applies:

Concept: *Electrons in a free atom can only have quantized values of total energy corresponding to discrete energy levels.*

Moreover, to completely account for the data from atomic spectra, it was necessary to assume that the electron possesses a spin. That is, the electron not only rotates around the nucleus but also around its own axis. This effect is represented by a fourth quantum number m_s which can assume one of two values, $\pm 1/2$ for an electron, depending on the direction of spin.

The four quantum numbers define an **electron state**. According to **Pauli's exclusion principle**, which can be justified from quantum-mechanical considerations, not more than one electron can be in the same state, that is, be described by the same set of four quantum numbers.

Energy Band Structure

When atoms are grouped together in a crystal, the wave functions of the electrons in the individual atoms interfere with one another. Consequently, the discrete energy levels of identical free atoms no longer coincide but are grouped together into energy bands separated by forbidden energy gaps. A similar situation is encountered, for example, if two oscillators of the same frequency of oscillation are coupled together; the mutual interference between the two oscillators shifts the frequency of oscillation of one oscillator upward, that of the other oscillator downward, so that the two frequencies of oscillation form a "band."

The number of energy levels in an atomic energy band is very large, of the order of the number of atoms in a crystal, the density of atoms being typically about 10^{23} atoms/cm^3. According to Pauli's exclusion principle, not more than two electrons, having opposite spins, can occupy the same energy level, that is, have the same value of total energy.

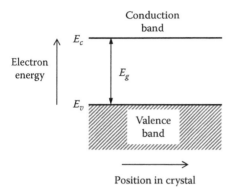

FIGURE 2.1.5
Energy band structure of semiconductor at 0K showing a filled valence band and an empty conduction band.

Concept: *At 0K, atoms in a crystal are at rest, and electrons occupy the lowest possible energy levels, subject to Pauli's exclusion principle.*

Note that at 0K electrons are not at rest. If they were, they would all have the same zero energy, which violates Pauli's exclusion principle. Because they occupy the lowest possible energy states, all valence electrons are bound to their parent atoms at 0K. The band that represents the energies of these electrons is the **valence band**. The next higher energy band is the **conduction band**, which is separated from the valence band by an energy gap E_g. In the case of semiconductors at 0K, the valence band is completely filled and the conduction band is completely empty, as illustrated in the energy band diagram of Figure 2.1.5.

The following should be noted concerning energy band diagrams: (1) The horizontal axis represents position in the crystal. The vertical axis represents total electron energy, that is, potential and kinetic, the total energy increasing upward on the diagram. (2) Conventionally, the highest electron energy is zero, corresponding to the potential energy of an electron at rest at an infinite distance from a positively charged nucleus. Hence, values of energy levels on the diagram are negative. Normally, energy differences are of interest and not absolute values. (3) The highest energy level in the valence band is denoted by E_v and the lowest energy level in the conduction band by E_c. The width of the gap separating these two bands is $E_g = E_c - E_v$; $E_g \cong 1.17$ eV for Si and $E_g \cong 1.52$ eV for GaAs at 0K.

Energy band diagrams are very helpful in understanding electric conduction in crystals and are used to characterize insulators, semiconductors, metals, and semimetals (Section ST2.4).

Exercise 2.1.7

If electrons in the valence band are to absorb energy from photons of green light, of 570 nm wavelength, what should be the minimum width of the air gap?
Answer: 2.18 eV. ∎

Exercise 2.1.8

If 1 in 10^{10} of the valence band electrons in Si at 0K gain sufficient energy to jump to the conduction band, what will be the density of conduction electrons, assuming the density of atoms in Si to be $5 \times 10^{22}/\text{cm}^3$?
Answer: $2 \times 10^{13}/\text{cm}^3$. ∎

2.2 Electric Conduction in Semiconductors

Electrons in a Periodic Potential

At $T > 0K$, some of the valence electrons acquire sufficient energy to break away from their parent atoms and circulate randomly in the crystal. They become **conduction electrons**, because in the presence of an applied electric field they can acquire a small drift velocity that is superposed on their random thermal motion, thereby supporting the flow of electric current. Their energies are now in the conduction band (Figure 2.1.5). The atoms that have lost valence electrons become positively charged ions.

When atoms are brought together in a crystal, the potential energy profiles for an electron in the vicinity of each positively charged nucleus (Section ST2.3) superimpose. Figure 2.2.1 illustrates the potential energy profile for a hypothetical, one-dimensional crystal, where positive charges denote nuclei of crystal atoms. Except at the ends of the crystal, the potential energy function $U(x)$ is periodic, of periodicity equal to the atomic spacing a. Strictly speaking, the potential energy function should include not only the interaction of a conduction electron with the positive ions but also with all other conduction electrons. This latter interaction is very difficult to account for and is weak compared to the periodic, electron–ions interaction.

The behavior of an electron in the periodic potential of the crystal is described by the solution to Schrödinger's equation, which takes the form:

$$\psi(x) = v_k(x)e^{jkx} \tag{2.2.1}$$

where $v_k(x)$ is also a periodic function of periodicity a. This is **Bloch's theorem**, and the wave function of the form of Equation 2.2.1 is a **Bloch function**. If $\psi(x)$ from Equation 2.2.1 is substituted in Equation 2.1.5, we obtain the equation that is satisfied by $v_k(x)$:

$$\frac{d^2 v_k(x)}{dx^2} + 2jk\frac{dv_k(x)}{dx} + \left[\frac{8\pi^2 m}{h^2}(E - U(x)) - k^2\right]v_k(x) = 0 \tag{2.2.2}$$

Recall that in the cases of a free electron and an electron in an infinite potential well, the solution to Schrödinger's equation gave a continuous parabolic E vs. k relation for the former case, and, for the latter case, a series of points on the parabola that are spaced at intervals

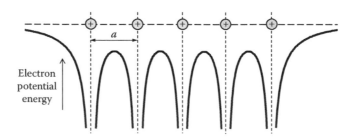

FIGURE 2.2.1
Potential energy profile for a row of atoms.

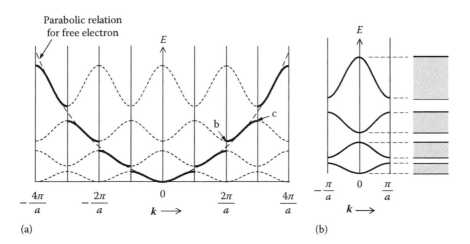

FIGURE 2.2.2
(a) Variation of electron energy with the wave vector in a single-dimensional crystal; (b) equivalent representation based on the periodicity in (a).

of π/L (Figure 2.1.3). Similarly, the solution to Equation 2.2.2 is a series of E vs. k curves, as indicated by the solid curves in Figure 2.2.2a, which also shows the aforementioned parabolic relation. The solid curves represent the allowed energies for various values of k. Thus, for $\frac{2\pi}{a} \leq k \leq \frac{3\pi}{a}$, the allowed energies are those in the range b to c. Since k is related to momentum by Equation 2.1.9, it is to be expected that higher energies are associated with larger values of k.

Two features of these curves are of immediate interest: (1) the discontinuity in E at values of k that are integral multiples of π/a, and (2) the periodicity of $2\pi/a$ in k, as illustrated by the thin dashed curves. For example, if the energy curve b to c is displaced by $-2\pi/a$ so that b lies on the vertical axis, it is still the same energy variation but for values of k that are reduced by $2\pi/a$. Because of this periodicity, it is customary to draw a reduced diagram for k between $-\pi/a$ and $+\pi/a$, as in Figure 2.2.2b, since such a diagram contains all the relevant information about the variation of E with k for all allowed energies. In a real crystal, the shapes of the E vs. k curves can be quite complicated and are different for $k's$ in different directions in the crystal. The range $-\frac{\pi}{a} \leq k \leq \frac{\pi}{a}$ is the first **Brillouin zone** in one dimension. The ranges $-\frac{2\pi}{a} \leq k \leq -\frac{\pi}{a}$ and $\frac{\pi}{a} \leq k \leq \frac{2\pi}{a}$ are the second Brillouin zone, and so on. Figure 2.2.2 again shows the formation of energy bands separated by gaps. Since the energy bands consist of discrete energy levels, the E vs. k curves are in fact not continuous but are made up of closely spaced points.

The formation of energy bands in connection with Figure 2.1.5 was attributed to interference between the wave functions of the discrete energy levels of free atoms as the atoms were brought close together in a crystal. In Figure 2.2.2, the same energy bands can be interpreted in terms of restrictions on wave travel in a crystal. The gaps occur at $k = \pm n\pi/a$, the condition for **Bragg reflection**, when traveling waves representing electron motion interfere with those reflected at the discontinuities due to the presence of crystal atoms. There is no wave propagation under these conditions because of the formation of standing waves (Section ST2.5).

Effective Mass

In the presence of an applied electric field, an electron in a crystal experiences a force F_{ext} and picks up energy from the field. The electron is represented as a wave packet having a group velocity $u_g = \dfrac{d\omega}{dk}$. From Equation 2.1.1, $\omega = 2\pi\nu = 2\pi E/h$. Differentiating,

$$u_g = \frac{d\omega}{dk} = \frac{2\pi}{h}\frac{dE}{dk} \tag{2.2.3}$$

The rate at which the electron picks up KE from the electric field is:

$$\frac{dE}{dt} = \frac{d}{dt}\left(\frac{1}{2}mu_g^2\right) = mu_g\frac{du_g}{dt} = F_{ext}u_g = F_{ext}\frac{2\pi}{h}\frac{dE}{dk} \tag{2.2.4}$$

assuming that m is constant, where $F_{ext} = mdu_g/dt$. But,

$$\frac{dE}{dt} = \frac{dk}{dt}\frac{dE}{dk} \tag{2.2.5}$$

Comparing Equations 2.2.4 and 2.2.5:

$$F_{ext} = \frac{h}{2\pi}\frac{dk}{dt} = \frac{d}{dt}\left(\frac{hk}{2\pi}\right) \tag{2.2.6}$$

For a classical particle subjected to a force F_{ext}, the equation of motion is:

$$F_{ext} = \frac{d}{dt}(p) \tag{2.2.7}$$

Equations 2.2.6 and 2.2.7 are of the same form. The quantity $hk/2\pi$ can therefore be considered as the **crystal momentum** of the electron.

We can go a step further and conclude that there should be associated with this momentum an **effective mass** m_e^* such that $m_e^* u_g = hk/2\pi$. Substituting Equation 2.2.3:

$$m_e^* = \left(\frac{h}{2\pi}\right)^2\frac{k}{dE/dk} \tag{2.2.8}$$

For a free electron, $E = \dfrac{h^2}{8\pi^2 m}|k|^2$ (Equation 2.1.7), so $m_e^* = m_e$.

Concept: *Replacing a true mass by an effective mass allows treating the electron as a free particle that obeys classical mechanics when subjected to an external force, such as an electric field.*

In other words, the effective mass accounts for the forces experienced by the electron in the crystal. As a free particle, *a conduction electron of effective mass m_e^* has the same potential energy everywhere in the crystal and is not subject to any force other than an external force.*

Hole Conduction

It is seen from Equation 2.2.8 that, in general, m_e^* is a function of k and can be larger or smaller than m_e depending on the value of $\dfrac{k}{dE/dk}$. Near the top or bottom of an energy band, the E vs. k curve can be approximated by a parabola, as in the case of a free electron. Under these conditions, $E = bk^2$, where b is a constant, so that $\dfrac{k}{dE/dk} = 2b$. This means that m_e^* is constant, and Equation 2.2.7 can then be written as $F = m_e^* \dfrac{du_g}{dt} = m_e^* a$, where a is the acceleration of the wave packet. From Equation 2.2.3:

$$a = \frac{du_g}{dt} = \frac{2\pi}{h}\frac{d}{dt}\left(\frac{dE}{dk}\right) = \frac{2\pi}{h}\frac{dk}{dt}\frac{d}{dk}\left(\frac{dE}{dk}\right) = \frac{2\pi}{h}\frac{dk}{dt}\frac{d^2E}{dk^2} \tag{2.2.9}$$

Substituting $\dfrac{dk}{dt} = \dfrac{2\pi}{h}F_{\text{ext}}$ from Equation 2.2.6:

$$a = \left[\left(\frac{2\pi}{h}\right)^2 \frac{d^2E}{dk^2}\right]F_{\text{ext}} \tag{2.2.10}$$

The quantity inside the square brackets must be $1/m_e^*$. Thus:

$$m_e^* = \left(\frac{h}{2\pi}\right)^2 \frac{1}{d^2E/dk^2} \tag{2.2.11}$$

Unlike Equation 2.2.8, Equation 2.2.11 does not explicitly involve k and shows that m_e^* depends on the curvature of the E vs. k curves. Near the bottom of an energy band, where the E vs. k curve is concave upward (Figure 2.2.2b), d^2E/dk^2, and hence m_e^*, is positive. On the other hand, the E vs. k curve is concave downward near the top of the energy band, so d^2E/dk^2, and hence m_e^*, is negative.

How is a negative m_e^* to be interpreted? Consider a one-dimensional motion of an electron in the positive x-direction. If a force F is applied in this direction, an electron of positive mass accelerates in the positive x-direction. When the mass is negative, a is negative and the electron decelerates. Such behavior cannot be explained by classical mechanics, but a "plausible" wave-mechanical explanation can be given in terms of the wave packet. If such a wave packet corresponds to electron energies near the top of the energy band, then although the individual waves making up the wave packet travel in the positive x-direction at phase velocities, their phases, amplitudes, and frequencies can be such that their superposition, that is the wave packet, decelerates rather than accelerates in the direction of the applied force.

To see how an electron of negative mass contributes to current flow, Figure 2.2.3 illustrates the directions of the force exerted, the velocity acquired, and the current when an electric field is applied in the positive x-direction to: (1) a positive charge of positive mass, (2) an electron of positive mass, and (3) an electron of negative mass. It is seen that an electron of negative mass acquires velocity in the same direction as that of a positive charge of positive mass but contributes to current in the opposite direction because of its negative charge.

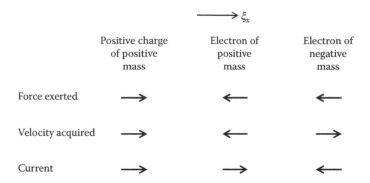

FIGURE 2.2.3
Effect of the electric field on particles of positive or negative charges and masses.

When valence electrons in a nearly filled valence band are subjected to an electric field, they can gain kinetic energy from the field and move to available higher energy levels in the valence band, giving rise to a current, although their motion is restricted to the covalent bonds of the crystal atoms. If the valence band is filled, they cannot ordinarily gain sufficient additional energy from the electric field to move to the conduction band by breaking away from their parent atoms. The electric field strength required for this would cause electrical breakdown of the material, as explained in Section 3.2, Chapter 3, resulting in a large, uncontrolled current. It follows that a filled valence band does not contribute to current flow under normal conditions.

Let the energy levels of the valence band be labeled $1 \ldots N_v$ in the order of increasing energy, and let us assume that electrons in the levels $1 \ldots N_{vm+}$ have positive effective mass in the presence of an applied electric whereas electrons in the levels $(N_{vm+} + 1) \ldots N_v$ have negative effective mass. The current density due to all the electrons in the filled valence band is:

$$J'_{vb} = -\frac{2|q|}{\Lambda} \sum_{i=1}^{N_{vm+}} \vec{u}_i - \frac{2|q|}{\Lambda} \sum_{i=N_{vm+}+1}^{N_v} \vec{u}_i = 0 \qquad (2.2.12)$$

where Λ is the volume of the crystal, and the factor 2 accounts for the fact that two electrons, one of each spin, occupy a given energy level. Equation 2.2.12 is simply of the form of current density as the product: charge × velocity × (number of electrons)/volume. Assuming that the electric field is in the positive x-direction, the u_i's for electrons having positive mass will be in the negative x-direction, as indicated by the arrow over the velocity symbol, and will, therefore, have negative values. The current due to these electrons is represented by the first term on the RHS of Equation 2.2.12 and will have a positive numerical value, that is, it will be in the direction of the electric field. The u_i's for electrons having negative mass will be in the positive x-direction, as indicated by the arrow over the velocity symbol, in accordance with Figure 2.2.3. The second term on the RHS of Equation 2.1.12 will have a negative numerical value that cancels out the first term.

Let us assume, for simplicity, and without invalidating the essence of the argument, that the levels in the valence band corresponding to a positive electron mass are filled, whereas the levels corresponding to a negative electron mass are empty. The current due to the applied electric field will then be represented by the first term on the RHS of Equation 2.2.12:

$$J_{vb} = -\frac{2|q|}{\Lambda} \sum_{i=1}^{N_{vm+}} \bar{u}_i \tag{2.2.13}$$

But this current is the negation of the second term on the RHS of Equation 2.2.12, so that:

$$J_{vb} = +\frac{2|q|}{\Lambda} \sum_{i=N_{vm+}+1}^{N_v} \bar{u}_i \tag{2.2.14}$$

J_{vb} according to Equation 2.2.14 is the same as the current due to positive charges $+|q|$ having a positive mass and moving in the same direction as the electric field. Hence,

Concept: *The contribution of a nearly filled valence band to current flow is as if the band is devoid of electrons, and the initially empty levels are occupied by positive charges, each having the same magnitude of charge as an electron and a positive mass.*

These positive charges are referred to as **holes** and arise because of empty energy levels or "holes" near the top of the valence band. *Like electrons, an effective mass m_h^* can be assigned to a hole, which allows treating a hole as a free particle that obeys classical mechanics when subjected to an external force.* Because holes are positively charged, hole energies increase downward on the energy band diagram. The valence band of a semiconductor at $T \gg 0K$ can be considered devoid of electrons, with the empty energy levels near the top of the band occupied by holes, the lowest hole energy being E_v. When they acquire additional kinetic energy, holes move to the lower levels in the valence band.

Density of Electron States

Of primary interest is the concentration of current carriers in a semiconductor at a given temperature. This concentration is calculated as the product of: (1) the number of energy levels that can be occupied by electrons or holes, and (2) the probability that a given level is occupied, the probability depending on the energy of the given level and the temperature.

The density of electron states at a given energy E is denoted by $N(E)$. This means that $N(E)dE$ is the number of electron states per unit volume having energy between E and $E + dE$. The number of electron states $N'(E)$ per unit volume with energy less than or equal to E is then:

$$N'(E) = \int_0^E N(E)dE \quad \text{or} \quad N(E) = \frac{dN'(E)}{dE} \tag{2.2.15}$$

$N'(E)$ can be calculated quite simply by considering free electrons in a cubical potential well of side L, with infinite potential energy outside the well. From Example 2.1.1 the energy levels in a one-dimensional well are $E = \frac{h^2 n^2}{8m_e L^2}$ J $= \frac{h^2 n^2}{8m_e L^2 |q|}$ eV. For a three-dimensional

well, similar relations hold for the other two dimensions, so
that the energy levels in eV are given by:

$$E = \frac{h^2}{8m_e L^2 |q|} \left(n_x^2 + n_y^2 + n_z^2\right) \qquad (2.2.16)$$

where $n_x, n_y,$ and n_z are positive integers that account for the
quantization of energy in the x-, y-, and z-directions, respectively. The lowest energy state is that having $n_x = n_y = n_z = 1$,
and the energy increases by one level each time $n_x, n_y,$ or n_z is
incremented by 1. This can be represented by cubes of unit
sides in a three-dimensional space having axes $n_x, n_y,$ and n_z
(Figure 2.2.4).

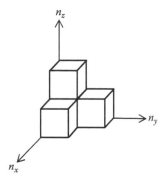

For a given E, $n_x^2 + n_y^2 + n_z^2 = \text{constant} = r^2$, which describes
a sphere of radius r centered at the origin. The number of
energy levels having energy less than or equal to E is then
the number of unit cubes inside an octant (that is, one-eighth of the volume) of the sphere
of radius r, since $n_x, n_y,$ and n_z can only take positive values. The volume of the octant is

$\frac{1}{6}\pi r^3 = \frac{\pi}{6}\left(\frac{8m_e L^2}{h^2}E|q|\right)^{\frac{3}{2}}$. The number of energy levels per unit volume of material is:

$$N'_L(E) = \frac{\pi}{6}\left(\frac{8m_e |q|}{h^2}\right)^{\frac{3}{2}} E^{\frac{3}{2}} \qquad (2.2.17)$$

where the L subscript is added to distinguish electron energy levels from electron states.
When the number of energy levels is large, the separation between allowed energy levels is
small, so E can be considered continuous. Equation 2.2.17 is differentiated to give:

$$N_L(E) = \frac{dN'_L(E)}{dE} = \frac{\pi}{4}\left(\frac{8m_e |q|}{h^2}\right)^{\frac{3}{2}} E^{\frac{1}{2}} \qquad (2.2.18)$$

For conduction electrons in a solid, the potential barriers at the surface can be considered
virtually infinite, if the solid is undisturbed. This is because if an electron attempts to
escape from the solid, it leaves behind a positive charge that attracts the electron back,
unless the electron is of exceptionally high energy. Moreover, the effect of the periodic
potential in the crystal can be accounted for by using the effective mass, as explained
previously. Replacing m_e by m_e^*, and bearing in mind that each energy level can accommodate two electrons, one of each spin, the density of electron states per unit volume in the
conduction band is:

$$N_e(E) = 4\pi\left(\frac{2m_e^* |q|}{h^2}\right)^{\frac{3}{2}}(E - E_c)^{\frac{1}{2}} \qquad (2.2.19)$$

where the energy E of the electron in the box in Equation 2.2.18 corresponds to the energy
$E - E_c$ of the electron in the conduction band in Equation 2.2.19. $N_e(E)$, the number
of energy states per unit volume per unit energy difference, increases with E, but at a
decreasing rate.

The density function for holes is obtained from Equation 2.2.18 by replacing m_e by m_h^* and E by $E_v - E$:

$$N_h(E) = 4\pi \left(\frac{2m_h^* |q|}{h^2} \right)^{\frac{3}{2}} (E_v - E)^{\frac{1}{2}} \tag{2.2.20}$$

where E is the *electron energy* of a given energy level in the valence band, so that $E \le E_v$.

Fermi–Dirac Distribution

The probability that a particle has a certain energy E is given by a **distribution function** $f(E)$ that depends on the system of the particles in question. For electrons subject to Pauli's exclusion principle, $f(E)$ is given by the **Fermi–Dirac distribution** (Section ST2.6) expressed as:

$$f(E) = \frac{1}{1 + e^{\frac{E - E_F}{kT}}} \tag{2.2.21}$$

where
  k is Boltzmann's constant in eV/K
  E_F is a constant known as the **Fermi Level**

$f(E)$ is plotted in Figure 2.2.5 as a function of $E - E_F$ for various T. At $T = 0K$, $f(E) = 1$ for $E < E_F$, and $f(E) = 0$ for $E > E_F$. This means that at 0K all allowed energy levels of energy less than E_F are occupied, and all levels of energy greater than E_F are empty. As T is raised, electrons of highest energy acquire additional energy and are able to move to higher energy levels. Electrons of low energies cannot move to higher levels because these are already occupied.

For $T > 0K$, $f(E) = 1/2$ at $E = E_F$, and $f(E)$ is symmetrical with respect to this point, since $f(E_F + \Delta E) = 1 - f(E_F - \Delta E)$ (Exercise 2.2.1). If $E - E_F \gg kT$,

$$f(E) \cong e^{-(E - E_F)/kT} \tag{2.2.22}$$

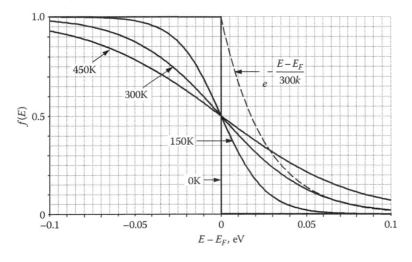

FIGURE 2.2.5
Variation of the Fermi–Dirac distribution function with energy and temperature.

which is another distribution function, the Maxwell–Boltzmann distribution. For $(E - E_F) >$ 0.08 eV, the deviation between the two distributions is less than 5% at 300K (Figure 2.2.5). The Maxwell–Boltzmann distribution applies to particles that are sufficiently far apart so that their wave functions do not interact to any significant extent, like the molecules of a dilute gas. When the probability of occupancy of an energy level is low, as at the tail end of the Fermi–Dirac distribution, the restriction of not more than two electrons per energy level is no more relevant and the concentration of carriers is low enough so that their wave functions do not interact to any significant extent. Hence, the Maxwell–Boltzmann distribution can be used instead of the Fermi–Dirac distribution under these conditions, which considerably facilitates mathematical analysis.

Exercise 2.2.1

Verify that $f(E_F + \Delta E) = 1 - f(E_F - \Delta E)$ for any $\Delta E = E - E_F$.   ■

Exercise 2.2.2

Calculate the percentage difference between the Fermi–Dirac and Boltzmann distributions for (a) $(E - E_F) = 2kT$; (b) $(E - E_F) = 3kT$; (c) $4kT$.
   Answers: (a) 13.5%; (b) 5%; (c) 1.8%.   ■

2.3 Intrinsic and Extrinsic Semiconductors

Crystal Structure

The most important semiconductor materials at present are the crystalline forms of the element silicon (Si) and the compound gallium arsenide (GaAs). The atoms in a crystal are regularly arranged in a periodic structure, the **crystal lattice**. The basic unit that periodically repeats in the crystal lattice is the **unit cell**. Silicon, germanium, and gallium arsenide have a diamond structure consisting of a tetrahedron in which each atom is surrounded by four equidistant atoms at the corners of the tetrahedron, as illustrated in Figure 2.3.1. The diamond structure is part of the unit cell, which is a face-centered cubic lattice with four additional atoms inside the cube. In the case of Si, the cube edge is 0.543 nm, the *lattice constant*. The edge of the smaller cube of Figure 2.3.1 is half the lattice

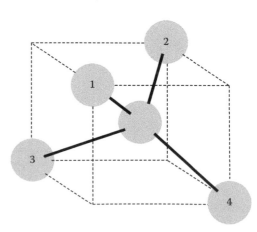

FIGURE 2.3.1
Tetrahedral crystal lattice.

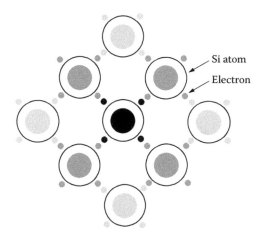

Si atom

Electron

FIGURE 2.3.2
Two-dimensional representation of a silicon crystal illustrating the sharing of valence electrons between each atom and its four closest neighbors.

constant, that is, approximately 0.272 nm. Bearing in mind that the radius of the Si atom is approximately 0.118 nm, it can be readily calculated that the Si atoms occupy only about 4.3% of the volume of the unit cell (Problem 2.2.1).

The tetrahedral structure can be conveniently represented in two dimensions, as shown in Figure 2.3.2 for Si. Every atom has four valence electrons, each of which is shared with one of its four closest neighbors in a **covalent bond**. Effectively, this provides each atom with a stable outer shell of eight electrons. Because of this sharing, the valence electrons are not bound to any particular atom but are continuously exchanged between atoms, thereby producing a strong attractive force that is mainly responsible for holding the crystal together.

GaAs has a tetrahedral structure in which each atom of one element has four closest neighbors of the other element. In a given sample, half the atoms are Ga and half are As. Ga and As atoms can be imagined to alternate along each diagonal row of atoms in the two-dimensional representation corresponding to that of Figure 2.3.2. Whereas, bonding in Si crystals is purely covalent and is largely covalent in GaAs crystals, there is some ionic bonding in GaAs because of the different number of valence electrons in Ga and As.

Intrinsic Semiconductor

This is a semiconductor without any impurities. The energy band diagram of an intrinsic semiconductor at 0K has a completely filled valence band and a completely empty conduction band, separated by a relatively narrow energy gap, as illustrated in Figure 2.1.5. When the temperature is raised above 0K, electrons from the valence band acquire sufficient energy to move to the conduction band. Every time an electron moves from the valence band to the conduction band, an *electron–hole pair is generated* (Figure 2.3.3). A conduction electron of total energy $E > E_c$ has a potential energy E_c, as discussed previously in connection with effective mass, the additional energy $E - E_c$ being KE. Similarly a hole of total energy E possesses potential energy E_v and KE $E_v - E$, where E_v and E for a hole are expressed in terms of the corresponding electron energies.

Exercise 2.3.1

If the atomic density of Si is 5×10^{22} atoms/cm^3 and $n_i = 10^{10}$/cm^3 at 300K. How many Si atoms contribute, on average, 1 electron–hole pair?
  Answer: 1 in 5×10^{12}. ∎

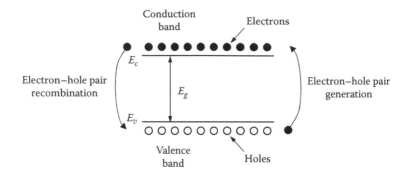

FIGURE 2.3.3
Electron–hole pair generation and recombination.

The movement of electrons between the valence and conduction bands is dynamic, occurring continually in both directions. A conduction electron may be "captured" by an unsatisfied covalent bond and becomes a valence electron, losing energy in the process. On the energy band diagram, the electron drops from the conduction band to the valence band, and an electron and a hole are said to have *recombined* (Figure 2.3.3). On average, an electron or a hole exists for a time τ before recombination, τ being the mean **lifetime** of the carrier. Unless otherwise indicated, the term "electron" will henceforth refer to a conduction electron.

Concept: *In an intrinsic semiconductor, the concentrations of electrons and holes are equal and are denoted by the* **intrinsic concentration** n_i. *This concentration increases with temperature and is larger, at a given temperature, for materials having a smaller E_g.*

Electrons and holes have the same concentration n_i in an intrinsic semiconductor because they are generated in pairs and recombine in pairs. n_i increases with temperature as more valence electrons acquire sufficient energy to move to the conduction band. The smaller E_g, the larger is the number of electrons that have sufficient energy, at a given temperature, to move to the conduction band. E_g decreases slightly with temperature (Problem P2.1.2), because electrons acquire additional energy at higher temperatures, so that it becomes easier for them to move to the conduction band. At 300K, E_g is approximately 1.12 eV for Si and 1.42 eV for GaAs. n_i at 300K is approximately $10^{10}/\text{cm}^3$ for Si and $2 \times 10^6/\text{cm}^3$ for GaAs.

Table 2.3.1 lists some properties of intrinsic Si and GaAs at 300K. Mobility and lifetime are discussed in Sections 2.6 and 2.7, respectively. In comparing Si and GaAs, it should be noted that Si has many desirable physical properties, such as intrinsic concentration and carrier mobilities which result in devices having good electrical properties. It can be readily oxidized to form SiO_2, which is an excellent and easily controlled insulator. The comparable electron and hole mobilities in Si allow complementary transistor circuits (CMOS) featuring greatly reduced power dissipation, leading to a tremendous increase in packaging density and reduction in manufacturing costs, as will be described later. For these reasons Si-based CMOS is the dominant semiconductor technology at present. Moreover, continued improvements in transistor materials and structure further enhance the dominant role of Si.

TABLE 2.3.1

Some Properties of Intrinsic Si and GaAs at 300K

	Si	GaAs
Density of atoms or molecules per cm^3	5×10^{22}	4.42×10^{22}
Energy gap, E_g, eV	1.12	1.42
Intrinsic concentration per cm^3	$\cong 10^{10}$	$\cong 2 \times 10^6$
Carrier mobility cm^2/Vs		
Electrons	1430	9200
Holes	480	320
Minority carrier lifetime, s	$\cong 2.5 \times 10^{-3}$	$\cong 10^{-8}$

Nevertheless, GaAs has some properties that are important for particular applications. It is widely used in light-emitting devices. Compared to Si, GaAs has a much smaller intrinsic concentration. This is advantageous for ICs that might be exposed to radiation, as in satellites, as well as for high-speed applications. The speed advantage of GaAs is further enhanced by its very high electron mobility, but its hole mobility is much lower. This, in addition to the fact that there is no GaAs-based insulator that possesses the many desirable properties of SiO$_2$, has made it impractical to implement complementary structures in GaAs. As a result, power dissipation is relatively high, and because of the higher density of crystal defects due to the difficulty of growing single GaAs crystals, the packaging density is lower than in CMOS. Other than in light-emitting devices, GaAs is mainly used in very high-speed and high-frequency applications.

Extrinsic Semiconductors

The operation of semiconductor devices depends on having carriers predominantly either holes or electrons. This is achieved by **doping**, that is, adding a carefully controlled concentration of certain impurity atoms, usually in the range of 1 impurity atom per 10^4–10^8 crystal atoms. The semiconductor becomes **extrinsic**, either *n*-type or *p*-type. The predominant carriers are the **majority carriers**, whereas carriers of the other type become **minority carriers**.

In an *n*-type semiconductor, the dopant is pentavalent, such as phosphorus or arsenic, having one electron more than the four needed for the covalent bonds. The ionization energy is therefore low, approximately 50 meV in a Si crystal (Example SE2.4), so that it takes little energy for the electron to break away from its parent atom and become a conduction electron, leaving a positive ion behind (Figure 2.3.4a). Because it donates an

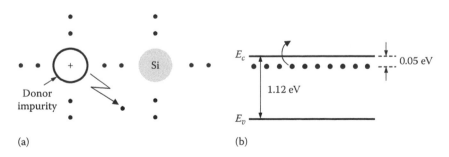

(a) (b)

FIGURE 2.3.4
(a) Donor impurity atom in silicon crystal; (b) energy-band diagram.

electron, this type of impurity is a **donor impurity**. Through associations of the letter n, it is easy to remember that in an n-type semiconductor, impurity atoms are do*n*ors, and majority carriers are *n*egative charges.

In the energy band diagram (Figure 2.3.4b), the donor impurity introduces a narrow donor energy band, having as many electron states as there are donor atoms in the crystal, and located about 0.05 eV below the edge of the conduction band. At 0K, the band is filled, which means that the extra electrons are bound to the impurity atoms. The valence band will also be filled and the conduction band empty, as in an intrinsic semiconductor. As the temperature is raised, some of the additional electrons jump the 0.05 eV gap into the conduction band and become conduction electrons, which corresponds to their breaking away from the donor atoms and becoming free to circulate in the crystal. At room temperature, practically all the donor atoms are ionized, and $n_{no} \cong N_D$, where n_{no} is the equilibrium concentration of electrons in the n-type semiconductor and N_D is the concentration of donor atoms, which is ordinarily much greater than n_i. In comparison, the concentrations of electrons and of holes due to the intrinsic carrier generation are very small at room temperature, but at high enough temperatures, the intrinsic mechanism predominates (Figure 2.3.5). In the temperature range that is usually of interest in a semiconductor operation (about 250K–450K), the concentration of electrons is practically constant at N_D, and is much greater than the concentration of holes.

An analogous situation exists for holes in a p-type semiconductor. The dopant is trivalent, such as boron or gallium, so that an electron is missing from one of the covalent bonds. It takes little energy for an electron from a crystal atom to attach to the impurity atom, making it a negative ion (Figure 2.3.6a). Because it accepts an electron, this type of

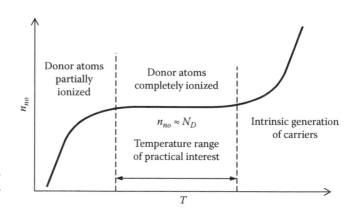

FIGURE 2.3.5
Temperature dependence of the majority carrier concentration in n-type semiconductor.

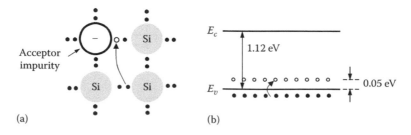

FIGURE 2.3.6
(a) Acceptor impurity atom in silicon crystal; (b) energy-band diagram.

impurity is an **acceptor impurity**. Through associations of the letter p, it is easy to remember that in a p-type semiconductor, impurity atoms are acce*p*tors, and majority carriers are *p*ositive charges.

In the energy band diagram (Figure 2.3.6b), the acceptor impurity introduces an acceptor energy band, having as many electron states as there are acceptor atoms in the crystal, and located about 0.05 eV above the edge of the valence band. At 0K, the band is empty, but as the temperature is raised, some of the valence electrons of the crystal atoms jump the 0.05 eV gap into the empty acceptor states, leaving holes in the valence band. At room temperature, practically all the acceptor atoms are ionized, and $p_{po} \cong N_A$, where p_{po} is the equilibrium concentration of holes in the p-type semiconductor and N_A is the concentration of acceptor atoms, which is ordinarily much greater than n_i. As for n-type semiconductors, the concentration of holes is practically constant at N_A over the temperature range of interest and is much greater than the concentration of electrons.

Concept: *Over the temperature range of practical interest, the type and concentration of current carrier in an extrinsic semiconductor are determined by the type of dopant and its concentration.*

In other words, the current carriers in an extrinsic semiconductor can be predominantly positive or negative electric charges whose concentrations can be varied over a wide range. This gives great flexibility to semiconductor devices and underlies their practical importance.

2.4 Electrochemical Potential

A characteristic feature of semiconductors is that the current can be a drift current due to an applied electric field, as in metals, or a diffusion current due to differences in carrier concentrations in different parts of the semiconductor. There is no diffusion current in metals, because the conductivity is so high that any concentration gradients result in large currents that immediately eliminate these gradients. The analysis of systems having both electric potential and concentration gradients is greatly facilitated by the concept of electrochemical potential.

Expression for Chemical Potential

Atomic particles, such as atoms, electrons, or holes that are free to move in a given medium, will move from regions of high concentration to regions of low concentration. A region that contains a given particle at a higher average concentration than adjacent regions will gradually expand due to random thermal motion of the particle. If a drop of a water-soluble dye, for example, is deposited in a glass of water, the dye spreads by diffusion until it is uniformly distributed throughout the glass. When this happens, the same number of dye particles, on average, will move per unit area per unit time in any specified direction.

Since the net motion of free particles can be considered quite generally to occur from regions of higher potential energy to regions of lower potential energy, a type of potential energy, referred to as **chemical potential energy**, is associated with concentration. That is, regions of high concentration of a given particle are at a higher chemical potential energy than regions of lower concentration. Just as electric potential is electric potential energy per particle, a **chemical potential** can be defined as chemical potential energy per particle. As in the case of electric potential, differences in chemical potential are normally of interest and not the absolute value of the potential.

To derive the expression for chemical potential, consider a one-dimensional flow of particles under the influence of a concentration gradient. The flux M, in the number of particles per unit area per unit time passing a given plane that is oriented perpendicularly to the direction of flow, is given by **Fick's law**:

$$M = -D\frac{dC}{dx} \tag{2.4.1}$$

where
  C is the concentration at x
  D is a **diffusion constant** that depends on the resistance to the motion of particles in the medium

If C decreases with x ($dC/dx < 0$), then M is in the positive x-direction ($M > 0$), since the particles move in this direction, from regions of high concentration to regions of low concentration. A negative sign is required in Equation 2.4.1 to make D positive.

Flux is also related quite generally to concentration and velocity u as follows:

$$M = Cu \tag{2.4.2}$$

Mobility μ is defined as the magnitude of drift velocity per unit electric field:

$$\mu = \left|\frac{u_{\text{drift}}}{\xi_x}\right| = |q|\frac{u_{\text{drift}}}{F_{\text{drift}}} \tag{2.4.3}$$

where the force on an electron or a hole due to an electric field ξ_x is $F_{\text{drift}} = q\xi_x$, q being the electronic charge, including the sign. For particles of positive mass, as is true of both electrons and holes, u_{drift} is in the direction F_{drift}, so that μ is a positive quantity. μ is expected to depend on the effective mass of the particle, so that the smaller the effective mass, the larger is the velocity acquired per unit driving force, and the greater is the mobility (Section ST2.7).

In the presence of a concentration gradient alone, particles can be considered subject to some "diffusion force" F_{diff} and acquire a velocity u_{diff}, just as in the case of drift in the presence of an electric field. The velocity acquired per unit driving force is independent of the nature of the force, depending only on the particle and the medium. It follows from Equation 2.4.3 that:

$$\frac{u_{\text{diff}}}{F_{\text{diff}}} = \frac{u_{\text{drift}}}{F_{\text{drift}}} = \frac{\mu}{|q|} \tag{2.4.4}$$

Clearly, diffusion and mobility are related, for both depend on the ease with which a particle moves in a given medium. This relationship is given by the **Nernst–Einstein relation**:

$$D = \mu\frac{k_J T}{|q|} = \mu V_T \tag{2.4.5}$$

where
  k_J is Boltzmann's constant in J/K
  T is the absolute temperature
  V_T is the thermal voltage

Note that the dimensions of μ are in m²/Vs, and the dimensions of D are m²/s.
  Eliminating D, μ, u_{diff}, and M between the preceding equations, bearing in mind that for diffusion, u in Equation 2.4.2 should be replaced by u_{diff}, gives:

$$F_{\text{diff}} = -\frac{k_J T}{C}\frac{dC}{dx} = -\frac{d}{dx}\left[k_J T \ln\frac{C}{C_r} + \kappa'_{Jr}\right] \tag{2.4.6}$$

where C_r and κ_r' are arbitrary constants.
  A driving force is quite generally the negative of a potential energy gradient. The quantity in square brackets can therefore be identified as the chemical potential energy per particle, κ', which is the chemical potential. Thus, for a given concentration C:

$$\kappa'_J = k_J T \ln\frac{C}{C_r} + \kappa'_{Jr} \tag{2.4.7}$$

where κ'_J is the chemical potential in J/particle and κ'_{Jr} is the chemical potential of some reference concentration C_r, since $\kappa'_J = \kappa'_{Jr}$ when $C = C_r$. Including κ'_{Jr} is in keeping with the fact that potential energy is generally measured with respect to some arbitrary reference.

Exercise 2.4.1

Verify the statement: "A driving force is quite generally the negative of a potential energy gradient," in the case of: (a) electric field; (b) gravitational field. ∎

Exercise 2.4.2

If the chemical potential of electrons in an intrinsic semiconductor is taken as zero, what is the chemical potential of electrons in an n-type semiconductor of donor concentration $N_D = 10^{15}/\text{cm}^3$ at 300K?
  Answer: 4.67×10^{-20} J or 0.29 eV. ∎

Expression for Electrochemical Potential

If the particle is electrically charged, and is located in a region where the concentration is C and the electric potential is V, the algebraic sum of the chemical potential energy and the electric potential energy is the electrochemical potential energy. The electrochemical potential energy per particle is the **electrochemical potential** κ. It follows that:

$$\kappa_J = k_J T \ln\frac{C}{C_r} + q(V - V_r) + \kappa_{Jr} \tag{2.4.8}$$

where the r subscript refers to a reference state whose concentration is C_r, whose electric potential is V_r, and whose electrochemical potential is κ_{Jr}. As noted previously, κ_{Jr} can be assigned any arbitrary value, because it is the differences in potential that are of interest and not the absolute values of potential. Note that all terms in Equation 2.4.8 are in J/particle; hence, the J subscript.

Exercise 2.4.3

If the *n*-type semiconductor of Exercise 2.4.2 is at a voltage of 1 V with respect to the intrinsic semiconductor, what is the electrochemical potential of an electron in the *n*-type semiconductor? If the semiconductor were *p*-type of the same doping level, what would be the electrochemical potential of a hole?

    Answers: -0.71 eV; 1.29 eV.   ∎

Generalized Ohm's Law

When electrons or holes flow under the influence of both a concentration gradient and an electric potential gradient, the current density is:

$$J = J_{\text{diff}} + J_{\text{drift}} \tag{2.4.9}$$

where

    J_{diff} is the diffusion current density due to the concentration gradient
    J_{drift} is the drift current density due to the electric field

The drift component is the familiar current in electric circuits that obeys Ohm's law. When current flow and electric field are not uniform,

$$J_{\text{drift}} = \sigma \xi_x \tag{2.4.10}$$

where σ is the conductivity. Conductivity is related to mobility, since the higher the mobility of current carriers, the larger is the velocity per unit applied force and the larger is the current. From Equation 2.4.2, $J_{\text{drift}} = qM_{\text{drift}} = qCu_{\text{drift}}$. Substituting for u_{drift} from Equation 2.4.3 gives $J_{\text{drift}} = \mu C |q| \xi_x$. Comparing with Equation 2.4.10,

$$\sigma = |q| \mu C \tag{2.4.11}$$

By definition, diffusion flux M is related to diffusion current density by $M = J_{\text{diff}}/q$. Substituting in Equation 2.4.1:

$$J_{\text{diff}} = -Dq \frac{dC}{dx} \tag{2.4.12}$$

Or, from Equations 2.4.5 and 2.4.11:

$$J_{\text{diff}} = -\frac{\sigma}{q} \frac{k_J T}{C} \frac{dC}{dx} \tag{2.4.13}$$

The sign of J_{diff} should be noted. If $dC/dx < 0$, the particle moves in the positive x-direction. The sign of J_{diff} is then the same as that of q. Combining Equations 2.4.10 and 2.4.13:

$$J = -\frac{\sigma}{q} \frac{k_J T}{C} \frac{dC}{dx} - \sigma \frac{dV}{dx} \tag{2.4.14}$$

$$= -\frac{\sigma}{q} \frac{d}{dx} \left[k_J T \ln \frac{C}{C_r} + q(V - V_r) + \kappa_{Jr} \right] = -\frac{\sigma}{q} \frac{d\kappa}{dx} \tag{2.4.15}$$

Equation 2.4.15 is known as **generalized Ohm's law** because it applies when current is due to both diffusion and drift. In the absence of a concentration gradient, it reduces to the familiar $J = -\dfrac{\sigma}{q}\dfrac{d[q(V - V_r)]}{dx} = -\dfrac{\sigma}{q}\dfrac{qdV}{dx} = \sigma\xi_x$ (Equation 2.4.10).

When considering electrons and holes, it is convenient to express energies, including chemical and electrochemical potentials, in eV. Dividing both sides Equation 2.4.7 by $|q|$:

$$\kappa' = V_T \ln\frac{C}{C_r} + \kappa'_r \tag{2.4.16}$$

where $V_T = kT$ and each term is in eV/particle. Equation 2.4.8 becomes for the holes:

$$\kappa_h = V_T \ln\frac{C}{C_r} + (V - V_r) + \kappa_r \tag{2.4.17}$$

where $(V - V_r)$ is in volts, numerically equal to eV, and the remaining terms are in eV/particle. For electrons:

$$\kappa_e = kT \ln\frac{C}{C_r} - (V - V_r) + \kappa_r \tag{2.4.18}$$

Equation 2.4.15 becomes for holes and electrons:

$$J = -\sigma_h \frac{d\kappa_h}{dx} = \sigma_e \frac{d\kappa_e}{dx} \tag{2.4.19}$$

Concept: *The driving force for electric current is, in general, the electrochemical potential gradient. In the absence of a concentration gradient, the electrochemical potential gradient is the same as the electric potential gradient, that is, the negative of the electric field.*

Exercise 2.4.4

Verify that when C is constant, Equation 2.4.15 reduces to Equation 2.4.10.   ■

Exercise 2.4.5

It appears from Equation 2.4.15 that the net current could be zero if J_{diff} and J_{drift} are equal but opposite. Derive the condition for this to occur.

Answer: $V - V_r = -\dfrac{kT}{q}\ln\dfrac{C}{C_r}$.   ■

State of Equilibrium

Equilibrium conditions play an essential role in the discussion of semiconductor systems, so the definition of a state of equilibrium should be clearly understood.

Definition: *Under equilibrium conditions, the temperature and pressure in the system, and the electrochemical potential of every species of freely moving particles, are equalized throughout the system, do not change with time, and the system is not disturbed in any way.*

The definition implies that no work is done in moving a freely moving particle from one part of the system to another. Hence,

Concept: *In a system at equilibrium, there can be no net flow of freely moving particles, which implies no current flow.*

A net flow of particles from one part of the system to another can only occur because of differences in temperature, pressure, or electrochemical potential. These differences are precluded by the definition of a state of equilibrium. The converse statement, however, is not true, that is, if there is no net flow of freely moving particles, then it cannot be concluded that the system is at equilibrium (see Example 3.3.1, Chapter 3).

Example 2.4.1 Electric Field in Nonuniformly Doped Semiconductor

Given an *n*-type semiconductor slab, oriented along the *x*-axis and with $x = 0$ at one end of the slab. The semiconductor is doped so that the impurity concentration varies exponentially with *x* as $N_D(x) = N_D(0)e^{-\alpha x}$. It is required to determine $\xi(x)$ at equilibrium assuming that $N_D(x) \gg n_i$ over the range of *x* involved, so that electron concentration equals $N_D(x)$.

SOLUTION

$J_{ediff} = D_e|q|dN_D(x)/dx$ and $J_{edrift} = |q|N_D(x)\mu_e\xi(x)$. At equilibrium, the electrochemical potential of electrons is equalized throughout the slab, so there cannot be a net current flow, which means that $J_{ediff} + J_{edrift} = 0$. Hence, $-\alpha D_e N_D(0)e^{-\alpha x} + N_D(0)e^{-\alpha x}\mu_e\xi(x) = 0$. This gives $\xi(x) = D_e\alpha/\mu_e = V_T\alpha$. $\xi(x)$ is in the positive *x*-direction, because electrons tend to diffuse in this direction, which causes a diffusion current in the negative *x*-direction. To counteract this current, $\xi(x)$ has to be in the positive *x*-direction. Note that for this exponential doping profile $\xi(x)$ is constant, depending only on α and *T*.

Example 2.4.2 Conductivity of Intrinsic Silicon

Based on the data in Table 2.3.1, it is required to determine the conductivity of intrinsic Si at 300K.

SOLUTION

$\sigma = \mu n|q|$ (Equation 2.4.11), where μ is the mobility of the particle. Electrons and holes have the same magnitude of charge, concentration n_i, but different mobilities. Hence, $J = \sigma_e\xi + \sigma_h\xi = \sigma_i\xi$, where $\sigma_i = (\mu_e + \mu_h)n_i|q| = (1430 + 480) \times 10^{10} \times (1.6 \times 10^{-19}) = 3.06 \times 10^{-6}$ S/cm. This corresponds to a resistivity of 3.3×10^5 Ω cm, compared to a resistivity for copper of 1.7×10^{-6} Ω cm. It is seen that intrinsic Si is a very poor conductor.

Fermi Level and Electrochemical Potential

The Fermi level E_F (Equation 2.2.21) is interpreted as: (1) the energy level below which all allowed levels are occupied at $T = 0$K and above which all levels are empty, or (2) the energy level that has a probability of 0.5 of being occupied at $T > 0$K, provided it is an allowed level, that is, it is not in a forbidden energy gap. It also follows from the derivation of the Fermi–Dirac distribution that E_F is of the nature of an electrochemical potential

(Section ST2.6). But since a state of equilibrium is explicitly assumed in the derivation of the Fermi–Dirac distribution, E_F is defined only under equilibrium conditions. On the other hand, electrochemical potential is defined under both equilibrium and nonequilibrium conditions. For this reason some authors refer to the electrochemical potential under nonequilibrium conditions as a "quasi-Fermi" level.

But if E_F is an equilibrium electrochemical potential, what equilibrium potential does it in fact represent? It is argued in Section ST2.8 that:

Concept: *E_F can be considered to be the equilibrium electrochemical potential of both electrons and holes in an intrinsic or an extrinsic semiconductor. This means that in any semiconductor system at equilibrium, E_F is a constant level throughout the system.*

The constancy of E_F in a semiconductor at equilibrium is almost invariably invoked in the analysis of these systems.

2.5 Carrier Concentrations in Semiconductors

By definition of the density of states for electrons (Equation 2.2.19), $N_e(E)dE$ is the number of electron states per unit volume having energies between E and $E + dE$ in the conduction band. If this is multiplied by $f(E)$, the probability that the energy level is occupied (Equation 2.2.21), then $N_e(E)f(E)dE$ is the number of electrons per unit volume having energies between E and $(E + dE)$ eV. Integrating this over all energies in the conduction band gives the equilibrium concentration of electrons n_o. Thus:

$$n_o = \int_{E_c}^{E_{c(top)}} N_e(E)\, f(E) dE \qquad (2.5.1)$$

where $E_{c(top)}$ is the energy at the top of the conduction band. The expression for p_o, the equilibrium concentration of holes, is analogous, except that $N_h(E)$ (Equation 2.2.20) has to be multiplied by $1 - f(E)$, the probability that the level of energy E is empty:

$$p_o = \int_{E_{v(bot)}}^{E_v} N_h(E)(1 - f(E)) dE \qquad (2.5.2)$$

where $E_{v(bot)}$ is the energy at the bottom of the valence band.

Two simplifications can be made in the evaluation of the integrals in Equations 2.5.1 and 2.5.2: (1) $E_{c(top)}$ and $E_{v(bot)}$ are identified with $+\infty$ and $-\infty$, respectively. This is justified by the fact that, for electrons, $f(E)$ decreases rapidly toward zero in the conduction band, unless E_F lies deep in this band. Similar considerations apply for holes in the valence band. (2) Equation 2.2.22 is used for $f(E)$. It turns out that as long as $E_c - E_F$, or $E_F - E_v$, is greater than about $2kT$, the values of the *integrals* in Equations 2.5.1 or 2.5.2 using Equation 2.2.22 are in excellent agreement with those using Equation 2.2.21. Semiconductors in which E_F is at least $2kT$ away from the edges of the conduction and valence bands are **nondegenerate**, and the Maxwell–Boltzmann distribution applies. Semiconductors in

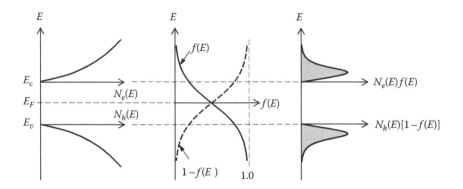

FIGURE 2.5.1
Derivation of carrier concentrations from the density of states and the Fermi–Dirac distribution in an intrinsic semiconductor.

which E_F is within $2kT$ from the edges of the conduction and valence bands, or falls within these bands, are **degenerate**, in which case the Fermi–Dirac distribution must be used to evaluate the integrals of Equations 2.5.1 and 2.5.2. At room temperature, the semiconductor is degenerate if the doping level exceeds about $10^{19}/cm^3$ (Exercise 2.5.2). Unless explicitly stated otherwise, nondegeneracy is assumed in what follows.

The integration is shown graphically in Figure 2.5.1 for an intrinsic semiconductor. $f(E)$ is drawn with E_F located in the middle of the energy gap, as will be justified later. $f(E)$ has a tail that extends into the conduction band, whereas $1 - f(E)$ has a tail that extends into the valence band. Each tail, when multiplied by the density of states in the corresponding band, and the product integrated over the band, gives the density of carriers in that band.

Substituting for $N_e(E)$ from Equation 2.2.19 and for $f(E)$ from Equation 2.2.21 in Equation 2.5.1,

$$n_o = \int_{E_c}^{\infty} 4\pi \left(\frac{2m_e^* |q|}{h^2} \right)^{\frac{3}{2}} (E - E_c)^{\frac{1}{2}} e^{-(E-E_F)/kT} dE$$

The integral evaluates to (Problem P2.2.8):

$$n_o = N_c e^{-(E_c - E_F)/V_T}, \quad \text{where } N_c = 2 \left(\frac{2m_e^* k_J T \pi}{h^2} \right)^{\frac{3}{2}} \tag{2.5.3}$$

N_c is the **effective density of states of the conduction band**. It is as if a number of states N_c per unit volume were concentrated at energy E_c. The exponential $e^{-(E_c - E_F)/V_T}$ represents, according to Equation 2.2.22, the probability that these states are occupied. Multiplying this probability by N_c gives n_o. If m_e^* is expressed in kg, k_J in J/K, and h in Js, N_c is in m^{-3}.

The integration of Equation 2.5.2 gives:

$$p_o = N_v e^{-(E_F - E_v)/V_T}, \quad \text{where } N_v = 2 \left(\frac{2m_h^* k_J T \pi}{h^2} \right)^{\frac{3}{2}} \tag{2.5.4}$$

TABLE 2.5.1

Densities of States in Si and GaAs

	Si	GaAs
N_c (300K)	$2.86 \times 10^{19}/\text{cm}^3$	$4.7 \times 10^{17}/\text{cm}^3$
N_v (300K)	$3.10 \times 10^{19}/\text{cm}^3$	$7.0 \times 10^{118}/\text{cm}^3$
m_e^*/m_e (density of states)	1.09	0.0705
m_h^*/m_e (density of states)	1.15	0.426

N_v is the **effective density of states of the valence band**. Because of uncertainties in the values of effective masses, different values of N_c and N_v are found in the literature. A set of these values for Si and GaAs at 300K is given in Table 2.5.1 together with the associated effective masses of electrons and holes based on Equations 2.5.3 and 2.5.4.

E_F can be eliminated from the expressions for n_o and p_o by multiplying them together:

$$n_o p_o = N_c N_v e^{-E_g/V_T} \tag{2.5.5}$$

where $E_g = (E_c - E_v)$ eV is the width of the energy gap.

In an intrinsic semiconductor, $n_o = p_o = n_i$. By applying Equation 2.5.5 to an extrinsic and to an intrinsic semiconductor, it follows that in any semiconductor at equilibrium:

$$n_o p_o = n_i^2 \tag{2.5.6}$$

assuming that E_g is the same in the extrinsic and intrinsic semiconductors, which is the case in nondegenerate semiconductors, in which the dopant concentration is not too high.

Carrier Generation and Recombination

An alternative derivation of Equation 2.5.6 is instructive. Carrier generation and recombination can be considered as a "reversible reaction":

$$\text{electron} + \text{hole} \rightleftarrows \text{recombined electron–hole pair} \tag{2.5.7}$$

The reaction to the left is that of electron–hole pair generation. Its rate at equilibrium, denoted by $G_o(T)$, is dependent on T alone, and is independent of the concentrations of impurity atoms, electrons, or holes, since these are normally much less than that of the concentration of crystal atoms. The reaction to the right is that of recombination. For **direct recombination**, that is, when an electron drops directly from the conduction band to the valence band, the rate can be expressed as $R_o(T)n_o p_o$, where $R_o(T)$ is a function of T and n_o and p_o are the equilibrium concentrations of electrons and holes, respectively. The rate of recombination is proportional to the product $n_o p_o$, in accordance with the **law of mass action**, because the larger this product, the greater is the chance of an encounter between an electron and a hole, and the greater is the recombination rate. At equilibrium, the rates of generation and recombination are equal, because n_o and p_o do not vary with time. Hence,

$$G_o(T) = R_o(T)n_o p_o \tag{2.5.8}$$

For an intrinsic semiconductor at the same temperature as a given extrinsic semiconductor, $G_o(T)$ and $R_o(T)$ are the same, and $n_o = p_o = n_i$, which gives:

$$G_o(T) = R_o(T)n_i^2 \qquad (2.5.9)$$

Comparing Equations 2.5.8 and 2.5.9 gives Equation 2.5.6.

If the equilibrium concentration of one type of carrier is known, the concentration of the other carrier can be determined from Equation 2.5.6. Consider, for example, Si that is doped with a donor impurity at a level of 1 in 10^6. Since the density of crystal atoms is $5 \times 10^{22}/\text{cm}^3$ (Table 2.3.1), $n_{no} = N_D = 5 \times 10^{16}/\text{cm}^3$, assuming that all donor impurities are ionized. At $T = 300\text{K}$, $n_i = 10^{10}/\text{cm}^3$, so $n_{no} \gg n_i$. From Equation 2.5.6, the minority carrier concentration p_{no} is 2×10^3, which is much smaller than n_i. It is seen that when n_{no} becomes larger than n_i, because of donor impurity atoms, p_{no} becomes less than n_i. This is known as **minority carrier suppression**, and is a direct consequence of Equation 2.5.6. It is important to keep in mind that:

Concept: *For a typical n-type semiconductor at room temperature: $n_{no} \gg n_i \gg p_{no}$, whereas for a typical p-type semiconductor at room temperature: $p_{po} \gg n_i \gg n_{po}$.*

It may be noted that the "reaction" of Equation 2.5.7 is similar to the ionization of water, which from basic physical chemistry is:

$$[\text{OH}^-] + [\text{H}^+] \rightleftarrows \text{H}_2\text{O} \qquad (2.5.10)$$

For pure water, which is analogous to an intrinsic semiconductor, the concentrations of OH^- and H^+ are equal. The rates of ionization and recombination are given by expressions similar to those in Equation 2.5.9. The rate of ionization is independent of the concentrations of OH^- and H^+ because these concentrations are normally much less than that of water. If the concentration of one of these ions is increased by adding acid or base, the concentration of the other ion is depressed below its value in pure water, as in minority carrier suppression.

Intrinsic Semiconductor

In an intrinsic semiconductor, Equation 2.5.5 becomes:

$$n_i^2 = N_c N_v e^{-E_g/V_T} \qquad (2.5.11)$$

which can be expressed as:

$$n_i = KT^{3/2}e^{-E_g/2V_T} \qquad (2.5.12)$$

where $K = \sqrt{N_c N_v}/T^{3/2}$, independently of T, and depends only on the material, neglecting the weak variation of effective mass with T. Using the values of N_c and N_v in Table 2.5.1, and $E_g = 1.1245$ eV for Si and 1.4225 eV for GaAs (Problem P2.1.2) gives $n_i = 1.07 \times 10^{10}/\text{cm}^3$ for Si, and $2.04 \times 10^{10}/\text{cm}^3$ for GaAs, at $T = 300\text{K}$. Because of T in the argument of the exponential in Equation 2.5.11, n_i *rapidly increases with temperature.*

Position of Fermi Level

The position of the Fermi level in the energy gap can be determined by equating the expressions from Equations 2.5.3 and 2.5.4:

$$N_c e^{-(E_{ci}-E_F)/V_T} = N_v e^{-(E_F-E_{vi})/V_T}$$

where the i subscript has been added to E_c and E_v to emphasize that the intrinsic case is being considered. It follows that:

$$E_F = \frac{1}{2}(E_{ci} + E_{vi}) + \frac{V_T}{2}\ln\frac{N_v}{N_c} = \frac{1}{2}(E_{ci} + E_{vi}) + \frac{3}{4}V_T\ln\frac{m_h^*}{m_e^*} \qquad (2.5.13)$$

At $T=0K$, $V_T=0$ and E_F is in the middle of the energy gap. As T increases, E_F moves slightly away from the middle of the gap, because $m_e^* \neq m_h^*$. In the case of S_i, for example (Table 2.5.1), the second term on the RHS of Equation 2.5.13 is approximately 1 meV at $T=300K$, so that E_F is only very slightly above the middle of the energy gap.

Since E_F in an intrinsic semiconductor can be considered to lie in the middle of the gap, it is convenient to use it as a reference energy level $E_{Fi} = \frac{1}{2}(E_{ci} + E_{vi})$. Equation 2.5.3 for an intrinsic semiconductor can then be written as $n_i = N_c e^{-(E_c-E_{Fi})/V_T}$. Dividing Equation 2.5.3 by this relation so as to eliminate N_c gives:

$$n_o = n_i e^{(E_F-E_{Fi})/V_T} \qquad (2.5.14)$$

Similarly, Equation 2.5.4 for an intrinsic semiconductor can be written as $n_i = N_v e^{-(E_{Fi}-E_v)/V_T}$. Dividing Equation 2.5.4 by this relation so as to eliminate N_v gives:

$$p_o = n_i e^{(E_{Fi}-E_F)/V_T} \qquad (2.5.15)$$

Note that in an n-type semiconductor, $E_F > E_{Fi}$, so that $n_o > n_i$. Similarly, in a p-type semiconductor, $E_F < E_{Fi}$, so that $p_o > n_i$.

n-Type Semiconductor

For an n-type semiconductor, Equation 2.5.6 is:

$$n_{no}p_{no} = n_i^2 \qquad (2.5.16)$$

To determine n_{no} and p_{no}, another relationship is provided by the condition that the crystal as a whole must be electrically neutral at equilibrium since it is not disturbed in any way. Thus:

$$n_{no} = p_{no} + N_D^+ \qquad (2.5.17)$$

where N_D^+ is the concentration of ionized donor impurities. Solving for n_{no}:

$$n_{no} = \frac{N_D^+}{2} + \sqrt{n_i^2 + \left(\frac{N_D^+}{2}\right)^2} \qquad (2.5.18)$$

At room temperature, the donor impurities are almost completely ionized, so $N_D^+ \cong N_D$ (Example SE2.2). The concentration of donor atoms N_D is normally such that $N_D \gg n_i$. It follows from Equation 2.5.18 that $n_{no} \cong N_D$ under these conditions. Note that in Equation 2.5.17 p_{no} is the same, in value and sign, as the ionized crystal atoms, and n_{no} is due to both crystal and dopant ions (Problem P2.2.9).

The position of the Fermi level in the energy gap can be found from Equation 2.5.3, assuming that $n_{no} \cong N_D$ and the semiconductor is nondegenerate. Thus:

$$E_{Fn} = E_{cn} - V_T \ln \frac{N_c}{N_D} \qquad (2.5.19)$$

where E_{Fn} denotes E_F in the *n*-type semiconductor. Alternatively, from Equation 2.5.14:

$$E_{Fn} = E_{Fi} + V_T \ln \frac{n_{no}}{n_i} \qquad (2.5.20)$$

E_{Fn} is closer to E_c, as illustrated in Figure 2.5.2 for the case for an *n*-type semiconductor. The donor density of states is shown in the figure on the left. These states do not enter directly in the integration shown in the figure on the right. Their effect is included in the position of E_{Fn}. It is seen that $n_{no} \gg p_{no}$ at room *T*. Note that the concentration of electrons in the conduction band is maximum close to the edge of the conduction band.

p-Type Semiconductor

The case of a *p*-type semiconductor is analogous. Equations 2.5.16 and 2.5.17 become:

$$p_{po} n_{po} = n_i^2 \qquad (2.5.21)$$

$$p_{po} = n_{po} + N_A^- \qquad (2.5.22)$$

where N_A^- is the concentration of ionized acceptor impurities. Solving for p_{po}:

$$p_{po} = \frac{N_A^-}{2} + \sqrt{n_i^2 + \left(\frac{N_A^-}{2}\right)^2} \qquad (2.5.23)$$

At room temperature $N_A^- \cong N_A$, and normally, $N_A \gg n_i$. Hence, $p_{po} \cong N_A$.

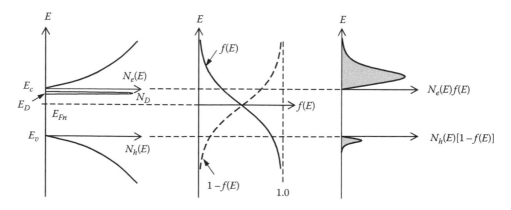

FIGURE 2.5.2
Derivation of carrier concentrations from the density of states and the Fermi–Dirac distribution in an *n*-type semiconductor.

The position of the Fermi level in the energy gap can be found from Equation 2.5.4, assuming that $p_{po} \cong N_A$ and that the semiconductor is nondegenerate:

$$E_{Fp} = E_{vp} + V_T \ln \frac{N_v}{N_A} \qquad (2.5.24)$$

Alternatively, from Equation 2.5.15:

$$E_{Fp} = E_{Fi} - V_T \ln \frac{p_{po}}{n_i} \qquad (2.5.25)$$

E_{Fp} is closer to E_v, as illustrated in Figure 2.5.3 for the case for an p-type semiconductor. Moreover, the maximum concentration of holes is close to the edge of the valence band. The general case, where the conductor is degenerate and impurity atoms are not completely ionized, is more complicated and is outlined in Section ST2.9.

The variation of the E_F with T in an extrinsic semiconductor is of interest. Consider an n-type semiconductor. At $T = 0K$, the donor level E_D is filled and the conduction band is empty. According to the interpretation of E_F as the level below which all allowed energy levels are filled at $T = 0K$, this level must lie above E_D. It can be shown (Example SE2.3) that in fact it lies midway between E_C and E_D. At a sufficiently high T, the semiconductor becomes essentially intrinsic, so E_F must be close to the middle of the energy gap.

Conclusion: *The position of E_F in the energy gap is indicative of the relative concentrations of electrons and holes in a semiconductor. When these concentrations are equal, as in an intrinsic semiconductor, E_F is essentially in the middle of the gap. In an n-type semiconductor, where electrons are majority carriers, E_F is closer to the conduction band. In a p-type semiconductor, where holes are majority carriers, E_F is closer to the valence band.*

Example 2.5.1 Fermi Level in Extrinsic Semiconductor

It is required to determine the Fermi levels in a p-type semiconductor having $N_A = 10^{18}/cm^3$ and in an n-type semiconductor having $N_D = 10^{15}/cm^3$.

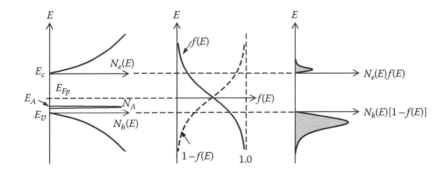

FIGURE 2.5.3
Derivation of carrier concentrations from the density of states and the Fermi–Dirac distribution in a p-type semiconductor.

SOLUTION

In the case of the *n*-type semiconductor, Equation 2.5.19 gives $E_{cn} - E_{Fn} = 0.026 \ln(2.86 \times 10^{19}/10^{15}) = 0.27$ eV, which is considerably larger than $2kT = 0.052$ V. For the *p*-type semiconductor, Equation 2.5.24 gives $E_{Fp} - E_{vn} = 0.026 \ln(3.10 \times 10^{19}/10^{18}) = 0.09$ eV, which is larger than $2kT$, although the acceptor impurity concentration is on the high side.

Example 2.5.2 Energy Bands in a *pin* Structure

A silicon crystal is doped so that it is *p*-type at one end, *n*-type at the other end, and intrinsic in the middle. It is required to sketch how the edges of the conduction and valence bands vary along the crystal at equilibrium.

SOLUTION

Imagine for the moment that the energy bands of the three regions are laid side by side so that the edges of the conduction and valence bands are aligned (Figure 2.5.4). As explained previously, E_F must be the same throughout the crystal. To make this happen, energy levels must be lowered by $(E_{Fn} - E_{Fin})$eV in the *n*-region, and raised by $(E_{Fip} - E_{Fp})$eV in the *p*-region. This implies that the electric potential energies of electrons are lowered by $(E_{Fn} - E_{Fin})$eV in the *n*-region, and raised by $(E_{Fip} - E_{Fp})$ eV in the *p*-region; conversely for holes. If Equations 2.5.20 and 2.5.25 are written as $E_{Fn} - E_{Fin} = V_T \ln(n_{no}/n_i)$, and $E_{Fip} - E_{Fp} = V_T \ln(p_{po}/n_i)$, respectively, the aforementioned changes in electric potential energy exactly compensate for the changes in chemical potential energy, so that the electrochemical potential of electrons and holes is the same throughout, as it must be at equilibrium.

The complete energy band diagram is illustrated in Figure 2.5.5. All energy levels are lowered by $(E_{Fn} - E_{Fin})$ eV in the *n*-region and raised by $(E_{Fip} - E_{Fp})$ eV in the *p*-region, so as to make $E_F = E_{Fn} = E_{Fp}$ the same level throughout. Since changes in electric potential energy are reflected as changes in voltage, then in going from the intrinsic to the *n*-region, the change in electric potential energy is $E_{cno} - E_{cio} = E_{Fin} - E_{Fn}$. When energies are expressed in eV, the change in potential energy of an electron is the negative of the change in voltage. Thus, $E_{cno} - E_{cio} = E_{Fin} - E_{Fn} = -(V_{no} - V_{io}) = (V_{io} - V_{no}) = V_{in} = -V_{ni}$, where V_{no} and V_{io} are the electric potentials of the *n*-region and the intrinsic region, respectively, with respect to some arbitrary reference, and $V_{ni} > 0$ is the electric potential of the *n*-region with respect to the intrinsic region. Hence, $E_{Fn} - E_{Fin} = V_{ni}$, as indicated in the figure. Equation 2.5.14 becomes, for electrons:

$$n_{no} = n_i e^{(E_{Fn}-E_{Fin})/V_T} = n_i e^{V_{ni}/V_T} \tag{2.5.26}$$

so that $n_{no} > n_i$.

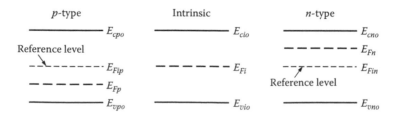

FIGURE 2.5.4
Figure for Example 2.5.2.

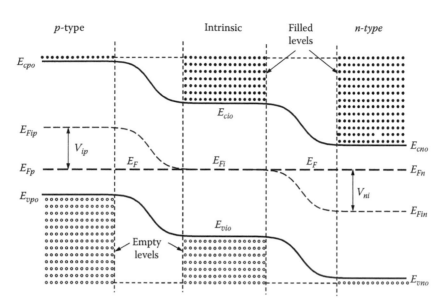

FIGURE 2.5.5
Figure for Example 2.5.2.

Let us consider holes in the *n*-region. In going from the intrinsic to the *n*-region, the change in the electric potential energy of holes is $-E_{vno} - (-E_{vio}) = E_{vio} - E_{vno} = E_{Fn} - E_{Fin}$, since the electric potential energy levels for the holes are the negatives of those for electrons. The corresponding change in the electric potential is $V_{no} - V_{io} = V_{ni}$. It follows that, for holes, $E_{Fin} - E_{Fn} = -V_{ni}$. Equation 2.5.15 becomes:

$$p_{no} = n_i e^{(E_{Fin} - E_{Fn})/V_T} = n_i e^{-V_{ni}/V_T} \tag{2.5.27}$$

so that $p_{no} < n_i$. It follows, of course, from Equations 2.5.26 and 2.5.27 that $n_{no} p_{no} = n_i^2$.

It is seen that, compared to the intrinsic region, electrons in the *n*-region have a higher chemical potential energy but an equally lower electric potential energy, so that their electrochemical potential is the same in both regions. Conversely, holes in the *n*-region have a lower chemical potential energy, compared to the intrinsic region, but an equally higher electric potential energy, so that their electrochemical potential is the same in both regions. The required change in the electric potential energy of electrons and holes in the *n*-region is brought about by having the *n*-region at a positive voltage, V_{ni}, with respect to the intrinsic region. The fact that the edge of the conduction band is lower in the *n*-region than in the intrinsic region can be interpreted to mean that electrons are more concentrated in the *n*-region than in the intrinsic region. This is because at the highest filled energy level above E_{cio} in the intrinsic region, and at a given *T*, more levels are filled in the *n*-region than in the intrinsic region, as illustrated diagrammatically in Figure 2.5.5. Similarly, the fact that the edge of the valence band is lower in the *n*-region than in the intrinsic region can be interpreted to mean that holes are less concentrated in the *n*-region than in the intrinsic region, since at the highest filled energy level below E_{vno} in the *n*-region, and at a given *T*, more levels are empty in the intrinsic region than in the *n*-region.

An exactly analogous argument shows that the intrinsic region is at a positive voltage V_{ip} with respect to the *p*-region. The edges of the conduction and valence bands vary smoothly from one region to the other in Figure 2.5.5 in a manner that depends on the type of junction, that is, whether the impurity concentration changes abruptly or in a specified, gradual manner.

Equations 2.5.26 and 2.5.27 are a form of **Boltzmann relations**. They derive very simply from equating the change in chemical potential to the change in electric potential.

Exercise 2.5.1

Verify the voltage relations between the p-type and intrinsic semiconductors in Figure 2.5.5. ■

Exercise 2.5.2

Determine N_D and N_A that will bring E_F to approximately $2kT$ below E_c or above E_v.
  Answers: $N_D = N_c/e^2 \cong 0.44 \times 10^{19}/cm^3$, $N_A = N_v/e^2 \cong 0.25 \times 10^{19}/cm^3$. ■

A semiconductor may contain both donor and acceptor impurities, in which case it is described as a **compensated semiconductor**. The behavior of the semiconductor will then depend on the net concentration of these impurities. Thus, the semiconductor behaves as an n-type if $N_D > N_A$, as p-type if $N_D < N_A$, and as intrinsic if $N_D = N_A$. Such semiconductors are encountered in practice in several situations. For example, the substrate of a semiconductor device may be a lightly doped n-type material. To add a p-type region, a sufficient concentration of acceptor impurities is added to compensate for the donor impurities present and result in a p-type material having a net desired concentration of acceptor impurities. Another situation is where there is a small but residual, unknown concentration of donor impurities that cannot be reduced for technological reasons. To produce a high-resistivity intrinsic material, a controlled concentration of acceptor impurities is added. An important difference, however, between a compensated semiconductor and an uncompensated semiconductor is that the effect of dopants on carrier mobility and lifetime (Equations 2.6.1, 2.6.2, and 2.7.2) depends on the *total* concentration of impurities and not on the net concentration of donor or acceptor impurities.

2.6 Carrier Mobility

Variation with Dopant Concentration

As its name implies, mobility is a measure of how easily an atomic particle moves in a medium under the influence of some applied force (Section 2.4). Mobility depends on temperature, on the effective mass of the particle, and on impediments to the motion of the particle in a given medium. In the intrinsic semiconductor, mobility is limited by **lattice scattering,** which refers to the interactions between the electron waves describing the motion of electrons in the crystal, and crystal vibrations, that is, crystal ions vibrating about their rest positions. In extrinsic semiconductors, **ionized impurity scattering** due to dopants further impedes the movement of carriers. For our purposes, we will consider electron and hole mobilities in Si at 300K in cm^2/Vs to be related to the *total* dopant concentration N of donor and acceptor impurities in particles/cm^3 as follows, independently of the types of dopants (Arora et al., 1982):

$$\mu_e = 130 + \frac{1300}{1 + 7 \times 10^{-18}N} \tag{2.6.1}$$

$$\mu_h = 65 + \frac{415}{1 + 3.75 \times 10^{-18}N} \tag{2.6.2}$$

Variation with Temperature

The variation of mobility with temperature is more complicated. In the intrinsic semiconductor, as the temperature increases, the amplitude of vibration of lattice atoms about their mean positions increases, the probability of scattering increases, and mobility decreases with increasing temperature. In the presence of ionized impurities, carrier velocities are small at relatively low temperatures, which means that the time of the passage of carriers past the ionized impurities is longer than at higher temperatures, so that the deflection of carriers is increased, and the mobility decreases with decreasing temperature. The net effect of these opposing mechanisms is a decrease of mobility with temperature over the temperature range normally of interest. An approximate, empirical expression for the variation of mobility with temperature over the range 300–400K is (Green, 1990):

$$\mu(T) = \frac{\mu(300)}{(T/300)^{2.5}} \tag{2.6.3}$$

Electron mobility in GaAs is considerably larger than in Si (Table 2.3.1). GaAs devices that depend for their operation on electron mobility will respond more quickly to changes in applied voltages and will therefore have a faster response than comparable Si devices.

The decrease of mobility with temperature has a somewhat unexpected effect. Whereas the conductivity of an intrinsic semiconductor increases with temperature, because of the predominant effect of increase in carrier concentrations, the conductivity of an extrinsic semiconductor generally *decreases* with temperature because of the reduction of mobility at a relatively constant concentration of carriers, as in metals (Problem P2.2.20).

Exercise 2.6.1

Verify that the temperature coefficient $\dfrac{1}{\mu(T)} \dfrac{d\mu(T)}{dT}$ is given by:

$$\frac{1}{\mu(T)} \frac{d\mu(T)}{dT} = -\frac{2.5}{T} \tag{2.6.4} \quad \blacksquare$$

2.7 Carrier Recombination

Whenever carrier concentrations in a semiconductor exceed their equilibrium values, as when a semiconductor is subjected to a pulse of radiation, for example, the rate of recombination increases above its equilibrium level so as to bring the concentrations back to their equilibrium values. Whenever an electron and a hole recombine both energy and momentum must be conserved, as will be discussed in more detail in Section 3.4, Chapter 3. In a direct-bandgap semiconductor such as GaAs, recombination is mainly direct, as described in

Section 2.5. That is, an electron in the conduction band drops directly into the valence band, conserving momentum and emitting the energy difference E_g as radiation.

In indirect-bandgap semiconductors such as Si, however, recombination occurs mainly via **traps** and surface states. Traps can be due to metallic impurities, which are always present in practice and introduce energy levels in the forbidden gap, just like donor and acceptor impurities, but located closer to the center of the energy gap. Gold in Si and copper in GaAs are effective in providing traps that increase the rate of recombination, and therefore reduce carrier lifetime.

Indirect recombination through a trap of energy E_t is illustrated in Figure 2.7.1. An electron in the conduction band (1) may drop into the energy level E_t (2). It may then drop into the valence band (3), annihilating the electron–hole pair. In effect, the sequence (1) to (3) represents electron–hole recombination through an intermediate energy level E_t. Once in the trap (2), there is some probability that the electron may return to the conduction band. The closer E_t is to E_c, the larger is this probability. That is why, for effective recombination, E_t should not be close to E_c. Moreover, the reverse sequence can also take place: an electron can move from the valence band to the energy level E_t and then into the conduction band, which constitutes indirect generation. For this reason, traps are also referred to as **generation–recombination centers**. The recombination rate in indirect recombination is derived in Section ST2.10.

Crystal imperfections also introduce energy levels in the forbidden gap. The imperfections may be due to atoms missing from crystal locations or due to the distortion of the crystal because of mechanical strain. The discontinuity at the surface of a crystal can be regarded as an extreme form of crystal imperfection. In effect, the crystal atoms right at the surface do not have neighboring atoms, beyond the surface, with which they can share electrons. This leaves so-called dangling bonds that are capable of capturing electrons from the bulk of the crystal at energy levels in the forbidden gap. The density of these energy levels, or **surface states**, can be controlled to a considerable extent by an appropriate choice of crystal orientation and the treatment of the surface during fabrication (Section 4.8, Chapter 4). Adsorbed ions or molecules, or mechanical damage to surface layers, can also introduce surface states.

Another type of recombination that becomes important at the high carrier concentrations is **Auger recombination**, in which the energy and momentum balance of an electron–hole recombination is transferred to a third particle, which could be an electron or a hole. This third particle then loses its energy to lattice interactions.

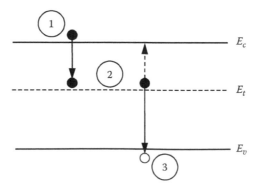

FIGURE 2.7.1
Indirect recombination through a trap of energy in the forbidden band.

Minority Carrier Lifetime

Let τ_x be the mean lifetime of a given type of carrier, which could be a minority or a majority carrier. Then $(1/\tau_x)$ is the average rate at which individual carriers recombine. Multiplying this by the carrier concentration gives the overall rate of recombination:

$$\left(\frac{1}{\tau_x}\right) n_x = R_x(T) \quad \text{or} \quad \tau_x = \frac{n_x}{R_x(T)} \tag{2.7.1}$$

where
  n_x is the carrier concentration
  $R_x(T)$ is the rate of recombination per unit volume

It is seen that τ_x decreases with an increased $R_x(T)$, as expected.

In an extrinsic semiconductor, the concentration of majority carriers greatly exceeds that of minority carriers, so that the rate of recombination is limited by the lifetime and concentration of minority carriers. This lifetime decreases with increased doping level in an extrinsic semiconductor, for all the recombination mechanisms mentioned previously. The higher the doping level, the higher is the concentration of majority carriers, the smaller is the concentration of minority carriers, and the smaller is the minority carrier lifetime (Equation 2.7.1). A simplified empirical relationship between minority carrier lifetime and the *total* dopant concentration N of donor and acceptor impurities at 300K is:

$$\tau \cong \frac{2.5 \times 10^{-3}}{1 + [N/(3.5 \times 10^{15})]^{1.33}} \tag{2.7.2}$$

As for temperature dependence, minority carrier lifetime increases with temperature for both direct and indirect recombinations. This is reasonable, because as the temperature is raised, the minority carrier concentration increases relative to the majority carrier concentration, so that minority carriers become less overwhelmed by majority carriers, and their lifetime increases. In Auger recombination, however, the mobility decreases with temperature.

In general,

Concept: *High mobilities and short minority carrier lifetimes in a semiconductor device are conducive to fast switching and high-frequency operation.*

It should be noted that minority carrier lifetime plays an important role in the operation of semiconductor devices such as *pn* junction diodes and bipolar junction transistors (BJTs). On the other hand, other semiconductor devices, such as Schottky diodes and field-effect transistors (FETs) are majority-carrier operated. The response time for majority carriers depends on the **dielectric relaxation time** τ_D, which equals ε/σ, where ε is the permittivity of the semiconductor and σ is its conductivity (Section ST3.4, Chapter 3). Because τ_D is generally much smaller than the minority carrier lifetime, majority-carrier-operated devices are in principle faster than minority-carrier-operated devices, *other factors being equal*. This latter proviso is critical, because, as will be demonstrated later, BJTs are in fact faster than FETs.

Summary of Main Concepts and Results

- Wave/quantum mechanics is the more general and rigorous form of mechanics. It reduces to classical mechanics when considering large-scale phenomena, or situations where the dimensions of the system are very large compared to the De Broglie wavelength, so that the variation of potential energy over distances comparable to this wavelength is quite negligible. If this is not the case, classical mechanics breaks down, because it does not take into account the effect of the means of observation on what is being observed.

- In wave mechanics, the wave function $\psi(x,t)$ satisfies Schrödinger's equation, where $|\psi(x,t)|^2$ is the probability of finding the particle under consideration at a certain location at a given time. A probability wave packet can be set up that is composed of a superposition of wave functions subject to the boundary conditions of the problem. In general, the individual wave functions in a wave packet travel at phase velocities. Their envelope travels at the group velocity, which corresponds to the velocity of the particle.

- Because of the probabilistic interpretation of the location of a particle in wave mechanics, there is a finite probability that a particle is able to penetrate, or tunnel through, a potential energy barrier of finite height and width. The probability of tunneling increases rapidly as the barrier height or width decreases.

- When atoms come together in a crystal, the discrete energy levels of free atoms form energy bands separated by forbidden gaps. The highest energy band is the conduction band, the next lower energy band being the valence band.

- At 0K electrons occupy the lowest energy levels available. In semiconductors, the valence band is filled at 0K and the conduction band is empty, the bandgap being relatively narrow.

- An electron state in a free atom is defined by four quantum numbers. According to Pauli's exclusion principle, not more than one electron can be in the same state, or not more than two electrons, one of each spin, can have the same energy.

- Electrons in the periodic electric potential of the crystal can be in stationary states because of Bragg reflections. Electrons in stationary states do not propagate, and the wave vectors correspond to forbidden energies in the gaps between energy bands.

- Electrons occupying energy levels near the top of the valence band have negative effective mass. In the presence of an electric field, a nearly filled valence band behaves as if it were devoid of electrons and the initially empty levels were occupied by positive charges having positive mass, these positive charges being referred to as holes.

- The concept of effective mass accounts for the interaction of an electron with the rest of the crystal and allows an electron or a hole in the crystal to be treated like a free particle that obeys classical mechanics when subjected to an external force.

- Freely moving, charged atomic particles posses chemical potential energy, because of concentration, and electric potential energy in the presence of electric potential. The algebraic sum of these two potential energies is the electrochemical potential energy. The electrochemical potential energy per particle is the electrochemical potential.

- The Fermi–Dirac distribution gives the probability that a given energy level is occupied. The product of this probability and the density of states is the carrier concentration.

- The Fermi level can be identified with the equilibrium electrochemical potential of electrons or holes in an intrinsic or an extrinsic semiconductor. Its location determines the equilibrium concentrations of electrons and holes. It is the same level throughout a system at equilibrium.

- The diffusion and drift currents are due to the flow of current carriers under the influence of concentration gradients and electric potential gradients, respectively. The total current is proportional to the electrochemical potential gradient as the net driving force, leading to a generalized ohm's law.

- Equilibrium is the state in which temperature, pressure, and electrochemical potential of freely moving particles are all equalized throughout the system, do not change with time, and the system is not disturbed in any way. There is no net flow of particles at equilibrium.

- In an intrinsic semiconductor, electrons and holes have equal concentrations. This intrinsic concentration is highly temperature sensitive and decreases as the band-gap increases.

- Extrinsic semiconductors have carefully controlled concentrations of dopant. If the dopant atom is pentavalent, it is a do*n*or type dopant and the semiconductor is *n*-type, in which the majority carriers are *n*egatively charged particles, or electrons. If the dopant atom is trivalent, it is an acce*p*tor type impurity and the semiconductor is *p*-type, in which the majority carriers are *p*ositively charged particles, or holes.

- Over the temperature range of practical interest, and for typical concentrations of dopants, the concentration of majority carriers in an extrinsic semiconductor is very nearly equal to the concentration of impurity atoms. This fact underlies the practical importance of semiconductor devices, since the type and concentration of the current carrier is completely determined by the type and concentration of dopants added.

- In an extrinsic semiconductor at equilibrium, the product of the concentrations of majority carriers and minority carriers equals the square of the intrinsic concentration.

- In an intrinsic or extrinsic semiconductor at equilibrium, the rate of electron–hole pair generation must be equal to the rate of electron–hole recombination. Carriers have a mean lifetime between generation and recombination.

- Mobility and minority carrier lifetime decrease with increasing concentration of dopants. Mobility decreases with increasing temperature over the temperature range of interest, whereas the minority carrier lifetime increases.

- It is generally true that high mobilities and short carrier lifetimes in a semiconductor device are conducive to fast switching and high-frequency operation.

Learning Outcomes

- Articulate some of the basic concepts that govern electric conduction in semiconductors, including diffusion and drift, carrier concentrations, carrier mobility, and lifetime.

Supplementary Examples and Topics on Website

SE2.1 Calculation of Atomic Densities of Si and GaAs. These densities are calculated from atomic/molecular mass and densities.

SE2.2 Calculation of Percentage Ionization of Impurities. The percentage ionization is calculated at 300K for phosphorous and aluminum in Si.

SE2.3 Fermi Level in an n-Type Semiconductor. The Fermi level is calculated in a nondegenerate n-type semiconductor as a function of temperature and donor impurity concentration, taking into account incomplete ionization.

SE2.4 Ionization Energy Using Bohr's Model. The ionization energy of donor atoms is estimated using Bohr's model of the hydrogen atom.

ST2.1 De Broglie Wavelength. Illustrates the validity of classical mechanics by calculating the De Broglie wavelength for: (a) a mass of 1 kg falling through 100 m, (b) an electron accelerated through 300 V over a distance of 1 cm, and (c) an electron in a hydrogen atom.

ST2.2 Wave Packets. Illustrates the concept of a wave packet and derives the expression for the group velocity of waves having slightly different frequencies. It is shown that the group velocity of a wave packet representing a free electron moving with a velocity u in a region of constant potential energy is the same as u.

ST2.3 Quantum Numbers. Discusses in more detail the interpretation of quantum numbers.

ST2.4 Grouping of Energy Levels into Energy Bands. Illustrates the formation of energy bands in a hypothetical crystal and in the case of Si.

ST2.5 Bragg Reflection and the Formation of Standing Waves. Illustrates the formation of standing waves when incident and reflected waves interfere, leading to Bragg reflection.

ST2.6 Fermi–Dirac Distribution. Derives this distribution and shows that the Fermi level is of the nature of an electrochemical potential.

ST2.7 Mobility. The expression for mobility is derived in terms of effective mass and momentum relaxation time.

ST2.8 Interpretation of the Fermi Level. Argues that the Fermi level can be interpreted as: (1) a reference level for the electric potential energy of both electrons and holes in an intrinsic semiconductor at equilibrium, and (2) the electrochemical potential of electrons and holes in an intrinsic or an extrinsic semiconductor at equilibrium. It is also shown that $n_{xo}p_{xo} = n_i^2$ for an extrinsic semiconductor that is not too heavily doped.

ST2.9 Carrier Concentrations in Degenerate Semiconductor. Outlines the analysis of the general case of a degenerate, compensated semiconductor in which the unknowns are n_o, p_o, N_D^+, N_A^-, and E_F.

ST2.10 Rate of Indirect Recombination. Derives the expression for this rate.

Problems and Exercises

P2.1 Basic Physical Concepts

P2.1.1 In Figure 2.3.2b, $|\psi(x)|^2 = 0$ at intermediate values of x. Does this mean that the particle does not cross such locations?

P2.1.2 Why does E_g become narrower as T increases? The variation of E_g with T can be considered as $E_g(T) = E_g(0) - \alpha T^2/(T+\beta)$ eV, where T is the absolute temperature, and the values of the various quantities are as follows:

	$E_g(0)$	α	β
Si	1.17	4.73×10^{-4}	636
GaAs	1.519	5.405×10^{-4}	204

P2.1.3 E_g in insulators is too wide to allow valence electrons to jump to the conduction band by acquiring energy from photons of light in the visible spectrum. Do you expect a pure insulator without imperfections to be transparent or opaque? What about metals?

P2.1.4 Sketch Equations ST2.5.2 and ST2.5.3 at $t=0$, $T/8$, $T/4$, $3T/8$, $T/2$, $5T/8$, $3T/4$, and $7T/8$ assuming $k = \pi/a$. Note that if there are many reflection sites equally separated by a distance that satisfies the condition for Bragg reflection, and which produce reflections either in accordance with Equation ST2.5.2 or Equation ST2.5.3, waves traveling at the same speed in opposite directions produce standing waves.

P2.1.5 Are electrons in stationary states in the crystal at rest? Explain in terms of the probabilistic interpretation of electron motion. Note that the interpretation applies to an electron of zero KE at the edge of the conduction band.

P2.1.6 In ionic systems, electrochemical potential is defined as J/mol rather than J/particle or eV/particle. What would be the expression for electrochemical potential for ions of valence z?

P2.1.7 Extend Figure 2.2.3 to include the case of positive charges having negative mass.

P2.1.8 A radio transmitter radiates 10 kW at 100 MHz. How many photons are radiated/s?

P2.1.9 Consider an energy level that is 0.5 eV below the Fermi level. What is the probability of its being occupied, or empty at: (a) 100K, (b) 300K, (c) 500K?

P2.1.10 Repeat Problem P2.1.9 for an energy level that is 0.5 eV above the Fermi level.

P2.1.11 If a state of energy E has a probability K of being occupied, where $E - E_F \gg kT$, what is the probability of occupation of a state of energy $E + kT$?

P2.2 Semiconductors

P2.2.1 The unit cell of the diamond structure consists of: (1) 8 atoms at the corners of a cube, (2) 6 atoms at the centers of each face of the cube, and (3) 4 atoms within the body of the cube, all these individual atoms being part of a tetrahedral structure as in Figure 2.3.1. (a) Considering each side of the cube to be 0.543 nm, calculate the density of Si atoms; (b) Assuming a Si atom to be a sphere of 0.118 nm radius, what fraction of the cube volume is filled with Si atoms?

P2.2.2 How do you expect a GaAs semiconductor to behave if: (a) one in every 10^{17} Ga atoms is replaced by a Si atom? (b) one in every 10^{17} As atoms is replaced by a Si atom?

P2.2.3 Calculate the position of E_F in intrinsic GaAs at $T=300$K using the data of Table 2.3.1.

P2.2.4 If E_g increases by 50% from 1.12 eV, what is the effect on n_i at 300K?

P2.2.5 The Fermi level in a Si sample at 300K is located 0.3 eV above the middle of the energy gap. Determine the equilibrium concentrations of electrons and holes.

P2.2.6 Determine the position of E_F with respect to E_v at 300K in Si having $N_A = 5 \times 10^{16}/\text{cm}^3$, assuming the impurity atoms are completely ionized.

P2.2.7 For what value of N_D/n_i is a 1% error introduced in assuming $n_{no} = N_D$ in Equation 2.5.18?

P2.2.8 Verify the integration leading to Equation 2.5.3 by substituting $x = (E - E_c)/kT$ and noting that $\int_0^\infty x^{1/2}e^{-x}dx = \sqrt{\pi}/2$.

P2.2.9 Consider the electroneutrality Equation 2.5.17. Show that the equation is valid if the positively charged Si crystal ions that have lost valence electrons due to the intrinsic carrier generation are taken into account. Assume that all donor impurities are ionized.

P2.2.10 A Si sample at 300K has $N_D = 1.0 \times 10^{16}/\text{cm}^3$ and $N_A = 1.1 \times 10^{16}/\text{cm}^3$. Determine n_o and p_o at 300K assuming that the impurities are completely ionized.

P2.2.11 Neglecting the variation of E_g and of effective mass with T, apply Equation 2.5.12 to determine n_i at 200K and at 400K in Si assuming $n_i = 10^{10}$ at 300K.

P2.2.12 If a semiconductor is doped at a concentration of 1 in 10^8, at what temperature will the intrinsic concentration be equal to that of the impurity atoms? What will be the majority carrier concentration? Assume that E_g remains constant at 1.12 V.

P2.2.13 Repeat Problem P2.2.12 for a doping level of 1 in 10^5.

P2.2.14 Using Equation 2.6.3, and the values in Table 2.3.1, determine electron and hole mobilities in intrinsic Si at 200K and 400K.

P2.2.15 Determine electron and hole mobilities in an extrinsic Si semiconductor doped at a concentration of 1 in 10^7 with: (a) donor impurities; (b) acceptor impurities.

P2.2.16 Consider an *n*-type Si that is doped at a concentration of 1 in 10^6. Assuming complete ionization of the dopant at $T = 300K$, compare the semiconductor to intrinsic Si with respect to: (a) the concentration of electrons; (b) the concentration of holes; and (c) conductivity, using Equations 2.6.1 and 2.6.2.

P2.2.17 A *p*-type semiconductor at equilibrium has a resistivity of 0.5 Ω cm at 300K. Assuming complete ionization of the dopant, determine: (a) the concentration of acceptor impurities that must be added to double the conductivity; (b) the concentration of donor impurities that must be added to reduce the conductivity to one-tenth of its original value. How do the electron and hole mobilities compare with those in the original sample?

P2.2.18 The diffusion constant, which is the diffusion current density per unit concentration gradient, is related to mobility by the Nernst–Einstein relation (Equation 2.4.5). Determine the diffusion constants of electrons and holes in: (a) intrinsic Si at 300K; (b) intrinsic Si at 400K; (c) *n*-type Si at 300K doped at a concentration of 1 in 10^7; (d) case (c) at 400K.

P2.2.19 Assume that in a bar of *n*-type silicon doped at a concentration of 1 in 10^7, a hole concentration gradient of $5 \times 10^6 \times p_{no}/\mu\text{m}$ is maintained at $T = 300K$. (a) Express the concentration as function of x, with $x = 0$ at the point of largest concentration, and determine the hole diffusion current density. (b) Express the chemical potential as function of x and verify Equation 2.4.19. Note that σ in this equation is that due to holes only.

P2.2.20 (a) Determine the change in σ of intrinsic Si at equilibrium between 300K and 400K, referring to Example 2.4.2, Problem P2.2.11, and Problem P2.2.14. (b) Repeat for Si having $N_D = 10^{17}/\text{cm}^3$. Note that whereas the conductivity of intrinsic Si increased with T, that of extrinsic Si decreases with T over the temperature range considered because of decrease in mobility.

P2.2.21 Consider silicon that is doped with As ($E_D = 0.048$ eV) at a concentration of $10^{16}/cm^3$. Assume that the semiconductor temperature range of interest is from 250K to 450K. Determine: (a) the percentage ionization of the donor impurity atoms at 250K (refer to Examples SE2.2 and SE2.3); (b) the intrinsic concentration at 450K. Calculate n_{no} at these two temperatures.

P2.2.22 In Example SE2.3, what should be the position of E_F to give a percentage ionization of 99.9% at 300K for: (a) donor impurities; (b) acceptor impurities? Are the semiconductors degenerate?

P2.2.23 Repeat Example SE2.3 for a *p*-type semiconductor.

3

pn Junction and Semiconductor Diodes

The purpose of this chapter is to explain the behavior of various types of semiconductor diodes, based on the physical principles enunciated in Chapter 2. The *pn* junction is a basic constituent of a wide variety of semiconductor devices, not only diodes but also metal–oxide–semiconductor transistors and bipolar junction transistors, covered in Chapters 5 and 6, respectively. It is highly important, therefore, to acquire a good physical understanding of the behavior of the *pn* junction.

The *pn* junction is first considered at equilibrium, which serves as a foundation for the discussion of the *pn* junction under forward or reverse bias. The behavior of the *pn* junction, and some associated phenomena, are explored in considerable detail, with a strong emphasis on fundamental physical principles. Mathematical derivations and topics that are more advanced are left to the Supplementary Examples and Topics on the website.

The discussion then shifts to some special types of semiconductor diodes. The electrical properties of a *pn* junction can be altered through irradiation by light, as in photodiodes and photocells. The reverse process of light emission by a *pn* junction occurs in light-emitting diodes (LEDs). The current–voltage characteristic of the tunnel diode has an unusual negative resistance part that is useful for some applications. Metal–semiconductor contacts are not only unavoidable in semiconductor devices, but are also the main structure in some special-purpose diodes and transistors. The chapter ends with a discussion of the hetero-junction, which is an important development that enhances the performance of the bipolar junction transistor.

Learning Objectives

❖ To be familiar with:

• The terminology associated with semiconductor diodes of various types.

❖ To understand:

• The basic physical principles underlying the operation of the *pn* junction.
• The capacitances associated with the *pn* junction and their implications.
• The effects of temperature on the *pn* junction diode.
• Breakdown phenomena in the reverse-biased *pn* junction diode.
• The effect of light on semiconductors and on the *pn* junction.
• The basic principles of operation of photodiodes, solar cells, LEDs, and tunnel diodes.
• The basic principles underlying the behavior of metal–metal contacts, metal–semiconductor contacts, and heterojunctions.

3.1 The *pn* Junction at Equilibrium

Junction Potential

A ***pn* junction** is formed in a crystal, part of which is *n*-type and part is *p*-type (Figure 3.1.1), as is explained in Chapter 4. It is assumed, for simplicity, that the two regions are uniformly doped and that the metallurgical boundary between them is abrupt. Electrons and holes can move freely between the *n*- and *p*-sides, as these are part of the same crystal.

How is equilibrium established? Consider the case of conduction electrons first. Because these electrons are more concentrated on the *n*-side than on the *p*-side, they diffuse down their concentration gradient. As they do so, however, they make the *p*-side negatively charged with respect to the *n*-side. An electric potential difference is thereby established that tends to drive electrons in the opposite direction, from the *p*-side to the *n*-side. At equilibrium, these two tendencies are equal and opposite. *Electrons have a higher chemical potential energy on the n-side, compared to the p-side, but a lower electric potential energy, so that their electrochemical potential is the same on both sides of the junction, as it must be at equilibrium* (Section 2.4, Chapter 2). Thus, as electrons move from one side to the other at equilibrium:

$$\Delta(\text{Chemical potential energy}) + \Delta(\text{Electric potential energy}) = 0 \qquad (3.1.1)$$

In going from the *n*-side to the *p*-side, for example, $\Delta(\text{Chemical potential energy/electron}) = \kappa'_{epo} - \kappa'_{eno} = V_T \ln(n_{po}/n_{no})$ eV, and $\Delta(\text{Electric potential energy/electron}) = -(V_{po} - V_{no})$ eV, where the change in a given quantity is taken as its value at the destination minus its value initially. Substituting in Equation 3.1.1 and rearranging,

$$V_{npo} = V_{no} - V_{po} = V_T \ln \frac{n_{no}}{n_{po}} \qquad (3.1.2)$$

where V_{npo} is the **junction potential** or the **built-in potential**.

An exactly analogous situation applies for holes. They tend to diffuse down their concentration gradient, from the *p*-side to the *n*-side. But this tendency is opposed by the electric potential difference between the two sides. At equilibrium, *holes have a higher chemical potential energy on the p-side than on the n-side but a lower electric potential energy,*

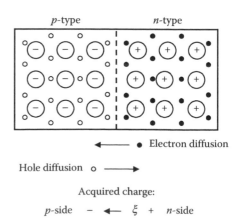

FIGURE 3.1.1
Establishment of equilibrium in a *pn* junction.

so that their electrochemical potential is the same on both sides. In going from the *n*-side to the *p*-side, for example, Δ(Chemical potential energy/hole) $= \kappa'_{hpo} - \kappa'_{hno} = V_T \ln(p_{po}/p_{no})$eV, and Δ(Electric potential energy/hole) $= (V_{po} - V_{no}) = -V_{npo}$ eV. This gives:

$$V_{npo} = V_{no} - V_{po} = V_T \ln \frac{p_{po}}{p_{no}} \tag{3.1.3}$$

Equating the two expressions for V_{npo} gives $n_{no}p_{no} = n_{po}p_{po}$, which follows from the basic equation at equilibrium, $n_o p_o = n_i^2$ (Equation 2.5.6, Chapter 2).

Conclusion: *Equilibrium is established in a pn junction, for both electrons and holes, by having the n-side at a positive voltage V_{npo} with respect to the p-side, where,*

$$V_{npo} = V_T \ln \frac{p_{po}}{p_{no}} = V_T \ln \frac{n_{no}}{n_{po}} = V_T \ln \frac{p_{po} n_{no}}{n_i^2} \cong V_T \ln \frac{N_A N_D}{n_i^2} \tag{3.1.4}$$

The approximation on the RHS applies at room temperature with the dopant atoms completely ionized, and their concentrations much larger than n_i (Section 2.5, Chapter 2). This last expression is usually the most convenient for calculating V_{npo}.

Exercise 3.1.1

Derive the expression for V_{npo} from Equations 2.4.17 and 2.4.18, Chapter 2 by setting $\kappa_{hp} = \kappa_{hn}$ and $\kappa_{ep} = \kappa_{en}$. ∎

Exercise 3.1.2

Given a *pn* junction having $N_A = 10^{16}/\text{cm}^3$ and $N_D = 10^{15}/\text{cm}^3$. (a) Determine p_o and n_o at 300K, stating any assumptions made; (b) Determine V_{npo}; (c) what is the electric potential energy of an electron on the *p*-side relative to the *n*-side at equilibrium? What is the corresponding chemical potential energy and the electrochemical potential energy?
   Answers: (a) $p_{po} = 10^{16}/\text{cm}^3$, $n_{po} = 10^4/\text{cm}^3$, $n_{no} = 10^{15}/\text{cm}^3$, and $p_{no} = 10^5/\text{cm}^3$, based on the assumptions of: (i) complete ionization of impurity atoms at 300K, and (ii) N_A, $N_D \gg n_i$; (b) 0.66 V; (c) 1.06×10^{-19} J, -1.06×10^{-19} J, 0. ∎

Depletion Approximation

Assuming an abrupt junction (Figure 3.1.2a) and $N_A^- > N_D^+$ so that $p_{po} > n_{no}$, the energy-band diagram of a *pn* junction is shown in Figure 3.1.2b. As explained in Example 2.5.2, Chapter 2, E_F is the same throughout, being closer to the edge of the conduction band on the *n*-side, and to the edge of the valence band on the *p*-side. In the region on either side of the metallurgical junction, the edges of the conduction and valence bands rise smoothly from their levels on the *n*-side to their levels on the *p*-side. Similarly, the majority and minority carrier concentrations change smoothly in this region from their values on one side to their values on the other side (Figure 3.1.2c). This region of adjustment of voltage and concentration between the two sides is the **depletion region**. Where E_F intersects the midgap level E_{Fi}, the concentrations of electrons and holes is n_i. In general, this does not occur at the metallurgical junction (Example 3.1.1).

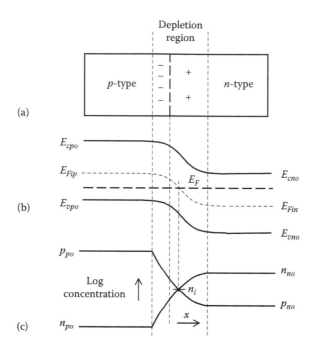

FIGURE 3.1.2
pn junction at equilibrium: (a) depletion region; (b) energy-band diagram; (c) carrier concentration profile.

Where there is a voltage difference, there must be charge separation. In the case of the *pn* junction, there is a net positive charge on the *n*-side, with an equal and opposite net negative charge on the *p*-side. These net charges arise in the depletion region on both sides of the junction (Figure 3.1.2a). According to the depletion approximation, the concentrations of carriers in the depletion region on either side of the junction is assumed to be negligibly small compared to the concentration of dopant atoms. The justification for this is the exponential variation of concentrations with energy levels (Equations 2.5.14 and 2.5.15, Chapter 2). If $N_D = 10^{16}/cm^3$, for example, $E_F - E_{Fi} = 0.357$ eV on the *n*-side. A decrease of $E_F - E_{Fin}$ in the depletion region to half this value reduces n_o by a factor of 1000.

If carrier concentrations in the depletion region are neglected, and *assuming complete ionization of impurity atoms*, the charge concentrations per unit volume are $|q|N_D$ on the *n*-side and $|q|N_A$ on the *p*-side. The electric field $\xi(x)$ is zero outside the depletion region, where the majority carrier concentrations are high, and is directed in the negative *x*-direction in the depletion region. Considering $x = 0$ to be at the metallurgical junction, let the edge of the depletion region be at $x = W_n$ on the *n*-side, and at $x = -W_p$ on the *p*-side (Figure 3.1.3a).

For $x \geq 0$, $\xi(x)$ is from Gauss's law:

$$\xi(x) = \frac{1}{\varepsilon}\int_{W_n}^{x} |q|N_D dx = \frac{|q|N_D}{\varepsilon}(x - W_n)$$

(3.1.5)

$$\text{At } x = 0, \quad \xi(0) = -\frac{|q|N_D}{\varepsilon}W_n.$$

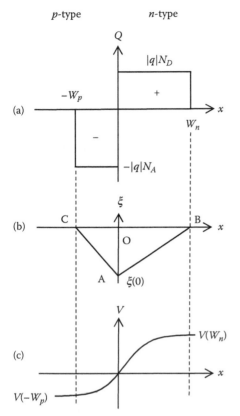

FIGURE 3.1.3
Depletion approximation: (a) charge distribution; (b) variation of electric field; and (c) variation of voltage.

For $x \leq 0$,

$$\xi(x) = \frac{1}{\varepsilon} \int_{-W_p}^{x} -|q|N_A dx = -\frac{|q|N_A}{\varepsilon}(x + W_p) \qquad (3.1.6)$$

At $x = 0$, $\xi(0) = -|q|N_A/\varepsilon W_p$. $\xi(x)$ has its largest magnitude at $x = 0$ (Figure 3.1.3b). Since $\xi(0)$ is the same, $N_D W_n = N_A W_p$, as to be expected, because the total charge per unit area is of the same magnitude, but opposite in sign, on the two sides of the junction. It follows that:

$$\frac{W_n}{W_p} = \frac{N_A}{N_D} \qquad (3.1.7)$$

If $N_A > N_D$, then $W_n > W_p$, that is,

Concept: *The depletion region extends more into the less heavily doped side.*

Using $\xi(x) = -dV(x)/dx$, Equation 3.1.5 gives:

$$V(x) = -\frac{|q|N_D}{\varepsilon} \int_{0}^{x} (x - W_n)dx + V(0) = \frac{|q|N_D}{\varepsilon}\left(xW_n - \frac{x^2}{2}\right), \quad 0 \leq x \leq W_n \qquad (3.1.8)$$

assuming $V(0) = 0$, for convenience (Figure 3.1.3c). At $x = W_n$, $V(W_n) = |q|N_D W_n^2/2\varepsilon = \frac{1}{2} W_n |\xi(0)|$, which is the area of triangle OAB, as expected. Equation 3.1.6 gives:

$$V(x) = \frac{|q|N_A}{\varepsilon} \int_0^x (x + W_p)dx + V(0) = \frac{|q|N_A}{\varepsilon}\left(xW_p + \frac{x^2}{2}\right), \quad -W_p \leq x \leq 0 \qquad (3.1.9)$$

At $x = -W_p$, $V(-W_p) = -|q|N_A/2\varepsilon W_p^2 = -\frac{1}{2}W_p|\xi(0)|$; $|V(-W_p)|$ is the area of triangle OAC.

W_n and W_p are unknown. Equation 3.1.7 is one relation between them. The other relation is obtained from the fact that $V(W_n) - V(-W_p) = V_{npo}$, the known junction potential. Let $W_d = W_n + W_p$. Then, from Equation 3.1.7,

$$W_n = \frac{N_A}{N_A + N_D}W_d, \quad \text{and} \quad W_p = \frac{N_D}{N_A + N_D}W_d \qquad (3.1.10)$$

From Equations 3.1.8 through 3.1.10,

$$V_{npo} = \frac{|q|}{2\varepsilon}\left(N_D W_n^2 + N_A W_p^2\right) = \frac{|q|}{2\varepsilon}W_d^2\frac{N_A N_D}{N_A + N_D}$$

or,

$$W_d = \sqrt{\frac{2\varepsilon}{|q|}V_{npo}\left(\frac{1}{N_A} + \frac{1}{N_D}\right)} \qquad (3.1.11)$$

Exercise 3.1.3

Calculate the junction potential in a Si *pn* junction diode if $N_A = 10^{16}/\text{cm}^3$, $N_D = 10^{15}/\text{cm}^3$, and:
(a) $T = 300\text{K}$; (b) $T = 500\text{K}$. Assume that $n_i = 10^{10}$ at 300K, all impurities are ionized, and E_g varies with T as in P2.1.2, Chapter 2.
   Answers: (a) 0.64 V; (b) 0.2 V.   ∎

Exercise 3.1.4

Calculate W_d for the *pn* junction of Exercise 3.1.3 at 300K assuming $\varepsilon = 12 \times 8.85 \times 10^{-14}$ F/cm for Si.
   Answer: 0.967 μm.   ∎

Equilibrium is dynamic; that is, electrons continually move from the *n*-side to the *p*-side, through diffusion, and from the *p*-side to the *n*-side, through drift under the influence of the voltage difference; similarly for holes. These movements constitute currents, which, because of the relatively large carrier concentrations, are extremely large, and can exceed tens of thousands of A/cm^2 at room temperature, depending on the doping level (Example 3.1.1). However, over any given time interval at equilibrium, the diffusion current for electrons, or holes, is equal in magnitude, on the average, but opposite in direction to the corresponding drift current.

Example 3.1.1 Equilibrium Diffusion and Drift Currents across a *pn* Junction

Consider an abrupt silicon *pn* junction with $N_A = 10^{16}/\text{cm}^3$ and $N_D = 10^{14}/\text{cm}^3$. It is required to determine the equilibrium drift and diffusion currents, and the carrier concentrations in the depletion region, assuming the depletion approximation and a temperature of 300K.

SOLUTION

Step 1: Calculation of equilibrium carrier concentrations and junction potential. $p_{po} \cong N_A = 10^{16}/cm^3$, $n_{no} \cong N_D = 10^{14}/cm^3$. From Equation 3.1.4, the equilibrium junction potential is $V_{npo} = 0.026 \ln(10^{16} \times 10^{14}/10^{20}) \cong 0.6$ V.

Step 2: Calculation of the extent of the depletion region and the electric field strength at the metallurgical junction. The total width of the depletion region is given by Equation 3.1.11. Substituting $\varepsilon = 12 \times 8.85 \times 10^{-14}$ F/cm, where the relative permittivity of Si is taken as 12

$$W_d = \sqrt{\frac{2 \times 12 \times 8.85 \times 10^{-14} \times 0.6}{1.6 \times 10^{-19}}}(10^{-14} + 10^{-16}) = 2.8 \times 10^{-4} \text{ cm} \equiv 2.8 \text{ } \mu m.$$ From the

relations of Equation 3.1.10, W_n and W_p are: $W_p = 10^{14} \times 2.8/(10^{14} + 10^{16}) = 0.0277$ μm, and $W_n = 10^{16} \times 2.8/(10^{14} + 10^{16}) = 2.773$ μm. The electric field strength at the

junction $\xi(0)$ can be calculated from Equation 3.1.5 as $\xi(0) = -\dfrac{|q|N_D}{\varepsilon} W_n =$

$-\dfrac{1.6 \times 10^{-19} \times 10^{14} \times 2.773 \times 10^{-4}}{12 \times 8.85 \times 10^{-14}} = -4.178 \times 10^3$ V/cm, and is directed in the negative

x-direction, from the *n*-side toward the *p*-side.

Step 3: Calculation of equilibrium hole diffusion and drift current densities. If generation or recombination of carriers in the depletion region is neglected, current densities do not vary in the depletion region and are most conveniently calculated at $x = 0$. The diffusion current density at any x is given by Equation 2.4.12, Chapter 2: $J_{h(diff)} = -D_h|q|dp_o(x)/dx$. To determine $p_o(x)$, we note that in going from the bulk *p*-region to any point $x \leq 0$, the change in hole chemical potential energy is $k_J T \ln p_o(x)/p_{po}$, while the change in hole electric potential energy is $|q|(V(x) - V_{po})$. Hence, $k_J T \ln p_o(x)/p_{po} + |q|(V(x) - V_{po}) = 0$. This gives:

$$p_o(x) = p_{po}e^{|q|(V_{po} - V(x))/V_T} \tag{3.1.12}$$

Differentiating with respect to x and substituting $\xi(x) = -dV(x)/dx$, we obtain $\dfrac{dp_o(x)}{dx} = \dfrac{p_o(x)\xi(x)}{V_T}$, and $\left.\dfrac{dp_o(x)}{dx}\right|_{x=0} = \dfrac{p_o(0)\xi(0)}{V_T}$. Thus:

$$J_{h(diff)} = \frac{-D_h|q|p_o(0)\xi(0)}{V_T} \tag{3.1.13}$$

Substituting for D_h from the Nernst–Einstein relation (Equation 2.4.5, Chapter 2),

$$J_{h(diff)} = -|q|\mu_h p_o(0)\xi(0) = -J_{h(drift)} \tag{3.1.14}$$

where $J_{h(drift)}(0) = \mu_h|q|p_o(0)\xi(0)$, from Equations 2.4.10 and 2.4.11, Chapter 2. From Equation 3.1.12, $p_o(0) = p_{po}e^{V_{po}/V_T}$, where from Equation 3.1.9, $V_{po} = -\dfrac{1}{2}W_p|\xi(0)| = -0.5 \times (0.0277$ μm) $\times (4.178 \times 10^5$ V/m) $\equiv -5.79$ mV. This gives $p_o(0) = 10^{16} \times e^{-5.79/25.85} = 8 \times 10^{15}/cm^3$. Taking $\mu_h = 400$ cm^2/(Vs), $J_{h(diff)} = 400 \times 1.6 \times 10^{-19} \times 8 \times 10^{15} \times (-4.178 \times 10^3) = 2.14 \times 10^3$ A/cm^2. This current increases considerably with the doping level. For example, if $N_A = 10^{18}/cm^3$ and $N_D = 10^{16}/cm^3$, $J_{h(drift)} \cong 8.7 \times 10^5$ A/cm^2 (Problem P3.1.22). The calculation of equilibrium electron drift and diffusion currents can be carried out in an exactly analogous manner and is left as a problem (Problem P3.1.21).

Step 4: Calculation of carrier concentrations in the depletion region. $p_o(x)$ is given by Equation 3.1.12. The concentration of electrons as a function of x can be determined in a similar manner. Thus, in going from any point $x \geq 0$ to the bulk *n*-region, the change in electron chemical potential energy is $kT \ln(n_{no}/n_o(x))$, while the change in electron electric potential energy is $-|q|(V_{no} - V(x))$. It follows that:

$$n_o(x) = n_{no}e^{(V(x)-V_{no})/V_T} \qquad (3.1.15)$$

where $V(x)$ is given by Equation 3.1.8, $0 \leq x \leq W_n$.

It will be noted that at the metallurgical junction ($x = 0$), the semiconductor is effectively p-type, because $p(0) = 8 \times 10^{15}/\text{cm}^3 > n_i = 10^{10}/\text{cm}^3$. The semiconductor is effectively intrinsic in the n-region. It is of interest to determine the value of $x = x_i$ at which this occurs. From Equation 3.1.8, $V_{no} = \frac{1}{2}W_n|\xi(0)| = 0.579$ V, and from Equation 3.1.15, $V(x_i) = 0.579 + 0.026 \times \ln(10^{10}/10^{14}) = 0.34$ V. From Equation 3.1.8, $V(x_i) = (|q|N_D/\varepsilon) \times \left(x_iW_n - \frac{x_i^2}{2}\right)$. Substituting and solving for x_i gives $x_i = 8 \times 10^{-4}$ μm, which is only very slightly into the n-side. At this x_i, $E_F = E_{Fi}$, that is, E_F is in the middle of the energy gap.

The assumptions made in Example 3.1.1 are well justified in practice under normal conditions. The more exact analysis, outlined in Section ST3.1, is not difficult but awkward.

Exercise 3.1.5

Using Equations 3.1.4, 3.1.12, and 3.1.15, show that $p_o(x)n_o(x) = n_i^2$.   ■

3.2 The Biased *pn* Junction

The *pn* Junction as a Rectifier

If an external voltage is applied across the *pn* junction, the junction becomes *biased*. If the p-side is made positive with respect to the n-side, the junction is *forward biased*, whereas if the p-side is made negative with respect to the n-side, the junction is *reverse biased*. It is helpful to remember that the p-side is positive with respect to the n-side under forward (spelled with **ph**!) bias. Because the *pn* junction is disturbed by the bias, it is no longer at equilibrium.

A forward bias opposes the junction potential, which reduces the electric potential energy difference relative to the chemical potential energy difference, for both electrons and holes. The diffusion–drift balance is upset in favor of diffusion, causing a *net diffusion current* in the forward direction, that is, in the direction of the drop in applied voltage and in electrochemical potential. The current is due to majority carriers, that is, holes moving from the p-side to the n-side and electrons moving from the n-side to the p-side.

A reverse bias augments the junction potential. The electric potential energy difference is increased relative to the chemical potential energy difference, so the diffusion–drift balance is upset in favor of drift. There will, therefore, be a *net drift current* in the reverse direction, but still in the direction of the drop in applied voltage and in electrochemical potential. This current is carried by minority carriers, that is, by holes moving from the n-side to the p-side and by electrons moving from the p-side to the n-side. These relations are summarized in Table 3.2.1.

Concept: *In a biased pn junction, the forward current is due to majority carriers, whereas the reverse current is due to minority carriers. Since the majority carrier concentration vastly exceed the minority carrier concentration, the forward current is much larger than the reverse current. The pn junction therefore behaves as a rectifier.*

TABLE 3.2.1

Potential Energy Relations and Current under Bias

State	Difference in Electric Potential Energy Relative to Difference in Chemical Potential Energy	Diffusion–Drift Balance in Favor of	Direction of Current	Current Carriers
Equilibrium	Same	Neither	No current	—
Forward bias	Reduced	Diffusion	$p \to n$	Majority
Reverse bias	Increased	Drift	$n \to p$	Minority

Width of Depletion Region

The results of Example 3.1.1 lead to a fundamental assumption in the simplified analysis of the *pn* junction:

Concept: *Because the equilibrium diffusion and drift currents across a pn junction are much larger than the current that ordinarily flows even under heavy forward bias, the biased junction is never far from equilibrium.*

This concept is extensively applied in the simplified analysis of the *pn* junction. For example, Equation 3.1.11, based on the depletion approximation, is generalized to:

$$W_d = \sqrt{\frac{2\varepsilon}{|q|} \left(V_{npo} - v_D^*\right) \left(\frac{1}{N_A} + \frac{1}{N_D}\right)} \tag{3.2.1}$$

where v_D^* is, strictly speaking, the voltage across the junction and differs from the terminal voltage v_D because of the voltage drop due to the bulk resistance (Section 1.1, Chapter 1). In the forward direction, $v_D > v_D^* > 0$ and $v_D^* < V_{npo}$. If $v_D^* = V_{npo}$, the drift current is zero and the diffusion current will be extremely large. In fact, the diode will be damaged by overcurrent and overheating long before v_D^* approaches V_{npo}. Nevertheless, under forward bias, $v_D^* > 0$, and the net voltage across the junction is reduced from V_{npo} to $(V_{npo} - v_D^*)$, so that less charge in the depletion region is required to support this voltage. Since dopant concentrations are fixed, less charge can only result from a narrower width of the depletion region (Figure 3.1.3).

Conversely, under reverse bias, the current is small and the voltage drop due to the bulk resistance is small. $v_D^* \cong v_D = -v_R$, where v_R is the reverse bias. The net voltage across the junction is increased from V_{npo} to $(V_{npo} + v_R)$, and more charge in the depletion region is required to support this voltage, which means that the depletion region becomes wider. Hence,

Concept: *The width of the depletion region decreases under forward bias and increases under reverse bias.*

Charge Distributions and Currents under Bias

Under forward bias, electrons and holes diffuse across the junction from the region where they are majority carriers to the region where they are minority carriers. Minority carriers are said to be **injected** by forward bias. At the edges of the depletion region, the

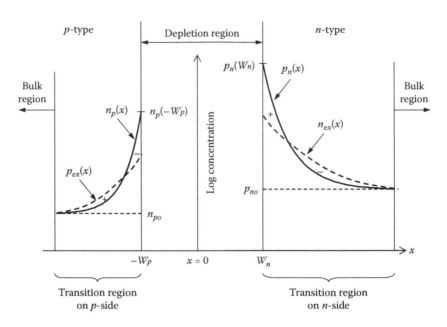

FIGURE 3.2.1
pn junction under forward bias.

concentrations of these carriers increase exponentially with forward bias, in accordance with the following relations, known as the **law of the junction**:

$$p_n(W_n) = p_{no}e^{v_D/V_T} \quad \text{and} \quad n_p(-W_p) = n_{po}e^{v_D/V_T} \tag{3.2.2}$$

where v_D is the forward bias across the junction (Figure 3.2.1). Equations 3.2.2 can be derived quite readily from the electrochemical potential difference of the carriers (Section ST3.5). Thus, neglecting the small electric field outside the depletion region, the change in hole electrochemical potential in going from the *p*-side through the transition region on the *n*-side is $V_T \ln(p_{no}/p_n(W_n))$ eV. In going through the external circuit, the change in hole electrochemical potential is $-v_D$ eV. Equating these expressions gives the law of the junction on the *n*-side.

Injected minority carriers recombine with majority carriers on either side, their concentration decreasing exponentially with distance at a rate determined by the **diffusion length**, L_h or L_e, for holes and electrons, respectively. The diffusion length is analogous to the time constant in the case of exponential variation with time and equals $\sqrt{D\tau}$ (Section ST3.4), where D is the diffusion constant of minority carriers and τ is the minority carrier lifetime. The larger the diffusion constant, the more mobile is the particle, and the further it can travel in a given time. The larger the lifetime, the longer will the particle exist before recombination. Typically, D is of the order of μm, whereas τ is in the range of ns to μs.

The net charge due to injected minority carriers attracts majority carriers, which tend to neutralize the distribution of minority carriers, as illustrated in Figure 3.2.1, where $n_{ex}(x)$ and $p_{ex}(x)$ are the respective *excess* majority carrier concentrations above their equilibrium values. The distribution of majority carriers is, however, flatter than the exponential distribution of the minority carriers, leaving equal net charges of opposite polarity. These net charges are required to produce the electric field necessary for supporting the distribution of majority carriers. Thus, the distribution of minority carriers is maintained by injection from the opposite side, whereas the distribution of majority carriers is maintained by the relatively small electric field that arises from the difference between the distributions of minority and majority carriers.

At distances sufficiently removed from the junction, the presence of the junction is not significantly felt, and the concentrations of majority carriers and minority carriers approach their equilibrium values. The region on either side of the junction, beyond the depletion region, where the concentration of minority carriers is significantly above its equilibrium value, is the **transition region.** The region, on either side of the junction, beyond the transition region, where carrier concentrations are substantially at their equilibrium values, is the **bulk region** (Figure 3.2.1). Since minority carriers are injected across the junction into the transition region, through the depletion region, the concentrations of holes and electrons in the depletion region are higher than at equilibrium. This means that there will be a net recombination of carriers in this region.

In the steady state, a continuous flow of majority carriers on a given side toward the junction is required for two purposes: (1) to replace majority carriers lost to recombination with minority carriers, both in the transition and depletion regions on the given side; and (2) to inject the carriers that would cross the depletion region and become minority carriers on the other side. The majority carriers are supplied through the external circuit that applies the forward bias, as illustrated in Figure 3.2.2. Electrons flow from the negative terminal of the dc supply, through the cathode terminal, and into the bulk region, toward the junction. On the *p*-side, holes flow in bulk region toward the junction and have to be replaced by the external circuit. Since the metallic conductor of the external circuit does not ordinarily support the flow of holes, electron–hole pairs must be generated at the metal–semiconductor on the *p*-side, so that holes flow into the bulk region and electrons flow in the external circuit. On the *n*-side, the metal–semiconductor contact is only required to allow the flow of electrons.

Similar considerations apply to reverse bias, with all currents reversed and generation replacing recombination (Figure 3.2.3). At the edges of the depletion region, minority carrier concentrations are depressed below their equilibrium values, in accordance with Equations 3.2.2 with $v_D < 0$. Instead of being injected across the junction into the depletion and transition regions, minority carriers are now **extracted** from these regions. This extraction of minority carriers is due to the dominance of the electric potential energy difference under reverse bias, so the electric field, which is in the negative *x*-direction, sweeps holes from the *n*-side to the *p*-side, and electrons from the *p*-side to the *n*-side.

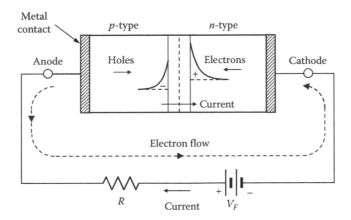

FIGURE 3.2.2
Flow of charges for a *pn* junction under forward bias.

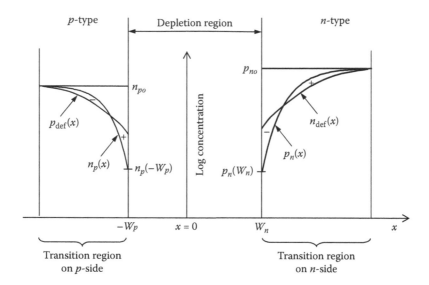

FIGURE 3.2.3
pn junction under reverse bias.

The depression of the concentrations of minority carriers is accompanied by deficiencies in the concentrations of majority carriers $n_{def}(x)$ and $p_{def}(x)$ below their equilibrium values (Figure 3.2.3). Since carrier concentrations in the depletion and transition regions are below equilibrium levels, there is a net generation of carriers. Electrons extracted from the transition region on the *p*-side, and those generated in the depletion and transition regions on the *n*-side, flow in the bulk regions away from the junction and out of the cathode terminal (Figure 3.2.4); similarly for holes. At the metal–semiconductor contacts on the *p*-side, electrons and holes recombine, so that electron inflow at the anode terminal is converted to flow of holes toward this terminal. Because the rate at which minority carriers are extracted is limited, the reverse current I_S quickly saturates with reverse bias.

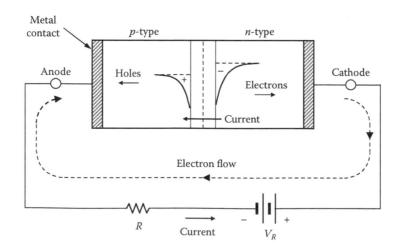

FIGURE 3.2.4
Flow of charges for a *pn* junction under reverse bias.

Current distributions under forward bias and reverse bias are discussed qualitatively in Section ST3.2, where the drift and diffusion components of the carrier current in are identified in each of the three regions on either side. It should be noted that:

Concept: *The forward current of the pn junction diode is essentially a recombination current, and the reverse current is essentially a generation current.*

All electrons that flow on the *n*-side toward the junction under forward bias eventually recombine with holes in the transition and depletion regions on the two sides; similarly for holes that flow on the *p*-side toward the junction. In an analogous manner, the reverse current is a generation current. Because the forward and reverse currents are recombination and generation currents, respectively, minority carriers play an essential role in the operation of the *pn* junction diode. Exposure to radiation, as may occur in space applications, can have a detrimental effect on operation, because it increases the rate of generation, and hence the rate of recombination.

Current–Voltage Relation

Based on some simplifying assumptions, the diode current–voltage relation can be derived from i_{Dh} and i_{De}, the hole and electron current, respectively. The relations are (Equations ST3.5.10 and ST3.5.11):

$$i_D = i_{Dh} + i_{De} = I_S(e^{v_D/V_T} - 1) \tag{3.2.3}$$

where

$$I_S = A_J|q|\left(\frac{D_h p_{no}}{L_h} + \frac{D_e n_{po}}{L_e}\right) = A_J|q|\left(\frac{D_h}{L_h N_D} + \frac{D_e}{L_e N_A}\right)n_i^2 \tag{3.2.4}$$

$$i_{Dh} = \frac{A_J|q|D_h p_{no}}{L_h}(e^{v_D/V_T} - 1) = \frac{A_J|q|D_h n_i^2}{L_h N_D}(e^{v_D/V_T} - 1) \tag{3.2.5}$$

$$i_{De} = \frac{A_J|q|D_e n_{po}}{L_e}(e^{v_D/V_T} - 1) = \frac{A_J|q|D_e n_i^2}{L_e N_A}(e^{v_D/V_T} - 1) \tag{3.2.6}$$

A_J being the area of the junction.

Under forward bias, the effect of recombination in the depletion region is generally not negligible, particularly in Si diodes when the forward bias is relatively low. It is shown in Section ST3.6 that the current due to recombination in the depletion region is:

$$i_D = I_S(e^{v_D/2V_T} - 1) \tag{3.2.7}$$

where the emission coefficient η is 2. At small enough forward currents, the current component due to recombination predominates and Equation 3.2.7 applies. At large forward currents, the concentration of injected minority carriers at the edge of the transition region can become comparable to the majority carrier concentration, or can even exceed it. In such a state of **high-level injection**, η is also theoretically equal to 2 (Section ST3.7). As mentioned in Section 1.1, Chapter 1, η varies with the diode current, but is normally considered 2 for a discrete Si *pn* junction diode and 1 for an IC diode.

In a **short-base diode**, the length of one side, say the n-side, is short compared to the diffusion length of holes on this side. The advantages of a shorter side are a larger current and a faster turnoff time because of reduced stored charge (Section ST3.8).

Charge–Current Relation

The diode current can be expressed in terms of the stored minority carrier charge on either side. As argued in connection with Equation 2.7.1, Chapter 2, $1/\tau_h$ is the average rate at which individual holes recombine in the transition region on the n-side. If the number of excess holes stored on this side is N_h, the rate at which holes recombine is N_h/τ_h. The rate at which hole charge is lost to recombination is $qN_h/\tau_h = Q_h/\tau_h$, where Q_h is the excess hole charge stored on the n-side. Under steady-state conditions, the stored charge is replenished by the hole current i_{Dh} so that $i_{Dh} = Q_h/\tau_h$. Similarly, the charge stored on the p-side is replenished by the electron current $i_{De} = Q_e/\tau_e$. The total diode current is $i_D = i_{Dh} + i_{De}$, which gives:

$$i_D = \frac{Q_h}{\tau_h} + \frac{Q_e}{\tau_e} \tag{3.2.8}$$

The charge–current relation is also derived in Section ST3.5. The total charge stored on both sides of the junction is $Q_{\text{diff}} = Q_h + Q_e$, where $Q_h = i_{Dh}\tau_h$ and $Q_e = i_{De}\tau_e$. Thus:

$$Q_{\text{diff}} = \tau_h i_{Dh} + \tau_e i_{De} = \tau_T i_D \tag{3.2.9}$$

where τ_T is the **mean transit time** and can be readily expressed in terms of Equations 3.2.3, 3.2.5, 3.2.6, and the relation $L = \sqrt{D\tau}$ (Problem P3.1.16).

Normally, $N_A \gg N_D$ so as to control the breakdown voltage, as discussed later. This makes $i_{Dh} \gg i_{De}$ (Equations 3.2.5 and 3.2.6), so that $i_D \cong Q_h/\tau_h$.

pn Junction Capacitances

Capacitance is a circuit parameter that accounts for energy stored in the electric field arising from separated electric charges. In a linear capacitor, the charge is directly proportional to the voltage difference between the two plates of the capacitor. In the more general case encountered in *pn* junctions, charge varies nonlinearly with voltage difference, so an incremental capacitance is defined at a given operating point as the ratio of a small change in charge to a small change in the associated voltage difference. Under dc conditions, voltages are constant, and capacitors behave as open circuits. Under time-varying conditions, the charge changes with time, which constitutes a current through the capacitor. The higher the frequency, the smaller is the time during which a given change of charge takes place and the larger is the current.

In a *pn* junction at equilibrium, a positive charge exists in the depletion region on the n-side, with an equal negative charge in the depletion region on the p-side (Figure 3.1.2). As explained in connection with Equation 3.2.1, the charge in the depletion region decreases with forward bias and increases with reverse bias. Hence, a change Δv_D produces a change $-\Delta Q_{dp}$ in the charge on either side of the junction. A **depletion capacitance** C_{dp} is defined as:

$$C_{dp} = -\frac{dQ_{dp}}{dv_D} = \frac{dQ_{dp}}{dv_R}\bigg|_{v_R = V_R} \tag{3.2.10}$$

To make C_{dp} positive, it is defined as the derivative of Q_{dp} with respect to $v_R = -v_D$, since Q_{dp} increases with v_R. C_{dp} is evaluated at a particular value V_R. The charge Q_{dp} at any given diode voltage can be readily derived using the depletion approximation for an abrupt junction (Figure 3.1.3). Thus, $Q_{dp} = |q|A_J N_D W_n = |q|A_J W_d N_A N_D/(N_A + N_D)$, where A_J is the area of the junction and W_d is given by Equation 3.2.1 with $v_R = -v_D$. Applying Equation 3.2.10,

$$C_{dp} = \frac{C_{dp0}}{\sqrt{1 + V_R/V_{npo}}} \qquad (3.2.11)$$

where C_{dp0} is the depletion capacitance at equilibrium, when $V_R = 0$. Because Q_{dp} varies as the square root of $(V_{npo} + V_R)$ (Equation 3.2.1), dQ_{dp}/dv_R, and hence C_{dp}, decrease with increasing V_R, while Q_{dp} increases. From Equation SE3.1.3, Example SE3.1, C_{dp0} is given by:

$$C_{dp0} = A_J \sqrt{\frac{\varepsilon_s|q|}{2V_{npo}} \left(\frac{N_A N_D}{N_A + N_D}\right)} \qquad (3.2.12)$$

For a nonabrupt junction, Equation 3.2.11 can be written as:

$$C_{dp} = \frac{C_{dp0}}{\left(1 + V_R/V_{npo}\right)^m} \qquad (3.2.13)$$

where m depends on the grading of the junction. m is $1/2$ for an abrupt junction and $1/3$ for a linearly graded junction (Example SE3.1). It is shown in Example SE3.1 that for abrupt and linearly graded junctions C_{dp} is that of a parallel-plate capacitor of area A_J and a silicon dielectric of thickness W_d:

$$C_{dp} = \frac{\varepsilon A_J}{W_d} \qquad (3.2.14)$$

Exercise 3.2.1
Assume that the diode of Exercise 3.1.2 has an abrupt junction of 2000 μm^2 area. If $C_{dp} = 0.4$ pF at $V_R = 5$ V, determine C_{dp0} per unit area.
    Answer: 0.59 fF/μm^2.   ■

    In the biased junction, there is also charge in the transition region. Under forward bias, there is an excess charge on either side of the junction due to injected minority carriers. The charge is partly counterbalanced by a flatter distribution of majority carriers, leading to charge separation and the establishment of an electric field that maintains the distribution of the majority carriers (Figure 3.2.1). A **diffusion capacitance** is defined under these conditions as $C_d = dQ_{\text{diff}}/dv_D$. From Equation 3.2.9, and using Equation 1.1.3, Chapter 1,

$$C_d = \tau_T \frac{di_D}{dv_D} = \frac{\tau_T}{r_d} = \frac{\tau_T}{\eta V_T} i_D \qquad (3.2.15)$$

If $N_A \gg N_D$, $i_{Dh} \gg i_{De}$ (Equations 3.2.5 and 3.2.6), so $Q_{\text{diff}} \cong \tau_h i_{Dh}$ and $C_d \cong \tau_h di_D/dv_D = \tau_h/r_d$.

Concepts: *(1) The diffusion capacitance increases rapidly with forward bias and is negligible for reverse bias. (2) The depletion capacitance also increases with forward bias and* *decreases with reverse bias.* (3) *In a forward-biased diode, the diffusion capacitance is dominant, whereas in a reverse-biased diode the depletion capacitance is dominant.*

Exercise 3.2.2

Estimate C_d at $i_D = 10$ mA if $\tau_h = 100$ ns, assuming $\eta = 2$ and $V_T = 0.026$ V.
    Answer: 19.2 nF.   ∎

Application Window 3.2.1 Varactors

Varactors are reverse-biased diodes that are used to provide a voltage-controlled capacitance. Such a capacitance can vary the resonant frequency of an *RLC* circuit, as illustrated in Figure 3.2.5. The resonant frequency is determined by L and the varactor capacitance, which depends on the reverse bias. The reverse bias depends in turn on R_v, which is usually the voltage-controlled resistance of a transistor. The nominal capacitance of varactors is generally in the range of tens of pFs and can be varied over a fraction of the nominal value. Varactors are used for electronic tuning in radio and television receivers.   ∎

Diode transients are discussed mainly qualitatively in Section ST3.3. Of particular importance is the reverse switching transient that occurs when a diode is switched from forward bias to reverse bias. Before the diode can assume this latter state, the stored charge due to injected minority carriers must be removed. Manufacturer's data sheets quote a **reverse recovery time** t_{rr}, which is the total time taken for the diode to recover to a specified reverse current, or to a specified reverse resistance, under specified conditions. t_{rr} is typically in the nanosecond range.

Temperature Effects

From Equation 3.2.4, the temperature dependence of I_S is due to n_i^2, the diffusion constants, and the minority carrier lifetimes. From Equation 2.5.12, Chapter 2, n_i^2 varies with temperature as $T^3 e^{-E_g/kT}$. The diffusion constant varies with temperature as μT (Equation 2.4.5, Chapter 2). As discussed in Section 2.6, Chapter 2, mobility and minority carrier lifetime are temperature dependent. The temperature variation of I_S can be expressed as:

$$I_S = M T^b e^{-E_g/kT} \tag{3.2.16}$$

where M is independent of temperature, and b can assume values between 2.5 and 5.5. Neglecting the small variation of E_g with T (Problem P2.1.2, Chapter 2), taking logarithms of both sides of Equation 3.2.16, and differentiating with respect to T:

$$\frac{1}{I_S}\frac{dI_S}{dT} = \frac{d(\ln I_S)}{dT} = \frac{b}{T} + \frac{E_g}{kT^2} \tag{3.2.17}$$

Over the temperature range of practical interest, the contribution of the first term is small compared to the second term, irrespective of b. Substituting $E_g = 1.12$ eV for Si and $k = 8.62 \times 10^{-5}$ eV/K gives $\dfrac{1}{I_S}\dfrac{dI_S}{dT} \cong$ 0.15 for Si at 300K. This means that I_S increases by 15% per °C. It approximately doubles for every 5°C rise in temperature, since

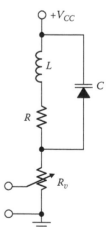

FIGURE 3.2.5
Figure for Application Window 3.2.1.

$(1.15)^5 \cong 2$. However, I_{leak} (Equation 1.1.2, Chapter 1) does not increase as rapidly with temperature. As a rule of thumb, the reverse current I_R is considered to double for every 10°C rise in temperature. Thus, if I_R is 1 nA at a junction temperature of 25°C, it becomes $2^5 = 32$ nA at a junction temperature of 75°C.

Under forward bias, we can consider temperature effects under conditions of constant diode voltage or constant diode current. For constant diode voltage, it follows from Equation 1.1.1, Chapter 1, neglecting unity in comparison with the exponential, that:

$$\frac{1}{i_D}\frac{di_D}{dT} = \frac{1}{i_D}\frac{d\left(I_S e^{v_D/\eta kT}\right)}{dT} = \frac{1}{I_S}\frac{dI_S}{dT} - \frac{v_D}{\eta kT^2} \tag{3.2.18}$$

If $v_D = 0.7$ V, $\eta = 1$, and $T = 300$K, the second term evaluates to 0.09, compared to 0.15 for the first term. Hence, i_D increases by about 6% per °C, that is, it doubles every approximately 12°C. If $\eta = 2$, i_D doubles approximately every 7°C.

If i_D is assumed constant,

$$\frac{dv_D}{dT} = \frac{v_D}{T} - \eta kT\left(\frac{1}{I_S}\frac{dI_S}{dT}\right) \tag{3.2.19}$$

At $T = 300$K, and $v_D = 0.7$ V, $\frac{dv_D}{dT} \cong -1.6$ mV/°C if $\eta = 1$, and $\frac{dv_D}{dT} \cong -5$ mV/°C if $\eta = 2$. As a rule of thumb, v_D is considered to change at a rate of approximately -2 mV/°C, based on measurements at i_D of about 1 mA. The forward i–v characteristic therefore changes with temperature as illustrated in Figure 1.4.6, Chapter 1.

Junction Breakdown

As the reverse bias is increased, the reverse current $I_S + I_{leak}$ increases slowly at first, due to I_{leak}. Eventually, at a reverse voltage BV the reverse current increases rapidly with a small increase in the reverse voltage (Figure 1.4.1, Chapter 1). The part of the i–v characteristic where this occurs is the **breakdown region**. Breakdown can occur according to two distinct mechanisms that differ in the magnitude of BV and in its temperature dependence.

One breakdown mechanism is **avalanche breakdown**. If the electric field ξ in the depletion region is large enough, electrons can acquire sufficient KE from the electric field so that when they "collide" with atoms in the crystal they break covalent bonds, thereby generating electron–hole pairs. The released electrons can in turn receive sufficient KE from the electric field so that when they "collide" with crystal atoms, they generate more electron–hole pairs and so on. The generated electrons and holes contribute to the reverse current, which therefore increases very rapidly, like an avalanche.

The process is diagrammatically illustrated in Figure 3.2.6. Electron 1 has potential energy E_{cp} at the edge of the depletion region on the p-side and no KE as a classical particle. In terms of wave mechanics, the electron wave is a standing wave. As it moves into the depletion region, the electron is subjected to ξ and gains KE, as indicated by the difference between its energy level and the corresponding energy of the edge of the conduction band. If it gains sufficient energy after a distance d_{coll}, which on the average equals the mean free path between "collisions," it generates an electron–hole pair, denoted by the number 2. Electrons 1 and 2 are shown at the edge of the conduction band, just after "collision," and hole 2 at the edge of the valence band. This is tantamount to assuming that electron 1 loses on "collision" all the KE it had gained, and that electron 2 and hole 2 are

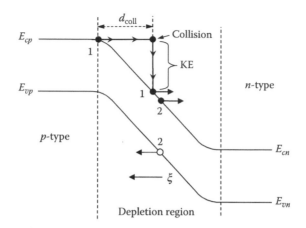

FIGURE 3.2.6
Diagrammatic illustration of avalanche breakdown.

released with no KE as classical particles. If ξ is strong enough, both electrons 1 and 2 may gain sufficient KE, so that, after traveling an average distance d_{coll}, each can cause the generation of an electron–hole pair, and so on. Holes can also cause ionizing "collisions" but at a higher reverse voltage. In practice, avalanche breakdown due to electrons occurs before hole collisions can make a significant contribution.

It is seen from this simple illustration that for avalanche breakdown to occur, the KE gained from ξ must be greater than E_g, because this is the minimum energy required for a valence electron in a covalent bond to break away from its parent atom and become a conduction electron. Moreover, the width of the depletion region must be greater than several mean free paths, so that avalanche multiplication occurs. Using the depletion approximation, it can be shown that (Problem P3.1.8):

$$BV = \xi_{brk}^2 \frac{\varepsilon}{2|q|} \left(\frac{1}{N_D} + \frac{1}{N_A} \right) \tag{3.2.20}$$

where ξ_{brk} is the breakdown value of the electric field. For a given ξ_{brk}, BV varies inversely with the doping levels. The larger the doping levels, the narrower is the depletion region, the larger is $|\xi(0)|$ (Figure 3.1.3b), and the smaller is the voltage required to cause avalanche breakdown. ξ_{brk} increases somewhat with dopant concentration (Problem 3.1.8) because of the increased probability of "collisions." This effectively reduces the mean free path of carriers, so that a larger ξ_{brk} is required for avalanche breakdown. ξ_{brk} is in the range of about 3×10^5 to 10^6 V/cm, in both Si and GaAs, for a concentration range of 10^{15} to 10^{18}/cm^3. For a given ξ_{brk},

Concept: *If one side of a pn junction diode is much less heavily doped than the other, the breakdown voltage can be closely controlled by varying the dopant concentration in the less heavily doped side.*

For example, if $N_D \ll N_A$, $BV \cong \xi_{brk}^2 \varepsilon / 2|q|N_D$.

The other breakdown mechanism is **zener breakdown**. If the doping levels of both sides are high, the width of the depletion region is relatively small (Equation 3.2.1). At a sufficiently high reverse voltage, empty energy levels in the conduction band on the n-side

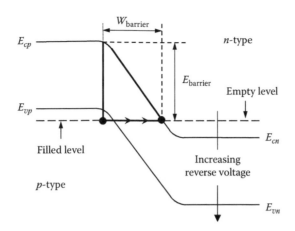

FIGURE 3.2.7
Diagrammatic illustration of zener breakdown.

of the junction become aligned with filled energy levels in the valence band on the *p*-side, as illustrated in Figure 3.2.7. Electrons in the valence band encounter an energy barrier in trying to move to the conduction band, as shown in heavy lines in the figure. If the width of the barrier $W_{barrier}$ is small enough, in the range of nanometers, an electron in the valence band can "tunnel" through the barrier to the conduction band (Example 2.1.2, Chapter 2), thereby creating an electron–hole pair. As the reverse voltage v_R is increased, the net effect is that $W_{barrier}$ decreases. Since the probability of tunneling increases exponentially as the barrier width decreases, the current increases rapidly with the reverse voltage once tunneling occurs.

The critical factor that determines which type of breakdown occurs is the doping level. If the doping level is sufficiently high, the zero-bias width of the depletion region is small. Tunneling occurs at values of v_R in the range of 2.5–5 V. The depletion region is not wide enough to cause avalanche breakdown. At lower doping levels, the depletion region is wider, which is not conducive to tunneling. Instead, avalanche breakdown occurs at v_R in the range of 7 V to a few kilovolts. For v_R in the range 5–7 V, breakdown can occur by both mechanisms. Breakdown diodes are referred to as "zener diodes" irrespective of the breakdown mechanism, because it was believed at one time that breakdown occurred only by the zener mechanism.

The effect of temperature is different in the two types of breakdown. In zener breakdown, a rise in temperature reduces the width of the energy gap, because the average energy of electrons increases with temperature. This reduces the barrier height so that breakdown occurs at a lower v_R. The zener breakdown voltage therefore has a negative temperature coefficient. On the other hand, an increase in temperature reduces the mean free path of the carriers, because of increased amplitude of oscillations of atoms in the crystal. A larger electric field, and hence a larger voltage, is required to impart sufficient energy to the carriers to initiate avalanche breakdown. The avalanche breakdown voltage therefore has a positive temperature coefficient. When breakdown is due to both mechanisms, the temperature coefficient can be very small. Moreover, diodes can be connected in the forward direction in series with zener diodes having a positive temperature coefficient so as to reduce the overall temperature coefficient, but at the expense of increasing the incremental resistance.

A diode can operate safely in the breakdown region if adequate measures are taken to dissipate the heat generated. If not, the temperature can rise sufficiently to cause irreversible damage (Application Window 11.1.1, Chapter 11).

3.3 Semiconductor Photoelectric Devices

Concept: *Electron–hole pairs are generated in a semiconductor by light of frequency ν_{rad} if the photon energy $h\nu_{\text{rad}}$ exceeds the bandgap energy E_{gJ}.*

When $h\nu_{\text{rad}} > E_{gJ}$, where h is Planck's constant, and E_{gJ} is the bandgap energy in joules, electrons in the valence band acquire sufficient energy from incident photons to move to the conduction band, thereby generating electron–hole pairs. This is the basis of operation of semiconductor photoelectric devices.

Photoconductive Cell

Basically, **the photoconductive cell** is a thin film of semiconductor that is exposed to light. The conductivity of the semiconductor increases with the rate of carrier generation, and hence with light intensity. For a given applied voltage, the current through the semiconductor therefore increases with the light intensity (Example 3.3.1 and Application Window 3.3.1).

Example 3.3.1 Photoconductive Current

It is required to derive an expression for the photoconductive current in a semiconductor film subjected to incident radiation.

SOLUTION

If the power of the incident radiation is P_{rad}, then the number of incident photons/s is $P_{\text{rad}}/h\nu$. The additional number of electron–hole pairs generated per unit time per unit volume is:

$$G_x = \zeta \left(\frac{P_{\text{rad}}}{h\nu} \right) \frac{1}{\Lambda} \qquad (3.3.1)$$

where
  ζ is an efficiency parameter defined as the number of electron–hole pairs generated per incident photon
  Λ is the volume

It is assumed that the film is thin enough so that electron–hole pairs are generated uniformly by the radiation throughout the volume Λ.

Under steady-state conditions, $G_x = R_x(T)$, the rate of recombination. From Equation 2.7.1, Chapter 2, the excess hole and electron concentrations are, respectively, $G_x\tau_h$ and $G_x\tau_e$. The total concentrations are $p = G_x\tau_h + p_o$ and $n = G_x\tau_e + n_o$, where p_o and n_o are the equilibrium concentrations of holes and electrons, respectively. The conductivity is then,

$$\sigma = |q|(p\mu_h + n\mu_e) \tag{3.3.2}$$

If G_x is large enough so that $p = n$, then because the rate of recombination is the same for both under steady-state conditions, $\tau_h \cong \tau_e$.

Note that when carrier concentrations reach steady values, the semiconductor is in a steady state but not at equilibrium. Although pressure, temperature, and electrochemical potential of carriers are equalized throughout the system and do not vary with time, the semiconductor is disturbed by the radiation so that carrier concentrations are higher than their equilibrium values, which means that the semiconductor is not at equilibrium.

Photodiode

When a biased *pn* junction is exposed to light, the increase in the rate of generation of carriers in the depletion and transition regions modifies the $i-v$ relation to (Section ST3.5):

$$i_D = I_S(e^{v_D/\eta V_T} - 1) - I_{SS} \tag{3.3.3}$$

where

$$I_{SS} = A_J |q| G_x (W_d + L_h + L_e) \tag{3.3.4}$$

G_x is given by Equation 3.3.1, W_d is the width of the depletion region, and L_h and L_e are the diffusion lengths of holes and electrons, respectively.

It is seen from Equation 3.3.3 that the $i-v$ characteristic is shifted away from the origin, as illustrated in Figure 3.3.1. In the third quadrant, the reverse current under reverse bias increases markedly with the light intensity. The diode now operates as a **photodiode**, as illustrated in Figure 3.3.2a, which also shows the photodiode symbol. Figure 3.3.2b depicts the load-line construction, with the characteristics in the third quadrant redrawn so that the diode reverse voltage and current are in the positive directions. The parameter of the characteristics is the total reverse current $I_S + I_{SS} \cong I_{SS}$. If I_{SS1} is the reverse current for a given light intensity, P is the operating point, v_{PD} is the voltage across the diode, and $V_{SRC} - v_{PD}$ is the voltage across R_L. If the light intensity increases so that the reverse current is I_{SS2}, the operating point moves to Q. The voltage across the diode decreases, while that across R_L increases.

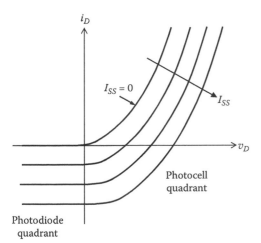

FIGURE 3.3.1
Effect of light on the $i-v$ characteristic of a *pn* junction diode.

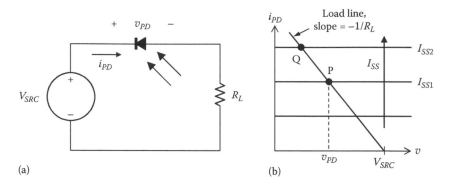

FIGURE 3.3.2
(a) Circuit and symbol of a photodiode; (b) load-line construction.

Photocell

Operation in the fourth quadrant of Figure 3.3.1 results in a **photocell**, or solar cell, which converts solar energy to electric energy. The basic photocell circuit is illustrated in Figure 3.3.3a, which also shows the photocell symbol. The $i-v$ characteristic in the fourth quadrant is redrawn as a source characteristic in Figure 3.3.3b with a line of slope R_L representing the load resistance. Since the photocell acts as a power source, it is important to have maximum power delivered to the load. The condition for maximum power transfer, derived in Section SE3.2, is that the magnitude of the slope of the source characteristic at the operating point is equal to R_L.

It is instructive to consider qualitatively, and in terms of electrochemical potential, how a photocell forward voltage is developed under open-circuit conditions in the presence of light. At equilibrium, the chemical potential energy difference of holes between the p-side and the n-side is proportional to $\ln(p_{po}/p_{no})$ (Equation 2.4.16, Chapter 2), and that of electrons is proportional to $\ln(n_{no}/n_{po})$. In the presence of light, electrons and holes are generated in equal numbers. If the concentration of holes increases by Δp, the chemical potential energy difference of holes becomes $\ln \dfrac{p_{po}+\Delta p}{p_{no}+\Delta p}$. Since $p_{po} \gg p_{no}$, the effect is to reduce the chemical potential energy difference for holes; similarly for electrons. The electric potential energy difference therefore predominates. This means that in the

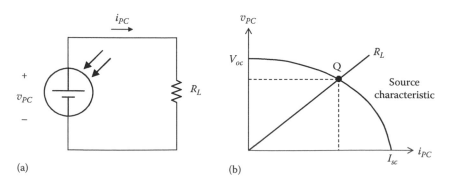

FIGURE 3.3.3
(a) Circuit and symbol of a photocell; (b) source characteristic and load resistance.

depletion region, where the electric field exists and is directed from the *n*-side to the *p*-side, generated holes are swept to the *p*-side and electrons to the *n*-side. The *p*-side becomes positively charged with respect to the *n*-side and an open-circuit forward voltage V_{oc} is developed. If i_D is set to zero in Equation 3.3.3,

$$V_{oc} = \eta V_T \ln\left(\frac{I_{SS}}{I_S} + 1\right) \qquad (3.3.5)$$

At normal levels of radiation, the short-circuit current I_{sc} of the photocell ($v_D = 0$) is practically equal to I_{SS}, as can be seen from Figure 3.3.1.

Application Window 3.3.1 Semiconductor Photoelectric Devices

Semiconductor photoelectric devices are examples of transducers that convert radiation energy to electric energy. The material of a photoconductive cell could be lightly doped *n*-type silicon, cadmium sulfide, or lead sulfide. The material is chosen not only for good sensitivity to light, but also for its ability to dissipate power, and for its **spectral response**, that is, the variation of its light sensitivity with the wavelength of light. Depending on the application, the photoconductive cell may be required to have its maximum sensitivity to visible light of a particular frequency, for example, or to infrared radiation. The conductance of most photoconductive cells varies almost linearly with light intensity. The change in resistance could be quite considerable, from many MΩs in the dark to a few hundreds of ohms in bright illumination. Photoconductive cells are used in some light meters to measure light intensity.

The current of a photodiode diode (Figure 3.3.2) can increase several thousand-folds from darkness to bright illumination. Practical photodiodes have a *pin* structure, in which a relatively wide intrinsic region is interposed between the *p*- and *n*-type regions (Figure 3.3.4). This increases the width of the depletion region, and hence I_{SS} (Equation 3.3.4). However, the speed of response is reduced because of the additional time it takes the carriers to drift through the intrinsic region under the influence of the electric field. The width of this region is therefore a compromise between photosensitivity and speed of response. The diode is operated at a reverse bias close to breakdown, resulting in two advantages: (1) the high electric field, which appears mostly in the intrinsic region, increases the drift velocity, and improves the speed of response; and (2) the current is increased because of avalanche multiplication.

The basic structure of a silicon photocell is illustrated in Figure 3.3.5. It is important to have a high efficiency of conversion of solar energy to electric energy. To capture more of the incident light, the exposed surface is coated with antireflective material. Placing the contacts on the sides would provide maximum exposure of the surface of the cell to incident light, but is avoided because current will then have to flow laterally through the relatively high resistance of the semiconductor in order to reach the contacts. This increases the source resistance of the cell and hence the power loss. Instead, the top contact is in the form of a grid that allows most of the light to reach the semiconductor while shortening the current path. Even then, the efficiency remains low; the power delivered is only about 15% of the incident radiation power. The main cause of inefficiency is the narrow band of wavelengths that is effective in generating electron–hole pairs. Light of energy less than E_g cannot

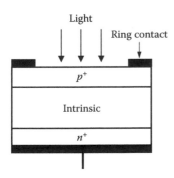

FIGURE 3.3.4
Figure for Application Window 3.3.1.

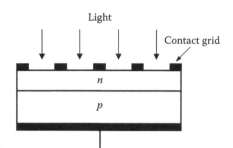

FIGURE 3.3.5
Figure for Application Window 3.3.1.

generate electron–hole pairs, whereas with light of greater energy, the excess energy above E_g appears as heat. The efficiency is significantly improved in **multijunction cells** having two or more layers of material with different bandgaps. The material nearer the surface has a wider bandgap and absorbs high-energy photons, whereas the low-energy photons are absorbed by the lower layers of material having a narrower bandgap. The current delivered by the cell is increased by having a large cell area. In a solar panel, photocells are connected in series-parallel so as to increase the output voltage and current.

A **charged-coupled device** (CCD) is another important example of a semiconductor photoelectric device. Basically, it consists of an array of metal–oxide–semiconductor (MOS) capacitors (Section ST4.3, Chapter 4) that are coupled by *pn* channels. The image to be captured is projected onto the exposed array through a lens. Carriers are generated in each capacitor structure, or picture element (**pixel**), in proportion to the intensity of the light received. By applying a sequence of pulses to the gates, which constitute one set of electrodes of the capacitors, the stored charge in a row of capacitors can be passed from one end of the row to the other and read out sequentially for each capacitor in the row. The process is repeated for all the rows in the array. The information that is read is digitized and processed to provide a two-dimensional reconstruction of the original image. Color can be reproduced in a number of ways. The best, but the most expensive, scheme uses three CCD arrays, one for each of the three primary colors (red, green, and blue). CCD arrays can have more than 6 million elements, which gives high spatial resolution, and are about 100 times more sensitive than photographic film, because they are much more efficient in capturing incident light. This makes them particularly suitable for astronomical telescopes. CCD arrays are widely used in digital, still and video cameras, scanners, optical character recognition (OCR), and machine vision for robots.

3.4 Light-Emitting Diodes

It was pointed out earlier that the diode forward current is accounted for by the recombination of carriers, which implies release of energy as electrons and holes recombine. It may be thought that energy is released as photons of radiation of energy $h\nu = E_{gJ}$, where ν is the frequency of the radiation and E_{gJ} is the bandgap energy is joules. However, this is not the case in all semiconductors, because,

Concept: *When an electron and a hole combine, energy and momentum must be conserved.*

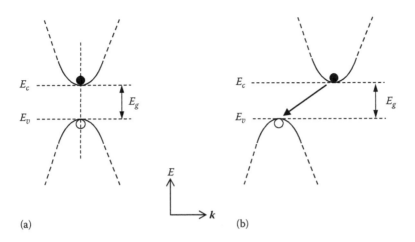

FIGURE 3.4.1
Diagrammatic illustration of energy-band diagrams for: (a) a direct-bandgap semiconductor, and (b) an indirect-bandgap semiconductor.

To see how conservation of momentum affects the type of recombination that takes place, consider the E vs. k diagram introduced in Section 2.2, Chapter 2, for a hypothetical one-dimensional case (Figure 2.2.2, Chapter 2). In a real, three-dimensional crystal, these diagrams have different forms for different directions of k, but all forms for a given semiconductor show separation between conduction bands and valence bands. Based on these E vs. k diagrams, semiconductors can be classified as **direct-bandgap** or **indirect-bandgap** semiconductors. In direct-bandgap semiconductors, such as GaAs, the lowest energy of the conduction band occurs at the same value of k as the highest energy of the valence band (Figure 3.4.1a), unlike the case of indirect-bandgap semiconductors, such as Si (Figure 3.4.1b).

In direct-bandgap semiconductors, an electron in the conduction band can drop into an empty level in the valence band, with little or no change in k and hence in its momentum. Conservation of momentum is easily satisfied and a photon of wavelength hc/E_{gJ} is emitted. Direct-bandgap semiconductors are therefore efficient emitters of light through recombination and are suitable as a substrate for LEDs and semiconductor lasers. They can also efficiently absorb photons of light having energies greater than E_g.

If an electron in an indirect-bandgap semiconductor is to transfer directly from the conduction band to the valence band, then because k at the energy minimum of the conduction band is different from that at the energy maximum of the valence band, the right amount of momentum must be exchanged in interactions with crystal vibrations so as to conserve momentum. Such interactions are rare. Hence, recombination is much more likely to take place via traps that introduce energy levels in the bandgap (Figure 2.7.1, Chapter 2). In this case, energy and momentum are exchanged with the crystal, the energy difference appearing as heat.

Application Window 3.4.1 Light-Emitting Diodes

Photons of visible light having a wavelength in the range of about 460 nm (blue) to 770 nm (red), have corresponding energies, hc/wavelength, that are approximately in the range 2.7 eV (blue) to 1.6 eV (red), which is the range of gap energies obtained with suitable semiconductor materials. Lower gap energies down to about 1.1 eV can give rise to photons of wavelength in the infrared region, which is useful in some practical applications.

The substrates of practical LEDs are gallium arsenide (GaAs), gallium phosphide (GaP), gallium-arsenide-phosphide (GaAsP), gallium nitride (GaN), or indium gallium nitride (InGN). The substrate and type of dopant added to create the necessary *pn* junction determine the wavelength of emitted light, which is nearly monochromatic, that is, consists of a single frequency, and can be in the blue, green, yellow, orange, or red regions of the electromagnetic spectrum. White LEDs are blue LEDs having a phosphor that fluoresces and emits white light when irradiated by blue light, which is very similar to what happens in a fluorescent lamp.

Special precautions are taken to maximize the fraction of generated light that is emitted through the exposed surface of the diode. Photons that are not emitted are reabsorbed resulting in the generation of electron–hole pairs. LEDs have a diode offset voltage V_{D0} that is typically in the range 1.2–2 V and normally operate with forward currents of 20–50 mA.

LEDs are useful as indicators of various types, such as mains, or power-on indicators, and tuning indicators. LEDs are widely used as light sources in many applications, ranging from flashlights to instrumentation systems utilizing optical fibers. Compared to incandescent and fluorescent lamps, they have the following advantages:

1. They are rugged and have a lifetime of at least 100,000 h if they are not overheated. Moreover, their life expectancy is not shortened by frequent switching on and off, as is the case with incandescent lamps and with fluorescent lamps, particularly at low temperatures.

2. They do not generate much heat and are easy to control and dim.

3. They have a faster turn-on time, which is important in some applications such as brake lights on cars. Because of elimination of the time for heating a filament, LEDs turn on about 0.18 s faster than incandescent lamps. At 80 km/h, this gives an additional braking distance of about 4 m.

4. Because they do not normally need replacement, LEDs can be soldered in place, thereby avoiding the deterioration of contacts in corrosive atmospheres.

Light is also emitted due to electron–hole recombination in an organic LED (OLED). The OLED, whose structure and operation are diagrammatically illustrated in Figure 3.4.2, consists of an anode, a cathode, and two layers of organic polymer semiconductors: (1) the emissive layer, made of a polymer such as polyfluorene, that transports electrons from the adjacent cathode, and (2) a conductive layer, made of a polymer such as polyaniline, that transports holes. When the anode is made positive with respect to the cathode, electrons are emitted from the cathode and collected at the anode, so the emissive layer becomes negatively charged with electrons and the conductive layer positively charged with holes. Electrons and holes are attracted toward each other and recombine in the emissive layer,

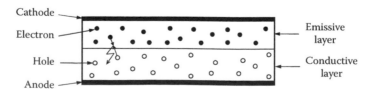

FIGURE 3.4.2
Figure for Application Window 3.4.1.

close to the boundary, because holes have a higher mobility than electrons in organic semiconductors. The energy released on recombination in the emissive layer is in the visible spectrum, resulting in emission of light through either a transparent anode or a transparent cathode. The color of the emitted light depends on the polymer in the emissive layer and its side chains.

OLEDs can be produced by low-cost manufacturing processes, such as ink-jet or silk-screen printing on suitable substrates, which can be flexible or rollable. Compared to LCD panels, they do not need a backlight, consume less power, and have a larger range of color, contrast, and viewing angle. They are used for screen displays on cellular phones, media players, car radios, digital cameras, and in lightweight TV displays that can be hung on a wall. Their useful life is, however, limited, as they degrade after several years.

Exercise 3.4.1

The luminous intensity I_{lum} of an LED is almost directly proportional to the forward diode current. The relation for a given LED is: $I_{lum} = 0.05 I_D$, where I_D is in mA and I_{lum} is in millicandela (mcd). The LED has a forward voltage drop of 1.8 V and is connected in series with a resistor R across a 5 V supply. Determine R such that the luminous intensity is 2 mcd.

Answer: 80 Ω. ∎

3.5 Tunnel Diode

Although its practical importance has declined, the tunnel diode is of fundamental interest, as will be demonstrated in this section.

In the tunnel diode, the doping levels of the p- and n-sides are so high that tunneling occurs even for small forward or reverse bias. The energy-band diagram at equilibrium is illustrated in Figure 3.5.1a, where κ_{hpo} and κ_{eno} denote, respectively, the electrochemical potentials of majority carriers, which are the same as E_F at equilibrium (Section 2.4, Chapter 2). Because of the high doping, E_F is no longer in the energy gap, and the donor and acceptor impurity levels broaden to bands that overlap the edges of the conduction and valence bands, respectively. At equilibrium, energy levels are filled to the same level on both sides. Tunneling occurs in both directions with equal probabilities, so the net tunneling current is zero.

When a small reverse bias is applied, energy levels on the n-side, including κ_{eno}, are lowered by v_R eV (Figure 3.5.1b). The energy levels of the highest-energy electrons in the valence band on the p-side are now opposite to empty energy levels in the conduction band on the n-side. The probability of electron tunneling is therefore larger from the p-side to the n-side than in the opposite direction. A relatively large reverse current flows (Figure 3.5.2a, region 1).

When a small forward bias is applied, net tunneling occurs in the opposite direction (Figure 3.5.2b), giving rise to an equally large forward current (Figure 3.5.2a, region 2). The i–v characteristic thus has a high slope in the vicinity of the origin, equivalent to a relatively small resistance of a few tens of ohms or less. With increasing forward bias, the forward tunneling current increases until all the energy levels in the conduction band on the n-side are opposite empty levels in the valence band on the p-side (Figure 3.5.2c). At this point, the

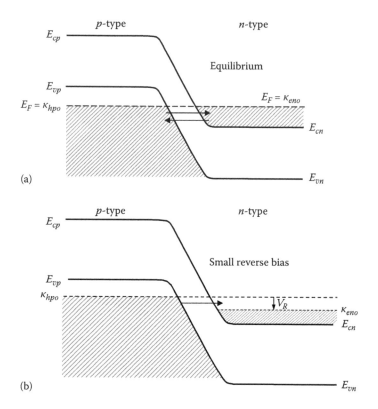

FIGURE 3.5.1
Energy-band diagram of a tunnel diode, (a) at equilibrium and (b) for small reverse bias.

tunneling current is at its maximum (Figure 3.5.2a, point 3). If the forward bias is increased further, the tunneling current decreases in magnitude (Figure 3.5.2d), which gives the negative resistance part of the characteristic (Figure 3.5.2a, region 4). Tunneling stops when the energy levels of the conduction band on the n-side are opposite the forbidden gap (Figure 3.5.2e). While tunneling takes place in the forward direction, there is also the normal, but still small, pn junction current due to injection of minority carriers. When tunneling ceases, this latter current builds up rapidly, as in a regular pn junction diode (Figure 3.5.2a, region 5).

The tunnel diode therefore has the composite i–v characteristic illustrated in Figure 3.5.3, which also shows the symbol for the tunnel diode. There is a peak point (I_p, V_p), a valley point (I_v, V_v), and a forward voltage V_F, at which the current again equals I_p. If a variable current source is applied to the diode, then as the current increases from zero, the i–v characteristic is swept by the intersection point with a horizontal line, as shown in Figure 3.5.3. Part OP of the characteristic is first traced. If the current is increased just beyond I_p, the diode voltage switches abruptly from V_p to V_F. Further increase of current causes the voltage to increase along part FS. If the current is reduced, path SV is followed. A small decrease in current below I_v causes the voltage across the diode to switch from V_v to a small value. The negative resistance part of the characteristic is not traced but a switching action occurs when the voltage across the diode changes suddenly. If a variable voltage source is applied, then the entire characteristic can be swept, including the negative resistance region, as illustrated by the vertical line in Figure 3.5.3. When a load line of

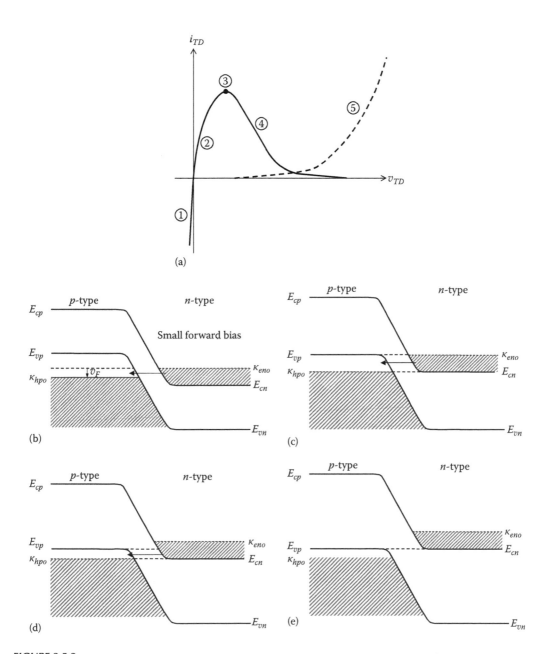

FIGURE 3.5.2
(a) Tunneling current of a tunnel diode; (b) to (e) show the energy-band diagrams for increasing forward bias; (b) for a small forward bias; (c) for maximum tunneling current; (d) for reduced tunneling current; and (e) for zero tunneling current.

slope $-1/R$ is superposed on an i–v characteristic having a negative resistance part, it is instructive to examine the stability of operating points in different parts of the characteristic (Section ST3.9).

Tunnel diodes are usually made from GaAs, which gives a high ratio of I_p/I_v. Typical parameters are $V_p \cong 0.15$ V, $V_v \cong 0.5$ V, $V_F \cong 1$ V, $I_p \cong 2$ mA, and $I_v \cong 0.1$ mA. Tunneling

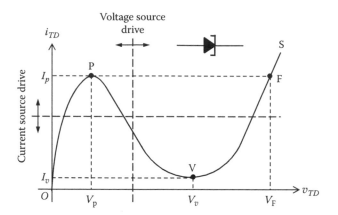

FIGURE 3.5.3
i–v characteristic of tunnel diode.

takes place at virtually the speed of light, with very little capacitance associated with the junction. Tunnel diodes are therefore suitable for high-frequency operation, as switches and as negative resistance oscillators (Example SE3.4). The peak point (I_p, V_p), which depends only on tunneling, is quite insensitive to temperature. On the other hand, the valley point (I_v, V_v) depends on the normal forward characteristic of a *pn* junction diode and is temperature sensitive.

3.6 Contacts between Dissimilar Materials

Contacts between dissimilar materials are ubiquitous. The *pn* junction is between two dissimilar semiconductors that are in turn connected to external elements by semiconductor–metal contacts. A different metal is almost invariably used to connect these metal contacts to the diode terminals, which introduces yet another type of contact between dissimilar materials, this time a metal–metal contact.

   When dissimilar materials are in contact, some general principles apply.

Concepts:

1. *Under equilibrium conditions, the Fermi level E_F, being the equilibrium electrochemical potential of current carriers, must be the same throughout.*

2. *The alignment of E_F is brought about by a contact potential, which is the junction potential in the case of the pn junction. The contact potential lowers electron energy levels on the positively charged side, with respect to those on the negatively charged side, so as to align E_F.*

3. *The contact potential is associated with a net charge on either side of the contact. In semiconductors, the net charge is a space charge in a depletion region of the semiconductor. In metals, the net charge is a surface charge, because the high conductivity precludes a space charge.*

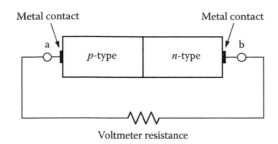

FIGURE 3.6.1
Measurement of voltage between the two terminals of a *pn* junction diode at equilibrium using a current-actuated voltmeter.

Given that a contact difference of potential exists between dissimilar materials, would it be possible to measure such differences of potential by means of an ordinary, current-actuated voltmeter? To answer this question, consider the voltmeter connected to diode terminals at a and b, as in Figure 3.6.1, where the voltmeter is represented by its resistance, which can be very large but is finite nevertheless. Clearly, no current flows in the circuit at equilibrium; the electrochemical potential of current carriers, which are electrons in the external copper wires, is the same throughout. In other words, there is no driving force to drive current through the circuit.

The preceding argument emphasizes that a current-actuated voltmeter actually measures electrochemical potential difference, since this is fundamentally the driving force for current. In the absence of differences of concentrations of current carriers, that is, differences in chemical potential of carriers, the electrochemical potential difference is the same as the electric potential difference. In principle, contact potential can be measured by an electrometer-type voltmeter, whose indication depends on forces between charges in capacitor structures.

Another implication is for Kirchhoff's voltage law, which fundamentally depends on conservation of energy (Sabah 2008, p. 50). Since the equilibrium contact potential balances the chemical potential, it does not enter into conservation of energy in terms of currents and voltages. It follows that, in the presence of concentration gradients of current carriers, Kirchhoff's voltage law applies to electrochemical potential differences and not to voltages, as discussed in connection with Equation ST5.6.7, Section ST5.6, Chapter 5, for the JFET.

Metal–Metal Contacts

If an electron attempts to escape from a metal, it leaves behind a net positive charge $+|q|$ that attracts the electron back to the metal. An electric potential barrier therefore exists at the surface of the metal, which must be surmounted by the electron if it is to escape from the metal.

The energy-band diagram of a metal is illustrated in Figure 3.6.2a. At 0K, E_F is the highest energy level of conduction electrons. The zero energy level is theoretically at infinity and is that of a free electron at rest. In practice, zero energy is that of an electron that just escapes from the metal with zero KE. The **work function** ϕ_W of the metal is the minimum energy that must be possessed by an electron at 0K to enable it to escape from the metal. It is the energy difference between E_F and the zero level outside the metal.

Consider two metals 1 and 2 in contact at equilibrium (Figure 3.6.2b). Let the work functions of the two metals be ϕ_{W1} and ϕ_{W2} eV, where $\phi_{W1} < \phi_{W2}$. Because of the

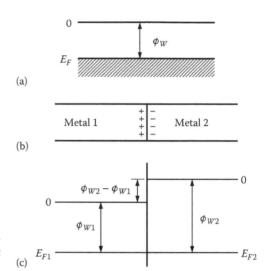

FIGURE 3.6.2
Contact between two metals of different work functions.
(a) Energy-band diagram of a metal at 0K; (b) metal–
metal contact at equilibrium; (c) energy-band diagram at
equilibrium.

difference in work functions, more electrons are able to leave metal 1 to metal 2 than in the opposite direction. As a result, metal 2 acquires a net negative charge, whereas metal 1 acquires an equal and opposite positive charge. The resulting difference in electric potential opposes the flow of electrons from metal 1 to metal 2. At equilibrium, the electrochemical potential of electrons in the two metals is the same, and E_F is aligned (Figure 3.6.2c). There will be a contact potential in volts of $(\phi_{W2} - \phi_{W1})$, which is that required to raise the energy levels of metal 2 with respect to those of metal 1 so as to align E_F. The charges on metals 1 and 2 are surface charges.

Metal–Semiconductor Contacts

It will be argued in this section, in a highly simplified manner that is intended mainly to highlight some of the basic principles involved, that metal–semiconductor contacts are of two general types: rectifying and nonrectifying. A practical example of a rectifying contact is the Schottky diode, considered at the end of this section. Nonrectifying contacts, or ohmic contacts, are required for making connections between semiconductors and external elements.

Consider a metal of high work function in contact with an n-type semiconductor having a lower work function (Figure 3.6.3a). Initially, there will be a net flow of electrons from the n-type semiconductor to the metal. At equilibrium, the metal will have a negative surface charge, the semiconductor an equal, positive space charge. In the semiconductor, the space charge arises from a reduced concentration of conduction electrons leaving an excess of positively charged donor ions. The edge of the conduction band is bent upward, away from the Fermi level (Figure 3.6.3b) so as to reduce the concentration of electrons in the conduction band, as required to produce the space charge. Alternatively, it may be argued that the body of the semiconductor, sufficiently far from the contact surface, is at a positive potential with respect to the metal, because of the positive charge in depletion region. This reduces the electric potential energy of electrons and shifts electron energy levels in the semiconductor downward in order to align E_F. As the contact surface is approached, less positive charge is enclosed between the given point and the contact surface. The potential of the semiconductor becomes less positive. Electron energy levels are therefore raised as the contact surface is approached.

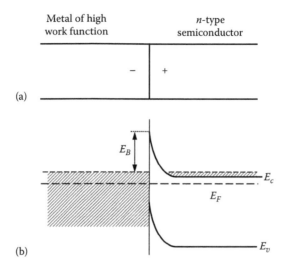

(a)

(b)

FIGURE 3.6.3
Contact between an *n*-type semiconductor and a metal of higher work function. (a) Metal-semiconductor contact; (b) energy-band diagram.

The upward bending of the conduction band introduces at the interface an energy barrier to electrons moving across the contact in either direction. The barrier is normally too wide for any significant tunneling to occur, so only electrons possessing energies greater than the barrier height are able to cross the barrier. Electron energies are distributed according to the Fermi–Dirac statistical distribution function (Section 2.2, Chapter 2), which can be approximated by a Maxwell–Boltzmann distribution for higher-energy electrons. Thus, if E_B is the nominal height of the energy barrier for conduction electrons, the number of electrons having energies of at least E_B is proportional to e^{-E_B/V_T}. The electron current at equilibrium, in either direction, can therefore be expressed as Be^{-E_B/V_T}, where B is some proportionality constant.

If the semiconductor is made negative with respect to the metal by a bias voltage v_B, electron energies in the semiconductor are raised, which reduces the barrier height on the semiconductor side by v_B V (Figure 3.6.4a); conversely, if the semiconductor is made positive with respect to the metal (Figure 3.6.4b). In both cases, the barrier, as seen from the side of the metal is unchanged, so electron flow from the metal to the semiconductor is the same as at equilibrium. On the semiconductor side, the number of electrons having energies that enable them to surmount the barrier is proportional to $e^{-(E_B-v_B)/V_T}$, where this number is larger than the equilibrium value for $v_B > 0$ (Figure 3.6.4a), and is smaller when

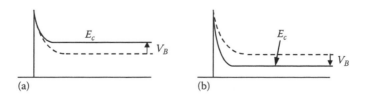

(a)   (b)

FIGURE 3.6.4
Energy barrier of the contact of Figure 3.6.3 under forward bias (a) and under reverse bias (b).

$v_B < 0$ (Figure 3.6.4b). If we subtract from this the electron flow from the metal to the semiconductor, the net electron flow from the semiconductor to the metal is proportional to $e^{-E_B/kT}\left(1 - e^{v_B/V_T}\right)$. The current, in the conventional sense, from the metal to the semiconductor can be expressed as:

$$i_{BD} = I_{B0}\left(e^{v_B/V_T} - 1\right) \tag{3.6.1}$$

Equation 3.6.1 is of the same form as Equation 3.2.3 for a *pn* junction diode and shows that a contact between a metal and an *n*-type semiconductor of lower work function has rectifying properties. The preceding discussion neglects the effect on the barrier height of charge in surface states. The density of these surface states is very much influenced by the crystal orientation, by the metal used, and by the treatment of the semiconductor surface (Section 4.8, Chapter 4).

Consider next the case of a metal–semiconductor contact in which the metal has a lower work function than the semiconductor (Figure 3.6.5a). There will be a net flow of electrons from the metal to the *n*-type semiconductor, with the result that the semiconductor gains electrons, whereas the metal loses electrons. At equilibrium, the metal will have a positive surface charge, the semiconductor an equal but negative space charge. In the semiconductor, the volume space charge arises from an increased concentration of conduction electrons. The edge of the conduction band is therefore bent downward, toward the Fermi level (Figure 3.6.5b), which increases the concentration of electrons in the conduction band, as required. The barrier will be small, or virtually nonexistent. The contact is therefore nonrectifying, or ohmic.

Similarly, a contact between a *p*-type semiconductor and a metal of lower work function is rectifying, whereas a contact between such a semiconductor and a metal of higher work function is nonrectifying. The Schottky diode is based on a metal–Si semiconductor contact, whereas the metal–semiconductor field-effect transistor (MESFET) is based on a metal–GaAs semiconductor contact (Section 5.7, Chapter 5).

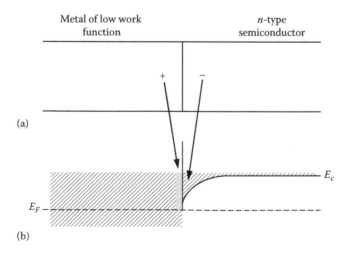

FIGURE 3.6.5
Contact between an *n*-type semiconductor and a metal of lower work function. (a) Metal-semiconductor contact; (b) energy-band diagram.

Schottky Diode

A rectifying contact is formed in this diode between a metal and an *n*-type semiconductor of lower work function. In a Si Schottky diode, the metal used, which is usually platinum or nickel, reacts with Si to produce a silicide (Section 4.3, Chapter 4), which has the energy-band structure of a metal and makes the diode properties insensitive to the type of cleaning treatment of the semiconductor surface prior to application of the metal. The diode is forward biased by making the metal positive with respect to the semiconductor, as in a *pn* junction diode. Figure 3.6.6 shows the symbol for the Schottky diode.

Anode (metal)

Cathode (*n*-type semiconductor)

FIGURE 3.6.6
Symbol of a Schottky diode.

The *i–v* characteristic of the Schottky diode is of the form given by Equation 3.6.1. The saturation current I_{B0} is, however, several orders of magnitude larger than that of a *pn* junction diode having the same junction area. This also reduces the diode offset voltage to approximately 0.3 V (Exercise 1.2.1, Chapter 1). The reverse breakdown voltage of the Schottky diode is also smaller than that of a comparable *pn* junction diode. Nevertheless, Schottky diodes have some distinct advantages. They do not depend on minority carrier injection for their operation. The absence of minority carrier storage means that the Schottky diode can switch substantial currents at very high speeds. Another consequence is insensitivity to radiation, which increases the rate of generation of carriers and therefore affects devices that depend for their operation on minority carriers. Schottky diodes are extensively used in Schottky transistor–transistor logic (TTL) digital circuits (Section 13.6, Chapter 13) to limit transistor saturation.

3.7 Heterojunction

A heterojunction is a *pn* junction formed between two dissimilar semiconductors, which are chosen to have different bandgap energies E_g but nearly equal lattice constants, that is, dimensions of the unit cell. The near-equality of lattice constants ensures crystal continuity, so there are no unsatisfied bonds at the interface that would trap carriers and therefore impede their motion between the two semiconductors. In the Si heterojunction of practical importance, the *n*-type semiconductor is Si, having a bandgap of 1.12 eV, whereas the *p*-type semiconductor is made up of silicon-germanium (SiGe) having 15–20% Ge, which reduces the bandgap to 0.8–0.9 V (Section 4.8, Chapter 4). In GaAs heterojunctions, the *p*-type semiconductor is GaAs, having a bandgap of 1.42 eV, and the *n*-type semiconductor is usually GaAsAl having a bandgap of approximately 1.6–1.8 eV, depending on the concentration of Al. The Si heterojunction is considered in what follows.

We will start by applying the depletion approximation (Section 3.1) to the heterojunction, as shown in Figure 3.7.1. In comparing this figure with Figure 3.1.2 for a homojunction, charge neutrality of the semiconductor as a whole requires, as in the case of a homojunction that:

$$N_D W_n = N_A W_p \tag{3.7.1}$$

At the junction, the electric displacement $D = \varepsilon\xi$ is continuous, so that:

$$\varepsilon_n \xi_n(0) = \varepsilon_p \xi_p(0) \tag{3.7.2}$$

where ε_n and ε_p are the permittivities of the n-type and p-type semiconductors, respectively. As the relative permittivity of Ge is 16.2, compared to 11.9 for Si, $\varepsilon_p > \varepsilon_n$, so that $|\xi_p(0)| < |\xi_n(0)|$. The electric field is discontinuous at the junction, as shown in Figure 3.7.1b. In accordance with Gauss's law, the magnitude of the electric field at the junction on either side equals the area of the corresponding charge rectangle in Figure 3.7.1a divided by the permittivity. Thus,

$$|\xi_n(0)| = \frac{|q|N_D}{\varepsilon_n} W_n, \quad \text{and} \quad |\xi_p(0)| = \frac{|q|N_A}{\varepsilon_p} W_p \qquad (3.7.3)$$

It is seen that Equation 3.7.3 is consistent with Equations 3.7.1 and 3.7.2.

Considering, for convenience, the voltage at the junction to be zero, the voltage of either semiconductor beyond the depletion region is given by the area of the corresponding ξ-x triangle:

$$V_n = \frac{|q|N_D}{2\varepsilon_n} W_n^2, \quad \text{and} \quad V_p = \frac{|q|N_A}{2\varepsilon_p} W_p^2 \qquad (3.7.4)$$

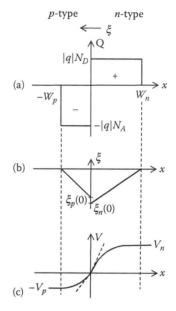

FIGURE 3.7.1
(a) Charge distribution in a *pn* heterojunction; (b) variation of the electric field; and (c) variation of voltage.

It follows from Equations 3.7.4 that:

$$W_n = \sqrt{\frac{2\varepsilon_n}{|q|N_D}} V_n, \quad \text{and} \quad W_p = \sqrt{\frac{2\varepsilon_p}{|q|N_A}} V_p \qquad (3.7.5)$$

Using Equations 3.7.1 and 3.7.4,

$$\frac{V_n}{V_p} = \frac{\varepsilon_p N_A}{\varepsilon_n N_D} \qquad (3.7.6)$$

The voltage variation is shown in Figure 3.7.1c. The voltage at the junction is continuous, because a discontinuity in voltage is generally due to an electric double layer at the interface. This is a surface sheet of charge, of infinitesimal thickness, having positive charges on one side and equal and opposite negative charges on the other. Since atomic bonding is assumed to be continuous across the interface, there are no surface charges. However, the slope of the voltage curve is discontinuous at the interface because of the discontinuity in the electric field.

The equilibrium junction potential V_{npo} is $V_n + V_p$. From Equation 3.7.6,

$$V_n = \frac{\varepsilon_p N_A}{\varepsilon_p N_A + \varepsilon_n N_D} V_{npo}, \quad \text{and} \quad V_p = \frac{\varepsilon_n N_D}{\varepsilon_p N_A + \varepsilon_n N_D} V_{npo} \qquad (3.7.7)$$

Substituting in the relations of Equation 3.7.5,

$$W_n = \sqrt{\frac{2\varepsilon_n \varepsilon_p N_A}{|q|N_D(\varepsilon_p N_A + \varepsilon_n N_D)}} V_{npo} \qquad (3.7.8)$$

$$W_p = \sqrt{\frac{2\varepsilon_n \varepsilon_p N_D}{|q|N_A(\varepsilon_p N_A + \varepsilon_n N_D)}} V_{npo} \tag{3.7.9}$$

The total width of the depletion region is $W_d = W_n + W_p$. Hence,

$$W_d = \sqrt{\frac{2\varepsilon_n \varepsilon_p}{|q|(\varepsilon_p N_A + \varepsilon_n N_D)} \frac{(N_A + N_D)^2}{N_A N_D}} V_{npo} \tag{3.7.10}$$

If $\varepsilon_n = \varepsilon_p = \varepsilon$, Equation 3.7.10 reduces to Equation 3.1.11.

The energy-band diagrams of the two semiconductors in isolation are shown in Figure 3.7.2a. When the two semiconductors form a junction, E_F is aligned as usual by having the n-side at a more positive potential V_{npo} with respect to the p-side. This is reflected in Figure 3.7.2b by having the zero energy level on the n-side lowered by V_{npo} with respect to the p-side, where all energy levels are expressed in electron volts.

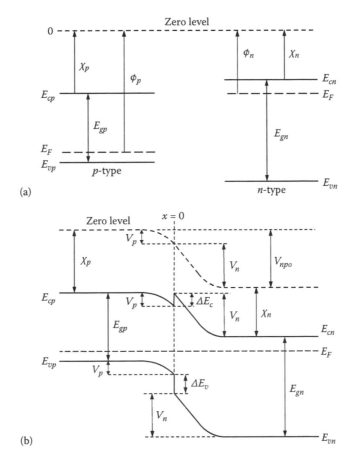

FIGURE 3.7.2
Energy-band diagrams of an n-type semiconductor and a p-type semiconductor, (a) in isolation and (b) when forming a pn heterojunction.

To align E_F,

$$V_{npo} = \phi_p - \phi_n \qquad (3.7.11)$$

$V_{npo} = V_p + V_n$, where V_p is the voltage between the zero level on the *p*-side and the junction at $x = 0$, and V_n is the voltage between the junction and the zero level on the *n*-side.

In a homojunction, where E_g is the same on both sides of the junction, the edges of the conduction and valence bands on the *n*-side are also lowered by V_{npo} with respect to those on the *p*-side when E_F is aligned (Figure 3.1.2). In the heterojunction, the gaps differ by ΔE_g expressed as:

$$\Delta E_g = E_{gn} - E_{gp} = \Delta E_c + \Delta E_v \qquad (3.7.12)$$

The division of ΔE_g between ΔE_c and ΔE_v is governed by the **electron affinity**, defined as the energy χ eV needed to move an electron from the edge of the conduction band E_c to the zero level. It follows from Figure 3.7.1b that:

$$\Delta E_c = \chi_p - \chi_n \qquad (3.7.13)$$

If $\chi_p > \chi_n$, as in Figure 3.7.1b, a discontinuity ΔE_c is introduced in the conduction band, as shown in the figure. A discontinuity ΔE_v is also introduced in the valence band to give E_{gn} in accordance with Equation 3.7.12. Note that if $\chi_p = \chi_n$, then $\Delta E_c = 0$; there is no discontinuity in the conduction band, but a discontinuity of ΔE_g occurs in the valence band.

In each semiconductor,

$$p_{po}n_{po} = n_{ip}^2, \quad \text{and} \quad n_{no}p_{no} = n_{in}^2 \qquad (3.7.14)$$

From Equation 2.5.5, Chapter 2,

$$n_{ip}^2 = N_{cp}N_{vp}e^{-E_{gp}/kT}, \quad \text{and} \quad n_{in}^2 = N_{cn}N_{vn}e^{-E_{gn}/kT} \qquad (3.7.15)$$

It follows that:

$$\frac{n_{ip}^2}{n_{in}^2} = \frac{p_{po}n_{po}}{n_{no}p_{no}} = \frac{N_{cp}N_{vp}}{N_{cn}N_{vn}}e^{\Delta E_g/kT} \qquad (3.7.16)$$

Equating the change in chemical potential of holes to the change in electric potential,

$$kT \ln \frac{p_{po}}{p_{no}} = E_{vp} - E_{vn} = V_{npo} + \Delta E_v \qquad (3.7.17)$$

Similarly for electrons,

$$kT \ln \frac{n_{no}}{n_{po}} = E_{cp} - E_{cn} = V_{npo} - \Delta E_c \qquad (3.7.18)$$

Subtracting Equation 3.7.18 from Equation 3.7.17 gives:

$$kT \ln \frac{p_{po} n_{po}}{n_{no} p_{no}} = \Delta E_g \qquad (3.7.19)$$

Equation 3.7.19 is consistent with Equation 3.7.16, because Equations 3.7.17 and 3.7.18 do not account for any changes in the effective masses of electrons and holes on the two sides. If these effective masses are the same, $N_{cp} = N_{cn}$ (Equation 2.5.3, Chapter 2) and $N_{vp} = N_{vn}$ (Equation 2.5.4, Chapter 2). Equation 3.7.16 then reduces to Equation 3.7.19.

The importance of heterojunctions is mainly in their use as the base–emitter junctions in bipolar junction transistors (Section 6.4, Chapter 6).

Summary of Main Concepts and Results

- In a *pn* junction at equilibrium, electrons (holes) have a higher (lower) chemical potential energy on the *n*-side than on the *p*-side but a lower (higher) electric potential energy, so that their electrochemical potential is the same on both sides. The electric potential energy difference results in an equilibrium junction potential. At equilibrium, the electron (hole) diffusion current is equal and opposite to the electron (hole) drift current. These currents are so large that the biased junction is never far removed from equilibrium.

- Under forward bias, the drift–diffusion balance is upset in favor of diffusion. Minority carriers are injected, and the forward current of the diode is a recombination current that supplies the carriers that recombine in the depletion and transition regions on both sides of the junction.

- Under reverse bias, the drift–diffusion balance is upset in favor of drift. Minority carrier extraction takes place, and the reverse current of the diode is essentially a generation current that is carried by minority carriers and which removes the carriers generated in the depletion and transition regions on both sides of the junction.

- Since the majority carrier concentration vastly exceeds the minority carrier concentration, the forward current is much larger than the reverse current. The *pn* junction therefore behaves as a rectifier.

- A *pn* junction has two types of capacitance: a depletion capacitance associated with charges in the depletion regions, and a diffusion capacitance associated with charges due to injection or extraction of minority carriers in the transition regions. Both capacitances increase with the forward bias and decrease with the reverse bias. Under forward bias, the diffusion capacitance is dominant, whereas under reverse bias, the depletion capacitance is dominant. The capacitances introduce a finite reverse recovery time when the diode is switched from forward bias to reverse bias.

- The effect of temperature on a *pn* junction diode is to cause I_R to approximately double per 10°C rise in T, and v_D to decrease by about 2 mV/°C at constant i_D.

- In a heavily doped *pn* junction diode, breakdown occurs at a reverse voltage of 2.5–5 V due to zener breakdown. Electrons in the valence band of the *p*-semiconductor tunnel through the energy gap and become conduction electrons in the *n*-semiconductor. *BV* has a negative temperature coefficient. In a less heavily doped *pn* junction diode, breakdown occurs at a reverse voltage between about 7 V and a few kilovolts due to avalanche breakdown. Electrons and holes acquire sufficient KE between "collisions" with crystal atoms to cause ionizing "collisions" with these atoms. *BV* has a positive temperature coefficient.

- The effect of light on semiconductors is to increase the rate of generation of carriers, and hence the electrical conductivity. In a *pn* junction diode, the reverse current is greatly increased–an effect used in photodiodes. The forward *i–v* characteristic is displaced in the direction of positive forward voltage leading to a photocell that converts light energy to electric energy.

- Energy released when electrons and holes recombine during forward conduction of a *pn* junction diode falls within the infrared and visible parts of the spectrum. However, only in direct recombination, which occurs in direct gap semiconductors such as GaAs, is useful radiation emitted. This forms the basis of operation of LEDs. In indirect gap semiconductors, such as Si, recombination takes place mainly via traps or surface states having energies within the forbidden gap. Energy is released as heat.

- In a tunnel diode, doping levels are so high that electron tunneling takes place for small forward or reverse bias. The tunneling current produces a negative resistance region in the forward *i–v* characteristic, which can be used for switching purposes or to construct high-frequency oscillators.

- When dissimilar materials are in contact at equilibrium, the electrochemical potential of current carriers, and E_F, must be the same throughout. The alignment of E_F is brought about by a contact potential between the two materials. The contact potential is associated with a net charge distribution that is a surface charge in metals and a volume charge in semiconductors.

- A current-actuated voltmeter fundamentally measures differences in electrochemical potential. In the absence of differences in chemical potential, the voltmeter measures differences in electric potential. In the presence of chemical potential differences, KVL applies to electrochemical potential differences.

- The work function of a metal is the minimum energy that must be imparted to an electron at 0K to enable it to escape from the metal. A contact between a metal of high work function and an *n*-type semiconductor is rectifying, as in a Schottky diode. On the other hand, a contact between a metal of low work function and an *n*-type semiconductor is nonrectifying, or ohmic.

- A heterojunction is formed between *n*-type Si and *p*-type SiGe that contains 15%–25% Ge. The energy gaps between the conduction and valence bands in the two semiconductors are no longer equal, which can be utilized in improving the performance of bipolar junction transistors.

Learning Outcomes

- Articulate the physical basis for rectification in a *pn* junction diode, junction capacitances, junction breakdown, effect of light on a *pn* junction, emission of light by a *pn* junction, and operation of the tunnel diode, Schottky diode, and heterojunction.

Supplementary Examples and Topics on Website

SE3.1 Capacitances of Abrupt and Graded Junctions. Determines the depletion capacitances of abrupt and linearly graded junctions.

SE3.2 Maximum Power Output from Solar Cell. Derives the condition for maximum power transfer from a solar cell.

SE3.3 LED Mains Indicator. Considers three alternative designs for such an indicator.

SE3.4 Tunnel Diode Oscillator. Analyzes the operation of a tunnel diode oscillator.

ST3.1 Exact Analysis of *pn* Junction at Equilibrium. Indicates how N_A^+, N_D^-, n_{no}, p_{no}, p_{po}, and p_{no}, as well as $V(x)$, $n_o(x)$, and $p_o(x)$ in the depletion region, can be derived without making the usual simplifying assumptions. Applies Poisson's equation and Boltzmann's relations to obtain an expression for the Debye length.

ST3.2 Diffusion and Drift Currents under Forward Bias and Reverse Bias. Illustrates the division of electron and hole currents into drift and diffusion components under forward bias and reverse bias.

ST3.3 Diode Transients. Discusses diode transients in terms of the diode capacitances and the charge concentration profile.

ST3.4 Continuity Equation. Derives the continuity equation and applies it to minority carrier injection across a *pn* junction; shows that the diffusion length can be interpreted as the mean distance through which a minority carrier travels before it recombines. The continuity equation is used to derive the dielectric relaxation time and its relation to the Debye length.

ST3.5 Quantitative Analysis of *pn* Junction. States, in terms of electrochemical potential, the assumptions made in the simplified analysis of the *pn* junction and derives the diode current.

ST3.6 Recombination Current of a *pn* Junction Diode. Derives the expression for the recombination current, which dominates at low forward currents, and which has $\eta = 2$.

ST3.7 High-Level Injection in a *pn* Junction Diode. Shows that $\eta = 2$ for high-level injection.

ST3.8 Short-Base Diode. Derives the expression for current in a short-base diode, in which the length of one side is small compared to the diffusion length of minority carriers on that side.

ST3.9 Stability of Operating Points on *i–v* Characteristic with Negative Resistance Part. Derives the criterion for stability in terms of the slopes of the load line and the different parts of an *i–v* characteristic having a negative resistance part.

Problems and Exercises

P3.1 *pn* Junction Diode

P3.1.1   Determine V_{npo} for a *pn* junction in which the doping concentration is 1 in 10^6 on the *n*-side and is: (a) 1 in 10^6, (b) 1 in 10^8 on the *p*-side.

P3.1.2   Repeat Problem P3.1.1 with the *n*-side and *p*-side interchanged.

P3.1.3   What would you expect V_{npo} to be at $T = 0$? Justify your answer.

P3.1.4 Compare V_{npo} for Si and GaAs pn junctions having the same N_A and N_D at 300K.

P3.1.5 Consider a Si pn junction at 300K having $N_A = 10^{16}/cm^3$ and $N_D = 10^{14}/cm^3$. Assuming that the junction is abrupt and that the depletion approximation applies, determine the effect of multiplying N_D by 10 on: (a) W_n and W_p; (b) the maximum electric field.

P3.1.6 Repeat Problem P3.1.5 with $N_A = 10^{15}/cm^3$ and N_D of: (a) $10^{13}/cm^3$, (b) $10^{14}/cm^3$. What can you conclude from the results of this problem and those of Problem P3.1.5?

P3.1.7 N_A and N_D on the two sides of a pn junction are varied but their product is kept constant at $10^{32}/cm^6$. For what N_A and N_D is the width of the depletion region W_d, a minimum?

P3.1.8 (a) Derive Equation 3.2.20. (b) A Si pn diode has $N_A = 10^{18}/cm^3$. Determine N_D that would give a breakdown voltage of 50 V, assuming that in the range of $N_D = 10^{16}/cm^3$ to $10^{17}/cm^3$, ξ_{brk} varies with N_D as $\xi_{brk} = 4.45 \times 10^5 + 4\log_{10}(N_D/10^{16})$ V/cm.

P3.1.9 Consider a pn junction diode with $N_A > N_D$. Compare the magnitudes of the hole and electron currents under: (a) forward bias, (b) reverse bias.

P3.1.10 If I_R doubles every 10°C, what is the percentage increase per °C. By what factor does the current changes if T increases by: (a) 15°C, or (b) by an arbitrary change of ΔT °C.

P3.1.11 If I_S varies by 15% per °C and I_R varies by 10% per °C, what percentage of I_R is I_S, assuming that I_{leak} does not change with T? If the variation of I_R is measured at $I_R = 1$ nA and a reverse voltage of 10 V, what is the leakage resistance of the diode?

P3.1.12 Consider a Si pn junction diode having $I_S = 10$ nA, $\eta = 2$, and operating at a forward voltage of 0.7 V at 300K. If the temperature increases by 50K, what is the change in: (a) I_S, (b) i_D? Use Equation 3.2.16 for I_S assuming $b = 3$ and $E_g = 1.1.2$ eV, and use Equation 1.1.1, Chapter 1, for i_D.

P3.1.13 A pn junction diode has a capacitance of 1.05 pF at a reverse voltage of 3 V and 0.72 pF at a reverse voltage of 9 V. If $V_{npo} = 0.75$ V, determine: (a) the grading coefficient m, (b) C_{dp0}.

P3.1.14 Are the capacitances due to charges of stored minority carriers on the two sides of a forward-biased pn junction, in series or in parallel? Justify your answer.

P3.1.15 A pn junction that is reverse biased at 5 V has $I_S = 10$ pA, $V_{npo} = 0.7$ V, and $C_{dp0} = 2$ pF. If the reverse voltage is suddenly changed by 100 mV, what is the transient change in the charge flowing through the diode? What percentage of the total charge that passes through the diode in 100 μs does this represent? Assume that the junction is abrupt.

P3.1.16 Using Equations 3.2.5 and 3.2.6, express the mean transit time τ_T in terms of the diode parameters.

P3.1.17 A Si pn diode has $N_A = 10^{17}/cm^3$, $N_D = 10^{15}/cm^3$, $\mu_e = 1200$ cm²/Vs, $\mu_h = 400$ cm²/Vs, $\tau_e = 2$ μs, and $\tau_h = 8$ μs. Calculate the hole and electron current densities at a forward voltage of 0.68 V at 300K.

P3.1.18 Repeat Problem P3.1.17 for GaAs, assuming $\mu_e = 8000$ cm²/Vs, $\mu_h = 350$ cm²/Vs, $\tau_e = 1$ ns, and $\tau_h = 10$ ns.

P3.1.19 Consider a Si pn diode at 300K having $N_A = 10^{17}/cm^3$ and $N_D = 10^{15}/cm^3$. (a) Determine v_D at which the concentration of holes at the edge of the transition region on the n-side becomes comparable with the concentration of majority carriers on this side, say one-fifth. This can be considered the limit of low-level injection. At this voltage, (b) what is the corresponding ratio of the electron concentration at the edge of the transition region on the p-side to the concentration of majority carriers on this side? (c) What is the width of the depletion region on each side?

P3.1.20 The hole diffusion length in a p^+n diode is 2 μm. In a p^+n diode under forward bias, the ratio of the hole diffusion current to the electron diffusion current at the edge of the transition region on the n side is 50. Determine this ratio at a distance L_h from the edge of the transition region, assuming that the total diffusion current is the same.

P3.1.21 Calculate the equilibrium electron drift and diffusion currents in Example 3.1.1.

P3.1.22 Repeat Example 3.1.1 if $N_A = 10^{18}/\text{cm}^3$ and $N_D = 10^{16}/\text{cm}^3$.

P3.1.23 Sketch the voltage variations across the junction in Example 3.1.1 using the depletion approximation.

P3.1.24 The junction of a *pn* Si diode has an area of 10^5 μm^2, $N_A = 10^{18}/\text{cm}^3$, and $N_D = 10^{16}/\text{cm}^3$. Determine (a) C_{dp0} and (b) C_{dp} at a reverse bias of 10 V.

P3.1.25 A *pn* junction Si diode has $\tau_h = 100$ ns and $\eta = 1.6$. Determine C_d at i_D of: (a) 1 mA, (b) 10 mA.

P3.2 Special-Purpose Diodes

P3.2.1 Is an abrupt or a graded junction more desirable for a varactor diode? Assume that a varactor diode has an abrupt junction, a C_{dp0} of 5 pF, and a V_{npo} of 0.8 V. It is required to resonate with a 1 μH inductor at 100 MHz. Determine the required reverse bias.

P3.2.2 A photocell has $I_S = 1$ nA and $I_{SS} = 100$ mA. Subject to the simplifying assumptions made in Section 3.3, determine: (a) I_{sc}; (b) V_{oc}; (c) v_{PC} for maximum power transfer; and (d) the maximum power available.

P3.2.3 A simplified equivalent circuit of a photocell based on Equation 3.3.3 is shown in Figure P3.2.3. Given two banks of photocells, one having an open-circuit voltage of 0.68 V and a short-circuit current of 1.4 A, the other having an open-circuit voltage of 0.52 V and a short-circuit current of 1 A. Determine the open-circuit voltage and the short-circuit current when the photocell banks are connected in: (a) parallel, and (b) series. Assume that $\eta V_T = 0.04$ V for all the photocells.

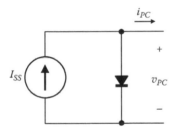

FIGURE P3.2.3

P3.2.4 Based on Equations 3.3.5, 3.2.4, and 2.5.11, Chapter 2, how will V_{oc} of a photocell vary with E_g? Using the qualitative argument in terms of electrochemical potential difference in a photocell, what is the theoretical maximum V_{oc}?

P3.2.5 Show that for a photocell, $v_{PC(\text{max})}$, the voltage for maximum power output, satisfies the equation: $v_{PC(\text{max})} = V_{oc} - \eta V_T \ln(1 + v_{PC(\text{max})}/\eta V_T)$.

P3.2.6 The *i–v* characteristic of a tunnel diode can be expressed as $i_D = I_p(v_D/V_p)e^{(1-v_D/V_p)} + I_S e^{v_D/V_T}$, where the first term is due to tunneling and the second term is due to normal diode action. Neglecting the latter term, show that the negative incremental resistance R is given by $\dfrac{1}{R} = -\dfrac{I_p}{V_p}\left(\dfrac{v_D}{V_p} - 1\right)e^{(1-v_D/V_p)}$. Determine: (a) the largest magnitude of the incremental resistance for a tunnel diode having $V_p = 0.15$ V, $I_p = 2$ mA, and $I_S = 0.1$ nA; (b) the corresponding i_D.

P3.2.7 I_S of a Schottky diode can be expressed as $I_{B0} = K_{SB}T^2 e^{-E_B/V_T}$. If $E_B = 0.6$ V and $v_D = 0.4$ V, by how much does v_D change, at constant i_D, for a 1°C rise in T above 300K?

P3.2.8 The forward current of both a *pn* junction Si diode and a Schottky diode is given by: $I_S e^{v_D/V_T}$. (a) What should be the ratio of the saturation currents if the forward voltage drop of the Schottky diode is 0.3 V less than that of the *pn* junction Si diode? (b) If the saturation current density of the Schottky diode is 5000 times that of the Si *pn* junction diode, what should be the ratio of the areas?

P3.2.9 Consider a Si–SiGe heterojunction having $N_D = 10^{16}/cm^3$, $N_A = 10^{18}/cm^3$, and $V_{npo} = 1.58$ V. Determine V_n, V_p, W_n, and W_p, at equilibrium, assuming ε is 12 for Si and 13 for SiGe.

P3.2.10 Repeat Problem P3.2.9 for a reverse bias of 10 V, replacing V_{npo} by $(V_{npo} - v_D)$.

P3.3 Miscellaneous

P3.3.1 An n-type semiconductor slab of 3 mm×3 mm section and 10 μm depth has $N_D = 10^{15}/cm^3$ and is irradiated at a power density of 0.4 W/cm². If $E_g = 2$ eV, what is the maximum wavelength of light that can be detected? Determine the conductivity at this wavelength if $\mu_e = 800$ cm²/Vs, $\mu_h = 300$ cm²/Vs, $\tau_e = \tau_h = 10^{-5}$ s, and $\zeta = 10\%$.

P3.3.2 Express Equation ST3.4.9 as $p_n(x) - p_{no} = A \cosh(x/L_h) + B \sinh(x/L_h)$. Determine A and B subject to the general boundary conditions of Equations ST3.8.1 and ST3.8.2. Show that when $(x/L_h) \ll 1$, $p_n(x) - p_{no} = (p_n(0) - p_{no}) \times \left(1 - \left(\coth\dfrac{W_n}{L_h}\right)\dfrac{x}{L_h}\right)$, which is the same as Equation ST3.8.5 when W_n/L_h is small. Deduce that the hole current density can be expressed as $J_h(x) = \dfrac{|q|D_h n_i^2}{L_h N_D} \coth\dfrac{W_n}{L_h}(e - 1)$.

P3.3.3 Show that Equation ST3.8.4 reduces to Equation ST3.4.10 when $W_n \gg L_h$.

P3.3.4 Show that if $N_A \gg N_D$ in a short-base diode, $C_d r_d = W_n^2/2D_h$.

4

Semiconductor Fabrication

After having discussed the basic properties of semiconductors and the various types of *pn* junction diodes, and before considering transistors, it is opportune at this stage to familiarize the reader with the nature and some basic features of semiconductor fabrication. This makes the description of the structure of semiconductor devices, particularly transistors of various types, more realistic and concrete. The description of fabrication processes in what follows applies specifically to Si, not to GaAs. This is justified, not only by the introductory nature of this book, but also by the fact that Si is currently by far the most popular material for microelectronic fabrication. Moreover, GaAs fabrication processes are of essentially the same nature as those for Si, but differ in some important details, so that a discussion of Si fabrication processes will provide the reader with a good understanding of semiconductor fabrication in general.

The chapter begins with a brief description of the preparation of a Si wafer, which acts as the substrate for all the ICs fabricated from the wafer. Various patterning processes are then considered that are involved in the transfer of the desired circuit design to appropriate structural patterns in a number of levels of the IC. This is followed by deposition processes, which are concerned with the controlled addition of substances to the wafer, including doping. The discussion of fabricating processes ends with packaging.

Semiconductor fabrication is illustrated with the fabrication of simple devices, namely, resistors, *pn* junction diodes, and capacitors, followed by the fabrication of CMOS and bipolar junction transistors (BJTs), including BiCMOS. Although MOSFETs and BJTs have not yet been considered, the reader is expected to refer to these sections while reading Chapters 5 and 6, respectively. The chapter ends with some special topics related to semiconductor fabrication and operation, namely, SiGe technology, crystal defects, and the Si–SiO$_2$ interface.

Finally, it should be mentioned that Si-based microelectronic fabrication techniques are increasingly being used to produce devices for nonelectronic applications, such as microelectromechanical systems (**MEMS**). These exploit the mechanical properties of materials used in Si IC fabrication, such as single-crystal Si, polycrystalline Si, SiO$_2$, and silicon nitride (Si$_3$N$_4$). Further developments and extensions of microelectronic fabrication techniques have paved the way to **nanotechnology**, in which materials and structures exhibit at this scale new characteristic that are not encountered in structures of larger dimensions.

Learning Objectives

❖ To be familiar with:

- Terms used in conjunction with semiconductor fabrication.
- The nature of the major processes involved in semiconductor fabrication.

❖ To understand:

- The significance of the basic steps in the preparation of silicon wafers, patterning processes, deposition processes, and packaging.
- The basic structure of some simple semiconductor devices such as resistors, *pn* junction diodes, and capacitors, as well as the basic structure of CMOS transistors and BJTs.

4.1 Preparation of Silicon Wafer

High-Purity Silicon

In IC fabrication, a number of component assemblies known as **dice**, whose singular is a die, are formed on a **wafer**, which is a thin slice of a large Si crystal. A usable **die** is the IC before it is packaged. The packaged die, or IC, is also known as a **chip**.

The starting point in the preparation of a wafer is natural SiO_2, or quartzite, which is a relatively pure form of sand. To extract Si, quartzite is subjected to high temperatures of about 2000°C in an arc furnace in the presence of a source of carbon, such as SiC. This reduces SiO_2 and yields **metallurgical grade silicon** of 90%–99% purity, which is still far from the degree of purity required. Because the dopant density in semiconductors is usually in the range of 10^{15}–$10^{19}/cm^3$, the concentration of unwanted impurities should be much lower, of the order of $10^{13}/cm^3$. Since the concentration of Si atoms is about $5 \times 10^{22}/cm^3$ (Table 2.3.1), the degree of purity required is nearly one impurity atom for every 5 billion Si atoms. Obtaining such high purity was the first challenge in the fabrication of Si semiconductors.

To achieve the required high purity, metallurgical grade Si is pulverized and treated with HCL to form trichlorosilane ($SiHCl_3$). As this is liquid at room temperature, it can be distilled to a high degree of purity. The purified $SiHCl_3$ is then reduced by hydrogen to produce highly purified **electronic grade silicon**. The silicon at this stage is polycrystalline (polysilicon), or **poly**, in the form of small crystals having many defects.

Crystal Growth

Single crystals for IC fabrication are almost invariably grown by the **Czochralski growth** technique. The silicon is melted in a fused silica crucible in an inert gas (usually argon) atmosphere at a temperature of about 1500°C, just above the melting point of Si. A chemically etched seed crystal, about 10 cm long, 0.5 cm in diameter, and having the desired crystal orientation is lowered into contact with the melt. The seed is pulled up at a controlled rate of a few mm/h while both the seed and crucible are rotated at a slowspeed in opposite directions. As the seed is pulled up, the melted silicon attaches to the growing seed and solidifies, becoming part of a single crystal having the same crystal orientation as the seed crystal. The fully grown crystal, referred to as a **boule** or **ingot**, is 1–2 m in length, and of a diameter determined by the temperature and pulling speed. Boron or phosphorus is commonly added to the melt to obtain *p*-type or *n*-type boules, respectively, of specified resistivity.

The three possible crystal orientations can be defined with respect to the tetrahedral structure of Figure 4.1.1, referred to in Section 2.3, Chapter 2. The orientation is <100> if the surface of the crystal is parallel to the plane ABCD, <110> if the surface of the crystal is parallel to the plane AFGD, and <111> if the surface of the crystal is parallel to the plane EDG.

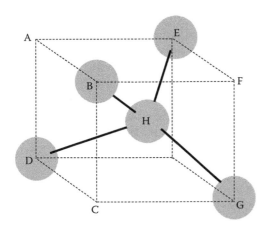

FIGURE 4.1.1
Crystal orientations of a tetrahedral structure.

Wafer Production

The top and bottom parts of the boule are cut off, and the boule is ground to a near-final diameter. Flat regions may be ground along the length of a boule to identify crystal orientation and semiconductor type. The boule is then dipped in a chemical etching solution to remove the damage caused by mechanical grinding. The boule is sliced into circular wafers 0.6–1.3 mm thick using a wire impregnated with diamond particles. Wafer thickness generally increases with diameter, for more rigidity. The wafers are polished on one surface by a **chemical mechanical process** (CMP) in which the wafer is rotated over a polishing pad that is continually supplied with an abrasive chemical solution, or slurry, of progressively finer grain. Polishing is due to both fine mechanical abrasion as well as chemical action. As a final step, the wafers are chemically cleaned to give a mirrorlike finish that is virtually free from imperfections.

Wafer diameter has increased over the years. Wafers with 200 mm diameters were introduced in 1991, those with 300 mm diameters were introduced in 2001, and those with 450 mm diameters are planned to be introduced around 2012. Increasing the wafer diameter increases the number of chips of a given size that can be produced from the wafer, which reduces the cost per chip, assuming that the costly upgrade to the larger wafer diameter can be offset by a larger volume of chips produced. A 300 mm diameter wafer can supply more than 50,000 small chips, like some op amps, and a 450 mm diameter wafer would carry about twice the number of chips as a 300 mm diameter wafer.

Oxidation

When the surface of the wafer is exposed to air at room temperature, Si reacts readily with O_2 to produce SiO_2, which is a form of glass and is a very good insulator. Rather than having a layer of SiO_2 form on the surface of a wafer in an uncontrolled manner, an oxide layer of controlled thickness is grown on the wafer. To oxidize the surface, it is exposed at a temperature of 1000°C–1200°C to oxygen (dry oxidation) or to steam (wet oxidation). The oxide grows more rapidly in the wet process, which is therefore used to grow thick oxide layers, but the oxide is not as dense and does not have as good mechanical and electrical properties (Section 4.8). An important feature of the oxidation process is that as the oxide film grows, new oxide is formed, not at the exposed SiO_2 surface, but at the Si–SiO_2 interface, because the rate of diffusion of oxygen in SiO_2 is much higher than that of Si. The Si–SiO_2 interface is thereby protected from contamination. The thickness of the oxide ranges between about 0.1 μm (dry oxidation) and 1 μm (wet oxidation), depending on the application. Oxidation consumes Si from the wafer and adds to its thickness. If the thickness of the oxide is t_{ox}, the thickness of the wafer is increased by about $0.55t_{ox}$, and the original thickness of Si in the wafer is reduced by $0.45t_{ox}$. Following oxidation, the wafer is usually **annealed** by heating in an inert gas atmosphere, as this reduces the positive charge that inevitably forms in the oxide near the Si–SiO$_2$ interface (Section 4.8).

The oxide can serve several purposes, such as protection, electrical isolation of different structures on the chip, acting as a functional dielectric, or as a barrier that allows the introduction of dopants in predefined areas of the chip, as described later.

4.2 Patterning Processes

Preparation of Masks

Patterning refers to the transfer of the circuit design to a Si wafer that has been appropriately treated. The starting point is, of course, the design of a circuit that operates in the desired manner. Generally, this involves PSpice simulations at various stages in the design to validate the design and ensure its proper operation. The physical realization of the circuit design in the form of a chip consists of building up the chip layer by layer, each layer involving a number of fabrication steps. A complex IC can have more than 10 layers, with 30 or more major fabrication steps involving something like 200 operations or process steps. Each major fabrication step is governed by a **mask**, which is a two-dimensional pattern that is transferred to the wafer through a lithographic process, as described later (Figure 4.2.1).

A layout computer program generates the required series of masks from the circuit design, subject to certain layout design rules that are dictated by the fabrication technology used. The design rules can be generally divided into two categories: one category is concerned with spatial resolution in a given level and defines the minimum allowed widths of lines and their spacing; the other category is concerned with alignment between successive layers. The design rules are based on manufacturing variability such as

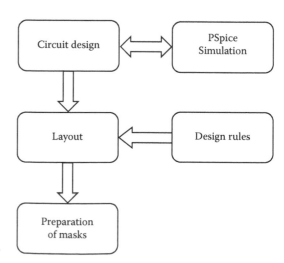

FIGURE 4.2.1
Work flow diagram for preparation of masks.

dimensional tolerances, registration errors in the alignment of different masks, and variations in the timing and duration of various processes.

The **minimum feature size** is the minimum resolvable metal line widths of the interconnections. It is also the minimum *half-pitch*, or separation between adjacent metal lines. In a 45 nm technology, for example, the minimum feature size is 45 nm.

Lithography

Lithography is the process of transferring the physical patterns of the masks to the Si wafer in order to select parts of the wafer that are to be exposed to fabrication steps such as oxidation, ion implantation, diffusion, etc., as will be described later. Currently, optical lithography, or **photolithography** using ultraviolet (UV) light, is the main lithographic process used in IC fabrication. Literally, photolithography means "using light to write on stone (lithos in Greek)." Figure 4.2.2 illustrates a simple mask that can be used as a first step in producing an *n*-well in a *p*-substrate. The *n*-well can in principle be used to fabricate a *pn* junction diode or an IC resistor, or to form the body of a metal–oxide–semiconductor (MOS) transistor, as will be explained later. The well is rectangular in shape, defined by the inner, white rectangle in top view. As illustrated in cross-sectional view, the mask is up to about 7 mm thick and consists of a quartz layer with chromium backing that has been removed in the area that defines the well.

The first step in lithography is chemical mechanical polishing (CMP), as previously described, which planarizes the surface to make it optically flat so that small features can be more accurately reproduced. The entire surface of the wafer is then coated with a light-sensitive emulsion, or **photoresist**. The surface must be clean and dry for good adherence of the photoresist. To dry the surface, the wafer is placed in an oven at 150°C–200°C for at least 15 min. A quantity of photoresist is then deposited on the surface of the wafer, and the wafer spun rapidly at 2000–6000 rpm to cover the surface of the wafer with a uniform film of photoresist about 1 μm thick depending on the speed (Figure 4.2.3). A pre-exposure bake, or softbake, as specified by the photoresist manufacturer, removes the solvent from the photoresist and improves its adhesion to the wafer. The wafer is then subjected to UV light through the mask, as diagrammatically illustrated in Figure 4.2.3. The UV light is blocked by the opaque chromium backing, but where the chromium backing has been removed, the UV light is transmitted through the transparent quartz layer and reacts with the photoresist.

Photoresists are of two types: positive and negative. Positive photoresist is composed of long-chain molecules that are broken into small chains by the UV light, whereas negative

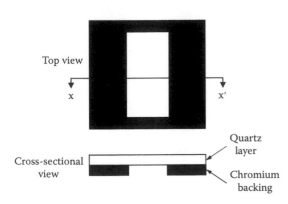

FIGURE 4.2.2
Illustration of a simple lithographic mask.

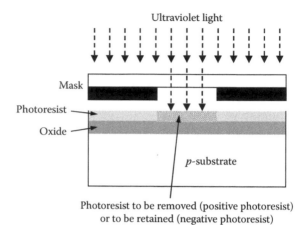

FIGURE 4.2.3
Illustration of lithographic process.

Photoresist to be removed (positive photoresist)
or to be retained (negative photoresist)

photoresist consists of small-chain molecules that are polymerized into long chains by the UV light. When either photoresist is developed, the short-chain photoresist is removed, leaving the long-chain photoresist. Consequently, the positive photoresist that remains after development conforms to the areas of chromium backing, and the exposed surface of the wafer conforms to the clear pattern of the mask. The opposite is true of negative photoresist. In the case of the n-well of Figure 4.2.3, a positive photoresist should be used with the mask shown. A positive photoresist is more commonly used in IC fabrication because of its better resolution, although using both types of photoresists can reduce the number of masks required.

After exposure to UV light, the photoresist is developed, usually by immersing the wafer in the developing solution. The non-polymerized photoresist is softened by the developer and is subsequently washed away. A hardbake often follows, to improve the adhesion of the polymerized photoresist and its resistance to subsequent etching processes. The surface of the wafer becomes as in Figure 4.2.4a. Before any doping processes can be performed, the oxide layer has to be etched away from the exposed surface of the wafer, as illustrated in Figure 4.2.4b, and discussed in the next subsection. Note that some etching occurs under the resist. The photoresist is then removed, usually using a solvent such as acetone.

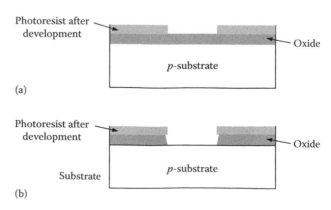

FIGURE 4.2.4
Surface of wafer (a) after removal of the photoresist and (b) after etching of the oxide.

Before describing etching, there are some important points to note concerning photolithography. During exposure to UV radiation, the mask could be in contact with the wafer, in close proximity to it, as in Figure 4.2.3, or the mask pattern could be projected from a mask onto a wafer that is many centimeters away, using an optical system. Having the mask in physical contact with the wafer is simple and reproduces the image on the mask as 1:1 with minimal blurring. However, repeated contact can damage the mask due to scratching of the chromium or the sticking of photoresist to it. Damage to the mask can be avoided by having the mask spaced some 10–50 μm from the wafer. Resolution is degraded, however, because of the diffraction of light as it passes through small apertures, or openings, in the mask, if the dimensions of the apertures are not large compared to the wavelength of light used. As a result of diffraction, some light spreads as fringes or bands into what would have been shadow regions had the light propagated in straight lines through the mask. The edges of these small openings would therefore appear blurred. For these reasons, a projection system is almost invariably used nowadays.

A mask may include patterns for the whole wafer, or it may contain the pattern for a single die or several dice, in which case it is referred to as a **reticle**. The reticle is projected onto the wafer, usually with a 5:1 or 10:1 reduction. Since the pattern of the reticle is 5 or 10 times larger than the pattern on the wafer, the fabrication of the reticle becomes easier. A stepper, referred to as an **optical stepper**, is used to step the wafer so that the entire array of circuits is patterned. A reticle can be used on several thousand wafers before it has to be replaced. The maximum dimensions of a die are limited by the size of the imaging field of the optical system. The maximum die dimensions are currently about 30 mm on each edge.

Of crucial importance is the alignment of successive reticles so that the circuit patterns would be properly placed relative to one another in different layers on the chip. The wafer and reticles for different layers have sets of alignment marks, which, before every exposure for a given layer, must be properly aligned relative to one another. In VLSI fabrication, the alignment of each individual die is done automatically under computer control by imaging and pattern-recognition systems. The error in the alignment of masks for successive levels is the **registration error**.

The smallest resolution possible with optical lithography is fundamentally limited by diffraction phenomena. To achieve finer resolution, UV light of shorter wavelengths must be used, such as deep UV light of 248, 193, or 157 nm wavelength. The shorter wavelengths necessitate different light sources, materials for the reticle and its backing, and photoresists. The quartz of the mask is replaced by CaF, and an UV-emitting laser is used as a source instead of a mercury lamp. The laser scans the surface of the wafer, through the reticle, rather than project the light over the whole surface to be exposed, before the wafer is stepped. Resolution-enhancement techniques (RETs) and chemically amplified resists have been developed to obtain resolutions better than those dictated by lens optics. Resolution has been enhanced by immersion, that is, having the UV light reach the wafer through a film of high-purity water instead of air. This changes the refractive index and increases the numerical aperture of the lens by about 40%. With these improvements, minimum feature sizes of 25 nm have been realized in the laboratory using deep UV photolithography. For finer resolution, other technologies may have to be used, such as electron-beam lithography, currently used to produce the masks for photolithography; ion-beam lithography; extreme ultraviolet (EUV) lithography of 13.5 nm wavelength; or X-ray lithography. Note, however, that not all the layers in the fabrication of a given chip may require the highest resolution. Some of the theoretical and practical aspects of lithography are elaborated in Section ST4.2.

Etching

Etching is the removal from the wafer of unwanted substances, which could be Si, SiO_2, metal, or photoresist. Etching can be chemical, physical, or a combination of both, the etching processes being classified as wet or dry.

In wet etching, a liquid chemically erodes the material to be removed. Wet etching is used extensively for cleaning and for removal of thin films, as in the development of the photoresist. However, it is not appropriate for critical fabrication steps for two reasons: (1) contamination by etching chemicals or their by-products, and (2) difficulty in controlling the extent of removal in specific directions, with consequent encroachment on the dimensions of small features and the degradation of resolution.

Dry etching, on the other hand, uses a plasma, which is an electric discharge in a partially ionized gas. A plasma may contain charged atoms or molecules in the form of positive or negative ions, and may also contain **radicals**, that is, dissociated atomic or molecular fragments having unsatisfied bonds, which makes them highly reactive. A plasma is initiated in a gas at low pressure by a voltage between two electrodes that is sufficiently high to cause electrical breakdown due to ionization of the gas and the creation of free electrons and positive ions. Electrons travel to the anode and positive ions to the cathode, where they release electrons by secondary emission because of the high energies of the ions. The secondary electrons travel to the anode, colliding with gas atoms along the way, thereby causing more ionization. The discharge becomes a self-sustaining plasma. Because in IC fabrication some of the materials on one or both electrodes can be insulating, such as SiO_2, radio frequency (rf) plasmas are used in these cases at a frequency of 13.56 MHz. Plasma density can also be increased in various ways.

Three general types of dry etching are used in IC fabrication. In what is referred to simply as plasma etching, a specific gas or gas mixture such as carbon tetrafluoride (CF_4) and oxygen, is used at a relatively high vacuum pressure of about 0.5 mm of mercury. The chemistry involved is rather complicated, but basically, the reactive components are adsorbed to the surface to be removed, react with surface atoms, become desorbed or are nudged away by other components of the plasma, and are removed out of the etch chamber by the vacuum system. Plasma etching has high selectivity, which means that the rate of removal of the desired material is much higher than that of other materials that are not to be removed.

In another method of dry etching, known as ion milling or sputter etching, no chemical reactions are involved. A plasma at low pressure in an inert gas, such as argon, generates ions that strike the surface at moderate energy and physically eject atoms from the surface. The advantages of ion milling is that it is highly directional, that is *anisotropic*, because the ions follow the direction of the electric field. Anisotropy is important for minimizing lateral etching compared to depth etching. Ion milling can be applied to a wide variety of materials, which implies, however, that it is not very selective.

The third method that of **reactive ion etching** (RIE), is a combination of both methods and involves adding certain chemicals to the Si surface, such as fluorocarbon polymers. This method is extensively used, particularly for etching narrow, deep structures, because it combines the high selectivity of plasma etching with the high directionality of ion milling.

4.3 Deposition Processes

Deposition processes, including doping, are concerned with the controlled addition of substances to the Si wafer in order to control the electrical properties of various parts of the wafer. In common with the patterning processes discussed in Section 4.2, the deposition processes are **planar**, that is, they are implemented from the top surface of the wafer.

Ion Implantation

In this process, dopant ions of high energy bombard the surface and become embedded within a given penetration depth from the surface. Commonly used dopants are boron for p-type semiconductors, and phosphorus or arsenic for n-type semiconductors.

The ion source is a gaseous compound such as BF_3, PH_3, or AsH_3. The gas is passed through an arc chamber where it is broken up into ionized components. To select a particular ionic species, the ionized gas is accelerated by an electric field and subjected to a magnetic field perpendicular to the direction of flow. Because ions are deflected according to their charge/mass ratio, the electric and magnetic fields can be adjusted so that only the desired ionic species will emanate as a beam. The beam is focused by means of electrostatic lenses and accelerated to a high energy that is usually in the range 20–200 keV but could be as high as several MeV in some cases. Regions that are not to be implanted with ions are masked off with a sufficiently thick layer of oxide, silicon nitride, polysilicon, or photoresist, which cannot be penetrated by the beam. The narrow ion beam is scanned across the wafer, implanting ions over unmasked as well as masked regions that are subsequently removed.

The implant dose, which determines the doping concentration, is controlled by the beam intensity and can be precisely adjusted by measuring the beam current. The concentration profile of implanted ions depends mainly on the mass of the ions and their energies. The depth of penetration can be accurately controlled by the beam energy and is usually between 0.1 and 1 μm, although very shallow or very deep profiles are difficult to control. Because of the penetration of bombarding ions, the concentration of these ions peaks not at the surface but some distance into the semiconductor, resulting in what is known as a **retrograde profile**. Ion implantation is widely used, because it is highly controllable and does not require a high temperature. Its major disadvantage is that ion bombardment causes local damage to the crystal by dislocating crystal atoms (Section 4.8). Ion implantation is followed by annealing, that is, heating the wafer to a temperature of about 1000°C for a predetermined time in an inert gas atmosphere. The objective is twofold: (1) to repair crystal damage by allowing the crystal to reform where crystal atoms have been dislocated, and (2) to allow the implanted ions to occupy lattice sites, which they must if they are to act as a dopant. This is known as **impurity activation**.

Diffusion

Diffusion is used to introduce specified impurities to depths in the range of few tenths to several μm. Diffusion proceeds in two steps: **predeposition**, which deposits the required impurity at a high concentration at the surface, and **drive-in**, which allows the deposited impurity to diffuse to the required depth. The final impurity concentration is maximum at, or very close to, the surface and decreases with depth into the semiconductor.

In modern practice, the impurity is deposited very near the surface by ion implantation, but when a high doping concentration is required, vapor deposition at the surface is used. An inert gas, such as nitrogen, is bubbled through a liquid compound of the dopant to produce a vapor of the liquid compound. Typically, BBr_3 and $POCl_3$ are used to deposit boron and phosphorus, respectively. Oxygen is added to the mixture of nitrogen and the vapor of the liquid compound, and the gases are passed to the furnace containing the wafer. When they come in contact with the surface of the wafer, the gases react to release the dopant at the surface. The drive-in step occurs at a higher temperature of about 1100°C and allows the dopant to diffuse below the surface to a depth that depends on the temperature and the time allowed.

Substances can diffuse in a crystal in two ways: if they do not bond readily with crystal atoms, they end up in the interstitial spaces between crystal atoms and do not contribute to doping. Examples are gold, copper, nickel, and zinc. The other diffusion mode is substitutional, where the impurity, which acts as a dopant, occupies the position of a crystal atom. It may diffuse by occupying vacant sites in the crystal, or by exchanging places with a crystal atom. Boron and phosphorus have the advantage of high solubility and a high rate of diffusion in Si, whereas their rate of diffusion in SiO_2 is very low, so that the oxide forms an effective barrier to diffusion.

Chemical Vapor Deposition

In chemical vapor deposition (CVD), chemical reactions involving gases and vapors at the surface of the wafer lead to deposition of material on this surface. A variety of materials can be deposited in this manner, including silicon, polysilicon, dielectrics such as SiO_2 or silicon nitride, metals such as tungsten and copper, and barrier metals such as titanium nitride and tungsten nitride. Depending on the material to be deposited, CVD can take place at atmospheric pressure (APCVD), at a low pressure of 0.25–10 mm of mercury (LPCVD), or can be plasma enhanced (PECVD). An important advantage of CVD is its good **step coverage**, meaning a good coverage of both the vertical and horizontal surfaces of a steplike geometry on the surface of the wafer. Such geometries become more prominent with continuing reduction in minimum feature size.

When the temperature is raised to around 1000°C, so that the reactions are driven by heat energy, the process is referred to as thermal CVD, vapor-phase epitaxy, or **epitaxy**. An important feature of silicon epitaxy is that it allows the growth of a doped layer of crystalline silicon on a single-crystal substrate, so that the layer becomes part of the single crystal, with the doping level in the grown layer being _less_ than that in the substrate. This is not possible with ion implantation or diffusion, which result in regions having higher dopant concentrations. In a typical epitaxial growth system for growing doped silicon, a gas mixture of silane (SiH_4), hydrogen as a diluent, and gaseous dopant compounds such as PH_3, AsH_3, or B_2H_6 is passed through the furnace. A relatively thick layer of up to few tens of μm can be grown in this manner, and abrupt _pn_ junctions can be formed, compared to the graded junctions formed by diffusion processes.

At a lower temperature of about 600°C, the deposited silicon does not integrate with the single crystal of the substrate, but becomes a layer of polycrystalline silicon. This layer can then be heavily doped to produce polysilicon for the gates of MOSFETs and for interconnections.

Metallization

The first metallization that takes place is in making low-resistance ohmic contacts to terminal regions of various devices. Al is extensively used for this purpose because it has a low resistivity, adheres well to Si, and is easy to evaporate and etch. Being trivalent, it forms a heavily doped p^+ contact region. To make a metallic contact to p-type Si, for example, Al is deposited on the Si surface and the wafer is heated to about 400°C, so that the Al forms a thin, highly doped p^+ region. Al has a relatively low-work function (Section 3.6, Chapter 3) and electrons can tunnel through the thin barrier from the p^+ semiconductor to Al under forward bias and in the opposite direction under reverse bias (Figure 3.2.2, Chapter 3). To make a metallic contact with n-type Si, a heavily doped n^+ region is formed, and the heated aluminum then makes a p^+n^+ junction through which electrons can tunnel (Section 3.5, Chapter 3).

The common method of depositing metal is **sputtering** in one of its several variants. In this process, a disk of the metal to be deposited is placed in close proximity to the wafer in a chamber filled with argon gas at low pressure. The metal disk, or target, is made the cathode and a voltage is applied that is sufficiently high to ionize the gas. The gas ions strike the target and vaporize metal atoms, which coat the whole surface of the wafer. The unwanted metal is then selectively etched, leaving the desired interconnection pattern.

A second form of metallization is the interconnections at the same level between various components (such as different transistors) or between regions of the same component (such as gate–drain or base–collector shorts in diode-connected transistors). Aluminum, alloyed with a small percentage of Si, has also been very popular for metallization. However, continued miniaturization has increased the importance of the electrical properties of interconnections (Section 12.8, Chapter 12). The reduction of the width and thickness of interconnections increases the resistance, so that Cu, whose resistivity is about 1.7 μΩ-cm compared to 2.7 μΩ-cm for Al, is being increasingly used for smaller-size interconnections. An added advantage is that small-size Al interconnections are susceptible to electromigration (Section 5.3, Chapter 5), whereas Cu is less susceptible to this phenomenon. However, copper interconnections are more difficult to work with. Doped polysilicon is also used in some situations for making metallic connections.

A third form of metallization is the connections between metallic conductors in adjacent levels of the wafer. The connection is made by a small plug or stud, also known as a **via**. Copper plugs are used with copper conductors, but tungsten plugs are used with aluminum conductors because of electromigration problems with Al.

A fourth form of metallization is concerned with the external, off-chip connections of the die. Square metal pads of several tens of μm side, known as **bonding pads**, are provided on the chip for this purpose. The metal of the pads is almost invariably Al. Connections to these pads are described in Section 4.4.

Clean Room Environment

All fabrication processes are carried out in ultraclean environments in which the temperature and humidity are precisely controlled. Of particular importance is dust control through stringent filtering of air in IC fabrication facilities. Dust particles on the surface of the wafer or mask may behave as opaque objects during lithography, thereby interfering with the circuit pattern transferred to the wafer. Dust particles on the surface of the wafer can interfere with crystal growth, resulting in unoccupied crystal sites, or they may lodge in the gate oxide, causing partial or complete shorting. A metallic conductor of submicron size can be damaged by a dust particle.

Dust count is specified differently in the English and metric systems. A Class N clean room in the English system has a maximum of N particles of 0.5 μm diameter or larger per cubic foot of air. In the metric system, a Class M X clean room has a maximum of

10^X particles of 0.5 μm diameter or larger per cubic meter of air. Thus, a Class 100 clean room would have a maximum count of 100 particles of 0.5 μm diameter or larger per cubic foot of air, which is equivalent to a count of nearly 3531 per cubic meter of air. This gives $X = \log_{10}(3531) = 3.55$, which corresponds to a Class M 3.55 in the metric system. An IC fabrication facility should be Class 100 or cleaner, which is typically at least four orders of magnitude lower than that of ordinary room air and at least an order of magnitude lower than that of a hospital operating room.

4.4 Packaging

When all fabrication steps have been performed, the individual dice are tested electrically using an automatic probing station to make sure that they comply with the required performance specifications. Bad circuits are marked, and the wafer is cut into separate dice by a diamond-coated blade, the operation being referred to as **dicing**. The good dice are attached to a plastic or ceramic substrate or to a metal lead frame to which the external pins of the package are connected. A fine gold wire, known as a **bond wire**, is bonded to the Al bonding pad by a combination of heat, pressure, and ultrasonic vibration, which is used to break the oxide layer that inevitably forms on the Al surface. The other end of the gold wire is similarly bonded to the metal lead frame of the package. In an alternative arrangement, known as the **flip-chip process**, the die is provided with aluminum pads on which blobs of solder are deposited, resulting in **solder bumps**. The die is flipped over, face down, and the solder bumps are mated with the metal pads on the package. Heating to about 250°C causes the solder to flow, making a very good connection between the chip and the package. The flip-chip process eliminates the resistance and inductance of the bond wires and allows the connection pads to cover the entire chip area, rather than just the periphery, as is the case with bond wires, which results in a high pin density and better heat removal. However, thermal fatigue of the solder bumps may cause failure of the connections with time, unless special measures are taken. Moreover, the use of common, lead-based solder is prohibited in some countries because of the environmental hazards of lead.

Package Types

A traditionally popular type of package is the **through-hole technology** (THT) package having metal leads or pins that are inserted into holes in the circuit board and soldered in place. THT packages may be **dual-in-line** packages (DIPs) having metal leads on two long sides of the rectangular package, or **quad-in-line** packages (QIPs) having metal leads along the four sides of the rectangular package, or **pin-grid array** (PGA) packages having an array of pins on the bottom of the package. The trend toward higher densities of ICs having a larger number of pins, faster data rates, and greater power dissipation has curbed the popularity of THT packages.

An increasingly popular type of package is the **surface-mount-technology** (SMT) package having metal leads with flat ends parallel to the bottom surface of the package and which are soldered to the surface of the circuit board. Compared to THT packages, SMT packages are smaller, lighter, and more resistant to mechanical shock. Several versions of SMT are in use. Some have leads on two opposite sides, some on all four sides, and some have the flat end bent inward to fit under the package. Another version is the

ball grid array (BGA) package having a grid of solder bumps on the bottom of the package for attachment to the circuit board.

Both THT and SMT packages may house several chips stacked on top of one another in what are known as **multichip modules** (MCMs). These are becoming increasingly popular as they result in a denser packing of chips per unit surface area of the circuit board.

Packages not only protect the chip mechanically but should also conduct heat so as to limit the temperature rise of the chip due to power dissipation in normal operation. The least expensive packages are molded over the die and lead frame using thermosetting plastics. These packages do not completely protect the chip from contamination by water or other environmental contaminants during operation despite encapsulating the circuit with glass, polyamide, or silicon nitride. Full protection is provided by the more expensive **hermetically sealed** packages commonly used in aerospace and military applications. These packages have metal, ceramic, or metal–ceramic enclosures with glass seals. The packaging material and glass seal should have matched thermal expansion coefficients so as to reduce thermal stresses on the seal.

After packaging, the chip is electrically tested to ensure that it meets the specifications. Electrical testing is sometimes done at an elevated temperature, a process referred to as **burn-in** or **accelerated testing**, which identifies chips that may be subject to early failure.

4.5 Fabrication of Simple Devices

This section illustrates how some of the basic processes discussed in Sections 4.2 and 4.3 can be used to fabricate some fairly simple devices, mainly semiconductor resistors, *pn* junction diodes, and semiconductor capacitors.

Resistors

In principle, a simple type of resistor can be made from the *n*-well of Figure 4.2.4. After the oxide has been etched, an *n*-well is produced by ion implantation followed by diffusion, or by diffusion through vapor deposition. The oxide is reformed on the surface of the chip, and then etched again at the two ends of the well for the contact. An n^+ region is formed, usually by ion implantation, and Al is deposited to form the contacts, as illustrated diagrammatically in cross section in Figure 4.5.1. Resistance values obtained in this manner are in the range of few kΩs. A lower value of resistance can be obtained by increased doping of the *n*-well. For higher values of resistance, a zigzag pattern can be used. A resistor formed in this manner has the following disadvantages: (1) its tolerance can exceed 20%, although different resistors can be matched to within a few percent, and (2) the

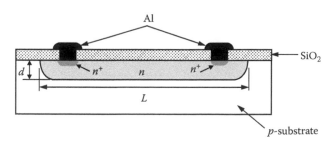

FIGURE 4.5.1
A simple IC resistor.

resistor is usually isolated from the rest of the chip by reverse biasing the *pn* junction formed between the *n*-well and the *p*-substrate. This, however, introduces a parasitic capacitance and makes the resistance voltage-dependent because of the variation of the width of the depletion layer with voltage. A better resistor can be formed by a polysilicon layer that is isolated from the rest of the chip by SiO_2.

The resistance of structures on chips, including interconnections, is usually specified in terms of the **sheet resistance**, denoted as $\Omega/\square$ (ohms/square) and having the dimensions of ohms. Recall that the resistance of a block of length *L*, width *W*, and thickness *d* is:

$$R = \frac{\rho}{d} \times \frac{L}{W} \tag{4.5.1}$$

where ρ is the resistivity in ohm-(unit length). The sheet resistance, R_{sq}, is defined as:

$$R_{sq} = \frac{\rho}{d} \tag{4.5.2}$$

and takes into account the variation of concentration of the dopant with the depth *d* as well as the variation of carrier mobility with dopant concentration. The total resistance is given by:

$$R = R_{sq} \times \frac{L}{W} \tag{4.5.3}$$

pn Junction Diodes

A *pn* junction diode can be formed in an *n*-well process by forming p^+ and n^+ regions, as illustrated in Figure 4.5.2a. The p^+ region is the anode, the *pn* junction being formed between this region and the *n*-well. As mentioned in connection with diode-connected transistors, IC diodes of better performance are formed by shorting together the base and collector metallization of a BJT (Section 6.3, Chapter 6).

Discrete, planar diodes are often made starting with an n^+ substrate, on which an epitaxial *n*-layer is grown, which allows accurate control of the doping of the *n*-region of

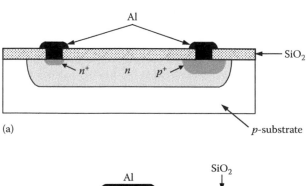

FIGURE 4.5.2

(a) An IC *pn* junction diode, and (b) a discrete, planar *pn* junction diode.

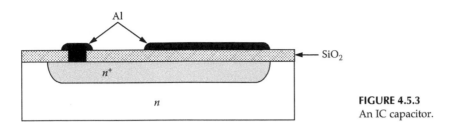

FIGURE 4.5.3
An IC capacitor.

the diode. A *p*-region is then implanted and contacts are made to both sides of the chip (Figure 4.5.2b).

Capacitors

Capacitors can be formed by reverse-biased *pn* junctions or by MOS structures in which the metal is one plate of the capacitor, the semiconductor is the other plate, and SiO_2 is the dielectric. In both of these types, the capacitance is voltage dependent. The voltage dependency of an MOS capacitor (Section ST4.3) can be reduced by implanting on the surface of the substrate an n^+ layer that acts as the other plate of the capacitor, as illustrated in Figure 4.5.3.

A better capacitor can be formed by two polysilicon layers separated by a thin layer of SiO_2, the polysilicon layers being themselves isolated from the rest of the chip by SiO_2. The tolerance of IC capacitors can be about 1%, but capacitors can be matched to within 0.1%.

4.6 CMOS Fabrication

The objective of this section is to describe the basic processes of CMOS fabrication, the details of which may vary depending on the intended application for the chip. This section should be read in conjunction with Section 5.2, Chapter 5, describing MOSFET operation.

CMOS ICs can be fabricated starting with a *p*-substrate, on which *n*-channel MOSFETs are formed, with *p*-channel MOSFETs formed in **n-wells** in the *p*-substrate. Alternatively, the start can be an *n*-substrate on which *p*-channel MOSFETs are formed, with *n*-channel MOSFETS in **p-wells**. Better performance is achieved in a **twin-well** construction in which each type of transistor has its own well, usually in an *n*-substrate on which a low-resistivity epitaxial *n*-layer is grown. The *n*-wells and *p*-wells are then formed in this layer, each type of well having the required dopant concentration. Individual NMOS or PMOS transistors can, of course, be fabricated in the same manner as the corresponding CMOS transistor. In all cases, the substrate is preferably of <100> crystal orientation, as this orientation has the lowest interface trap density.

Transistor Isolation

The isolation of adjacent transistors on a chip so as to eliminate unwanted coupling between them through the substrate has an important bearing on the packing density. In normal operation, the channel, source, and drain regions of a MOSFET are surrounded by a depletion region, which effectively isolates the MOSFET from the rest of the IC. To be effective, the depletion region should be relatively wide, which severely limits the packing density. Moreover, if the drain of one transistor, say D_A, is adjacent to the source of another transistor, say S_B, on the same IC, with metallic interconnections running over the surface oxide between D_A and S_B, then if the voltages between these three regions are of appropriate magnitude and polarity, a parasitic MOSFET is formed having the metallic interconnections as gate and a conducting channel between D_A and S_B. To eliminate this possibility, the threshold of the parasitic transistor is made sufficiently high by: (1) creating a **channel stop**, or chanstop, between transistors by increasing the surface doping level in the substrate between transistors using ion implantation, which increases the electric field required to induce a channel in this region; and (2) increasing the oxide thickness over this parasitic channel, which increases the voltage required to produce the larger electric field. The relatively thick oxide between the active areas of isolated transistors is referred to as the **field oxide** (FOX) and is usually grown by a process of **local oxidation** (LOCOS).

The first step in the LOCOS process is to grow a thin layer of SiO_2 followed by a thicker layer of Si_3N_4. The thin oxide layer, called an oxide pad, reduces the stress in the Si substrate that occurs during oxidation because of the difference in thermal expansion coefficients between Si and Si_3N_4, and because of the increasing volume of the growing oxide. A mask is used to pattern the nitride, and both the photoresist and nitride are removed from the areas where the oxide is to be made thick. The doping level of the substrate in these regions is increased by ion implantation so as to produce the channel stops. The photoresist is stripped from the nitride layer and the wafer is oxidized. The nitride acts as a barrier to oxidation, so that the oxide grows thicker only in areas not covered by the nitride. The areas covered by the nitride define the active regions of the transistors. The nitride is subsequently removed from these regions to create the wells. Although extensively used in technologies having minimum feature sizes of 0.5 μm or larger, LOCOS alone is not appropriate for smaller feature sizes because of encroachment of the oxide into active regions, which limits the packing density.

In technologies of smaller feature sizes, trench isolation is used, which can be shallow trench isolation (STI) or deep trench isolation. STI is illustrated in Figure 4.6.1a, which shows the pad SiO_2 and Si_3N_4 layers previously mentioned, with a shallow trench less than 1 μm in depth. The trench walls are implanted with dopants so as to form chanstops that will also act as sidewalls for the *p*-well on one side and the *n*-well on the other side. The trench is filled with CVD-grown oxide and an optically flat surface is produced by CMP, the Si_3N_4 serving as a polish stop. The thickness of the oxide in the trench and the chanstops effectively isolates the transistors and prevents the formation of a parasitic MOSFET. However, STI does not eliminate latchup in CMOS (Section ST6.6, Chapter 6). A deep trench less than 1 μm wide and about 5 μm deep eliminates latchup (Figure 4.6.1b) but is considerably more difficult to fabricate. As illustrated in Figure 4.6.1b, the trench is deeper than the ensuing wells.

The process of making a trench, overfilling it with material, and then grinding the surface flat is known as a **Damascene process**, after the process used in medieval times by craftsmen in Damascus to inlay gold or silver in sword handles and sheaths.

Transistor Formation

To form the wells, the Si_3N_4 layer is removed, leaving the pad oxide. The *p*-well region is patterned by photolithography and etching, followed by several boron ion implantations to achieve the desired retrograde profile. The *n*-well is then similarly formed using phosphorus ion implantation. The wells are usually about 2 μm deep. Special implants are made to adjust the thresholds of the two transistors. The next major step is to form the

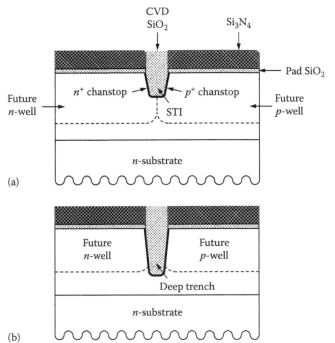

FIGURE 4.6.1
(a) Chanstops and a shallow trench; (b) chanstops and a deep trench.

gate stack comprising the gate oxide and the polysilicon gate electrode. A high-quality thin oxide of thickness appropriate to the technology used is thermally grown to serve as the gate oxide. In modern transistors, the gate oxide is nitrided, by oxidizing in the presence of NO or by using N_2O, to increase the breakdown voltage and the dielectric constant. Oxidation is followed by a blanket LPCVD deposition of polysilicon over the whole surface. The main reason poly is used as the gate electrode rather than Al is that the poly can withstand the high temperatures required to activate the source and drain implants. The gate and its interconnections are patterned, and undesired poly and oxide are removed. A series of steps then follow resulting in the structure diagrammatically illustrated in Figure 4.6.2. The following features should be noted: (1) The source and drain have more lightly doped regions, the **source/drain extensions**, also referred to as

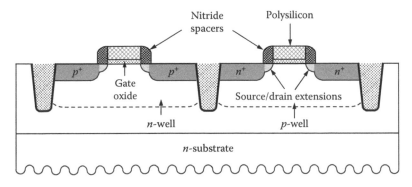

FIGURE 4.6.2
Diagrammatic structure of the CMOS transistors.

lightly doped drain (LDD), which extend into the channel and provide a resistive transition between the channel and the highly doped source/drain regions. This spreads the drain–channel voltage drop over a longer distance and reduces the electric field, thereby attenuating some short-channel effects, particularly hot-carrier injection into the oxide. (2) Before the source and drain are implanted, nitride spacers are formed on the gate. The polysilicon and spacers act as a barrier to the implanted ions, thereby leading to **self-alignment** of the source and drain, in the sense that the channel length is defined by the gate regions that had already been formed. In the older technology, where Al was used for the gate, the gate was formed after the source and drain, using a separate mask. Registration errors in the alignment of the different masks resulted in a larger overlap between the gate on one side and the source and the drain on the other, thereby creating relatively large parasitic capacitances between the gate and the source/drain regions.

The next series of steps are concerned with metallization of the source and drain. Al was originally used for contacts but became unsuitable as dimensions kept shrinking and the source and drain became much shallower. The problem with Al is that Si is partially soluble in Al at the annealing temperatures used. The migration of Si into Al allows Al to encroach into Si in the form of spikes that dangerously reduce the thickness of already shallow source/drain regions. Al encroachment can be reduced by adding Si to Al, which reduces the amount of wafer Si that dissolves in the Al contact. Another measure is to introduce a barrier metallic compound, such as TiN, which has low contact resistance with Si and does not react with Al.

Because of the problem of Al spiking, a **silicide** is used nowadays as a contact metal. Silicides are compounds of Si and a refractory metal, such as $TiSi_2$, $TaSi_2$, $MoSi_2$, WSi_2, $PtSi_2$, or $CoSi_2$. Silicides have the energy band structure of a metal and a reasonably low **specific contact resistance** in the range of 10–25 $\mu\Omega$-cm^2 compared to 2.7 $\mu\Omega$-cm^2 for aluminum, where contact resistance is the specific contact resistance divided by contact area. The product of the reaction with polysilicon is a **polycide**, and a self-aligned silicide is referred to a **salicide**. To form the salicide, Ti or Co is deposited over the entire surface of the wafer and the wafer is heated to a relatively low temperature so that the metal reacts with Si in the source and drain areas to form a silicide, and with the polysilicon to form a polycide. The metal does not react with the nitride spacers or with the oxide over the trenches so that the silicide is self-aligned in the required areas without the need for a masking operation. The unreacted metal is etched away.

The next step is to grow a SiO_2 layer that will act as an electrical insulator between the silicides/polycides formed so far on the surface of the wafer and the first layer of metal interconnections. A layer of CVD-grown oxide is deposited over the whole wafer, a photolithography operation is performed that defines the contact windows, and the oxide is etched away at these locations. The contact openings are filled with tungsten and the surface is planarized, as illustrated in the cross section of Figure 4.6.3. The tungsten studs to the gate poly are not shown because they are not, in general, in the same plane as those of the sources and drains. Tungsten is used to avoid electromigration problems with Al studs and because of its good adhesion to silicide/polycide and to Al in the interconnect. Before tungsten is applied, a layer of TiN is sometimes added to promote adhesion of tungsten to SiO_2 and to act as a barrier between tungsten and the source/drain.

The next step is to add the first metal layer, referred to as metal1, or M1. When Al is used, it is applied, often with some Cu or Ti added to reduce electromigration. The connections in this layer are patterned as required. Another oxide layer is grown and patterned for the tungsten vias between metal1 layer and metal2, as mentioned previously under "Metallization."

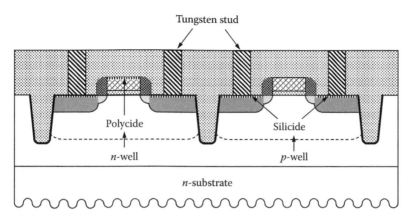

FIGURE 4.6.3
The transistors of Figure 4.6.2 after addition of contacts to the next layer.

The preceding operations are repeated for subsequent metal layers. The upper layers of metals are primarily used for power connections, clock, and global signals and are normally thicker than the lower layers because they have to carry heavier currents. The lower metal layers are primarily for connections between transistors or transistor groups. Because vias have an appreciable resistance, they are kept as short as possible, and several vias may be paralleled to reduce resistance. Where Cu is used instead of Al, both Cu vias and interconnects are formed by a dual-Damascene process. The advantage of this process is that it does not require metal etching. This is particularly important for Cu because of the lack of a good etch process for Cu.

After all the metal layers have been formed, a final passivation layer is added, usually in the form of doped glass or Si_3N_4. The purpose of this layer is to protect the chip from contaminants, such as water or ionic salts, as well as from abrasion during probe testing and packaging. Finally, openings are made for bond pads by photolithography and etching of the passivation layer.

It should be mentioned that the gate insulation is so thin and has such a high resistance (normally greater than about 10^9 Ω) that it may be readily damaged by static electricity created by handling the chip or transistor. The layout of the I/O pad is usually such that the gates of transistors connected to I/O pads, and which are therefore liable to be damaged by electrostatic discharge, are protected by diodes, as illustrated in Figure 4.6.4. These diodes are implemented by providing appropriate n^+ and p^+ implants in the substrate and transistor wells. As long as the gate voltage is not less than zero or more than V_{DD}, both diodes are nonconducting and do not affect normal operation. Should the gate voltage fall below 0, or rise above V_{DD}, by more than the diode cut-in voltage of about 0.5 V, one of the diodes will conduct and clamp the gate voltage to a safe value. A similar arrangement is used in discrete devices.

It should be noted that in the preceding description separate NMOS and PMOS transistors were assumed, as in Figures 4.6.2 and 4.6.3. In some circuits, a terminal of one transistor is connected to a terminal of another transistor.

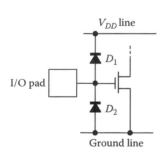

FIGURE 4.6.4
Diode protection of transistor gates.

In the CMOS inverter (Section 5.5, Chapter 5), for example, the drains of the NMOS and PMOS transistors are connected together. The two drains are therefore *merged* together in one structure during fabrication. This merging of structures in a given layer is implemented whenever possible so as to minimize the inactive area on the chip, which can also improve performance by eliminating the parasitic resistance and capacitance.

Silicon-on-Insulator Technology

In silicon-on-insulator (SOI) technology, each transistor is completely encased by SiO_2. Transistor fabrication begins with a Si wafer having at its surface an oxide layer underneath a thin Si layer. Such a structure can be produced in several ways. In the **im**plantation of **o**xygen (SIMOX) process, oxygen ions are implanted, producing a nearly amorphous layer having about twice as many oxygen as Si atoms. This is followed by a high temperature anneal that repairs the crystal damage caused by the implantation and forms a buried oxide (BOX) layer that is 50–200 nm thick, with a Si layer above it of approximately the same thickness.

Alternative processes rely on wafer bonding. In the **e**pitaxial **l**ayer **tran**sfer (ELTRAN) process, a wafer, referred to as the device wafer, is anodized to create two distinct, porous Si layers of different porosities. A high-quality epitaxial layer of Si is then grown above the porous layers followed by thermal oxidation to create a SiO_2 layer (Figure 4.6.5a). Another wafer, referred to as the handle wafer, is bonded to the device wafer by pressing them together at a high temperature. The bonded wafers are then separated between the two

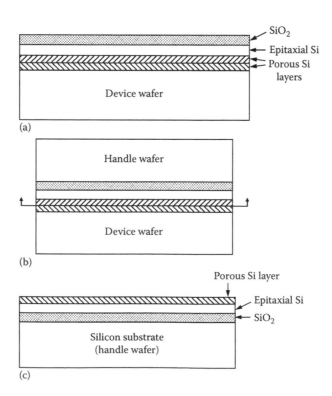

FIGURE 4.6.5

(a) Preparation of a device wafer in SOI technology; (b) a handle wafer bonded to the device wafer; (c) preparation of silicon substrate.

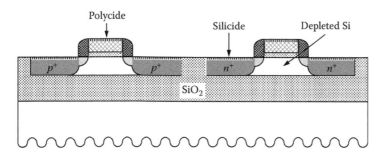

FIGURE 4.6.6
Diagrammatic structure of SOI CMOS transistors.

porous layers, as illustrated in Figure 4.6.5b. The handle wafer, which is used as the Si substrate for forming transistors, has a porous Si layer on top of an epitaxial Si layer, with a buried SiO_2 layer (Figure 4.6.5c). The porous silicon layer is removed, the source/drain are formed in the epitaxial layer, and the transistor is fabricated as previously described, but with a buried SiO_2 layer beneath the transistor.

The resulting CMOS structure is very similar to that previously considered (Figure 4.6.2), except that each transistor is encased in SiO_2 insulation (Figure 4.6.6). When the thickness of the Si layer in which the transistors are formed is small, that is, about 50 nm, the Si layer under the channel is fully depleted at zero bias by the built-in junction potentials between the source/drain and the Si under the channel. The transistor is said to be of the **fully depleted** (FD) type. If the silicon layer is thicker, that is, more than 100 nm, the Si under the channel is partially depleted, and the transistor is said to be of the **partially depleted** (PD) type. Although PD transistors are easier to manufacture, the trend is to adopt FD transistors for the more advanced technologies.

4.7 Fabrication of Bipolar Junction Transistors

Operation of the BJT is described in Section 6.1, Chapter 6.

Basic BJT Structure

The basic structure of an *npn* BJT is diagrammatically illustrated in Figure 4.7.1. Starting with a lightly doped *p*-substrate, the first step is to form a buried n^+ layer that reduces the resistance in the current path to the collector. The steps involved are the thermal growing of a thick oxide layer, of 0.5–1 μm depth, then opening a window in the oxide, implanting As ions, followed by annealing to repair the damage caused by ion implantation and to activate the implanted impurities. The oxide that is formed during annealing is removed and an epitaxial *n*-layer is grown over the n^+ layer. Epitaxy is here indispensable for forming a layer of reduced doping concentration over the n^+ layer, and which will form the collector region. The next step is to grow a thick LOCOS layer that reaches all the way to the n^+ buried layer for lateral isolation between adjacent transistors. The transistor is isolated from the *p*-substrate by the n^+p junction between the n^+ buried layer and the substrate. Boron-doped p^+ channel stops are formed under the isolation oxide to prevent the formation of an E-NMOS transistor consisting of the *p*-substrate, the oxide layer, and overlying metallic connections and n^+ regions on either side.

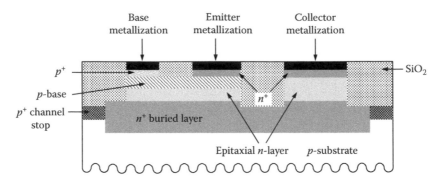

FIGURE 4.7.1
Diagrammatic structure of *npn* BJT.

The *n*-collector region is now almost formed, except for the contact region. The next major step is to implant boron ions to form the *p*-base. This is followed by As ion implant to form the n^+ emitter and the n^+ collector contact region. The p^+ base contact region is then formed by boron ion implantation, followed by the metallization of the contact.

It should be noted that the depth of the collector region under the base is that of the *n*-epitaxial layer less the base diffusion and the updiffusion from the n^+ buried layer. Arsenic is used in the n^+ buried layer to minimize this updiffusion. For analog applications, which require higher breakdown voltages, the depth of the *n*-epitaxial layer is relatively large and the doping level is relatively low, compared to digital applications, which require lower voltages for switching purposes. Not shown in Figure 4.7.1 is a sinker implant that is used to connect the n^+ buried layer to the n^+ collector contact so as to lower the collector resistance.

In Figure 4.7.1, the structure comprising the emitter, the base, and the collector region directly beneath the emitter is referred to as the *intrinsic device*. The region from the intrinsic device to the base contact is the *extrinsic base*, and the region from the intrinsic device to the collector terminal is the *extrinsic collector*. The extrinsic base adds to the base resistance, whereas the extrinsic collector adds to the collector–substrate capacitance. Both of these parasitic elements degrade the high-frequency response of the transistor.

Modified BJT Structures

One of the improvements to the basic BJT structure is to use SOI substrates. This not only reduces the collector junction capacitance but increases the breakdown voltage as well. However, because thermal conductivity of SiO_2 is much lower than that of Si, not as much power can be dissipated in SOI devices for the same junction temperature.

The parasitic base resistance and collector–substrate capacitance are reduced in the self-aligned, double-poly structure, illustrated diagrammatically in Figure 4.7.2. After the buried n^+ layer, the *n*-epitaxial layer, and the SiO_2 isolation are formed, a p^+ poly layer, heavily doped with boron, is deposited and then patterned to open a window for the active regions of the base and the emitter. A thermal oxide is grown over the etched structure to form the p^+ poly vertical sidewalls that determine the spacing between the base and emitter contacts. Because boron is highly soluble in Si, it outdiffuses from the heavily doped p^+ poly during the thermal oxidation and forms the extrinsic base regions. The intrinsic base region is then formed by implantation of boron ions. The intrinsic and extrinsic base regions are self-aligned in this step because of the sidewall spacers.

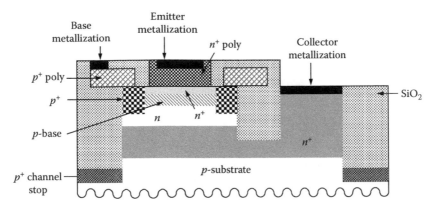

FIGURE 4.7.2
Improved, double-poly structure of *npn* BJT.

The second poly layer is then deposited and implanted with As or P. Outdiffusion from this n^+ poly forms the n^+ emitter region. Finally, a Pt film is deposited and heated to form PtSi polycide over some of the emitter and base poly regions. The resulting structure has reduced base resistance, collector–substrate capacitance, and base–collector capacitance.

A *pnp* transistor can be formed by the same self-alignment process discussed in connection with Figure 4.7.2 using complementary semiconductors. It is also possible to combine such a *pnp* transistor with the self-aligned *npn* transistor of Figure 4.7.2.

A **lateral-type** *pnp* transistor, as opposed to the vertical-type *npn* transistors considered so far, is formed by the structure of Figure 4.7.3 using the same manufacturing process as *npn* transistors, with the lightly doped *n*-type epitaxial layer constituting the base. As a result, the collector/base depletion layer extends mostly into the base, so the base must be made wider, which reduces current gain and unity gain frequency, and makes the recombination current a larger fraction of the base current (Example 6.1.1, Chapter 6). Lateral *pnp* transistors are also compatible with MOSFET fabrication. The emitter and collector regions could be the source and the drain of a PMOS transistor whose *n*-well, with an added contact, would act as the base.

BiCMOS

Most BiCMOS technologies aim at using the same CMOS processes as much as possible. The *n*-well CMOS construction is well suited for incorporating *npn* BJTs, as this basically involves the addition of a *p*-base diffusion region. Starting with a p^- substrate, an n^+

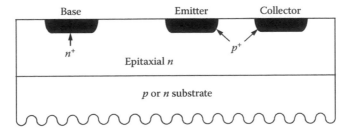

FIGURE 4.7.3
Lateral-type *pnp* BJT.

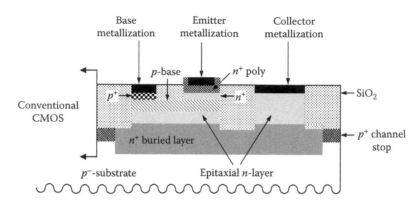

FIGURE 4.7.4
npn BJT in BiCMOS fabrication.

buried layer is implanted for reducing collector series resistance (Figure 4.7.4). Arsenic is usually used in this layer because its low diffusion rate minimizes outdiffusion from this layer. An *n*-type epitaxial layer is then deposited for further processing. *n*-wells are formed for both the PMOS and BJT transistors at a doping level of about $10^{17}/cm^3$, which is found to give good performance for both types of transistors. A series of high-dose *n*-type implants in parts of the *n*-well establishes a connection between the collector contact region and the buried layer. A boron ion implant forms the *p*-base. A polysilicon emitter is used to establish the emitter contact.

4.8 Miscellaneous Topics

SiGe Technology

Adding Ge to Si is an important technological development that can enhance the performance of both MOSFETs and BJTs. Ge is a group IV element, like Si, but has a larger atomic weight, and hence is a bigger atom. Its E_g of about 0.7 eV is smaller than that of Si, because valence electrons, being farther away from the nucleus, require less energy to move to the conduction band. When incorporated in a Si crystal, Ge replaces some Si atoms but does not dope the crystal because its outermost shell has four electrons, just like Si.

The effect of adding Ge at concentrations that are usually between 15% and 25% is twofold: (1) SiGe has a bandgap of approximately 0.8–0.9 eV, depending on the concentration of Ge, compared to 1.12 for Si. The difference in bandgap energies dramatically increases current gain in the heterojunction bipolar transistor (HBT) (Section 6.4, Chapter 6). (2) The Si crystal is strained, with resultant increase in carrier mobility. Using strained Si in a MOSFET channel increases the speed of the transistor (Section 5.3, Chapter 5).

The common method of straining Si basically consists of growing a thin epitaxial layer of Si on a SiGe crystal. The dimensions of the unit cell of the SiGe crystal are about 1% larger than those of the Si crystal. In maintaining continuity across the interface by satisfying the atomic bonds, the Si crystal becomes strained in the plane parallel to the interface, for example, the *xy* plane. As the Si crystal is strained in tension in the *xy* plane, it is compressed in the *z*-direction, as illustrated in Figure 4.8.1a. However, if the Si layer is too thick, the stress is relieved during subsequent annealing, and the Si crystal assumes its

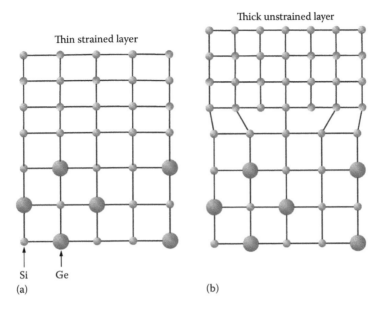

FIGURE 4.8.1
(a) A thin strained layer of Si on SiGe; (b) strain relieved in a thicker *Si* layer.

normal dimensions, leaving unsatisfied bonds, or misfit dislocations, at the interface (Figure 4.8.1b). The critical thickness of the Si layer decreases as the concentration of Ge in the underlying substrate is increased. For concentrations normally used, the critical thickness is about 100 nm.

Crystal Defects

As crystal defects can have a substantial impact on the performance of semiconductors, it is important to appreciate their nature and effects. It should be borne in mind that crystal boundaries are in themselves a major departure from an ideal crystal of infinite extent, because atoms at the surface of the crystal do not have four neighbors to form covalent bonds with (Section 2.3, Chapter 2), leaving dangling bonds that can trap current carriers (Section 2.7, Chapter 2).

Crystal defects are normally classified into four categories: point defects, line defects, area defects, and volume defects. Point defects are localized, as illustrated in Figure 4.8.2a. Examples are: (1) a vacant lattice site (A); (2) a self-interstitial in which a lattice atom occupies a space between lattice positions (B); (3) a **Frenkel defect** (C), in which a lattice atom is displaced into the interatomic space, creating both a vacant site and a self-interstitial; (4) a substitutional defect, in which an impurity atom occupies a lattice site (D), as in the case of dopants; and (5) an interstitial defect, in which an impurity atom occupies a space between lattice atoms (E). Defects A to C involving only the crystal atoms are **intrinsic defects**, whereas defects D and E are **extrinsic defects**. Vacancies and self-interstitials occur, with a small probability, in an otherwise perfect crystal due to thermal excitation at nonzero temperatures. They also propagate through the crystal, particularly at high temperatures. When a vacancy is created (A), not all the bonds may be broken, so that some electrons are left behind. Substitutional impurities (E) introduce energy levels in the

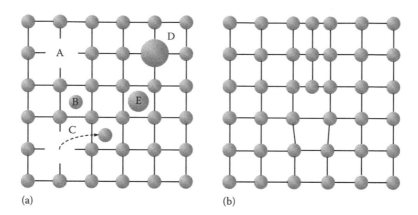

(a)   (b)

FIGURE 4.8.2
(a) Examples of point crystal defects and (b) crystal dislocation.

bandgap, acting as generation–recombination centers (Section 2.7, Chapter 2). Point defects are important in ion implantation and diffusion processes.

Line defects occur in one dimension, most commonly as a **dislocation**, in which an additional line of atoms is inserted between two other lines, as illustrated diagrammatically in Figure 4.8.2b. It is seen that some bonds are stressed in expansion and some in compression, which affects carrier mobilities and energy levels. Line defects can occur as a result of *agglomeration* of point defects, or as a result of stress that may be induced by uneven heating or mechanical pressure, or by ionic bombardment, or by a high concentration of substitutional impurities. Dislocations can also move through the crystal.

Area defects are discontinuities in two dimensions, such as a **stacking fault**, which is an extra plane of atoms rather than an extra line, as in a line fault. The atomic spacing is irregular in two dimensions but is regular in the third. Area defects can also occur as a change in the crystal orientation across a plane, as happens at grain boundaries between crystals. Volume defects have irregularity in three dimensions and usually occur as precipitates due to a decrease in the solubility of an impurity, because of temperature changes, for example.

Si–SiO$_2$ Interface

The structure of SiO$_2$ is basically that of four oxygen atoms at the corners of a tetrahedron, with a Si atom at the center (Figure 4.8.3). The Si valence shell of 8 is satisfied by covalent

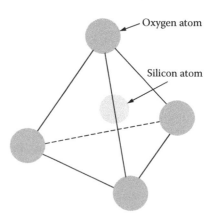

Oxygen atom

Silicon atom

FIGURE 4.8.3
Structure of SiO$_2$.

bonds with the four oxygen atoms at the corners of the tetrahedron. If each oxygen atom is part of two tetrahedra, then each oxygen atom is bonded to two Si atoms, the valence of oxygen will also be satisfied, and the oxygen atoms are referred to as **bridging sites**. Quartz has such a regular, lattice structure in which all oxygen atoms are bridging sites. In thermally grown SiO_2, some of the oxygen atoms are nonbridging, that is, they are not common to two tetrahedra and are bonded to one Si atom only. The larger the fraction of bridging sites, the denser and more cohesive is the oxide. Dry oxides have a much larger fraction of bridging sites than wet oxides.

The Si–SiO_2 interface is basically between a regular Si crystal and an essentially amorphous SiO_2. Hence, not all the bonds of the Si crystal atoms will be satisfied, which creates low-energy traps that can capture and release electrons and holes, resulting in flicker noise (Section 6.5, Chapter 6). The mean charge held in these traps depends on the voltage across the SiO_2 layer. The trap density of states can be reduced in two ways: (1) making the surface plane the <100> orientation (Figure 4.1.1), which has the lowest density of surface atoms of about 6.8×10^{14} atoms/cm^2; and (2) low-temperature hydrogen annealing at about 450°C. Hydrogen satisfies dangling bonds at the interface and in the body of the oxide.

Apart from the charge due to surface states, a fixed positive charge exists in a thin, transition layer of about 3 nm thickness at the Si–SiO_2 interface due to excess Si ions that have broken away from the Si crystal. This charge is also lowest for the <100> orientation and is effectively reduced by postoxidation, high-temperature annealing at about 1200°C in an inert atmosphere of nitrogen or argon.

In addition to the aforementioned charges at the interface, fixed and mobile charges can exist in the body of the oxide. The fixed charges can be positive or negative and are due to defects in the SiO_2 or due to hot carriers, resulting from a high electric field at the drain, and which get trapped in the oxide. The mobile charges are due to contamination during fabrication by Na^+ and K^+, which have a large mobility in SiO_2, particularly at temperatures exceeding about 100°C. These charges move in the oxide under the influence of an applied voltage.

Electric charges in the oxide or at the Si–SiO_2 interface affect the threshold of MOS transistors. If the charge is not affected by voltage and causes only a small shift in the threshold, then it is not much of a problem. Much more serious are the charges affected by the applied bias voltage, such as the charge due to mobile ions or due to hot carriers. These can have a detrimental effect on the long-term reliability of the transistor.

Summary of Main Concepts and Results

- Silicon of high purity is obtained by distillation of $SiHCl_3$, a liquid compound of silicon.
- A large single crystal, or ingot, is grown from a seed crystal having the desired crystal orientation. The ingot is then cut into wafers which are highly polished on one surface using a CMP.
- SiO_2 is thermally grown using dry or wet processes. SiO_2 grows at the Si–SiO_2 interface.
- The major fabrication steps depend on masks that transfer the required layout patterns to the wafer using a photolithographic process followed by etching. Etching could be wet or dry, using a plasma. RIE utilizes a plasma in conjunction with suitable chemicals.

- Ion implantation is extensively used for doping. It allows close control at a relatively low temperature but causes local damage to the crystal. The damage is repaired by a high-temperature anneal, which also activates the dopant ions by making them occupy lattice sites.

- Diffusion is used to introduce specific impurities to depths of several μm, following an initial deposition by ion implantation or by vapors of compounds of the required dopant. Subsequent heating allows the impurities to diffuse to the required depths.

- CVD is used to deposit many different types of materials. An important advantage of CVD is good step coverage. A form of CVD is silicon epitaxy, which allows the growth of a less heavily doped layer of crystalline Si over wafer Si, the epitaxial layer becoming part of the same crystal.

- Metallization includes the formation of low-resistance metallic contacts to terminal regions of devices, metallic interconnections at the same level of the chip and between different levels, as well as bonding pads for external connections. Polysilicon, aluminum, and silicides are used for low-resistance contacts, aluminum and copper are mainly used for interconnections, tungsten vias are used with aluminum interconnections, whereas copper vias are used with copper interconnections, and aluminum is used for bonding pads.

- IC packages of various forms are used to connect the chip to the circuit board, protect it mechanically, and conduct the heat dissipated in the chip. Hermetic sealing protects the chip from environmental contaminants such as water and ionic compounds.

- Simple devices such as resistors, *pn* junction diodes, and capacitors, albeit of not very exacting specifications, can be based on a basic well structure in a complementary semiconductor substrate.

- With the increasing density of transistors, reverse-biased *pn* junctions could no longer be used to isolate transistors and prevent the formation of the parasitic MOS transistors. STI, deep trench isolation, or SOI provide effective isolation.

- Fully depleted, SOI transistors are an important development for submicron CMOS technologies.

- Reduction of parasitic base resistance, collector resistance, base–collector capacitance, and collector–substrate capacitance are important considerations in advanced structures for the BJT.

- Bipolar transistors can be integrated into CMOS fabrication to produce BiCMOS ICs.

- SiGe technology can be used to enhance the performance of both BJTs and MOSFETs.

- Charge in the gate oxide and at the $Si–SiO_2$ interface affects the threshold of the MOSFETs and can have a detrimental long-term effect on the reliability of the transistor.

Learning Outcomes

- Articulate the basic processes involved in semiconductor fabrication and their essential features.

Supplementary Examples and Topics on Website

ST4.1 Doping Concentration in Czochralski Growth. Derives the expression for the doping concentration in the boule as a function of the solidification fraction in Czochralski growth.

ST4.2 Some Theoretical and Practical Aspects of Lithography. Elaborates some topics on lithography, including optical considerations, resolution-enhancement techniques, and nonoptical lithography.

ST4.3 MOS Capacitor. Derives the basic relations of the MOS capacitor.

Problems and Exercises

P4.1.1 A Czochralski-grown boule is to be doped with phosphorus at a concentration of $10^{16}/\text{cm}^3$, where the **equilibrium segregation coefficient** $k_0 = C_s/C_m$ is 0.35, C_s and C_m are, respectively, the concentrations of dopant in the solid crystal and melt. This means that as the melt is converted to a solid crystal, the dopant concentration in the melt, and hence in the crystal increases. It can be shown (Equation ST4.1.6) that $C_s = k_0 C_0 (1 - M_s/M_0)^{k_0 - 1}$, where M_s/M_0 is the fraction of the melt that has solidified. (a) What should be the mass of phosphorus added to a 50 kg melt of Si to give initially the required concentration in the solid crystal? Assume that the density of Si is 2.53 g/cm^3 and that the atomic weight of phosphorus is 31 g/mol. (b) What is the dopant concentration in the solid crystal by the time that half the melt has solidified?

P4.1.2 If an SiO_2 layer 10 nm thick is grown by thermal oxidation, what is the thickness of Si consumed? Assume that Si has a molecular weight of 28.9 g/mol and a density of 2.33 g/cm^3, while the corresponding figures for SiO_2 are 60.1 g/mol and 2.21 g/cm^3. (Hint: calculate the volume of 1 mol from the figures given.)

P4.1.3 Assume that the surface-trap and fixed charge at an Si–SiO_2 interface is one electronic charge per 10^4 Si atoms, the density of atoms in the $<100>$ orientation being 6.8×10^{14} atoms/cm^2. Determine the shift in the threshold if $t_{ox} = 10$ nm. (Hint: calculate the charge/unit area).

P4.1.4 CMP is to be used to remove a layer of material 1 μm thick that is on top of a "stop layer" that is 0.01 μm thick. If the rates of removal of the top and stop layers are 0.15 and 0.02 μm/min, respectively, how much polishing time should be allowed?

P4.1.5 A 300 mm diameter wafer is exposed for 1 min to a laminar airflow of 25 m/min in a class M 3 clean room. How many dust particles of 0.5 μm diameter or larger will be deposited on the wafer?

P4.1.6 **Fick's diffusion equation,** $\dfrac{\partial C(x,t)}{\partial t} = D \dfrac{\partial^2 C(x,t)}{\partial x^2}$, where $C(x, t)$ is the concentration of the dopant and D is its diffusion constant at the given temperature, applies for low dopant concentrations. The solution to this equation subject to the boundary conditions that a fixed amount S of the dopant is deposited at the semiconductor surface, and that $C(\infty, t) = 0$, is $C(x,t) = \dfrac{S}{\sqrt{\pi Dt}} e^{-x^2/4Dt}$. If As is deposited from the gaseous phase at a surface concentration of 2×10^{14} atoms/cm^2, determine the concentration of As at the surface and at a depth of 1 μm after 2 min of diffusion at 1200°C, assuming that the diffusion constant is 2.6×10^{-13} cm^2/s.

P4.1.7 The length of an n-well is 10 times its width. Calculate the resistance of this n-well assuming that its sheet resistance is 2 k$\Omega/\square \pm 20\%$.

P4.1.8 Titanium silicide has a resistivity of 15 $\mu\Omega$-cm. Determine the thickness that will give a sheet resistance of 0.5 $\Omega/\square$.

P4.1.9 An IC resistor 100 μm long and 8 μm wide has a sheet resistance of 1.2 k$\Omega/\square$. If the two end contacts are equivalent to 1.5 $\square$, determine the total resistance.

P4.1.10 Two parallel Cu wires of 0.5 μm $\times$ 0.5 μm cross section and 1 mm length are separated by a SiO$_2$ dielectric 0.5 μm wide. Determine: (a) the resistance of each wire, and (b) the capacitance between the wires neglecting fringe effects. Assume the resistivity of copper is 1.7 $\mu\Omega$-cm and the dielectric constant of SiO$_2$ is 3.9.

P4.1.11 A 0.5 mm long, 0.25 μm wide metal line is separated from the substrate below it by a 1 μm layer of SiO$_2$. Determine the capacitance between the metal and the substrate neglecting fringing effects and assuming the dielectric constant of SiO$_2$ is 3.9.

P4.1.12 The inductance of an IC spiral inductor of N turns and outer radius r is approximately given by $L \cong \mu_0 N^2 r$, where μ_0 is the permeability of free space. Determine r that is required to give $L = 15$ nH with 25 turns.

5

Field-Effect Transistors

Chapters 5 and 6 are concerned with the common types of transistors–the active devices used in electronic circuits mainly for signal amplification and switching. Transistors fall into two general classes: **field-effect transistors** (FETs), of which there are several types, and **bipolar junction transistors** (BJTs), with each type having its own advantages and disadvantages. As a switch, the transistor is either fully conducting or nonconducting, whereas as an amplifier the transistor is in an intermediate state of conduction that allows amplification of voltage, or current, or both. In all cases, it is important to have a good understanding of the basic operating principles of the various types of transistors and their voltage–current characteristics.

Analog electronics is principally concerned with the use of transistors as amplifying devices, although some switching operations may be included. *Digital electronics*, on the other hand, is mostly concerned with the use of transistors as switches in logic gates and in other digital elements, although some nonswitching applications may be included.

The general approach in Chapters 5 and 6 is to analyze the operation of an idealized device, based on physical principles, and derive its voltage–current characteristics and small-signal equivalent circuit. Effects that significantly affect device behavior are then considered, including secondary effects that are important under certain conditions. Mathematical treatment is kept to a minimum; derivations of transistor equations and other mathematical relations are left to Supplementary Examples and Topics on the website.

The reader may find the many assumptions made in transistor operation rather disconcerting. Nevertheless, as is true of engineering models in general, the assumptions are well justified under normal conditions. It is important, therefore, to appreciate the nature of the assumptions underlying different models, and not draw unwarranted conclusions based on incorrect assumptions. Particular attention is paid to explaining the limitations of derived models.

The operation of various types of FETs is considered in this chapter, beginning with a general discussion of the essential attribute of an amplifier and the various possible modes of operation of a generic, hypothetical amplifying device.

Learning Objectives

❖ To be familiar with:

- The symbols, terminology, and basic structure of FETs.
- Various secondary effects and some more advanced concepts that influence the behavior of FETs.

❖ To understand:

- The basic operation of an active element as an amplifying device, a voltage controlled resistance, a switch, and a transmission gate.
- The basic operation of enhancement-type and depletion-type, n-channel and p-channel, metal–oxide–semiconductor field-effect transistors (MOSFETs) in terms of the basic square-law model, and the resulting transfer and output characteristics.
- The voltage and current relations in a basic MOSFET amplifier circuit for small-signal operation, and the derivation of the small-signal equivalent circuit.
- The effect of channel-length modulation on MOSFET characteristics.
- The various capacitances that influence the behavior of the MOSFET at high frequencies.
- The different i–v characteristics of diode-connected MOSFETs.
- The basic operation of the CMOS amplifier, transmission gate, and inverter.
- The basic operation of the junction field-effect transistor (JFET) and the resulting transfer and output characteristics as well as the small-signal equivalent circuit.
- The basic operation of the metal–semiconductor field-effect transistor (MESFET).

5.1 Amplifiers

An electronic amplifier is essentially a three-terminal circuit having an input terminal, an output terminal, and a terminal that is common between input and output (Figure 5.1.1). An input signal is applied to the input terminal, and the output signal is applied to some kind of load, which for simplicity can be considered purely resistive, but which could, in general, be reactive or another amplifying stage.

Definition: *An amplifier amplifies power, so that the output signal power is considerably larger than the input signal power:*

$$v_O(-i_O) > v_I i_I \qquad (5.1.1)$$

where $v_I i_I$ is the instantaneous power delivered to the amplifier, and $v_O(-i_O)$ is the instantaneous power delivered to the load.

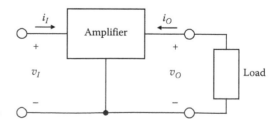

FIGURE 5.1.1
Electronic amplifier connected to a load.

The following should be noted concerning this inequality:

1. The inequality strictly applies to the average, or real, power under steady-state conditions. Under transient conditions, when internal capacitances are being charged, for example, the output power can be small.

2. In order to satisfy the inequality, either v_O must be sufficiently larger than v_I (corresponding to voltage amplification) or $-i_O$ must be sufficiently larger than i_I (corresponding to current amplification), or both. The latter condition prevails in most amplifier circuits.

3. Energy must be conserved in an amplifier circuit, despite inequality 5.1.1. This implies that some auxiliary supply, which usually consists of one or more **dc bias supplies**, must be associated with the amplifier so that energy is conserved in the circuit as a whole.

Hypothetical Amplifying Device

In order to appreciate some basic concepts concerning amplifiers, consider the hypothetical device illustrated in Figure 5.1.2 consisting of a capacitor-like structure having upper and lower plates as well as side plates. The upper plate is connected to the control terminal c, whereas the lower plate, connected to terminal d, is grounded. Terminal a is connected to a source v_x, and terminal b is connected to a load R_L.

The material between the two plates is assumed to have some very special properties. It does not conduct current between the plates connected to terminals c and d, so that i_C is always zero. When $v_C \geq 0$, the concentration of electrons is high, so that the material acts as an excellent conductor between the plates connected to a and b, the resistance r_{ab} being practically zero. But when $v_C < 0$, electrons are repelled and r_{ab} increases as v_C becomes more negative. At a sufficiently negative v_C, $r_{ab} \to \infty$.

The device of Figure 5.1.2 can be operated in one of four modes:

1. As a **voltage-controlled resistance**. This gives $v_O = v_x R_L / (r_{ab} + R_L)$, where r_{ab} can be varied between zero and infinity by varying v_C between zero and a sufficiently negative value.

2. As a **switch**. If v_x is a dc supply, the dc current through the load can be switched on and off by varying v_C between a positive value ($r_{ab} = 0$) and a sufficiently negative value ($r_{ab} \to \infty$).

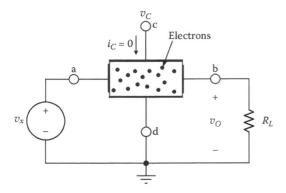

FIGURE 5.1.2
Hypothetical amplifying device.

3. As a **transmission gate**. If v_x is a time-varying signal, v_C can control or "gate" the transmission between the signal source v_x and load. If $v_C \geq 0$, $r_{ab} = 0$, and the signal v_x is applied to the load. But if v_C is sufficiently negative, $r_{ab} \to \infty$, and no signal is applied to the load.

4. As an **amplifier**. In this case, terminal a is grounded, so that the device becomes three-terminal. A small-signal ac source v_{src} in series with a dc bias V_{BB} is applied to input terminal c, which ensures that the total voltage v_I applied to the upper plate is always negative (Figure 5.1.3a). The load R_L is connected to output terminal b in series with a dc supply V_{CC}. When $v_{src} = 0$, $v_I = -V_{BB}$. A steady current I_L flows and a steady voltage $R_L I_L$ is developed across R_L (Figure 5.1.3b). When $v_{src} \neq 0$, $v_I = -V_{BB} + v_{SRC}$, and i_L varies in phase with v_{src}. That is, when v_I becomes less negative, i_L increases in magnitude, and conversely. When i_L increases, the voltage drop $v_L = R_L i_L$ increases and v_O decreases, since $v_O + v_L = V_{CC}$. The overall effect is that the signal v_{src} effectively *modulates* the current through the device, causing a substantial variation of the voltage across the load due to signal variations. If the ac input signal is $v_i = V_{im}\sin\omega t$, then $i_l = I_{lm}\sin\omega t$, $v_l = V_{lm}\sin\omega t$, and $v_o = -V_{om}\sin\omega t$, where $V_{om} = V_{lm}$. The signal power delivered to the load is $V_{lm}^2/2R_L$. The power expended by the signal source v_{src} is ideally zero, because $i_C = 0$. The power amplification is theoretically infinite. Since it essentially generates signal power, an amplifying device is an **active** device.

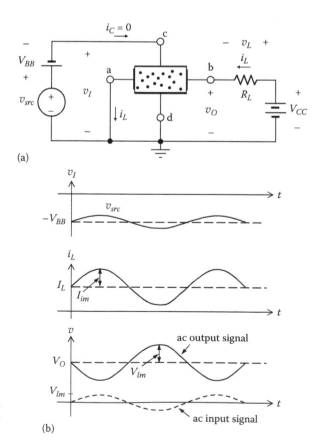

FIGURE 5.1.3
(a) Connection of the device of Figure 5.1.2 as an amplifier and (b) voltage and current waveforms.

Concept: *In an amplifier, the input signal essentially modulates, at a low power level, the output current, at a high power level.*

As will be demonstrated in due course, the output current is modulated in some way by the input signal in both FETs and BJTs.

Exercise 5.1.1

An ideal transformer of primary-to-secondary turns ratio a is a two-port device whose v–i relations at the input and output are given by $\dfrac{v_O}{v_I} = a$ and $\dfrac{i_O}{i_I} = -\dfrac{1}{a}$, following the voltage and current designations of Figure 5.1.1. If $a > 1$, voltage is amplified. If $a < 1$, current is amplified. Is the ideal transformer an amplifier?

Answer: No, because $v_O(-i_O) = v_I i_I$, that is, power is not amplified. ∎

5.2 Basic Operation of the MOSFET

MOSFETs are of two types: enhancement-type or depletion-type, each of which can be either *n*-channel (NMOS), or *p*-channel (PMOS). Where it is necessary to differentiate between the four types, we will refer to them as E-NMOS, D-NMOS, E-PMOS, and D-PMOS, where the E- and D-prefixes denote enhancement and depletion types, respectively.

Structure

Figure 5.2.1 diagrammatically illustrates the basic structure of an E-NMOS transistor. The source and drain are highly doped (n^+) regions that are formed in a *p*-type Si **body**. A gate electrode is separated from the semiconductor by a thin layer of silicon dioxide (SiO_2), which is an excellent insulator. For this reason, MOSFETs are sometimes referred to as **insulated-gate field-effect transistors** (IGFETs). Metallic contacts are made to the source, drain, and gate, and are brought out as terminals in discrete devices.

Henceforth, the term "body" will be used in this book to denote the semiconductor that borders the source and drain in a particular transistor. In contrast, the **substrate** is the

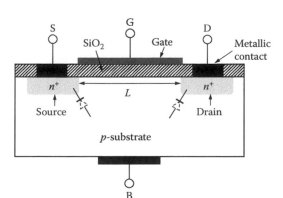

FIGURE 5.2.1
Diagrammatic illustration of the structure of an E-NMOS transistor.

base semiconductor that is common to all the transistors on the chip. The body is the substrate itself if the source and drain are formed directly in the substrate. However, the body could be a well that is formed in the substrate for a single transistor, or could be the semiconductor encased by insulation in silicon-on-insulator (SOI) technology (Section 4.6, Chapter 4). Note, however, that the B terminal (Figure 5.2.1) is connected to the substrate.

In early MOSFETs, the gate electrode was made of aluminum, and the structure at the gate (metal–oxide–semiconductor) gave the transistor its name. In later technologies, the gate metal was replaced by polysilicon, which is a heavily doped polycrystalline silicon composed of many crystals, in contrast to the single-crystal structure of the silicon chip from which the rest of the transistor is made (Section 4.6, Chapter 4). However, there has been a return once more to metal gates in the most recent technologies, as will be mentioned later in Section 5.3.

The separation between the source and drain constitutes a channel of length L, which can be in the range 1 μm to several μms in long-channel devices, or much less than 1 μm in short-channel devices. The width W of the channel in the direction normal to the plane of the paper typically ranges between approximately $1.5L$ to several hundred times L. The thickness of the oxide t_{ox} is typically about $L/50$. Thus, $t_{ox} \cong 20$ nm for $L = 1$ μm. The trend has been to decrease these dimensions as much as possible in order to pack a larger number of transistors per unit area of the Si chip, and to increase switching speeds, as discussed later under short-channel effects.

The structure of Figure 5.2.1 is symmetrical; the drain and source can be interchanged without affecting device characteristics. In normal operation, the *pn* junctions between the body and source or drain are *never* forward biased, as discussed later under body effect (Section 5.3).

Operation of Enhancement-Type MOSFETs

The dc gate current is normally extremely small, typically of the order of 10^{-15} A, and is neglected. With no gate bias applied ($v_{GS} = 0$), two *pn*-junction diodes are effectively connected back-to-back between drain and source, resulting in a very high resistance between these two regions (typically $> 10^{10}$ Ω). If $v_{DS} > 0$ is applied, the drain current i_D is practically zero.

If $v_{GS} > 0$ and is small, an electric field is directed from the positively charged gate electrode into the *p*-type body. Holes are repelled, so that the electric field lines end on negatively charged acceptor ions. As v_{GS} is increased further, the field becomes strong enough to attract electrons, primarily from the source and drain regions, so that the electric field lines end on both negatively charged acceptor ions and conduction electrons. An *n*-channel is thereby formed in a thin surface layer underneath the gate, between the source and drain (Figure 5.2.2a). This induced *n*-channel is an **inversion layer**, because it effectively inverts the *p*-semiconductor into an *n*-type-like semiconductor in the channel. If a small v_{DS} is applied, a current i_D flows that is essentially directly proportional to v_{DS} (Figure 5.2.2b). It is important to note that i_D does not flow until the *n*-channel is formed, which requires a minimum $v_{GS} = V_{tn}$, where V_{tn} is the **threshold voltage** of the E-NMOS transistor. As at any *pn* interface, a depletion layer is formed around the channel, source, and drain even with zero bias across this interface. For small v_{DS}, the channel and the depletion layer underneath it are essentially of uniform depth.

Increasing v_{GS} beyond V_{tn} enhances the *n*-channel and hence i_D for a given v_{DS}, which gives the enhancement-type transistor its name. As v_{DS} increases, with the body and source

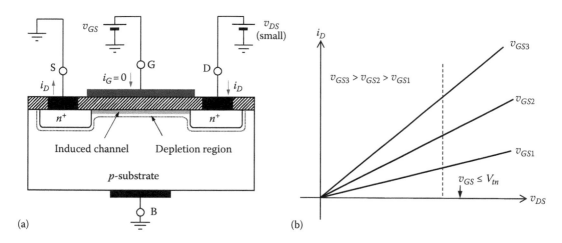

FIGURE 5.2.2
(a) E-NMOS transistor when drain–source voltage is small and (b) the resulting current–voltage characteristic.

voltages constant, the drain–body voltage increases, so that the depletion region below the drain becomes deeper than below the source (Figure 5.2.3a).

v_{DS} appears as a voltage drop along the channel, due to the flow of i_D. If we consider a plane xx′ transversely across the channel, the voltage in the channel at this plane is the same, because no current flows transversely in the channel. The voltage v_x at xx′ with respect to the source is the voltage drop in the channel from xx′ to the source, due to i_D in the channel. The voltage $v_{Gx} = v_{GS} - v_x$ is the voltage between the gate and body at xx′. v_{Gx} determines the concentration of electrons in the channel at xx′, and hence the depth of the channel at xx′. At the source, $v_x = 0$ and $v_{Gx} = v_{GS}$. At the drain, v_x reaches its maximum value of v_{DS}, so that $v_{Gx} = v_{GS} - v_{DS}$. For a given i_D, therefore, the channel is tapered, its depth decreasing from source to drain.

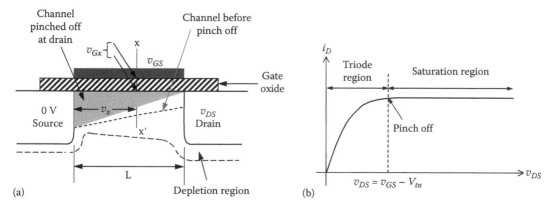

FIGURE 5.2.3
(a) The channel at increased drain–source voltage and (b) the corresponding variation of drain current.

As i_D increases, v_x increases at a given xx' and v_{Gx} decreases. The channel depth decreases and its resistance correspondingly increases. i_D no longer increases linearly with v_{DS}, as with small v_{DS}, but increases at a progressively slower rate (Figure 5.2.3b). When,

$$v_{GS} - v_{DS} = V_{tn} \tag{5.2.1}$$

the channel depth at the drain becomes almost zero, and the channel is **pinched off**. Further increase in v_{DS} theoretically has no effect on i_D, because of the ideally infinite incremental resistance at the pinch-off point, so that a change in v_{DS} causes no change in i_D. The region of the $i_D - v_{DS}$ characteristics in which i_D remains nominally constant is the **saturation region**. The region in which i_D increases with v_{DS}, at constant v_{GS}, is the **triode region**, because of its resemblance to a similar region in the characteristics of a vacuum-tube triode. The following should be noted:

1. When v_{DS} increases beyond pinch off in Equation 5.2.1, $v_{GS} - v_x = V_{tn}$ is satisfied at $v_x < v_{DS}$, so that pinch off takes place closer to the source, that is, the effective channel length is reduced (Figure 5.2.4). The region between the drain and the pinch-off point is the **pinch-off region**. As $v_{GS} - v_x < V_{tn}$ in this region, the region is depleted of carriers, so that the depletion region near the drain expands into the pinch-off region. Electrons flow in this depleted pinch-off region under the influence of the electric field, which is directed from the drain toward the channel.

2. When the channel is pinched off, i_D is not reduced to zero. This is because, where the channel is sufficiently deep, nearer the source, significant electron concentration supports current flow, but conservation of current requires that i_D be the same all along the channel. i_D in the channel is a drift current that is proportional to the product of electron concentration and velocity u, where $|u| = \mu_e|\xi|$; $|u|$ is directly proportional to the magnitude of the electric field $|\xi|$ as long as the mobility μ_e is constant. Hence, where the electron concentration decreases, the electric field and velocity increase to keep the same current flowing.

    A hydraulic analogy is helpful. Consider a pump that drives Q liters of water per minute through a pipe of uniform cross section and length L (Figure 5.2.5a). The pressure at the inlet of the pipe is P_a, and the pressure gradient P_a/L being uniform along the pipe. Suppose that the pipe is replaced by another pipe that is constricted over a length x. In order to have the same flow Q, the pressure at the inlet has to be increased to P_b, analogous to a higher drain–source voltage. In the constricted region, water must flow faster in order to keep the same flow through the smaller cross section. The pressure gradient $\Delta P/x$, analogous to the electric field, is

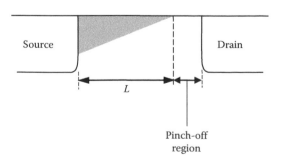

FIGURE 5.2.4
The pinch-off region at the drain and reduced channel length when the drain–source voltage is increased in the saturation region.

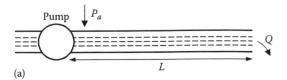

(a)

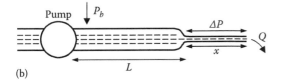

(b)

FIGURE 5.2.5
Hydraulic analogy of current flow in the channel, before pinch off (a) and after pinch off (b).

correspondingly larger than in the nonconstricted region. Similarly, in the pinched-off region of the channel, fewer electrons move faster under the influence of a larger electric field.

3. The reduction in the effective channel length L as v_{DS} increases beyond its value that causes pinch off at the drain (Figure 5.2.4) is **channel-length modulation**. The effect of this modulation on i_D can be accounted for as described in Section 5.3, but is neglected in the simplified analysis of transistor behavior, which means that as v_{DS} is increased in the saturation region, i_D is assumed to remain constant at its value when pinch first occurs (Figure 5.2.3b).

Current–Voltage Relations

The dependence of i_D on v_{DS} and v_{GS} in the triode region is derived in Section ST5.1. According to the approximation of the square-law model (Equation ST5.1.24),

$$i_D = \mu_e C_{ox}^* \frac{W}{L} \left[(v_{GS} - V_{tn})v_{DS} - \frac{1}{2}v_{DS}^2 \right] \tag{5.2.2}$$

where
μ_e is electron mobility
$C_{ox}^* = \varepsilon_{ox}/t_{ox}$ is the gate capacitance per unit area (the asterisk denotes capacitance per unit area), where ε_{ox} and t_{ox} being the permittivity and thickness of the oxide layer, respectively

$\mu_e C_{ox}^* = k_n'$ is the **process transconductance parameter** and is determined by the manufacturing process and the material properties of the transistor. $k_n = k_n'(W/L)$ is the **device transconductance parameter**, and W/L is the **aspect ratio**. W/L is used to obtain different values of i_D, for the same v_{GS} and v_{DS} in transistors having the same $\mu_e C_{ox}^*$, L, and V_{tn}.

For small v_{DS}, the squared term in Equation 5.2.2 can be neglected. i_D then varies linearly with v_{DS} for a given $(v_{GS} - V_{tn})$, and with $(v_{GS} - V_{tn})$ for a given v_{DS} (Figure 5.2.2b). The effective drain–source resistance r_{DS} can be expressed under these conditions as (see also Example SE5.1):

$$r_{DS} = \frac{v_{DS}}{i_D} = \frac{1}{k_n(v_{GS} - V_{tn})} \tag{5.2.3}$$

Exercise 5.2.1

Given an E-NMOS transistor having the following parameters: $t_{ox} = 0.1$ μm, $\varepsilon_{ox} = 3.9 \times 8.85 \times 10^{-14} = 3.45 \times 10^{-13}$ F/cm, $\mu_e = 800$ cm²/Vs, $W/L = 60$ μm/1 μm, and $V_{tn} = 2$ V. Determine: (a) k'_n; (b) k_n; (c) r_{DS} for $v_{GS} = 3$, 5, and 7 V.

Answers: (a) 27.6 μA/V²; (b) 331.2 μA/V²; (c) 3.02 kΩ, 1.01 kΩ, 604 Ω.   ∎

i_D saturates when $v_{DS(sat)} = v_{GS} - V_{tn}$ (Equation 5.2.1). Substituting this value of v_{DS} in Equation 5.2.2 gives for the saturation value of i_D:

$$i_{D(sat)} = \frac{1}{2}k_n(v_{GS} - V_{tn})^2, \quad v_{GS} \geq V_{tn} \tag{5.2.4}$$

Equation 5.2.4, plotted in Figure 5.2.6a, is the **transfer characteristic** of the transistor, because it relates i_D on the drain side to v_{GS} on the gate side.

The **output characteristics** are a family of curves that relate i_D to v_{DS} for various values of v_{GS} (Figure 5.2.6b). For a given $v_{GS} > V_{tn}$, i_D is related to v_{DS} in the triode region by Equation 5.2.2. The saturation value of i_D (Equation 5.2.4) is then extended into the saturation region, because of the assumption that once pinch off occurs, i_D becomes independent of v_{DS}, depending only on v_{GS}. The boundary between the triode and saturation regions, (dashed in Figure 5.2.6b), can be derived by substituting $v_{GS} - V_{tn} = v_{DS(sat)}$ in Equation 5.2.2 or 5.2.4 to obtain:

$$i_{D(sat)} = \frac{1}{2}k_n v^2_{DS(sat)} \tag{5.2.5}$$

The behavior of the MOSFET at low frequencies is completely described by the output characteristics, because i_D, v_{DS}, and v_{GS} are the only circuit variables involved in the operation of the transistor. The transfer characteristic is not an independent characteristic.

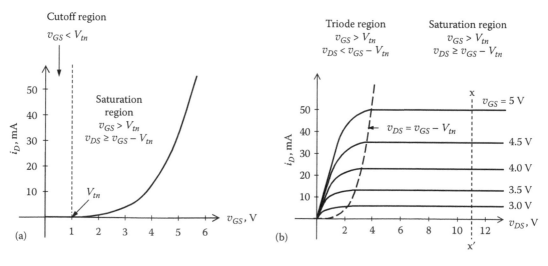

FIGURE 5.2.6
Transfer characteristic (a) and output characteristics (b) of an E-NMOS transistor.

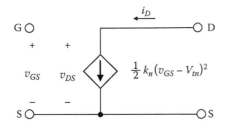

FIGURE 5.2.7
Large-signal equivalent circuit of a MOSFET in the saturation region.

It can be obtained from the output characteristics as the relation between i_D and v_{GS} at any value of v_{DS} in the saturation region, such as that corresponding to line xx' in Figure 5.2.6b.

For operation in the saturation region, and according to the assumptions made, the transistor can be replaced by the equivalent circuit of Figure 5.2.7. The gate terminal is open circuited, and i_D is given by a dependent current source in accordance with Equation 5.2.4. The MOSFET behaves as a voltage-controlled dependent current source having a quadratic voltage dependence. Note that this is a large-signal equivalent circuit, in contrast to the small-signal, linearized equivalent circuit discussed later (Figure 5.2.14).

Exercise 5.2.2

Determine $i_{D(\text{sat})}$ for the MOSFET of Exercise 5.2.1 for the three values of v_{GS} mentioned.
  Answers: 165.6 μA, 1.49 mA, 4.14 mA. ∎

The basic symbol for an E-NMOS transistor is shown in Figure 5.2.8a. The broken line on which the arrow terminates denotes the channel of an enhancement type transistor, the arrow being in the forward direction of the *pn* junction between the *p*-body and the

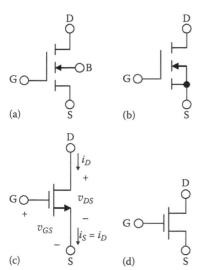

FIGURE 5.2.8
(a)–(d) illustrate various symbols for an E-NMOS transistor.

channel. The gate terminal is drawn closer to the source terminal. The symbol in Figure 5.2.8b is similar but assumes that the body is connected to the source. It can be simplified to that of Figure 5.2.8c, where the arrow is at the source and in the direction of actual current flow at this terminal. The symbol in Figure 5.2.8d is often used in digital circuits, without reference to the source or drain. The circuit is drawn so that the upper terminal is at a positive voltage with respect to the lower terminal, which means that the upper terminal is that of the drain.

p-Channel MOS Transistor

A *p*-channel enhancement-type MOSFET (E-PMOS) is of complementary structure, having an *n*-type body and p^+ source and drain regions. Making the gate sufficiently negative with respect to the body repels electrons and induces a *p*-channel between source and drain. If the drain is made negative with respect to the source, a hole current flows from source to drain. This operation is completely analogous to that of an E-NMOS transistor, but with the roles of electrons and holes interchanged. Equations 5.2.2 and 5.2.4 apply but with hole mobility μ_h replacing electron mobility, and the process transconductance parameter defined as $k'_p = \mu_h C^*_{ox}$. The symbol for an E-PMOS transistor is the same as that of an E-NMOS transistor shown in Figure 5.2.8a through c but with the direction of the arrows reversed. The symbol corresponding to that of Figure 5.2.8d is shown in Figure 5.2.9a, with a bubble at the gate to indicate that the gate voltage polarity is reversed with respect to that of an E-NMOS transistor. The circuit is drawn so that the upper terminal is at a positive voltage with respect to the lower terminal, and would, therefore, correspond to the source.

If the same positive directions of i_D, v_{DS}, and v_{GS} are assigned for an E-PMOS transistor, as for an E-NMOS transistor (Figure 5.2.9b), then all these variables will have negative numerical values. To use positive numerical values, the assigned positive direction of i_D is reversed, so that it flows into the source terminal in the direction of the arrow (Figure 5.2.9c), and new terminal voltages are defined as $v_{SG} = -v_{GS}$ and $v_{SD} = -v_{DS}$. If the threshold is considered as $\overline{V}_{tp} = -V_{tp} > 0$, Equations 5.2.2 and 5.2.4 will then apply in the form given, with v_{SD}, v_{SG}, and $\overline{V}_{tp}$ replacing v_{DS}, v_{GS}, and V_{tn}, respectively.

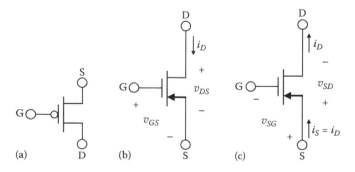

FIGURE 5.2.9
(a)–(c) illustrate symbols and polarities for an E-PMOS transistor.

In CMOS (Section 5.5), PMOS transistors are combined with NMOS transistors in the same circuit. Hole mobility is only about one-third of electron mobility in Si, and even less in GaAs (Table 2.3.1, Chapter 2), which means that, for the same magnitudes of terminal voltages, the current in a PMOS transistor is about one-third that in an NMOS transistor of the same W and L, and r_{DS} is approximately three times as much. For the same current and resistance values, the W/L ratio of the NMOS transistor is about one-third that of the PMOS transistor. For this W/L ratio and the same L, the NMOS transistor occupies nearly one-third of the area of an equivalent PMOS transistor and can operate at faster speeds because of its smaller dimensions.

Exercise 5.2.3

An E-PMOS transistor has the same dimensions as the E-NMOS transistor of Exercise 5.2.2 but with $\mu_h = 350\,\text{cm}^2/\text{Vs}$. Determine: (a) k'_p; (b) k_p; and (c) $i_{D(\text{sat})}$ for $v_{SG} = 3$, 5, and 7 V, assuming $\overline{V}_{tp} = 2$ V. Note the similarity of the voltage values to those of the E-NMOS transistor and compare the currents to those of Exercise 5.2.2.

Answers: (a) 12.08 μA/V^2; (b) 144.9 μA/V^2; and (c) 72.45 μA, 652 μA, and 1.81 mA. ∎

Small-Signal Operation

Consider an E-NMOS transistor in the circuit of Figure 5.2.10. Although, in practice, a MOSFET is biased differently to a desired operating point (Section 8.1, Chapter 8), the circuit is convenient for illustrating some basic concepts concerning small-signal operation.

Assume to begin with, that $v_{src} = 0$, $v_{GS} = V_{GG} > V_{tn}$, and v_{DS} is large enough so that a saturation dc drain current I_D flows. In the drain circuit, KVL under dc conditions gives:

$$V_{DS} + R_D I_D = V_{DD} \tag{5.2.6}$$

Equation 5.2.6 is the constraint placed on I_D and V_{DS} by the external drain circuit. In addition, I_D and V_{DS} are related by the output characteristics, which can be formally expressed as $i_D = f(v_{DS}, v_{GS})$. For dc conditions, with $v_{GS} = V_{GG}$, this relation becomes:

$$I_D = f(V_{DS}, V_{GG}) \tag{5.2.7}$$

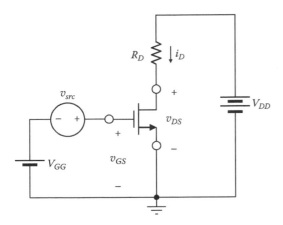

FIGURE 5.2.10
Basic circuit for small-signal analysis of an E-NMOS amplifier.

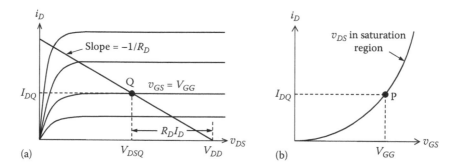

FIGURE 5.2.11

(a) Load line construction for determining the quiescent operating point Q on the output characteristics and (b) corresponding operating point on the transfer characteristic.

For a given V_{GG} and V_{DD}, the two unknowns I_D and V_{DS} are completely determined by Equations 5.2.6 and 5.2.7. Graphically, a load line can be drawn on the output characteristics from the point $(V_{DD}, 0)$ having a slope $-1/R_D$ (Figure 5.2.11a). The intersection of this line with the particular characteristic having $v_{GS} = V_{GG}$ defines the quiescent operating point Q having coordinates of V_{DSQ} and I_{DQ}, because at this point both Equations 5.2.6 and 5.2.7 are satisfied. Q corresponds to $P(I_{DQ}, V_{GG})$ on the transfer characteristic (Figure 5.2.11b).

When v_{src} is applied, the instantaneous v_{GS} is given by:

$$v_{GS} = V_{GG} + v_{gs} \qquad (5.2.8)$$

where $v_{gs} = v_{src}$. As v_{gs} varies with time, the operating point traces a path on the transfer characteristic determined by the time variation of v_{gs}, as illustrated in Figure 5.2.12, where it is assumed for the sake of argument, that v_{gs} is a sinusoidal function of time of amplitude V_{gm}. The corresponding i_d can, in principle, be determined graphically, point by point from the transfer characteristic, or by substituting Equation 5.2.8 in Equation 5.2.4 to obtain:

$$i_D = \frac{1}{2}k_n(V_{GG} + v_{gs} - V_{tn})^2 = \frac{1}{2}k_n(V_{GG} - V_{tn})^2 + k_n(V_{GG} - V_{tn})v_{gs} + \frac{1}{2}k_n v_{gs}^2 \qquad (5.2.9)$$

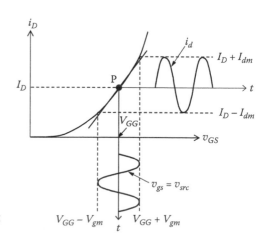

FIGURE 5.2.12

Small-signal variations around the quiescent operating point on the transfer characteristic.

The first term on the RHS is the dc component I_D resulting from V_{GG}, with $v_{gs} = 0$. The second term is a component of i_D that is directly proportional to v_{gs}. The last term involves v_{gs}^2 and arises from the quadratic nature of Equation 5.2.4. In small-signal operation, it is assumed that the term in v_{gs}^2, and any terms in higher powers of v_{gs} that may be present, are negligibly small compared to the term that is linear in v_{gs}. This is tantamount to assuming that v_{gs} is small enough in magnitude so that the transfer characteristic in the vicinity of P can be approximated by the tangent to the curve at P. Equation 5.2.9 can then be expressed as:

$$i_D = I_D + i_d \tag{5.2.10}$$

where

$I_D = \dfrac{1}{2}k_n(V_{GG} - V_{tn})^2$ is the dc component of i_D

$i_d = k_n(V_{GG} - V_{tn})v_{gs} = k_n(V_{GG} - V_{tn})v_{src}$ is the small-signal component, which is directly
proportional to v_{src}

Equation 5.2.10 underlies an important concept in linearized, small-signal analysis.

Concept: *In a linear system, currents and voltages of different frequencies do not interact, so that each frequency component can be considered separately.*

In other words, linearization implies superposition, which allows considering the dc and ac components independently of one another and adding them algebraically to obtain the total instantaneous values in accordance with Equations 5.2.9 and 5.2.10.

More formally, we can regard i_D in the saturation region as a function of v_{GS}. Then $i_D(V_{GG}) = I_D$ at the quiescent point, and $i_D(V_{GG} + \Delta v_{GS})$ is the value of i_D when v_{GS} is varied by a small increment away from V_{GG}; $i_D(V_{GG} + \Delta v_{GS})$ can be expressed as the sum of the first two terms of a Taylor series (Appendix A) in the neighborhood of the operating point. Thus:

$$i_D(V_{GG} + \Delta v_{GS}) = I_D(V_{GG}) + \left.\frac{di_D}{dv_{GS}}\right|_{v_{GS} = V_{GG}} \Delta v_{GS} \tag{5.2.11}$$

or

$$i_D(V_{GG} + \Delta v_{GS}) - I_D(V_{GG}) = \Delta i_D = \left.\frac{di_D}{dv_{GS}}\right|_{v_{GS} = V_{GG}} \Delta v_{GS} \tag{5.2.12}$$

We identify Δv_{GS} with the small-signal $v_{gs} = v_{src}$ and Δi_D with the small-signal i_d. From

Equation 5.2.4, $\left.\dfrac{di_D}{dv_{GS}}\right|_{v_{GS} = V_{GG}} = k_n(V_{GG} - V_{tn})$. Equation 5.2.12 gives $i_d = k_n(V_{GG} - V_{tn})v_{gs}$,

as in Equation 5.2.9. This is to be expected because $\left.\dfrac{di_D}{dv_{GS}}\right|_{v_{GS} = V_{GG}}$ is the slope of the tangent

at P. According to Figure 5.2.12, i_d is v_{gs} multiplied, point by point, by the slope of the tangent at P.

The slope $\dfrac{di_D}{dv_{GS}}\Big|_{v_{GS}=V_{GG}}$ at any point on the transfer characteristic whose voltage intercept is V_{GG} defines the **transconductance** g_m of the MOSFET:

$$g_m = \frac{di_D}{dv_{GS}}\Big|_{v_{GS}=V_{GG}} = k_n(V_{GG} - V_{tn}) \tag{5.2.13}$$

so that,

$$i_d = g_m v_{gs} \tag{5.2.14}$$

If $v_{src} = V_{gm}\cos\omega t$, then $i_d = I_{dm}\cos\omega t$ where $I_{dm} = g_m V_{gm}$; v_{GS} varies between $V_{GG} + V_{gm}$ and $V_{GG} - V_{gm}$, whereas i_D varies between $I_D + I_{dm}$ and $I_D - I_{dm}$ on both the transfer and output characteristics (Figure 5.2.13). As the operating point moves between Q_1 and Q_2, the small-signal v_{ds} is given by the horizontal excursions of Q between Q_1 and Q_2. In effect, the small-signal variation i_d is "reflected" off the load line to give the small-signal variation $v_{ds} = \Delta v_{DS}$.

KVL must also be satisfied in the drain circuit for instantaneous values. Thus:

$$v_{DS} + R_D i_D = V_{DD} \tag{5.2.15}$$

For small-signal variations $\Delta V_{DD} = 0$, because V_{DD} is a constant supply voltage and has no small-signal component. In other words, the dc supply is considered to have zero source impedance, so that any small-signal current that flows through this supply will not cause a small-signal voltage across the supply. Hence, $\Delta v_{DS} + R_D \Delta i_D = 0$, or:

$$v_{ds} = -R_D i_d \tag{5.2.16}$$

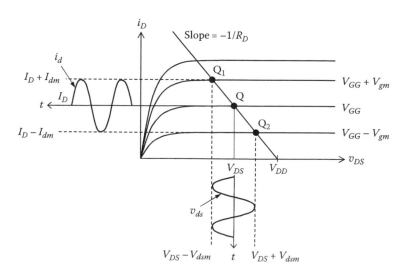

FIGURE 5.2.13

Small-signal variations around the quiescent operating point on the output characteristics.

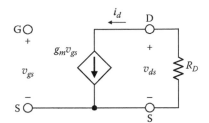

FIGURE 5.2.14
Small-signal equivalent circuit assuming a constant i_D in the saturation region.

The small-signal variation in v_{ds} is in antiphase with i_d, because an increase in i_d increases the voltage drop across R_D; v_{DS} decreases, and $\Delta v_{DS} = v_{ds}$ is therefore negative. Equation 5.2.16 also follows from the fact that i_d is reflected by the load line of negative slope to give v_{ds}.

Equations 5.2.14 and 5.2.16 can be interpreted in terms of a *small-signal equivalent circuit* of the transistor that relates v_{gs}, i_d, and v_{ds} (Figure 5.2.14). Note that the assigned positive directions of these quantities are the same as v_{GS}, i_D, and v_{DS}, respectively, in the large-signal equivalent circuit (Figure 5.2.7) and in the transistor symbol (Figure 5.2.8c). Moreover, because an increase in v_{gs} increases i_d, the arrow of the dependent current source is in the direction from drain to source. Because the dc supply is assumed to have ideally zero source impedance, *the dc supply is replaced by a short circuit in the small-signal equivalent circuit.*

From Figure 5.2.14, $v_{ds} = -R_D i_d$, as in Equation 5.2.16. The small-signal voltage gain can be derived by substituting for i_d from Equation 5.2.14 in Equation 5.2.16 to obtain:

$$v_{ds} = -g_m R_D v_{gs} \qquad (5.2.17)$$

from which it follows that the small-signal voltage gain v_{ds}/v_{gs} is $-g_m R_D$.

Equation 5.2.17 can also be derived in accordance with the aforementioned concept on independence of components of different frequencies in a linear system. Substituting for instantaneous values v_{DS} and i_D in Equation 5.2.15 gives:

$$(V_{DS} + v_{ds}) + R_D(I_D + i_d) = V_{DD} \qquad (5.2.18)$$

Equation 5.2.18 is a mixture of dc and small-signal quantities. The dc and ac quantities can each be equated separately to give Equations 5.2.6 and 5.2.16, respectively.

The small-signal equivalent circuit is the same for a PMOS transistor, in which the polarities of all currents and voltages are reversed. The reversal does not affect the relations between the small-signal quantities. Thus, in Figure 5.2.15 for a PMOS transistor,

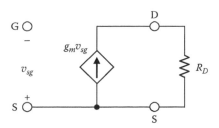

FIGURE 5.2.15
Small-signal equivalent circuit of a PMOS transistor.

increasing v_{SG} increases the current from source to drain, so that the polarity of the dependent source is from source to drain. Setting $v_{gs} = -v_{sg}$ in Figure 5.2.15 reverses the polarity of the current source in terms of v_{gs}, which gives Figure 5.2.14.

Exercise 5.2.4

If two of the three quantities i_D, v_{DS}, and v_{GS} are given, formulate a general procedure for determining whether a MOSFET is cut off, or is operating in the triode region or in saturation. ∎

Exercise 5.2.5

Consider the E-NMOS transistor of Exercise 5.2.2 with $v_{GS} = 4$ V and v_{DS} varying from zero to 10 V. Derive the i–v relation in the triode region and determine $v_{DS(sat)}$ and $i_{D(sat)}$.
   Answers: In the triode region, $i_D = 331.2v_{DS}(2 - v_{DS}/2)$; $v_{DS(sat)} = 2$ V, $i_{D(sat)} = 0.66$ mA.   ∎

Exercise 5.2.6

Repeat Exercise 5.2.5 for the E-PMOS transistor of Exercise 5.2.3 with $v_{SG} = 4$ V and v_{SD} varying from zero to 10 V.
   Answers: $i_D = 144.9v_{SD}(2 - v_{DS}/2)$ μA; $v_{SD(sat)} = 2$ V, $i_{D(sat)} = 0.29$ mA.   ∎

The power relations in small-signal, linear operation are discussed in Section ST5.2, where it is shown that the average power delivered by the supply is not changed in the presence of an applied signal and that the ac power delivered to the load comes from a reduction in the power dissipated in the transistor. A T-model is derived from Figure 5.2.14 in Section ST5.3.

5.3 Secondary Effects in MOSFETs

Channel-Length Modulation

This is an important effect that significantly modifies the output characteristics. As explained in connection with Figure 5.2.4, when v_{DS} increases beyond pinch off, the pinch-off point moves toward the source, thereby reducing the channel length by ΔL (Figure 5.3.1). Since the saturation value of i_D is inversely proportional to the effective channel length (Equation 5.2.4), i_D is no longer independent of v_{DS} but increases with v_{DS} in the saturation region. The resulting channel-length modulation is evidently more

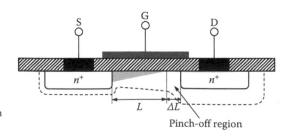

FIGURE 5.3.1
Reduction of the effective channel length with v_{DS} in the saturation region.

pronounced, in proportion to channel length, in shorter channels. It is to be expected that as the channel becomes shorter with increasing v_{DS}, its resistance decreases and the drain current increases.

Channel-length modulation is investigated quantitatively in Section ST5.1, where it is shown that the effect is to make i_D increase in the saturation region at a rate λv_{DS}, where λ is a constant for a given transistor and is typically in the range of 0.005–0.04 V^{-1}. It would be thought that because channel-length modulation does not occur in the triode region, i_D in this region is still given by Equation 5.2.2. However, this causes a problem, because according to Equation 5.2.2, $\left.\dfrac{\partial i_D}{\partial v_{DS}}\right|_{v_{GS}} = 0$ at the boundary between the triode and saturation regions ($v_{DS} = v_{DS(\mathrm{sat})}$), whereas the effect of channel-length modulation is to make the slope $\lambda v_{DS(\mathrm{sat})}$ nonzero at this boundary (Equation ST5.1.29, Section ST5.1). In SPICE level-1 model, based on the square-law approximation, i_D is considered to be given by (Appendix B):

$$i_D = \frac{1}{2}k_n(v_{GS} - V_{tn})^2(1 + \lambda v_{DS}) \text{ (saturation region)} \tag{5.3.1}$$

$$i_D = k_n\left[(v_{GS} - V_{tn})v_{DS} - \frac{1}{2}v_{DS}^2\right](1 + \lambda v_{DS}) \text{ (triode region)} \tag{5.3.2}$$

This makes the slope of the i_D vs. v_{DS} characteristic continuous across the boundary between the triode and saturation regions (Problem P5.1.11). The expression for $i_{D(\mathrm{sat})}$ is obtained by substituting $v_{DS(\mathrm{sat})} = v_{GS} - V_{tn}$ in Equation 5.3.1 or 5.3.2:

$$i_{D(\mathrm{sat})} = \frac{1}{2}k_n v_{DS(\mathrm{sat})}^2(1 + \lambda v_{DS(\mathrm{sat})}) \tag{5.3.3}$$

According to Equation 5.3.1, $i_D = 0$ when $v_{DS} = -1/\lambda$, irrespective of v_{GS}, which means that if the straight-line output characteristics in the saturation region are extended backward, they will all intersect the v_{DS} axis at $-1/\lambda$. Channel-length modulation also affects the transfer characteristic, which is no longer a single curve, independent of v_{DS}, but becomes a family of curves, each for a constant v_{DS}, in accordance with Equation 5.3.1 (Figure 5.3.3b).

From Equation 5.3.1, the output characteristics in the saturation region will have a slope g_{ds} that increases with $v_{DS(\mathrm{sat})} = v_{GS} - V_{tn}$ and hence with v_{GS} (Figure 5.3.3a):

$$g_{ds} = \left.\frac{\partial i_D}{\partial v_{DS}}\right|_{v_{GS}} = \frac{\lambda}{2}k_n v_{DS(\mathrm{sat})}^2 = \frac{\lambda i_{D(\mathrm{sat})}}{1 + \lambda v_{DS(\mathrm{sat})}} \cong \lambda i_{D(\mathrm{sat})} \tag{5.3.4}$$

In terms of the output resistance $r_o = 1/g_{ds}$, an approximate expression for r_o is:

$$r_o = \frac{1}{g_{ds}} \cong \frac{1}{\lambda i_{D(\mathrm{sat})}} = \frac{V_A}{i_{D(\mathrm{sat})}} \cong \frac{V_A}{I_D} \tag{5.3.5}$$

where $V_A = 1/\lambda$ and $I_D \cong i_{D(\mathrm{sat})}$. r_o can be related to L by substituting for k_n in Equation 5.3.4:

$$r_o = \frac{1}{g_{ds}} = \frac{2}{\lambda k_n v_{DS(\mathrm{sat})}^2} = \frac{2L}{\lambda \mu_e C_{ox} W v_{DS(\mathrm{sat})}^2} \tag{5.3.6}$$

From Equation ST5.1.28 (Section ST5.1), $\lambda \propto 1/L$. Hence, with μ_e and C_{ox} constant,

$$r_o \propto \frac{L^2}{W v_{DS(sat)}^2} \tag{5.3.7}$$

Since $\left.\dfrac{\partial i_D}{\partial v_{DS}}\right|_{v_{GS}}$ is no longer zero, we can regard i_D as a function of two variables, that is, $i_D = i_D(v_{GS}, v_{DS})$. A Taylor series expansion (Appendix A) then gives:

$$\Delta i_D = \left.\frac{\partial i_D}{\partial v_{DS}}\right|_{v_{GS}} \Delta v_{DS} + \left.\frac{\partial i_D}{\partial v_{GS}}\right|_{v_{DS}} \Delta v_{GS} \tag{5.3.8}$$

where the partial derivative means that the independent variable that is not involved in the differentiation is kept constant. This is indicated, for emphasis, by the variable next to the vertical bar. The derivatives, which represent small variations, are taken at the quiescent point Q. $\left.\dfrac{\partial i_D}{\partial v_{GS}}\right|_{v_{DS}}$ is evaluated at constant v_{DS} and around $v_{GS} = V_{GG}$. This is g_m in accordance with Equation 5.2.13, where it was assumed that the transfer characteristic is independent of v_{DS} in the saturation region. Whereas on the transfer characteristic g_m is the slope at the operating point, it is on the output characteristics the limit of $\dfrac{\Delta i_D}{\Delta v_{GS}}$, along a vertical line of constant v_{DS}, as the small changes become infinitesimal. In terms of small-signal quantities, Equation 5.3.8 becomes:

$$i_d = g_{ds} v_{ds} + g_m v_{gs} \tag{5.3.9}$$

Consequently, the small-signal equivalent circuit of Figure 5.2.14 is modified to that shown in Figure 5.3.2. The transistor now has a finite output resistance $r_o = 1/g_{ds}$ in the saturation region. It follows from this figure that $v_{ds} = -(r_o \| R_D) g_m v_{gs}$, or:

$$\frac{v_{ds}}{v_{gs}} = -g_m(r_o \| R_D) = -g_m \frac{r_o R_D}{r_o + R_D} \tag{5.3.10}$$

$$= -g_m r_o, \quad R_D \gg r_o, \quad \text{or}$$

$$= -g_m R_D, \quad R_D \ll r_o$$

It should be emphasized that whereas the dependent current source is $g_m v_{gs}$, directed from drain to source, $i_d = g_m v_{gs}$ only when r_o is large compared with R_D. Equation 5.3.10 is interpreted graphically in Section ST5.4.

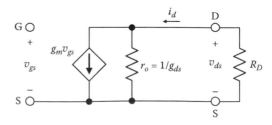

FIGURE 5.3.2
Small-signal equivalent circuit taking into account channel-length modulation.

Transconductance

If the effect of channel-length modulation is included, the definition of transconductance is modified from that of Equation 5.2.13 to:

$$g_m = \left.\frac{\partial i_D}{\partial v_{GS}}\right|_{\substack{v_{DS} \\ v_{GS}=V_{GS}}} = k_n(V_{GS} - V_{tn})(1 + \lambda V_{DS}) = \frac{2I_D}{V_{GS} - V_{tn}} \tag{5.3.11}$$

where g_m is evaluated for V_{GS} and V_{DS} at a quiescent operating point Q in the saturation region, and I_D, the dc drain current at Q, is given by Equation 5.3.1 with $v_{GS} = V_{GS}$ and $v_{DS} = V_{DS}$. Substituting for $(V_{GS} - V_{tn})$ in Equation 5.3.11 from the expression for I_D (Equation 5.3.1),

$$g_m = \sqrt{2k_n I_D(1 + \lambda V_{DS})} \cong \sqrt{2k_n I_D} \tag{5.3.12}$$

The transconductance is an important figure of merit for the transistor, because it is a measure of how effectively the input signal modulates the drain current. It is seen from Equation 5.3.12 that g_m varies as the square root of I_D, at constant V_{DS}. If we consider $g_m r_o$, it follows from Equations 5.3.5 and 5.3.12 that:

$$g_m r_o \cong \frac{\sqrt{2k_n I_D}}{\lambda I_D} = \frac{\sqrt{2k_n}}{\lambda \sqrt{I_D}} \tag{5.3.13}$$

Increasing I_D increases g_m but *decreases* $g_m r_o$, which is the maximum or open-circuit voltage gain. The gain $g_m R_D$, however (Equation 5.3.10), increases with g_m and hence I_D.

Simulation Example 5.3.1 E-NMOS Characteristics from Transistor Parameters

Given an E-NMOS transistor having the following parameters: $t_{ox} = 0.05$ μm, $\varepsilon_{ox} = 3.9 \times 8.85 \times 10^{-14} = 3.45 \times 10^{-13}$ F/cm, $\mu_e = 580$ cm^2/Vs, $V_{tn} = 1$ V, $W/L = 20$ μm/2 μm, and $\lambda = 0.015$ V^{-1}. It is required to determine the characteristics of the device.

SOLUTION

(a) $k'_n = 580$ cm^2/Vs $\times 3.45 \times 10^{-13}$ F/cm $\times (1/5 \times 10^{-6}$ cm) $\cong 40$ μA/V^2; $k_n = k'_n W/L = 400$ μA/V^2.

(b) r_{DS} at small v_{DS} (Equation 5.2.3): $r_{DS} = 1/400(5 - 1) \equiv 625$ Ω, when $v_{GS} = 5$ V.

(c) Boundary between the triode and saturation regions: For a given v_{GS}, $v_{DS(sat)}$ is determined from the relation $v_{DS(sat)} = v_{GS} - V_{tn}$, and $i_{D(sat)}$ can then be calculated from Equation 5.3.3. Corresponding values of v_{GS}, $v_{DS(sat)}$, and $i_{D(sat)}$ are indicated in Table 5.3.1.

TABLE 5.3.1

v_{GS}, v_{DS}, and i_D at Edge of Saturation

v_{GS}, V	3	4	5	6	7	8	9
$v_{DS(sat)}$, V	2	3	4	5	6	7	8
$i_{D(sat)}$, mA	0.824	1.881	3.392	5.375	7.848	10.829	14.336

(d) The output characteristics can be plotted using Equations 5.3.1 and 5.3.2.

(e) The transfer characteristics can be plotted from Equation 5.3.1 for various values of v_{DS} in the saturation region.

To simulate the output and transfer characteristics, choose from the BREAKOUT library part MbreakN3, which is an E-NMOS transistor having its body and source terminals connected together. The schematic is entered with just the transistor and a VDC source connected between the gate and source and another VDC source connected between the drain and source. Click on the transistor symbol to select it, then right click and choose Edit PSpice Model. The window of the PSpice Model Editor shows only: model Mbreakn NMOS, without any parameters. Enter the parameters of the transistor under consideration as follows, after the word "NMOS": Vto = 1, kp = 40u, lambda = 0.15, and exit the PSpice Model Editor after saving the file. In the Property Editor spreadsheet, enter 2u under L and 20u under W. To obtain the output characteristics, perform a DC Sweep analysis, sweeping the V_{DS} source from 0 to 16 V in increments of 1 mV, as linear primary sweep, and choose the V_{GS} source as secondary sweep, entering 3, 4, 5, 6, 7, 8, and 9 in the Value list field. The output characteristics are shown in Figure 5.3.3a.

To obtain the transfer characteristics, select the V_{GS} source for primary sweep between 0 and 10 V, and the V_{DS} source as secondary sweep at 8 and 16 V. The transfer characteristics are shown in Figure 5.3.3b. The two characteristics for $v_{DS} = 8$ V and $v_{DS} = 16$ V are close together.

(f) Assume that the transistor is connected in the circuit of Figure 5.2.10, with $V_{GS} = 5$ V, $V_{DD} = 15$ V, and $R_D = 1.5$ kΩ. The operating point Q can be determined graphically or by solving the load-line equation $V_{DS} + 1.5I_D = 15$ and the characteristic for $v_{GS} = 5$ V, which is $I_D = 3.2 + 0.048V_{DS}$ (Equation 5.3.1), assuming that the transistor is operating in saturation. This gives: $I_D = 3.66$ mA, $V_{DS} = 9.51$ V. Since $V_{DS} > V_{GS} - V_{tn}$, the assumption is justified. Otherwise, the load-line equation has to be solved with Equation 5.3.2 for operation in the triode region.

(g) Equation 5.3.4 gives for $V_{GS} = 5$ V, $g_{ds} = (\lambda/2)k_n(v_{GS} - V_{tn})^2 = 0.048$ mS, equivalent to 20.8 kΩ. From Equation 5.3.11, $g_m = 2 \times 3.66/(5 - 1) = 1.83$ mA/V. Note that using $g_m = \sqrt{2k_nI_D}$ (Equation 5.3.12) gives $g_m = 1.71$ mA/V.

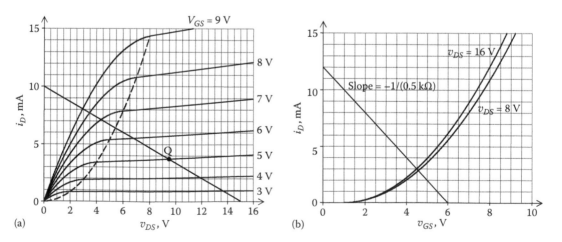

FIGURE 5.3.3
Figure for Simulation Example 5.3.1. (a) Output characteristics; (b) transfer characteristics.

Exercise 5.3.1

Consider the E-NMOS transistor of Exercise 5.2.1, with $V_{DD}=12$ V, $R_D=2$ kΩ, and $\lambda = 0.02$ V^{-1}. Determine $v_{DS(sat)}$, $i_{D(sat)}$, the quiescent values of I_D and V_{DS}, r_o, and g_m for $V_{GS}=4$ V. What happens if V_{GS} is increased to 8 V?

Answers: $v_{DS(sat)}=2$ V, $i_{D(sat)}=0.689$ mA, $I_D=0.8$ mA, $V_{DS}=10.40$ V, the transistor is in saturation, $r_o=72.6$ kΩ, $g_m=0.8$ mA/V. If V_{GS} is increased to 8 V, the transistor will be in the triode region. ∎

Overdrive Voltage

It is convenient when discussing transistor circuits to define an **overdrive voltage** as:

$$V_{ov} = v_{GS} - V_{tn} = v_{DS(sat)} \tag{5.3.14}$$

As its name implies, V_{ov} is a measure of the excess v_{GS} beyond threshold. Ignoring channel-length modulation, it follows from Equation 5.2.4 that:

$$V_{ov} = \sqrt{\frac{2I_D}{k_n'(W/L)}} = \frac{g_m}{k_n'(W/L)} \tag{5.3.15}$$

Increasing V_{GS} increases I_D, g_m, and V_{ov}. For given I_D and k_n', V_{ov} is reduced by increasing W/L.

In terms of V_{ov}, it follows from Equations 5.3.6 and 5.3.11, ignoring channel-length modulation, that:

$$g_m r_o \cong (k_n V_{ov}) \frac{2}{\lambda k_n V_{ov}^2} = \frac{2}{\lambda V_{ov}} \tag{5.3.16}$$

Temperature Effects

The effect of a rise in temperature is to decrease V_{tn} by about 1 mV/°C (Section ST5.1), which tends to increase i_D for given bias conditions (Equation 5.3.1); however, μ_e and hence k_n', decrease with temperature (Equation 2.6.3, Chapter 2). The overall effect is that i_D increases with temperature at low currents, but decreases with temperature at large currents.

Breakdown

Avalanche breakdown can occur at high v_{DS} in the reverse-biased junction between the drain and the body, or in the pinch-off region of the channel (Figure 5.3.1). Typically, breakdown occurs in the pinch-off region first, when the electric field becomes so high that electrons acquire sufficient kinetic energy to initiate avalanche multiplication, as in a reverse-biased *pn* junction (Section 3.2, Chapter 3). A large current flows between the drain and the body (Figure 5.3.4), and the gate effectively loses control over the drain current. Avalanche-generated electrons are swept to the drain, whereas avalanche-generated holes are swept into the body, resulting in current through the body terminal. The larger v_{GS}, the larger is the current i_D available for avalanche multiplication in the channel, so that breakdown is initiated at a smaller v_{DS}.

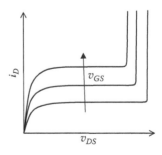

FIGURE 5.3.4
Modification of MOSFET output characteristics because of breakdown.

Body Effect

The **body effect** is the variation of I_D with a voltage between the source and body. It does not occur: (1) if the body is directly connected to the source, as in discrete MOSFETs and in MOSFETs formed in a well in the substrate, and (2) in fully depleted SOI MOSFETs (Figure 5.3.13). The body effect can occur when the source and drain are formed directly in the substrate. In order to ensure that the junctions between the substrate and MOSFETs on the chip are never forward biased, a *p*-substrate is connected to the most negative voltage supply on the chip, usually ground, whereas an *n*-type substrate is connected to the most positive supply. If the source of an E-NMOS transistor is at a voltage above ground, a reverse bias exists between the source and the substrate, which increases the width of the depletion region of the *pn* junction between the substrate and source-channel-drain regions, the depletion region being narrowest at the source and widest at the drain. The increase in width of the depletion region occurs almost totally in the lightly doped substrate and increases in turn the exposed charge due to dopant atoms in the substrate, which have the same sign of charge as the carriers in the channel. Hence,

Concept: *For a given $|v_{GS}|$, the charge on the gate is constant, which means that the sum of the induced charge in the channel and the exposed charge in the depletion region is constant. If the charge in the depletion region increases, the charge in the channel must decrease, and conversely.*

As the source-to-substrate voltage v_{SB} increases in an E-NMOS transistor, the charge in the channel decreases, and i_D is reduced for given v_{GS} and v_{DS}. This means that the threshold has increased (Equation 5.3.1). The threshold is given by (Equation ST5.1.18):

$$V_t = V_{t0} + \gamma\left(\sqrt{2|\varphi_F| + |V_{SB}|} - \sqrt{2|\varphi_F|}\right) \tag{5.3.17}$$

where
$V_{SB} > 0$ is the voltage of the source with respect to the substrate
V_{t0} is the threshold for $V_{SB} = 0$
$\gamma = \sqrt{2|q|N_A \varepsilon_S}/C_{ox}$ is the **body-effect parameter**

$|\phi_F|$, usually about 0.3 V, is a constant for a given transistor at a given temperature and is equal to the junction potential of a substrate-intrinsic junction. In terms of absolute values, Equation 5.3.17 applies equally well to E-NMOS and E-PMOS transistors, where for an E-NMOS transistor, V_{SB} is positive and $V_t = V_{tn}$, whereas for an E-PMOS transistor, V_{SB} is negative, $V_{t0} = \overline{V}_{tp0}$, and $V_t = \overline{V}_{tp}$.

Simulation Example 5.3.2 Body Effect under dc Conditions

It is required to simulate the effect of body–source voltage on the transfer characteristic, assuming the same transistor parameters as in Example 5.3.1, with $\gamma = 0.3$ V$^{1/2}$ and $2\phi_F = 0.6$ V.

SIMULATION

The schematic is shown in Figure 5.3.5 using the E-NMOS transistor Mbreakn from the BREAK-OUT library. In the PSpice model editor, add gamma = 0.3 phi = 0.6. V_{DS} is set to 16 V, v_{GS} is

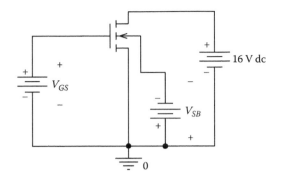

FIGURE 5.3.5
Figure for Simulation Example 5.3.2.

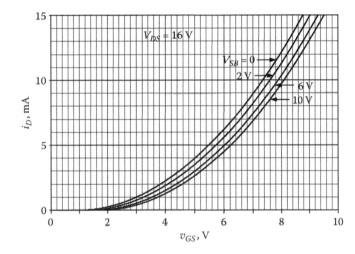

FIGURE 5.3.6
Figure for Simulation Example 5.3.2.

swept between 0 and 10 V, with $V_{SB} = 0$, 2, 6, and 10 V. The transfer characteristics are shown in Figure 5.3.6, the corresponding thresholds being: 1, 1.25, 1.54, and 1.74 V, respectively.

The bias supply to which the substrate is connected normally acts as a short circuit for time-varying signals, as explained for V_{DD} in connection with Figure 5.2.14, and the substrate is said to be *ac grounded*. If the source is grounded, there is no ac signal between the source and substrate, and the small-signal equivalent circuit of the transistor is not affected, *except for a change in g_m because of the change in V_{tn}*. If the source is not grounded for time-varying signals, a signal voltage appears between the source and substrate, and the small-signal equivalent circuit has to be modified accordingly, as discussed in Example 5.3.3.

Example 5.3.3 Small-Signal Equivalent Circuit Including Body Effect

It is required to determine how the small-signal equivalent circuit is modified because of the body effect.

SOLUTION

Consider the circuit of Figure 5.3.7, in which the substrate is connected to a negative supply V_{BB}. Resistors R_1, R_2, and R_S and capacitor C are used for dc biasing, as explained later in

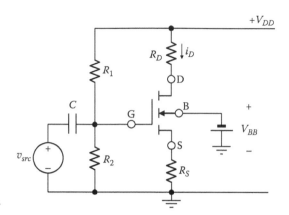

FIGURE 5.3.7
Figure for Example 5.3.3.

Section 8.1, Chapter 8. In this circuit $V_{SB} = R_S I_D - V_{BB}$, where V_{BB} has a negative value. Because i_D has a signal component due to the applied ac signal, a time-varying signal v_{bs} appears between the substrate and the source, which affects i_d. To take this into account, we consider i_D to be a function of the three variables, v_{DS}, v_{GS}, and v_{BS}. Equation 5.3.8 becomes:

$$\Delta i_D = \frac{\partial i_D}{\partial v_{DS}}\bigg|_{\substack{V_{GS}\\V_{BS}}} \Delta v_{DS} + \frac{\partial i_D}{\partial v_{GS}}\bigg|_{\substack{V_{DS}\\V_{BS}}} \Delta v_{GS} + \frac{\partial i_D}{\partial v_{BS}}\bigg|_{\substack{V_{GS}\\V_{DS}}} \Delta v_{BS} \tag{5.3.18}$$

In terms of small signals, Equation 5.3.18 can be written as:

$$i_d = g_{ds}v_{ds} + g_m v_{gs} + g_{mb}v_{bs} \tag{5.3.19}$$

where, $g_{ds} = \frac{\partial i_D}{\partial v_{DS}}\big|_{\substack{V_{GS}\\V_{BS}}} = \frac{1}{r_o}$, $g_m = \frac{\partial i_D}{\partial v_{GS}}\big|_{\substack{V_{DS}\\V_{BS}}}$, and $g_{mb} = \frac{\partial i_D}{\partial v_{BS}}\big|_{\substack{V_{GS}\\V_{DS}}}$ is the **body transconductance**.
According to Equation 5.3.19, the small-signal equivalent circuit of the transistor becomes as in Figure 5.3.8. In effect, the body terminal becomes the terminal of a second gate that draws a negligible current under normal conditions because of the reverse-biased source–substrate junction. Note that in terms of v_{bs}, the current source is directed from drain to source, like the current source due to v_{gs}. If the substrate becomes more positive with respect to the source, v_{BS} increases, the body effect is reduced, and i_d increases. The effect is like an increase in v_{gs}.
  If channel-width modulation is neglected, the relation between g_m and g_{mb} is determined as (Equation ST5.1.19):

$$g_{mb} = g_m \frac{\gamma}{2\sqrt{2|\phi_F| + |V_{SB}|}} = \chi g_m \tag{5.3.20}$$

where $\chi = \gamma/2\sqrt{2|\phi_F| + |V_{SB}|}$. If $\gamma = 0.3$ V$^{1/2}$, $|\phi_F| = 0.3$ V, and $V_{SB} = 5$ V, then $\chi = 0.06$. In most MOSFETs, χ is in the range 0.1–0.3.

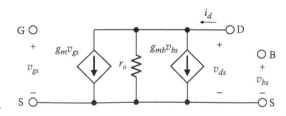

FIGURE 5.3.8
Figure for Example 5.3.3.

Capacitances

The internal capacitances of a MOSFET arise because structures, such as the gate-oxide-semiconductor and the substrate-drain/source depletion regions, store electric charges. These internal capacitances critically affect the speed of operation of the MOSFET, because the stored charges change with voltage variations, which implies in turn, a flow of current associated with these changes. If the circuit cannot supply these currents with minimal voltage drops, the high-frequency performance and switching speed of the MOSFET will be compromised.

Consider a MOSFET operating in the triode region, with a small v_{DS} that results in a channel of nearly uniform depth (Figure 5.3.9). The following capacitances can be identified:

1. The gate-channel capacitance C_{gc} that we had identified earlier as WLC_{ox}^* and which is considered to be divided equally between gate–source and gate–drain capacitances when the transistor is in the triode region.

2. An additional **overlap capacitance** C_{ol}, exists between the source or drain and gate, given by:

$$C_{ol} = C_{ov} + C_f \tag{5.3.21}$$

where $C_{ov} = WL_{ov}C_{ox}^*$ is due to some inevitable lateral diffusion of the source and drain regions to a distance L_{ov} under the gate (Section 4.6, Chapter 4), as illustrated in Figure 5.3.9, and C_f is a fringe capacitance between the source or drain and the sides of the gate (fringe capacitance is discussed in Section ST12.8, Chapter 12). Hence,

$$C_{gs} = C_{gd} = \frac{1}{2}WLC_{ox}^* + C_{ol} \quad \text{(triode region)} \tag{5.3.22}$$

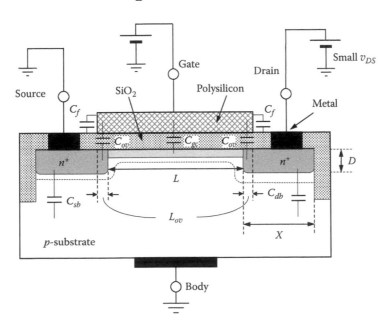

FIGURE 5.3.9
Capacitances of a MOSFET operating in the triode region.

3. The capacitances C_{sb} and C_{db} are the depletion capacitances of the reverse-biased source–body and drain–body *pn* junctions. The capacitances per unit area at zero bias, C_{sb0}^* and C_{db0}^*, are calculated using Equation 3.2.12, Chapter 3, with $N_D \gg N_A$, and assuming an abrupt junction:

$$C_{j0}^* = C_{sb0}^* = C_{db0}^* = \sqrt{\frac{\varepsilon_s |q| N_A}{2V_{npo}}} \tag{5.3.23}$$

The depletion capacitances C_{sb} and C_{db} are then obtained using Equation 3.2.13, Chapter 3:

$$C_{sb}, C_{db} = A_J \frac{C_{j0}^*}{\sqrt{1 + V_x/V_{npo}}} \tag{5.3.24}$$

where $V_x = V_{SB}$ for the source and $V_x = V_{DB}$ for the drain. The area $A_J = WX + WD$, where X is the length of the source or drain regions and D is the depth of these regions. The area WX is that between the bottom surface of the drain or source and the substrate, whereas WD is the area of the sidewall between the drain or source and the substrate on the channel side. It is assumed in deriving Equation 5.3.24 that the sidewall junction is abrupt. This is not strictly correct but the error introduced is small. In modern transistors, the other three sidewalls of the source or drain are with the oxide separating neighboring transistors. The capacitance due to these sidewalls is negligible because of the relative thickness of the oxide (Section 4.6, Chapter 4).

When the transistor operates in the saturation region, the channel is tapered and pinched off near the drain. Under these conditions (Equation ST5.5.4, Section ST5.5):

$$C_{gs} = \frac{2}{3} WLC_{ox}^* + C_{ol}, \quad C_{gd} = C_{ol} \quad \text{(saturation region)} \tag{5.3.25}$$

When the transistor is cut off with a sufficiently negative v_{GS}, holes are attracted under the gate oxide, resulting in a gate-body capacitance C_{gb} of maximum value WLC_{ox}^*. Thus,

$$C_{gb(\max)} = WLC_{ox}^*, \quad C_{gs} = C_{ol}, \quad C_{gd} = C_{ol} \quad \text{(cutoff)} \tag{5.3.26}$$

The MOSFET capacitances are included in the equivalent circuit of Figure 5.3.10. The following should be noted concerning these capacitances:

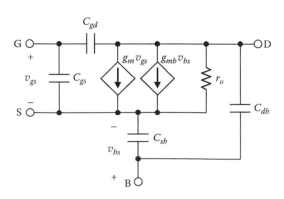

FIGURE 5.3.10
Small-signal equivalent circuit of a MOSFET, including capacitances and the body effect.

1. Except for C_{ol}, the capacitances are nonlinear, that is, voltage dependent. For large-signal operation, such as between saturation and cutoff, average values are often used. A more elaborate method is presented in Section ST13.2, Chapter 13.

2. The actual values of the capacitances in Figure 5.3.9 depend on the transistor size, structure, and method of fabrication. The capacitance of the gate/unit width of channel, $C'_{gb} = C^*_{ox}L = \varepsilon_{ox}L/t_{ox}$ is relatively independent of the size of the transistor, because t_{ox} is reduced by the same factor as L (Table 13.1.1, Chapter 13). $C^*_{ox}L$ has remained approximately constant at 1.5 fF/μm.

3. C_{gb} is maximum when V_{GS} is sufficiently negative to attract holes under the gate. When V_{GS} is only slightly negative, or is positive but less than V_{tn}, a depletion layer is formed under the channel, so that C_{gb} is C_{ox} in series with the channel-substrate depletion capacitance. At $V_{GS}=0$, $C_{gb} \cong C_{ox}/2$. The MOS capacitor is discussed in Section ST4.3, Chapter 4.

4. C_{gd} in Figure 5.3.10 provides a path from output to input, that is, changes in v_{ds} can affect v_{gs}, depending on the input circuit. We now have some reverse transmission (Section 7.1, Chapter 7), which is absent under dc conditions, with I_G being essentially zero.

Exercise 5.3.2

Calculate (a) C'_{gb}, assuming $t_{ox} = 0.02L$ and a relative permittivity of the oxide of 3.9; (b) C'_{ol}, the capacitance per unit width W, assuming $L_{ov} = 0.05L$, and $C'_f = 0.024$ fF/unit width W; (c) C_{gs} and C_{gd}, assuming the transistor operates in the saturation region and $W=5$ μm; (d) C_{sb} and C_{db}, assuming, $N_A = 10^{17}/cm^3$, $N_D = 10^{20}/cm^3$ in the source and drain, the relative permittivity of silicon is 11.9, $X = 1.5$ μm, $D = 0.25$ μm, $V_{SB} = 1$ V, and $V_{DS} = 2$ V.

Answers: (a) 1.73 fF/μm, independently of L; (b) 0.11 fF/μm, independently of L; (c) $C_{gs} = 6.30$ fF, $C_{gd} = 0.55$ fF; (d) $C_{sb} = 5.66$ fF, and $C_{db} = 4.01$ fF. ∎

Unity-Gain Bandwidth

If the source and substrate are effectively connected together for time-varying signals, $v_{bs} = 0$, C_{sb} in Figure 5.3.10 is shorted out and C_{db} effectively becomes a drain-source capacitance C_{ds}. The resulting simplified equivalent circuit of Figure 5.3.11 can be used to derive an important parameter for the MOSFET, namely, the **unity-gain bandwidth**, defined as the frequency at which the magnitude of the short circuit current gain from the gate to drain becomes unity. Experimentally, the drain is short-circuited to the source

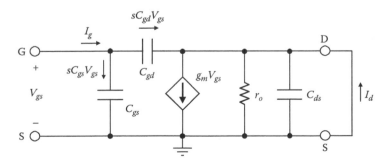

FIGURE 5.3.11
Simplified small-signal equivalent circuit of a MOSFET used for calculating the unity-gain bandwidth.

for the ac signal by connecting a large capacitor between them, while preserving dc bias conditions. It is seen that:

$$I_g = s(C_{gs} + C_{gd})V_{gs} \qquad (5.3.27)$$

At the output node,

$$I_d = g_m V_{gs} - sC_{gd}V_{gs} \cong g_m V_{gs} \qquad (5.3.28)$$

because $sC_{gd} \ll g_m$ at the frequencies of interest. Eliminating V_{gs} between these two equations,

$$A_i = \frac{I_d}{I_g} = \frac{g_m}{s(C_{gs} + C_{gd})} \qquad (5.3.29)$$

and

$$|A_i| = \frac{g_m}{\omega(C_{gs} + C_{gd})} \qquad (5.3.30)$$

$|A_i| \to \infty$ as $\omega \to 0$, because I_G is practically zero. $|A_i| = 1$ at a frequency ω_T given by:

$$\omega_T = \frac{g_m}{C_{gs} + C_{gd}} \qquad (5.3.31)$$

where ω_T is the unity-gain bandwidth. The larger ω_T, the larger is the response to high frequencies or to rapidly changing signals, that is, the faster is the response of the transistor. Modern MOSFETs have a unity-gain frequency $f_T = \omega_T/2\pi$ in the range of tens of GHz.

When the MOSFET operates in the saturation region, $C_{gs} = \frac{2}{3}WLC_{ox}^*$ and $C_{gd} \cong 0$ (Equation 5.3.25), neglecting C_{ol}. Substituting for C_{gs} in Equation 5.3.31,

$$f_T = \frac{3g_m}{4\pi WLC_{ox}^*} \qquad (5.3.32)$$

Substituting $g_m = k_n(V_{GS} - V_t) = \mu_e C_{ox}^*(W/L)V_{ov}$,

$$f_T = \frac{3}{4\pi}\mu_e \frac{V_{ov}}{L^2} \qquad (5.3.33)$$

f_T is increased by having large μ_e and V_{ov} and small L. This is to be expected because a large μ_e and V_{ov} mean a larger drain current, while a large μ_e and small L reduce the transit time of carriers through the channel. However, a smaller L and a larger V_{ov} lead to a small r_o (Equation 5.3.7 with $V_{ov} = v_{DS(sat)}$) and hence a reduced low-frequency gain (Equation 5.3.10). Moreover, a large $v_{DS(sat)}$ means reduced voltage swing in both analog (Figure 5.2.13) and digital circuits, as will be discussed later.

A figure of merit for the MOSFET is the gain $\times f_T$ product (GFT). Multiplying f_T (Equation 5.3.32) by $g_m r_o$ (Equation 5.3.10), substituting $r_o = 1/\lambda i_{D(sat)}$ (Equation 5.3.4), $g_m \cong \sqrt{2k_n I_D}$ (Equation 5.3.12), and considering $I_D \cong i_{D(sat)}$,

$$GFT \cong \frac{3\mu_e}{2\pi L^2 \lambda} \qquad (5.3.34)$$

From Equation ST5.1.28 (Section ST5.1), $\lambda \cong 1/L$. Hence,

$$\text{GFT} \propto \frac{\mu_e}{L} \tag{5.3.35}$$

Again, it appears advantageous to have a small L. However, it will be shown later that a number of factors collectively referred to as **short-channel effects** come into play when the channel is too short. In particular, velocity saturation of carriers reduces mobility, so that μ_e/L becomes nearly constant and the GFT does not significantly improve. In fact, for analog applications, which are primarily concerned with gain, bandwidth, and a high r_o of the transistor, L must not be reduced to the point where short-channel effects significantly compromise analog circuit performance. This occurs for L of about 0.5–1 μm.

In digital circuits on the other hand, the primary concerns are high packing density, fast speed of operation, and low power density, that is, power dissipated per unit area of the chip. Reducing L not only increases the packing density but speeds up the response as well, because of the reduction of capacitances, which are proportional to area. Capacitance has a major impact on the switching behavior of the MOSFET, as discussed in Chapters 12 and 13. Hence, the trend in digital circuits has been a steady reduction in channel length. The recent technologies are designated in terms of the **minimum feature size**, or minimum dimensions. These terms refer to the minimum resolvable metal line widths of the interconnections and to the *half-pitch*, or separation between adjacent metal lines, which determines how closely the transistors are packed. The minimum dimension is also the nominal channel width, although channel length can be considerably smaller than this dimension.

In 2007, 45 nm technology started replacing the 65 nm technology, and by 2010, 32 nm technology is expected to go into production. However, shortening the gate increases power dissipation by reducing the threshold voltage, as mentioned later, which in turn increases the leakage current when the MOSFET is off (Section 13.1, Chapter 13). Hence, in some cases gate length is slightly increased in logic paths that do not need high speed in a given chip.

Short-channel transistors are also referred to as **submicron transistors** and those of very short channels as **deep submicron transistors**.

Short-Channel Effects

Shortening the channel, while keeping other dimensions constant, degrades performance mainly due to a significant longitudinal electric field along the channel (Figure 5.3.12a). This longitudinal field, which is neglected in the gradual channel approximation that leads

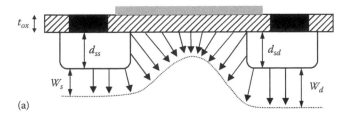

(a)

(b)

FIGURE 5.3.12
Effect of shortening the channel (a) without scaling other dimensions of the transistor, and (b) with scaling.

to Equation 5.2.2, is due to v_{DS} along the channel and to the lateral penetration into the channel of the electric field associated with source–body and drain–body voltages.

In principle, it should be possible to obtain long-channel behavior with shorter channels if measures are taken to maintain the validity of the gradual channel approximation. One obvious measure is to reduce the oxide thickness, as this increases the transverse electric field for a given gate-channel voltage. However, at oxide thicknesses less than about 1.5 nm, electrons can tunnel through the oxide, resulting in gate current. Moreover, the gate oxide begins to break down at about 10 MV/cm, or 1 V/nm, so that the gate voltage across a 1.5 nm gate oxide layer should be limited to 1.5 V for reliable long-term operation. Another measure is to decrease the depth of the source and drain regions as well as the depth of the depletion region underneath them (Figure 5.3.12b). The effect is twofold: (1) it makes the transverse field more nearly perpendicular to the surface, as assumed in the gradual channel approximation, and (2) it reduces the lateral penetration, into the channel region, of electric field lines emanating from the source and drain. However, too shallow drain and source regions increase the series, parasitic resistance, and reduce the drain current. The upshot is that in modern short-channel devices the effect of the longitudinal field is appreciable, which invalidates the gradual channel approximation made in deriving the i–v characteristics of the transistor (Section ST5.1).

Reducing all dimensions of the transistor by the same factor is referred to as **scaling**. The effects of scaling on transistor parameters that are relevant to digital circuits are indicated in Table 13.1.1, Chapter 13. Because of the nonlinear variation of many important parameters affecting the behavior of short-channel devices, and the empirical description of these relations, PSpice simulation is usually resorted to in the design of circuits involving these transistors.

Carrier Velocity Saturation and Hot Carriers

The longitudinal field accelerates carriers in the channel, thereby increasing their drift velocity u_{drift}. At low electric fields $u_{drift} = -\mu_e \xi$ for an electron, where μ_e is electron mobility (Equation 2.4.3, Chapter 2) and is substantially independent of ξ. However, quantum mechanical considerations indicate that at high electric fields u_{drift} saturates at its **scattering-limited** value u_{sat}, which is nearly that of the thermal velocity of electrons ($\sim 10^7$ cm/s). In Si this happens at electric fields of about 10^5 V/cm, which are encountered in short-channel devices. Velocity saturation obviously limits the increase of drain current with v_{DS} at high values of v_{DS}. The variation of i_D with ($v_{GS} - V_{tn}$), which is the transfer characteristic shifted by V_{tn}, becomes linear instead of quadratic as in Equation 5.3.3, and saturation occurs at lower v_{DS} (Section ST5.1). The variation of i_D with v_{GS} in the saturation region becomes (Equation ST5.1.47):

$$i_D = u_{sat} W C_{ox}(v_{GS} - V_{tn} - V_{DS(sat)})$$ (5.3.36)

and

$$g_m = \left. \frac{\partial i_D}{\partial v_{GS}} \right|_{\substack{v_{gs}=V_{GD} \\ i_D=I_D}} = u_{sat} C_{ox} W$$ (5.3.37)

According to Equation 5.3.37, g_m depends only on W, for a given technology and constant u_{sat}. In fact, u_{sat} does not remain constant but increases somewhat with v_{GS} or v_{DS}.

The saturation voltage becomes (Equation ST5.1.48):

$$v_{DS(\text{sat})} = \frac{(v_{GS} - V_{tn})\xi_c L}{v_{GS} - V_{tn} + \xi_c L} \tag{5.3.38}$$

where $\xi_c \cong 10^5$ V/cm is the critical electric field at which velocity saturates. It is seen that $v_{DS(\text{sat})} = v_{GS} - V_{tn}$ of the long-channel model is multiplied by $\xi_c L/(v_{GS} - V_{tn} + \xi_c L)$, which is less than 1.

Additional effects of the high electric field are due to **hot carriers**. In n-channel devices, these are high-velocity electrons in the pinch-off region near the drain. Because mobility decreases with temperature, electrons of high velocity and reduced mobility are effectively at a high temperature; hence the term "hot carrier." These high-velocity electrons can cause ionizing collisions with crystal atoms, though not leading to avalanche multiplication as in avalanche breakdown. The electrons generated by these collisions flow to the drain, where they increase i_D, whereas the holes flow to the body, resulting in a drain-to-body current. Moreover, some of these electrons may tunnel through to the conduction band of the oxide layer, where they are collected by the gate as gate current. Or, they may get trapped in the oxide, leading to a gradual increase in V_{tn} and the consequent degradation of the long-term reliability of the transistor.

Reduced Output Resistance and Threshold

Channel-length modulation has a relatively larger effect in shorter channels, which reduces r_o (Equation 5.3.7). In addition, the following effects reduce both r_o and V_{tn}: (1) Because of the longitudinal field at the source and drain, less of the electric field lines due to the gate–source voltage terminate in the depletion region and more in the channel, thereby increasing i_D and decreasing V_{tn}. Moreover, this causes i_D to increase with v_{DS}, which implies a reduction in r_o. (2) As mentioned previously, current flows between the drain and body due to the ionizing collisions of hot carriers. In the presence of the body effect, the flow of this current through the substrate resistance, increases the voltage of the substrate adjoining the source and reduces V_{SB}. This decreases V_{tn} (Equation 5.3.17) and increases the drain current, which again reduces r_o.

Other Effects

At high electric fields in short channel devices, the drain depletion region can extend all the way to the source, effectively reducing the channel length to zero and resulting in a large drain current. This phenomenon is referred to as **punchthrough**.

Moreover, the following effects, which are inconsequential in long-channel devices, become significant in deep submicron devices and can adversely affect the long-term reliability of the circuit:

1. As channel length approaches molecular dimensions, the position of Si atoms in the channel can affect channel resistance and heating. Note that the radius of a Si atom is about 0.118 nm and the side of the cube in Figure 2.3.1, Chapter 2, is 0.2715 nm.

2. **Electromigration**, or the displacement of metal atoms from their regular positions due to the momentum of current carriers, can be serious as the width of interconnections becomes very small and the current density exceeds a certain level that depends on the metal. Atoms accumulate in some places and voids form in other

places. Voids can lead to open circuits, whereas accumulations can lead to short circuits between neighboring lines on the same level or in adjacent levels of the chip.

3. Stress in various interconnection layers of the chip during fabrication can cause voids in these layers leading to breaks or to short circuits.

4. Slow degradation in the very thin gate oxide due to incipient breakdown and trapping of high-energy electrons.

Improved Performance

In modern SOI technology, each transistor is encased in a layer of insulator, usually silicon dioxide, as diagrammatically illustrated in Figure 5.3.13 (Section 4.6, Chapter 4). This not only simplifies chip layout and processing but results in a higher packing density and improved performance. The insulating layer effectively isolates adjacent transistors, thereby reducing internal capacitances and leakage current. Reduction of capacitances improves switching speeds because the same currents can charge and discharge the smaller capacitances at faster rates. For a given switching speed, reduction of capacitances reduces the power dissipated in charging and discharging these capacitances. Reduction of leakage currents reduces the quiescent power dissipation. Short-channel effects are also less pronounced, because the source and drain regions extend all the way to the buried part of the insulator layer, which reduces the longitudinal electric field due to the drain–substrate and source–substrate depletion regions (Figure 5.3.6a). SOI technology eliminates latchup due to a parasitic *pnpn* structure (Section ST6.6, Chapter 6). Fully depleted SOI transistors (Section 4.6, Chapter 4) have no body effect and are less affected by radiation because of the small volume of Si between the source and drain.

Strained Si technology (Section 4.8, Chapter 4) has been used since the 90 nm process to enhance speed. Straining affects the electrical properties of the crystal. It distorts the lattice and changes the spatial distribution of the electric potential due to crystal ions and electrons in the crystal. Recall that the forces exerted by a crystal on a conduction electron are accounted for by the effective mass of the electron. When the crystal is distorted, these forces are modified in a way that reduces the effective mass of conduction electrons. A smaller effective mass means higher mobility (Example ST2.7, Chapter 2). Another effect

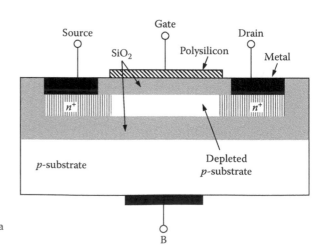

FIGURE 5.3.13
Diagrammatic illustration of the structure of a
silicon-on-insulator MOSFET.

of distortion of the crystal lattice is to redistribute energy levels in the conduction band so that it becomes more difficult for electrons to move to higher energy levels, which means less variability in their kinetic energies and hence velocities. The fact that it is easier for conduction electrons to move in certain directions with less variability in their velocities means that in the presence of an electric field, the mean scattering time between collisions increases. The drift velocity therefore increases, which again leads to higher mobility. Similar considerations apply to holes. The result is that electron mobility is doubled in strained Si grown over SiGe with 25% Ge. The improvement in hole mobility is less, but increases more rapidly than electron mobility at higher strain levels. The increased mobility is utilized in MOSFETs by having the SiGe doped in the same way as the substrate in a normal transistor, with the strained Si layer constituting the channel.

Metal gates in combination with gate insulation of high dielectric constant, the so-called **high-k dielectrics**, have been used in the 45 nm and later technologies to enhance speed by up to 40%. A "figure of merit" for the improvement in switching speed is the ratio of the turn-on current of the channel to its leakage current when the transistor is off. This ratio is typically 700 μA/1 nA per μm for a polysilicon gate and about 1000 μA/1 nA per μm for the metal gate/high-k stack. The materials used for the stack and its method of fabrication vary between different manufacturers. Aluminum is used in some processes, but if the metal is to withstand the high temperature required to activate the dopants in the source and drain (Section 4.6, Chapter 4), a metal having a high melting point, such as molybdenum is used. The high-k dielectric could be a thin layer of hafnium dioxide on top of silicon dioxide, or a layer of dysprosium oxide on top of silicon oxynitride. For best performance, the work function of the metal has to be tuned differently for NMOS and PMOS transistors by some form of doping of the metal.

Application Window 5.3.1 Thin-Film Transistors

For better quality and stability of the image on an LCD display, every pixel of each of the three primary colors should have its own transistor switch. Because of the large area of the LCD panel, it is impractical to use conventional MOSFETs, having a common, single-crystal substrate. Instead, thin-film transistors (TFTs) of hydrogenated amorphous silicon Si (a-Si:H) are formed on a glass substrate. Because the silicon is noncrystalline, the motion of electrons is severely impeded, particularly by unsatisfied bonds of Si atoms, which capture conduction electrons. Hydrogenating the silicon provides hydrogen atoms that satisfy some of these bonds. Even then, electron mobility is very low, typically less than 1 cm^2/Vs. However, the main consideration is neither gain nor speed, but rather a low leakage current when the transistor is off, so that the charge on the pixel does not leak too rapidly.

The structure of a TFT is diagrammatically illustrated in Figure 5.3.14. The structure is inverted, so that the metal gate is at the bottom. Above the gate is an insulation layer of SiO$_2$ or SiN, then a layer of undoped (intrinsic) i-a-Si:H that forms the channel. The reason

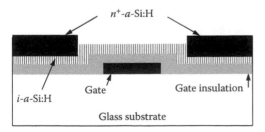

n$^+$-a-Si:H

i-a-Si:H   Gate   Gate insulation

Glass substrate

FIGURE 5.3.14
Diagramatic illustration of the structure of a thin-film transistor.

for the inverted structure is twofold: (1) it makes it easier to form the n^+-a-Si:H source and drain, and (2) it avoids damage to the channel if it is formed before the dielectric layer.

Other TFT structures are possible. Polysilicon on quartz can be used in a conventional, top-gate structure, which greatly increases electron mobility. TFTs can be combined with OLEDs in flexible displays (Application Window 3.4.1, Chapter 3). Interestingly, using high-k dielectrics for gate insulation allows the gate to be placed on the side of the channel, rather than on top of it, which simplifies fabrication.

5.4 Depletion-Type MOSFETs

In a depletion-type, n-channel MOSFET (D-NMOS), donor impurities are implanted at the semiconductor surface, so that a channel exists between the source and drain even with $v_{GS} = 0$. Any $v_{DS} > 0$ causes electrons to flow from source to drain, as in an E-NMOS transistor. To reduce i_D, v_{GS} should be made negative so as to repel electrons from the channel and therefore deplete it of current carriers; hence the designation depletion type. At a sufficiently negative v_{GS}, equal to the threshold V_{tnD}, electron concentration, and hence i_D, become almost zero.

The behavior of the D-NMOS transistor is exactly analogous to that of an E-NMOS transistor and is described by the same equations, except that V_{tnD} has a negative value. Figure 5.4.1 shows the characteristics of a D-NMOS transistor that is identical to the E-NMOS transistor of Figure 5.2.6 having $V_{tn} = 1$ V, except that $V_{tnD} = -3$ V. The boundary between the saturation and triode regions on the output characteristics (Figure 5.4.1b) is still defined by $v_{DS(sat)} = v_{GS} - V_{tnD}$ but is shifted to more positive v_{DS} because V_{tnD} is negative. When $v_{GS} = 0$, $v_{DS(sat)} = -V_{tnD} = \overline{V}_{tnD} = 3$ V. Note that the transistor could also be operated in the enhancement mode, with $v_{GS} > 0$, as long as the rated power dissipation of the transistor is not exceeded.

A value of i_D that is of special interest in a D-NMOS transistor is $I_{DSS} = i_{D(sat)}$ for $v_{GS} = 0$ (Figure 5.4.1a). It follows from Equation 5.2.4, neglecting channel-length modulation, that:

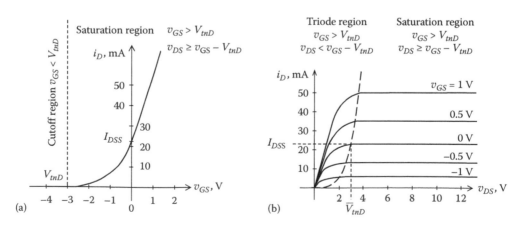

FIGURE 5.4.1
Depletion-type MOSFET: (a) transfer characteristic; (b) output characteristics.

$$I_{DSS} = \frac{1}{2} k_n V_{tnD}^2 \tag{5.4.1}$$

In terms of I_{DSS}, i_D in the saturation region is given by:

$$i_D = I_{DSS}\left(1 + \frac{v_{GS}}{V_{tnD}}\right)^2 (1 + \lambda v_{DS}) \tag{5.4.2}$$

The symbols for a D-NMOS transistor are illustrated in Figure 5.4.2. The main difference between the symbols of Figure 5.4.2a and b, and those of Figure 5.2.8a and b for an E-NMOS transistor, is that the vertical line representing the channel is solid rather than broken, which signifies the presence of a channel with no bias applied. This same feature is symbolized in Figure 5.4.2c and d by a thick vertical line. The transconductance is:

$$g_m = \frac{di_D}{dv_{GS}} = 2\frac{I_{DSS}}{V_{tnD}}\left(1 + \frac{v_{GS}}{V_{tnD}}\right)(1 + \lambda v_{DS}) \tag{5.4.3}$$

As in the case of the enhancement-type transistor, a depletion-type p-channel MOS (D-PMOS) transistor can be constructed having an n-type body and p^+ source and drain regions, with acceptor impurities implanted at the surface of the semiconductor to create a p-channel between source and drain with no bias applied. Operation is completely analogous to that of a D-NMOS transistor, but with the roles of electrons and holes interchanged. The symbol for a D-PMOS transistor is the same as that of the D-NMOS transistor shown in Figure 5.4.2 but with the direction of the arrow reversed in Figure 5.4.2a through c. A bubble is placed on the gate in Figure 5.4.2d and the upper and lower terminals become the source and drain, respectively. If the same positive directions of i_D, v_{DS}, and v_{GS} are assigned for a D-PMOS transistor, as for a D-NMOS transistor, then all these quantities will have negative numerical values. To use positive numerical values in the equations, then as in the case of an E-PMOS transistor, the assigned positive direction of i_D is reversed, and new terminal voltages are defined as $v_{SG} = -v_{GS}$ and $v_{SD} = -v_{DS}$. Note that the threshold of a D-PMOS transistor is positive.

Figure 5.4.3 summarizes the voltage polarities in MOSFETs. The shaded triangle depicts the normal operating range of v_{GS}. The direction of the v_{GS} arrow is that of increasing conduction, so that the width of the triangle signifies the degree of conduction. The following generalizations can be made: (1) V_t is of the same polarity as v_{DS} in enhancement-type devices and is of opposite polarity in depletion-type devices; (2) in all cases, the device becomes more conducting, that is, $|i_D|$ increases, as v_{GS} changes toward v_{DS}; (3) in both n-channel and p-channel devices, current carriers, which are the majority carriers of the source and drain, flow from source to drain.

The relation that defines the edge of the saturation region for all MOSFETs is:

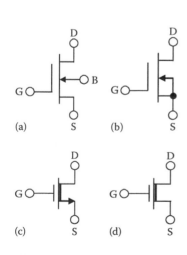

FIGURE 5.4.2
(a)–(d) illustrate various symbols for a D-NMOS transistor.

$$v_{DS(sat)} = v_{GS} - V_t \tag{5.4.4}$$

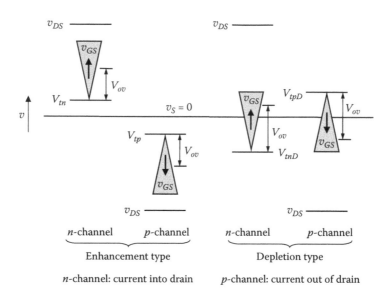

FIGURE 5.4.3

Relative polarities of terminal voltages for various types of MOSFETs.

TABLE 5.4.1

Relations for the Four Types of MOSFETs

MOSFET	V_t	Saturation Voltage	Saturation Current, $i_{D(\text{sat})}$	Threshold Modification by Body Effect
E-NMOS	>0	$v_{DS(\text{sat})} = v_{GS} - V_{tn}$	$\frac{1}{2}k_n v_{DS(\text{sat})}^2(1 + \lambda v_{DS(\text{sat})})$	$V_{tn} = V_{tn0} + \gamma(\sqrt{2|\phi_F| + |V_{SB}|} - \sqrt{2|\phi_F|})$
E-PMOS	<0	$v_{SD(\text{sat})} = v_{SG} - \overline{V}_{tp}$	$\frac{1}{2}k_p v_{SD(\text{sat})}^2(1 + \lambda v_{SD(\text{sat})})$	$\overline{V}_{tp} = \overline{V}_{tp0} + \gamma(\sqrt{2|\phi_F| + |V_{SB}|} - \sqrt{2|\phi_F|})$
D-NMOS	<0	$v_{DS(\text{sat})} = v_{GS} + \overline{V}_{tnD}$	$\frac{1}{2}k_n v_{DS(\text{sat})}^2(1 + \lambda v_{DS(\text{sat})})$	$\overline{V}_{tnD} = \overline{V}_{tnD0} - \gamma(\sqrt{2|\phi_F| + |V_{SB}|} - \sqrt{2|\phi_F|})$
D-PMOS	>0	$v_{SD(\text{sat})} = v_{SG} + V_{tpD}$	$\frac{1}{2}k_p v_{SD(\text{sat})}^2(1 + \lambda v_{SD(\text{sat})})$	$V_{tpD} = V_{tpD0} - \gamma(\sqrt{2|\phi_F| + |V_{SB}|} - \sqrt{2|\phi_F|})$

Equation 5.4.4 is adapted to the four types of MOSFETs as indicated in Table 5.4.1. These relations can be derived from first principles (Exercise 5.4.1) and can be summarized as follows: (1) v_{DS} and v_{GS} are used for NMOS transistors, whereas v_{SD} and v_{SG} are used for PMOS transistors, so that all these terms are positive; (2) a positive value of threshold in Equation 5.4.4 is used for all transistors, which means using $\overline{V}_{tp} = -V_{tp}$ for E-PMOS transistors and $\overline{V}_{tnD} = -V_{tnD}$ for D-NMOS transistors; (3) For enhancement devices, the sign of the threshold term in Equation 5.4.4 is negative, whereas for depletion devices, the sign of the threshold term is positive. These signs follow readily from Equation 5.4.4. In the case of an E-PMOS transistor, for example, if both sides of the equation are negated, it becomes $-v_{DS(\text{sat})} = -v_{GS} + V_{tp}$. Changing to $v_{SD(\text{sat})}$ and v_{SG} gives $v_{SD(\text{sat})} = v_{SG} + V_{tp}$. The relation in Table 5.4.1 follows by substituting $\overline{V}_{tp} = -V_{tp}$. Note the variation in threshold due to the body effect. In all cases, the body effect reduces the drain current for given gate–source and drain–source voltages, as explained earlier, which means that the magnitude of V_{ov} must decrease. The reduction in the magnitude of V_{ov} implies an increase in the threshold in enhancement devices, and a decrease in the magnitude of the

threshold in depletion devices, in accordance with the sign of the threshold term in Table 5.4.1 (see Example SE5.2). It is seen that in n-channel MOSFETs (E-NMOS and D-NMOS), the body effect shifts the threshold in the positive direction, whereas in p-channel MOSFETs (E-PMOS and D-PMOS), the threshold is shifted in the negative direction.

Exercise 5.4.1

Derive the relations in the third column of Table 5.4.1 from first principles by setting the total voltage drop, from gate to source to drain, equal to the threshold. ■

Exercise 5.4.2

Given a D-NMOS transistor having a device transconductance parameter k_n of 400 μA/V^2 and a threshold of -3 V. Determine I_{DSS}.
    Answer: 1.8 mA. ■

Exercise 5.4.3

By setting $v_{GS}=0$ in Equation 5.3.1, show that $I_{DSS} = \dfrac{1}{2}k_n V_{tn}^2$ in the absence of channel-length modulation. Note that since this is proportional to V_{tn}^2, it applies equally well to enhancement-type devices, where I_{DSSE} denotes I_{DSS} and I_{DSSE}/V_{tn}^2 simply becomes another way of writing $\dfrac{1}{2}k_n$. Show that for an enhancement-type device, Equation 5.3.1 can then be written as:

$$i = \frac{I_{DSSE}}{V_{tn}^2}(v_{GS} + \overline{V}_{tn})^2(1+\lambda v_{DS}) = I_{DSSE}\left(1 + \frac{v_{GS}}{\overline{V}_{tn}}\right)^2(1+\lambda v_{DS}) \tag{5.4.5}$$

Diode Connection

In some applications, the MOSFET is connected as a two-terminal device. In such a **diode connection**, it is natural to connect together the gate and source of a depletion-type transistor, leaving the drain as the other terminal, as illustrated in Figure 5.4.4 for an n-channel transistor. The i–v relation is the same as the output characteristic for $v_{GS}=0$ in Figure 5.4.1b. Once the transistor reaches saturation, $i=I_{DSS}$, as given by Equation 5.4.1. Beyond saturation, and taking into account channel-length modulation, it follows from Equation 5.3.1 that:

$$i = I_{DSS}(1+\lambda v) \tag{5.4.6}$$

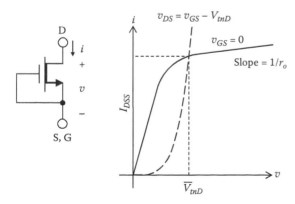

FIGURE 5.4.4
Diode connection of D-PMOS transistor.

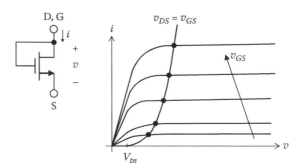

FIGURE 5.4.5
Diode connection of E-NMOS transistor.

In enhancement-type devices, the gate cannot be connected directly to the source, as the transistor is cut off when $v_{GS} = 0$. Although it is possible, in principle, to apply a suitable bias between gate and source, this is inconvenient in practice, and the device is no longer two terminal. The gate is therefore connected to the drain, as illustrated in Figure 5.4.5 for an n-channel transistor. In the saturation region, $v_{DS} > v_{GS} - V_{tn}$. If v_{GS} is replaced by v_{DS}, this inequality becomes $v_{DS} > v_{DS} - V_{tn}$, which is always true, because $V_{tn} > 0$. Hence, once $v_{DS} = v_{GS}$ exceeds V_{tn}, the transistor is always in saturation. The locus of points $v_{DS} = v_{GS}$ on the output characteristics is the $i–v$ relation for the diode-connected transistor, and is in fact, the transfer characteristic superposed on the output characteristics, neglecting channel-length modulation. If this effect is taken into account, the locus is given by Equation 5.3.1 with $v_{DS} = v_{GS} = v$.

It is seen that the major difference between diode-connected enhancement-type transistors and depletion-type transistors is in their incremental resistance, which is the reciprocal of the slope of the respective $i–v$ relation at a given operating point. For an enhancement-type transistor, the incremental output resistance is $(1/g_m)\|r_o \cong 1/g_m$ (Exercise 5.4.4) and is $1/g_m$ if channel-width modulation is neglected ($r_o \to \infty$), in which case the i_D vs. v_{DS} relation is the same as the transfer characteristic. Whereas $1/g_m$ is typically about 1 kΩ (Example 5.3.1), the incremental output resistance for a depletion-type transistor is r_o, and is typically at least few tens of kΩ.

Exercise 5.4.4

(a) Consider the small-signal equivalent circuit of Figure 5.3.2 with the gate connected to the drain, which makes the current of the dependent source $g_m v_{ds}$. Deduce that the source can be replaced by a resistance $1/g_m$, so that the small-signal resistance of the diode-connected transistor is $(1/g_m)\|r_o$. (b) Determine di/dv in Equation 5.3.1, where $v = v_{DS} = v_{GS}$, and show that it is equivalent to $g_m + 1/r_o$. ∎

5.5 Complementary MOSFETs

CMOS technology, which involves fabricating both NMOS and PMOS transistors on the same chip (Section 4.6, Chapter 4), is at present the dominant technology for digital ICs. It offers many advantages over using NMOS or PMOS transistors alone, as explained in this section.

CMOS Amplifier

In a CMOS amplifier, the drain load for a given MOSFET is a complementary MOSFET. Figure 5.5.1a shows an example, where a diode-connected D-PMOS transistor Q_p is the drain load of an E-NMOS transistor Q_n. In ICs, the substrate of Q_p would normally be connected to V_{DD}, whereas that of Q_n would be grounded. There is no body effect in this case.

The v_O–v_I relation, or voltage transfer characteristic (VTC), assuming zero output current, can be readily derived. From KVL in the output circuit, $v_{SDp} + v_{DSn} = V_{DD}$, where $v_{DSn} = v_O$. This is of the same form as Equation 5.2.15, with $R_D i_D$ replaced by v_{SDp}. Hence, the same type of graphical construction can be used to determine the operating point, with the load line replaced by a "load curve" (Figure 5.5.1b). In this figure, the $v_{SGp} = 0$ output characteristic of Q_p is drawn "backward" with V_{DD} as the origin, and the output characteristics of Q_n are labeled with v_I instead of v_{GS}. The intersection of the output characteristic for a particular v_I with the Q_p load curve defines the operating point P, because at this point, i_D is the same in both transistors, assuming zero load current, and $v_{SDp} + v_{DSn} = V_{DD}$.

It is seen from Figure 5.5.1b that for $v_I < V_{tn}$, Q_n is cut off, $i_D = 0$, and $v_{SDp} = 0$, so that $v_O = V_{DD}$ (Figure 5.5.1c). When $v_I = V_{tn}$, Q_n begins to conduct, and the operating point is at A with Q_n in saturation and Q_p at the beginning of the triode region. When v_I increases so that the operating point moves to B, Q_p is at the edge of the saturation region, $v_{SDp} = V_{tpD}$ with $V_{SGp} = 0$, and $v_O = V_{DD} - V_{tpD}$, where $V_{tpD} > 0$. A further small increase in v_I causes a large change in v_O, because the operating point P is the intersection of two nearly

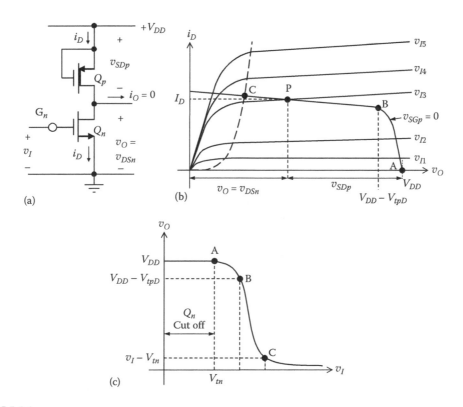

FIGURE 5.5.1
CMOS amplifier having a diode-connected D-PMOS transistor load: (a) circuit diagram; (b) graphical construction based on output characteristics; and (c) transfer characteristic.

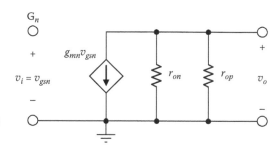

FIGURE 5.5.2
Small-signal equivalent circuit of the CMOS amplifier of
Figure 5.5.1.

horizontal lines. At C, Q_n is at the edge of the triode region, $v_{DS} = v_{GS} - V_{tn}$, or $v_O = v_I - V_{tn}$.
For large v_I, v_O is small.

Exercise 5.5.1

Determine the states of the two transistors in Figure 5.5.1a for the different parts of the transfer
characteristic in Figure 5.5.1c.
   Answers: $v_I < v_{IA}$: Q_n off, Q_p in triode region; $v_{IA} < v_I < v_{IB}$: Q_n in saturation, Q_p in triode
region; $v_{IB} < v_I < v_{IC}$: both transistors in saturation; $v_I > v_{IC}$: Q_n in triode region, Q_p in
saturation.   ∎

For small-signal operation, the equivalent circuit is shown in Figure 5.5.2. Q_p is
replaced by its incremental output resistance r_{op} at the operating point. The small-signal
voltage gain is:

$$\frac{v_o}{v_i} = -g_{mn}(r_{on}\|r_{op}) \tag{5.5.1}$$

The gain is largest in the steepest part of the VTC, when both transistors are in saturation.
   What are the advantages of using a complementary transistor in the circuit of Figure 5.5.1a
rather than a resistor? It is clear from Figure 5.5.1b that Q_p allows a larger r_{op}, and hence
a larger gain (Equation 5.5.1), but with a much smaller dc voltage drop than is possible with a
fixed resistor. The dc voltage drop across Q_p can be in the range of few volts to less than 10 V,
whereas r_{op} is at least few tens of kΩ. Suppose that $r_{op} = 20$ kΩ, and that a resistance R_D of this
value is used to give the same small-signal voltage gain. Assume that in the CMOS case,
$V_{DD} = 12$ V, $I_D = 2$ mA, and $V_O = V_{DD}/2 = 6$ V at the operating point. When $R_D = 20$ kΩ is
used, the voltage drop across R_D is 40 V, and V_{DD} has to be 46 V instead of 12 V. Hence,

Concept: *A diode-connected transistor load provides a higher voltage gain with a lower
dc voltage drop than is possible with a linear resistor.*

Another advantage is that to obtain a value of $R_D = r_{op}$ in an IC would require a much
larger chip area than the PMOS transistor.

Exercise 5.5.2

Assume that in Figure 5.5.1a an E-PMOS transistor is used as the load with proper biasing to give
a high r_o. (a) Draw a suitable biasing circuit for this transistor. (b) How will the voltage gain
change if the gate is connected to the drain? (c) How will v_O change for $v_I \le V_{tn}$? (d) How will the
VTC be different if an E-NMOS transistor is used instead of the E-PMOS transistor assuming that
in both of these transistors are matched to the E-NMOS driver?

Answers: (a) Possible biasing schemes: (i) resistive voltage divider, (ii) an identical diode-connected PMOS supplied by a current source, with the two gates joined together, which gives a current mirror (Section 8.1, Chapter 8). (b) The voltage gain becomes $g_{mn}r_{on}\|r_{op}\|(1/g_{mp})$ instead of $g_{mn}(r_{on}\|r_{op})$. (c) $v_O = V_{DD} - \overline{V}_{tp}$ instead of V_{DD}. (d) No difference. ∎

A possible limitation of the circuit of Figure 5.5.1a is that the $i-v$ relation of Q_p is limited to a single output characteristic, namely that of $v_{SGp} = 0$. A current mirror (Section 8.1, Chapter 8) provides more flexibility by allowing the use of other output characteristics of Q_p.

Simulation Example 5.5.1 MOS Amplifier with Diode-Connected D-PMOS Transistor

It is required to simulate the v_O-v_I characteristic of an amplifier consisting of an E-NMOS transistor with a D-PMOS load. Assume that $V_{DD} = 12$ V, $k'_n = 40$ μA/V^2, $(W/L)_n = (10$ μm/1 μm$)$, $k'_p = 16$ μA/V^2, $(W/L)_p = (25$ μm/1 μm$)$, $V_{tn} = V_{tpD} = 2$ V, and $\lambda_n = \lambda_p = 0.015$V^{-1}.

SIMULATION

MbreakN3 from the BREAKOUT library is used for the E-NMOS transistor and MbreakP3D for the D-PMOS transistor, both having the substrate connected to the source. The first simulation is based on the same circuit as in Figure 5.5.1a. After entering the schematic, select the E-NMOS transistor, right click on its symbol, and choose Edit/PSpice Model. In the window of the PSpice Model Editor enter in the first line after NMOS: Kp = 40u Vto = 2 lambda = 0.015. In the Property Editor spreadsheet, enter 1u under L and 10u under W. Repeat the same procedure for the D-PMOS transistor, entering kp = 16u Vto = 2 lambda = 0.015 in the PSpice Model editor, and 1u under L and 25u under W in the Property Editor spreadsheet. In the simulation profile, select DC Sweep, and vary the input dc source between 0 and 8 V. Simulation results are shown in Figure 5.5.3.

The two transistors are matched, since $k_n = k_p$ and $V_{tn} = V_{tpD}$. The $v_{SGp} = 0$ output characteristic of the PMOS transistor corresponds to $V_{ov} = 2$ V. Hence, the corresponding output characteristic of the NMOS transistor has $v_{GSn} = v_I = 2 + V_{tn} = 4$ V. The intersection of the two characteristics is at $V_{DD}/2 = 6$ V. Analytically, $(k_n/2)(v_I - V_{tn})^2(1 + \lambda V_{DD}/2) = (k_p/2)(0 + V_{tpD})^2(1 + \lambda V_{DD}/2)$. With $k_n = k_p$ and $V_{tpD} = V_{tn}$, this gives $v_I = 4$ V, $i_D = 0.87$ mA. In Figure 5.5.3, the midpoint of the straight part of the characteristic is at $v_I = 4$ V and $v_O = 6$ V.

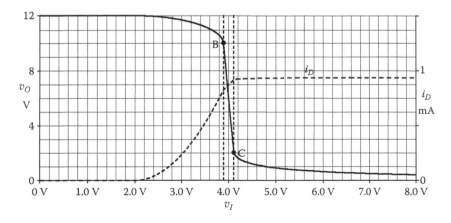

FIGURE 5.5.3
Figure for Simulation Example 5.5.1.

The D-PMOS transistor enters saturation when $v_O = V_{DD} - V_{tpD} = 10$ V, corresponding to $v_I = 3.89$ V and $i_{Dp(sat)} = 0.824$ mA (point B of Figure 5.5.3). This agrees with Equation 5.3.3, which gives: $i_{Dp(sat)} = \frac{1}{2}k_p v^2_{SDp(sat)}(1 + \lambda v_{SDp(sat)}) = 0.2(2)^2(1 + 0.015 \times 2) = 0.824$ mA. With this value of $i_{Dp(sat)}$, Equation 5.3.1 gives $v_I = v_{GSn} = 3.89$ V.

The E-NMOS transistor leaves saturation when $v_{DSn(sat)} = v_I - V_{tn}$. To find $v_{O(sat)}$, we equate the saturation currents of the two transistors: $\frac{1}{2}k_n v^2_{DSn(sat)}(1 + \lambda v_{DSn(sat)}) = \frac{1}{2}k_p$ $4[1 + \lambda(12 - v_{DSn(sat)})]$. This gives $v_{DSn(sat)} = 2.11$ V, $v_I = 4.11$ V and $i_{Dn(sat)} = 0.919$ mA, in agreement with the values for point C in Figure 5.5.3. The difference between $i_{Dn(sat)}$ and i_{Dp} $_{(sat)}$, which is 0.095 mA, is due to the nonzero slope of the $v_{SGp} = 0$ output characteristic, which is $0.2(2)^2 \times 0.015(10 - 2.11) = 0.095$ mA.

For $v_I > 4.11$ V, the E-NMOS transistor is in the triode region, v_O drops to a low value, v_{SDp} varies little, and the current stays nearly constant. The limiting value of i_D is when $v_O \cong 0$. The saturation current of the PMOS transistor is $0.2 \times 4(1 + 0.015 \times 12) = 0.944$ mA.

The small-signal voltage gain for any v_I is the slope of the VTC characteristic at the given v_I. The slope can be plotted by selecting Add Trace and entering $-D(V(VO))$. The largest magnitude of this slope occurs at point B (Problem P5.2.18) and is approximately 38.3. To apply Equation 5.5.1, we have, from Equation 5.3.11, $g_{mn} = 2 \times 0.824/1.89 = 0.872$ mA/V. From Equation 5.3.4, $r_{op} = (1 + 0.015 \times 2)/(0.015 \times 0.824) \equiv 83.3$ kΩ. From Equation 5.3.1, $g_{dsn} = 0.2(1.89)^2 \times 0.015$ and $r_{on} = 1/(0.015 \times 0.2 \times (1.89)^2) \equiv 93.3$ kΩ. The small-signal voltage gain is $0.872(83.3 \| 93.3) = 38.4$.

CMOS Transmission Gate

Figure 5.5.4a shows a CMOS transmission gate using two complementary MOSFETs. The source and drain are not marked, since either terminal a or b of Q_n or Q_p can act as source, the other as drain, because of symmetry of the transistors. It is assumed that the control voltage v_C could be at either +5 V or −5 V, and that the input v_I is in the range of ±5 V. The substrates of the E-PMOS and E-NMOS transistors are connected to +5 V and −5 V, respectively.

When $v_C = -5$ V, and $-5 \text{ V} \le v_I \le 0$, Q_n cannot be conducting. This is because if Q_n is on, current would flow from ground, through the load and the transistor, to the input, making b_n the drain and a_n the source. The source voltage would equal v_I, which makes the gate-to-source voltage between zero and −5 V. Under these conditions, Q_n must be off.

With $v_C = -5$ V, and $0 \le v_I \le 5$ V, Q_n again cannot be conducting. This is because if Q_n is on, current would flow from the input, through the transistor and load, to ground, making

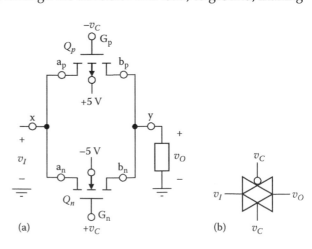

FIGURE 5.5.4
CMOS transmission gate: (a) circuit; (b) symbol.   (a)

a_n the drain and b_n the source. The source voltage would also equal v_I, neglecting the voltage drop in the transistor compared to that of the load, which makes the gate-to-source voltage between -5 and -10 V. Q_n must again be off.

A similar argument can be applied to show that Q_p is also off (Exercise 5.5.3). In fact, one may argue, on the basis of complementarity, that if $v_{Gn} = -5$ V cuts off the NMOS transistor, then for the same range of v_I, the PMOS transistor must also be cut off when $v_{Gp} = +5$ V. It follows that when $v_C = -5$ V, both transistors are off for inputs -5 V $\leq v_I \leq +5$ V. Theoretically, the transmission gate as a whole behaves as an open switch, and the load is effectively disconnected from the input v_I. In practice, the resistance between terminals x and y when both transistors are off is very much larger than the load resistance, so $v_O \cong 0$.

Consider next the case when $v_C = v_{Gn} = +5$ V and $v_{Gp} = -5$ V. If -5 V $\leq v_I \leq 0$, then as argued previously on the basis of the direction of current flow, a_n is the source and b_n the drain. This makes $v_{GSn} = 10$ V when $v_I = -5$ V, and $v_{GSn} = 5$ V when $v_I = 0$. Q_n will be turned hard on, as V_{GSn} is considerably larger than V_{tn}. The transistor is in the triode region, v_{DSn} is small compared to v_O, and $v_O \cong v_I$. As for Q_p, assume, for the sake of argument, that $V_{tp} = -2$ V and that $v_I = -1$ V. Since Q_n is on, b_p will be slightly positive with respect to a_p by the small voltage v_{DSn}. With $v_{Gp} = -5$ V, Q_p gate will be more negative than a_p or b_p by about 4 V, which is more than the threshold. Q_p is therefore conducting, with b_p acting as the source and a_p as the drain. If v_I goes more negative than -3 V, say -4 V, then because Q_n is on, the voltage of b_p is about -4 V. The gate to source voltage of Q_p is now less than the threshold and this transistor will be off.

An analogous argument shows that when $0 \leq v_I \leq 5$ V, Q_p is always on, but Q_n is on as long as $v_I < 5 - V_{tn}$ and is off when $v_I > 5 - V_{tn}$. The states of the two transistors are summarized in Figure 5.5.5, for the case $V_{tn} = +2$ V, and $V_{tp} = -2$ V. The width of the triangle is proportional to the magnitude of v_{GS}, so that the wider the triangle the more conducting is the respective transistor. As long as $v_C = +5$ V, a low-resistance path exists between the input and output terminals, through one or both transistors. Note that the transmission gate is symmetrical, or bilateral, that is, the input and output are interchangeable. This is reflected in the transmission gate symbol (Figure 5.5.4b), showing two superimposed triangles pointing in opposite directions.

Exercise 5.5.3

Verify that in the transmission gate Q_p is off when $v_C = -5$ V. ∎

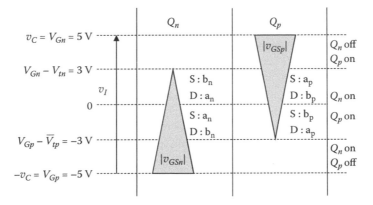

FIGURE 5.5.5
States of conduction of the transistors in a CMOS transmission gate.

It may be argued that either Q_n or Q_p alone can be used as a transmission gate. Whereas this is true, the CMOS combination has two advantages compared to using say Q_n alone: (1) In order to allow transmission of v_I in the range of ± 5 V, v_C must be $(5 + V_{tn})$. In the CMOS combination, v_C is conveniently equal to the positive limit of v_I. (2) As v_I increases from -5 V, Q_n becomes less conducting, and its on-resistance increases (Example 5.5.2). This not only decreases the magnitude of v_O, because of the voltage divider effect between the on-resistance and the load, but introduces phase shift as well when the load is not a pure resistance. In the CMOS transmission gate, the on-resistance is the parallel on-resistances of both transistors. As one transistor becomes less conducting, the other transistor becomes more conducting. Hence, the combined on-resistance stays more nearly constant than that of either transistor alone. This, in turn, makes the attenuation and phase shift more nearly independent of signal level.

Simulation Example 5.5.2 CMOS Transmission Gate

In a CMOS transmission gate $k_n = k_p = 400$ µA/V^2, $V_{tn} = -V_{tp} = 2$ V, and $R_L = 40$ kΩ. It is required to determine v_O and the resistance of the switch, when conducting, for the following values of v_I: (a) 0, (b) -1 V, (c) -2 V, (d) -3 V, and (e) -5 V, neglecting the body effect. It is also required to simulate the operation of the transmission gate, with and without the body effect.

ANALYSIS

When the transmission gate conducts, with $v_I \leq 0$, a_n and b_p in Figure 5.5.4a are sources, a_p and b_n are drains. The drain–source voltage is small, as long as the switch resistance r_{xy} is small compared to the load resistance. Equation 5.2.3 can be used to calculate r_{DSn} and r_{DSp}:

$$r_{DSn} = \frac{1}{k_n(v_{GSn} - V_{tn})}, \quad r_{DSp} = \frac{1}{k_n(v_{SGp} - \overline{V}_{tp})}, \quad r_{xy} = r_{DSn} \| r_{DSp} \qquad (5.5.2)$$

(a) If $v_I = 0$, there is no voltage in the drain–source circuit to drive current. Hence, $v_O = 0$. From Equation 5.5.2, $r_{DSn} = 1/(0.4(5 - 2)) = 0.83$ kΩ, $r_{DSp} = 1/(0.4(5 - 2)) = 0.83$ kΩ, $r_{xy} = 0.42$ kΩ.

(b) If $v_I = -1$ V, $v_{GSn} = 5 - (-1) = 6$ V, $v_{GSp} = v_O - (-5) = v_O + 5$ V. Hence, $r_{DSn} = 1/(0.4(6 - 2)) = 0.625$ kΩ, $r_{DSp} = 1/(0.4(v_O + 5 - 2)) = 2.5/(v_O + 3)$ kΩ, and $r_{xy} = 2.5/(v_O + 7)$. But $v_O = -1 \times 40/(40 + r_{xy})$. Solving for v_O and r_{xy} gives: $v_O = -0.99$ V and $r_{xy} = 0.416$ kΩ. Note that the assumption of small drain–source voltage is justified. Since $v_{DSn} < v_{GSn} - V_{tn}$ and $v_{DSn} < v_{GSn} - \overline{V}_{tp}$, both transistors operate in the triode region. $r_{DSp} = 1.24$ kΩ, which is larger than r_{DSn} because Q_n is more conducting than Q_p. The load current is $i_L = v_O/40$ kΩ $\equiv 24.7$ µA.

(c) If $v_I = -2$ V, $v_{GSn} = 5 - (-2) = 7$ V, $r_{DSn} = 1/(0.4(7 - 2)) = 0.5$ kΩ, $v_{GSp} = v_O + 5$ V, $r_{DSp} = 2.5/(v_O + 3)$ kΩ, $r_{xy} = 2.5/(v_O + 8)$, and $v_O = -2 \times 40/(40 + r_{xy})$. Solving for v_O and r_{xy} gives: $v_O = -1.98$ V and $r_{xy} = 0.413$ kΩ; $r_{DSp} = 2.38$ kΩ. Note that as r_{DSn} decreased from 0.625 to 0.5 kΩ, r_{DSp} increased from 1.24 to 2.38 kΩ, but r_{xy} stayed almost constant.

(d) If $v_I = -3$ V, $v_{GSn} = 5 - (-3) = 8$ V, $r_{DSn} = 1/(0.4(8 - 2)) = 0.417$ kΩ, Q_p is now just off. Hence, $r_{xy} = 0.417$ kΩ and $v_O = -3 \times 40/(40 + r_{xy}) = -2.97$ V.

(e) If $v_I = -5$ V, $v_{GSn} = 5 - (-5) = 10$ V, $r_{DSn} = 1/(0.4(10 - 2)) = 0.313$ kΩ. Hence, $r_{xy} = 0.313$ kΩ, and $v_O = -5 \times 40/(40 + r_{xy}) = -4.96$ V. Note that as long as both transistors are on, r_{xy} does not change much. It decreases when Q_p is off and Q_n becomes more conducting.

As can be seen from Figure 5.5.5, the situation for $v_I > 0$ is analogous, with the roles of the E-NMOS and E-PMOS transistors interchanged.

SIMULATION

The schematic is entered as in Figure 5.5.4a, the transistor parameters being as in Example 5.3.2. v_I is a VDC source that is swept between -5 V and $+5$ V. The plot of v_O vs. v_I is a straight line passing through the origin, with $v_O = -4.96$ V for $v_I = -5$ V and $v_O = +4.96$ V for $v_I = +5$ V. The body effect makes no noticeable difference to the v_O vs. v_I characteristic because of the large R_L (Problem 5.2.23).

Exercise 5.5.4

Include the body effect for the case $v_I = 0$ in Example 5.5.2, assuming $\gamma = 0.3$ $V^{1/2}$ and $2|\phi_F| = 0.6$ V. What complicates taking the body effect into account when $v_I \neq 0$?

Answers: $r_{DSn} = r_{DSp} = 0.99$ kΩ; the source voltage of one of the transistors is the unknown output voltage v_O, so that $|V_{SB}|$, and hence V_t of the transistor will be a function of v_O.

CMOS Inverter

The CMOS inverter consists of an E-NMOS transistor and an E-PMOS transistor connected between V_{DD} and ground as in Figure 5.5.1a but with the gates of both transistors connected to v_I, which assumes one of two values: a low value, $v_{IL} \cong 0$, and a high value $v_{IH} \cong V_{DD}$. When v_I is low, v_O is high, and when v_I is high, v_O is low; hence, the name inverter. The inverter is a basic circuit in digital electronics, and its analysis provides a good indication of the performance of the logic family to which the inverter belongs. The CMOS inverter is considered in detail in Section 13.1, Chapter 13.

5.6 Junction Field-Effect Transistor

Structure

The basic structure of an n-channel JFET is illustrated in longitudinal section in Figure 5.6.1. Essentially, it consists of a slab of semiconductor having terminals at both ends, labeled source (S) and drain (D), and p^+ regions on opposite faces, to which a gate terminal (G) is made. The region between source and drain forms the channel through which electrons, the

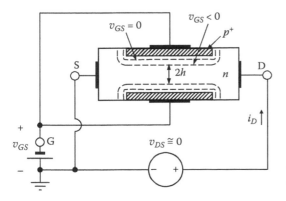

FIGURE 5.6.1
Diagrammatic illustration of the structure of an n-channel JFET.

majority carriers, can flow under the influence of an applied voltage. The source and drain terminals can be interchanged, as the structure is symmetrical.

Operation

Let us assume, to begin with, that the source, drain, and gate are all at zero voltage, that is, $V_{GS} = v_S = 0$. The pn junction formed between the n-channel and the p^+ regions on both sides is at equilibrium. A depletion layer of small width extends mainly into the channel, because the p^+ regions are much more heavily doped. If the gate is made negative with respect to the source ($v_{GS} < 0$), the pn junction becomes reverse biased, and the depletion region widens, reducing the channel depth $2h$. The channel between drain and source thus behaves like a variable resistance r_{DS} controlled by v_{GS}. The value of this resistance is simply $L/\sigma A$, where L is the channel length, $A = 2h \times W$, W being the width of the channel, and σ is the conductivity given by $|q|N_D\mu_e$. For a sufficiently negative $v_{GS} = V_{th}$, and in the absence of any current in the channel, the depletion layer extends all the way across the channel. V_{th} is the threshold voltage, and no current can flow under these conditions because the channel is depleted of carriers.

Example 5.6.1 JFET Threshold

It is required to calculate V_{th} for an n-channel JFET having $N_D = 10^{15}/\text{cm}^3$, $N_A = 10^{18}/\text{cm}^3$, and a channel depth of 4 μm in the absence of a depletion region ($\varepsilon_s = 1.05 \times 10^{-12}$ F/cm).

SOLUTION

In a pn junction having $N_D \ll N_A$, the width of the depletion region on the n-side is, from Equation 3.1.11, Chapter 3, $W_{dn} = \sqrt{2\varepsilon_s V_{np}/|q|N_D}$, where V_{np} is the effective voltage of the n side with respect to the p side. At equilibrium, with no externally applied voltage, $V_{np} = V_{npo}$ and is a positive quantity. When the gate-to-source voltage is V_{th}, and is a negative quantity, V_{np} becomes $V_{npo} - V_{th}$, that is, the effective reverse voltage across the junction becomes $V_{npo} + |V_{th}|$. If the depth of the channel when there is no depletion region is $2h$ and the gate is on both sides of the channel (Figure 5.6.2a), the channel is depleted, or pinched off, when $W_{dn} = h$. This gives

$$V_{np} = h^2 |q|N_D/2\varepsilon_s \tag{5.6.1}$$

Substituting numerical values, $V_{np} = \dfrac{(2 \times 10^{-4})^2 \times 1.6 \times 10^{-19} \times 10^{15}}{2 \times 1.05 \times 10^{-12}} = 3.05$ V. From Equation 3.1.3, Chapter 3, $V_{npo} = 0.026 \ln \dfrac{10^{15} \times 10^{18}}{10^{20}} = 0.78$ V. Hence, $V_{th} = V_{npo} - V_{np} = -2.9$ V.

V_{np} at pinch off is referred to as the **pinch-off voltage** and denoted by V_P.

Exercise 5.6.1

Consider the n-channel JFET of Example 5.6.1, with $v_{GS} = 0$. (a) Calculate the width of the depletion region in the channel; (b) channel resistance, assuming a channel area of 40 μm², a length of 10 μm, and $\mu_e = 800$ cm²/Vs.
  Answers: (a) 1 μm; (b) 19.5 kΩ.   ■

   Consider next the case where a constant, negative $v_{GS} > V_{th}$ is applied. As v_{DS} increases from zero, i_D increases linearly with v_{DS}, because r_{DS} stays almost constant (region I of the

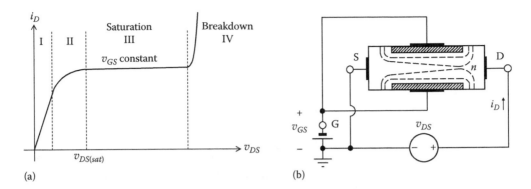

FIGURE 5.6.2
(a) JFET current–voltage relation and (b) channel profile for various values of drain–source voltage.

i–v characteristic in Figure 5.6.2a). With further increase in v_{DS}, and hence i_D, the voltage drop along the channel becomes appreciable, so the reverse bias at the drain increases by $r_{DS}i_D$, the voltage drop along the channel. The channel therefore narrows near the drain (Figure 5.6.2b), and r_{DS} is increased. This means that i_D does not increase as fast with v_{DS} as when v_{DS} is small (region II of the *i–v* characteristic in Figure 5.6.2a). If v_{DS} is increased further, the channel becomes pinched off at the drain, as happens in a MOSFET. i_D now flows through a small, constricted area under the influence of a larger electric field, as explained with the aid of the hydraulic analogy (Figure 5.2.5). Ideally, i_D stays constant as v_{DS} increases, giving a saturation region III of the *i–v* characteristic (Figure 5.6.2a). With higher v_{DS}, the constriction of the channel spreads toward the source, as in the MOSFET, thereby reducing the channel length and increasing i_D somewhat, which is analogous to channel-length modulation in a MOSFET. At a large enough v_{DS}, the drain–gate reverse voltage becomes high enough to cause avalanche breakdown at the drain, so i_D increases sharply (a region IV of the *i–v* characteristic in Figure 5.6.2a). Saturation occurs when $v_{GS} - v_{DS} = V_{th}$, so that in the saturation region, $v_{DS} \geq v_{GS} - V_{th}$. Thus, if $V_{th} = -4$ V, for example, and $v_{GS} = -3$ V, saturation occurs at $V_{DS(sat)} = 1$ V.

The output characteristics of a JFET are illustrated in Figure 5.6.3a and are of the same shape as the i_D–v_{DS} curve of Figure 5.6.2a. Since the saturation current is maximum when

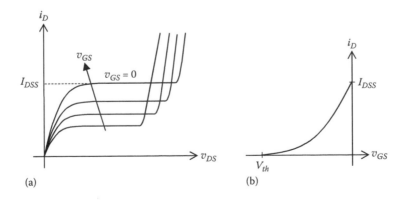

FIGURE 5.6.3
(a) JFET output characteristics and (b) JFET transfer characteristic for a particular value of v_{DS}.

$v_{GS} = 0$ and decreases for $v_{GS} < 0$, the JFET is a depletion-type transistor. It is not operated in an enhancement mode, as in the case of a MOSFET, because $v_{GS} > 0$ can only widen the channel to a small extent by reducing the width of the depletion region at equilibrium. If v_{GS} becomes positive by more than the cut-in voltage of the pn junction, a forward current flows that provides a low resistance path between the gate and source, and transistor action is abolished. Normally, the gate current is very small under reverse bias, of the order of 1 nA or less. Because it is the reverse current of a pn junction, it approximately doubles for every 10°C rise in temperature.

Breakdown occurs when the reverse bias across the junction is large enough to initiate avalanche multiplication. At the onset of breakdown, the more negative v_{GS}, the larger is the maximum electric field in the depletion region at the drain end of the channel, and the *smaller* is v_{DS} required to initiate breakdown.

Current–Voltage Relation

The *i–v* characteristic is derived in Section ST5.6, where it is shown that, subject to some approximations, the drain current in the saturation region is given by the same expression as Equation 5.4.2 for a depletion-type MOSFET:

$$i_D = I_{DSS}\left(1 - \frac{v_{GS}}{V_{th}}\right)^2 (1 + \lambda v_{DS}) \tag{5.6.2}$$

where V_{tD} has been replaced by V_{th}. Despite the assumptions made in deriving Equation 5.6.2, it is found experimentally that this relation is closely approximated.

It follows that the output conductance $g_{ds} = 1/r_o$ is given by:

$$g_{ds} = \left.\frac{\partial i_D}{\partial v_{DS}}\right|_{v_{GS}} = \lambda I_{DSS}\left(1 - \frac{V_{GS}}{V_{th}}\right)^2 \cong \lambda I_D \tag{5.6.3}$$

The transconductance $g_m = \left.\frac{\partial i_D}{\partial v_{GS}}\right|_{\substack{v_{DS} \\ v_{GS}=V_{GG}}}$ is:

$$g_m = \frac{2I_D}{|V_{th}|(1 - V_{GS}/V_{th})} \tag{5.6.4}$$

where g_m is evaluated at a quiescent point corresponding to the dc values I_D and V_{GS}. If we substitute for $(1 - V_{GS}/V_{th})$ from Equation 5.6.2,

$$g_m = \left(\frac{2I_{DSS}}{|V_{th}|}\right)\sqrt{\frac{I_D}{I_{DSS}}(1 + \lambda V_{DS})} \cong \left(\frac{2I_{DSS}}{|V_{th}|}\right)\sqrt{\frac{I_D}{I_{DSS}}} \tag{5.6.5}$$

The small-signal equivalent circuit of the JFET is the same as that of the MOSFET (Figure 5.3.2). C_{gs} and C_{ds} are in this case depletion capacitances that depend on the reverse bias across the pn junction and are typically, in the ranges of 1–3 and 0.1–0.5 pF, respectively.

A *p*-channel JFET is possible, in which the p^+ and n regions of the *n*-channel JFET are replaced by n^+ and p regions, respectively, so that all voltages and currents are reversed in polarity. The symbols for *n*-channel and *p*-channel JFETs are shown in Figure 5.6.4a and b,

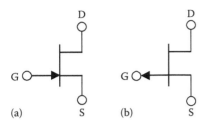

FIGURE 5.6.4
Symbol for (a) n-channel JFET; (b) p-channel JFET.

(a)   S   (b)   S

respectively. The arrow is in the forward direction of the gate–channel pn junction and is drawn at the same level as the source.

Because of the absence of an oxide layer at the gate, and the consequent gate voltage drop across this layer, the drain current is generally more effectively controlled than in a comparable MOSFET, resulting in a higher g_m. The capacitances are generally smaller, compared to a MOSFET of the same overall dimensions, because the gate can be made shorter than the channel and the depletion layer is wider than the oxide thickness. On the other hand, the gate current is many orders of magnitude larger than in a MOSFET. Overall, the JFET does not have decisive advantages over the MOSFET that warrant its use in ICs. It is mostly used in discrete form as an amplifier, as a switch, and as a current regulator diode.

Exercise 5.6.2

It can be shown that $I_{DSS} = \dfrac{\mu_e \varepsilon_s}{h} \dfrac{W}{L} V_{th}^2$ (Equation ST5.6.14). Determine: (a) I_{DSS} for a transistor having $\mu_e = 800$ cm^2/Vs, $\varepsilon_s = 1.05 \times 10^{-12}$ F/cm, $h = 2$ μm, $W/L = 200$, and $V_{th} = -3$ V; (b) g_m for $V_{GS} = 0$ and V_{DS} that will cause pinch off; (c) I_D and g_m for $V_{GS} = -1$ V and $V_{GS} = -2$ V, assuming $\lambda = 0$; and (d) r_o at $V_{GS} = -1$ V and $V_{GS} = -2$ V, assuming $\lambda = 0.01$ V^{-1}.

Answers: (a) 7.56 mA; (b) 5.04 mA/V, 3 V; (c) 3.36 mA, 3.36 mA/V, 0.84 mA, 1.68 mA/V; (d) 29.8 kΩ, 119 kΩ. ∎

Application Window 5.6.1 Current Regulator Diode

A JFET is combined with an IC source resistance R_S to form a two-terminal device, a **current regulator diode** (Figure 5.6.5a) having the symbol shown. The i_F–v_F curve is illustrated in Figure 5.6.5b. i_F is substantially constant over a voltage range from a fraction of a volt or so to a maximum of about 100 V, limited by breakdown. Nominally constant values of i_F range between few hundred μAs and few mAs, the incremental output resistance being typically several hundred kΩ to over 1 MΩ. The device is useful for regulating current in a given circuit, that is, keeping it nearly constant, and as a current limiter.

The general shape of Figure 5.6.5b can be readily explained. At very low v_F and i_F, the curve nearly follows the transistor output characteristic for $v_{GS} = 0$. As v_F and i_F increase, v_{GS} becomes more negative, because it equals $-R_S i_F$, and the i_F–v_F curve is determined by the transistor output characteristics in the triode region. When v_F is sufficiently high so that $v_{DS} \geq v_{GS} - V_{th}$, the transistor is in saturation, and i_F remains substantially constant.

It would appear that R_S could be eliminated altogether, with the gate connected directly to the source. Under these conditions, $I_F = I_D$ for $v_{GS} = 0$, which is nearly equal to I_{DSS}, and the output resistance is r_o. The disadvantage is that I_{DSS} can vary considerably between

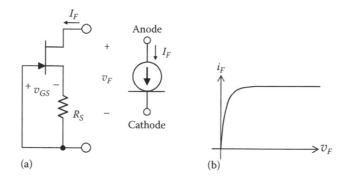

(a)   (b)

FIGURE 5.6.5
Figure for Application Window 5.6.1. (a) Circuit and symbol of current regulator diode; (b) current-voltage characteristic.

individual JFETs of the same type. R_S provides series–series negative feedback that keeps i_F more nearly constant and increases the output resistance (Section 7.3, Chapter 7). That the feedback is negative can be easily verified by considering what happens if v_F is increased, so that i_F tends to increase. The voltage drop $R_S i_F$ increases. But this makes v_{GS} more negative, which opposes the change in i_F, and therefore keeps it more nearly constant. Since i_F is kept more nearly constant when v_F is increased, the effective output resistance appears larger.

The effect of the negative feedback can be analyzed with the help of the small-signal equivalent circuit (Figure 5.6.6a). A dependent current source $g_m v_{gs}$ appears in parallel with r_o, between the drain and source, where $v_{gs} = -R_S i_f$. The gate is assumed to be open circuited through the transistor. Although the circuit can be easily analyzed to determine

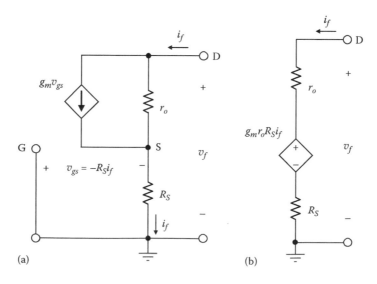

(a)   (b)

FIGURE 5.6.6
Figure for Application Window 5.6.1. (a) Small-signal equivalent circuit; (b) current source transformed to a voltage source.

the relation between v_f and i_f, it is more instructive to transform the current source in parallel with r_o to a voltage source (Figure 5.6.6b), whose polarity has been reversed to account for the negative sign relating v_{gs} and i_f. The voltage of the dependent source is now a voltage drop in the direction of current and is proportional to the current through the source. The source can be replaced by a resistance $g_m r_o R_S$. It follows that the total incremental output resistance r_{out} is

$$r_{out} = \frac{v_f}{r_f} = r_o + R_s(1 + g_m r_o) \tag{5.6.6}$$

Note if the gate is connected to the source, rather than to ground, through a battery that maintains the same gate–source dc bias, there is no feedback through R_S, and the incremental output resistance is $r_o + R_S$. The additional term $g_m r_o R_S$ is due to the feedback.

As a numerical example, consider a JFET having $V_{th} = -3$ V, $I_{DSS} = 10$ mA, and $R_S = 1$ kΩ. The operating point can be determined graphically as the intersection of the line $V_{GS} = -R_S i_F$ with the transfer characteristic (Exercise 5.6.3). Analytically, V_{GS} and i_F can be determined by substituting $V_{GS} = -R_S i_F$ in Equation 5.6.2, with $i_D = i_F$, neglecting λ. This gives $\left(\frac{R_S}{V_{th}}\right)^2 i_F^2 - \left(\frac{1}{I_{DSS}} - \frac{2R_S}{V_{th}}\right) i_F + 1 = 0$. Substituting numerical values and solving, gives: $i_F = 5.15$ and 1.75 mA. The larger value should be discarded because it makes $v_{GS} = -5.15$ V $< V_{th}$. It can be readily shown that if I_{DSS} ranges between 5 and 15 mA, i_F varies between 1.41 and 1.93 mA, respectively. The variation in i_F is proportionately much less than that of I_{DSS}.

From Equation 5.6.5, $g_m = 2.8$ mA/V. If $r_o = 300$ kΩ, $R_S(1 + g_m r_o) \cong 2.8 \times 300 = 840$ kΩ, so r_{out} becomes about 1.14 MΩ.

Simulation Example 5.6.2 JFET Output Resistance

It is required to investigate, through simulation, the behavior of the 2N38918 JFET as a current regulator diode (Application Window 5.6.1).

SIMULATION

The schematic circuit of the current regulator is shown in Figure 5.6.7a, using for the JFET part number J2N3819 from the EVAL library. $R_S = 885$ Ω so as make $I_D \cong 2$ mA. The VAC source applies a dc voltage of 12 V and an ac signal of 1 V. A current printer measures the ac current i_d as 1.129 μA, using AC Sweep at 1 kHz. Hence, $r_{out} = 1$ V/1.129 μA $= 886$ kΩ.

To apply Equation 5.6.6, we need g_m and r_o at $I_D = 2$ mA. These are obtained using the schematic of Figure 5.6.7b. With $V_{GS} = 0$ and 12 V applied between the drain and source, $I_{DSS} = 11.96$ mA. With $V_{GS} = -1.773$ V, which makes $I_D = 2$ mA, $i_d = 4.388$ μA, so that $r_o = 228$ kΩ. From the PSpice model of the JFET, $V_{th} = -3$ V. Substituting in Equation 5.6.5, $g_m = 3.26$ mA/V. Applying Equation 5.6.6 gives $r_{out} = 887$ kΩ, in agreement with the measurement.

It may be thought that applying a 1 V ac signal violates small-signal conditions. This is not the case, because PSpice applies the ac signal to a linearized circuit, which makes the results independent of the magnitude of the ac signal.

Exercise 5.6.3

Plot Equation 5.6.2 for the JFET of Application Window 5.6.1 and determine i_F from the intersection with a line of slope $-1/R_S$, where $R_S = 1$ kΩ. ∎

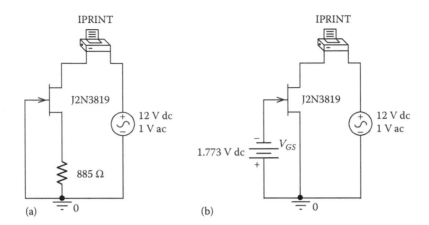

FIGURE 5.6.7
Figure for Simulation Example 5.6.2. Schematic circuit of current regulator (a), and for determining r_o (b).

Exercise 5.6.4

Draw the small-signal equivalent circuit of Figure 5.6.5a with a voltage v_i applied between gate and ground and derive Equation 5.6.6 from Thévenin's equivalent circuit between drain and ground, as v_{oc}/i_{sc}.

5.7 Metal–Semiconductor Field-Effect Transistor

The MESFET is a device based on GaAs technology. The advantages of GaAs over Si were noted in connection with Table 2.3.1, Chapter 2. GaAs has a much lower intrinsic concentration than Si ($2 \times 10^6/\text{cm}^3$ vs. $10^{10}/\text{cm}^3$) and an electron mobility that is more than six times that in Si at electric fields less than few kV/cm. However, at electric fields of the order of 10^5V/cm, the mobility becomes almost inversely proportional to the electric field and the saturation electron velocity is about the same in both GaAs and Si. Higher electron mobility means increased conductivity due to electrons, so that for the same input voltages, GaAs devices whose operation depends on electron current have higher output currents, and hence higher g_m. However, hole mobility is less in GaAs than in Si, which means that the advantages of higher mobility apply only to GaAs devices whose operation depends on electron current, and as long as the electric field acting on the conduction electrons is not very high.

Because of the low intrinsic carrier concentration, undoped GaAs has a high resistivity ($\sim 10^8$ Ω-cm) and is considered semi-insulating. This is an important advantage of GaAs technology, because it simplifies the isolation of transistors on the same chip from one another and reduces the internal capacitances encountered in Si MOSFETs. These capacitances are associated not only with the substrate–source and substrate–drain pn junctions but also with the capacitances of the metallic interconnections between devices, because in Si devices these interconnections are isolated from the substrate by a layer of silicon dioxide that has a relatively high dielectric constant. The speed advantage of GaAs devices arises from two factors: (1) smaller internal capacitances, and (2) increased output currents due to higher electron mobilities, which allows faster charging and discharging of internal

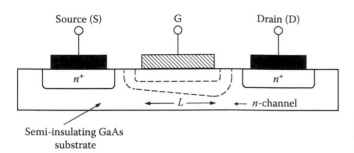

Semi-insulating GaAs
substrate

FIGURE 5.7.1
Diagrammatic illustration of the structure of a MESFET.

and load capacitances. This makes GaAs devices useful as amplifiers at frequencies in the GHz range and in very-high-speed digital ICs.

Structure

Figure 5.7.1 illustrates the MESFET in longitudinal section. The n-channel is formed between the n^+ source and the n^+ drain. The gate metal used is usually tungsten or platinum, which reacts with Si to form a silicide (Section 4.3, Chapter 4). The length L of the channel and its width W (normal to the plane of the page) are determined by the dimensions of the gate electrode.

Operation

Operation of the MESFET is analogous to that of the JFET. When no voltages are applied, a depletion region exists in the channel, associated with the equilibrium potential of the metal–semiconductor contact, as in the Schottky diode (Section 3.6, Chapter 3). The depletion region expands when a negative, reverse voltage is applied between the gate and channel, which reduces the width of the channel and increases its resistance. At a threshold of -0.5 to -3 V, the channel is completely pinched off, when no drain current is flowing. The flow of drain current increases the reverse bias at the drain end, so the channel becomes tapered and narrower at the drain end. The output characteristics are similar in shape to those of the JFET. If the MESFET has a long channel ($L > 10$ μm or so), the same $i–v$ relation applies as the JFET, namely Equation ST5.6.4, Section ST5.6. However, high-speed MESFETs have short channels (L in the range of 0.2–1 μm) in order to capitalize on the speed advantage of GaAs technology. In these devices the onset of current saturation is due to velocity saturation (Section 5.3) rather than channel pinch off.

MESFETs are normally depletion-type devices. In an enhancement-type MESFET the channel depth is small enough so that the channel is pinched off by the equilibrium depletion layer, and no current flows when $v_{GS} = 0$. A positive v_{GS} narrows the depletion layer and allows current flow. However, v_{GS} cannot be made more positive than about 0.4 V, because that would forward bias the gate-channel Schottky diode and cause gate current to flow, thereby providing a low resistance path between the gate and source and abolishing transistor action. The gate input voltage range is thus limited, and manufacturing tolerances have to be tight. The symbols for depletion and enhancement MESFETs are shown in Figure 5.7.2a and b, respectively.

The variation of drift velocity with the electric field is rather complicated. Also, the longitudinal electric field in short-channel devices cannot be neglected compared with the transverse electric field, as discussed for the short-channel MOSFET. Consequently, it becomes practically impossible to derive analytical expressions that model the behavior

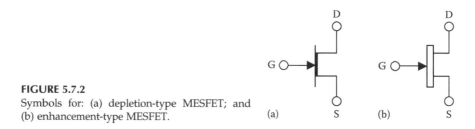

FIGURE 5.7.2
Symbols for: (a) depletion-type MESFET; and
(b) enhancement-type MESFET.

of the MESFET. Therefore, empirical models of varying complexities are used for analysis and simulation.

High-Mobility Devices

An improvement over the simple MESFET is the **modulation-doped field-effect transistor** (MODFET), also known as **high-electron-mobility transistor** (HEMT), illustrated in longitudinal section in Figure 5.7.3. A layer of heavily doped n-type AlGaAs is formed underneath the gate, which together with the gate, source, and drain, form a MESFET structure. However, a very thin layer of undoped AlGaAs separates the heavily doped layer from the GaAs substrate. AlGaAs has a wider bandgap than GaAs. Hence, although the n^+-AlGaAs provides the conduction electrons, these electrons preferentially move to the GaAs, where their energies in the conduction band are lower because of the narrower energy gap. What is referred to as a **two-dimensional (2-D)electron gas** is formed in the GaAs boundary region, the current due to these electrons being controlled by the gate voltage. Electron mobility is considerably higher in the undoped GaAs, because electrons are not scattered by any impurity atoms (Section 2.6, Chapter 2). The purpose of the thin undoped AlGaAs layer is to enhance the isolation between the donor impurities of the n^+-AlGaAs layer and the 2-D electron gas.

The electron mobility advantage becomes more marked at lower temperatures, because mobility increases at lower temperatures due to less scattering by crystal ions. Electron mobilities exceeding 100,000 cm^2/V in the 2-D electron gas have been observed at 77K. As noted earlier, higher mobility means larger transconductance and faster operation. Moreover, less scattering means less noise, which is also an important advantage.

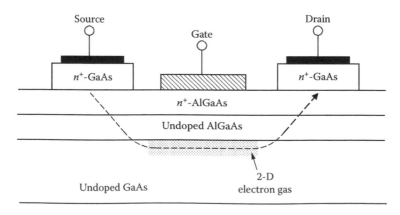

FIGURE 5.7.3
Diagrammatic illustration of the structure of a MODFET.

Summary of Main Concepts and Results

- In an enhancement-type MOSFET, v_{GS} of the same polarity as v_{DS} creates a conduction channel between the source and drain. In a depletion-type MOSFET, a relatively large current I_{DSS} flows at $v_{GS} = 0$. v_{GS} of polarity opposite to v_{DS} must be applied to reduce the drain current.

- The output characteristics of FETs have four regions: (1) the cutoff region, where the output current is negligibly small; (2) the triode region where the output current is appreciable but the output voltage is small; (3) the saturation region where the output current increases relatively slowly with the output voltage; and (4) the breakdown region, where the output current increases rapidly with only a small increase in output voltage. Channel-length modulation causes the output current to increase with the output voltage in the saturation region.

- In small-signal operation, the device input and output characteristics are considered linear for small variations about the dc, or quiescent, operating point; dc conditions and small-signal variations can then be treated independently. The small-signal variations can be represented by an equivalent circuit consisting of ideal resistors and capacitors and a current source at the output that is controlled by the input voltage.

- The transconductance of a transistor is a measure of how effectively the input voltage controls the output current. The transconductance is relatively small in MOSFETs, because of the gate oxide layer. JFETs and MESFETs have higher transconductance because of increased effectiveness of the reverse-biased gate–channel junction in modulating the output current.

- A rise in temperature increases the drain current in MOSFETs at low current levels but decreases it at high current levels.

- In IC MOSFETs, the body effect, due to a voltage difference between the source and substrate, affects the threshold under dc conditions. In addition, if the substrate is not at ac ground, it acts as a second gate and adds to the transconductance.

- CMOS circuits have important advantages in both analog and digital applications. A CMOS amplifier having a diode-connected depletion-type MOSFET load gives higher gain at a lower supply voltage, and occupies a smaller area on the chip than one that uses the same driver transistor but with a passive resistive load. A CMOS transmission gate allows having control voltages at the two extremes of signal voltages and provides a resistance in the conducting state that is more nearly constant for different inputs.

- A JFET, in combination with a source resistance, is a useful two-terminal current regulator.

- The GaAs, n-channel MESFET is potentially a faster device than the MOSFET because of the higher electron mobility in GaAs and reduced internal capacitances.

Learning Outcomes

- Articulate the basic operation of various types of MOSFETs, the JFET, and the MESFET as well as some of the important secondary effects.

Supplementary Examples and Topics on Website

SE5.1 Drain Current for Small Drain–Source Voltages. Derives, from first principles, the expression for i_D when v_{DS} is small.

SE5.2 Body Effect in MOSFETs. Simulates the body effect in the four types of MOSFETs and derives the corresponding thresholds.

SE5.3 Single-Transistor Equivalent of CMOS Pair. Derives a single-transistor equivalent for a CMOS pair operating in the saturation mode under large-signal conditions.

SE5.4 Linear Transconductance Amplifier. Derives a transconductance amplifier, based on the equivalent CMOS pair, which is ideally linear for large signals.

ST5.1 Basic Operation of MOSFETs. Derives the expressions for the threshold voltage, the relation between transconductance and body transconductance, and the current–voltage relation in a MOSFET, including the effect of channel-length modulation. Discusses sub-threshold operation of the MOSFET.

ST5.2 Power Relations in Small-Signal Sinusoidal Operation. Investigates power relations in a transistor circuit assuming linear operation.

ST5.3 T-Model of MOSFET. Derives the T-model of a MOSFET.

ST5.4 Graphical Interpretation of Small-Signal Voltage Gain. Gives a graphical interpretation of Equation 5.3.10.

ST5.5 Gate–Source Capacitance in Saturation. Calculates C_{gs} in the saturation region due to the pinched-off channel.

ST5.6 Junction Field-Effect Transistor. Derives the voltage–current relation for the JFET and shows how it can be approximated by a square law.

Problems and Exercises

P5.1 Operation and Parameters of FETs

P5.1.1   Identify the types of MOSFETs that can have a threshold of ± 2 V and a drain current of ± 1 mA in any of the four sign combinations, where a positive drain current is considered to flow into the drain.

P5.1.2   An E-NMOS transistor having $k_n = 0.4$ mA/V^2 and $V_{tn} = 1$ V is to be operated as a switch controlled by a gate-to-source voltage that can assume one of two values: 0 or v_C. If the voltage drop across the switch when closed is not to exceed 0.25 V while conducting 1 mA, what should be the minimum value of v_C?

P5.1.3   An E-NMOS transistor having $k_n = 50$ μA/V^2, $V_{tn} = 2$ V, and $\lambda = 0$ is used for small signals to provide a linear resistance in the range 10–100 kΩ. What range of v_{GS} is required? If $v_{GS} = 4$ V, at what v_{DS} is the departure from linearity 5%?

P5.1.4   Show that in the triode region, $g_{m(\text{triode})} = \left.\dfrac{\partial i_D}{\partial v_{GS}}\right|_{v_{DS}} = k_n v_{DS}$ and $g_{ds} = \left.\dfrac{\partial i_D}{\partial v_{DS}}\right|_{v_{GS}} = k_n (v_{GS} - V_{tn} - v_{DS})$, $v_{GS} \geq V_t$. Explain why $g_{m(\text{triode})}$ is small for small v_{DS}. Note that g_{ds} at small v_{DS} equals g_m in the saturation region for the same v_{GS}.

P5.1.5 An E-NMOS transistor is to conduct in the saturation mode a current of 100 mA at a minimum v_{DS} of 2 V. If $t_{ox} = 0.02$ μm and $V_{tn} = 1$ V, what should be the W/L ratio of the transistor, assuming $\mu_e = 600$ cm^2/Vs? What would be the W/L ratio if the transistor were E-PMOS having $\mu_h = 250$ cm^2/Vs? Neglect channel-length modulation.

P5.1.6 Given an E-NMOS transistor having $V_{tn} = 1$ V and $V_{GS} = 3$ V. Determine v_{DS} at which i_D is: (a) 0.5 times, and (b) 0.25 times its saturation value. Neglect channel-length modulation.

P5.1.7 An E-PMOS transistor has $k'_n = 20$ μA/V^2, $W/L = 10$, $v_{GS} = -2$ V, $V_{tp} = -0.8$ V, and $\lambda = 0.05$ V^{-1}. Determine $v_{SD(sat)}$, $i_{D(sat)}$, and i_D when the magnitude of v_D exceeds $v_{SD(sat)}$ by 1 V.

P5.1.8 An E-NMOS transistor has $k'_n = 90$ μA/V^2, $W/L = 10$, $v_{GS} = 2$ V, $V_{tn0} = 1$ V, $v_{DS} = v_{DS(sat)} + 1$ V, $\lambda = 0.02$ V^{-1}, $\gamma = 0.5$ V$^{1/2}$, and $V_{SB} = 0.5$ V. Determine g_m, g_{mb}, and r_o.

P5.1.9 Explain how the breakdown voltage varies with v_{GS} and v_{DS} in a MOSFET (Figure 5.3.4 and in a JFET (Figure 5.6.3a). What is the voltage that contributes to breakdown in the gate–channel junction of a JFET?

P5.1.10 Consider the E-NMOS transistor of Example 5.3.1 with $\lambda = 0$, $V_{tn} = 1$ V, and $v_{GS} = 5$ V. If the temperature increases from 300 to 350K, what is the change in $I_{D(sat)}$ if: (a) μ_e remains constant and V_{tn} decreases by 2 mV/°C; (b) V_{tn} remains constant and μ_e varies in accordance with Equation 2.6.3; (c) both V_t and μ_e vary with temperature as in (a) and (b).

P5.1.11 Show that when i_D is given by Equations 5.3.1 and 5.3.2, $\left.\dfrac{\partial i_D}{\partial v_{DS}}\right|_{v_{GS}}$ is continuous at the boundary between the triode and saturation regions.

P5.1.12 A JFET having $V_{th} = -2$ V, $I_{DSS} = 5$ mA, and $\lambda = 0.02$ V^{-1} is used as a current regulator with $R_S = 1$ kΩ. Determine i_F, g_m, r_o of the transistor and r_{out} of the regulator.

P5.1.13 For the JFET of Example 5.6.1, determine: (a) v_{GS} that reduces the channel width by one-half, if $v_{DS} = 0$; and (b) v_{GS} that causes pinch off when $v_{DS} = 3$ V.

P5.1.14 Determine I_{DSS} for the JFET of Problem P5.1.13, assuming $W = 1$ mm and $L = 10$ μm.

P5.1.15 For the JFET of Problem P5.1.13, determine $I_{D(sat)}$ and $v_{DS(sat)}$ for $v_{GS} = 0.5 V_{th}$.

P5.2 *i–v* Relations of FETs

P5.2.1 Given an E-NMOS transistor with the source grounded, a fixed voltage source $V_{GG} > V_{tn}$ between the gate and source and a variable current source I_{SRC} between the drain and source, the current direction being that of the drain current. Is this a valid connection? Explain your answer. Simulate with PSpice.

P5.2.2 Consider Problem P5.2.1 with the variable current source I_{SRC} connected between source and ground, the voltage source $V_{GG} > V_{tn}$ connected between gate and ground, and the drain grounded. Is this a valid connection? Explain your answer. Simulate with PSpice.

P5.2.3 Consider an E-NMOS transistor in Figure 5.2.10, with $v_{src} = 0$ and given R_D and V_{DD}. Assume V_{GG} increases from zero. Determine when the transistor is at the edge of the saturation region. What is the limit of V_{DS} as V_{GG} becomes very large? What is the effect of including a resistance R_{SG} in series with V_{GG}? Neglect channel-length modulation.

P5.2.4 An E-NMOS transistor has $\lambda = 0.02$ V^{-1} and operates in the triode region with $v_{DS} = 0.2$ V. If $i_D = 50$ μA at $v_{GS} = 2$ V, and 100 μA at $v_{GS} = 3$ V, determine V_{tn} and k_n.

P5.2.5 Consider the E-NMOS transistor of Example 5.3.1 at the quiescent point. How will r_o change if: (a) V_{tn} is increased from 1 to 2 V; (b) W is doubled; (c) L is doubled?

P5.2.6 Consider the E-NMOS transistor of Example 5.3.1 at the quiescent point. How will g_m change if: (a) W is doubled; (b) L is doubled?

P5.2.7 Given a MOSFET having $k = 0.4$ mA/V^2, $|V_t| = 2$ V, and $\lambda = 0$. The source is grounded, and the gate voltage is 1 V beyond threshold, in the conducting direction. The drain voltage is varied from zero to 5 V in the same direction as the gate voltage. Sketch the transfer and output characteristics if the transistor is: (a) E-PMOS, (b) D-NMOS, or (c) D-PMOS. For the PMOS transistors, label the characteristics in terms of v_{SG}, v_{SD}, i_{SD}, and $\overline{V}_{tp}$. Simulate with PSpice.

P5.2.8   Given an E-PMOS transistor having $V_G = 0$ and $V_D = -1$ V. If $k_p = 0.2\,\text{mA/V}^2$, $V_{tp} = -2$ V, and $\lambda = 0.02$ V^{-1}, determine i_D if: v_S is (a) 1 V; (b) 3 V; or (c) 5 V. Neglect the body effect.

P5.2.9   A D-NMOS transistor has $k_n = 1\,\text{mA/V}^2$, $V_{tnD} = -2$ V, and $\lambda = 0$, with its gate connected to one of the other terminals. Sketch the i–v relation as v is varied from -5 V to $+5$ V. Identify the regions of operation, noting that the MOSFET is a symmetrical device. Simulate with PSpice.

P5.2.10 Repeat Problem P5.2.9 for a D-PMOS transistor having $k_p = 1\,\text{mA/V}^2$ and $V_{tp} = 2$ V.

P5.2.11 Consider the E-NMOS transistor of Example 5.3.1. If the gate is connected to the drain, what would be the value of voltage across the device if the current is 1 mA or 3 mA? Solve the cubic equation using MATLAB® and compare with the voltage values assuming $\lambda = 0$. Determine the incremental resistance at the two current values.

P5.2.12 An E-NMOS transistor having $k_n = 0.5$ mA/V^2, $V_{tn} = 2$ V, and $\lambda = 0$ is biased as in Figure 5.3.7, with $R_S = 0.5$ kΩ and $V_{DD} = 12$ V. It is desired to have $i_D = 2$ mA and $V_D = 9$ V with respect to ground, with the transistor in saturation. Determine all resistance values and g_m, neglecting the body effect. If the transistor is replaced in the same circuit with one having $k_n = 0.75\,\text{mA/V}^2$ and $V_{tn} = 2$ V, determine the new value of i_D. Simulate with PSpice.

P5.2.13 Repeat Problem P5.2.12 with $R_S = 2$ kΩ. How does this compare with the results of Problem P5.2.12? How do you explain the difference?

P5.2.14 Draw the circuit diagram that is exactly analogous to that of Problem P5.2.12 using a matched E-PMOS transistor with $V_{DD} = -12$ V. Determine the quiescent conditions if the substrate is grounded and the transistor has $\gamma = 0.3$ $\text{V}^{1/2}$.

P5.2.15 It is desired to use three diode-connected MOSFETs in series across a 12 V supply to obtain voltage division at 4 and 8 V. Bearing in mind the desirability of having a low output resistance, which transistor type should be used? If $|V_{tn}| = 2$ V, $i_D = 1$ mA, and $\lambda = 0$, determine k of the transistors. What is the output resistance at the voltage taps? Simulate with PSpice.

P5.2.16 A D-PMOS transistor has $k'_p = 100$ $\mu\text{A/V}^2$ and $V_{tpD} = 2$ V. Assume $\lambda = 0$. If I_{DSS} of 10 mA is required, what should be W/L? If the gate is joined to the source and the transistor is connected in series with a 1 kΩ across a supply V_{CC} V, what should be the minimum value of V_{CC} for the transistor to conduct in the saturation region?

P5.2.17 Explain how the operation of the CMOS amplifier of Figure 5.5.1 would be affected if $|V_{SB}| \neq 0$, but sources and substrates are ac grounded. Simulate with PSpice and compare the VTCs with and without body effect, assuming the transistors have $k = 0.5$ mA/V^2, $V_t = 1$ V, $\lambda = 0.05$ V^{-1}, $|V_{SB}| = 2$ V, $\gamma = 0.5$ $\text{V}^{1/2}$, and $|\varphi_F| = 0.3$ V.

P5.2.18 By equating the saturation currents of the two transistors in Example 5.5.1 and taking derivatives with respect to v_I show that the largest magnitude of the slope is at point B. How do you interpret this result? Compare the calculated values of the slope at points B and C with the values from the simulation.

P5.2.19 In the CMOS amplifier of Example 5.5.1, determine v_O, if: (a) $v_I = 3.5$ V; (b) $v_I = 4.5$ V. Compare with the simulation values.

P5.2.20 Repeat the simulation of Example 5.5.1 for the CMOS amplifier assuming: (a) $V_{tn} = 1$ V, and (b) transistors are of equal dimensions. Compare with analysis.

P5.2.21 Simulate the current variation in Example 5.5.1 with an E-NMOS load and interpret all the results, as was done for the D-PMOS load.

P5.2.22 Given a CMOS transmission gate with matched transistors having $k_n = k_p = 0.4\,\mathrm{mA/V^2}$, $V_{tn} = \overline{V}_{tp} = 1.5$ V, $\lambda = 0$, $v_C = 5$ V, $R_L = 20$ kΩ, and no body effect. Determine v_O and the switch resistance for: (a) $v_I = 5$ V; (b) $v_I = 2.5$ V; (c) $v_I = 0$. Simulate with PSpice.

P5.2.23 Let $R_L = 100$ Ω in Example 5.5.2. Simulate the v_O vs. v_I characteristic for: (a) gamma = 0 (no body effect); (b) gamma = 0.3; and (c) gamma = 0.5, and compare v_O at $v_I = \pm 5$ V.

P5.2.24 The load in Example 5.5.2 is a 100 pF capacitor. The input suddenly changes from +5 V to −5 V while $v_C = 5$ V. (a) Determine the initial capacitor current, assuming the capacitor is initially charged to 5 V. (b) What are the states of the two transistors during the discharge, and when do the transistors change state? Simulate with PSpice.

P5.2.25 A single E-NMOS transistor is to be used as a transmission gate with $-5 \leq v_I \geq 5$ V and the substrate connected to a −5 supply. It is desired to have a minimum v_{GS} of 1 V above threshold when the switch is on, so that the transistor conducts a sufficient current. Determine the range of gate control signal required, taking the body effect into account and neglecting any load impedance. Assume $k_n = 0.4\,\mathrm{mA/V^2}$, $V_{tn0} = 1$ V, $\gamma = 0.6\,\mathrm{V^{1/2}}$, and $|\varphi_F| = 0.3$ V.

P5.2.26 In the circuit of Figure P5.2.26, $k_n = 0.1\,\mathrm{mA/V^2}$, $\lambda = 0$, and $V_{tn} = 1$ V. Determine (a) V_{GS1} if $I_{D1} = 0.1$ mA and (b) I_{D2}.

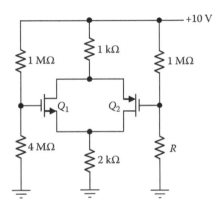

FIGURE P5.2.26

P5.2.27 The two transistors in Figure P5.2.27 are identical and have $k_n = 1\,\mathrm{mA/V^2}$ and $V_{tn} = 1$ V. If $V_I = 2$ V, determine V_O.

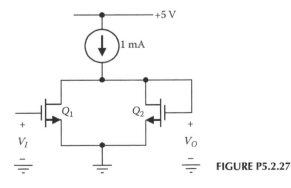

FIGURE P5.2.27

P5.3 Miscellaneous

P5.3.1 Neglecting channel-length modulation, show that Equation 5.3.12 for the g_m of a MOSFET is the same as Equation 5.6.5 for a JFET.

P5.3.2 Consider an E-NMOS transistor and an E-PMOS transistor having the same dimensions, with $t_{ox} = 0.1$ μm, $\mu_e = 1000$ cm^2/Vs, $\mu_h = 500$ cm^2/Vs, $W/L = 20$, $V_{tn} = 1$ V, and $V_{tp} = -1.5$ V, determine V_{teq} and K_{eq} of the composite transistor defined in SE5.3.

P5.3.3 Verify Equation ST5.6.8.

P5.3.4 Verify Equation ST5.6.11.

P5.3.5 Verify Equation ST5.6.12.

In the following problems, refer to Section ST5.1.

P5.3.6 Given an E-NMOS transistor having $t_{ox} = 15$ nm, $N_A = 10^{15}$/cm^3, $N_{Dg} = 10^{20}$/cm^3, and $Q_{ox} = 10^{10}$/cm^2. It is required to determine V_{tn} assuming: (a) $V_{SB} = 0$, or (b) $V_{SB} = 5$ V.

P5.3.7 Consider an E-PMOS transistor forming a CMOS pair with that of Problem P5.3.6 using the same technology, that is, having $t_{ox} = 15$ nm, $N_{Dg} = 10^{20}$/cm^3, $Q_{ox} = 10^{10}$/cm^2, and $N_D = 5 \times 10^{16}$/cm^3. Determine V_{tn}, assuming: (a) $V_{BS} = 0$, or (b) $V_{BS} = 5$ V.

P5.3.8 Determine the concentration of impurities that must be implanted at the surfaces of the transistors of Problems P5.3.6 and P5.3.7 so that the thresholds of the two transistors are of equal magnitude but of opposite signs, with $V_{SB} = 0$.

P5.3.9 An E-NMOS transistor has $t_{ox} = 10$ nm and $\gamma = 0.2$ V$^{1/2}$. Determine N_A of the substrate.

P5.3.10 Determine the surface charge density of the transistor of Example ST5.1.1 when $V_{GS} = 2$ V, with the acceptor impurities implanted.

P5.3.11 Determine the change in threshold in the transistor of Example ST5.1.1 when the temperature is raised to 400K.

6

Bipolar Junction Transistor

The chapter explains the operation of the bipolar junction transistor (BJT). It is shown, to begin with, that the BJT is an amplifying device that basically amplifies power and voltage, albeit in a manner that is quite different from that of FETs. An important difference between BJTs and FETs is that the input current in the BJT is not negligible. In fact, in one of the BJT configurations, the common-base (CB) configuration, the output current is slightly less than the input current. However, considerable current amplification occurs in other BJT configurations.

The general approach in analyzing the behavior of the BJT is the same as that applied to the MOSFET, namely, to consider first the behavior of an idealized device based on physical principles, and derive its voltage–current relations and small-signal equivalent circuit. Effects that significantly influence device behavior are then considered, including secondary effects that become important under certain operating conditions. Mathematical treatment is kept to a minimum; derivations of transistor equations and other mathematical relations are left to Supplementary Examples and Topics on the website.

Analogous to channel-length modulation in FETs, base-width modulation in BJTs makes the incremental output resistance finite, rather than ideally infinite. Moreover, because of base-width modulation, the output of a BJT has a small effect on the input, even under dc conditions–an example of reverse transmission. High-frequency performance is limited by the effects of stored charges, which are modeled by incremental capacitances, and by physical considerations, such as the time it takes current carriers to move through the base of the BJT. An interesting, recent variation on the conventional BJT is the heterojunction bipolar transistor (HBT), which affords additional design flexibility and improved performance. BJT fabrication is discussed in Section 6.4.7, Chapter 6.

Learning Objectives

❖ To be familiar with:

- The symbols, terminology, and basic structure of the BJT.
- Various effects and some more advanced concepts that influence the behavior of the BJT.
- Noise considerations in transistors.

❖ To understand:

- The basic operation of the BJT, its current–voltage characteristics in the CB and common-emitter (CE) configurations, and the corresponding small-signal equivalent circuits.
- The effect of base-width modulation on BJT characteristics.

- The various capacitances that influence the behavior of the BJT at high frequencies.
- The basic Ebers–Moll model of the BJT and its interpretation.
- The behavior of the BJT in the saturation mode.
- The basic operation of the HBT.

6.1 Basic Operation of the BJT

Consider the n^+pn structure illustrated in Figure 6.1.1 consisting of a p-region sandwiched between two n-regions, where the n^+, p, and n regions are the **emitter** (E), **base** (B), and **collector** (C), respectively. Superficially, the structure resembles two pn junctions connected back-to-back. The crucial difference, however, is that the base region is very thin so that the two pn junctions are "coupled" in a special way.

Assume that the base-emitter pn^+ junction is forward biased and the base-collector pn junction is reverse biased, with the small-signal voltage $v_{src} = 0$ to begin with. A forward current flows due to electrons injected from emitter to base and holes injected from base to emitter, as in any forward-biased pn junction diode, bearing in mind that the electron current is in the opposite direction to electron flow. Because the emitter is much more heavily doped than the base, the electron concentration in the emitter is much larger than the hole concentration in the base. The forward electron emitter current I_{eE} at the base-emitter junction (BEJ) is therefore much larger than the hole current I_{hE} (Equations 3.2.5 and 3.2.6, Chapter 3). Moreover, in a normal pn junction diode, the p- and n-regions are wide compared to the diffusion lengths of injected minority carriers, so that these carriers are eventually lost to recombination with majority carriers on the receiving side. In the BJT, the base is made narrow compared to the diffusion length of carriers injected into the base. Consequently, a small fraction $I_{e(rec)}$ of the electron current I_{eE} is lost to recombination in the base; most of the electrons cross the base region and arrive at the reverse-biased

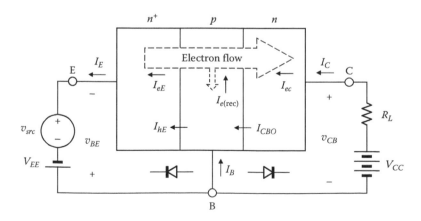

FIGURE 6.1.1
Basic operation of an *npn* BJT.

base-collector junction (BCJ). In a reverse-biased *pn* junction, the reverse current is a net drift current under the influence of the electric field in the depletion region. Electrons arriving at the edge of the base depletion region of the BCJ are, therefore, swept by the electric field into the collector region resulting in an electron current I_{eC} in this region. In addition to this current, there is the small, reverse current I_{CBO} of a reverse-biased *pn* junction.

To see how the BJT acts as an amplifier, the following should be noted:

1. Because the BEJ is forward biased, the $i_E - v_{BE}$ relation is of the same shape as that of a forward-biased diode (Figure 1.1.2, Chapter 1). The bias voltage V_{EE} defines an operating point P on the $i_E - v_{BE}$ curve (Figure 6.1.2a). A small change in the base-emitter voltage $\Delta v_{BE} = v_{src}$ produces a relatively large change in emitter current $\Delta i_E = \Delta v_{BE}/r_e$, where r_e is the small, incremental forward resistance of the BEJ once I_E becomes appreciable.

2. It is seen from the preceding discussion that $I_E \cong I_{eE} \cong I_{eC} \cong I_C$. In terms of instantaneous values, $i_E \cong i_C$ and $\Delta i_E \cong \Delta i_C$. That is, a change in emitter current produces an almost equal change in collector current, and the small-signal current gain is only slightly less than unity.

3. i_C is almost independent of the reverse base-collector bias, because all electrons arriving at the collector junction are swept by the dominant electric field anyway. Theoretically, increasing the reverse bias does not increase the collector current because there are no more electrons to sweep. The output characteristic for a given I_E is ideally a straight horizontal line whose intercept I_C is nearly equal to I_E (Figure 6.1.2b). This means that a relatively large load resistance R_L can be connected in series with the collector, with a suitable value of V_{CC} to maintain I_C. The operating point Q is determined in the usual manner by a load line construction. A change $\Delta i_E \cong \Delta i_C$ can therefore cause a relatively large change Δv_{CB} in the collector-base voltage, as long as the BCJ remains reverse biased. Hence,

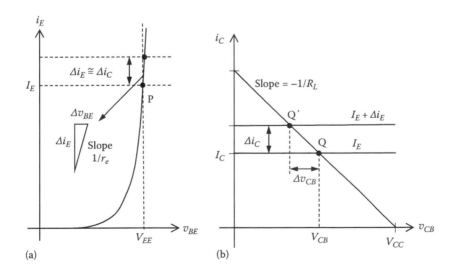

FIGURE 6.1.2
Input characteristic (a) and output characteristic (b) of the transistor of Figure 6.1.1.

$\Delta v_{CB} = -R_L \Delta i_C$, the minus sign accounting for the decrease in v_{CB} due to Δi_C and the resulting voltage drop $R_L \Delta i_C$ across R_L. R_L could be in the range of few tens of kΩ, whereas r_e is in the range of few tens of ohms. The small-signal voltage gain $\Delta v_{CB}/\Delta v_{BE}$ is therefore $-R_L/r_e$, which could be quite large. As the small-signal current gain is nearly unity, the power gain would be large and the BJT acts as an amplifier.

Concept: *In a BJT, an appreciable current through a reverse-biased pn junction can be controlled by small voltage variations across a forward-biased pn junction, resulting in substantial power amplification.*

The complete output characteristics are shown in Figure 6.1.3, from which it is seen that the BJT can operate in a number of different modes:

1. **Active mode** in which the BEJ is biased into forward conduction ($v_{BE} > V_{\gamma E}$, the cut-in voltage of the BEJ), whereas the BCJ is not forward biased, but v_{CB} is less than the breakdown voltage of the BCJ.
2. **Cutoff mode** in which the BCJ and the BEJ are not forward biased. Only very small currents flow under these conditions, and the transistor is cut off. The BCJ reverse current I_{CBO} referred to in Figures 6.1.1 and 6.1.3 is defined for $i_E = 0$, that is, with the emitter terminal open circuited (hence the O subscript). I_{CBO} is ideally the saturation current of the BCJ. In practice, the reverse current of the BCJ is I_{CBO} plus a leakage component, as in the case of the pn junction, and approximately doubles for every 10°C rise in temperature.
3. **Saturation mode** in which both the BEJ and the BCJ are forward biased beyond the cut-in voltage V_γ, but with I_C flowing into the collector. In this region, once v_{BC} exceeds $V_{\gamma C}$, which is normally 0.4–0.5 V, electrons are injected from collector to base. These electrons oppose the flow of electrons from base to collector, and the collector current decreases, eventually becoming zero at a sufficiently large forward bias v_{BC}. It is to be expected that the larger i_E, the larger is the forward bias v_{BC} that is required to reduce i_C to zero (Figure 6.1.3).

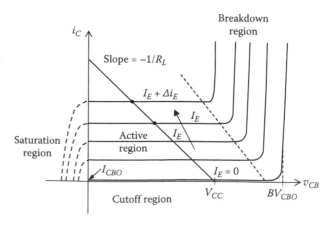

FIGURE 6.1.3
Output characteristics of the transistor of Figure 6.1.1.

TABLE 6.1.1

Modes of Transistor Operation

Mode	Is the Junction Forward Biased to beyond Cut-In?		I_C
	BEJ	**BCJ**	
Active	Yes	No	>0
Saturation	Yes	Yes	>0
Cutoff	No	No	$\cong 0$
Inverse active	No	Yes	<0

Whereas the active mode is the normal mode for operating the BJT as an amplifier, the BJT as a switch operates in the cutoff or saturation modes. In principle, it is also possible to operate the BJT in the **inverse active mode**, in which the roles of emitter and collector are interchanged between the n^+ and n regions. The four possible modes of transistor operation are summarized in Table 6.1.1. Other modes, namely, an inverse saturation mode and the base acting as anode of two forward-biased diodes, are also possible (Problem P6.1.2).

Common-Base dc Current Gains

Referring to Figure 6.1.1, the dc base current can be expressed as:

$$I_B = I_{hE} + I_{e(\text{rec})} - I_{CBO} \tag{6.1.1}$$

The ratio of I_{eE} to I_E is the **emitter injection efficiency** γ. Thus,

$$\gamma = \frac{I_{eE}}{I_E} = \frac{I_{eE}}{I_{hE} + I_{eE}} \cong 1 \tag{6.1.2}$$

As mentioned earlier, γ is made close to unity by having the emitter much more heavily doped than the base, so that $I_{eE} \gg I_{hE}$ (Example 6.1.1).

The ratio of I_{eC} to I_{eE} is the **base transport factor** δ. Thus,

$$\delta = \frac{I_{eC}}{I_{eE}} = \frac{I_{eC}}{I_{eC} + I_{e(\text{rec})}} \tag{6.1.3}$$

δ is made close to unity by having a thin base and a large collector junction area. The **dc CB current gain** α_F is defined as:

$$\alpha_F = \gamma\delta = \frac{I_{eC}}{I_E} \cong 1 \tag{6.1.4}$$

The designation CB refers to the fact that the base terminal is common to input and output. The F subscript denotes the forward active mode. From Figure 6.1.1,

$$I_C = I_{eC} + I_{CBO} \tag{6.1.5}$$

Substituting from Equation 6.1.4 in Equation 6.1.5,

$$I_C = \alpha_F I_E + I_{CBO} \qquad (6.1.6)$$

Concept: *The dc CB current gain α_F is only slightly less than unity, being the product of two quantities γ and δ, each of which is in turn slightly less than unity.*

Typical Structure

A typical structure of an IC *npn* transistor is shown in Figure 6.1.4. Connection to each region is made through a contact between metal and a highly doped semiconductor (Section 4.3, Chapter 4). If the collector consisted of the low-doped *n*-region only, the emitter current flowing to the collector encounters considerable resistance, which degrades the performance of the transistor, as clarified later. An n^+ buried layer is therefore included that, together with the n^+ sinker layer, provide a low-resistance path to the collector, thereby considerably reducing the collector resistance. The n^+ buried layer also isolates the transistor from the *p*-substrate below, whereas lateral isolation from neighboring devices is provided in modern transistors by SiO_2 barriers.

Note that, unlike the FETs previously discussed, the BJT is not a symmetrical structure. First, the emitter is much more heavily doped than the base, to make $I_{eE} \cong I_E$, and the collector is less heavily doped than the base, in order to control the breakdown voltage of the BCJ, as discussed later. Second, the area of the BCJ is considerably larger than that of the BEJ in order to have the BCJ collect practically all the electrons injected into the base, which reduces $I_{e(rec)}$, and makes $I_{eC} \cong I_{eE}$. Hence, interchanging the emitter and collector leads to poor performance and is avoided in analog circuits (Example 6.1.1). But in some digital circuits, the transistor operates in the inverse active mode under certain conditions (Section 13.6, Chapter 13), and this mode is also relevant to modeling transistor behavior (Section 6.3, Chapter 6).

The transistor of Figure 6.1.4 is an *npn* transistor. In contrast, a *pnp* transistor has a complementary structure, in which the roles of electrons and holes are interchanged, so that all corresponding voltages and currents are reversed. The circuit symbols for both

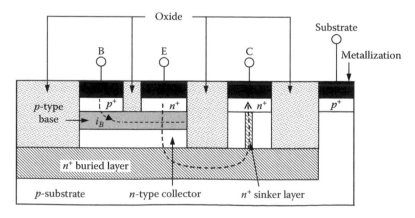

FIGURE 6.1.4
Diagrammatic illustration of the structure of an *npn* IC transistor.

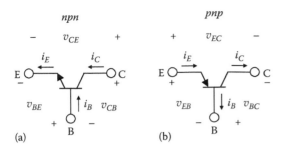

FIGURE 6.1.5
Terminal voltages and currents for an *npn* transistor (a), and a *pnp* transistor (b).

types of transistor are shown in Figure 6.1.5, where the arrow at the emitter is in the actual direction of emitter current. The assigned positive directions of currents and voltage drops indicated for both transistors are actual directions in the active region and would have positive numerical values. An *npn* transistor has a larger α_F than a *pnp* transistor of the same dimensions and doping levels because of higher electron mobility compared to hole mobility (Example 6.1.1).

Common-Emitter Configuration

Concept: *Interchanging the input and common terminals of the BJT in the CB configurations allows both current and voltage amplification.*

The resulting CE configuration is illustrated in Figure 6.1.6. The polarities indicated are actual polarities for an *npn* transistor in the active mode. Considering dc values to begin with, KCL gives:

$$I_E = I_C + I_B \tag{6.1.7}$$

Substituting for I_E from Equation 6.1.7 in Equation 6.1.6 and collecting terms,

$$I_C = \frac{\alpha_F}{1 - \alpha_F} I_B + \frac{1}{1 - \alpha_F} I_{CBO} \tag{6.1.8}$$

$\alpha_F/(1 - \alpha_F)$ in Equation 6.1.8 is the **dc CE current gain**, denoted by β_F. Note that as $\alpha_F \cong 1$, β_F is a relatively large number and is very sensitive to variations in α_F. For example, if α_F varies between 0.99 and 0.995, that is, a variation of about 0.5%, β_F varies between 99 and 199, a variation of about 100% (Problem P6.1.5).

When the base is open circuited, $I_B = 0$ and $I_C = I_{CEO}$, where,

$$I_{CEO} = \frac{1}{1 - \alpha_F} I_{CBO}, \quad I_B = 0 \tag{6.1.9}$$

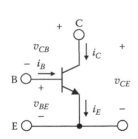

FIGURE 6.1.6
Terminal voltages and currents of an *npn* transistor in the CE configuration.

Equation 6.1.8 becomes:

$$I_C = \beta_F I_B + I_{CEO} \tag{6.1.10}$$

It would seem that I_{CEO} is much larger than I_{CBO} because of the $1/(1 - \alpha_F)$ factor. However, α_F at this low current is considerably less than unity, as explained later.

It is instructive to consider in terms of transistor action why $I_{CEO} = I_{CBO}/(1 - \alpha_F)$ when the base is open circuited (Figure 6.1.8). The BCJ is reverse biased and I_{CBO} flows from collector to base. But because I_{CEO} flows out of the emitter, the BEJ must be forward biased to allow this current to flow. By transistor action, the collector component of I_{CEO} is $\alpha_F I_{CEO}$. KCL applied to collector or base gives $I_{CEO} = I_{CBO} + \alpha_F I_{CEO}$, and Equation 6.1.9 follows.

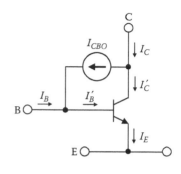

FIGURE 6.1.7
Figure for Exercise 6.1.3.

Exercise 6.1.1

Verify that $\alpha_F = \dfrac{\beta_F}{\beta_F + 1}$ and that $\dfrac{1}{1 - \alpha_F} = \beta_F + 1$.   ∎

Exercise 6.1.2

Given a transistor having $\alpha_F = 0.99$, $I_{CBO} = 100$ nA, and $I_B = 50$ μA. Calculate: (a) β_F; (b) I_C; (c) I_E; (d) I_{CEO} assuming that when $I_B = 0$, β_F has one-tenth of its value in (a); what is the corresponding α_F? (see Problem P6.1.6).

Answers: (a) 99; (b) 9.91 mA; (c) 10.01 mA; (d) 1.09 μA, 0.908.   ∎

Exercise 6.1.3

Show that the effect of I_{CBO} can be modeled by a current source as in Figure 6.1.7, where the transistor is free of a BCJ leakage current so that $I'_C = \alpha_F I_E$ and $I'_C = \beta_F I'_B$. Prove that Equations 6.1.6 and 6.1.8 apply.   ∎

Exercise 6.1.4

Assume that the transistor in Figure 6.1.9 has $V_{BE} = 0.7$ V and a very large β_F so that $I_B = 0$ and $I_C = I_E$. Determine (a) collector-to-emitter voltage; (b) transistor current; (c) in which mode is the transistor operating? (d) What happens if the 13.4 kΩ resistor is replaced by a 3.3 kΩ resistor? Verify by determining V_{BE}.

Answers: (a) 6.7 V; (b) 0.5 mA; (c) active mode, because V_{BE}, $V_{CE} > 0$; (d) transistor cuts off.   ∎

Example 6.1.1 α_F and β_F from Transistors Parameters

Given an *npn* transistor having the following physical parameters: $N_{DE} = 5 \times 10^{17}/\text{cm}^3$, $N_{AB} = 10^{16}/\text{cm}^3$, $N_{DC} = 10^{15}/\text{cm}^3$, effective base width $W_B = 1$ μm, minority-carrier lifetimes $\tau_{eB} = 4 \times 10^{-7}$ s, $\tau_{hE} = 10^{-9}$ s, $\tau_{hC} = 5 \times 10^{-7}$ s in base, emitter, and collector, respectively, it is

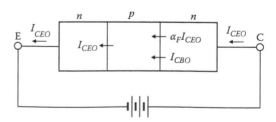

FIGURE 6.1.8
Currents in an *npn* transistor with open-circuited base.

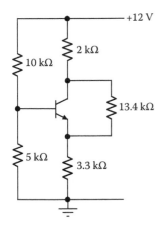

+12 V

2 kΩ

10 kΩ

13.4 kΩ

5 kΩ

3.3 kΩ

FIGURE 6.1.9
Figure for Exercise 6.1.4.

required to derive the currents α_F and β_F: (a) in the active mode, (b) in the inverse active mode, and (c) α_F and β_F for a *pnp* transistor having the same dimensions, doping levels, and minority-carrier lifetimes. Estimate the three components of base current, assuming $n_i = 10^{10}/\text{cm}^3$, $V_{BE} = 0.7$ V, and area of BEJ and BCJ $A_J = 0^{-4}$ cm². Note that in the basic analysis of the BJT, a planar structure is assumed in which the BEJ and BCJ are parallel planes of equal area. Although real transistors do not have this structure, the general conclusions of the basic analysis still apply.

SOLUTION

It is necessary first to calculate some of the physical parameters of the transistors. From Equations 2.6.1 and 2.6.2, Chapter 2, the mobilities are:

$$\mu_{eB} = 1345 \text{ cm}^2/\text{Vs}, \quad \mu_{hE} = 209 \text{ cm}^2/\text{Vs}, \quad \mu_{hC} = 478 \text{ cm}^2/\text{Vs}$$

From these, we can derive the diffusion constants using the Nernst–Einstein relation $D = \mu kT \cong 0.02585\mu$ at 300K (Equation 2.4.5, Chapter 2). This gives:

$$D_{eB} = 34.77 \text{ cm}^2/\text{s}, \quad D_{hE} = 5.40 \text{ cm}^2/\text{s}, \quad D_{hC} = 12.36 \text{ cm}^2/\text{s}$$

The diffusion lengths are calculated from the relation $L = \sqrt{D\tau}$, as:

$$L_{eB} = 3.73 \times 10^{-3} \text{ cm}, \quad L_{hE} = 0.735 \times 10^{-4} \text{ cm}, \quad L_{hC} = 2.486 \times 10^{-3} \text{ cm}$$

(a) $\dfrac{W_B}{L_{eB}} = \dfrac{10^{-4}}{36.09 \times 10^{-4}}$. From Equations ST6.1.23 and ST6.1.24, Section ST6.1,

$\delta \cong \dfrac{1}{1 + (W_B/L_{eB})^2/2}$ for small $\dfrac{W_B}{L_{eB}}$, and $\gamma \cong \dfrac{1}{1 + D_{hE}W_BN_{AB}/D_{eB}L_{hE}N_{DE}}$. Substituting numerical values, $\delta \cong 0.9996$ and $\gamma = 0.9958$, it follows that $\alpha_F = 0.9954$ and $\beta_F = \dfrac{\alpha_F}{1 - \alpha_F} \cong 216$.

(b) If the transistor is operated in the inverse active mode, it would appear that δ stays the same because it depends on W_B and L_{eB}. However, because in practice the collector has a much larger area than the emitter, a smaller fraction of the electrons injected from collector to base would be able to reach the emitter. A more realistic value of δ_R may be about 0.8. γ_R is obtained by replacing emitter quantities by corresponding collector quantities. Thus, $\gamma_R \cong \dfrac{1}{1 + D_{hC}W_BN_{AB}/D_{eB}L_{hC}N_{DC}} = 0.87$. It follows that $\alpha_R = \delta_R\gamma_R \cong 0.7$ and $\beta_R \cong 2.3$.

(c) From Equations 2.6.1 and 2.6.2, Chapter 2, the mobilities are:

$$\mu_{hB} = 465 \text{ cm}^2/\text{Vs}, \quad \mu_{eE} = 419 \text{ cm}^2/\text{Vs}, \quad \mu_{eC} = 1421 \text{ cm}^2/\text{Vs}$$

The diffusion constants are:

$$D_{hB} = 12.02 \text{ cm}^2/\text{s}, \quad D_{eE} = 10.83 \text{ cm}^2/\text{s}, \quad D_{eE} = 36.73 \text{ cm}^2/\text{s}$$

The diffusion lengths are

$$L_{hB} = 2.193 \times 10^{-3} \text{ cm}, \quad L_{eE} = 1.04 \times 10^{-4} \text{ cm}, \quad L_{eC} = 4.285 \times 10^{-3} \text{ cm}$$

It follows that $\dfrac{W_B}{L_{hB}} = \dfrac{10^{-4}}{21.93 \times 10^{-4}}$ and $\delta = 0.9990$. Interchanging donor and acceptor impurities, and electrons and holes, $\gamma \cong \dfrac{1}{1 + D_{eE}W_B N_{DB}/D_{hB}L_{eE}N_{AE}} = 0.9830$. Hence, $\alpha_F = 0.9820$ and $\beta_F = \dfrac{\alpha_F}{1 - \alpha_F} \cong 55$.

This example underscores the dramatic difference in current gains of a transistor operated in the active and in the inverse active modes. It also demonstrates how differences in electron and hole mobilities affect the current gains in otherwise identical *npn* and *pnp* transistors.

Let us calculate the components of the base current in Equation 6.1.1. From Equation 3.2.4, Chapter 3, $I_{CBO} = A_J |q| \left(\dfrac{D_{hC}}{L_{hC}N_{DC}} + \dfrac{D_{eB}}{L_{eB}N_{AB}} \right) n_i^2 = 9.4 \times 10^{-13}$ A. From Equation ST6.1.1, Section ST6.1, $I_{hE} = \dfrac{|q| D_{hE} A_J n_i^2}{L_{hE} N_{DE}} e^{V_{BE}/V_T} = 1.35 \times 10^{-4}$ A. From Equations ST6.1.21 and ST6.1.22, $I_{e(rec)} = I_{eE} - I_{eC} = \dfrac{|q| D_{eB} A_J n_i^2}{L_{eB} N_{AB}} e^{V_{BE}/V_T} \left(\dfrac{\cosh (W_B/L_{eB}) - 1}{\sinh (W_B/L_{eB})} \right) = 1.15 \times 10^{-5}$ A. I_{CBO} is thus negligible compared to the other components, and about $1.35 \times 100/(1.35 + 0.115) \cong 92\%$ of the base current is due to I_{hE}. In other words, $I_B = I_{e(rec)}(1.35 + 0.115)/0.115 \cong 13 \times I_{e(rec)}$. In most IC *npn* transistors, $I_{e(rec)}$ is only a small fraction of I_B. At the other extreme are lateral *pnp* transistors (Section 4.7, Chapter 4), where most of I_B can be due to $I_{e(rec)}$. Such transistors have a low β_F, approximately in the range of 10–50, and a relatively low unity-gain bandwidth (Equation 6.2.38).

By reducing the base width to a sufficiently low value, and having a high ratio of emitter-to-base doping levels, super-beta transistors having β_F of several thousands are possible. Too thin a base, however, makes the transistor susceptible to punchthrough, as discussed later, so that such transistors can be used only in some special circuits (Section 10.5, Chapter 10). Another way of increasing β_F to several thousands is to have a reasonably narrow base that makes δ close to unity and increase γ by having a heterojunction for the BEJ, as discussed in Section 6.4.

Exercise 6.1.5

Calculate α_F and β_F if: (a) $W_B = 5$ μm; (b) $N_{AB} = 10^{17}/\text{cm}^3$.
 Answers: (a) 0.9945, 181; (b) 0.99, 99.   ∎

Small-Signal Current Gains

Formally, the output characteristics of Figure 6.1.3 can be expressed in terms of the functional relation: $i_C = f(i_E, v_{CB})$. From a Taylor series expansion in the neighborhood of the operating point (Appendix A):

$$\Delta i_C = \left. \frac{\partial i_C}{\partial i_E} \right|_{v_{CB}} \Delta i_E + \left. \frac{\partial i_C}{\partial v_{CB}} \right|_{i_E} \Delta v_{CB} \tag{6.1.11}$$

The second term on the RHS is the slope of the CB output characteristics at constant i_E and is ideally zero if base-width modulation, considered later, is neglected. The first term is the limit of $\Delta i_C/\Delta i_E$ at constant v_{CB}, as the changes become infinitesimally small. It defines the **small-signal, short-circuit, common-base current gain** α. At constant v_{CB}, $v_{cb} = 0$, which means that the collector and base are short-circuited for the ac small-signal.

In practice, this is achieved by connecting a large capacitor between base and collector, so that the two terminals are short-circuited by the capacitor for ac signals *without disturbing dc bias conditions*.

If I_E and I_C in Equation 6.1.6 are replaced by their instantaneous values i_E and i_C, and both sides of the equation are differentiated with respect to i_E bearing in mind that I_{CBO} is independent of i_E,

$$\alpha = \left.\frac{\partial i_C}{\partial i_E}\right|_{v_{CB}} = \alpha_F + \left.\frac{\partial \alpha_F}{\partial i_E}\right|_{v_{CB}} i_E \tag{6.1.12}$$

If α_F is independent of i_E, then:

$$\alpha = \left.\frac{\partial i_C}{\partial i_E}\right|_{v_{CB}} = \alpha_F \tag{6.1.13}$$

In fact, α_F can be considered independent of I_E over a fairly wide range of I_E. However, α_F decreases both at low and high I_E. The decrease at low I_E is mainly due to electron–hole recombination in the depletion region of the BEJ. This recombination current is an appreciable fraction of I_E at low values of I_E so that less of I_E is injected into the base, which reduces γ and hence α_F. A large I_E implies **high-level injection,** so that the minority-carrier concentration in the base becomes comparable to that of majority carriers. In order to compensate for the increase in minority carrier concentration, the majority-carrier concentration rises above its equilibrium value. As a result, α_F is reduced in two ways: (1) The increase in majority-carrier concentration is akin to increasing the doping level in the base, which reduces γ, because as noted earlier, γ is made close to unity by having the doping level in the emitter much larger than that of the base (Example 6.1.1). (2) The increased carrier concentrations increase the recombination rate by increasing the probability of recombination of minority carriers, which in turn reduces the base transport factor δ. Moreover, both of these effects are aggravated by "crowding" of the emitter current at high injection levels, because of curvature of the BEJ in practice, leading to a nonuniform distribution of current across the BEJ. Because current at the emitter terminal is the average emitter current over the BEJ, emitter current crowding implies that the aforementioned effects of a high concentration of majority carriers in the base become manifest at a lower value of emitter current.

Neglecting the variation of α_F with i_E, Equation 6.1.11 becomes:

$$i_c = \alpha i_e \tag{6.1.14}$$

In the CE configuration, i_C can be expressed as the functional relation: $i_C = f(i_B, v_{CE})$. From a Taylor series expansion in the neighborhood of the operating point (Appendix A):

$$\Delta i_C = \left.\frac{\partial i_C}{\partial i_B}\right|_{v_{CE}} \Delta i_B + \left.\frac{\partial i_C}{\partial v_{CE}}\right|_{i_B} \Delta v_{CE} \tag{6.1.15}$$

The second term is the slope of the output characteristics for constant i_B. Neglecting the effects of base-width modulation, considered later, this slope is ideally zero as in the CB configuration. The first term is the limit of $\Delta i_C/\Delta i_B$ at constant v_{CE}, as the changes become infinitesimally small. It defines the **small-signal, short-circuit, CE current gain** β. As in the case of α, constant v_{CE} implies that the collector and emitter are short-circuited by a large capacitor for the ac small-signal only.

Replacing I_B and I_C in Equation 6.1.10 by the corresponding instantaneous values i_B and i_C,

$$i_C = \beta_F i_B + (1 + \beta_F) I_{CBO} \tag{6.1.16}$$

Differentiating with respect to i_B at constant v_{CE}:

$$\beta = \left.\frac{\partial i_C}{\partial i_B}\right|_{v_{CE}} = \beta_F + (i_B + I_{CBO}) \left.\frac{\partial \beta_F}{\partial i_B}\right|_{v_{CE}} \tag{6.1.17}$$

Whereas α_F could be considered constant over a wide range of transistor currents, there is much less justification for neglecting the variation of β_F with transistor currents because small variations in α_F are greatly amplified in β_F. As explained in connection with α_F, β_F decreases at low currents and at high currents. It follows that in using β_F or β, the appropriate value should be used for a given I_B or I_C because, strictly speaking, β_F and β cannot be considered constant over a wide range of currents. Bearing this in mind, Equation 6.1.15 can be written as:

$$i_c = \beta i_b \tag{6.1.18}$$

Simulation Example 6.1.2 Variation of BJT Current Gains

It is desired to examine the variation of dc and small-signal current gains of a popular *npn* transistor, the 2N2222, using PSpice simulation.

SIMULATION

The part name of the transistor is Q2N2222 in the EVAL Library. For the CB configuration, a 15 V dc source is connected between collector and base. A dc current source of the same polarity as the emitter current is connected between emitter and base. The current source is swept in logarithmic sweep from 1 μA to 1 A at 20 points/decade. After the simulation is run, the traces selected are IC(Q1)/I(I1) for α_F and D(IC(Q1)) for α, where I1 is the current source and Q1 is the transistor. The traces are shown in Figure 6.1.10a. α_F varies between 0.988 and a maximum of 0.9954 (at $i_E = 22.4$ mA) over a current range of 4.08 μA to 802 mA, a variation of only about 0.75% over more than 5 decades of current. The difference between α and α_F is small up to a few hundred mAs.

The simulation of the CE configuration is similar. The input current source is applied to the base and is swept over the range 10 nA to 100 μA. The x-axis is changed to IC(Q1), the β_F trace is obtained as IC(Q1)/IB(Q1), and the β trace as 1/D(IB(Q1)). The traces are shown in Figure 6.1.10b. β_F varies between 80 and a maximum of 210 (at $i_C = 23.6$ mA) over a current range of 3.7 μA to 800 mA, a variation of more than 150%. The difference between β and β_F is about 20 in the lower part of the current range and about 40 in the range of few hundred mAs.

It should be noted that β_F was defined in connection with Equation 6.1.10 as $\dfrac{\alpha_F}{1 - \alpha_F}$, α was defined as $\left.\dfrac{\partial i_C}{\partial i_E}\right|_{v_{CB}}$ (Equation 6.1.12), and β was defined as $\left.\dfrac{\partial i_C}{\partial i_B}\right|_{v_{CE}}$ (Equation 6.1.15). So what is the relationship between α and β? This relationship can be derived from the basic KCL relation (Figure 6.1.1) $i_C + i_B = i_E$. Differentiating with respect to i_C at constant v_{CE} gives:

$$1 + \left.\frac{\partial i_B}{\partial i_C}\right|_{v_{CE}} = \left.\frac{\partial i_E}{\partial i_C}\right|_{v_{CE}} \tag{6.1.19}$$

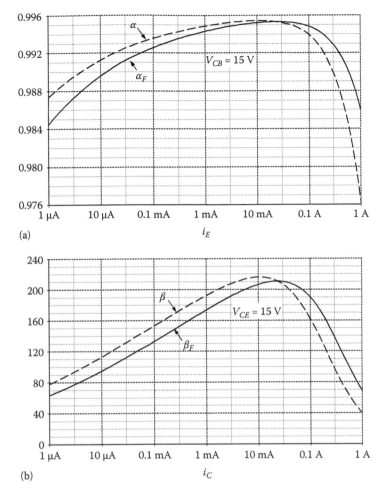

FIGURE 6.1.10
Figure for Example 6.1.2. (a) Variation of α and α_F with i_E; (b) variation of β and β_F with i_C.

The differential term on the LHS is $1/\beta$, whereas the term on the RHS is $1/\alpha$ if constant v_{CE} is identified with constant v_{CB}. This equivalence is usually made for the BJT in the active mode. From KVL, $v_{CE} = v_{CB} + v_{BE}$. In normal operation, v_{BE}, being the forward voltage drop across a conducting *pn* junction is nearly constant at about 0.7 V. Changes in v_{BE} are small compared to v_{CB}, which means that if v_{CB} is constant, v_{CE} can be considered constant, and conversely (Exercise 6.1.5). With this assumption, Equation 6.1.19 gives:

$$1 + \frac{1}{\beta} = \frac{1}{\alpha} \quad \text{or} \quad \beta = \frac{\alpha}{1 - \alpha} \tag{6.1.20}$$

which is the same as that between α_F and β_F.

The current gains for a BJT are summarized in Table 6.1.2 and apply to both *npn* and *pnp* transistors. For practical purposes, it is convenient to make following assumption:

TABLE 6.1.2

Summary of BJT Current Gain Relations (Neglecting
Base-Width Modulation)

dc Values	Small-Signal Values
$\alpha_F = \gamma\delta = \dfrac{I_{eC}}{I_E}$ (by definition)	$\alpha = \left.\dfrac{\partial i_C}{\partial i_E}\right\|_{v_{CB}} = \alpha_F$ (α_F independent of I_E)
$I_C = \alpha_F I_E$ (negligible I_{CBO})	$i_c = \alpha i_e$ (α_F independent of I_E)
$\beta_F = \dfrac{\alpha_F}{1 - \alpha_F}$ (by definition)	$\beta = \left.\dfrac{\partial i_C}{\partial i_B}\right\|_{v_{CE}} = \dfrac{\alpha}{1 - \alpha}$ (constant v_{CE} implies constant v_{CB})
$I_C = \beta_F I_B$ (negligible $(\beta_F + 1)I_{CBO}$) (β_F evaluated at operating point)	$i_c = \beta i_b$ (β evaluated at operating point)
$I_B = (1 - \alpha_F)I_E = I_E/(1 + \beta_F)$ (follows from preceding relations)	$i_b = (1 - \alpha)i_e = i_e/(1 + \beta)$ (follows from preceding relations)

Assumption: *For many practical purposes, it is convenient to assume that $\alpha = \alpha_F$ and $\beta = \beta_F$. Nevertheless, the nature of the assumptions made must be kept in mind.*

Exercise 6.1.6

Assume that v_{BE} varies between 0.6 and 0.8 V from cutoff to saturation. Determine the percentage change in v_{CE} if v_{CB} is maintained constant at: (a) 5 V; (b) 15 V. Repeat if i_B varies between 50 μA and 0.5 mA in Figure 6.2.4b, assuming that the curve for $v_{CE} = 1$ V applies.

Answers: For 0.2 V variation: (a) 3.6%; (b) 1.3%. If i_B varies between 50 μA and 0.5 mA as shown in Figure 6.2.4b, v_{BE} varies between about 0.7 and 0.77 V. Then (a) 1.25%; (b) 0.45%. ∎

Small-Signal Equivalent Circuits

The small-signal equivalent circuit can be readily derived from the preceding description of operation of the BJT. Considering the CB configuration first, it follows from Figure 6.1.2a that $\Delta v_{BE} = r_e \Delta i_E$, where r_e is the reciprocal of the slope of the i_E–v_{BE} characteristic. In the small-signal equivalent circuit it is conventional to consider positive input current as entering the input terminal, in the direction of the small-signal voltage drop from the input terminal to the common terminal. Hence, if we define $i_e = -\Delta i_E$ and $v_{eb} = -\Delta v_{EB}$ (Figure 6.1.11), $v_{eb} = r_e i_e$, where r_e is still positive and represents the small-signal incremental resistance as seen at the emitter terminal.

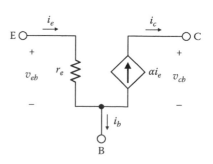

FIGURE 6.1.11
Small-signal equivalent circuit of a BJT in the CB configuration.

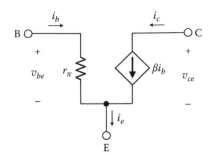

FIGURE 6.1.12
Small-signal equivalent circuit of a BJT in the CE configuration.

Equation 6.1.14 can be represented by a current source αi_e directed from base to collector (Figure 6.1.11). With α positive and i_e flowing inward, i_c is considered positive when flowing out of the collector terminal. Physically, an increase in v_{eb} reduces the base-emitter forward voltage and the current i_E flowing out of the emitter (Figure 6.1.1). This means that i_e, flowing *inward*, is increased ($i_e = -\Delta i_E$). By transistor action, i_C, flowing into the collector terminal is reduced. That is, i_c flowing out of the collector terminal in Figure 6.1.10 is increased.

In the CE configuration, i_B is a function of v_{BE} only; it is independent of v_{CE} in the absence of base-width modulation. Hence, $\Delta i_B = (di_B/dv_{BE})\Delta v_{BE}$. In terms of small-signal quantities, this can be written as $i_b = v_{be}/r_\pi$, where r_π is the reciprocal of the slope of the i_B-v_{BE} input characteristic and appears as an input resistance in Figure 6.1.12. It is seen that

$$r_\pi = \frac{v_{be}}{i_b} = \frac{i_e}{i_b}\frac{v_{be}}{i_e} = (\beta + 1)r_e \text{ (Table 6.1.2). In other words, } r_\pi \text{ is the incremental input resist-}$$

ance of the BEJ *referred to the base side*.

Note that i_b is directed inward at the input terminal, in the direction of the voltage drop v_{be}. Physically, an increase in i_B in Figure 6.1.1 implies an increase in both i_C and i_E in the directions indicated. Hence, i_c is also directed inward in Figure 6.1.12. On the output side, Equation 6.1.18 is represented by a current source βi_b directed from collector to emitter.

It is shown in Example 6.1.3 that the CE equivalent circuit of Figure 6.1.12 can be derived from the CB equivalent circuit of Figure 6.1.11 by simple circuit transformation.

Example 6.1.3 Transformation of CB to CE Small-Signal Equivalent Circuit

It is required to transform the CB equivalent circuit of Figure 6.1.11 to the CE equivalent circuit of Figure 6.1.12.

SOLUTION

As the input current i_e is directed inward in Figure 6.1.11, the first step is to reverse the directions of currents and to redraw the circuit as in Figure 6.1.13a, with reversed polarity of the dependent current source. The modified circuit is the **T-model** of the BJT. In this circuit, $v_{be} = r_e i_e$. But because $i_e = (\beta + 1)i_b$ (Table 6.1.2), it follows that $v_{be} = r_e(\beta + 1)i_b$. In other words, v_{be} could equally well result from i_b flowing in a resistance $r_\pi = (1 + \beta)r_e$, as in Figure 6.1.13b. The controlled current source is then reconnected between the collector and the emitter as the common terminal, so that only i_b flows in r_π. The value of the current source is similarly expressed in terms of i_b as $\alpha i_e = \alpha(1 + \beta)i_b = \beta i_b$. Reconnecting the current source preserves the relation $i_e = i_c + i_b$ but changes the voltage across the source from v_{cb} to v_{ce}. This is acceptable, because the voltage across an ideal current source is determined by the rest of the circuit and does not affect the source current. Redrawing the circuit gives Figure 6.1.12.

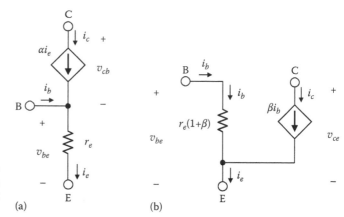

FIGURE 6.1.13

Figure for Example 11.6.3. (a) Figure 6.1.11 redrawn with all currents reversed; (b) circuit modified so that if flows in input resistor.

Exercise 6.1.7

Transform the circuit of Figure 6.1.12 to that of Figure 6.1.11.   ∎

6.2 Secondary Effects in BJTs

Base-Width Modulation

Base-width modulation, or the **Early effect**, is an important phenomenon that is analogous to channel-length modulation in FETs. Its primary effect is that it results in a nonzero slope of the output characteristics, that is, a finite output resistance, as in the case of FETs. However, in the BJT, base-width modulation has a small secondary effect that makes the output voltage affect the input side. This secondary effect is absent in the MOSFET because under dc conditions the input circuit is open ($I_G \cong 0$) so that changes in v_{DS} or i_D do not affect the input. Both the primary and secondary effects arise from the change in the width of the depletion layer of the BCJ with reverse bias, as occurs in a reverse-biased *pn* junction diode.

The BJT is analyzed in detail in Section ST6.1. For now, it suffices to note that in a *pn* junction diode, the concentration of injected minority carriers falls exponentially with distance at a rate determined by the diffusion length of these carriers (Figure 3.2.1, Chapter 3). Because the base of a BJT is narrow compared to the diffusion length, the exponential variation can be considered almost linear over its initial part. The concentration of injected electrons in the base of an *npn* transistor can thus be represented as in Figure 6.2.1. At the edge of the depletion region of the BEJ in the base ($x = 0$), the electron concentration is $n_p(0) = n_{po}e^{v_{BE}/V_T}$, whereas at the edge of the depletion region of the BCJ in the base the electron concentration is negligibly small. The electron current in the base is a diffusion current of density $|q|D_{eB}dn_p(x)/dx$, where D_{eB} is the diffusion constant of electrons in the base (Equation 2.4.12, Chapter 2). Assuming a linear concentration profile, $dn_p(x)/dx \cong n_p(0)/W_B$. We therefore have:

$$i_{eE} = \frac{|q|A_J D_{eB}}{W_B} n_{po}e^{v_{BE}/V_T} = I_S e^{v_{BE}/V_T} \cong i_E = i_C/\alpha_F \cong i_C \qquad (6.2.1)$$

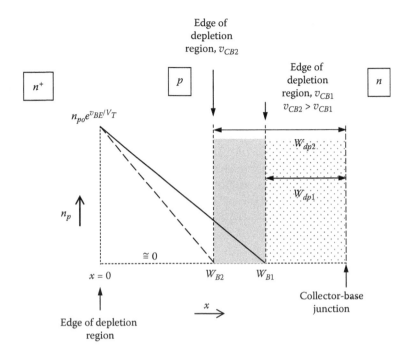

FIGURE 6.2.1
Illustration of base-width modulation.

where A_J is the effective area of the emitter junction. Strictly speaking, the magnitude of the slope of the concentration profile at the edge of the depletion region on the base side of the BEJ is proportional to i_{eE}, whereas the magnitude of the slope at the edge of the depletion region on the base side of the BCJ is slightly less and proportional to i_{eC}. The difference between these currents is $i_{e(rec)}$ and is proportional to the area under the curve, which represents the stored charge due to the injected electrons (Problem P6.3.1). Hence,

Concept: *If the base current is neglected, the concentration profile of minority carriers in the base is linear, the slope of the profile being proportional to the emitter or collector current, and the area under the profile being proportional to the minority carriers stored in the base.*

The approximations made in using Equation 6.2.1 are also examined in Example ST6.1.2, Section ST6.1. Note that W_B, the effective width of the base region, is the actual base width less than the widths of the depletion regions of the BEJ and the BCJ in the base. Although i_C, according to Figure 6.2.1 is in the negative x-direction and should therefore have a negative sign, this sign is omitted from Equation 6.2.1 to make i_C positive in conformity to the usual convention of considering the inward collector current positive.

It is seen from Equation 6.2.1 that i_C is inversely proportional to W_B. Thus, if v_{CB} increases from v_{CB1} to v_{CB2}, the width of the depletion region in the base increases from W_{dp1} to W_{dp2}, and the effective base width decreases from W_{B1} to W_{B2} (Figure 6.2.1). i_C therefore increases with v_{CB}; this is the primary effect of base-width modulation. The secondary effect of base-width modulation, which is neglected in Equation 6.2.1 is the variation of $i_{e(rec)}$ with effective base width, because of the change in area under the concentration profile.

The BJT output characteristics that are analogous to those of FETs are the CE output characteristics, with v_{BE} as parameter, illustrated in Figure 6.2.2 in exaggerated form. Compared to the CB characteristics (Figure 6.1.2), not only do they have a larger slope, but they are also shifted to the right by v_{BE}, because $v_{CE} = v_{CB} + v_{BE}$. The saturation region occurs at small values of v_{CE} because in this region, both the BEJ and BCJ are forward biased in the range of 0.6–0.8 V, but with $v_{BC} < v_{BE}$, mainly because the BCJ is larger in area than the BEJ (Figure 6.1.4). Hence, $v_{CE(sat)} = v_{BE} - v_{BC} \cong 0.2$ V. Note that the region where the output current increases rapidly with the output voltage is the triode region in the FET and the saturation region in the BJT, whereas the region where the changes in output current are small is the saturation region in the FET and the active region in the BJT.

The variation of i_C with v_{CE} at constant v_{BE} can be expressed as:

$$\left.\frac{\partial i_C}{\partial v_{CE}}\right|_{v_{BE}} = \left.\frac{\partial i_C}{\partial W_B}\right|_{v_{BE}} \left.\frac{\partial W_B}{\partial v_{CE}}\right|_{v_{BE}} \tag{6.2.2}$$

From Equation 6.2.1, $\left.\dfrac{\partial i_C}{\partial W_B}\right|_{v_{BE}} = -\dfrac{i_C}{W_B}$. Hence,

$$\left.\frac{\partial i_C}{\partial v_{CE}}\right|_{v_{BE}} = \frac{-i_C}{W_B} \left.\frac{\partial W_B}{\partial v_{CE}}\right|_{v_{BE}} \tag{6.2.3}$$

In the saturation region, where the BCJ is forward biased, base-width modulation is negligible. If the quantities on the RHS of Equation 6.2.3 are evaluated at the edge of active region, just as the transistor comes out of saturation, with the values of i_C and W_B denoted by $I_{C(sat)}$ and $W_{B(sat)}$, respectively, Equation 6.2.3 can be expressed as:

$$\left.\frac{\partial i_C}{\partial v_{CE}}\right|_{v_{BE}} = \frac{-i_{C(sat)}}{W_{B(sat)}} \left.\frac{\partial W_B}{\partial v_{CE}}\right|_{v_{BE}} \tag{6.2.4}$$

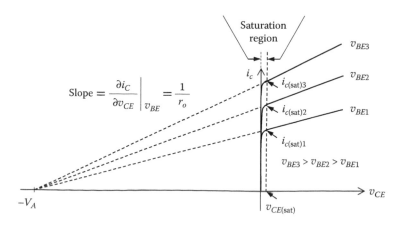

FIGURE 6.2.2
Output characteristics of a BJT in the CE configuration with v_{BE} as parameter.

Note that $\dfrac{\partial W_B}{\partial v_{CE}}\bigg|_{v_{BE}} p < 0$, because W_B decreases with v_{CE}, so that $\dfrac{\partial i_C}{\partial v_{CE}}\bigg|_{v_{BE}} > 0$ and increases with v_{BE}, at a constant v_{CE}, due to the increase in $i_{C(sat)}$ (Equation 6.2.1).

Since $\dfrac{\partial i_C}{\partial v_{CE}}\bigg|_{v_{BE}}$ is proportional to $i_{C(sat)}$ (Equation 6.2.4), and $i_{C(sat)}$ increases exponentially with v_{BE} (Equation 6.2.1), this means that as v_{BE} increases, the output characteristics for constant v_{BE} have: (1) larger $I_{C(sat)}$, and (2) slope more in the active region (Figure 6.2.2). However, the ratio $i_{C(sat)}\bigg/\left(\dfrac{\partial i_C}{\partial v_{CE}}\bigg|_{v_{BE}}\right) = -W_{B(sat)}\dfrac{\partial v_{CE}}{\partial W_B}\bigg|_{v_{BE}}$ is nearly independent of $i_{C(sat)}$.

Hence, if the straight parts of the output characteristics in the active region are extended backward to negative voltages, they would all intersect the voltage axis at a point $-V_A$ (Figure 6.2.2). V_A is the **early voltage** and is a positive quantity, typically in the range of 50–100 V. It is analogous to $1/\lambda$ in the case of MOSFETs (Equation 5.3.1, Chapter 5) and is given by:

$$V_A = -W_{B(sat)}\dfrac{\partial v_{CE}}{\partial W_B}\bigg|_{v_{BE}} \tag{6.2.5}$$

The output characteristics in the active region at constant v_{BE} can then be expressed as:

$$i_C = i_{C(sat)}\left(1 + \dfrac{v_{CE} - v_{CE(sat)}}{V_A}\right), \quad v_{CE} \geq v_{CE(sat)} \tag{6.2.6}$$

$$\cong i_{C(sat)}\left(1 + \dfrac{v_{CE}}{V_A}\right), \quad v_{CE} \gg v_{CE(sat)}$$

Equation 6.2.6 implies that $i_C = 0$ at $v_{CE} = v_{CE(sat)} - V_A \cong -V_A$, independently of v_{BE}.

The input characteristics of the CE configuration are affected by base-width modulation. When $v_{CE} = 0$, the $i_B - v_{BE}$ relation has the general shape of the i–v relation of a forward-biased *pn* junction (Figure 6.2.3a). With v_{CE} greater than about 0.8 V, the BCJ becomes reverse biased, and the secondary effect of base-width modulation comes into play. As v_{CE} increases at constant v_{BE}, the minority-carrier stored charge in the base decreases because the effective base width is reduced. $i_{e(rec)}$, and hence i_B, decrease (Equation 6.1.1). The decrease in i_B at constant v_{BE} means that the input characteristics for constant v_{CE} are

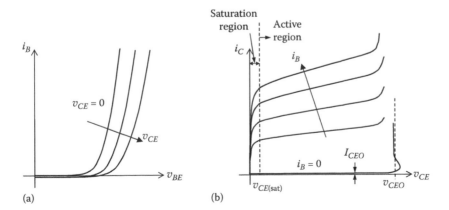

FIGURE 6.2.3
Input characteristics (a) and output characteristics (b) of a BJT in the CE configuration.

shifted to the right, as illustrated in Figure 6.2.3a in exaggerated form. The extent of this shift depends on how much the recombination current contributes to the base current (Example 6.1.1). Moreover, the characteristics for $v_{CE} > 0$ no longer pass through the origin. This is because when $v_{BE} = 0$, that is, the base is shorted to the emitter, and if $v_{CE} > 0$, a small negative base current flows (Example ST6.1.1, Section ST6.1).

The output characteristics of the CE configuration are conventionally plotted as a family of curves at constant i_B (Figure 6.2.3b) rather than constant v_{BE}. They have the same general shape as the constant v_{BE} curves of Figure 6.2.2 but with somewhat larger slopes in the active region (Section ST6.2).

Simulation Example 6.2.1 CE Characteristics

It is desired to simulate the characteristics of a 2N2222 transistor in the CE configuration.

SIMULATION

To obtain the output characteristics, a VDC source is applied between collector and emitter and an IDC source between base and emitter. The voltage source is swept, as primary sweep, between 0 and 20 V in steps of 0.01 V. The current source is swept, as secondary sweep, between 0 and 0.5 mA in steps of 50 μA. Figure 6.2.4a shows the resulting output characteristics. Considering the 0.5 mA characteristic, $i_{C(sat)} = 79.15$ mA, $v_{CE(sat)} = 0.35$ V, and $V_A = 73.1$ V.

The input characteristics are obtained by connecting a VDC source between base and emitter and sweeping it, as primary sweep, between 0 and 1 V. The collector-emitter voltage source is swept as secondary sweep using a list of values 0, 0.1, 1, 5, and 20 V. The resulting characteristics are shown in Figure 6.2.4b. With the voltage and current scales used in this figure, the input characteristics for $v_{CE} > 1$ V are too close together to be differentiated, and the reverse base current for very small v_{BE} and $v_{CE} > 0$ is too small to be seen. The change in input characteristics occurs in the saturation region, but once the BCJ becomes reverse biased i_B changes very little with v_{CE} because the base recombination current is a small fraction of i_B.

The saturation region is shown on an expanded scale in Figure 6.2.5. It is seen that the saturation voltage is between about 0.2 and 0.35 V for the current range considered.

Figure 6.2.6 illustrates the variation of β_F with v_{CE} and temperature. If v_{CE} increases at constant v_{BE}, i_C increases due to base-width modulation. To maintain the same i_C, v_{BE} should decrease in order to have the same slope of the concentration profile as before the increase in v_{CE}. The decrease in v_{BE} reduces i_B. In other words, the same i_C at a larger v_{CE} requires less i_B, which means that β_F has increased. The variation of β_F with temperature is discussed in the next section in connection with temperature effects.

Hybrid-π Equivalent Circuit

The hybrid-π, low-frequency, small-signal equivalent circuit of the transistor, shown in Figure 6.2.7, is derived in detail in Section ST6.2 and related to the effective output resistances of the CB and CE configurations. The term "hybrid" refers to the voltage-controlled current source $g_m v_{be}$, which relates a current on the output side to the input voltage. r_μ, shown dotted, is a relatively large resistance (Example 6.2.2) that accounts for the secondary effect of base-width modulation, and is usually omitted in simplified analysis. Unlike the effect of channel-length modulation in a MOSFET, r_μ introduces a conducting path between base and collector.

$g_m = \left. \dfrac{\partial i_C}{\partial v_{BE}} \right|_{v_{CE}}$ is the small-signal transconductance. Assuming a linear concentration profile of minority carriers in the base (Figure 6.2.1), g_m is derived from Equation 6.2.1 as:

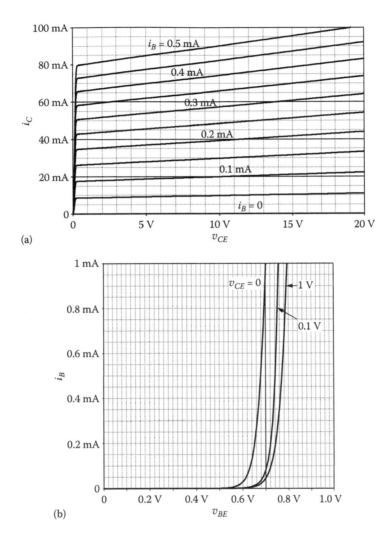

FIGURE 6.2.4
Figure for Example 6.2.1. (a) Output characteristics; (b) input characteristics.

$$g_m = \left.\frac{\partial i_C}{\partial v_{BE}}\right|_{v_{CE}} = \left.\frac{I_C}{V_T}\right|_{v_{CE}} \qquad (6.2.7)$$

The definition and interpretation of g_m are analogous to that for the FET. If a vertical line representing constant v_{CE} is drawn on the output characteristics of Figure 6.2.2, g_m is the limit of $\Delta i_C / \Delta v_{BE}$ as the changes become infinitesimally small. As in the FET, g_m is a measure of how effective is the input voltage, v_{BE} in this case, in controlling the output current. As $V_T \cong 0.026$ V at 300K, g_m for the BJT is much higher than that of the FET (Example 6.2.2). Moreover, g_m increases faster with I_C, being proportional to I_C, compared to $\sqrt{I_D}$ in the FET.

$r_\pi = \left.\dfrac{\partial v_{BE}}{\partial i_B}\right|_{v_{CE}}$, as previously defined, is the reciprocal of the slope of the input characteristics of Figure 6.2.4b at constant v_{CE}. If r_μ is omitted, it follows from Figure 6.2.7 that r_π is the incremental input resistance, irrespective of whether the output is open circuited

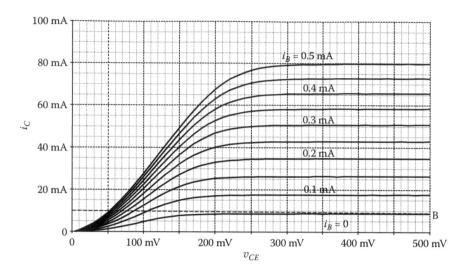

FIGURE 6.2.5
Figure for Example 6.2.1.

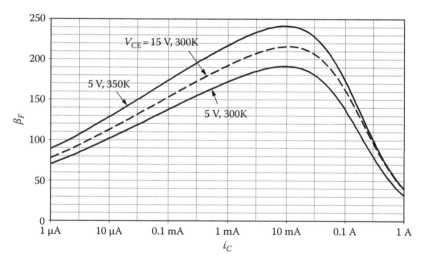

FIGURE 6.2.6
Figure for Example 6.2.1.

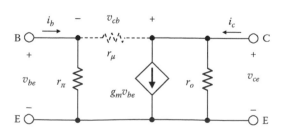

FIGURE 6.2.7
Hybrid-π, small-signal equivalent circuit of a BJT.

or short-circuited for the ac signal. A relation between g_m, r_π, and β can readily be derived. Thus,

$$\left.\frac{\partial i_C}{\partial v_{BE}}\right|_{v_{CE}} \left.\frac{\partial v_{BE}}{\partial i_B}\right|_{v_{CE}} = \left.\frac{\partial i_C}{\partial i_B}\right|_{v_{CE}} \quad \text{or} \quad g_m r_\pi = \beta \tag{6.2.8}$$

If r_μ is omitted, $v_{eb} = r_\pi i_b$ and $g_m v_{be} = g_m r_\pi i_b = i_b = \beta i_b$. The $g_m v_{be}$ VCCS can therefore be replaced by a βi_b CCCS (Figure 6.2.8), *but only with r_μ omitted.*

If $v_{be} = 0$ in Figure 6.2.7 and r_μ is neglected, $r_o = \left.\dfrac{\partial v_{CE}}{\partial i_C}\right|_{v_{BE}}$ is the reciprocal of the slope of the output characteristics with v_{BE} as parameter. Neglecting the secondary effect of base-width modulation ($r_\mu \to \infty$), which is not included in Equation 6.2.1, r_o is, from Equation 6.2.6:

$$r_o = \frac{V_A}{i_{C(sat)}} \tag{6.2.9}$$

Note that in the absence of r_μ, r_o is the incremental output resistance with the input either short-circuited to small signals ($v_{be} = 0$) or open circuited ($i_b = 0$), corresponding to the CE characteristics with constant v_{BE} or constant i_B, respectively. Typical values are in the ranges of a few hundred ohms to several kΩ for r_π, few tens of kΩ to several hundred kΩ for r_o, and hundreds of kΩ to several hundred mΩ for r_μ.

If g_m is evaluated at $i_C = i_{C(sat)}$, then combining Equations 6.2.7 and 6.2.9,

$$g_m r_o = \frac{V_A}{V_T} \tag{6.2.10}$$

Table 6.2.1 summarizes the expressions for small-signal parameters and their interrelations, the relations in bold being particularly useful in practice. The approximations involved in these relations are commonly made, as by assuming $I_C \cong I_{C(sat)}$. It should be emphasized

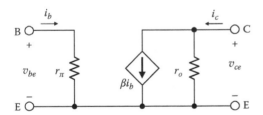

FIGURE 6.2.8
Replacement of the $g_m v_{be}$ current source by a βi_b source.

TABLE 6.2.1

Parameters of the CE Hybrid-π Equivalent Circuit

$$g_m = \frac{I_C}{V_T} = \frac{\beta}{r_\pi} = \frac{\alpha}{r_e} = \frac{1}{r_o}\frac{V_A}{V_T}$$

$$r_\pi = \frac{\beta}{g_m} = \frac{V_T}{I_B} = \beta\frac{V_T}{I_C} = \frac{V_T}{I_B} = (\beta + 1)r_e$$

$$r_o = \frac{V_A}{I_{C(sat)}} = \frac{1}{g_m}\frac{V_A}{V_T}$$

$$r_e = \frac{r_\pi}{\beta + 1} = \frac{V_T}{I_E} = \alpha\frac{V_T}{I_C} = \frac{\alpha}{g_m} \quad \text{and} \quad \frac{1}{r_e} = g_m + \frac{1}{r_\pi}$$

that for small signals $i_c = \beta i_b$ only when r_o is omitted (Figure 6.2.8), whereas for dc or instantaneous values, $i_C = \beta_F i_B$, as long as I_{CEO} can be neglected (Equation 6.1.10). This also illustrates the distinction that should, strictly speaking, be made between β and β_F.

Exercise 6.2.1

Verify all the relations in Table 6.2.1.   ∎

Example 6.2.2 Determination of Transistor Parameters

Given $i_{C(sat)} = 1$ mA, $\beta = 200$, and $V_A = 80$ V. Determine: (a) g_m, r_π, r_e, r_o; (b) r_μ; (c) input and output resistances.

SOLUTION

(a) From Equation 6.2.7, $g_m = 1/0.026 = 38.5$ mA/V. From Equation 6.2.8, $r_\pi = 200/38.5 = 5.2$ kΩ. From Equation 6.2.9, $r_o = 80/1 = 80$ kΩ. From Example 6.1.3 and Equation 6.2.8, $r_e \cong r_\pi/\beta \cong 1/g_m \cong 26\,\Omega$.

(b) $r_\mu = \left.\dfrac{\partial v_{BC}}{\partial i_B}\right|_{V_{BE}}$ and accounts for the effect of base-collector voltage on the recombination component $i_{e(rec)}$. It can be expressed as:

$$r_\mu = \left.\frac{\partial v_{BC}}{\partial i_B}\right|_{V_{BE}} = -\left.\frac{\partial v_{BC}}{\partial i_C}\right|_{V_{BE}} \left.\frac{\partial i_C}{\partial i_B}\right|_{V_{BE}} = \left.\frac{\partial v_{CE}}{\partial i_C}\right|_{V_{BE}} \left.\frac{\partial i_C}{\partial i_B}\right|_{V_{BE}} = r_o \left.\frac{\partial i_C}{\partial i_B}\right|_{V_{BE}} \qquad (6.2.11)$$

where the minus sign accounts for the decrease of i_C with v_{BC} (or increase with v_{CB}) due to the primary effect of base-width modulation, and the change in v_{CE} is the same as the change in v_{CB} at constant v_{BE}. Consider $\left.\dfrac{\partial i_C}{\partial i_B}\right|_{V_{BE}}$. β is defined as $\left.\dfrac{\partial i_C}{\partial i_B}\right|_{V_{CE}}$. In normal transistor operation at constant v_{CE}, the change in i_B is $\Delta i_B = \Delta i_{hE} + \Delta i_{e(rec)}$ (Equation 6.1.1). β is then given by $\Delta i_C/(\Delta i_{hE} + \Delta i_{e(rec)})$. When v_{BE} is constant, $\Delta i_{hE} = 0$ (Example 6.1.1) and $\left.\dfrac{\partial i_C}{\partial i_B}\right|_{V_{BE}} \cong \dfrac{\Delta i_C}{\Delta i_{e(rec)}}$ and is considerably larger than β. Hence $r_\mu > \beta r_o$. How much larger depends on what fraction of i_B is $i_{e(rec)}$, as explained at the end of Example 6.1.1. If we assume that $i_B = 10 i_{e(rec)}$, then $r_\mu \cong 10 \times 200 \times 80 \equiv 160$ MΩ.

(c) Because r_μ is very much larger than r_π, r_π is effectively the input resistance, irrespective of whether the output is open circuited or short-circuited for small signals (Figure 6.2.7).

It may be noted that if $v_{be} = 0$, then the output resistance looking into terminal C in Figure 6.2.7 is $(r_o \| r_\mu)$. This means that the slope of the output characteristics at constant v_{BE} is $(r_o \| r_\mu)$ and not r_o as defined by $r_o = \left.\dfrac{\partial v_{CE}}{\partial i_C}\right|_{V_{BE}}$. The discrepancy is due to a small approximation made in the derivation of the circuit of Figure 6.2.7, as explained in Section ST6.2. It is also argued in Section ST6.2 that the output resistance at constant i_B is very nearly $r_o \| (r_\mu/\beta)$ and the output resistance for the CB configuration is $r_\mu \| \beta r_o$. If the secondary effect of base-width modulation is neglected, $r_\mu \to \infty$, and the output resistance of the CE configuration is r_o at constant v_{BE} or i_B, which means that the slopes of the output characteristics for constant v_{BE} or i_B are equal. Moreover, if $r_\mu \to \infty$, the output resistance of the CB configuration becomes βr_o and is the slope of the CB output characteristics.

Considering the numerical values derived above, it is seen that $(r_o \| r_\mu) = (80\text{ kΩ}\|160\text{ MΩ}) \cong 80$ kΩ; $(r_o \| r_\mu/\beta) = (80\text{ kΩ}\|800\text{ kΩ}) = 72.7$ kΩ; and $(r_\mu \| \beta r_o) = (160\text{ MΩ}\|16\text{ MΩ}) = 15.5$ MΩ.

Variation of i_C with v_{be}

It is instructive to derive an expression for i_c as a function of v_{be}. Equation 6.2.1 can be expressed in terms of dc values as $I_C = \dfrac{I_S}{\alpha_F} e^{V_{BE}/V_T}$, or in terms of instantaneous values as $i_C = \dfrac{I_S}{\alpha_F} e^{v_{BE}/V_T}$. Substituting $v_{BE} = V_{BE} + v_{be}$ gives:

$$i_c = \frac{I_S}{\alpha_F} e^{V_{BE}/V_T} e^{v_{be}/V_T} = I_C e^{v_{be}/V_T} \tag{6.2.12}$$

But $i_C = I_C + i_c$. It follows that:

$$i_c = I_C(e^{v_{be}/V_T} - 1) \tag{6.2.13}$$

Suppose, for argument's sake, that $i_C = 10$ mA for $v_{BE} = 0.7$ V. Let v_{BE} change by 1%, that is, $v_{be} = 7$ mV. Equation 6.2.13 gives for $V_T = 26$ mV, $i_c \cong 0.31 I_C$. That is, a change of only 7 mV changes i_C by about 31%, from 10 to 13.1 mA. It follows that:

Concept: *Because of the exponential variation of i_C with v_{be}, a small change in v_{be} causes a relatively large change in i_C, which gives the BJT a large g_m.*

h-Parameter Equivalent Circuit

Transistor parameters are often specified by the h parameters of two-port circuits (Section 7.1, Chapter 7), where h_{11}, h_{12}, h_{21}, and h_{22} are denoted as h_{ie}, h_{re}, h_{fe}, and h_{oe}, respectively. The relationships between these parameters and those of the hybrid-π equivalent circuit are derived in Section ST6.3 for the CE configuration and summarized in Table 6.2.2. If $r_\mu \to \infty$, $h_{re} \cong 0$, and the h-parameter equivalent circuit is that of Figure 6.2.8 with $r_\pi = h_{ie}$, $\beta = h_{fe}$, and $r_o = 1/h_{oe}$. The measurement of h parameters is discussed in Section ST6.4.

Temperature Effects

As in the case of a *pn* junction diode, the effect of a rise in temperature is to make the reverse current of the BCJ approximately double for every 10°C rise in temperature, and to make v_{BE} decrease by about 2 mV/°C at constant current. Usually, v_{BE} opposes the voltage driving i_B at the input, so that the decrease in v_{BE} increases i_B and hence i_C. A rise in temperature also increases β_F (Figure 6.2.6) and hence i_C for a given i_B, because of the effect

TABLE 6.2.2

Relations between h Parameters and Hybrid-π Parameters

h Parameter	Hybrid-π Parameter
h_{fe}	β
h_{ie}	$(r_\pi \| r_\mu) \cong r_\pi$
h_{oe}	$\dfrac{1}{r_o} + \dfrac{\beta+1}{r_\pi + r_\mu} \cong \dfrac{1}{r_o}$
h_{re}	$\dfrac{r_\pi}{r_\pi + r_\mu} \cong \dfrac{r_\pi}{r_\mu}$

of the high doping level on the bandgap in the emitter. High doping levels decrease the bandgap due to several effects, including formation of an impurity energy band, rather than a discrete level, increased lattice deformation, and coulombic interactions of carriers at high concentrations. A reduced emitter bandgap makes ΔE_g negative in Equation 6.4.5 for a heterojunction, which reduces β_F but makes it increase with temperature. The temperature coefficient of β_F is typically about 0.7% per °C.

Breakdown

When avalanche multiplication sets in, Equation 6.1.6 can be expressed as:

$$i_C = M(\alpha_F i_E + I_{CBO}) \tag{6.2.14}$$

M in Equation 6.2.14 is an empirically-determined multiplication factor of the form:

$$M = \frac{1}{1 - (v_{CB}/BV)^n} \tag{6.2.15}$$

where
  v_{CB} is the reverse voltage across the BCJ
  BV is the breakdown voltage of this junction for a given i_C
  n assumes values between 3 and 6

As $v_{CB} \to BV$, $M \to \infty$, and i_C increases rapidly, being limited only by the external circuit. If the breakdown voltage in the CB configuration, with $i_E = 0$, is denoted by BV_{CBO} (Figure 6.1.2), Equation 6.2.15 becomes:

$$M = \frac{1}{1 - (v_{CB}/BV_{CBO})^n} \tag{6.2.16}$$

It is seen from Equation 6.2.14 that, for a given M, i_C increases with i_E, which means that a larger current contributes to avalanche multiplication so that breakdown sets in at lower values of v_{CB}. As illustrated in Figure 6.1.2, the larger i_E, the lower is the breakdown voltage.

For the CE configuration, breakdown is specified at $i_B = 0$. Substituting $i_E = i_C$ in Equation 6.2.14, and solving for i_C,

$$i_C = \frac{M}{1 - \alpha_F M} I_{CBO} \tag{6.2.17}$$

where M is given by Equation 6.2.16 with v_{CB} replaced by BV_{CEO}. This is because $v_{CE} \cong v_{CB}$, and breakdown occurs in the CE configuration when $v_{CE} = BV_{CEO}$. Thus,

$$M = \frac{1}{1 - (BV_{CEO}/BV_{CBO})^n} \tag{6.2.18}$$

From Equation 6.2.17, breakdown occurs when $\alpha_F M = 1$, or, from Equation 6.2.18, $\alpha_F = 1 - (BV_{CEO}/BV_{CBO})^n$. This gives:

$$BV_{CEO} = BV_{CBO} \sqrt[n]{1 - \alpha_F} \tag{6.2.19}$$

It is seen that BV_{CEO} is considerably less than BV_{CBO}. For example, if $n = 4$, and $\alpha_F = 0.99$, $BV_{CEO} = 0.32 BV_{CBO}$. The reason for this is that avalanche multiplication is amplified by transistor action in the CE configuration. As electron–hole pairs are generated in the depletion region of the BCJ, electrons are swept into the collector and holes into the base. To compensate for the positive charge in the base due to these holes, electrons are injected into the base from the emitter. But, by transistor action, a fraction $(1 - \alpha_F)$ of these electrons recombine with the holes swept into the base, whereas the remainder, which is α_F times, diffuses to the BCJ, where they increase the collector current and result in more multiplicative collisions. Breakdown therefore sets in at a lower reverse voltage across the BCJ. In practice, the effect is not as marked as Equation 6.2.19 suggests. This is because in actual transistor structures, as opposed to the plane-junction structures assumed in analysis, there are regions of the BCJ that are not adjacent to the emitter (Figure 6.1.4) and which, therefore, do not cause amplification of the avalanche current of the BCJ, as just described.

In a reverse-biased *pn* junction, the breakdown voltage is determined by the doping on the more lightly doped side (Section 3.2, Chapter 3 and Example 6.2.3). Hence, the collector is less heavily doped than the base in order to control BV_{CBO} and obtain values that are normally in the range of 100 V. In the case of the BEJ, the emitter is much more heavily doped in order to obtain an injection efficiency close to unity (Equation 6.1.2). The doping of the base therefore determines the BEJ breakdown voltage. As this voltage is normally less than about 8 V, zener breakdown plays a significant role (Section 3.2, Chapter 3).

Exercise 6.2.2

If $\alpha_F = 0.99$, determine BV_{CEO} in terms of BV_{CBO}, assuming (a) $n = 3$; (b) $n = 6$.
  Answers: (a) 0.22; (b) 0.46. ∎

Example 6.2.3 Determination of Breakdown Voltages

Given $N_{AB} = 10^{16}/cm^3$, $N_{DC} = 10^{15}/cm^3$, and a breakdown electric field ξ_{brk} of 2×10^5 V/cm, it is required to determine BV_{CBO} and BV_{CEO}, assuming that $n = 4$ in Equation 6.2.19 and $\beta = 50$ at the low value of I_{CEO}.

SOLUTION

According to the depletion approximation (Section 3.1, Chapter 3), the voltages V_p and V_n across the depletion region on the *p*-side and *n*-side, respectively, of a *pn* junction are from Equations 3.1.9 and 3.1.10, Chapter 3:

$$V_p = \frac{1}{2}W_p|\xi_{max}| \quad \text{and} \quad V_n = \frac{1}{2}W_n|\xi_{max}| \tag{6.2.20}$$

where ξ_{max} is the maximum electric field in the depletion region, which is at the metallurgical junction. The total reverse bias $V_R = V_p + V_n$. Hence,

$$\frac{V_n}{V_p + V_n} = \frac{1}{1 + W_p/W_n} \tag{6.2.21}$$

Using the relations of Equation 3.1.11, Chapter 3,

$$\frac{V_n}{V_p + V_n} = \frac{1}{1 + N_D/N_A} \tag{6.2.22}$$

If $N_D \ll N_A$, then $V_R \cong V_n$. Thus, if $N_D = 0.1N_A$, $V_R = 0.91\,V_n$. The breakdown voltage is therefore determined by the voltage across the depletion layer on the more lightly doped side.

With $N_D \ll N_A$, it follows from Equation 3.1.12 that $\xi_{max} = \sqrt{2|q|N_D(V_R + V_{npo})/\varepsilon_s}$. Identifying N_D with collector doping N_{DC}, V_R with BV_{CBO}, and neglecting V_{npo} compared to BV_{CBO}:

$$BV_{CBO} = \frac{\varepsilon_s \xi_{brk}^2}{2|q|N_{DC}} \tag{6.2.23}$$

Substituting numerical values gives $BV_{CBO} \cong 131$ V. The actual breakdown voltage is generally less than that given by Equation 6.2.23, because in real transistors, the BCJ is not planar but necessarily has some curvature at the edges. The electric field in these regions can be markedly higher than that in the planar parts of the junction, so breakdown occurs at lower voltages than predicted by Equation 6.2.23.

If we replace $(1 - \alpha_F)$ by its equivalent $1/(1 + \beta_F)$, Equation 6.2.19 becomes:

$$BV_{CEO} = \frac{BV_{CBO}}{\sqrt[4]{1 + \beta_F}} = \frac{131}{\sqrt[4]{51}} = 49 \text{ V}$$

Equation 6.2.19 can also be used to explain the "kink" in the CE characteristic for $i_B = 0$ at voltages near breakdown (Figure 6.2.3b). As BV_{CEO} is approached with only I_{CEO} flowing, α_F is small at this low current, as explained in connection with Equation 6.1.13, and BV_{CEO} as given by Equation 6.2.19 is relatively high. As breakdown sets in, the current increases so α_F increases, and BV_{CEO} decreases, causing the characteristic to bend back as shown. At higher collector currents, α_F assumes an almost constant value, and BV_{CEO} stays constant.

Punchthrough

This occurs when, with a sufficiently large reverse bias, the depletion region of the BCJ extends all across the base and merges with the depletion layer of the BEJ. Normally, avalanche breakdown occurs before this happens. But if the base is very thin or very lightly doped, punchthrough can occur. The base effectively ceases to have any control on the current. An increase of v_{CE} beyond the punchthrough value causes a large current to flow between collector and emitter, usually with catastrophic effects.

The depletion regions of the BEJ and the BCJ of an *npn* transistor at punchthrough are illustrated in Figure 6.2.9 together with the energy band diagram. The increase in current can be readily understood with reference to Figure 6.2.1. At punchthrough, the effective base width W_B becomes zero, so that, according to Equation 6.2.1 the diffusion current approaches infinity. Once in the base, electrons flow from the base to the collector under the influence of the large electric field due to the reverse bias.

BJT Capacitances

The effect of stored charges in a BJT can be simulated, for small changes, by incremental capacitances. Considering the diffusion capacitance of the forward-biased BEJ, and because the hole current injected into the emitter is small compared to the electron current injected into the base, the charge due to hole storage in the transition region on the emitter side of an *npn* transistor is neglected in comparison with the charge Q_{eB} due to electron storage in the base. $|Q_{eB}|$ can be readily related to i_C. If τ_F is the average time it takes an electron at the emitter end of the base to move to the collector end of the base, then the total charge

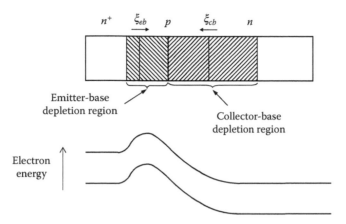

FIGURE 6.2.9
Depletion regions and energy band diagram of an *npn* transistor just before punchthrough.

transferred during this time is $|Q_{eB}|$, and the collector current is the charge transferred per unit time. Thus,

$$|Q_{eB}| = \tau_F i_C \qquad (6.2.24)$$

where i_C is a positive current flowing into the collector. Equation 6.2.24 is more rigorously derived in Section ST6.1. τ_F is the mean forward **transit time** through the base.
An incremental capacitance can be defined as:

$$C_{dE} = \left.\frac{\partial |Q_{eB}|}{\partial v_{BE}}\right|_{v_{CE}} = \left.\tau_F \frac{\partial i_C}{\partial v_{BE}}\right|_{v_{CE}} = \tau_F g_m = \tau_F \frac{I_C}{V_T} \qquad (6.2.25)$$

where τ_F is assumed to be independent of v_{BE} because electrons flow in the base predominantly by diffusion (Example ST6.1.4, Section ST6.1). C_{dE} appears between the base and emitter because it is defined with respect to changes in v_{BE}. To this capacitance must be added the capacitance C_{jE} that accounts for the change in the width of the depletion layer of the BEJ due to a change in v_{BE}. From Equation 3.2.6, Chapter 3,

$$C_{jE} = \frac{C_{jE0}}{\left(1 - V_{BE}/V_{jE0}\right)^m} \qquad (6.2.26)$$

where
C_{jE0} is the capacitance when $V_{BE} = 0$
V_{jE0} is the equilibrium junction potential
m depends on the grading of the BEJ and is typically between 0.2 and 0.5

As a rule of thumb, C_{jE} is approximately twice C_{jE0}. The total capacitance between base and emitter terminals is:

$$C_\pi = C_{dE} + C_{jE} \qquad (6.2.27)$$

At the reverse-biased BCJ the diffusion capacitance is negligible (Section 3.2, Chapter 3). The depletion capacitance C_{jC} is:

$$C_{jC} = \frac{C_{jC0}}{\left(1 + V_{CB}/V_{jC_o}\right)^m} \tag{6.2.28}$$

where
   C_{jC0} is the capacitance when $V_{CB} = 0$
   V_{jC_o} is the equilibrium potential of the BCJ

Because the charge stored in the base varies with v_{BC}, there is an additional capacitance $C_{b\mu}$ that must be considered and is given by:

$$C_{b\mu} = \left|\frac{\partial Q_{eB}}{\partial v_{BC}}\right|_{v_{BE}} = \left.\frac{\partial(\tau_F i_C)}{\partial v_{CB}}\right|_{v_{BE}} = \tau_F \left.\frac{\partial(i_C)}{\partial v_{CE}}\right|_{v_{BE}} = \tau_F \frac{I_C}{V_A} \tag{6.2.29}$$

where τ_F is assumed to be independent of v_{BC}, and hence of v_{CE} at constant v_{BE}, and Equation 6.2.9 was used, assuming $I_C = I_{C(\text{sat})}$. As explained earlier, electrons in the base flow by diffusion, and the dependence of τ_F on v_{BC} is small and mainly due to the variation in the width of the depletion region with v_{BC}. $\tau_F I_C$ can be substituted from Equation 6.2.25 to obtain:

$$C_{b\mu} = C_{dE} \frac{V_T}{V_A} \tag{6.2.30}$$

As V_T is about 0.026 V at 300K, whereas V_A is generally 50–100 V, $C_{b\mu} \ll C_{dE}$. The total capacitance between base and collector is then:

$$C_\mu = C_{jC} + C_{b\mu} \cong C_{jC} \tag{6.2.31}$$

The complete small-signal equivalent circuit is shown in Figure 6.2.10. A capacitance C_{cs} has been included to account for the capacitance between the collector and substrate, which in modern ICs is quite small (Figure 6.1.4). Typical values of C_μ are in the range of a fraction of a pF to a few pFs, and the corresponding range for C_π is a few pFs to a few tens of pFs.

   Figure 6.2.10 also shows "parasitic resistances" r_B, r_C, and r_E. As in the case of the bulk resistance of the *pn* junction diode, these include the resistance of the metallic contact and of the bulk region of the semiconductor, all the way to the active region of the base, collector, or emitter, respectively. The resistances generally vary with current in a complex manner depending on the structure of the BJT. Typically, r_B is in the range of a few tens to a few hundred ohms, r_C is in the range of a few ohms to a few tens of ohms, and r_E is a few ohms or less.

   Note that whereas reverse transmission is due to r_μ under dc conditions, it is due under ac conditions to the parallel combination of r_μ and C_μ.

Exercise 6.2.3

Deduce from Figure 6.2.1 that $|Q_{eB}| = \frac{1}{2} n_p(0) W_B A_J |q|$ and show that:

$$\frac{|Q_{eB}|}{i_C} = \tau_F = \frac{W_B^2}{2 D_{eB}} \tag{6.2.32}$$

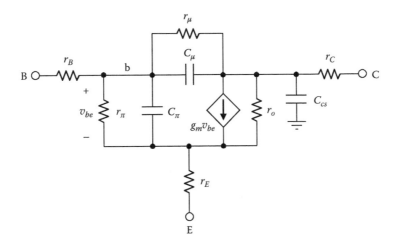

FIGURE 6.2.10
Small-signal equivalent circuit of the BJT including parasitic resistances and internal capacitances.

and,

$$C_{dE} = \frac{\partial |Q_{eB}|}{\partial v_{BE}}\bigg|_{v_{CE}} = \frac{1}{2V_T} W_B A_J |q| n_{po} e^{v_{BE}/V_T} \tag{6.2.33}$$

Unity-Gain Bandwidth

An important figure of merit of the BJT is the **unity-gain bandwidth**, determined using the small-signal equivalent circuit of Figure 6.2.11, with the collector ac short-circuited to the emitter. This equivalent circuit is that of Figure 6.2.10, ignoring r_μ and the parasitic resistances. It follows from Figure 6.2.11 that:

$$I_c(j\omega) = (g_m - j\omega C_\mu) V_\pi(j\omega) \tag{6.2.34}$$

and,

$$I_b(j\omega) = \left(\frac{1}{r_\pi} + j\omega(C_\pi + C_\mu) \right) V_\pi(j\omega) \tag{6.2.35}$$

Dividing and assuming that at the frequencies of interest $g_m \gg \omega C_\mu$,

$$\beta = \frac{I_c(j\omega)}{I_b(j\omega)} \simeq \frac{\beta_0}{1 + j\omega(C_\pi + C_\mu) r_\pi} \tag{6.2.36}$$

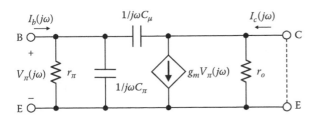

FIGURE 6.2.11
A simplified version of the equivalent circuit of Figure 6.2.10.

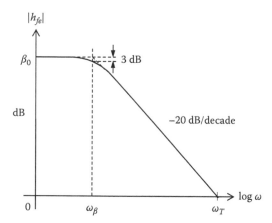

FIGURE 6.2.12
Frequency variation of the magnitude of the short circuit
CE current gain.

where $\beta_0 = g_m r_\pi$ is the value of β at zero frequency. It is usual to refer to β in Equation 6.2.36 as h_{fe}. A Bode plot of $|h_{fe}|$ is shown in Figure 6.2.12. The 3-dB cutoff frequency ω_β is given by:

$$\omega_\beta = \frac{1}{(C_\pi + C_\mu)r_\pi} \qquad (6.2.37)$$

The unity-gain bandwidth ω_T is the frequency at which $|h_{fe}| = 1$. From Equations 6.2.36 and 6.2.37, this occurs when $\beta_0 = \sqrt{1 + (\omega_T/\omega_\beta)^2}$. As $\beta_0 \gg 1$, it follows that $\omega_T \gg \omega_\beta$, so that:

$$\omega_T = \beta_0 \omega_\beta = \frac{\beta_0}{(C_\pi + C_\mu)r_\pi} = \frac{g_m}{(C_\pi + C_\mu)} \qquad (6.2.38)$$

As discussed in the case of the MOSFET (Equation 6.2.19, Chapter 6), the larger ω_T, the better is the "quality" of the transistor in terms of a high gain and a large bandwidth. ω_T represents the maximum theoretically attainable gain-bandwidth product of the amplifying stage using the given transistor. ω_T increases with g_m and decreases with the capacitances, as in the case of the MOSFET. C_μ is largely made up of the depletion capacitance of the BCJ and depends on the grading and levels of doping as well as on the reverse bias across the junction. C_π is made up of a depletion capacitance C_{jE} and a diffusion capacitance C_{dE}, which is usually considerably larger than C_{jE}, because the BEJ is forward biased. Equation 6.2.38 can be written as:

$$\frac{1}{\omega_T} \simeq \frac{C_{dE} + C_{jE} + C_{jC}}{g_m} = \tau_F + \frac{C_{jE} + C_{jC}}{g_m} \qquad (6.2.39)$$

where $C_{dE} = g_m \tau_F$ (Equation 6.2.25). Generally, $\tau_F > (C_{jE} + C_{jC})/g_m$ in the active region (Example 6.2.4) so that ω_T is dominated by $1/\tau_F$. It is seen that:

Concept: *The mean forward transit time τ_F is a critical parameter in determining the high-frequency performance of the BJT.*

Physically, if the signal frequency is high enough, the motion of electrons through the base cannot follow the signal and the current gain of the transistor falls off. Evidently, τ_F is reduced by having a small W_B and a large D_{eB}, in accordance with Equation 6.2.32.

At a given T, the diffusion constant is directly proportional to mobility (Equation 2.4.5, Chapter 2). As electron mobilities are approximately three times hole mobilities in Si, high-frequency BJTs are *npn*, rather than *pnp*. τ_F can be reduced by graded doping in the base, the doping concentration decreasing from the BEJ to the BCJ. At equilibrium, holes nearer the emitter in a *p*-type base are at a higher chemical potential, so they must be at a lower electric potential. This means that there will be a built-in electric field in the base, directed from collector to emitter, which accelerates electrons injected into the base and reduces τ_F. Typically, $f_T = \omega_T / 2\pi$ is in the range of a few hundred MHz to tens of GHz or more.

f_T decreases at high and low values of I_C. For intermediate values of I_C, where τ_F dominates in Equation 6.2.39, $f_T = 1/2\pi\tau_F$. At low values of I_C, the term $(C_{jE} + C_{jC})/g_m$ dominates in Equation 6.2.39 because g_m decreases with I_C (Equation 6.2.7), which decreases f_T as I_C decreases. At high values of I_C the effects of high-level injection, which decrease β_F, also decrease β_0, and hence f_T (Equation 6.2.38).

It is seen from the more complete circuit of Figure 6.2.10 that r_B plays an important role at high frequencies. It forms, in conjunction with the capacitances appearing at node b, a time constant that limits the high-frequency response (Section 8.3, Chapter 8). Recall that in an *npn* transistor, i_B provides the holes that: (1) are injected into the emitter under forward bias of the BEJ, and (2) recombine with some of the electrons injected into the base. The path of i_B is diagrammatically illustrated in Figure 6.1.4. r_B is the resistance encountered by i_B. If the base is made very thin so as to decrease τ_F, r_B increases and limits the high frequency response. Moreover, reducing the base width magnifies the effect of base-width modulation and lowers V_A. If the doping level in the base is increased so as to reduce the width of the depletion region in the base, and hence the effect of base-width modulation, the higher doping reduces the emitter injection efficiency, increases C_{jE0}, and reduces electron mobility in the base, which degrades the high-frequency performance. Hence in trying to maximize f_T, a compromise is made between all these factors. A heterojunction (Section 6.4) offers improved performance. It should be noted that realistic modeling of r_B at very high frequencies is complicated by the strictly distributed nature of r_B and C_μ. A simple, lumped-parameter representation may not be adequate in some cases.

r_C, in conjunction with any load capacitance, also limits the high frequency response. When transistors are cascaded, r_C adds to r_B of the following stage, again adversely affecting the high-frequency response. Having an n^+ buried layer (Figure 6.1.4) reduces r_C.

Example 6.2.4 Calculation of Unity-Gain Frequency

Given $I_C = 1.1$ mA, $\tau_F = 50$ ps, $C_{je} = 0.1$ pF, $C_{jC} = 0.02$ pF, and $V_A = 100$ V, calculate C_{dE}, $C_{b\mu}$, and f_T.

Solution

From Equation 6.2.7, $g_m = I_{C(sat)}/V_T = 1.1/0.026 = 42.3 \text{ mA/V}$. From Equation 6.2.25, $C_{dE} = \tau_B I_C / V_T = 50 \times 10^{-12} \times 1.1 \times 10^{-3}/0.026 \equiv 2.1 \text{ pF}$. Note that $C_{b\mu} = C_{dE}\dfrac{V_T}{V_A} = 2.1 \times 0.026/100 = 0.55 \text{ fF}$, which is much less than C_{jC}. In Equation 6.2.39, $(C_{jE} + C_{jC})/g_m = 0.12 \times 10^{-12}/42.3 \times 10^{-3} \equiv 2.8 \text{ pS}$, which is much smaller than τ_F. From Equation 6.2.39, $\omega_T = 1/52.8 \text{ pS}$, and $f_T = 3.0 \text{ GHz}$.

Simulation Example 6.2.5 High-Frequency Response

It is required to obtain by simulation the 3-dB and unity-gain frequencies of the 2N2222.

SOLUTION

The schematic is shown in Figure 6.2.13. An IAC source applies a 53.75 μA dc current and 1 μA ac current to the base. The dc current is adjusted to give a 10 mA dc collector current and a high β. The collector is ac short-circuited to ground by a 1 F capacitor. An ac sweep is performed over the frequency range 1 kHz to 1 GHz, and $20 \log_{10}[I(Q1{:}c)/I(Q1{:}b)]$ is plotted (Figure 6.2.14). From measurement on the plot, h_{fe} at low frequencies $= 45.75$ dB, a 3-dB cutoff frequency of 1.557 MHz, and a unity-gain frequency of 302.7 MHz are obtained.

From probe measurements at low frequencies, $i_c = (Q1{:}c) = 0.194$ mA and $v_{be} = V(Q1{:}b) = 0.528$ mV. It follows from these values that: $\beta = 0.194$ mA/1 μA $= 194$, $g_m = 0.194$ mA/0.528 mV $\equiv 367$ mA/V, and $r_{\pi} = 0.528$ mV/1 μA $= 528\,\Omega$. We can compare these figures with calculated values. Thus, $20 \log_{10}\beta = 45.76$ dB compared with 45.75 dB from the plot of Figure 6.2.14; $g_m = I_C/V_T = 10$ mA/0.026 V $= 385$ mA/V; $r_{\pi} = \beta/g_m = 194/367$ mA/V $\equiv 529\,\Omega$.

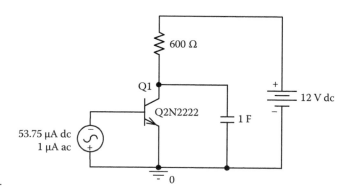

FIGURE 6.2.13
Figure for Example 6.2.5.

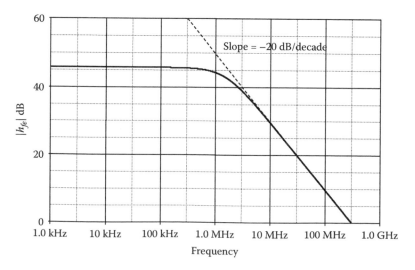

FIGURE 6.2.14
Figure for Example 6.2.5.

From the PSpice model of the 2N2222: $C_{jC} = 7.306$ pF, $C_{jE} = 22.01$ pF, and $\tau_F = 411.1$ ps. From Equation 6.2.25, $C_{dE} = \tau_B g_m = (411.1$ ps$) \times (385$ mA/V$) = 158$ pF so that $C_\pi = C_{dE} + C_{jE} = 158 + 22 = 180$ pF. Considering $C_\mu \cong C_{jC} = 7.31$ pF, it follows from Equation 6.2.37 that $f_\beta = 1/(2\pi(C_\pi + C_\mu)r_\pi) = 1/(2\pi \times (187$ pF$) \times 529) = 1.61$ MHz compared to a value of 1.557 MHz in Figure 6.2.14. From Equation 6.2.39, $\dfrac{1}{\omega_T} = 411 + \dfrac{(22 + 7.3)\text{ pF}}{385\text{ mA/V}} = 411 + 76.1 = 487$ ps. This gives $f_T = 1/(2\pi \times 487$ ps$) = 327$ MHz, compared to a value of 302.7 MHz in Figure 6.2.14.

Because PSpice performs the simulation on a linearized model, using an IAC of 1 A rather 1 μA gives the same values for β and r_π, although the simulation values of i_c and v_{be} would seem "unnatural." Note that an IPRINT printer can be used to measure i_c and a VPRINT printer to measure v_{be}, the AC Sweep being performed at a single frequency of, say, 1 kHz.

Application Window 6.2.1 Voltage Proportional to Absolute Temperature

A voltage proportional to absolute temperature (PTAT), based on Equation 6.2.1, is at the heart of today's digital thermometers. If two transistors are matched, except for their BEJ areas, and carry the same current, it follows from Equation 6.2.1 that:

$$V_{\text{PTAT}} = V_{BE2} - V_{BE1} = \frac{kT}{|q|} \ln \frac{A_1}{A_2} \tag{6.2.40}$$

where, if transistor Q_1 has a BEJ area A_1 larger than the BEJ area A_2 of Q_2, it will have a smaller V_{BE} for the same emitter current. The temperature coefficient of V_{PTAT} is:

$$\frac{dV_{\text{PTAT}}}{dT} = \frac{V_{\text{PTAT}}}{T} \tag{6.2.41}$$

A basic circuit for generating a V_{PTAT} is illustrated in Figure 6.2.15. Q_1 has 10 times the BEJ area of Q_2. A high-gain current amplifier A_I equalizes the currents in Q_1 and Q_2 by producing the appropriate $V_{BE2} - V_{BE1}$ across R_1. The V_{PTAT} at Q_1 base is $(V_{BE2} - V_{BE1}) \times (R_2/R_1)$, neglecting the base current of Q_1. This voltage is processed, converted to digital form, and displayed as temperature in °C or °F.

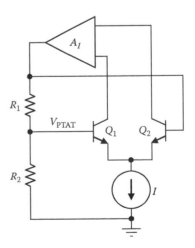

FIGURE 6.2.15
Figure for Application Window 6.2.1.

6.3 BJT Large-Signal Models

Ebers–Moll Model

The Ebers–Moll (EM) equations are extensively used to model the large-signal, nonlinear behavior of a BJT in all modes of operation. In their most basic form, they apply at low frequencies to an "ideal transistor," that is, one having uniform doping, abrupt, parallel-plane junctions, low-level injection, no recombination and generation in the depletion regions, and no base-width modulation. These equations are derived in Section ST6.1 and can be written in terms of instantaneous values for an *npn* transistor as:

$$i_C = \alpha_F I_{SE}(e^{v_{BE}/V_T} - 1) - I_{SC}(e^{v_{BC}/V_T} - 1) \tag{6.3.1}$$

$$\overline{i_E} = -I_{SE}(e^{v_{BE}/V_T} - 1) + \alpha_R I_{SC}(e^{v_{BC}/V_T} - 1) \tag{6.3.2}$$

where
  α_R is the dc CB current gain with the transistor in inverse active mode
  I_{SC} is the saturation current of the base–collector *pn* junction ($e^{v_{BC}}/V_T \ll 1$) with the base short-circuited to the emitter ($V_{BE} = 0$)
  I_{SE} is similarly defined as the saturation current of the base-emitter *pn* junction with the base short-circuited to the collector

The assigned positive direction of $\overline{i_E}$ is that flowing into the emitter junction, so $\overline{i_E} = -i_E$, as has been used for an *npn* transistor. The four parameters in these equations are not independent because:

$$\alpha_F I_{SE} = \alpha_R I_{SC} \tag{6.3.3}$$

(Section ST6.1). Equations 6.3.1 and 6.3.2 can be represented by the equivalent circuit of Figure 6.3.1, which is the static, **injection version** of the EM model. Diodes D_E and D_C represent the BEJ and BCJ, respectively, whereas the current sources account for transistor action due to the narrow base. The forward currents of the two diodes are:

$$i_{DE} = I_{SE}(e^{v_{BE}/V_T} - 1) \quad \text{and} \quad i_{DC} = I_{SC}(e^{v_{BC}/V_T} - 1) \tag{6.3.4}$$

In terms of i_{DE} and i_{DC}, Equations 6.3.1 and 6.3.2 become:

$$i_C = \alpha_F i_{DE} - i_{DC} \tag{6.3.5}$$

$$\overline{i_E} = \alpha_R i_{DC} - i_{DE} \tag{6.3.6}$$

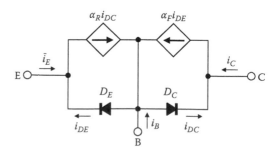

FIGURE 6.3.1
Injection version of the Ebers–Moll model.

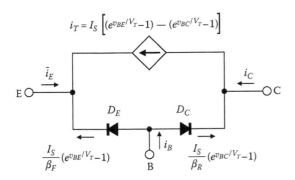

$$i_T = I_S \left[(e^{v_{BE}/V_T} - 1) - (e^{v_{BC}/V_T} - 1) \right]$$

FIGURE 6.3.2
Transport version of the Ebers–Moll model.

In the normal active region ($v_{BE} > 0$, $v_{BC} < 0$), $i_{DC} = -I_{SC}$, so that $i_C = \alpha_F i_{DE} + I_{SC}$. This is of the form of Equation 6.1.6, although i_{DE} is not the same as i_E, and I_{SC} is not the same as I_{CBO}. Also, $i_E = i_{DE} + \alpha_R I_{SC}$. The emitter current is thus the sum of the forward current of the emitter diode and the collector saturation current multiplied by α_R.

The injection version of Figure 6.3.1 can be transformed to the simpler, **transport version** of Figure 6.3.2 (Section ST6.1), in which $I_S = \alpha_F I_{SE} = \alpha_R I_{SC}$ and i_T is defined as:

$$i_T = \alpha_F I_{SE}(e^{v_{BE}/V_T} - 1) - \alpha_R I_{SC}(e^{v_{BC}/V_T} - 1) \tag{6.3.7}$$

$$= I_S(e^{v_{BE}/V_T} - 1) - I_S(e^{v_{BC}/V_T} - 1) = I_S(e^{v_{BE}/V_T} - e^{v_{BC}/V_T}) \tag{6.3.8}$$

$\beta_F = \alpha_F/(1 - \alpha_F)$ is the CE current gain in the normal active mode, and $\beta_R = \alpha_R/(1 - \alpha_R)$ is the corresponding current gain in the inverse active mode.

In Equation 6.3.7, $I_{SE}(e^{v_{BE}/V_T} - 1)$ is the diffusion current across the forward-biased BEJ, and $\alpha_F I_{SE}(e^{v_{BE}/V_T} - 1)$ is the fraction of this current that reaches the collector junction. Similarly, $I_{SC}(e^{v_{BC}/V_T} - 1)$ is the diffusion current across the forward-biased BCJ, and $\alpha_R I_{SC}(e^{v_{BC}/V_T} - 1)$ is the fraction of this current that reaches the emitter junction.

It follows from Figure 6.3.2 that the terminal currents can be expressed as:

$$i_E = -i_T - \frac{I_S}{\beta_F}(e^{v_{BE}/V_T} - 1) = -\frac{I_S}{\alpha_F}(e^{v_{BE}/V_T} - 1) + I_S(e^{v_{BC}/V_T} - 1) \tag{6.3.9}$$

$$i_C = i_T - \frac{I_S}{\beta_R}(e^{v_{BC}/V_T} - 1) = I_S(e^{v_{BE}/V_T} - 1) - \frac{I_S}{\alpha_R}(e^{v_{BC}/V_T} - 1) \tag{6.3.10}$$

$$i_B = -i_E - i_C = \frac{1}{\beta_F} I_S(e^{v_{BE}/V_T} - 1) + \frac{1}{\beta_R} I_S(e^{v_{BC}/V_T} - 1) \tag{6.3.11}$$

In the normal active mode $e^{v_{BC}/V_T} \ll 1$, so that these equations become:

$$i_E \cong \frac{I_S}{\alpha_F} e^{v_{BE}/V_T} - \frac{I_S}{\beta_F} \cong \frac{I_S}{\alpha_F} e^{v_{BE}/V_T} \tag{6.3.12}$$

$$i_C \cong I_S e^{v_{BE}/V_T} + \frac{I_S}{\beta_R} \cong I_S e^{v_{BE}/V_T} \tag{6.3.13}$$

$$i_B \cong \frac{I_S}{\beta_F} e^{v_{BE}/V_T} - I_S\left(\frac{1}{\beta_F} + \frac{1}{\beta_R}\right) \cong \frac{I_S}{\beta_F} e^{v_{BE}/V_T} \tag{6.3.14}$$

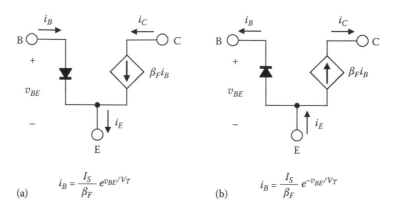

FIGURE 6.3.3
Approximate large-signal models for an *npn* transistors (a) and a *pnp* transistor (b).

The relations on the RHS of these equations are those commonly used to characterize the active mode. Equations 6.3.12 and 6.3.13 should be compared with Equation 6.2.1. In the active mode, the current through D_C is very small, so this branch may be open circuited and $i_T \cong i_C$. The circuit reduces to the approximate large-signal models sometimes used, as shown in Figure 6.3.3a and b for *npn* and *pnp* transistors, respectively. Note that whereas small-signal models are the same for complementary transistors, large-signal models are not.

To further illustrate the connection between the EM equations and the relations used in Section 6.1, it is shown in Example ST6.1.1, Section ST6.1, that the EM equations give $I_{CEO} = I_{CBO}/(1 - \alpha_F)$ and $I_{SC} = I_{CBO}/(1 - \alpha_F\alpha_R)$, where I_{SC} is the collector current with the base and emitter terminals short-circuited. The emitter currents are compared in Example ST6.1.2 and the equivalent circuits in Example ST6.1.3.

Saturation Mode

Figure 6.3.4a is the basic circuit of a BJT inverter, analyzed in detail in Section ST6.5. When $v_B = 0$, then $i_B \cong 0$, and $i_C = \dfrac{I_{CBO}}{1 - \alpha_F\alpha_R} < I_{CEO} \cong 0$ (Example ST6.1.1, Section ST6.1). The operating point Q_1 (Figure 6.3.4b) is practically on the voltage axis, with $v_{CE} \cong V_{DD}$. As v_B is increased, i_B begins to flow. The transistor is now in active mode, the operating point being Q_2, say, at the intersection of the $i_B = i_{B2}$ output characteristic and a load line of slope $-1/R_C$. With further increase in v_B, i_B and i_C increase, which decreases v_{CE} and v_{CB}. The operating point eventually moves to the edge of the saturation region (Q_3 in Figure 6.3.4b), where i_B, i_C, and v_{CE} have values of $i_{B(\text{sat})}$, $i_{C(\text{sat})}$, and $v_{CE(\text{sat})}$, respectively. At this point, the BCJ is forward biased to the forward cut-in voltage. Any further increase in v_B beyond $v_{B(\text{sat})}$ moves the transistor into saturation.

In the saturation region, which has been greatly exaggerated in Figure 6.3.4b for clarity, the individual characteristics for constant i_B are crowded together (Figure 6.2.4a). The forward voltage drop v_{BC} is slightly less than the forward voltage drop v_{BE} because the area of the BCJ is larger than that of the BEJ, and for a given forward current, the forward voltage across a *pn* junction is inversely proportional to the area of the junction (Equation ST3.5.10, Section ST3.5, Chapter 3). Hence, $v_{CE(\text{sat})} = v_{BE} - v_{BC}$ is normally in the range of 0.1–0.3 V. Thus, when $v_B = 0$, $v_{BE} \cong 0$, the transistor is practically cutoff and $v_{CE} \cong V_{CC}$.

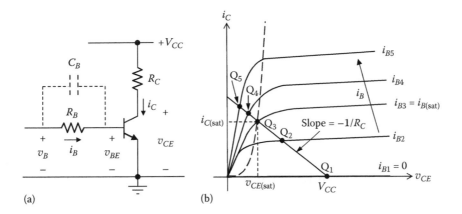

FIGURE 6.3.4
Basic circuit of a BJT inverter: (a) speed-up capacitor is shown dotted; (b) output characteristics with load line.

When v_B is sufficiently large, the transistor goes into saturation, and v_{CE} is small. The change in v_B is inverted at the collector, which gives the inverter its name.

The collector current in the saturation region is determined by the external circuit as:

$$i_C = \frac{V_{CC} - v_{CE(sat)}}{R_C} \qquad (6.3.15)$$

Because $v_{CE(sat)}$ is small, in the range of 0.1–0.3 V, and can only decrease very slightly as i_B increases (Figure 6.2.5), it follows from Equation 6.3.15 that i_C increases very little in the saturation region. The base current at the edge of saturation is:

$$i_{B(sat)} = \frac{v_{B(sat)} - v_{BE(sat)}}{R_B} \qquad (6.3.16)$$

If v_B is increased, i_B is determined by the base circuit as:

$$i_B \cong \frac{v_B - v_{BE(sat)}}{R_B} \qquad (6.3.17)$$

where $v_{BE(sat)}$ is considered to be about 0.1 V larger than its value in the active mode. The reason v_{BE} increases slightly with i_B in saturation is simply due to the shape of the i_B–v_{BE} characteristic. As i_B increases, with i_C staying almost constant, i_E increases with i_B, and the ratio i_C/i_B decreases. A β_{forced} is defined in the saturation region as:

$$\beta_{forced} = \frac{i_C}{i_B}, \quad i_B > i_{B(sat)} \qquad (6.3.18)$$

where $i_{B(sat)} = i_{C(sat)}/\beta_F$ at the edge of the saturation region.

Because in the saturation mode $v_{CE(sat)}$ is small and i_B is relatively large, the parasitic resistances r_B, r_C, and r_E may have to be considered. Figure 6.3.5 shows the equivalent circuit of an *npn* transistor in saturation, where r_E is omitted because it is usually quite

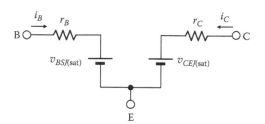

FIGURE 6.3.5
Equivalent circuit of a BJT in saturation.

small, and $v_{BEJ(sat)}$ and $v_{CEJ(sat)}$ are the voltage drops between base and emitter and collector and emitter, respectively, exclusive of the voltage drops in the parasitic resistances.

Neglecting the voltage drops in parasitic resistances, $v_{CE(sat)}$ can be determined from Equations 6.3.10 and 6.3.11 assuming that in the saturation mode $e^{v_{BE(sat)}/V_T} \gg 1$ and $e^{v_{BC(sat)}/V_T} \gg 1$; $e^{v_{BE(sat)}/V_T}$ and $e^{v_{BC(sat)}/V_T}$ can then be determined from the expressions for i_C and i_B, and $v_{CE(sat)} = v_{BE(sat)} - v_{BC(sat)}$ evaluated, with $\beta_{forced} = i_C/i_B$. This gives:

$$v_{CE(sat)} = V_T \ln \frac{1 + (\beta_{forced} + 1)/\beta_R}{1 - \beta_{forced}/\beta_F} \tag{6.3.19}$$

Example 6.3.1 Transistor in Saturation Mode

Determine $v_{CE(sat)}$ in Figure 6.3.4a if $V_{CC} = 12$ V, $R_C = 4.8$ kΩ, $v_B = 6$ V, and $R_B = 20$ kΩ, assuming $r_C = 50$ Ω, $r_E = 5$ Ω, $\beta_F = 200$, and $\beta_R = 1$.

Solution

i_C and i_B are determined from Equations 6.3.15 and 6.3.16, respectively, assuming for a start that $v_{CE(sat)} = 0.2$ V and $v_{BE(sat)} = 0.7$ V. This gives $i_C = (12 - 0.2)/4.8 = 2.46$ mA and $i_B = (6 - 0.7)/20 = 0.27$ mA. It follows from Equation 6.3.18 that $\beta_{forced} = 2.46/0.27 = 9.1$. Substituting in Equation 6.3.19 and adding the voltage drops in r_C and r_E: $v_{CE} = 0.026 \times \ln 11.63 + 2.46 \times 10^{-3} \times 50 + (2.46 + 0.27) \times 10^{-3} \times 5 = 0.064 + 0.123 + 0.014 = 0.201$ V which is close to the assumed value of 0.2 V, so no new iteration is needed.

Eliminating v_{BC} between the expressions for i_C and i_B in Equations 6.3.10 and 6.3.11, assuming $e^{v_{BE(sat)}/V_T} \gg 1$ and $e^{v_{BC(sat)}/V_T} \gg 1$, $v_{BE(sat)} = V_T \ln \dfrac{i_B}{I_S} \dfrac{1 + \beta_{forced} + \beta_R}{1 + (\beta_R + 1)/\beta_F}$, where $I_S = \dfrac{\beta_F}{\beta_F + 1} I_{SE}$ and $I_{SE} \cong \dfrac{|q|A_E D_{eB}}{W_B} n_{po}$. Substituting numerical values from Example 6.1.1 gives $v_{BE(sat)} \cong 0.6$ V, which does not significantly affect the value used for i_B.

The distribution of minority carriers in the base region in the saturation mode is illustrated in Figure 6.3.6. The concentrations of injected electrons at the edges of the depletion regions of the BEJ and BCJ are $n_p(0) = n_{po}e^{v_{BE}/V_T}$ and $n_p(W_B) = n_{po}e^{v_{BC}/V_T}$, respectively. Neglecting the recombination current in the base, the concentration profile of minority carriers is linear, represented by the line AC. Using Equation 6.2.1,

$$|q|A_J D_{eB} \frac{dn_p(x)}{dx} = \frac{|q|A_J D_{eB}}{W_B} n_{po}(e^{v_{BE}/V_T} - e^{v_{BC}/V_T}) \tag{6.3.20}$$

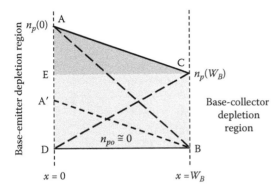

FIGURE 6.3.6
Electron concentration profile of a BJT in saturation.

When the transistor is just at the edge of saturation, corresponding to Q_3 in Figure 6.3.4b, the concentration profile is that of the dotted line A′B, with $i_C = i_{C(sat)}$ and $i_{B(sat)} = i_{C(sat)}/\beta_F$. The BCJ is not forward biased beyond the cut-in voltage and the injection of electrons from collector to base is negligible. The area of triangle A′BD in Figure 6.3.6 represents the charge stored in the base. If v_{BE} is increased, with i_C staying almost constant in saturation, as discussed earlier, the line A′B moves vertically upward to become AC. i_C is now the difference between the currents due to electrons injected from emitter to base (represented by the line AB) and electrons injected from collector to base (represented by the line CD). i_B is larger and equals $i_C/\beta_{\text{forced}}$ (Equation 6.3.18). The larger i_B causes more minority carriers to be stored in the base, corresponding to the additional area CBDE. This means that if the transistor is to be turned off, it would take longer to do so, because *more stored charge has to be removed from the base.*

The BJT as a Switch

The inverter circuit of Figure 6.3.4a is a basic circuit of the BJT as a switch that can turn on and off a load connected in series with the collector. When the transistor is saturated, corresponding to switch closure, an **offset voltage** remains across the switch equal to $v_{CE(sat)}$ in the range of 0.1–0.3 V. Turn-on and turn-off times of the BJT inverter are discussed in Section ST6.5. The turn-on time, or the time it takes the BJT to conduct from an initially cutoff state, is reduced by: (1) a large i_B to turn the transistor on, (2) a small $i_{C(sat)}$, where both of these conditions imply that the transistor saturates sooner, and (3) a small minority-carrier lifetime τ_{eB}; this is because $|Q_{eB}|/\tau_{eB}$ is the rate of recombination of the charge stored in the base (Equation 3.2.8, Chapter 3) and equals $i_{e(rec)}$, which is a component of i_B (Equation 6.1.1). Thus, for a given i_B and hence $i_{e(rec)}$, the smaller τ_{eB}, the smaller is the charge stored in the base. The turn-off time depends on the charge stored in the base. It is reduced by having: (1) a small i_B, and hence a lower degree of saturation, and (2) a small τ_{eB} so that the stored charge decays faster. The turn-on and turn-off times also depend on the external base circuit. For example, the turn-off time is reduced if the BEJ is reverse biased on switching so that the stored charge is removed more rapidly. A **speed-up capacitor** C_B across R_B (Figure 6.3.4a) considerably reduces turn-on and turn-off times. To appreciate the effect of C_B, assume that v_B is initially low, with C_B discharged through R_B. When v_B suddenly goes high, the capacitor voltage does not change instantaneously, so that v_B is applied across the BEJ causing a large i_B to flow through C_B, which charges rapidly to nearly v_B through the low resistance of the BEJ. This satisfies requirement (1) for reducing the turn-on time. R_B is chosen to satisfy requirement (2). When v_B returns to zero, the full voltage across C_B is applied as a reverse bias across the BEJ, thereby reducing the turn-off time. The value of C_B is normally chosen so that the time constant $C_B r_\pi$ is much smaller than the duration of the on or off periods, whichever is smaller.

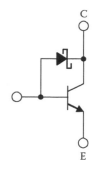

FIGURE 6.3.7
Schottky transistor.

Because of the adverse effects of saturation on switching speed, measures are taken in BJT digital logic families to prevent saturation. One approach, used in the Schottky transistor–transistor logic (TTL) family is to incorporate a Schottky diode in parallel with the BCJ (Figure 6.3.7). The diode conducts at about 0.3 V and effectively clamps v_{BC} at this voltage. As this is less than the cut-in voltage of the BCJ, the BCJ is prevented from conducting in the forward direction. The diode diverts current that would otherwise flow into the base if the transistor were allowed to saturate. The Schottky diode is itself majority-carrier operated and does not suffer from minority-carrier storage (Section 3.6, Chapter 3). Another approach is to use emitter-coupled logic (ECL) based on the differential pair (Section 13.7, Chapter 13).

Diode Connection

An IC transistor with the base connected internally to the collector (Figure 6.3.8) is often used in place of a *pn* junction diode, although it occupies slightly more surface area on the chip. The reason is that the IC transistor structure gives lower series resistance and faster switching time. From the expression for $\bar{i}_E$ (Equation 6.3.9) with $v_{BC}=0$ and $i=-\bar{i}_E$, it follows that:

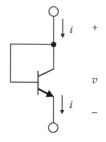

FIGURE 6.3.8
Diode connection of BJT transistor.

$$i = \frac{I_S}{\alpha_F}(e^{v/V_T} - 1) \cong I_S e^{v/V_T} \qquad (6.3.21)$$

The emission coefficient η in a diode-connected transistor is close to unity except at low and high current values. At low values of current, i_B varies nearly as $e^{v_{BE}/2V_T}$, because the recombination current in the depletion region of the BEJ varies with v_{BE} in this manner and is an appreciable fraction of the base current at low values of the base current (Section ST3.6, Chapter 3). At larger base currents the recombination current is swamped by the normal base current that varies with v_{BE} as e^{v_{BE}/V_T} in accordance with Equation 6.3.14. At high currents, i_C varies nearly as $e^{v_{BE}/2V_T}$ due to high-level injection (Example ST3.7, Chapter 3). These are the same considerations that reduce β at low and high values of current, respectively. An example of the use of a diode-connected transistor in log/antilog amplifiers is given in Example SE6.1.

Regenerative Pair

In the regenerative pair, a *pnp* and an *npn* transistor are connected in a positive feedback configuration (Figure 6.3.9). The circuit can be used to explain the operation of *pnpn* devices and the latching that can occur in complementary transistor circuits (CMOS) and BJT devices (Section ST6.6).

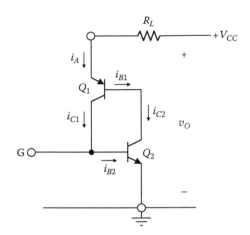

FIGURE 6.3.9
Regenerative pair.

Example 6.3.2 Regenerative Pair

It is required to investigate the basic operation of the regenerative pair of Figure 6.3.9.

ANALYSIS

Using Equation 6.2.14,

$$i_{C1} = M_1(\alpha_{F1} i_A + I_{CBO1}), \quad i_{C2} = M_2(\alpha_{F2} i_A + I_{CBO2})$$

$$i_{B1} = i_A - i_{C1}, \quad i_{B2} = i_A - i_{C2}, \quad i_{B1} = i_{C2}, \quad i_{B2} = i_{C1}$$

where M is the avalanche multiplication factor and the gate terminal G is assumed to be open. Solving for i_A:

$$i_A = \frac{M_1 I_{CBO1} + M_2 I_{CBO2}}{1 - \alpha_{F1} M_1 - \alpha_{F2} M_2} \qquad (6.3.22)$$

At low voltages, M_1 and M_2 are unity, and at low currents, α_{F1} and α_{F2} are small. Hence, $(\alpha_{F1} + \alpha_{F2})$ is small and i_A is small. However, if:

$$\alpha_{F1} M_1 + \alpha_{F2} M_2 = 1 \qquad (6.3.23)$$

then $i_A \to \infty$. The connection is that of positive feedback. Thus, an increase in i_{C1}, say, increases i_{C2}, which in turn increases i_{C1}, and so on. As a result, the transistors are driven hard into saturation. The condition of Equation 6.3.23 makes the gain in the positive feedback loop equal to unity. Thus, if α_F is replaced by $\alpha_F M$, Equation 6.1.8 becomes:

$$i_C = \frac{\alpha_F M}{1 - \alpha_F M} i_B + \frac{1}{1 - \alpha_F M} I_{CBO} = \beta' i_B + \frac{1}{1 - \alpha_F M} I_{CBO} \qquad (6.3.24)$$

Neglecting the second term on the RHS with respect to the first, $\dfrac{i_{C1}}{i_{B1}} = \beta_1' = \dfrac{\alpha_{F1} M_1}{1 - \alpha_{F1} M_1}$ and $\dfrac{i_{C2}}{i_{B2}} = \beta_2' = \dfrac{\alpha_{F2} M_2}{1 - \alpha_{F2} M_2}$. The loop gain, from the base of Q_2 to its collector and back through Q_1, is $\dfrac{i_{C2}}{i_{B2}} \times \dfrac{i_{C1}}{i_{B1}} = \beta_2' \beta_1'$. Substituting and equating this product to unity gives Equation 6.3.23.

Once the transistors go into saturation, the current is limited by the resistance in the external circuit. The transistors can be turned off either by interrupting the current externally, or by reducing the voltage across the transistors so that the current is reduced to make the α's sufficiently small. The transistors can be turned on by any mechanism that increases the current so that Equation 6.3.23 is satisfied (Section 11.10, Chapter 11).

Augmented Models

The basic EM model can be augmented by adding the capacitances and parasitic resistances used in the small-signal equivalent circuit of Figure 6.2.10. The effects of base-width modulation can be included by modifying the saturation current I_S (Equation 6.3.8) to become:

$$I_S = I_{S0}\left(1 - \frac{v_{BC}}{V_A}\right) \tag{6.3.25}$$

It is also possible to include the emission coefficients η that multiply V_T in the exponents involving v_{BE} and v_{BC}. The EM model, augmented as mentioned, is the simplest BJT model used by PSpice (Appendix B). In actual fact, PSpice uses a more complex model, the Gummel–Poon model, which includes the variation of current gain with collector current, due to the effects of recombination in the depletion region of the BEJ at low currents and the effects of high-level injection at high currents (Section ST6.1). However, when the parameters required by the Gummel–Poon model are not specified, PSpice uses the simpler augmented model described.

6.4 Heterojunction Bipolar Transistor

In a Si *npn* heterojunction bipolar transistor (HBT), the emitter is made of *n*-type Si, whereas the base is a thin layer of *p*-type SiGe (Section 4.8, Chapter 4). The major effect of the heterojunction is on the emitter injection efficiency γ. Thus,

$$\beta_F = \frac{\alpha_F}{1 - \alpha_F} = \frac{\gamma\delta}{1 - \gamma\delta} \cong \frac{\gamma}{1 - \gamma} \tag{6.4.1}$$

assuming that the base transport factor $\delta \cong 1$. From Equation ST6.1.24,

$$\gamma = \frac{1}{1 + D_{hE} W_B p_{no} / D_{eB} L_{hE} n_{po}} \tag{6.4.2}$$

Substituting for γ in Equation 6.4.1, with $n_{po} = n_{iB}^2/N_{AB}$ and $p_{no} = n_{iE}^2/N_{DE}$,

$$\beta_F = \frac{D_{eB}L_{hE}}{D_{hE}W_B} \frac{n_{iB}^2}{N_{AB}} \frac{N_{DE}}{n_{iE}^2} \qquad (6.4.3)$$

From Equation 3.7.17, Chapter 3, the ratio of the n_i's in the emitter and the base is:

$$\frac{n_{iB}^2}{n_{iE}^2} = \frac{N_{cB}N_{vB}}{N_{cE}N_{vE}} e^{\Delta E_g/kT} \qquad (6.4.4)$$

where

N_c and N_v refer to the effective densities of states in the conduction and valence bands, respectively

$\Delta E_g = E_{gE} - E_{gB} > 0$ is the difference in the bandgap energies between the emitter and base

Substituting in Equation 6.4.3, β_F can be written as:

$$\beta_F = \left(\frac{D_{eB}L_{hE}N_{cB}N_{vB}}{D_{hE}W_BN_{cE}N_{vE}} \right) \frac{N_{DE}}{N_{AB}} e^{\Delta E_g/kT} \qquad (6.4.5)$$

Assuming that the terms other than the exponential in Equation 6.4.5 are not much different from those in a homojunction transistor having $\Delta E_g = 0$, β_F is increased by the exponential factor. Thus β_F is multiplied by approximately 2300 for $\Delta E_g = 0.2$ eV and by approximately 11,000 for $\Delta E_g = 0.3$ eV. Physically, the difference in bandgap energies facilitates injection of electrons from emitter to base because conduction electrons have lower energies in the base, and opposes the injection of holes from base to emitter, thereby increasing γ. However, too large ΔE_g is undesirable because it accentuates the temperature dependence of β_F and enhances the spike in the conduction band at the emitter-base heterojunction (Figure 3.7.1b, Chapter 3). This spike constitutes a barrier that impedes the motion of electrons from emitter to base.

The exponential term in Equation 6.4.5 increases β_F when $\Delta E_g > 0$ but gives β_F a negative temperature coefficient. A high level of doping of the emitter of a normal, homojunction transistor makes ΔE_g slightly negative so that β_F increases with temperature.

For the same or somewhat larger β_F, the doping concentration N_{AB} in the base can be increased. The resulting increase in conductivity of the base has two desirable consequences: (1) reduction in r_B, which as explained earlier, improves the high-frequency response; (2) the depletion region on the base side of the BCJ becomes narrower, which reduces base-width modulation and increases the early voltage V_A as well as the punchthrough voltage. Alternatively, the doping concentration N_{DE} in the emitter can be reduced. This increases the width of the depletion region on the emitter side of the BEJ, which reduces C_{je}. It is seen that the HBT gives more flexibility in tailoring transistor performance to particular applications.

Having the base made of SiGe increases electron mobility in the base. Electron and hole mobilities are higher in Ge than in Si so that carrier mobilities in SiGe are higher than in Si, which reduces the base transit time (Equation ST6.1.35) and increases the unity-gain bandwidth (Equation 6.2.39). The base transit time can be decreased further by grading the concentration of Ge in the base. If this concentration increases from the emitter toward the collector, the bandgap becomes correspondingly narrower. The lowering of the energy of the edge of the conduction band means that the electric potential energy of conduction electrons in the base is reduced as one goes from emitter to collector. In other words, an electric field is created that accelerates electrons as they cross the base. An f_T in excess of 100 GHz can be achieved with Si HBTs.

6.5 Noise in Semiconductors

Noise in semiconductors should be briefly mentioned, although the subject is too specialized to be discussed here. Noise is important because it sets a limit to the maximum useful amplifier gain and to the smallest signals that can be handled by an electronic device. If the gain is sufficiently high, the output stages can be "swamped," or saturated, by noise from the input stages. If the signal is weak compared to the noise, that is, the signal-to-noise ratio is small, the signal cannot be usefully amplified, as the output will be mostly noise.

The types of noise encountered in semiconductor devices are:

1. **Thermal noise,** also referred to as **Johnson noise** or **Nyquist noise**, arises because of current fluctuations due to random motion of electrons. This type of noise is present in all conductors, whether metals or semiconductors. The conducting channels of junction field-effect transistors (JFETs) and MOSFETs exhibit this type of noise, as do the parasitic resistances of BJTs. Resistances such as r_π and r_o, are fictitious resistances introduced for modeling purpose only and do not exhibit this type of noise.

2. **Shot noise** arises because the flow of even a dc current through a space charge region, as in a *pn* junction, is not perfectly smooth but is subject to "pulsations" due to the discrete nature of current carriers, like a stream of gunshots hitting a target. The noise increases with the dc current. The base and collector currents of a BJT, as well as the small gate input current of a JFET exhibit this type of noise.

3. **Flicker noise,** or **$1/f$ noise**, is associated with a dc current flowing through an imperfect medium, such as a medium having impurities or crystal defects. In semiconductors it is mainly due to surface traps. The noise is generated when current carriers are held in the traps for some time and then released. This type of noise is significant in MOSFETs because of the surface states that occur at the semiconductor–oxide interface, and is usually negligible in BJTs and JFETs. Typically, *p*-channel devices have less $1/f$ noise than *n*-channel devices because holes are less likely to be trapped in surface states.

4. Other types of noise include **avalanche noise** and **burst noise**, or **"popcorn" noise**. Avalanche noise is due to avalanche multiplication, which occurs in short-channel MOSFETs (Section 5.3, Chapter 5) and in zener diodes having breakdown voltages larger than about 7 V (Section 3.2, Chapter 3). Burst noise is associated with the presence of heavy-metal impurities such as gold. These introduce additional generation-recombination centers (Section 2.7, Chapter 2), which cause random fluctuations in current.

Thermal noise is modeled by means of a noise voltage generator in series with the input, whereas shot noise, flicker noise, and burst noise are modeled by means of noise current generators in parallel with the input. Thus, noise behavior depends not only on the bandwidth of the circuit but also on the resistance of the signal source (see Problem P10.3.10).

6.6 Comparison of BJTs and FETs

Before comparing BJTs and FETs, it should be stressed that FETs are basically *unipolar* devices, in which operation depends essentially on one type of carrier: electrons in *n*-channel devices or holes in *p*-channel devices. In contrast, BJTs are so called because both types of carriers are involved in their operation. Although in an *npn* transistor the output current is mainly an electron current, storage of electrons in the base, as minority carriers, and the resulting electron–hole recombination play an important role in device operation.

BJTs and FETs can be compared in different respects. The manufacturing process of MOSFETs is comparatively simple, and a MOSFET typically occupies one-tenth of the chip area as a BJT. Functionally, the following comparison can be made between the two types of transistor:

1. The effective, active area of a BJT is that through which the emitter current flows. The active area of a MOSFET is restricted by the depth of the channel, which is small because of the limited penetration of the electric field into the semiconductor under the gate. Hence, the BJT structure can handle larger currents and therefore higher power signals. Although power MOSFETs have a different structure from low-power MOSFETs (Section 11.9, Chapter 11), they still cannot handle high powers like a BJT.

2. v_{BE} of the BJT is much more effective in controlling the output current than v_{GS} of a MOSFET, thereby resulting in a higher g_m. A change in v_{BE} of approximately 0.2 V is sufficient to drive the output current from practically zero to its maximum value. In a MOSFET, a considerably larger change in v_{GS} is needed to achieve the same effect because of the presence of the gate oxide. The reverse-biased *pn* junction of a JFET is more effective in controlling the output current that the gate of a MOSFET, which makes g_m of a JFET larger than that of a MOSFET but still less than that of a BJT. A higher transconductance means higher gain, for a given load resistance, and larger currents to charge and discharge parasitic capacitances, thereby reducing switching times.

3. BJTs offer faster *operating speeds*, where the operating speed is essentially determined by the carrier transit time through the base. In MOSFETs, the transit time of electrons through the channel is L/u_{sat}, where L is the length of the channel and u_{sat} is the scatter-limited velocity at high electric fields, which is close to the thermal velocity of about 10^7 cm/s. As a result, the maximum operating frequency is not as high as that possible with BJTs. That is why ECL is the fastest logic family in silicon (Section 13.7, Chapter 13).

4. The big advantage of the FET is the high input impedance, particularly of the MOSFET.

5. In switching applications, a FET does not have an offset voltage in the conducting state, whereas in a BJT, $v_{EC(sat)}$ is present across the closed switch. Moreover, a MOSFET maintains the conducting state without the need for significant input and output currents, and hence significant power, as in the BJT. Because the nonconducting state does not require power either, complementary MOSFETs require no power to maintain a logic state. Low power consumption, low manufacturing cost, and the possibility of reducing transistor size so as to pack a larger number on the chip, without significantly degrading performance, make the MOSFET, particularly CMOS technology, highly suitable for very-large-scale-integration (VLSI).

BiCMOS technology (Section 4.7, Chapter 4) combines many of the advantages of bipolar and CMOS technologies. Moreover, the insulated gate BJT (IGBJT) is a useful power device (Section 11.9, Chapter 11).

Summary of Main Concepts

- In a BJT, a small voltage across a low-resistance, forward-biased junction controls the current through a high resistance, reverse-biased junction resulting in considerable voltage and power amplification. Terminal voltages and output current in a CE *npn* transistor are analogous to those of an enhancement-type *n*-channel MOSFET.

- In the BJT output characteristics, the region where the output current is appreciable but the output voltage is small is the saturation region, whereas the region where the output current increases relatively slowly with the output voltage is the active region. Base-width modulation causes the output current to increase with the output voltage in the active region.

- In small-signal operation, small-signal variations can be represented by an equivalent circuit consisting of linear time-invariant (LTI) resistances and capacitances and a dependent current source at the output that is controlled by the input voltage, or by the input current if the input is isolated from the output.

- The BJT has higher transconductance than FET because the forward-biased BEJ is more effective in modulating the output current.

- A rise in temperature increases the collector current due to reduction in v_{BE} and increase in β_F.

- Diode-connected BJTs have smaller resistances, faster speeds of operation, and a more nearly logarithmic i–v relation than comparable pn junction diodes.

- Whereas FETs are unipolar devices that depend for their operation on one type of current carriers, BJTs are bipolar devices in which minority-carrier storage and recombination in the base play an important role in device operation. BJT turn-off is considerably slowed down in the saturation mode because of increased minority carrier storage in the base.

- Compared to MOSFETs, BJTs have the advantages of: (1) greater current and power handling capabilities, (2) a larger transconductance, and hence larger amplifier gains and, for a given change in input voltage, larger currents that can charge and discharge parasitic capacitances more rapidly, and (3) potentially faster operating speeds, as gauged by transit time through the base. MOSFETs have the advantages of: (1) negligible input current, (2) no offset voltage in the state of maximal conduction, (3) low steady-state power consumption in CMOS circuits, and (4) simpler and lower cost of manufacturing. These factors make the MOSFET, particularly CMOS technology, highly suitable for VLSI.

- An HBT offers considerable flexibility in tailoring transistor performance to particular applications. In particular, it is possible to increase β_F, reduce r_B, increase V_A, and reduce C_{je}.

Learning Outcomes

- Articulate the basic operation of the BJT as well as some of the important secondary effects.

Supplementary Examples and Topics on Website

SE6.1 Log/Antilog Amplifiers. Analyzes ideal op amp circuits that use the exponential variation of the *i–v* relation of a *pn* junction to construct logarithmic and antilogarithmic amplifiers.

ST6.1 Basic Analysis of BJT. Derives the Ebers–Moll equations, in both the injection and transport versions, and applies them to deriving: (1) the collector current when the base is open circuited or short-circuited to the emitter, (2) Equation 6.2.1, and (3) the small signal equivalent circuit. Expresses transistor operation in terms of charge-control relations and determines the transit time through the base.

ST6.2 Hybrid-π, Small-Signal BJT Equivalent Circuit. Derives the hybrid-π, low-frequency, small-signal equivalent circuit of a BJT.

ST6.3 h-Parameter, Small-Signal BJT Equivalent Circuit. Expresses the parameters of the hybrid-π, small-signal equivalent circuit in terms of the h parameters.

ST6.4 Measurement of Transistor h-parameters. Describes how transistor h-parameters can be measured using an ac signal while supplying the transistor input and output with a dc bias.

ST6.5 Analysis and PSpice Simulation of BJT Inverter. Investigates the steady-state and transient behavior of the BJT inverter ad simulates it using a 2N2222 transistor.

ST6.6 Parasitic Transistors. Shows how parasitic transistor structures in CMOS and BJT devices can behave as regenerative pairs that cause latchup.

ST6.7 Kirk Effect. Explains the Kirk effect, which causes a decrease in the collector current under high-level injection.

Problems and Exercises

P6.1 Operation and Parameters of BJTs

P6.1.1 Assume that an enhancement mode of operation is defined as one in which the output current is zero when the input terminals are shorted together, and the output current increases as the voltage of the input terminal moves toward that of the output terminal. How would this definition apply to *npn* or *pnp* BJTs in the CB or CE configurations?

P6.1.2 In Figure P6.1.2, $V_{BB}=0$ and V_{CC} and V_{EE} can assume values of -5 V, 0, and $+5$ V. Determine the transistor mode for all combinations of V_{CC} and V_{EE}.

P6.1.3 Repeat Problem P6.1.2 for a *pnp* transistor.

P6.1.4 When the base of an *npn* transistor is open circuited, $I_{CEO} = I_{CBO}/(1 - \alpha_F)$ (Equation 6.1.9). If the base is now shorted to the emitter, do you expect i_C to increase or decrease? Justify your answer based on a simple consideration of transistor action. When the base is shorted to the emitter, in which direction do you expect base current to flow, from basic circuit considerations?

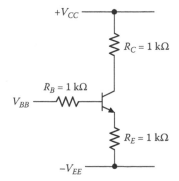

FIGURE P6.1.2

P6.1.5 Using the relation $(\beta + 1)(1 - \alpha) = 1$, show that if $\beta \gg 1$, $\Delta\beta/\beta \cong \beta(\Delta\alpha/\alpha)$.

P6.1.6 If in Exercise 6.1.2, I_{CEO} is calculated from I_{CBO} using $\beta_F = 99$ instead of 9.9, what is the percentage error in determining I_C from the relation: $I_C = \beta_F I_B + I_{CEO}$?

P6.1.7 Show that: (a) $I_B = \dfrac{I_E}{\beta_F + 1} - I_{CBO}$; (b) $I_E = \dfrac{\beta_F + 1}{\beta_F}(I_C - I_{CBO})$.

P6.1.8 Given a transistor having $\beta_F = 100$, $I_{CBO} = 1$ μA, and $I_B = 100$ μA at 300K. If I_B remains constant, determine the percentage increase in I_C at 375K, assuming that I_{CBO} doubles every 10°C and β_F increases by 0.7% per °C.

P6.1.9 Simulate: (a) the output characteristics of the 2N2222 transistor in the CB configuration for v_C in the range of 0–20 V and for I_E in the range of 0–100 mA, in steps of 10 mA; (b) the input characteristics for v_{BE} in the range of 0–0.8 A and for $v_{CB} = 0$, 1, and 20 V.

P6.1.10 The collector current of a transistor is 10 mA at $v_{BE} = 0.7$ V. What is the collector current if: (a) $v_{BE} = 0.65$ V; (b) the doping level in the base is reduced by a factor of 10; (c) the carrier mobility becomes one-third of its value.

P6.1.11 A transistor has $I_{C(sat)} = 10$ mA for a particular value of v_{BE}. Determine V_A if $i_C = 12.4$ mA at $v_{CE} = 12$ V, assuming $V_{CE(sat)} = 0.2$ V.

P6.1.12 A transistor has $g_m = 0.4$ A/V at 300K. At what value of I_C is g_m measured? Determine g_m at 350K if I_C increases by 15%.

P6.1.13 A transistor has $g_m = 0.4$ A/V and $\beta = 150$. Neglecting r_μ, determine the input resistance: (a) in the CE configuration; (b) in the CB configuration.

P6.1.14 In the transistor of Problem P6.1.13, determine r_o if $V_A = 80$ V.

P6.1.15 In the transistor of Problem P6.1.14, if $i_B = 5i_{e(rec)}$, estimate r_μ.

P6.1.16 In the transistor of Problem P6.1.15, determine the output resistance: (a) in the CE configuration, with i_B as parameter; (b) in the CB configuration.

P6.1.17 Determine the h parameters of the transistor in Problems P6.1.13 through P6.1.15.

P6.1.18 Using PSpice simulation, determine β_F, β, g_m, r_π, r_o, and V_A for the Q2N3904 at $I_C = 20$ mA and $V_{CE} = 20$ V.

P6.1.19 Repeat Problem P6.1.18 for the Q2N3906 *pnp* transistor in the EVAL library.

P6.1.20 Using the hybrid-π equivalent circuit (Figure 6.2.7), show that the small-signal output resistance of a diode connected BJT is $r_e \| r_o \cong r_e$.

P6.1.21 Repeat Problem P6.1.20 using the T-model (Figure 6.1.13a), including r_o.

P6.1.22 Determine BV_{CEO} in terms of BV_{CBO} if $n = 5$ and: (a) $\alpha_F = 0.99$; (b) 0.995.

P6.1.23 For the transistor of Example 6.2.3, determine W_B at which punchthrough occurs, assuming a reverse voltage of 100 V. Neglect the width of the depletion region of the BEJ.

P6.1.24 Derive Equation 6.3.19 from the EM equations.

P6.1.25 A transistor having $\beta_F = 100$ and $\beta_R = 2$ is to be operated as a switch. If $v_{CE(sat)}$ is equal to 0.2 V, with $i_C = 20$ mA, $r_C = 6$ Ω, and negligible r_E, what should be β_{forced}?

P6.1.26 Consider the expanded characteristics of Figure 6.2.5 with $I_B = 0.5$ mA. (a) Estimate r_o in the saturation region as $v_{CE(sat)}$ varies between 100 and 150 mV; (b) Determine $v_{CE(sat)}$ assuming $\beta_F = 150$, $\beta_R = 3$, and $\beta_{forced} = 15$.

P6.1.27 Using Equation 6.2.32, and assuming $W_B = 5$ μm and $D_{eB} = 30$ cm^2/s, calculate τ_F. Determine the stored charge in the base when $I_C = 10$ mA.

P6.1.28 If the transistor of Problem P6.1.27 operates at $I_C = 10$ mA and $C_{je} + C_{jc} = 0.2$ pF, determine f_T.

P6.1.29 Deduce from Equation ST6.1.34, Section ST6.1, that $\alpha_F = 1/(1 + \tau_F/\tau_{eB})$.

P6.1.30 Determine α_F and β_F in Example 6.1.1 if the base width doubled to 2 μm.

P6.1.31 If in Example 6.1.1 τ_{eB} is multiplied by 10, what is the effect on: (a) β_F; (b) $I_{e(rec)}/I_B$?

P6.1.32 If in Example 6.1.1 τ_{hE} is multiplied by 10, what is the effect on: (a) β_F; (b) $I_{e(rec)}/I_B$?

P6.2 *i–v* Relations of Bipolar Junction Transistors

P6.2.1 Given that $\beta = \dfrac{di_C}{di_B}\bigg|_{v_{CE}}$ and the variation of β with i_C, how do you expect the spacing of the BJT CE output characteristics for constant I_B to vary with increasing i_C at constant v_{CE}?

P6.2.2 In Figure P6.1.2, $V_{CC}=12$ V, $V_{EE}=-12$ V, $R_B=30$ kΩ, $R_C=1$ kΩ, $R_E=2$ kΩ, and $\beta=100$. Determine V_{BB} that will cut off the transistor with $V_{BE}=0.5$ V neglecting transistor currents in the cutoff state.

P6.2.3 Find V_{CE} in Problem P6.2.2 if $V_{BB}=0$ and $V_{BE}=0.7$ V. Simulate with PSpice.

P6.2.4 Consider Problem P6.1.2 with $V_{EE}=0$. Determine V_{BB} that will just bring the transistor to saturation, assuming $V_{BE}=0.7$ V. Simulate with PSpice.

P6.2.5 In Figure P6.2.5, $V_{CC}=12$ V, $V_{BB}=-12$ V, $R_C=2$ kΩ, $R_E=0$, $R_{B1}=20$ kΩ, and $R_{B2}=100$ kΩ. Determine V_B that will cut off the transistor with $V_{BE}=0.5$ V, neglecting transistor currents in the cutoff state.

P6.2.6 Consider Figure P6.2.5 with $R_E=100$ Ω and $\beta=100$. Determine V_B that will just bring the transistor to saturation, assuming $V_{BEA}=0.7$ V. Simulate with PSpice.

P6.2.7 Consider Figure P6.2.5 with $V_B=3.5$ V. Assuming $V_{BE}=0.7$ V, determine I_C if: (a) $\beta=100$; (b) $\beta=200$. What is the percentage variation of I_C when β is doubled? Simulate with PSpice.

P6.2.8 Repeat Problem P6.2.7 with $R_E=1$ kΩ. Simulate with PSpice. How do you explain the effect of larger R_E?

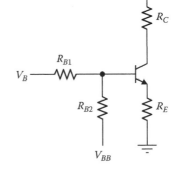

FIGURE P6.2.5

P6.2.9 The circuit of Figure P6.2.5, with $\beta=100$, $R_E=0$, $R_C=2$ kΩ, $V_{CC}=5$ V, and $V_{BB}=-5$ V, is to be operated as an inverter. Determine R_{B1} and R_{B2} so that when $v_B=0$, $v_{BE}=-1$ V but when $v_B=5$ V, $v_{BE}=0.8$ V assuming $\beta_{forced}=10$. Simulate with PSpice.

P6.2.10 With the values of R_{B1} and R_{B2} selected as in Problem P6.2.9, determine the output voltage levels for: (a) $v_B=0.8$ V; (b) $v_B=3.8$ V. What should be the minimum β to ensure saturation? Simulate with PSpice.

P6.2.11 In Figure P6.2.11, the voltage across the 1 kΩ resistor is 1 V and $V_{BE}=0.7$ V. Determine: (a) V_{CE}; (b) V_{CB}; (c) I_B; (d) I_C; (e) β_F.

P6.2.12 Repeat Problem P6.2.11 using a *pnp* transistor, replacing the +12 V supply with a −12 V supply, and label the actual directions and polarities of all currents and voltages.

P6.2.13 If the transistor in Figure P6.2.13 has $\beta_F=500$, determine the voltages at all the transistor terminals.

P6.2.14 Repeat Problem P6.2.13 with the *npn* transistor replaced by a *pnp* transistor having $\beta_F=100$ and with the +15 V supply replaced by a −15 V supply.

P6.2.15 Consider the inverter circuit of Figure 6.3.4a with $V_{CC}=5$ V, $R_C=1$ kΩ, and a resistor R_L connected between collector and ground. If $v_B=0$, what is the minimum value of R_L that will maintain the output voltage across R_L above 4 V? Transistor currents in the cutoff state may be neglected at these low values of resistances.

P6.2.16 Assume that in the inverter circuit of Figure 6.3.4a, i_C, β, v_B, and v_{BE} can vary over a range of values whose extremes may

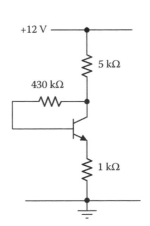

FIGURE P6.2.11

be denoted by "max" and "min" subscripts. Derive an expression for the maximum value of R_B that ensures saturation.

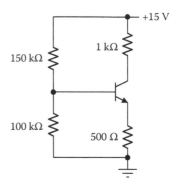

P6.2.17 Assume that in the inverter circuit of Figure 6.3.4a, $V_{CC} = V_B = 5$ V, $R_C = 1$ kΩ, $R_B = 20$ kΩ, $v_{CES} = 0.2$ V, and $\beta = 100$. Determine whether or not the transistor is saturated, and if so, β_{forced}. What is the minimum β that ensures saturation? Simulate with PSpice.

FIGURE P6.2.13

P6.2.18 Assume an inverter with circuit values as in Problem P6.2.17 connected at its output to a number N of identical inverters. What is the upper limit on N when the transistor is saturated? What is the upper limit when the transistor is cut off and all the inverters connected to the output are saturated with β_{forced} of 50?

P6.2.19 Assume that the inverter of Problem P6.2.17 is supplying the maximum number of identical inverters. Determine the power dissipated in the collector circuit when the transistor is: (a) saturated, and (b) cut off.

P6.2.20 Consider that a number of transistors in the circuit of Figure 6.3.4a are paralleled at their collectors so that they share the same R_C but have separate base resistors and inputs. What logic function is performed on the inputs? How would you convert this function to OR?

P6.2.21 Referring to Equations 6.3.1 and 6.3.2, show that

$$i_C = \alpha_F i_E - (1 - \alpha_F \alpha_R) I_{SC}(e^{v_{BC}/V_T} - 1)$$

$$i_E = -\bar{i}_E = \alpha_R i_C + (1 - \alpha_F \alpha_R) I_{SE}(e^{v_{BE}/V_T} - 1)$$

P6.2.22 Using the results of Problem P6.2.21, show that $v_{BE} = V_T \ln \left(1 + \dfrac{i_E - \alpha_R i_C}{(1 - \alpha_F \alpha_R) I_{SE}}\right)$.

P6.2.23 Show that if a transistor is diode-connected by joining the base to the emitter $i \cong \dfrac{I_S}{\alpha_R} e^{v/V_T}$, where $v = v_{BC}$ and i is the current flowing out of the collector.

P6.2.24 The parameters of the EB model are $I_S = 1$ pA, $\beta_F = 500$, and $\beta_R = 1$. Determine I_C and I_T if (a) $V_{BE} = 0.3$ V, $V_{CB} = 5$ V; (b) $V_{BE} = 0.65$ V, $V_{CB} = 5$ V; (c) $V_{EB} = 0.8$ V, $V_{BC} = 0.5$ V.

P6.2.25 An *npn* transistor has $\alpha_F = 0.98$ and $\alpha_R = 0.8$. When the emitter is open, the reverse current of the BEJ is 0.1 µA. Determine: (a) I_{CBO}; (b) the collector and base currents when the emitter is short-circuited to the base.

P6.2.26 Identifying I_{SE} in Example ST6.1.2 (Section ST6.1) with $|q|A_E D_{eB} n_{po}/W_B$ in Equation 6.2.1, determine I_{SE} using the values in Example 6.1.1. Calculate i_C if $v_{BE} = 0.7$ V.

P6.2.27 Determine β_F for a transistor having the same parameters as in Example 6.1.1, except that the BEJ is a heterojunction having $\Delta E_g = 0.1$ V. Assume the same effective densities of states in the valence and conduction bands in the base and the emitter.

P6.3 Miscellaneous

P6.3.1 Consider an exponential function Ae^{-ax}, where A and a are constants. Show that the area under the curve from $x = 0$ to x is proportional to the difference in magnitudes between the slopes at $x = 0$, and apply this to the concentration profile of minority carriers injected into the base.

P6.3.2 Implement Equation ST6.2.4, Section ST6.2, without disturbing Equation ST6.2.2 by isolating r_μ using a unity-gain follower whose input is connected to terminal C and the parallel branches and whose output is connected to r_μ. Show that under these conditions, the output resistance is: (a) r_o at constant v_{BE}; (b) $r_o||(r_\mu + r_\pi)/\beta \cong r_o||(r_\mu/\beta)$ at constant i_B, since the current flowing in at the C terminal is $\left(\dfrac{1}{r_o} + \dfrac{\beta}{r_\mu + r_\pi}\right) v_{ce}$; (c) $r_\pi + (1+\beta)r_o \cong \beta r_o$ at constant i_E since the current flowing in at the C terminal is $\left(\dfrac{1}{r_\pi + (1+\beta)r_o}\right) v_{cb}$.

P6.3.3 Consider a transistor turning off from a saturated state with $i_B = -I_{BL}$. Show that $t_S = \tau_{eB} \ln\left(\dfrac{|Q_{eB}(0)| + I_{BL}\tau_{eB}}{|Q_{eB(\text{sat})}| + I_{BL}\tau_{eB}}\right)$, where t_S is the time when $|Q_{eB}(t_S)| = |Q_{eB(\text{sat})}|$ (Section ST6.5). Note that if $I_{BL}\tau_{eB} \gg |Q_{eB}(0)|$, $t_S \to 0$.

7

Two-Port Circuits, Amplifiers, and Feedback

The purpose of the present chapter is to introduce some topics that are needed for proper understanding of amplifier circuits in the following chapters. As these circuits are basically two-port in nature, it is appropriate to discuss them within the framework of two-port circuit theory. Although this theory offers, in addition, some powerful methods for analyzing circuits, other than the conventional mesh-current and node voltage methods, these methods are rarely applied to electronic circuits.

The chapter begins by reviewing two-port circuit equations, parameters, and equivalent circuits. The two-port circuit description naturally leads to the representation of the four ideal amplifier types, namely, ideal voltage, current, transresistance, and transconductance amplifiers. It is then shown how a practical amplifier can be made to approach any of these ideal types by appropriate negative feedback connections. This also illustrates the great importance of negative feedback in conferring desirable characteristics on an amplifier circuit.

The chapter ends with a discussion of the ideal operational amplifier, in both the noninverting and inverting configurations, in terms of negative feedback and signal-flow diagrams. This illustrates some important features of operational amplifiers and the role that negative feedback plays in the improvement of some aspects of the performance of practical operational amplifiers, particularly input impedance, output impedance, response to extraneous signals, and frequency response.

Learning Objectives

❖ To be familiar with:

- The terminology and description of two-port circuits.
- The broad range of applications of operational amplifiers.

❖ To understand:

- The interpretation of the parameters of two-port circuits and their representation in terms of equivalent circuits.
- The interpretation of ideal controlled sources as two-port circuits and the resulting classification of the four ideal types of amplifiers.
- The treatment of feedback amplifiers as interconnected two-port circuits and the effects of the various forms of negative feedback on circuit behavior.
- The essential features of the noninverting and inverting operational amplifier configurations.
- The role negative feedback can play in the improvement of some aspects of amplifier performance, particularly input impedance, output impedance, response to extraneous signals, and frequency response.

7.1 Two-Port Circuits*

Two-port circuits have a pair of input terminals, or input ports, and a pair of output terminals, or output ports, as illustrated in Figure 7.1.1, where the subscripts 1 and 2 refer to the input and output, respectively. The assigned positive directions of voltages and currents are conventionally as shown in the figure. It is understood that a two-port circuit can contain passive circuit elements and dependent sources but no independent sources other than those applied at the input or output. The voltage and current variables are to be considered, in general, as Laplace transforms of the corresponding time functions. They reduce to dc variables under dc conditions and to phasors in the sinusoidal steady state.

Clearly, if a voltage V_1 is applied to a given two-port circuit of known parameters, such as a simple, resistive T-circuit, I_1, V_2, and I_2 are not determined unless another constraint is imposed, such as $I_2 = 0$ or a relation between V_2 and I_2. In general, two of the four terminal variables may be specified independently, in which case the other two variables are determined by the parameters of the given circuit. The two-port circuit can therefore be described in terms of two simultaneous equations. However, as four variables are involved, there are six ways of choosing two of these variables as independent, which gives rise to six sets of simultaneous equations, listed in Table 7.1.1. Each of these equations is written in terms of four coefficients, or **parameters**, that characterize the equation. Thus, one speaks of the z-parameter equations, the y-parameter equations, etc.

The sets of two equations in each row of Table 7.1.1 are inversely related, in the sense that the independent variables in one set of equations are the dependent variables in the other set. This makes the corresponding matrices of parameters inversely related. Thus,

$$
\begin{vmatrix} z_{11} & z_{12} \\ z_{21} & z_{22} \end{vmatrix} = \begin{vmatrix} y_{11} & y_{12} \\ y_{21} & y_{22} \end{vmatrix}^{-1},
$$

$$
\begin{vmatrix} z_{11} & z_{12} \\ z_{21} & z_{22} \end{vmatrix}^{-1} = \begin{vmatrix} y_{11} & y_{12} \\ y_{21} & y_{22} \end{vmatrix}
\tag{7.1.1}
$$

The other sets of parameters are similarly related.

Interpretation of Parameters

The parameters of two-port circuits can be interpreted in terms of voltage and current ratios under specified open-circuit or short-circuit terminations, as indicated in Table 7.1.2. Thus, z_{11} is the input impedance looking into port 1, with port 2 open-circuited, and y_{11} is the input admittance looking into port 1, with port 2 short-circuited. These interpretations provide a convenient means for the evaluation or measurement of the respective parameters.

As the same circuit variables are involved in the six sets of equations, the parameters in these equations must be related, as discussed in Section ST7.1.

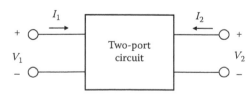

FIGURE 7.1.1
Two-port circuit and assigned positive direction.

* Sabah 2008, pp. 520–530.

TABLE 7.1.1

Two-Port Circuit Equations

$$V_1 = z_{11}I_1 + z_{12}I_2 \qquad I_1 = y_{11}V_1 + y_{12}V_2$$
$$V_2 = z_{21}I_1 + z_{22}I_2 \qquad I_2 = y_{21}V_1 + y_{22}V_2$$

$$V_1 = a_{11}V_2 - a_{12}I_2 \qquad V_2 = b_{11}V_1 - b_{12}I_1$$
$$I_1 = a_{21}V_2 - a_{22}I_2 \qquad I_2 = b_{21}V_1 - b_{22}I_1$$

$$V_1 = h_{11}I_1 + h_{12}V_2 \qquad I_1 = g_{11}V_1 + g_{12}I_2$$
$$I_2 = h_{21}I_1 + h_{22}V_2 \qquad V_2 = g_{21}V_1 + g_{22}I_2$$

TABLE 7.1.2

Interpretation of Two-Port Parameters

$$z_{11} = \frac{V_1}{I_1}\bigg|_{I_2=0}\ \Omega \quad z_{12} = \frac{V_1}{I_2}\bigg|_{I_1=0}\ \Omega \qquad y_{11} = \frac{I_1}{V_1}\bigg|_{V_2=0}\ S \quad y_{12} = \frac{I_1}{V_2}\bigg|_{V_1=0}\ S$$

$$z_{21} = \frac{V_2}{I_1}\bigg|_{I_2=0}\ \Omega \quad z_{22} = \frac{V_2}{I_2}\bigg|_{I_1=0}\ \Omega \qquad y_{21} = \frac{I_2}{V_1}\bigg|_{V_2=0}\ S \quad y_{22} = \frac{I_2}{V_2}\bigg|_{V_1=0}\ S$$

$$a_{11} = \frac{V_1}{V_2}\bigg|_{I_2=0} \quad a_{12} = -\frac{V_1}{I_2}\bigg|_{V_2=0}\ \Omega \qquad b_{11} = \frac{V_2}{V_1}\bigg|_{I_1=0} \quad b_{12} = -\frac{V_2}{I_1}\bigg|_{V_1=0}\ \Omega$$

$$a_{21} = \frac{I_1}{V_2}\bigg|_{I_2=0}\ S \quad a_{22} = -\frac{I_1}{I_2}\bigg|_{V_2=0} \qquad b_{21} = \frac{I_2}{V_1}\bigg|_{I_1=0}\ S \quad b_{22} = -\frac{I_2}{I_1}\bigg|_{V_1=0}$$

$$h_{11} = \frac{V_1}{I_1}\bigg|_{V_2=0}\ \Omega \quad h_{12} = \frac{V_1}{V_2}\bigg|_{I_1=0} \qquad g_{11} = \frac{I_1}{V_1}\bigg|_{I_2=0}\ S \quad g_{12} = \frac{I_1}{I_2}\bigg|_{V_1=0}$$

$$h_{21} = \frac{I_2}{I_1}\bigg|_{V_2=0} \quad h_{22} = \frac{I_2}{V_2}\bigg|_{I_1=0}\ S \qquad g_{21} = \frac{V_2}{V_1}\bigg|_{I_2=0} \quad g_{22} = \frac{V_2}{I_2}\bigg|_{V_1=0}\ \Omega$$

Example 7.1.1 Determination of *h*-Parameters

The following dc measurements were made on a two-port resistive circuit:

Port 2 open-circuited: $V_1 = 10$ mV, $I_1 = 50$ μA, $V_2 = 20$ V
Port 2 short-circuited: $V_1 = 40$ mV, $I_1 = 100$ μA, $I_2 = -1$ mA

It is required to find the *h*-parameters of the circuit.

SOLUTION

With port 2 short-circuited, $h_{11} = \frac{V_1}{I_1}\bigg|_{V_2=0} = \frac{40\text{ mV}}{100\text{ μA}} \equiv 400\ \Omega$, and $h_{21} = \frac{I_2}{I_1}\bigg|_{V_2=0} =$

$-\frac{1\text{ mA}}{100\text{ μA}} \equiv -10$. $h_{12} = \frac{V_1}{V_2}\bigg|_{I_1=0}$ and $h_{22} = \frac{I_2}{V_2}\bigg|_{I_1=0}$ cannot be obtained from the given measurements because these do not include the case of port 1 open-circuited. However, we can obtain the *a*-parameters from the given measurements. Thus, $a_{11} = \frac{V_1}{V_2}\bigg|_{I_2=0} =$

$\frac{10 \times 10^{-3}\text{ V}}{20\text{ V}} = 5 \times 10^{-4}$; $a_{12} = -\frac{V_1}{I_2}\bigg|_{V_2=0} = -\frac{40\text{ mV}}{-1\text{ mA}} = 40\ \Omega$; $a_{21} = \frac{I_1}{V_2}\bigg|_{I_2=0} = \frac{50\text{ μA}}{20\text{ V}} \equiv$

2.5×10^{-6} S; $a_{22} = -\frac{I_1}{I_2}\bigg|_{V_2=0} = -\frac{100 \times 10^{-3}\text{ mA}}{-1\text{ mA}} = 0.1$, and $\Delta a = a_{11}a_{22} - a_{12}a_{21} =$

$5 \times 10^{-4} \times 0.1 - 40 \times 2.5 \times 10^{-6} = -5 \times 10^{-5}$. From the relations between parameters

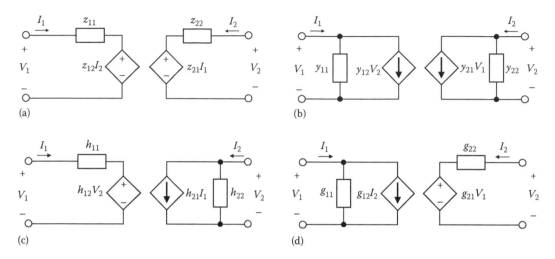

FIGURE 7.1.2
Equivalent circuits representing the two-port equations in terms of the z-parameters (a), y-parameters (b), h-parameters (c), and g-parameters (d).

(Section ST7.1): $h_{12} = \Delta a/a_{22} = -5 \times 10^{-5}/0.1 = -5 \times 10^{-4}$ and $h_{22} = a_{21}/a_{22} = 2.5 \times 10^{-6}/0.1 \equiv 25\ \mu S$.

Equivalent Circuits

The z-, y-, h-, and g-parameter equations can be represented in terms of the equivalent circuits shown in Figure 7.1.2a through d. The effect of the input on the output side, that is, the **forward transmission**, is expressed by the dependent source on the output side and its associated "21" subscript parameter. On the other hand, the effect of the output on the input side, that is, the **reverse transmission**, is expressed by the dependent source on the input side and its associated "12" subscript parameter. The remaining parameters can be readily interpreted in terms of these circuits. The "11" subscript parameter describes an input impedance or admittance when the dependent source on the input side is set to zero, which imposes an open-circuit or a short-circuit constraint on the output port. Similarly, the "22" subscript parameter describes an output impedance or admittance when the dependent source on the output side is set to zero, which imposes an open-circuit or a short-circuit constraint on the input port. In the case of the h-parameters, for example, if the output is short-circuited ($V_2 = 0$), it is seen from Figure 7.1.2c that the series-connected h_{11} is the input impedance under these conditions. Similarly, when the input is open-circuited ($I_1 = 0$), it is seen from Figure 7.1.2c that the shunt-connected h_{22} is the output admittance under these conditions.

Note that all the equivalent circuits of Figure 7.1.2 become three-terminal circuits if the lower input and output terminals are connected together.

7.2 Ideal Amplifiers

Consider next the form that an ideal amplifier would take as a two-port circuit. An ideal amplifier can be derived from each of the equivalent circuits of Figure 7.1.2 by setting all

TABLE 7.2.1

Ideal Amplifier Types

Parameter	Equations	Equivalent Circuit	Amplifier Type
z (Figure 7.1.2a) $z_{11}=0$; $z_{12}=0$, $z_{21}=R_m$; $z_{22}=0$	$v_I=0$ $v_O=R_m i_I$	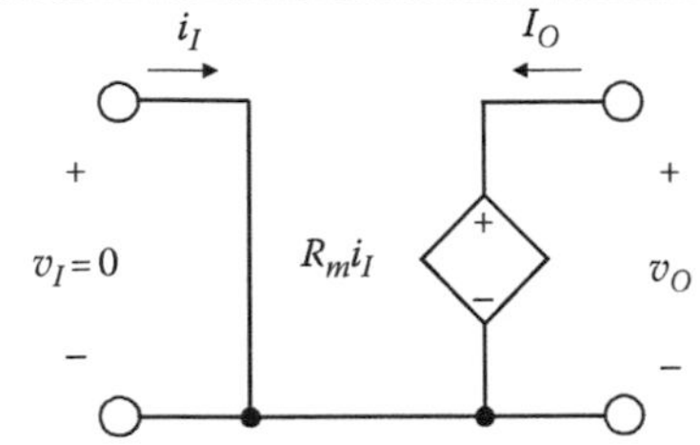	Ideal transresistance amplifier
y (Figure 7.1.2b) $y_{11}=0$; $y_{12}=0$, $y_{21}=G_m$; $y_{22}=0$	$i_I=0$ $i_O=G_m v_I$	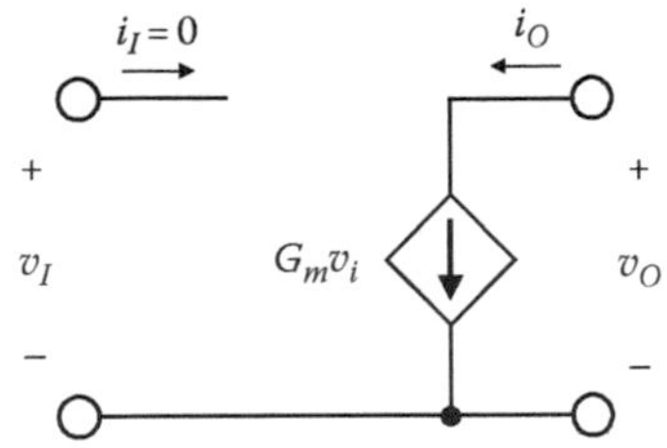	Ideal transconductance amplifier
h (Figure 7.1.2c) $h_{11}=0$; $h_{12}=0$, $h_{21}=A_i$; $h_{22}=0$	$v_I=0$ $i_O=A_i i_I$	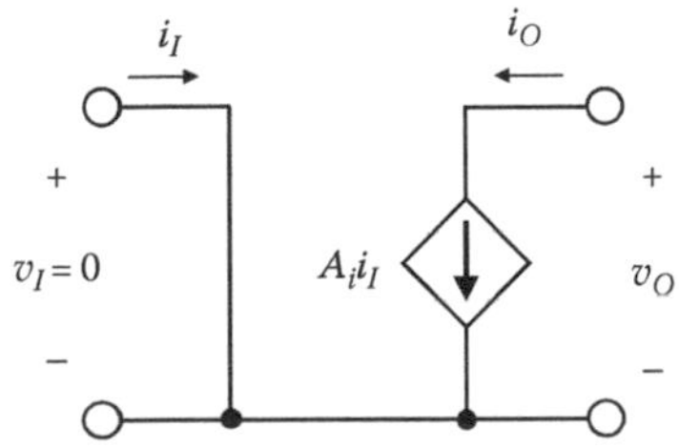	Ideal current amplifier
g (Figure 7.1.2d) $g_{11}=0$; $g_{12}=0$, $g_{21}=A_v$; $g_{22}=0$	$i_I=0$ $v_O=A_v v_I$	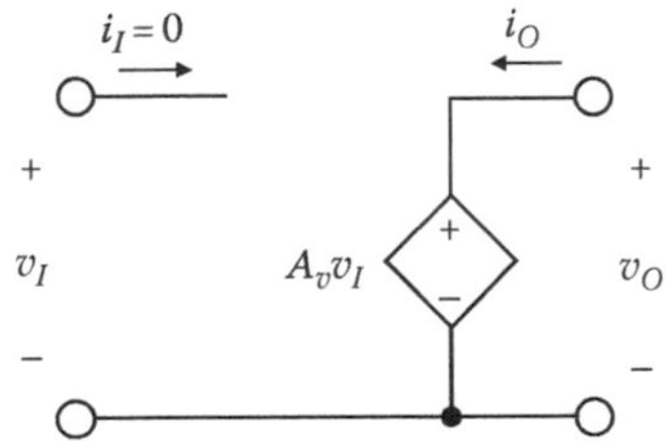	Ideal voltage amplifier

the coefficients in the corresponding two-port equations to zero, except the "21" parameter, which represents the gain of the amplifier, and is denoted by some constant. The "1" and "2" subscripts are replaced, respectively, by "I," for input, and "O," for output, as illustrated in Table 7.2.1.

The following should be noted concerning these ideal amplifiers:

1. That they are amplifiers is evident from the fact that the input power is zero, whereas the output power is finite. This makes the power amplification theoretically infinite.

2. The transresistance amplifier is so called because its output is a voltage and its input is a current, whereas the transconductance amplifier is so called because its output is a current and its input is a voltage.

3. In all cases, there is no reverse transmission, that is; the output does not affect the input. We will see in future chapters that in real amplifiers there is, in general, a small reverse transmission.

4. In transresistance and voltage amplifiers, the output is an ideal, dependent voltage source controlled by an input current or voltage.

5. In transconductance and current amplifiers, the output is an ideal, dependent current source controlled by an input current or voltage.

6. Where the controlling input is a voltage, as in transconductance and voltage amplifiers, and if the input circuit is represented by its Thevenin's equivalent, $i_I = 0$ ensures that v_I equals the Thevenin source voltage, irrespective of the Thevenin source impedance. In other words, the Thevenin source is ideally terminated.

7. Where the controlling input is a current, as in transresistance and current amplifiers, and if the input circuit is represented by its Norton's equivalent, $v_I = 0$ ensures that i_I equals the Norton source current, irrespective of the Norton source admittance. In other words, the Norton source is ideally terminated.

7.3 Negative Feedback*

A matter of great practical importance is the following: given an amplifier that is represented by any of the circuits of Figure 7.1.2, is it possible to make the amplifier approach any of the ideal amplifier types of Table 7.2.1? We will show that this is indeed possible under the following conditions:

1. Very high gain, or forward transmission
2. Negligible reverse transmission
3. An appropriate **negative feedback** connection
4. Negligible loading effect of the feedback circuit on the amplifier

As its name implies, feedback involves feeding part of the output back to the input. We can understand by the word "negative" at this stage that the fed back signal is of a polarity that opposes the applied input. A more general definition will be given later. It is necessary to specify that the feedback is negative, because it could alternatively be positive. But positive feedback serves quite different purposes that will be discussed in future chapters.

We will also show that negative feedback can be applied in one of four configurations, corresponding to the four ideal amplifier types of Table 7.2.1. The four feedback configurations derive from the fact that the output signal that is sampled could be a voltage or a current, whereas the signal that is fed back could also be a voltage or a current, leading to four possible combinations. But we must first decide on the form of the "nonideal" amplifier to which the negative feedback is going to be applied to make it "ideal." We note in this regard that when reverse transmission is neglected, the following applies to all four circuits of Figure 7.1.2:

---

* Sabah 2008, Section ST18.1, p. 755.

1. The input reduces to an input impedance z_i or an input admittance y_i that is independent of the output circuit.

2. At the output, we have either a dependent voltage source in series with an output impedance z_o, or a dependent current source in parallel with an output admittance y_o. Either output circuit can of course be transformed to the other.

3. Since the input consists of an impedance or an admittance, the controlling variable of the dependent source at the output can be considered as either the input current or the input voltage, since these two variables are related by the input impedance or admittance.

It follows that any of the representations referred to in the preceding items 1–3 can be used for the nonideal amplifier, whichever is most convenient for the purpose.

Series–Shunt Feedback

Figure 7.3.1 shows the z-equivalent circuit with zero reverse transmission, z_i at the input, and a dependent voltage source in series with z_o at the output. The dependent source is expressed in terms of the input voltage ε across z_i. Thus, $z_{21}I_{src} = z_{21}\varepsilon/z_i = A_{va}\varepsilon$, where A_{va} is the open-circuit voltage gain from the input to output of the *amplifier proper*, without feedback. The feedback circuit feeds a fraction $\beta_f V_o$ of the output back to the input as a voltage-controlled dependent voltage source, where β_f is the **feedback factor**. The feedback is negative because $\beta_f V_o$ is of a polarity that subtracts from the source input V_{src}, so that the net input to the amplifier proper is $\varepsilon = V_{src} - \beta_f V_o$. In the presence of negative feedback, the amplifier is referred to as a **feedback amplifier**. Note the change in notation in Figure 7.3.1 from instantaneous value symbols to Laplace transform symbols because the circuits being considered generally have impedances.

The negative feedback configuration of Figure 7.3.1 is **series–shunt feedback**, because current I_{src} flows in series through the amplifier input and $\beta_f V_o$ source. At the output, the feedback is shunt because the feedback circuit is connected in shunt across the output, thereby sampling the output voltage.

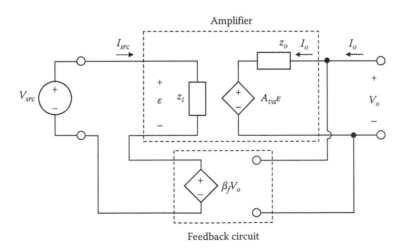

FIGURE 7.3.1
Ideal series–shunt, negative feedback configuration.

Ideally, the shunt feedback circuit draws no current from the output. Assuming no load at the output, $I_o = 0$, and $V_o = A_{va}\varepsilon$. At the input, $V_{src} = \varepsilon + \beta_f V_o$. Eliminating ε between these two relations gives:

$$\frac{V_o}{V_{src}} = \frac{A_{va}}{1 + \beta_f A_{va}} \tag{7.3.1}$$

$$\cong \frac{1}{\beta_f}, \quad \text{if } A_{va} \to \infty \tag{7.3.2}$$

The interpretation of Equation 7.3.2 is that if $A_{va} \to \infty$, then for any finite V_o, $\varepsilon \to 0$, since $\varepsilon = V_o / A_{va}$. KVL on the input side then gives $V_{src} = \beta_f V_o$ and Equation 7.3.2 follows. Note that with $\varepsilon \to 0$, $V_{src} = \beta_f V_o$ independently of I_o. Equation 7.3.2 underlies an important concept.

Concept: *In a high-gain amplifier with negative feedback, the overall gain of the feedback amplifier is determined by the feedback circuit.*

The practical implication is that if β_f is made to depend on ratios of resistances, for example, the overall gain $1/\beta_f$ can be precise and stable, irrespective of variations in the actual gain A_{va} of the amplifier proper. These variations may be due, for example, to changes in internal components caused by environmental factors or "aging," or due to changes in supply voltages, or due to differences between amplifiers of the same type because of manufacturing tolerances.

Let us look next at the input and output impedances. KVL at the input is $V_{src} = z_i I_{src} + \beta_f V_o$. Substituting for V_o from Equation 7.3.1, which is based on $I_o = 0$, and rearranging,

$$\frac{V_{src}}{I_{src}} = Z_{in} = z_i (1 + \beta_f A_{va}) \tag{7.3.3}$$

If $A_{va} \to \infty$, then $Z_{in} \to \infty$. This is to be expected, because if $A_{va} \to \infty$, then $\varepsilon \to 0$, as explained earlier, which means $I_{src} \to 0$ and $Z_{in} \to \infty$. The input behaves as an open circuit.

To determine Z_{out}, we note that the general definition of Z_{out} is Thevenin's impedance seen at the output terminals. Z_{out} can also be determined by applying a test voltage (current) source at the output, with $V_{src} = 0$ and determining the resulting source current (voltage). If a test source I_x is applied (Figure 7.3.2), it follows that $\varepsilon = -\beta_f V_x$ and $V_x = z_o I_x + A_{va}\varepsilon$. Eliminating ε and rearranging,

$$\frac{V_x}{I_x} = Z_{out} = \frac{z_o}{1 + \beta_f A_{va}} \tag{7.3.4}$$

As $A_{va} \to \infty$, $Z_{out} \to 0$. This follows from the fact that $\varepsilon \to 0$ when $A_{va} \to \infty$, which gives $\beta_f V_x = 0$ independently of I_o. In other words, the output behaves as a short circuit ($V_x = 0$). When A_{va} is finite, with V_{src} constant, then if I_o increases in Figure 7.3.1, due to a load for example, V_o increases because of the voltage drop in z_o. This increases the feedback signal $\beta_f V_o$, which decreases ε and $A_{va}\varepsilon$, and nearly restores V_o to its original value. Having nearly the same V_o despite changes in I_o is tantamount to saying that the

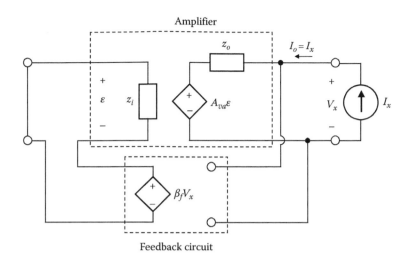

FIGURE 7.3.2
Application of a test current source I_x for determining output impedance.

effective output impedance seen at the output terminals is small. This feedback action can be generalized to an important concept.

Concept: *In a high-gain amplifier with negative feedback, the output variable that is sampled by the feedback circuit is maintained practically constant for a constant input, despite load variations.*

In the series–shunt feedback configuration of Figure 7.3.1, V_o is the output variable being sampled and is maintained nearly constant for a constant V_{src} despite variations in I_o.

   It is seen that a high-gain amplifier having $A_{va} \to \infty$ in series–shunt feedback has $Z_{in} \to \infty$ and $Z_{out} \to 0$. It is equivalent to the ideal voltage amplifier of Table 7.2.1, with $A_v = 1/\beta_f$. When A_{va} is finite, the effect of series–shunt feedback is to: (1) divide the gain of the amplifier by a factor $(1 + \beta_f A_{va})$ (Equation 7.3.1), (2) multiply the input impedance by the same factor $(1 + \beta_f A_{va})$ (Equation 7.3.3), and (3) divide the output impedance by the same factor $(1 + \beta_f A_{va})$ (Equation 7.3.4). Before proceeding to the other feedback configurations, we will clarify the significance of the term $(1 + \beta_f A_{va})$.

Signal-Flow Diagrams

The relationships between the variables in a feedback system can be conveniently represented diagrammatically by means of a **signal-flow diagram**, illustrated in Figure 7.3.3 for the series–shunt feedback configuration of Figure 7.3.1 with $I_o = 0$. The encircled variables are arranged in the order of signal flow, with the direction of this flow indicated by arrows joining the circles. Each arrow in the signal-flow diagram is labeled with a multiplying factor in accordance with the relationship between variables in the direction of signal flow. Thus, the arrow from ε to V_o is marked with A_{va} to indicate that $V_o = A_{va} \times \varepsilon$. The signal that is fed back is $\beta_f V_o$. The relation $\varepsilon = V_{src} - \beta_f V_o$ is represented by an arrow labeled 1 from V_{src} to ε and another arrow labeled -1 from $\beta_f V_o$ to ε. The convergence of the two lines on ε signifies summation. The negative feedback is represented in Figure 7.3.3 by a loop having a product of multipliers around the loop of $-\beta_f A_{va}$, where the term $-\beta_f A_{va}$ is the **loop gain**. This underscores an important concept.

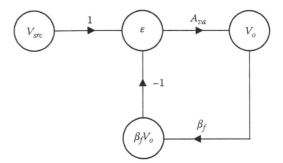

FIGURE 7.3.3
Signal-flow diagram for the feedback amplifier of
Figure 7.3.1.

Concept: *In a negative feedback system, the sign of the loop gain must be negative.*

A negative sign of the loop gain means that if a disturbance starts at any point in the loop and propagates around the loop, it must return to its point of origin with a phase opposite that with which it started, so that the disturbance is "squashed." This is, in fact, the more general definition of a negative feedback system. If the sign of the loop gain is positive, the system is a positive feedback system, the disturbance will be reinforced if the magnitude of the loop gain exceeds unity, leading to instability, as will be discussed later. Note that negative feedback can equally will be obtained by having the -1 in Figure 7.3.3 replaced by $+1$ and A_{va} replaced by $-A_{va}$.

In addition to showing the interrelations between the variables involved, signal-flow diagrams are useful for determining directly the relations between the variables, using some well-defined rules that are elaborated in books on control systems. It suffices for our purposes to use the following simple rule:

> **Rule:** *The value of a given variable equals the value of another variable upstream along the signal path, multiplied by the forward gain between the two variables, and divided by (1 − loop gain) of any feedback loop encountered anywhere along the path between the two variables, as long as the variables in the loop do not have additional inputs.* (7.3.5)

In going from V_{src} to V_O in Figure 7.3.3, for example, the forward gain is the product of all the factors on the arrows from V_{src} to V_O, which is $1 \times A_{va} = A_{va}$. Moreover, there is a feedback loop of gain $-\beta_o A_v$ in going from V_{src} to V_O, the gain around this loop being the loop gain $-\beta_o A_v$. The variables in the loop, ε and V_o, do not have inputs additional to those under consideration. Hence, according to Rule 7.3.5, $\dfrac{V_o}{V_{src}} = \dfrac{A_{va}}{1 + \beta_f A_{va}}$, which is the same as Equation 7.3.1. The justification of this result, of course, is that the relations between the variables in Figure 7.3.3 are precisely those leading to Equation 7.3.1. When additional inputs are present, superposition can be applied, as illustrated later in connection with Figure 7.4.18.

The signal-flow diagram can be also be used to derive Equations 7.3.3 and 7.3.4. Figure 7.3.4a shows the signal-flow diagram with the addition of the input current I_{src}. Applying Rule 7.3.5 in going from V_{src} to I_{src} gives $I_{src} = \dfrac{v_{src}}{z_i} \dfrac{1}{1 + \beta_f A_{va}}$, which is the same as Equation 7.3.3. Figure 7.3.4b shows the signal-flow diagram for determining Z_{out}. KVL at the output in Figure 7.3.2 is $V_x = A_{va}\varepsilon + z_o I_x$. Hence, the arrow from I_x to V_x is labeled z_o. Applying

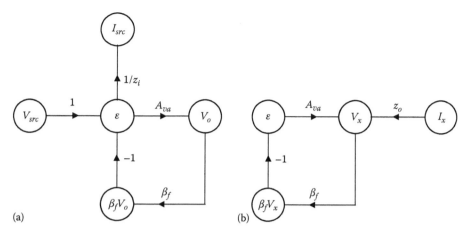

FIGURE 7.3.4
Signal-flow diagrams for determining input impedance (a), and output impedance (b) of the feedback amplifier of Figure 7.3.1.

Rule 7.3.5 to the signal-flow diagram of Figure 7.4.3 gives $V_x = \dfrac{z_o I_x}{1 + \beta_f A_{va}}$, which is the same as Equation 7.3.4.

Exercise 7.3.1

Determine Z_{out} in Figure 7.3.2 as the ratio $V_{o(oc)}/I_{sc}$ derived: (1) analytically, and (2) from the corresponding signal-flow diagrams.   ∎

Series–Series Feedback

Figure 7.3.5 illustrates a **series–series feedback** configuration in which the feedback signal is derived from the output current rather than the output voltage. Because the output current I_o is being sampled, it is more appropriate to represent the output as a current

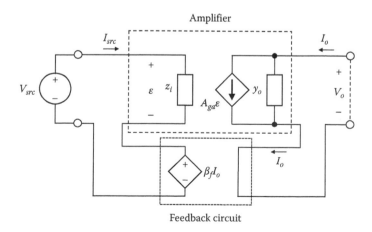

FIGURE 7.3.5
Ideal series–series, negative feedback configuration.

source $A_{ga}\varepsilon$ in parallel with an output admittance y_o, where g is added to the subscript to emphasize that A_{ga} has the dimensions of conductance. It is assumed that the output is short-circuited ($V_o = 0$), which is the ideal termination for a current drive. The series connection at the output implies that the same current I_o flows in the amplifier output and in the feedback circuit. Moreover, the series feedback circuit ideally presents a short circuit at the amplifier output so that the voltage across y_o is 0. From Kirchhoff's current law (KCL) at the output, $I_o = A_{ga}\varepsilon$. At the input, the feedback is series, as in the preceding case, so that $V_{src} = \varepsilon + \beta_f I_o$. Eliminating ε and rearranging,

$$\frac{I_o}{V_{src}} = \frac{A_{ga}}{1 + \beta_f A_{ga}} \tag{7.3.6}$$

$$\cong \frac{1}{\beta_f}, \quad \text{if } A_{ga} \to \infty \tag{7.3.7}$$

As in the preceding case, when A_{ga} is large, the overall gain I_o/V_{src} depends on the feedback circuit and not on the gain A_{ga}. The interpretation is that if $A_{ga} \to \infty$, then for any finite I_o, $\varepsilon \to 0$. KVL on the input side then gives $V_{src} = \beta_f I_o$ and Equation 7.3.7 follows.

To determine Z_{in}, KVL at the input is written as $V_{src} = z_i I_{src} + \beta_f I_o$. Substituting for I_o from Equation 7.3.6, which is based on $V_o = 0$, and rearranging,

$$\frac{V_{src}}{I_{src}} = Z_{in} = z_i(1 + \beta_f A_{ga}) \tag{7.3.8}$$

If $A_{ga} \to \infty$, then $Z_{in} \to \infty$. This is to be expected, because if $A_{ga} \to \infty$, then $\varepsilon \to 0$, which means $I_{src} \to 0$ and $Z_{in} \to \infty$. The input behaves as an open circuit.

To determine Y_{out}, a test source V_x is applied at the output, with $V_{src} = 0$ (Figure 7.3.6). $y_o V_x = I_x - A_{ga}\varepsilon$. Substituting $\varepsilon = -\beta_f I_x$ and rearranging,

$$\frac{I_x}{V_x} = Y_{out} = \frac{y_o}{1 + \beta_f A_{ga}} \tag{7.3.9}$$

As $A_{ga} \to \infty$, $Y_{out} \to 0$. Referring to Figure 7.3.5, the interpretation is that $\varepsilon \to 0$ when $A_{ga} \to \infty$, which gives $\beta_f I_o = 0$ when $V_{src} = 0$, independently of V_o. In other words, the

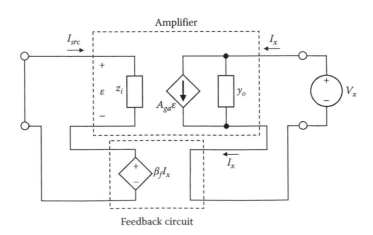

FIGURE 7.3.6
Application of a test voltage source V_x for determining the output admittance.

output behaves as an open circuit ($I_o = 0$). When A_{ga} is finite, with V_{src} constant, then if I_o increases, for example, in Figure 7.3.5, the feedback signal βI_o increases, which decreases ε and $A_{ga}\varepsilon$, and nearly restores I_o to its original value. The effective admittance at the output is therefore small; that is, the output appears as a current source. This is in accordance with the concept emphasized earlier that I_o, the output variable being sampled, is maintained practically constant by the negative feedback, for a constant V_{src}, despite load variations.

It is seen that a high-gain amplifier having $A_{ga} \to \infty$ in series–series feedback has $Z_{in} \to \infty$ and $Y_{out} \to 0$. The feedback amplifier is equivalent to the ideal transconductance amplifier of Table 7.2.1, with $G_m = 1/\beta_f$. When A_{ga} is finite, the effect of series–shunt feedback is to: (1) divide the gain of the amplifier by a factor $(1 + \beta_f A_{ga})$ (Equation 7.3.6), (2) multiply the input impedance by the factor $(1 + \beta_f A_{ga})$ (Equation 7.3.8), and (3) divide the output admittance by the factor $(1 + \beta_f A_{ga})$ (Equation 7.3.9).

Exercise 7.3.2

Derive Equations 7.3.6, 7.3.8, and 7.3.9 from corresponding signal-flow diagrams for the series–series feedback configuration. ∎

Exercise 7.3.3

Determine Y_{out} in Figure 7.3.6 as I_{sc}/V_{oc}. ∎

Shunt–Series Feedback

Figure 7.3.7 illustrates **shunt–series feedback**, in which the input is considered a current I_{src} and the feedback source $\beta_f I_o$ appears in shunt with the input. The source $\beta_f I_o$ subtracts from I_{src} so that the net input current to the amplifier proper is $\delta = I_{src} - \beta_f I_o$. At the output, with $V_o = 0$ as in the preceding case, $I_o = A_{ia}\delta$. Eliminating δ and rearranging,

$$\frac{I_o}{I_{src}} = \frac{A_{ia}}{1 + \beta_f A_{ia}} \tag{7.3.10}$$

$$\cong \frac{1}{\beta_f}, \quad \text{if } A_{ia} \to \infty \tag{7.3.11}$$

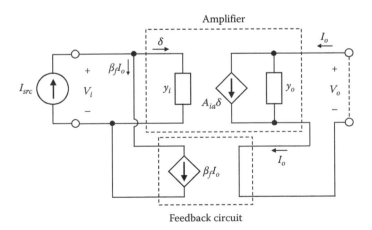

FIGURE 7.3.7
Ideal shunt–series, negative feedback configuration.

As in the preceding cases, when A_{ia} is large, the overall gain I_o/I_{src} depends on the feedback circuit and not on the gain A_{ia}. The interpretation is that if $A_{ia} \rightarrow \infty$, then for any finite I_o, $\delta \rightarrow 0$. KCL on the input side then gives $I_{src} = \beta_f I_o$ and Equation 7.3.10 follows.

To determine Y_{in}, KCL at the input is written as $I_{src} = y_i V_i + \beta_f I_o$. Substituting for I_o from Equation 7.3.10, which is based on $V_o = 0$, and rearranging,

$$\frac{I_{src}}{V_i} = Y_{in} = y_i(1 + \beta_f A_{ia}) \tag{7.3.12}$$

If $A_{ia} \rightarrow \infty$, then $Y_{in} \rightarrow \infty$. This is to be expected, because if $A_{ia} \rightarrow \infty$, then $\delta \rightarrow 0$, which means $V_i \rightarrow 0$ and $Y_{in} \rightarrow \infty$. The input behaves as a short circuit, which is the ideal termination for a current-source input.

If a test voltage source is applied at the output in Figure 7.3.7, but with the input open-circuited, $\delta = -\beta_f I_x$ and $I_x = y_o V_x + A_{ia} \delta$. This gives:

$$\frac{I_x}{V_x} = Y_{out} = \frac{y_o}{1 + \beta A_{ia}} \tag{7.3.13}$$

As $A_{ia} \rightarrow \infty$, $Y_{out} \rightarrow 0$, the interpretation being as explained previously.

It is seen that a high-gain amplifier having $A_{ia} \rightarrow \infty$ in shunt–series feedback has $Y_{in} \rightarrow \infty$ and $Y_{out} \rightarrow 0$. The feedback amplifier is equivalent to the ideal current amplifier of Table 7.2.1, with $R_m = 1/\beta_f$. When A_{ia} is finite, the effect of series–shunt feedback is to: (1) divide the gain of the amplifier by a factor $(1 + \beta_f A_{ia})$ (Equation 7.3.10), (2) multiply the input admittance by the factor $(1 + \beta_f A_{ia})$ (Equation 7.3.12), and (3) divide the output admittance by the factor $(1 + \beta_f A_{ia})$ (Equation 7.3.13).

Exercise 7.3.4

Derive Equations 7.3.10, 7.3.12, and 7.3.13 from corresponding signal-flow diagrams for the shunt–series feedback configuration.   ∎

Exercise 7.3.5

Determine Y_{out} in Figure 7.3.6 as I_{sc}/V_{oc}.   ∎

Shunt–Shunt Feedback

Finally, Figure 7.3.8 shows a **shunt–shunt feedback** configuration. With the same assumptions as before, KVL at the output gives $V_o = A_{ra} \delta$, with $I_o = 0$. At the input, $I_{src} = \beta_f V_o + \delta$. Substituting for δ and rearranging,

$$\frac{V_o}{I_{src}} = \frac{A_{ra}}{1 + \beta_f A_{ra}} \tag{7.3.14}$$

$$\cong \frac{1}{\beta_f}, \quad \text{if } A_{ra} \rightarrow \infty \tag{7.3.15}$$

As in the preceding cases, when A_{ra} is large, the overall gain V_o/I_{src} depends on the feedback circuit and not on the gain A_{ra}. The interpretation is that if $A_{ra} \rightarrow \infty$, then for any finite V_o, $\delta \rightarrow 0$. KCL on the input side then gives $I_{src} = \beta_f V_o$ and Equation 7.3.15 follows.

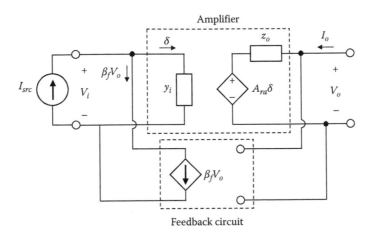

FIGURE 7.3.8
Ideal shunt–shunt, negative feedback configuration.

To determine Y_{in}, KCL at the input is written as $I_{src} = y_i V_i + \beta_f V_o$. Substituting for V_o from Equation 7.3.14, which is based on $I_o = 0$, and rearranging,

$$\frac{I_{src}}{V_i} = Y_{in} = y_i(1 + \beta_f A_{ra}) \tag{7.3.16}$$

If $A_{ra} \to \infty$, then $Y_{in} \to \infty$. This is to be expected, because if $A_{ra} \to \infty$, then $\delta \to 0$, which means $V_i \to 0$ and $Y_{in} \to \infty$. The input behaves as a short circuit.

If a test current source is applied at the output in Figure 7.3.8, with the input open-circuited, $\beta_f V_x = -\delta$ and $V_x = z_o I_x + A_{ra}\delta$. Eliminating δ and rearranging,

$$\frac{V_x}{I_x} = Z_{out} = \frac{z_o}{1 + \beta_f A_{ra}} \tag{7.3.17}$$

As $A_{ra} \to \infty$, $Z_{out} \to 0$, the interpretation being as explained previously.

It is seen that a high-gain amplifier having $A_{ra} \to \infty$ in shunt–shunt feedback has $Y_{in} \to \infty$ and $Z_{out} \to 0$. The feedback amplifier is equivalent to the ideal transresistance amplifier of Table 7.2.1, with $A_{ra} = 1/\beta_f$. When A_{ra} is finite, the effect of shunt–shunt feedback is to: (1) divide the gain of the amplifier by a factor $(1 + \beta_f A_{ra})$ (Equation 7.3.14), (2) multiply the input admittance by the factor $(1 + \beta_f A_{ra})$ (Equation 7.3.16), and (3) divide the output impedance by the factor $(1 + \beta_f A_{ra})$ (Equation 7.3.17).

Exercise 7.3.6

Derive Equations 7.3.14, 7.3.16, and 7.3.17 from corresponding signal-flow diagrams for the shunt–shunt feedback configuration. ∎

Exercise 7.3.7

Derive Z_{out} of the shunt–shunt feedback amplifier from Thevenin's equivalent circuit, as in Exercise 7.3.1. ∎

TABLE 7.3.1

Effect of Type of Feedback on Input and Output Impedances

Type of Feedback	Input Impedance	Output Impedance
Series–shunt	High	Low
Series–series	High	High
Shunt–series	Low	High
Shunt–shunt	Low	Low

The effect of the four negative feedback configurations on an amplifier can be summarized as follows:

Summary

1. *In all cases, negative feedback divides the gain of the amplifier by (1 – loop gain).*

2. *Series feedback, whether at the input or output, multiplies the corresponding impedance of the amplifier by (1 – loop gain).*

3. *Shunt feedback, whether at the input or output, divides the corresponding impedance of the amplifier by (1 – loop gain).*

4. *Shunt feedback at the output makes the output behave as a voltage source, whereas series feedback at the output makes the output behave as a current source.*

5. *Series feedback at the input is associated with voltage input and voltage feedback, whereas shunt feedback at the input is associated with current input and current feedback.*

The effects of the type of feedback on the input and output impedances are summarized qualitatively in Table 7.3.1.

7.4 Ideal Operational Amplifier*

Definition: *In an operational amplifier, or op amp, the output voltage is directly proportional to the difference between the voltages applied to its two input terminals.*

The op amp has an output terminal and two input terminals: a noninverting terminal denoted by the (+) sign, and an inverting terminal denoted by the (−) sign, as illustrated in the symbol for an op amp in Figure 7.4.1. According to the preceding definition,

$$v_O = A_v(v_P - v_N) = A_v\varepsilon \tag{7.4.1}$$

---

* Sabah 2008, pp. 714–723 and Example SE18.2, p. 756.

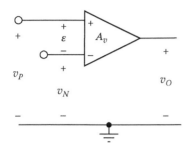

FIGURE 7.4.1
Operational amplifier.

where A_v is the **voltage gain** of the amplifier. The following should be noted concerning this definition:

1. Because the output is determined by the difference between the two inputs, the op amp is an example of a **differential amplifier**.

2. The two terminals are designated as noninverting and inverting, because v_O is of the same polarity as the input v_P applied to the noninverting terminal, and is of opposite polarity to the input v_N applied to the inverting terminal.

The ideal operational amplifier is a very useful idealization that can serve as an initial design step and which is also commonly invoked to illustrate some important circuit concepts. Many practical op amps approach the ideal in several respects, so the concept is not as farfetched as it may seem.

Concept: *An ideal op amp has the following properties*:

1. *The output is an ideal voltage source of voltage $v_O = A_v(v_P - v_N)$, with $A_v \to \infty$.*
2. *Like an ideal voltage source, the amplifier has zero output resistance and can deliver any output voltage or current, at any frequency.*
3. *Both inputs behave as open circuits.*
4. *The amplifier is free from any imperfections.*

The symbol for an ideal operational amplifier used in this book is that of Figure 7.4.1 but without indicating A_v. There are two basic op-amp configurations, noninverting and inverting, which are discussed next.

Noninverting Configuration

Consider an ideal op amp connected as shown in Figure 7.4.2. This is a series–shunt feedback configuration, where the feedback circuit is the voltage divider consisting of R_f and R_r connected in shunt across the output. The feedback signal taken from the output of

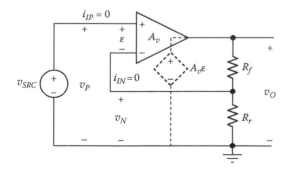

FIGURE 7.4.2
Noninverting operational amplifier.

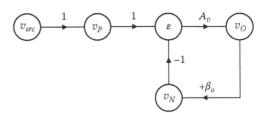

FIGURE 7.4.3
Signal-flow diagram for noninverting
operational amplifier.

the voltage divider is $\beta_o v_O$, where $\beta_o = R_r/(R_r + R_f)$ is the feedback factor, and the small o subscript refers to an operational amplifier. The feedback signal is applied to the inverting input so that it effectively subtracts from v_I applied to the noninverting input. It follows from Equation 7.3.2 that, with $A_v \to \infty$, that the overall voltage gain:

$$\frac{v_O}{v_{SRC}} = \frac{1}{\beta_o} = 1 + \frac{R_f}{R_r} \tag{7.4.2}$$

Because the sign of v_O is the same as that of v_{src}, the configuration is noninverting. To determine the voltage gain for finite A_v, we note that in Figure 7.4.2, $v_P = v_{SRC}$ and $v_N = \beta_o v_O$. Equation 7.4.1 is represented by the dependent source shown dotted. Substituting for v_P and v_N in Equation 7.4.1 and rearranging,

$$\frac{v_O}{v_{SRC}} = \frac{A_v}{1 + \beta_o A_v} \tag{7.4.3}$$

If $A_v \to \infty$, Equation 7.4.3 reduces to Equation 7.4.2. $\varepsilon = 0$ under these conditions, so that $v_P = v_N$. This is tantamount to having a **virtual short-circuit** between the two input terminals, without their being actually connected together. The fact that $v_P = v_N$ in an ideal op amp greatly simplifies the analysis of ideal op-amp circuits.

The voltage gain v_O/v_{SRC} cannot be less than unity. If $R_r = 1$ kΩ and $R_f = 4$ kΩ, for example, $\beta_o = 1/5$ and $v_O/v_I = 5$. A_v for a general purpose, low-frequency op amp is typically about 10^5. Substituting for β and A_v in Equation 7.4.3 gives $v_O/v_{SRC} = 4.99975$.

The signal-flow diagram for the noninverting configuration is illustrated in Figure 7.4.3. The relations between the variables conform to those of the circuit of Figure 7.4.2. Applying Rule 7.3.5 to the signal-flow diagram gives Equation 7.4.3.

Exercise 7.4.1

Determine V in Figure 7.4.4 so that no current flows in the 2 kΩ resistor.
    Answer: 10 V.   ∎

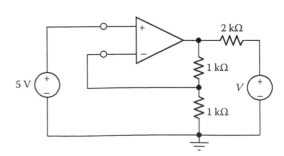

FIGURE 7.4.4
Figure for Exercise 7.4.1.

Unity-Gain amplifier

An important special case of the noninverting configuration is the **unity-gain amplifier**, also known as a **voltage follower**, in which $R_f = 0$, so that the output is directly connected to the inverting input (Figure 7.4.5). R_r across the output of the op amp becomes redundant and is removed. With $\beta_o = 1$, the overall gain is very nearly unity when A_v is high (Equation 7.4.3).

The unity-gain amplifier is extremely useful for *isolating* a source from a load. Isolation is desirable in many cases for the following reasons:

1. When the load resistance is much smaller than the source resistance, the signal at the load is considerably attenuated, and the power delivered to the load is limited. In Figure 7.4.6a, for example, the voltage v_O across the load is only $6 \times 0.1/1.2 = 0.5\,\mathrm{V}$ compared to an open-circuit source voltage of 6 V, and the power delivered to the load is 2.5 mW. With a unity-gain amplifier connected between the source and load (Figure 7.4.6b), the source current is ideally zero, independently of load variations. The voltage across the load is now 6 V, because the amplifier appears as a voltage source, and the power delivered to the load is 360 mW, assuming that the op amp is capable of delivering the required power.

2. Changes in the load do not affect the source current and hence the source voltage.

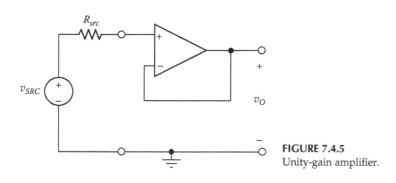

FIGURE 7.4.5
Unity-gain amplifier.

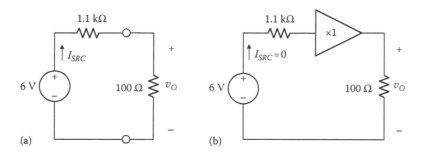

FIGURE 7.4.6
Effect of isolation of a load whose resistance is much smaller than that of the source. The load is connected to the source directly in (a) and through a unity-gain amplifier in (b).

3. In case of mismatch between the frequency dependence of the source and load impedances, the voltage across the load varies with frequency. Connecting a unity-gain amplifier having a high input impedance and a low output impedance overcomes this problem and makes the voltage across the load virtually independent of frequency.

Inverting Configuration

Figure 7.4.7a illustrates an op amp shunt–shunt feedback configuration. The noninverting input is connected to the common reference, which is usually grounded. With $v_P = 0$ and $\varepsilon_n = -\varepsilon = v_N$, Equation 7.4.1 gives $v_O = -A_v\varepsilon_n$. As $A_v \to \infty$, $\varepsilon_n \to 0$, which makes the inverting input a **virtual ground**: it is at ground potential without being actually grounded. Because of the virtual ground, no current flows in R_r and $i_r = i_{SRC}$. Connecting the output to the inverting input gives negative feedback, the feedback current i_f, being equal to $-v_O/R_f$, is proportional to v_O in accordance with shunt feedback at the output. Assume, for the moment, a finite input resistance between the inverting and noninverting inputs so that $i_{iN} \neq 0$. If v_O increases in the positive direction, i_f decreases, which increases i_{iN} and hence ε_n. This will make v_O more negative, thereby opposing the increase in v_O, because of the negative feedback. Note that when the op amp is ideal, the circuit of Figure 7.4.7a transforms a nonideal current source to an ideal voltage source at the op-amp output, the current source being ideally terminated in a short circuit.

Transforming the current source in parallel with R_r to a voltage source v_{SRC} in series with R_r, transforms the shunt–shunt transresistance amplifier to a voltage amplifier (Figure 7.4.7b). The input current is $i_r = v_{SRC}/R_r$ and is forced to flow through R_f, because the inverting input behaves as an open circuit. It follows that $i_r = i_f = -v_O/R_f$. Eliminating i_r between the two preceding relations,

$$\frac{v_O}{v_{SRC}} = -\frac{R_f}{R_r} \tag{7.4.4}$$

Because the sign of v_O is opposite to that of v_I, the configuration is inverting.

To analyze the circuit when A_v is finite, KVL at the input gives $v_{SRC} = R_r i_r + \varepsilon_n$. From KVL at the output: $\varepsilon_n = R_f i_r + v_O$. For the op amp, $v_O = -A_v\varepsilon_n$. Eliminating ε_n and i_r between these three equations and rearranging,

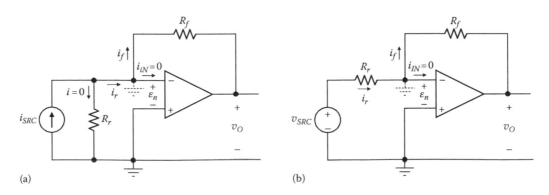

(a)   (b)

FIGURE 7.4.7
(a) Shunt–shunt operational amplifier connection; (b) inverting operational amplifier.

$$\frac{v_O}{v_{SRC}} = -\frac{R_f}{R_r}\frac{\beta_o A_v}{1+\beta_o A_v}$$ (7.4.5)

where $\beta_o = R_r/(R_r + R_f)$, as for the noninverting configuration. If $A_v \to \infty$, Equation 7.4.5 reduces to Equation 7.4.4.

An alternative and useful way of looking at the inverting configuration is to consider that a fraction of the input and a fraction of the output are superposed at the inverting input of the op amp. To determine the contribution ε_{n1} at the amplifier input due to v_{SRC} acting alone, we set $v_O=0$ by imagining the op amp to be modified internally so that $A_v=0$. This effectively connects the output end of R_f to the common reference through the short circuit of the ideal dependent voltage source (Figure 7.4.8a). It follows that $\varepsilon_{n1} = v_{SRC}\,R_f/(R_r + R_f)$. Next we imagine R_f to be disconnected from the amplifier output and connected to a voltage source v_O, with $v_{SRC}=0$ (Figure 7.4.8b). This gives $\varepsilon_{n2} = v_O\,R_r/(R_r + R_f) = \beta_o v_O$. In normal operation,

$$\varepsilon_n = \varepsilon_{n1} + \varepsilon_{n2} = \frac{R_f}{R_r + R_f}v_{SRC} + \frac{R_r}{R_r + R_f}v_O$$ (7.4.6)

In the ideal case, $\varepsilon_n=0$, and Equation 7.4.6 reduces to Equation 7.4.4.

Equation 7.4.6 can be represented on a signal-flow diagram as in Figure 7.4.9a. Note that the forward transmission $-A_v$ is negative. But the sign of the feedback path, from the output to the inverting input of the op amp is positive, so that the loop gain $-\beta_f A_v$, is negative, as it should be for a negative feedback system. Applying Rule 7.3.5 to this diagram gives Equation 7.4.5. We can make this signal-flow diagram formally look more like that of the noninverting amplifier, by changing ε_n to $\varepsilon=-\varepsilon_n$. This negates all the multiplying factors in Figure 7.4.9a leading to the signal-flow diagram of Figure 7.4.9b. The relation between v_O and v_{SRC} remains unchanged.

The signal-flow diagrams of the two configurations, as depicted in Figures 7.4.3 and 7.4.9b, respectively, differ only in the forward transmission from V_{SRC} to ε, this being unity in the noninverting configuration and $-R_f/(R_r + R_f)$ in the inverting configuration.

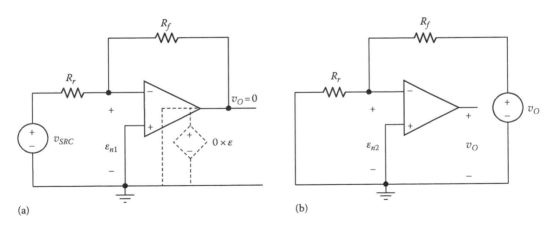

(a) (b)

FIGURE 7.4.8
Applying superposition to the inverting operational amplifier: (a) output set to zero; (b) input source set to zero.

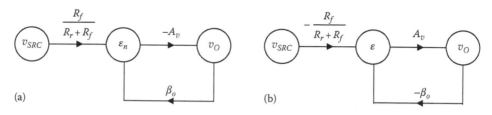

FIGURE 7.4.9
(a) Signal-flow diagram of the inverting configuration; (b) the signal-flow diagram with ε_n negated.

Exercise 7.4.2

Determine i_O in Figure 7.4.10.
   Answer: 2.5 mA.   ∎

Exercise 7.4.3

Show that in Figure 7.4.11, $i_L = -\dfrac{v_{SRC}}{R_{src}}\left(1 + \dfrac{R_1}{R_2}\right)$. Note that i_L is in this case larger than the current drawn from the source and is independent of R_L.   ∎

Exercise 7.4.4

Show that in Figure 7.4.12, $i_L = v_{SRC}/R$. Note that the voltage source is ideally terminated with an open circuit.   ∎

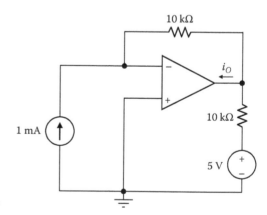

FIGURE 7.4.10
Figure for Exercise 7.4.2.

FIGURE 7.4.11
Figure for Exercise 7.4.3.

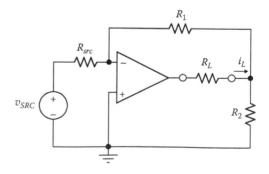

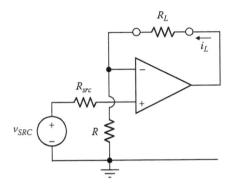

FIGURE 7.4.12
Figure for Exercise 7.4.4.

The fact that the inverting input of the op amp in the inverting configuration is a virtual ground is highly important and leads to many useful applications, the underlying concept being:

Concept: *When the inverting input is a virtual ground, the input current is determined solely by the input circuit. The input current is forced to flow through the circuit element connected between the inverting input and the output, irrespective of the value or nature of this element. The output voltage is then the negation of the voltage drop across the feedback element.*

In other words, the circuit element connected between the inverting terminal and output sees a current source whose value is the input current. This leads to a number of useful applications, such as adders, integrators, and differentiators, which are considered in detail in Chapter 10, taking into account op-amp imperfections. We will consider here integrators and differentiators based on op amps that are ideal except for finite gain. This will further illustrate some feedback principles that are needed in a later discussion.

Integrator

The basic circuit of an op-amp integrator is shown in Figure 7.4.13a and its signal-flow diagram in Figure 7.4.13b, based on Figure 7.4.9a. The relations for the integrator are

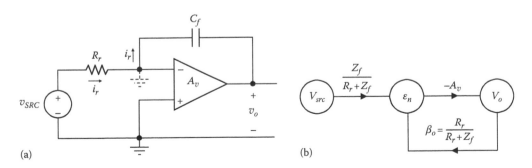

(a)   (b)

FIGURE 7.4.13
(a) Ideal op-amp integrator and (b) signal-flow diagram.

derived from those for the inverting amplifier of Figure 7.4.6 by replacing R_f with $Z_f = 1/j\omega C_f$. Equation 7.4.5 becomes, in Laplace transform notation:

$$\frac{V_o}{V_{src}} = -\frac{Z_f}{R_r}\frac{\beta_o A_v}{1 + \beta_o A_v} \tag{7.4.7}$$

where $\beta_o = R_r/(R_r + Z_f) = j\omega C_f R_r/(1 + j\omega C_f R_r)$. If $\beta_o A_v \to \infty$, Equation 7.4.5 reduces to:

$$\frac{V_o}{V_{src}} = -\frac{Z_f}{R_r} = -\frac{1}{j\omega C_f R_r} \tag{7.4.8}$$

which represents perfect integration in the frequency domain, with phase inversion, equivalent to $v_O = -\dfrac{1}{C_f R_r}\displaystyle\int v_{SRC}dt$ in the time domain. This relation readily follows from Figure 7.4.13a by writing $i_r = \dfrac{v_{SRC}}{R_r}$, $v_O = -\dfrac{1}{C_f}\displaystyle\int i_r dt$, and substituting for i_r.

If A_v is finite, $|\beta_o A_v| = A_v \omega C_f R_r \big/ \sqrt{1 + (\omega C_f R_r)^2}$. As long as $|\beta_o A_v| \gg 1$ in Equation 7.4.7, integration is near-perfect. At a sufficiently low ω, however, $|\beta_o A_v|$ becomes small enough to introduce appreciable error. Hence,

Concept: *Near-perfect integration down to very low frequencies requires very high values of amplifier gain.*

For example, if a low-frequency square wave is integrated to give a triangular wave, insufficient amplifier gain will make the sides of the triangular wave exponential rather than linear. Practical integrators are analyzed in Section 10.3, Chapter 10.

Differentiator

The basic circuit of an op-amp differentiator is shown in Figure 7.4.14a and its signal-flow diagram in Figure 7.4.14b, based on Figure 7.4.9a. The relations for the differentiator are derived from those for the inverting amplifier of Figure 7.4.6 by replacing R_r with $Z_r = 1/j\omega C_r$. Equation 7.4.5 becomes:

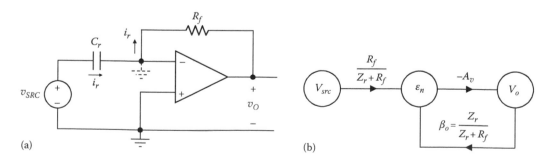

(a)   (b)

FIGURE 7.4.14
(a) Ideal op-amp differentiator, and (b) signal-flow diagram.

$$\frac{v_O}{v_{SRC}} = -\frac{R_f}{Z_r}\frac{\beta_o A_v}{1 + \beta_o A_v} \tag{7.4.9}$$

where $\beta_o = Z_r/(Z_r + R_f) = 1/(1 + j\omega C_f R_r)$. If $\beta_o A_v \to \infty$, Equation 7.4.9 reduces to:

$$\frac{v_O}{v_{SRC}} = -\frac{R_f}{Z_r} = -j\omega C_r R_f, \tag{7.4.10}$$

which represents perfect differentiation in the frequency domain, with phase inversion, equivalent to $v_O = -C_r R_f dv_{SRC}/dt$ in the time domain. This relation readily follows from Figure 7.4.14a by writing $i_r = C_r dv_{SRC}/dt$, $v_O = -R_f i_r$, and substituting for i_r.

If A_v is finite, $|\beta_o A_v| = A_v \Big/ \sqrt{1 + (\omega C_r R_f)^2}$. As long as $|\beta_o A_v| \gg 1$ in Equation 7.4.9, differentiation is near-perfect. At a sufficiently high ω, however, $|\beta_o A_v|$ becomes small enough to introduce appreciable error. Hence,

Concept: *Near-perfect differentiation up to very high frequencies requires very high values of amplifier gain.*

For example, if a square wave is differentiated, insufficient amplifier gain will give an initially sharp edge, coinciding with transitions of the square wave, followed by a slowly decaying exponential tail.

Exercise 7.4.5

Show that the circuit of Figure 7.4.15 is a noninverting integrator in which C_f is multiplied by the magnitude of the gain of the inverting amplifier. Note that the inverting input of the integrator is now a virtual ground. ∎

Example 7.4.1 Output and Input Impedances of Op-Amp Configurations

It is required to determine, for both the noninverting and inverting configurations: (a) the output resistance, assuming a finite output resistance of the op amp, and (b) the input resistance, assuming a finite input resistance of the op amp. The op amp is considered to have a finite gain in both cases.

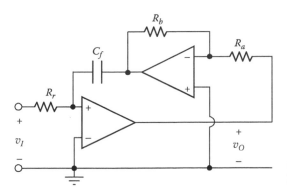

FIGURE 7.4.15
Figure for Exercise 7.4.5.

SOLUTION

(a) The output impedance of an op amp is essentially resistive and is denoted by R_o. Since the feedback is shunt type at the output in both configurations, the effect of the feedback on R_o is the same. This effect can be determined from the signal-flow diagram of Figure 7.4.16, which applies to both configurations, the only difference being in the value of k, as follows from Figures 7.4.3 and 7.4.9b.

If a test source I_x is applied at the output, as in Figure 7.3.2, KVL gives $V_x = A_v \varepsilon + R_o I_x$, for either op-amp configuration. V_x is related to I_x by R_o, as in Figure 7.3.4b. Applying Rule 7.3.5, $V_x = R_o I_x / (1 + \beta_o A_v)$. The effective output resistance in the presence of feedback is:

$$\frac{V_x}{I_x} = R_{out} = \frac{R_o}{1 + \beta_o A_v} \tag{7.4.11}$$

If $A_v \rightarrow \infty$, then $R_{out} \rightarrow 0$, as it should for shunt feedback at the output.

(b) Let the differential input impedance of the noninverting configuration be Z_i (Figure 7.4.17a). Assuming $Z_i \gg (R_f \| R_r)$, as is normally the case, the effect of I_{src} on the $R_r R_f$ voltage divider can be neglected. Hence, $\varepsilon = V_{src} - \beta_o V_o$, where $\varepsilon = Z_i I_{src}$, as depicted in the signal-flow diagram of Figure 7.4.17b, which is essentially the same as that of Figure 7.3.4a. It follows that $I_{src} = \frac{1}{Z_i} V_{src} \times \frac{1}{1 + \beta_o A_v}$, which gives $\frac{V_{src}}{I_{src}} = Z_{in} = Z_i (1 + \beta_o A_v)$. If $A_v \rightarrow \infty$, then $Z_{in} \rightarrow \infty$, as it should for series-feedback at the input.

For the inverting configuration, it follows from Figure 7.4.7a and b with Z_i connected at the input of the op amp, that Equation 7.4.6 is modified to $\varepsilon_n = R'_f V_{src} + \beta'_o V_o$, where

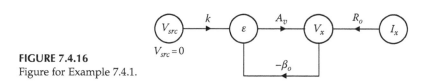

FIGURE 7.4.16
Figure for Example 7.4.1.

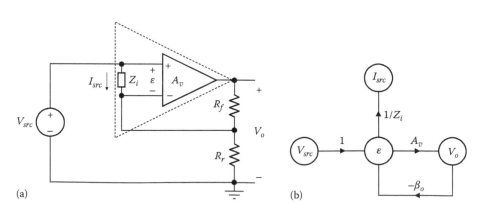

(a)   (b)

FIGURE 7.4.17
Figure for Example 7.4.1.

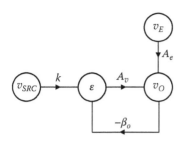

FIGURE 7.4.18
Signal-flow diagram for the inverting or noninverting configuration including an extraneous signal v_E.

$$R'_f = \frac{R_f \| Z_i}{R_r + R_f \| Z_i} \quad \text{and} \quad \beta'_o = \frac{R_r \| Z_i}{R_f + R_r \| Z_i}.$$ From Figure 7.4.6b, $V_{src} = \varepsilon_n + R_r I_r$ and $V_o = -A_v \varepsilon_n$. Substituting for V_o, eliminating ε_n, and rearranging,

$$\frac{V_{src}}{I_r} = Z_{in} = \frac{R_r}{1 - R'_f / (1 + \beta'_o A_v)} \qquad (7.4.12)$$

If $A_v \to \infty$, $Z_{in} = R_r$, as expected from the virtual ground at the inverting terminal.

Extraneous Signals

These signals may be due, for example, to: (1) interference from the surroundings, (2) nonlinear distortion, which generates harmonic frequencies in the amplifier, or (3) improper filtering of the rectified ac supply that provides the dc bias supplies of the op amp. Let us assume that an extraneous signal v_E appears at a point where the gain to the output is A_e (Figure 7.4.18). In this figure, $k = 1$ for the noninverting configuration and $k = -R_f / (R_r + R_f)$ for the inverting configuration. Applying superposition and Rule 7.3.5,

$$v_O = \frac{kA_v}{1 + \beta_o A_v} v_{SRC} + \frac{A_e}{1 + \beta_o A_v} v_E \qquad (7.4.13)$$

The extraneous signal at the output is A_e / kA_v relative to the input signal. Hence,

Concept: *In a negative feedback system, both the input and extraneous signals are attenuated by (1 − loop gain). The closer the source of the extraneous signal is to the output, the smaller is the gain A_e relative to A_v, and the smaller is the extraneous signal at the output, relative to the input signal.*

Simulation Example 7.4.2 Extraneous Signal

It is required to simulate the effect of an extraneous signal on the output of an op-amp feedback amplifier.

SIMULATION

The schematic circuit is that of the noninverting configuration with a 1 V, 1 kHz signal applied to the noninverting input and a 1 V, 10 kHz extraneous signal that appears as a perturbation of the output voltage of the op amp (Figure 7.4.19). PSpice has an "ideal op amp" part that is listed as OPAMP in the ANALOG library. The op amp has a default gain of 10^6 and default dc bias supplies of ± 15 V, both of which can be changed in the Property Editor spreadsheet.

Figure 7.4.20 shows the output of the op amp over a period of 0–5 ms. The 1 V, 1 kHz input appears at the output with a gain of 2, as expected. Since $\beta_o = 1/2$, and (1 − loop gain) is $(1 + 10^6/2)$, the 1 V, 10 kHz signal is attenuated by this factor so that its amplitude will be nearly 2 μV and is not visible in the figure.

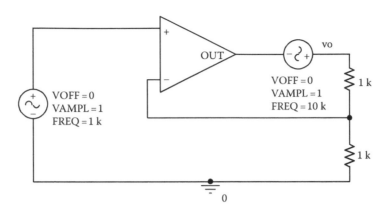

FIGURE 7.4.19
Figure for Simulation Example 7.4.2.

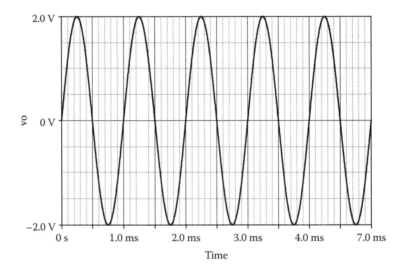

FIGURE 7.4.20
Figure for Simulation Example 7.4.2.

Gain-Bandwidth Product

Assume that A_v of the op amp is frequency dependent and is given by:

$$A_v(f) = \frac{A_{v0}}{1 + jf/f_c} \qquad (7.4.14)$$

where
  f_c is the 3-dB cutoff frequency
  A_{v0} is the gain at low frequencies

The asymptotic magnitude Bode plot (Appendix SC) is shown in Figure 7.4.21 as a plot of $20 \log_{10}|A_v(f)|$ in dB against $\log_{10} f$. The **gain-bandwidth product** (GB product),

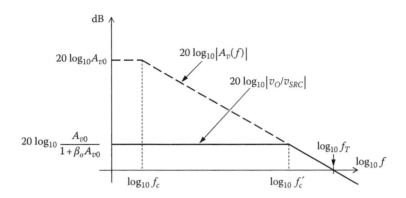

FIGURE 7.4.21
The effect of negative feedback on frequency response.

considered as $A_{v0}f_c$ (Exercise 7.4.7), is an important figure of merit of an amplifier; the larger this product, the better is the performance of the amplifier.

For the noninverting configuration, substituting for $A_v(f)$ from Equation 7.4.14,

$$\frac{v_O}{v_{SRC}} = \frac{A_v(f)}{1 + \beta_o A_v(f)} = \frac{A_{v0}}{1 + \beta_o A_{v0} + jf/f_c} = \frac{A_{v0}}{1 + \beta_o A_{v0}} \frac{1}{1 + jf/f_c(1 + \beta_o A_{v0})} \quad (7.4.15)$$

It is seen from Equation 7.4.15 that the 3-dB cutoff frequency is multiplied by the factor $(1 + \beta_o)$ to become $f_c' = f_c(1 + \beta_o A_{v0})$. In other words, the bandwidth is multiplied by $(1 + \beta_o A_{v0})$, but the dc gain A_{v0} is divided by the same factor and becomes $A_{v0}/(1 + \beta_o A_{v0})$. The magnitude of the gain $\left|\dfrac{A_v(f)}{1 + \beta_o A_v(f)}\right|$ is nearly $1/\beta_o$, independent of frequency as long as $|\beta_o A_{v0}| \gg 1$. The gain therefore remains substantially constant at $1/\beta_o$ as long as the magnitude of the loop gain is considerably larger than unity. In the asymptotic plot, the gain is shown as essentially $1/\beta_o$ all the way to f_c'. It must be kept in mind, however, that for the first-order response under consideration, the gain at the corner frequency f_c' is actually 3 dB less than its value at low frequencies, or $1/\sqrt{2}$ of this value. Note that for high frequencies both Equations 7.4.14 and 7.4.15 reduce to $\dfrac{A_{v0}}{jf/f_c}$, which means that both responses have the same high-frequency asymptote.

The GB in the presence of negative feedback is $\dfrac{A_{v0}}{1 + \beta_o A_{v0}} \times (1 + \beta_o A_{v0})f_c = A_{v0}f_c$, which is the same as for the op amp without the feedback. Thus,

Concept: *Negative feedback trades gain for bandwidth, without improving the GB product of an amplifier, which is an inherent characteristic of the amplifier.*

The GB product of a given amplifier circuit can be less than the theoretical maximum but cannot exceed this maximum (Exercise 7.4.8).

Exercise 7.4.6

Consider an op amp having $A_{v0} = 10^5$ and $f_c = 10$ Hz. What is the minimum β_o that makes the 3-dB cutoff frequency of the feedback amplifier at least 100 kHz?
    Answer: 0.1.   ■

Exercise 7.4.7

Show that the unity-gain frequency f_T in Figure 9.2.3 is given by $f_T = f_c \sqrt{A_{v0}^2 - 1} \cong A_{v0} f_c$, as to be expected from the GB product, since at unity gain, the bandwidth is expected to be $A_{v0} f_c$. If an op amp has $A_{v0} = 10^5$ and $f_T = 1$ MHz, determine f_c.
    Answer: 10 Hz.   ■

Exercise 7.4.8

Show that for the inverting configuration, the gain bandwidth product is $|k| A_{v0} f_c$, where $k = -R_f/(R_r + R_f)$.   ■

Simulation Example 7.4.3 Frequency Response

It is required to simulate the effect of a finite frequency response of an op amp, as well as the effect of feedback, on the output impedance of an op-amp feedback amplifier.

SIMULATION

An OPAMP from the ANALOG library is used in an inverting configuration having a gain of −1 (Figure 7.4.22). An *RC* lowpass circuit of 10 rad/s corner frequency is connected at the output of the op amp, and a 1 kΩ load connected at the output of the feedback amplifier. The gain of the op amp is the default value of 10^6.

    The frequency response, shown in Figure 7.4.23, gives a 3-dB cutoff frequency of 0.794 MHz. The 3-dB cutoff frequency of the *RC* circuit is 1.59 Hz. Multiplied by $(1 - \text{loop gain}) = \left(1 + \dfrac{10^6}{2}\right)$ gives a corner frequency of 0.796 MHz. The gain is −1 at low frequencies despite the 1 kΩ load and the 100 kΩ resistance in series with the output of the op amp, which means that the feedback amplifier has a low output resistance. The

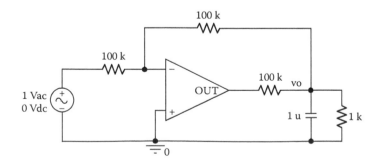

FIGURE 7.4.22
Figure for Simulation Example 7.4.3.

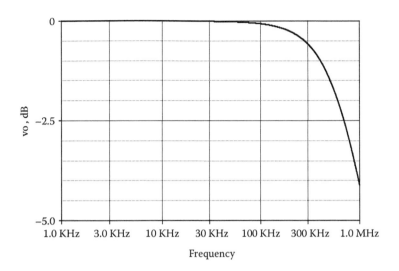

FIGURE 7.4.23
Figure for Simulation Example 7.4.3.

100 kΩ divided by (1 − loop gain) is only 0.2 Ω. This may be ascertained by replacing the input source by a short circuit and applying a 1 V VAC source at the output. If this source is denoted by V2, a plot of −V(V2:+)/I(V2) gives a low-frequency value of 0.2 Ω.

Summary of Main Concepts and Results

- A two-port circuit can be specified in terms of one of six sets of two simultaneous equations, each of which involves four parameters. In general, these parameters are nonzero and independent.

- The z-parameter, y-parameter, h-parameter, and g-parameter equations can be represented in terms of equivalent circuits in which forward and reverse transmissions are described by dependent sources on the output and input sides, respectively.

- The four ideal types of amplifiers are: ideal voltage, current, transresistance, and transconductance amplifiers. They follow from the g-, h-, z-, and y-parameter equations, respectively, by setting to zero all the parameters except the "21" parameters.

- An amplifier circuit can be made to approach any of the aforementioned ideal amplifier types under conditions of very high gain of the amplifier proper, a negligible reverse transmission, an appropriate negative feedback connection, and a negligible loading effect of the feedback circuit on the amplifier.

- In a high-gain amplifier with negative feedback, the overall gain of the feedback amplifier is determined by the feedback circuit.

- In a high-gain amplifier with negative feedback, the output variable that is sampled by the feedback circuit is maintained practically constant for a constant input, despite load variations.

- An ideal operational amplifier is an ideal voltage amplifier having a differential input.

- A unity-gain op amp is useful for isolation purposes.
- Near-perfect integration down to very low frequencies requires very high values of amplifier gain, whereas near-perfect differentiation up to very high frequencies requires very high values of amplifier gain.
- Negative feedback reduces distortion and increases the 3-dB cutoff frequency without affecting the maximum gain-bandwidth product of the op amp.

Learning Outcomes

- Derive the parameters of a two-port circuit.
- Articulate the benefits of negative feedback.
- Analyze ideal op-amp circuits.

Supplementary Topics and Examples on Website

ST7.1 Relations between Two-Port Parameters. Derives the relations between the six sets of two-port circuit parameters (Sabah 2008, pp. 521–524).

ST7.2 Two-Port Parameters of Linear and Ideal Transformers. Derives the two-port parameters for: (1) a linear transformer, including the case of perfect coupling, and (2) an ideal transformer. It is shown that in the former case the y-parameters do not exist, whereas in the latter case the z- and y-parameters do not exist (Sabah 2008, Example SE14.1, p. 554).

ST7.3 Ideal Transformer and Gyrator. Shows how consideration of lossless, two-port circuits that do not store energy naturally leads to the circuit characterization of an ideal transformer as well as another important device of theoretical and practical interest, namely, the **gyrator**, which converts capacitance to inductance, and conversely (Sabah 2008, Section ST14.5, p. 554).

Problems and Exercises

P7.1 Two-Port Circuits

P7.1.1 Determine the z- and y-parameters of the circuit of Figure P7.1.1 and verify matrix inversion (Equation 7.1.1). Note that $z_{12} = z_{21}$ and $y_{12} = y_{21}$. This defines a **reciprocal** two-port circuit that does not contain any dependent sources.

P7.1.2 Repeat Problem P7.1.1 for the h- and g-parameters. Note that reciprocity now implies that $h_{12} = -h_{21}$ and $g_{12} = -g_{21}$.

P7.1.3 Repeat Problem P7.1.1 for the a- and b-parameters. Note that reciprocity now implies that $\Delta a = a_{11}a_{22} - a_{12}a_{21} = 1$ and $\Delta b = b_{11}b_{22} - b_{12}b_{21} = 1$.

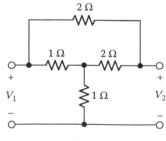

FIGURE P7.1.1

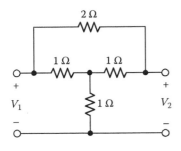

FIGURE P7.1.4

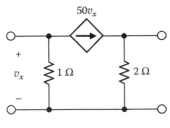

FIGURE P7.1.7

P7.1.4 Determine the z- and y-parameters of the circuit of Figure P7.1.4 and verify matrix inversion (Equation 7.1.1). Note that the circuit is now **symmetric**, that is, the input and output ports can be interchanged without affecting the voltage–current relations. In addition to the reciprocity relations, $z_{11} = z_{22}$ and $y_{11} = y_{22}$.

P7.1.5 Repeat Problem P7.1.4 for the h- and g-parameters. Note that symmetry now implies, in addition to the reciprocity relations, that $\Delta h = h_{11}h_{22} - h_{12}h_{21} = 1$ and $\Delta g = g_{11}g_{22} - g_{12}g_{21} = 1$.

P7.1.6 Repeat Problem P7.1.4 for the a- and b-parameters. Note that symmetry now implies, in addition to the reciprocity relations that $a_{11} = a_{22}$ and $b_{11} = b_{22}$. Verify these reciprocity relations from Table ST7.1.1.

P7.1.7 Determine the z- and y-parameters of the circuit of Figure P7.1.7. Note that the circuit is neither reciprocal nor symmetric.

P7.1.8 Repeat Problem P7.1.7 for the h- and g-parameters.

P7.1.9 Repeat Problem P7.1.7 for the a- and b-parameters.

P7.1.10 Determine the h-parameters of the high-frequency equivalent circuit of a bipolar junction transistor shown in Figure P7.1.10.

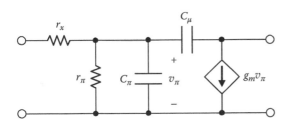

FIGURE P7.1.10

P7.2 Ideal Op-Amp Circuits

P7.2.1 If $v_{SRC} = 10$ V in Figure P7.2.1, determine R so that the current in R_L is 1 mA. Note that this current is independent of R_L, as long as the op-amp is not in saturation, which means that R_L sees a current source. If the output saturates at 15 V, what is the maximum R_L that can be used?

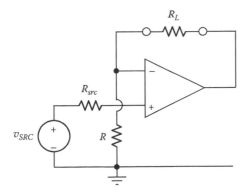

FIGURE P7.2.1

P7.2.2 In Figure P7.2.2 a T-circuit is used in the feedback path of the inverting configuration. Determine the voltage gain v_O/v_{SRC}. Note that, by providing a shunt path to ground through the 100 Ω resistor, the T-circuit simulates a large feedback resistor.

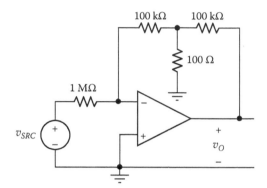

FIGURE P7.2.2

P7.2.3 Determine v_O in Figure P7.2.3 as a function of v_1, v_2, and v_3.

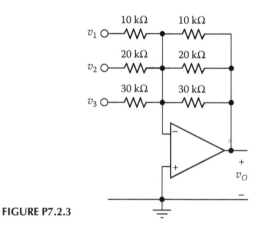

FIGURE P7.2.3

P7.2.4 In Figure P7.2.4, α is the fraction of the total resistance R_p of the potentiometer that appears between terminal a of the potentiometer and its slider. Show that $C_{in} = C/\alpha$.

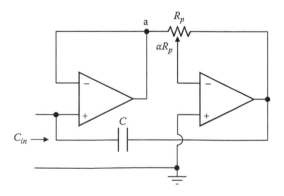

FIGURE P7.2.4

P7.2.5 Determine i_O in Figure P7.2.5. Simulate with PSpice.

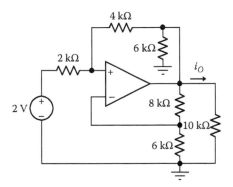

FIGURE P7.2.5

P7.2.6 Determine I_L in Figure P7.2.6.

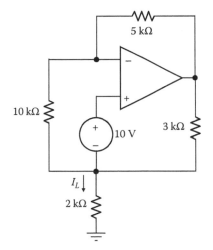

FIGURE P7.2.6

P7.2.7 Determine R_{in} in Figure P7.2.7.

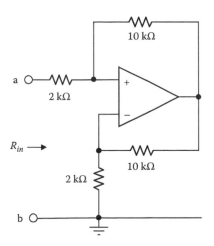

FIGURE P7.2.7

P7.2.8 Figure P7.2.8 illustrates a circuit that can be used with a strain gauge bridge. Show that $v_O = \dfrac{R}{R_o}\dfrac{\alpha}{(1+\alpha)(1+R_o/R)+1}V_{DC}$. Note that if $\dfrac{R_o}{R} \ll 1$ and $\alpha \ll 1$, then $v_O \cong V_{DC}\,\alpha R/2R_o$.

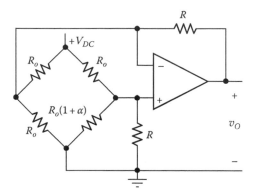

FIGURE P7.2.8

P7.2.9 Show that i_L in Figure P7.2.9 is given by $i_L = (v_2 - v_1)/R_2$. Note that the load sees a current source whose value is determined by the differential input $(v_2 - v_1)$.

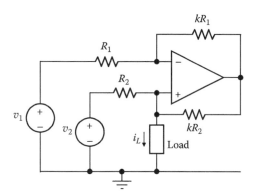

FIGURE P7.2.9

P7.2.10 Determine v_O in Figure P7.2.10. Simulate with PSpice.

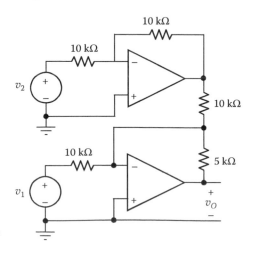

FIGURE P7.2.10

P7.2.11 Determine v_O in Figure P7.2.11. Simulate with PSpice.

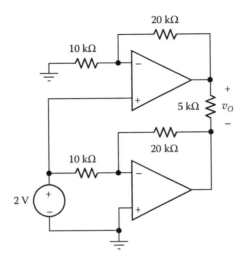

FIGURE P7.2.11

P7.2.12 Determine R in Figure P7.2.12 for zero I_{SRC}.

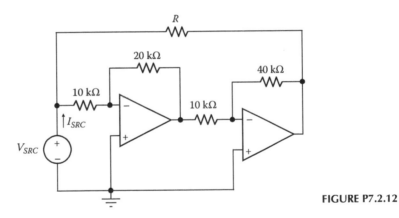

FIGURE P7.2.12

P7.2.13 Determine v_O in Figure 7.2.13. Simulate with PSpice.

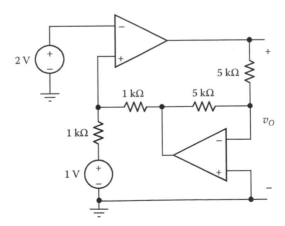

FIGURE P7.2.13

P7.2.14 Determine v_O in Figure P7.2.14. Simulate with PSpice.

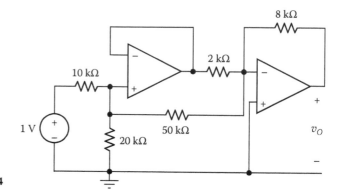

FIGURE P7.2.14

P7.2.15 In Figure P7.2.15, the switch is closed at $t=0$ with no initial energy storage in the capacitor and inductor. Determine v_O for $t \geq 0$. Simulate with PSpice.

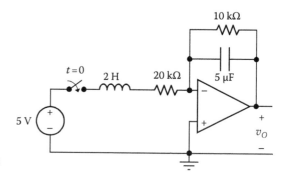

FIGURE P7.2.15

P7.2.16 Show that V_o in Figure P7.2.16 is given by $V_o = -\dfrac{1}{2}\dfrac{sC_fR_r}{sC_fR_r}\dfrac{sC_fR_f}{1+sC_fR_f}\dfrac{sC_rR_r}{1+sC_rR_r}V_{src}$. Note that if $C_fR_f = C_rR_r = \tau$, then $V_o = -\dfrac{1}{s\tau}\left(\dfrac{s\tau}{1+s\tau}\right)^2 V_{src}$.

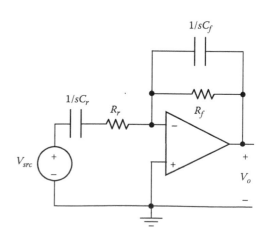

FIGURE P7.2.16

P7.2.17 The circuit of Figure P7.2.17 is referred to as a **charge amplifier** that can be used with capacitive transducers, where v_{src}, C_{src}, and R_{src} represent the transducer voltage, capacitance and resistance, respectively. R provides a dc path for the input bias current and should be large compared to the reactance of C at the lowest operating frequency.

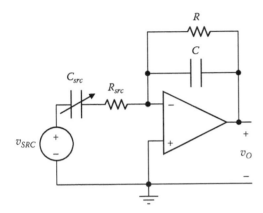

FIGURE P7.2.17

A change in either v_{src} or C_{src} changes the charge on C_{src}. This charge is transmitted to C, thereby causing a change in the output voltage. Ignoring R, show that $\Delta v_O = -\dfrac{\Delta C_{src}}{C} v_{SRC} - \dfrac{C_{src}}{C} \Delta v_{SRC}$. Note that this relation is not affected by a shunt capacitance at the virtual ground, which means that a long shielded cable can be used between the transducer and amplifier.

P7.2.18 Show that the circuit of Figure P7.2.18 acts as a gyrator (Section ST7.3). Simulate with PSpice.

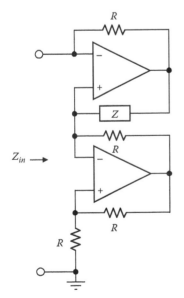

FIGURE P7.2.18

P7.2.19 Show that the circuit of Figure P7.2.19 acts as a gyrator (Section ST7.3). Simulate with PSpice.

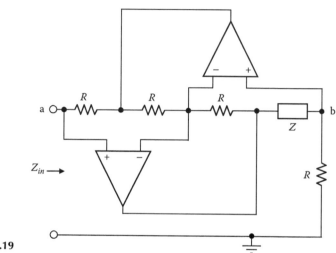

FIGURE P7.2.19

8

Single-Stage Transistor Amplifiers

After having considered the basic principles of operation of FETs and BJTs in the preceding chapters, the present chapter focuses on using transistors for amplifying small signals, with BJT and FET amplifiers integrated together in the discussion as much as possible. Amplification of large signals, as in the case of power amplifiers, is considered in Chapter 11.

In switching applications, the transistor is either cut off or fully turned on, whereas in amplifiers the quiescent operating point is chosen in the intermediate range of conduction on the basis of such requirements as maximum voltage gain, or high input resistance in case of BJTs, or small current drain from the power supply, which is particularly important in battery-operated equipment. After choosing the quiescent operating point, measures should be taken to ensure that this point is stabilized, that is, remains within specified limits despite variations in temperature, power supplies, or in transistor parameters because of inevitable manufacturing tolerances. The quiescent operating point is stabilized mainly by negative feedback in discrete amplifiers, and by some form of constant current source in IC amplifiers.

Depending on which of the three transistor terminals is common between input and output, three amplifier configurations are possible. Each of these configurations confers special advantages in terms of voltage gain, current gain, input and output resistances, and other features that are of importance in amplifier circuits. Of particular interest in amplifiers is the bandwidth, which is the frequency range of signals that can be amplified with a variation of gain within 3 dB. The reduced gain at high frequencies is fundamentally due to the effects associated with stored charges and to limitations on the speed of response of physical phenomena underlying transistor operation. An important measure of the performance of an amplifier at high frequencies is not the gain per se, but rather, the gain-bandwidth product.

Improved performance over that of a single-transistor amplifier can be achieved by using composite transistors consisting of two transistors that are directly connected together in a particular way. Several composite connections are considered and their salient features highlighted. Cascaded amplifying stages are discussed in the following chapter.

Learning Objectives

❖ To be familiar with:

• Terminology commonly used with amplifier circuits.

❖ To understand:

• The purpose and techniques of biasing BJTs and FETs in low-power amplifiers, both in discrete circuits and ICs.

- The essential features of the three basic amplifier configurations and the effects of an emitter/source resistor.
- The factors that determine the frequency response of an amplifier.
- How the three basic transistor amplifier configurations compare in terms of bandwidth and gain-bandwidth product.
- How some of the limitations of a single BJT or MOSFET can be overcome by using composite transistor connections, such as the CC-CE cascade, cascodes, Darlington pairs, and BiCMOS combinations.

8.1 Transistor Biasing

Concept: *The purpose of transistor biasing is to select an appropriate quiescent, or operating, point on the transistor output characteristics, and to stabilize this operating point despite variations in temperature, bias supplies, or transistor parameters, such as β_F or V_t, because of manufacturing tolerances. The stabilization entails minimizing the change in output current despite these variations.*

Biasing of Discrete Transistors

To appreciate the need for proper biasing, consider the basic CE amplifier of Figure 8.1.1a and the corresponding load-line construction on the output characteristics (Figure 8.1.1b). For best performance as an amplifier, the quiescent point Q is chosen so that: (1) the quiescent collector current I_{CQ} corresponds to a high value of β (Figure 7.1.9, Chapter 7), and (2) the distortion in the output signal is minimized for the desired magnitude of output. The latter condition implies that for small-signal amplification, the quiescent point should be located on the output characteristics where nonlinear distortion is small. Moreover, variations in temperature, bias supplies, or transistor parameters should keep

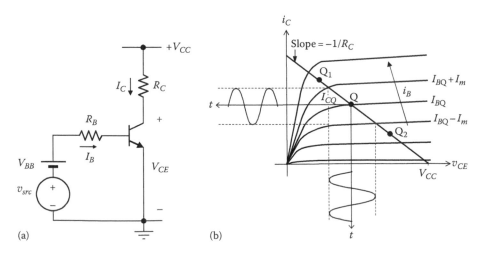

FIGURE 8.1.1
(a) Basic CE amplifier; (b) load-line construction on output characteristics.

the changes in the chosen quiescent point within acceptable limits. For example, if the operating point moves to Q_1 or Q_2 as in Figure 8.1.1b, where the output characteristics are not as equally spaced for equal changes in I_B as around Q, the output may be considerably distorted. The same applied signal can drive the transistor into the saturation region during part of the cycle (Q_1) or can cut it off (Q_2). If this happens, the signal will be severely distorted, the amplitude of the output signal reduced, and the gain decreased.

The quiescent operating point is stabilized in practice by keeping the quiescent output current, I_{CQ} in Figure 8.1.1, as nearly constant as possible. Ideally, this could be done by replacing R_C in Figure 8.1.1a by a current source I_{CQ}, which would have the additional advantage of making the load resistance for small-signal operation virtually infinite, thereby increasing the voltage gain. However, this may be precluded because of increased circuit complexity or because of the nature of the load. In practice, a nearly constant current is maintained in discrete transistor circuits using some form of negative feedback, whereas in ICs a current mirror circuit is used.

A biasing arrangement commonly used with discrete transistors is shown in Figure 8.1.2a. The voltage divider $R_1 R_2$ provides the required voltage bias at the base. R_E introduces series–series negative feedback that stabilizes the transistor output current I_C. Thus, an increase in I_C increases the voltage drop across R_E, which makes the emitter more positive with respect to the base, thereby reducing V_{BE} and opposing the increase in I_C. C is a **dc blocking capacitor** for dc isolation of the source from the amplifier input. If C is not included, the signal source, assumed to have a small source resistance, will act like a dc short circuit across R_2, which will make $V_{BE} \cong 0$ and cuts off the transistor. The value of C is made large enough so as not to significantly affect the circuit at the lowest signal frequency (Section 9.2, Chapter 9).

The dc analysis begins with converting the voltage divider in Figure 8.1.2a to its Thevenin's equivalent circuit (TEC, Figure 8.1.2b), where $V_{BB} = V_{CC} R_2 / (R_1 + R_2)$ and $R_B = R_1 \| R_2$. From KVL, $V_{BB} = R_B I_B + V_{BE} + R_E I_E$. Neglecting $I_{CBO}, I_B = I_C / \beta_F, I_E = I_C (\beta_F + 1)/\beta_F$, which gives:

$$I_C = \beta_F \frac{V_{BB} - V_{BE}}{R_E(\beta_F + 1) + R_B} \tag{8.1.1}$$

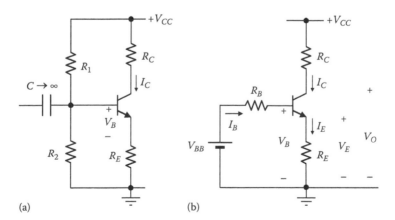

(a)   (b)

FIGURE 8.1.2
(a) Basic biasing circuit for discrete transistors; (b) Thevenin's equivalent circuit of voltage divider $R_1 R_2$.

According to Equation 8.1.1, I_C can change because of variation in V_{BE} or β_F. Having $V_{BE} \ll V_{BB}$ minimizes the effect of variation in V_{BE}. The effect of variation in β_F is minimized if:

$$R_E(\beta_F + 1) \gg R_B \tag{8.1.2}$$

This makes $I_C \cong \dfrac{\beta_F}{\beta_F + 1} \dfrac{V_{BB} - V_{BE}}{R_E}$, which is independent of β_F, for $\beta_F \gg 1$. In other words, when R_B is small, $I_E \cong (V_{BB} - V_{BE})/R_E$, independently of β_F, and $I_C = I_E\beta_F/(\beta_F + 1)$ is also independent of β_F, as long as β_F is large.

Exercise 8.1.1

Consider KVL in the base-emitter circuit of Figure 8.1.2b. Substitute $I_E = (I_C - I_{CBO})(\beta_F + 1)/\beta_F$ (Equation 7.1.6, Chapter 7) and $I_B = (I_C/\beta_F) - I_{CBO}(\beta_F + 1)/\beta_F$. Deduce that Equation 8.1.1 becomes $I_C = \beta_F \dfrac{V_{BB} - V_{BE}}{R_E(\beta_F + 1) + R_B} + (\beta_F + 1)\dfrac{I_{CBO}(R_E + R_B)}{R_E(\beta_F + 1) + R_B}$.   ∎

Design Example 8.1.1 Discrete Transistor Biasing Using Voltage Divider

It is desired to design the biasing circuit of Figure 8.1.2 to give $I_C = 2$ mA at which β_F is nominally 200. Consider $V_{CC} = 12$ V, the voltage drops across each of R_E and R_C to about 4 V, or one-third of V_{CC}, and $V_{BE} = 0.7$ V, independently of transistor current. The effect on I_C is to be determined for: (a) a 50°C rise in temperature, taking into account $I_{CBO} = 1$ nA, and (b) replacing the transistor with one having $\beta_F = 100$. The case is to be compared with that of $R_E = 0$.

SOLUTION

If $I_C = 2$ mA, then $I_B = I_C/\beta_F = 0.01$ mA, and $I_E = I_C + I_B = 2.01$ mA. $R_C = 4/2 = 2$ kΩ and $R_E = 4/2.01 = 1.99 \cong 2$ kΩ. To satisfy inequality 8.1.2, we may choose $R_B = 0.1(\beta_F + 1)R_E = 40.2$ kΩ. Applying KVL to the input circuit, $V_{BB} = 0.01 \times 40.2 + 0.7 + 4 = 5.1$ V. Dividing $R_B = R_1R_2/(R_1 + R_2)$ by $V_{BB} = 12R_2/(R_1 + R_2)$ gives $R_1 = 40.2 \times 12/5.1 = 94.6$ kΩ. Substituting in the expression for R_B, $R_2 = 69.9$ kΩ. Standard 5% tolerance resistors can be chosen so that $R_1 = 91$ kΩ and $R_2 = 68$ kΩ. These resistance values give $R_B = 38.92$ kΩ, $V_{BB} = 5.13$ V, $I_E = I_E = \dfrac{5.13 - 0.7}{2 + 38.92/201} = 2.02$ mA, $I_C = 2.01$ mA, and $I_B = 0.01$ mA. From KVL in the output circuit $V_{CC} = R_CI_C + V_{CE} + R_EI_E = I_C\left(R_C + \dfrac{\beta_F + 1}{\beta_F}R_E\right) + V_{CE}$. This gives $V_{CE} = 3.9$ V. Note that the resistance $R_C + R_E(\beta_F + 1)/\beta_F$ is the reciprocal of the slope of dc load line drawn on the transistor output characteristics.

(a) Consider first the effect of a temperature rise of 50°C on I_{CBO} alone. Assuming I_{CBO} doubles every 10°C, then it becomes $I_{CBOH} = 1 \times 2^5 = 32$ nA at 50°C. From the result of Exercise 8.1.1, the increase in I_C due to I_{CBO} is $\dfrac{I_{CBOH}(R_E + R_B)}{R_E + R_B/(\beta_F + 1)} = \dfrac{32 \times 40.92}{2 + 38.92/201} = 0.6$ μA, which represents an increase of only about 0.03% in I_C.

Next consider the effect of β_F acting alone. Assuming that β_F increases at a rate of 0.7%/°C, $\beta_{FH} = 200(1 + 50 \times 0.7/100) = 270$. From Equation 8.1.1, the new value of I_C is $I_{CH1} = 270(5.13 - 0.7)/(2 \times 271 + 38.92) = 2.06$ mA. Thus, I_C increases by about 2.5% due to the change in β_F alone. If V_{BE} decreases by 2 mV/°C, then $V_{BEH} = 0.7 - 50 \times 0.002 = 0.6$ V and $I_{CH2} = 200(5.13 - 0.6)/(2 \times 201 + 38.92) = 2.05$ mA. That is, I_C

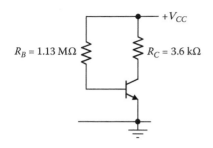

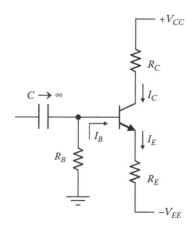

FIGURE 8.1.3
Figure for Design Example 8.1.1.

increases by about 2% due to the change in V_{BE} alone. The combined effects of the temperature rise is to make $I_{CH} = 2.11$ μA, which is an increase of about 5%.

If β_F becomes 100, $I_C = 100(5.13 - 0.7)/(2 \times 101 + 38.92) = 1.84$ mA, representing a decrease of about 8.5%. Note that $R_E(\beta_F + 1)$ is now $5R_B$ compared to about $10R_B$ when $\beta_F = 200$.

(b) If $R_E = 0$, R_2 can be dispensed with altogether, the biasing circuit becomes as shown in Figure 8.1.3. To give a base current of 0.01 mA, $R_B = (12 - 0.7)/0.01$ mA $= 1.13$ MΩ, and to give the same value of V_{CE} requires $R_C = (12 - 3.9)/2.01$ mA $= 4$ kΩ. On the input side, $I_B = (12 - V_{BE})/1.13 \times 10^3$ mA, so $I_C = \beta_F(12 - V_{BE})/1.13 \times 10^3$ mA. If $\beta_F = 270$ and $V_{BE} = 0.6$ V due to a temperature rise of 50°C, $I_C = 2.7$ mA, which is an increase of about 35%. If β_F changes to 100, I_C decreases by 50% to 1 mA.

Exercise 8.1.2

Assume that in Figure 8.1.2a, $R_E = 0$, $V_{CC} = 12$ V, $I_B = 0.01$ mA, $\beta_F = 200$, $R_C = 4$ kΩ, $R_1 + R_2 = 160$ kΩ, and the values of R_1 and R_2 are adjusted so that $V_{BE} = 0.7$ V. Verify that the transistor is driven into saturation by a temperature rise of 50°C and explain why. ∎

If both positive and negative power supplies are available, the biasing circuit of Figure 8.1.4 can be used. In the presence of a dc blocking capacitor, R_B provides a dc path for I_B. R_B cannot be set to zero because the ac input will then be shorted to ground and will not be applied to the transistor. KVL around the base-emitter circuit gives $R_B I_B + V_{BE} + R_E I_E = V_{EE}$. Neglecting I_{CBO}, substituting $I_B = I_C/\beta_F$, $I_E = I_C(\beta_F + 1)/\beta_F$, and solving for I_C:

$$I_C = \beta_F \frac{V_{EE} - V_{BE}}{R_E(\beta_F + 1) + R_B} \tag{8.1.3}$$

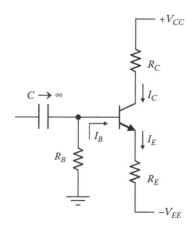

FIGURE 8.1.4
Biasing using a negative voltage supply.

This is the same as Equation 8.1.1 with V_{BB} replaced by a larger V_{EE}, which reduces the effect of a change in V_{BE}. Moreover, for the same I_E, R_E is larger, which reduces the effects of changes in β_F for the same R_B (Equation 8.1.2). The drain on V_{CC} due to R_1 and R_2 is eliminated at the expense of the additional power supply.

If V_{EE} and R_E are large, changes in R_C or V_{BE} will have little effect on I_E. This is equivalent to connecting a current source I_E in series with the emitter, which makes I_E independent of variations in temperature or in transistor parameters. $I_C = I_E \beta_F/(\beta_F + 1)$ and is only slightly affected by variations in β_F.

Negative feedback for bias stabilization can also be provided by connecting R_B to the collector (Figure 8.1.5). An increase in I_C, for example, reduces the collector voltage and hence I_B, which opposes the increase in I_C, as discussed in the following example.

Design Example 8.1.2 Transistor Biasing Using Shunt Feedback

Choose R_B and R_C in Figure 8.1.5 so that $I_C = 2$ mA and $V_{CE} = 3.9$ V, with $\beta_F = 200$, as in Example 8.1.1, and determine the effect of a temperature rise of 50°C and a change of β_F to 100.

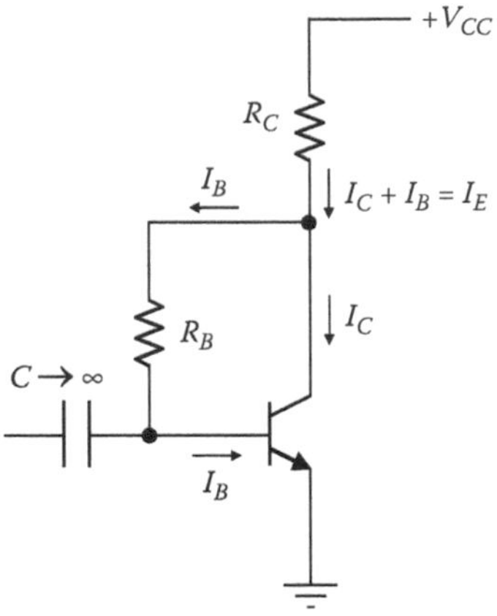

SOLUTION

Bearing in mind that the current in R_C is I_E, KVL gives $R_C I_E + R_B I_B + V_{BE} = V_{CC}$. Substituting $I_B = I_C/\beta_F$, $I_E = I_C(\beta_F + 1)/\beta_F$, and solving for I_C:

$$I_C = \beta_F \frac{V_{CC} - V_{BE}}{R_C(\beta_F + 1) + R_B} \qquad (8.1.4)$$

FIGURE 8.1.5
Figure for Design Example 8.1.2.

which is the same as Equation 8.1.1, but with V_{BB} replaced by V_{CC} and R_E replaced by R_C. To have $V_{CE} = 3.9$ V and $I_B = 0.01$ mA, $R_B = (3.9 - 0.7)/0.01 = 320\,\text{k}\Omega$ and $R_C = (12 - 3.9)/(0.01 \times 201) = 4.03\,\text{k}\Omega$. Choosing $R_B = 330$ kΩ and $R_C = 4$ kΩ gives $I_C = 1.99$ mA and $I_B = 0.01$ mA. Note that $(\beta_F + 1)R_C/R_B = 201 \times 4/330 = 2.44$.

If β_F increases to 270 (Example 8.1.1), $I_C = 270(12 - 0.7)/(4 \times 271 + 330) = 2.16$ mA. This is an increase of about 8%. If V_{BE} decreases to 0.6 V as before, I_C increases by less 1% to 2.01 mA. If β_F becomes 100, I_C decreases by about 23% to 1.54 mA. It is seen that overall, the performance of this biasing circuit is not as good as that of the voltage divider, but it does reduce the current drain on the power supply.

Exercise 8.1.3

It would appear from Equation 8.1.4 that in order to make I_C independent of β_F, $R_C(\beta_F + 1) \gg R_B$. Deduce that this would make $R_C I_C$ close to V_{CC} so that the transistor will be near saturation. ∎

In the circuits of Figures 8.1.2a, 8.1.4, and 8.1.5, the *npn* transistor can be replaced by an E-NMOS transistor. The MOSFET circuit of Figure 8.1.2a is analyzed in Section ST8.1, including the body effect, and that of Figure 8.1.5 is left as a problem (Problem P8.1.11). Although the effects of changes in temperature are generally not as marked in a MOSFET as in a BJT (Section 5.3, Chapter 5, and Section 6.2, Chapter 6), particularly in the region of operation where the drain current changes little with temperature, introducing negative feedback is advantageous in many respects (Section 7.3, Chapter 7). The effect of a source resistance in mitigating the effects of a change in threshold is illustrated in the following example.

Design Example 8.1.3 Effect of Source Resistance on Threshold Variations

An E-NMOS transistor having $k_n = 0.5$ mA/V^2 is to be biased in the two circuits of Figure 8.1.6 to $I_D = 1$ mA, $V_{DS} = 8$ V. The resistors are to be selected in both circuits, and it is required to determine the effect on I_D of a change in threshold from 2 to 1.8 V.

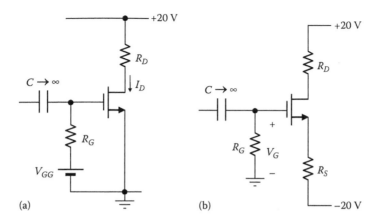

FIGURE 8.1.6
Figure for Design Example 8.1.3. E-NMOS biesing without a source resistor R_s(a), and with R_s(b).

SOLUTION

Neglecting channel-length modulation, $I_D = (k_n/2)(V_{GG} - V_{tn})^2$ in the circuit of Figure 8.1.6a. Substituting $I_D = 1$ mA and $V_{tn} = 2$ V gives $V_{GG} = 4$ V. It follows that $R_D = (20 - 8)/1 = 12$ kΩ. If V_{tn} changes to 1.8 V, $I_{DL} = 0.25(4 - 1.8)^2 = 1.21$ mA, an increase of 21%. V_{DS} is reduced to about 5.5 V. Since $I_G = 0$, R_G may be chosen to be 1 MΩ.

In the circuit of Figure 8.1.6b, $V_G = 0$. If $V_{GS} = 4$ V as before, the source voltage is -4 V, and $R_S = 16$ kΩ. The drain voltage is $+4$ V, and $R_D = 16$ kΩ. If V_{tn} changes to 1.8 V, I_{DL} and V_{GSL} are obtained by solving the equation of the transfer characteristic $I_{DL} = 0.25(V_{GSL} - 1.8)^2$ and KVL in the gate-source circuit, $V_{GSL} + 16I_{DL} = 20$. This gives $I_{DL} = 1.01$ mA and $V_{DSL} = 40 - 32 \times 1.01 = 7.68$ V. The change in I_D is only about 1% because of the negative feedback due R_S.

Current Mirror

Concept: *In a current mirror, a reference current applied to a diode-connected transistor establishes a voltage at the input of this transistor. If this voltage is applied to the input of an identical transistor, the output current of this transistor will be equal to, or will mirror, the reference current.*

BJT Current Mirror

Biasing IC transistors relies on a different philosophy altogether in order to avoid using multiple resistors for biasing and large capacitors for coupling signals. In ICs, these elements would take up a large chip area and the tolerance on their values would be rather high, typically around 20%. Instead, transistors are used in a **current mirror** circuit to provide constant-current biasing, which as we have seen earlier, is very effective in maintaining a constant output current.

The basic BJT current mirror is illustrated in Figure 8.1.7a, where Q_1 and Q_2 are matched transistors. Q_1 is diode connected in series with a resistor R. Neglecting base-width modulation to begin with, Equation 6.2.1, Chapter 6, gives:

$$\frac{I_{C2}}{I_{C1}} = e^{(V_{BE2} - V_{BE1})/V_T} = 1 \qquad (8.1.5)$$

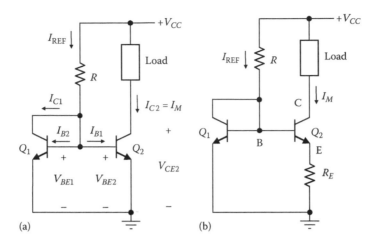

FIGURE 8.1.7
(a) Basic BJT current mirror; (b) Widlar current mirror.

because $V_{BE1} = V_{BE2} = V_{BE}$. From Figure 8.1.7a,

$$I_{REF} = \frac{V_{CC} - V_{BE}}{R} \tag{8.1.6}$$

and

$$I_{REF} = I_{C1} + I_{B1} + I_{B2} = I_M + 2I_B \tag{8.1.7}$$

where $I_{C1} = I_{C2} = I_M$ and $I_{B1} = I_{B2} = I_B$. Substituting $I_M = \beta_F I_B$, neglecting I_{CBO},

$$I_M = \frac{\beta_F}{\beta_F + 2} I_{REF} = \frac{\beta_F}{\beta_F + 2} \frac{V_{CC} - V_{BE}}{R} \tag{8.1.8}$$

If β_F is large, then $I_M \cong I_{REF}$, and if $V_{CC} \gg V_{BE}$,

$$I_M \cong I_{REF} \cong \frac{V_{CC}}{R} \tag{8.1.9}$$

Thus, V_{CC} and R set the value of I_{REF}, and I_M reflects, or "mirrors," this value. The physical interpretation is that if β_F is large, the base currents are negligible, which makes $I_{C1} = I_{REF}$. But because the transistors are identical and have the same V_{BE}, they must have the same collector current, neglecting base-width modulation.

It follows from Equation 8.1.8 that the effect of temperature variation on I_M is mainly due to variation in V_{BE} affecting I_{REF}. Changes in β_F due to temperature variation, or other causes, have little effect on I_M as long as β_F is large.

In Figure 8.1.7a, $V_{CB1} = 0$ and $V_{CE1} \cong 0.7$ V. However, V_{CE2} depends on the circuit to which Q_2 is connected and may vary over a wide range. In practice, this would affect I_{C2} because of base-width modulation. Adding an emitter resistor to Q_2 increases the effective output resistance, and therefore minimizes the variation of I_{C2} with V_{CE2}, while reducing R for a given I_M. The resulting circuit is the **Widlar current mirror** of Figure 8.1.7b (Example 8.1.4), also referred to as a **Widlar current source**, because of its enhanced output resistance.

Example 8.1.4 Comparison between Basic and Widlar Current Mirrors

A current $I_M = 50$ μA is to be derived from the basic and the Widlar current mirror, with $V_{CC} = 12$ V. It is required to compare R and r_{out} in both cases, assuming that at $I_M = 50$ μA, $r_o = 2$ MΩ, $r_\pi = 50$ kΩ, and $\beta = 100$, and that $V_{BE} = 0.7$ V at $I_C = 1$ mA.

SOLUTION

(a) Basic current mirror: Using Equation 8.1.5, $\dfrac{1}{0.05} = e^{(0.7 - V_{BE})/0.026}$, or $V_{BE} = 0.7 - 0.026 \times$ $\ln(20) = 0.62$ V. Hence, $R = (12 - 0.62)/0.05 = 228$ kΩ. $r_{out} = r_o = 2$ MΩ.

(b) Widlar current mirror: From Equation 6.2.1, Chapter 6, $V_{BE1} = V_T \ln(I_{REF}/I_S)$ and $V_{BE2} = V_T \ln(I_M/I_S)$. Hence, $V_{BE1} - V_{BE2} = V_T \ln(I_{REF}/I_M)$. But $V_{BE1} = V_{BE2} + R_E I_M$. It follows that $R_E = (V_T/I_M) \ln(I_{REF}/I_M)$. We may choose $I_{REF} = 1$ mA, in which case, $R = (12 - 0.7)/1 = 11.3$ kΩ, and $R_E = (0.026/0.05) \ln(1/0.05) = 1.56$ kΩ. It is advantageous to use lower resistance values in ICs because they would generally occupy a smaller chip area.

To determine the output resistance, we note that B is grounded through a diode connected to transistor Q_1, which presents a resistance of $r_e \| R \cong r_e$ to ground (Problem P6.1.20, Chapter 6). As the current in Q_1 is 20 times that in Q_2, r_e of Q_1 is one-twentieth r_e of Q_2. Because this is small compared with r_π of Q_2, we may consider B to be effectively grounded.

The small-signal equivalent circuit is shown in Figure 8.1.8. Let $R'_E = R_E \| r_\pi$ and transform the dependent current source $g_m v_{be}$ in parallel with r_o to a dependent voltage source $g_m r_o v_{eb}$, of reversed polarity, in series with r_o. The voltage source $g_m r_o v_{eb}$ in series with v_{eb} gives a voltage drop $(1 + g_m r_o) v_{eb} = (1 + g_m r_o) R'_E i$, where i is the current in the series combination. This voltage drop is therefore equivalent to a resistance $(1 + g_m r_o) R'_E$ through which i flows. It follows that $r_{out} = r_o + R'_E(1 + g_m r_o) \cong r_o(1 + g_m R'_E)$. Substituting numerical values, with $g_m = 100/50 = 2$ mA/V, gives $r_{out} = 2(1 + 2 \times 1.51) = 8$ MΩ.

Concept: *Although, strictly speaking, r_{out} is defined under small-signal conditions, it gives a useful indication of how much the output current varies with the output voltage.*

In Example 8.1.4, r_{out} was found to be 8 MΩ. This implies that if the output voltage at Q_2 collector changes by 2 V, the output current through the load changes by 0.25 μA.

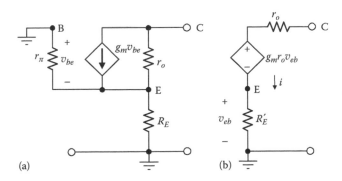

(a)   (b)

FIGURE 8.1.8
Figure for Example 8.1.4. (a) Small-signal equivalent circuit; (b) current source transformed to a voltage source.

Exercise 8.1.4

In Example 8.1.4, determine the following, for both the current mirrors of Figure 8.1.7: (a) the change in I_M if V_O (the collector voltage of Q_2) varies by 5 V, assuming that r_{out} applies over this voltage range; (b) R assuming that $\beta_F = 100$ and using Equation 8.1.8 with $V_{BE} = 0.7$ V.

Answers: (a) 2.5 μA and 0.63 μA; (b) 222 kΩ and 11.2 kΩ.   ■

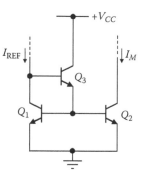

FIGURE 8.1.9
Figure for Exercise 8.1.5.

Exercise 8.1.5

Starting with I_M in Figure 8.1.9, express currents in terms of β_F and show that $\dfrac{I_M}{I_{REF}} \cong \left(1 - 2/\beta_F^2\right)$, compared to $\dfrac{I_M}{I_{REF}} \cong \left(1 - 2/\beta_F\right)$ for the basic current mirror. In effect, Q_3 supplies base current to Q_1 and Q_2, in a composite transistor type of connection.   ■

It is possible to use the same Q_r with a number of transistors in a **current-steering circuit** in order to derive multiple current sinks (Figure 8.1.10a). The term current sink rather than current source is used when I_M draws current from the load toward a lower voltage. If Q_1 to Q_3 are identical, then by analogy with Figure 8.1.7a, $I_{M1} = I_{M2} = I_{M3} = I_{REF}\beta_F/(\beta_F + 4)$, since I_{REF} supplies base current to four transistors. The currents I_{M1} to I_{M3} could be made different by having different emitter areas of Q_1 to Q_3 (Equation 6.2.1, Chapter 6). It is also possible to combine transistors Q_1 to Q_3 in a single device having a common emitter and base but multiple collectors of different areas (Figure 8.1.10b).

MOSFET Current Mirror

MOSFET current mirrors are analogous to their BJT counterparts, as illustrated in Figure 8.1.11 for a basic E-NMOS current mirror. Q_1 is diode connected, and is in saturation as long as $V_{DS1} = V_{GS} > V_{tn1}$ (Section 5.4, Chapter 5), which will always be satisfied if Q_1 is conducting. Q_2 is in saturation as long as $V_{DS2} \geq V_{GS} - V_{tn2}$. From Figure 8.1.11,

$$I_{D1} = I_{REF} = \frac{V_{DD} - V_{GS}}{R} \tag{8.1.10}$$

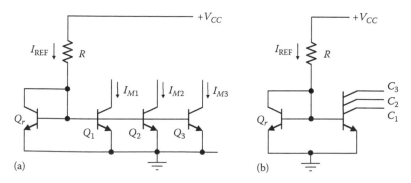

FIGURE 8.1.10
Current-steering circuit using: (a) separate transistors, (b) a single transistor with multiple collectors.

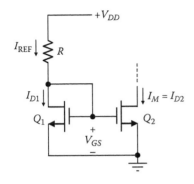

FIGURE 8.1.11
MOSFET current mirror.

Assuming that I_D of each transistor is given by Equation 5.3.1, Chapter 5, it follows that:

$$\frac{I_M}{I_{REF}} = \frac{I_{D2}}{I_{D1}} = \frac{k'_{n2}}{k'_{n1}} \frac{(W/L)_2}{(W/L)_1} \left(\frac{V_{GS} - V_{tn2}}{V_{GS} - V_{tn1}}\right)^2 \left(\frac{1+\lambda V_{DS2}}{1+\lambda V_{DS1}}\right)$$

(8.1.11)

If the transistors are identical in every respect and $V_{DS2} = V_{DS1}$, the current ratio is unity and $I_M = I_{REF}$. If the transistors differ only by their aspect ratios,

$$\frac{I_M}{I_{REF}} = \frac{(W/L)_2}{(W/L)_1}$$

(8.1.12)

This provides a convenient way of scaling I_M to be larger or smaller than I_{REF}.

Example 8.1.5 Basic MOSFET Current Mirror

In Figure 8.1.11, $V_{DD} = 5$ V, Q_1 and Q_2 are perfectly matched and have $k_n = 0.4$ mA/V^2, $V_{tn} = 1$ V, and $\lambda = 0.01$ V^{-1}. It is required to determine R and r_{out} for a nominal $I_M = 0.2$ mA.

SOLUTION

In order to determine R from Equation 8.1.10, we need V_{GS}. This can be determined from the transfer characteristic, neglecting channel-width modulation. Thus $I_{D1} = 0.2 = (0.4/2)$ $(V_{GS} - 1)^2$, so $V_{GS} = 2$, since, $V_{DS1} = V_{GS}$, $(1 + \lambda V_{DS1}) = 1.02$ and will not make a significant difference. From Equation 8.1.10, $R = (5 - 2)/0.2 = 15$ kΩ. The minimum output voltage that will maintain Q_2 in saturation is $V_{GS} - V_{tn} = 1$ V. $I_M = 0.2$ mA when $v_O = V_{DS2} = V_{DS1} = 2$ V. $r_{out} = r_o = 1/\lambda I_{DS2(sat)} \cong 1/0.01 \times 0.2 \equiv 500$ kΩ. If v_O increases by 2 V, the change in I_M is 2 V/500 k$\Omega \equiv 4$ μA.

In Figure 8.1.12 the diode-connected E-NMOS and E-PMOS transistors Q_1 and Q_3 of the current mirrors share the same resistor R for I_{REF}. Q_2 and Q_4, and any other transistors paralleled with them, provide current sources and sinks. The currents in the paralleled

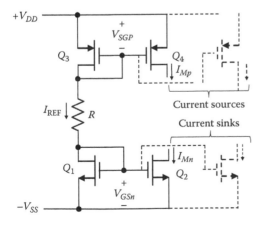

FIGURE 8.1.12
E-NMOS and E-PMOS current mirrors sharing the same R.

transistors could be made different by varying the widths of these transistors. The substrates of the n-channel and the p-channel transistors are connected to $-V_{SS}$, and $+V_{DD}$, respectively, which eliminates the body effect. If source resistors are added to the basic current mirror, the body effect modifies the thresholds and the effective transconductances of the transistors (Section 5.5, Chapter 5).

In comparing BJT and MOS current mirrors, it is seen that: (1) BJT current mirrors suffer from the effect of finite β_F, whereas MOS current mirrors may be subject to the body effect, as in the case of the Widlar current source; (2) r_{out} is generally higher for BJT current mirrors because V_A of BJTs is larger than $1/\lambda$ of MOSFETs for the same current; (3) the collector voltage of a BJT can swing to within $V_{CE(sat)}$ of about 0.2 V from the emitter voltage, whereas the drain-gate voltage of an E-NMOS transistor cannot be more negative than V_{tn}, if the transistor is to remain in saturation. More elaborate MOS current mirrors are discussed in Sections 8.4 and ST8.3.

8.2 Basic Amplifier Configurations

General Considerations

We will begin our analysis of basic transistor amplifiers with generalizations that are used later. An amplifier can be represented at its output terminals by a Thevenin equivalent circuit (TEC, Figure 8.2.1a) or a Norton equivalent circuit (NEC, Figure 8.2.1b) having a dependent source controlled, in general, by v'_{src}, where v'_{src} and R'_{src} are TECs of the input circuit. The following *small-signal parameters* are of interest:

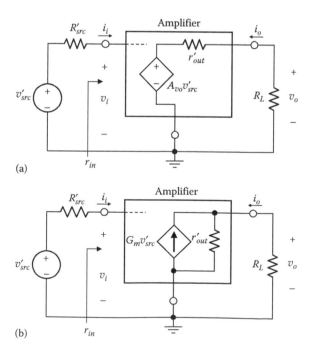

FIGURE 8.2.1
Representation of an amplifier output by a Thevenin-like equivalent circuit (a), or Norton-like equivalent circuit (b).

r_{in}: amplifier input resistance between input terminal and ground.

r_{out}: resistance looking into the output terminal of the amplifier proper, with $v'_{src} = 0$ and excluding all external resistances connected between the output terminal and ac ground.

r'_{out}: output resistance seen by the load, with $v'_{src} = 0$. r'_{out} is r_{out} in parallel with any resistance between the output terminal and the ac ground, other than the load, such as R_C in Figure 8.2.3.

R_{out}: total resistance between output terminal and ground, including the load resistance R_L. Thus, $R_{out} = R_L \| r'_{out}$.

A_v: voltage gain from input terminal to output, $A_v = v_o / v_i$.

A'_v: voltage gain from source to output, $A'_v = v_o / v'_{src}$.

A_{vo}: Open-circuit voltage gain from source to output.

A_i: current gain, $A_i = i_o / i_i$.

G_m: **equivalent transconductance** of the amplifier, $G_m = \left. \dfrac{i_o}{v'_{src}} \right|_{v_o=0}$, that is, G_m is the ratio of the short-circuit output current to the source voltage.

It is seen from Figure 8.2.1 that:

$$v_o = \frac{R_L}{r'_{out} + R_L} A_{vo} v'_{src} = G_m (r'_{out} \| R_L) v'_{src} = G_m R_{out} v'_{src} \tag{8.2.1}$$

The following relations apply between A_v, A'_v, and A_i (Exercise 8.2.1):

$$A'_v = \frac{r_{in}}{r_{in} + R'_{src}} A_v, \quad A_v = -\frac{R_L}{r_{in}} A_i, \quad \text{and} \quad A'_v = -\frac{R_L}{r_{in} + R'_{src}} A_i \tag{8.2.2}$$

The following should be noted in connection with the circuit of Figure 8.2.1:

1. In order to keep G_m positive, then for an inverting amplifier, the direction of the dependent current source $G_m v'_{src}$ is reversed, which makes $v_o = -G_m R_{out}$.

2. When $i_i = 0$, as when the input is applied to the gate of a MOSFET at low frequencies, $v'_{src} = v_i$, the output source can be expressed as $A_{vo} v_i$ or $G_m v_i$, where G_m and r_{out} are evaluated with $v_i = 0$. When $i_i \neq 0$, A_{vo}, G_m and, in general, r'_{out} depend on R'_{src}. If under these conditions, the output source is considered to be $A_{vo} v_i$ or $G_m v_i$, and G_m and r'_{out} are evaluated with $R'_{src} = 0$, then this implies that an ideal voltage source, of zero source resistance, is applied directly to the amplifier input.

3. If any of the output resistances is required with $v_i = 0$, it can be obtained from the expressions for the corresponding output resistances with $v'_{src} = 0$ by setting $R'_{src} = 0$ in these expressions.

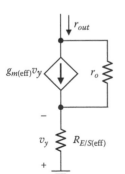

FIGURE 8.2.2
Circuit for determining r_{out}.

The circuit of Figure 8.2.2 is often encountered in amplifier circuits, where r_o is the output resistance of the transistor, $R_{E/S(eff)}$ is the effective resistance between emitter or source and ground, $g_{m(eff)}$ is the effective transconductance of the transistor, and v_y is v_{be} for an *npn* BJT and v_{gs} for an *n*-channel MOSFET transistor. Following the procedure applied to Figure 8.1.8b in transforming the dependent current source to a dependent voltage source controlled by the voltage across $R_{E/S(\text{eff})}$ and then representing this source by a resistance, r_{out} can be expressed as:

$$r_{out} = r_o + R_{E/S(\text{eff})} + g_{m(\text{eff})}r_o R_{E/S(\text{eff})} \tag{8.2.3}$$
$$\cong r_o(1 + g_{m(\text{eff})}R_{E/S(\text{eff})}), \quad \text{when } R_{E/S(\text{eff})} \ll r_o \tag{8.2.4}$$

$R_{E/S(\text{eff})}$ and $g_{m(\text{eff})}$ depend on the particular amplifier, as will be discussed later.

Exercise 8.2.1

Verify Equations 8.2.2.   ∎

Common-Emitter Amplifier

The CE configuration is widely used, because it provides both current and voltage gains as well as a moderately high input resistance. Its output resistance, however, is rather high, which attenuates the output voltage when the load resistance is relatively low.

A typical discrete, CE amplifier configuration is illustrated in Figure 8.2.3a using the biasing circuit of Figure 8.1.2a. The signal source is coupled to the base through a large dc blocking capacitor, as discussed earlier. The load R_L is similarly coupled to the collector through a large capacitor so as to isolate the load from the dc collector voltage. The large bypass capacitor C_E eliminates the effect of R_E at signal frequencies, as discussed later.

To perform small-signal analysis on the circuit of Figure 8.2.3a, the circuit is converted to the ac circuit of Figure 8.2.3b as follows: (1) The V_{CC} dc bias supply is replaced by a short-circuit, since its impedance is assumed to be zero. (2) The large coupling and emitter capacitors (C_{PI}, C_{PO}, and C_E) are replaced by short circuits, because their reactances at the signal frequencies are assumed to be negligibly small. (3) The circuit connected to the base is replaced by its TEC so that $v'_{src} = \dfrac{R_1 \| R_2}{R_{src} + R_1 \| R_2} v_{src}$ and $R'_{src} = (R_{src} \| R_1 \| R_2)$. (4) The transistor is considered as having an input resistance r_π between base and emitter

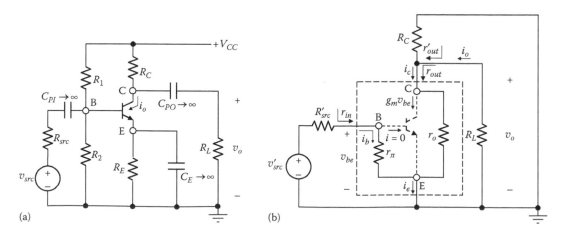

FIGURE 8.2.3
(a) Typical discrete CE amplifier circuit; (b) the circuit under ac conditions.

terminals and a resistance r_o between collector and emitter terminals, leaving an "ideal" BJT, shown dotted, that does not draw any base current and which behaves between collector and emitter as a dependent current source $g_m v_{be}$ directed from collector to emitter. This is essentially the same as invoking the small-signal equivalent circuit of Figure 7.2.7, Chapter 7, but highlights the role of the transistor in the circuit and facilitates performing small-signal analysis more or less by inspection.

It is evident from the circuit of Figure 8.2.3b that the input resistance r_{in} is:

$$r_{in} = \frac{v_{be}}{i_b} = r_\pi \tag{8.2.5}$$

To determine A_i, it follows from current division that $i_o = g_m v_{be}(r_o \| R_C \| R_L)/R_L$. Hence,

$$A_i = \frac{i_o}{i_b} = \frac{g_m v_{be}(r_o \| R_C \| R_L)/R_L}{v_{be}/r_\pi} = \beta \frac{r_o \| R_C \| R_L}{R_L} \tag{8.2.6}$$

If $R_L \ll r_o$ and R_C, the collector is effectively short circuited to ground for ac signals, and $A_i = \beta$, which agrees with the definition of β as the CE short-circuit current gain.

If $v'_{src} = 0$, then $v_{be} = 0$, $g_m v_{be} = 0$, and the source $g_m v_{be}$ does not contribute to r_{out}. Thus,

$$r_{out} = r_o \tag{8.2.7}$$

It is seen that $r'_{out} = r_o \| R_C$. If $r_o \to \infty$, then $r_{out} \to \infty$ and $r'_{out} \cong R_C$.

The voltage gain A_v from base to collector is:

$$A_v = \frac{v_o}{v_{be}} = \frac{-g_m v_{be}(r_o \| R'_L)}{v_{be}} = -g_m(r_o \| R'_L) = -g_m R_{out} \tag{8.2.8}$$

where $R'_L = (R_C \| R_L)$ and $R_{out} = r_o \| R_C \| R_L$. To evaluate G_m, we reverse the direction of the source $G_m v'_{src}$ in Figure 8.2.1b because the amplifier is inverting. Since $v_{be} = v'_{src} r_\pi/(r_\pi + R'_{src})$ and when the output is short circuited, $i_{sc} = g_m v_{be} = v'_{src} g_m r_\pi/(r_\pi + R'_{src})$, it follows from Figure 8.2.1 that:

$$G_m = \frac{i_{sc}}{v'_{src}} = \frac{g_m r_\pi}{r_\pi + R'_{src}} = \frac{\beta}{r_\pi + R'_{src}} \tag{8.2.9}$$

$$A'_v = G_m R_{out} = \frac{-\beta(r_o \| R'_L)}{r_\pi + R'_{src}} = A_v \frac{r_\pi}{r_\pi + R'_{src}} \tag{8.2.10}$$

in accordance with Equations 8.2.2. The following should be noted concerning Equation 8.2.10:

1. If $R'_{src} \gg r_\pi$, the input is essentially a current drive, so $i_b = v'_{src}/R'_{src}$. If the source $g_m v_{be}$ is represented as βi_b (Figure 6.1.12, Chapter 6), $v_o = -\beta(r_o \| R'_L)i_b = \frac{-\beta(r_o \| R'_L)}{R'_{src}} v'_{src}$, as in Equation 8.2.10.

2. On the other hand, if $R'_{src} \ll r_\pi$, the input is essentially a voltage drive, $v'_{src} = v_{be}$, and $A'_v = A_v = -g_m(r_o \| R'_L)$, as in Equation 8.2.8.

3. The maximum possible gain $A_{v(\max)}$ occurs when $R'_L \gg r_o$. Using Equation 6.2.10, Chapter 6:

$$A_{v(\max)} = -g_m r_o = -\frac{V_A}{V_T} \qquad (8.2.11)$$

4. The overall gain from v_{src}, rather than v'_{src}, to the output, is A'_v multiplied by $(R_1 \| R_2)/(R_{src} + R_1 \| R_2)$.

Common-Source Amplifier

In the case of the CS amplifier, the same circuit of Figure 8.2.3b applies but without r_π, which makes r_{in} and A_i theoretically infinite. Again, when $v_{src} = 0$, $g_m v_{gs} = 0$ and the dependent current source does not contribute to r_{out}. Thus,

$$r_{out} = r_o \qquad (8.2.12)$$

It follows that $r'_{out} = r_o \| R_D$, and $R_{out} = r_o \| R_D \| R_L$.

The voltage gain from gate to collector is:

$$A_v = -g_m(r_o \| R'_L) = -g_m(r'_{out} \| R_L) = -g_m R_{out} \qquad (8.2.13)$$

where $R'_L = R_D \| R_L$. To evaluate G_m, we note that when the output is short circuited, $i_{sc} = g_m v_{gs} = g_m v'_{src}$. It follows from Figure 8.2.1b, with the polarity of the source $G_m v'_{src}$ reversed, that:

$$G_m = g_m \qquad (8.2.14)$$

This result is in accordance with Equation 8.2.9, because for a MOSFET, β and r_π of a BJT both tend to infinity but their ratio remains finite and equal to g_m.

The maximum possible gain is $-g_m r_o$ (Equation 5.3.16, Chapter 5). Setting $V_A = 1/\lambda$,

$$A_{v(\max)} = -g_m r_o = -\frac{2V_A}{V_{ov}} \qquad (8.2.15)$$

Not only is V_A of the BJT generally higher than that of the MOSFET but $V_{ov}/2 \gg V_T$, so that $A_{v\max}$ of the BJT is normally much greater than that of the MOSFET.

Exercise 8.2.2

Verify that if $r_o \to \infty$, then $r_{out} \to \infty$, with $v_{src} = 0$ in Figure 8.2.3a, by applying a test source and determining the test current. ∎

Exercise 8.2.3

Verify Equations 8.2.2 for the CE and CS amplifiers. ∎

Common-Drain Amplifier

In some cases, it is desirable to have both a high-input resistance, so as not to overload a driving stage, and a low-output resistance that provides the load drive capability of a

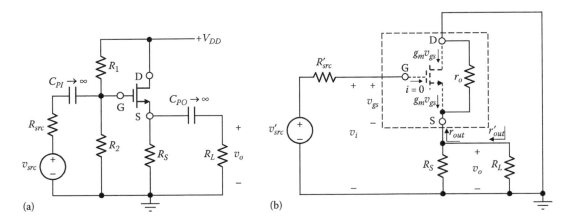

FIGURE 8.2.4
(a) An implementation of a CD amplifier; (b) the circuit under ac conditions.

voltage source. The common-drain (CD)/common-collector (CC) amplifiers fulfill this requirement.

Figure 8.2.4a illustrates a CD amplifier. Compared to a CS amplifier, R_D is zero and the output is taken directly from the source. The drain is thus common between input and output, which gives the circuit its name. The entire output voltage is fed back in series opposition to the input, which gives a series–shunt, negative feedback configuration. As discussed in Section 7.3, Chapter 7, such a configuration increases the input resistance and reduces the output resistance, while reducing the voltage gain. Although the gate input resistance of a MOSFET is theoretically infinite, with or without feedback, the output resistance is advantageously reduced.

The circuit for small-signal analysis, corresponding to Figure 8.2.4a, is shown in Figure 8.2.4b, where $v'_{src} = \dfrac{R_1 \| R_2}{R_{src} + R_1 \| R_2} v_{src}$ and $R'_{src} = (R_{src} \| R_1 \| R_2)$.

If the body effect is neglected for the moment, it is seen from Figure 8.2.4b that $v_{gs} = v_i - v_o$ and $v_o = g_m v_{gs} R'_S$, where $R'_S = r_o \| R_S \| R_L$. These relations can be represented by a signal-flow diagram (Figure 8.2.5). Applying Rule 7.3.5, Chapter 7, gives:

$$A_v = \frac{v_o}{v_i} = \frac{g_m R'_S}{1 + g_m R'_S} \tag{8.2.16}$$

If $g_m R'_S \gg 1$, $A_v \cong 1$, which means that the source voltage ideally follows the gate voltage. For this reason, the CD amplifier is also known as a **source follower**.

To account for the body effect, the small-signal equivalent circuit of Figure 5.3.8, Chapter 5, is applied to give the small-signal equivalent circuit of Figure 8.2.6a for the amplifier. The circuit can be readily converted to the form of Figure 8.2.1b, as shown in

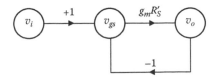

FIGURE 8.2.5
Signal-flow diagram of the CD amplifier, neglecting channel-length modulation.

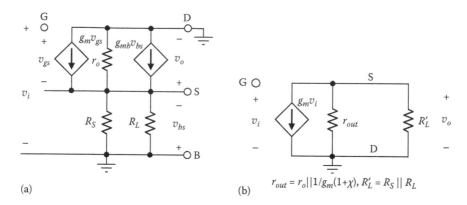

FIGURE 8.2.6
(a) Small-signal equivalent circuit of the CD amplifier taking channel-length modulation into account; (b) reduced small-signal equivalent circuit.

Figure 8.2.6b. R_S is paralleled with R_L to give R'_L. Since $v_{bs} = -v_o$, the source $g_{mb}v_{bs}$ becomes $g_{mb}v_o$ directed upward. Moreover, $v_{gs} = v_i - v_o$, so that $g_mv_{gs} = g_mv_i - g_mv_o$. Thus, the source g_mv_{gs} can be replaced by two sources: (1) a source g_mv_i directed downward in Figure 8.2.6a, and (2) a source g_mv_o directed upward, which can be combined with the $g_{mb}v_o$ source to give a source $g_m(1 + \chi)v_o$ directed upward, where $g_{mb} = \chi g_m$. Since this combined source is in the direction of a voltage droop v_o, it can be replaced by a resistance $1/g_m(1 + \chi)$ (Exercise 8.2.4). This resistance is then combined with r_o to give r_{out}. Thus:

$$r_{out} = r_o || [1/g_m(1 + \chi)] = \frac{r_o}{1 + g_m(1 + \chi)r_o} \tag{8.2.17}$$

$$\cong \frac{1}{g_m(1 + \chi)}, \quad \text{if } 1/g_m(1 + \chi) \ll r_o \tag{8.2.18}$$

Comparing Figure 8.2.6b with Figure 8.2.1b with $v_i = v'_{src}$, it follows that:

$$G_m = g_m \tag{8.2.19}$$

because when the source is shorted to the drain there is no body effect (Figure 8.2.26a). R_{out} is:

$$R_{out} = R_S || R_L || r_o || [1/g_m(1 + \chi)] = R'_S || 1/g_m(1 + \chi) = \frac{R'_S}{1 + g_m(1 + \chi)R'_S} \tag{8.2.20}$$

$$\cong \frac{1}{g_m(1 + \chi)}, \quad \text{if } 1 + g_m(1 + \chi)R'_S \gg 1 \tag{8.2.21}$$

From Figure 8.2.1, $A_v = A'_v = G_mR_{out}$. Hence,

$$A_v = g_mR_{out} = g_mR'_S || 1/g_m(1 + \chi) = \frac{g_mR'_S}{1 + g_m(1 + \chi)R'_S}$$

$$\cong \frac{1}{(1 + \chi)}, \quad \text{for } g_m(1 + \chi)R'_S \gg 1 \tag{8.2.22}$$

The following should be noted concerning the source follower:

1. The body effect reduces A_v because of the $(1+\chi)$ factor, thus degrading performance. Without the body effect, and assuming $1/g_m \ll R_S'$, $R_{out} \cong 1/g_m$ and $A_v \cong 1$.

2. In Equation 8.2.22, the $(1+\chi)$ factor multiplies g_m in the denominator only because r_{out} is affected by this factor but not the source $g_m v_i$ in Figure 8.2.6. The signal-flow diagram (Figure 8.2.5) is modified by multiplying the forward gain from v_{gs} to v_O by $(1+\chi)$ and dividing the forward gain from v_i to v_{gs} by this factor (Exercise 8.2.4).

3. In the absence of feedback, as when the ac input v_i is applied between gate and source (Figure 8.2.4b), rather than between gate and ground, $v_o = g_m R_S' v_{src}'$ and $R_{out} = R_S'$. The effect of the shunt negative feedback is to divide R_{out} and A_v by the factor $1 + g_m(1+\chi)R_S'$. This is $(1 - \text{loop gain})$ of the modified signal-flow diagram. The negative feedback reduces the gain by this factor because, for the same v_{gs}, v_i must be increased to counteract the feedback voltage across R_S'.

4. The resistance seen by R_L is $r_{out}' = r_{out} \| R_S$.

A CD amplifier with a current mirror active load is analyzed in Section ST8.4 and a super source follower of low output resistance is analyzed in Section ST8.5.

Exercise 8.2.4

(a) Verify that R_{out} can be derived by applying a test source v_x as in Figure 8.2.7. What is the effective resistance of the two current sources? (b) Modify Figure 8.2.5 to account for the body effect. ∎

Common-Collector Amplifier

The circuit for small-signal analysis of the CC amplifier, corresponding to Figure 8.2.4b, is shown in Figure 8.2.8, where r_π has been added to the input. This allows a simple derivation of the input resistance and the voltage gain.

KVL at the input gives $v_i = r_\pi i_b + R_E'[(\beta+1)i_b]$, where $R_E' = R_E \| R_L \| r_o$ (Figure 8.2.9a). This can equally be expressed as $v_i = r_\pi i_b + [(\beta+1)R_E']i_b$, where the $(\beta+1)$ factor is associated with R_E' rather than i_b (Figure 8.2.9b), which leads to an important interpretation in BJT circuits:

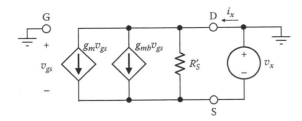

FIGURE 8.2.7
Determining the output resistance by applying a test source.

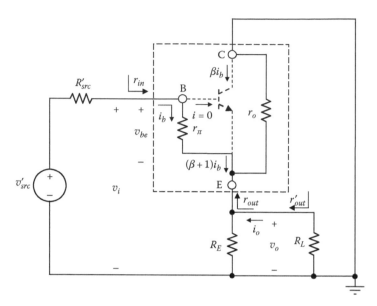

FIGURE 8.2.8
The CC amplifier under ac conditions.

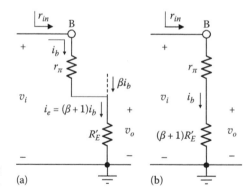

FIGURE 8.2.9
(a) i_e flowing through R'_E, as in Figure 8.2.9; (b) i_b flowing
through $(\beta+1)R'_E$ representing R'_E reflected to the base circuit. (a)   (b)

Concept: *A resistance in the emitter circuit can be reflected to the base circuit by multiplying it by $(\beta+1)$. Conversely, a resistance in the base circuit can be reflected to the emitter circuit by dividing it by $(\beta+1)$.*

It is seen from Figure 8.2.8b that:

$$r_{in} = \frac{v_i}{i_b} = r_\pi + (\beta+1)R'_E \cong (\beta+1)R'_E \cong r_\pi(1+g_m R'_E) \qquad (8.2.23)$$

The input resistance is increased by $(\beta+1)R'_E$, or multiplied by the factor $(1+g_m R'_E)$ because of the series–shunt negative feedback.

It also follows from voltage division in Figure 8.2.9b that:

$$A_v = \frac{v_o}{v_i} = \frac{(\beta + 1)R'_E}{r_{in}} = \frac{(\beta + 1)R'_E}{r_\pi + (\beta + 1)R'_E} \simeq \frac{g_m R'_E}{1 + g_m R'_E} \simeq 1 \qquad (8.2.24)$$

Because of this near-unity gain, the circuit is also known as an **emitter follower**. If the numerator and the denominator of Equation 8.2.24 are divided by r_π, $A_v \simeq \dfrac{g_m R'_E}{1 + g_m R'_E}$.

From Equation 8.2.2,

$$A'_v = \frac{v_o}{v'_{src}} = \frac{r_{in}}{r_{in} + R'_{src}} A_v = \frac{(\beta + 1)R'_E}{r_{in} + R'_{src}} = \frac{(\beta + 1)R'_E}{r_\pi + R'_{src} + (\beta + 1)R'_E} \qquad (8.2.25)$$

Applying current division at the output in Figure 8.2.8, the current gain is:

$$A_i = \frac{i_o}{i_b} = -(\beta + 1)\frac{R'_E}{R_L} \qquad (8.2.26)$$

If $R_L = 0$, then $R'_E/R_L = 1$, and the short-circuit current gain is $-(\beta + 1)$, as expected, the minus sign being due to the direction of i_o in Figure 8.2.8. When the emitter is short-circuited to ground, the short-circuit current flowing out of the emitter is $i_{sc} = (\beta + 1)i_b$, where, $i_b = v'_{src}/(r_\pi + R'_{src})$. Hence,

$$i_{sc} = \frac{\beta + 1}{r_\pi + R'_{src}} v'_{src} \qquad (8.2.27)$$

Comparing with Figure 8.2.1b, it is seen that:

$$G_m = \frac{\beta + 1}{r_\pi + R'_{src}} \qquad (8.2.28)$$

This is consistent with the result for the source follower because $G_m = g_m$ as β and r_π tend to infinity but their ratio is finite and equal to g_m.

R_{out} can be determined from Equations 8.2.18 and 8.2.25 since $A'_v = G_m R_{out}$ when the current source in Figure 8.2.1 is $G_m v'_{src}$. It follows that:

$$R_{out} = \frac{A'_v}{G_m} = R'_E \left\| \left(\frac{r_\pi + R'_{src}}{\beta + 1} \right) \right. \qquad (8.2.29)$$

$$\simeq \frac{R'_E}{1 + g_m R'_E}, \quad \text{for large } \beta \text{ and } r_\pi \qquad (8.2.30)$$

As $R'_E = R_E \| R_L \| r_o$, and $\dfrac{1}{R_{out}} = \dfrac{1}{R_E} + \dfrac{1}{R_L} + \dfrac{1}{r_o} + \dfrac{\beta + 1}{r_\pi + R'_{src}}$, then $\dfrac{1}{r_{out}} = \dfrac{1}{r_o} + \dfrac{\beta + 1}{r_\pi + R'_{src}}$, or:

$$r_{out} = r_o \left\| \left(\frac{r_\pi + R'_{src}}{\beta + 1} \right) \right. = r_o \left\| \left(r_e + \frac{R'_{src}}{\beta + 1} \right) \right. \qquad (8.2.31)$$

$$\simeq r_o \left\| \left(\frac{1}{g_m} + \frac{R'_{src}}{\beta + 1} \right) \right. \simeq \frac{r_o}{1 + g_m r_o} \simeq \frac{1}{g_m} \quad \text{when } \beta \text{ and } r_o \text{ are large} \qquad (8.2.32)$$

where $(r_\pi + R'_{src})/(\beta + 1)$ is $(r_\pi + R'_{src})$ reflected to the emitter. r_{out} can also be derived from v_{oc}/i_{sc} or by applying a test source between emitter and ground (Exercises 8.2.6 and 8.2.7).

Note that r_{in} is multiplied by nearly $(1 + g_m R'_E)$, while A_v and R_{out} are divided by nearly the same factor, all in accordance with what is to be expected of series–shunt feedback.

Exercise 8.2.5

(a) The βi_b current source in Figure 8.2.8 appears in parallel with R'_E. Transform this current source to a voltage source $\beta i_b R'_E$, which will be in series with r_π and R'_E, the current in the series combination being i_b. The voltage source $\beta i_b R'_E$, being a voltage drop in the direction of i_b, is equivalent to a resistance of $\beta i_b R'_E$. It follows that $r_{in} = r_\pi + \beta R'_E + R'_E$ as in Equation 8.2.23.

(b) Replace the βi_b source in Figure 8.2.8 by $g_m v_\pi$, where v_π, is the voltage across r_π. Transform this current source to a voltage source $g_m R'_E v_\pi$. Note that r_π in series with a voltage source that is proportional to v_π is equivalent to a resistance $r_\pi (1 + g_m R'_E)$. This is the same as Equation 8.2.23 when the substitution $\beta = g_m r_\pi$ is made and R'_E is added. ∎

Exercise 8.2.6

Derive Equation 8.2.29 by applying a test source v_x between emitter and ground. Deduce that in the absence of the "ideal" transistor, the resistance between emitter and ground is $R'_E || (r_\pi + R'_{src})$, and it becomes R_{out} in the presence of the transistor. Note also that reflection of $(r_\pi + R'_{src})$ accounts for the presence of the dependent current source of the transistor. ∎

Exercise 8.2.7

Derive r_{out} (Equation 8.2.31) from i_{sc} (Equation 8.2.27) and v_{oc} in Figure 8.2.8. ∎

Common-Source Amplifier with Source Resistor

If the source bypass capacitor (corresponding to C_E in Figure 8.2.3a) is omitted, the configuration becomes that of series–series, negative feedback, since the output is taken from the drain instead of the source. Because of the voltage drop across R_D, then for the same dc supply and a passive R_S, this resistor is generally smaller than that of CD amplifier. A_v will therefore be intermediate between that of the CS amplifier and the CD amplifier. However, r_{out} will now exceed that of the CS amplifier because of the series feedback due to R_S.

The small-signal equivalent circuit (Figure 8.2.10) is that of Figure 8.2.6a modified by moving R_L from source to ungrounded drain, with $R'_L = R_D || R_L$. We will first determine r_{out},

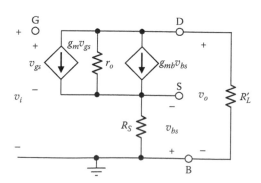

FIGURE 8.2.10
Small-signal equivalent circuit of the CS amplifier with unbypassed source resistor.

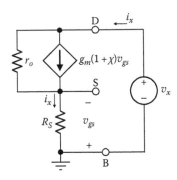

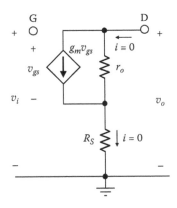

FIGURE 8.2.11
Determining the output resistance by applying a test source.

FIGURE 8.2.12
The small-signal equivalent circuit of Figure 8.2.10 under open circuit conditions.

the resistance looking into the drain with $v'_{src} = v_i = 0$. When the gate is grounded, $v_{gs} = v_{bs}$, so that the two sources can be combined in a single source $g_m(1 + \chi)v_{gs}$. If a test source v_x is applied between drain and ground (Figure 8.2.11), then since $v_{gs} = -R_S i_x$, the current source becomes $g_m(1 + \chi)R_S i_x$ of reversed polarity. Applying Equation 8.2.3, with $R_{E/S(eff)} = R_S$ and $g_{m(eff)} = g_m(1 + \chi)$:

$$r_{out} = r_o + R_S + g_m(1 + \chi)r_o R_S \cong r_o[1 + g_m(1 + \chi)R_S] \quad (8.2.33)$$

The resistance seen by the load is $r'_{out} = r_{out} \| R_D$.

We will determine A_{vo} in this case with the drain terminal open circuited, the small-signal circuit becoming as shown in Figure 8.2.12. No current flows through R_S, so there is neither a body effect nor a feedback signal. $v_i = v_{gs}$, the source $g_m v_{gs}$ flows through r_o, and $v_o = -g_m r_o v_i$. This gives $A_{vo} = -g_m r_o$. In Figure 8.2.1b, $r'_{out} = r_{out}$ on open circuit of the drain terminal, and $v_o = -G_m r_{out} v_i$, with the polarity of the $G_m v'_{src}$ reversed because the amplifier is inverting. Hence,

$$G_m = g_m \frac{r_o}{r_{out}} \cong \frac{g_m}{1 + g_m(1 + \chi)R_S} \quad (8.2.34)$$

When $R_S = 0$, $r_{out} = r_o$ and $G_m = g_m$ as expected.

The series–series negative feedback due to R_S has multiplied r_o by a factor of nearly $[1 + g_m(1 + \chi)R_S]$ and has divided the transconductance G_m by nearly the same factor. G_m is reduced because, for the same v_{gs}, v_i has to be increased to counteract the feedback voltage across R_S.

It follows from Equations 8.2.2 that:

$$A'_v = A_v = \frac{v_o}{v_i} = -G_m(r'_{out} \| R_L) = -G_m(r_{out} \| R'_L) = -G_m R_{out}$$

$$= -g_m \frac{r_o(r_{out} \| R'_L)}{r_{out}} = -\frac{g_m r_o R'_L}{R'_L + r_o + R_S + g_m(1 + \chi)r_o R_S}$$

$$\cong -\frac{g_m R'_L}{1 + g_m(1 + \chi)R_S}, \quad r_o \gg R'_L + R_S \quad (8.2.35)$$

where R_{out} is the total resistance to ground at the drain, including the effect of the transistor. If $R_S = 0$, $r_{out} = r_o$, and $A_v = -g_m(r_o \| R'_L)$, as expected.

Exercise 8.2.8

Ignore r_o and assume that g_m associated with R_S becomes $g_m(1 + \chi)$. Draw the signal-flow diagram for the CS amplifier with source resistance and deduce that $A_v = -\dfrac{g_m R'_L}{1 + g_m(1 + \chi)R_S} \cong -\dfrac{R'_L}{R_S}$ if $g_m(1 + \chi)R_S \gg 1$. Compare with Equation 8.2.35 with $r_o \to \infty$. ∎

Exercise 8.2.9

Since $A_{vo} = g_m r_o$, TEC at the output of the amplifier, with $R'_L \to \infty$, is a voltage source $A_{vo} v_i$ in series with r_{out}. Transform this to a current source, then add R'_L in parallel with r_{out} as in Figure 8.2.1, and show that Equation 8.2.35 results.   ∎

Common-Emitter Amplifier with Emitter Resistor

The circuit for the small-signal analysis of the CE amplifier with emitter resistor is shown in Figure 8.2.13. To derive r_{in}, the circuit can be redrawn as in Figure 8.2.14, where $R'_L = R_C \| R_L$ and the dependent current source has been transformed to the voltage source. From KVL around the mesh on the RHS, the current gain is:

$$A_i = \frac{i_o}{i_b} = \frac{i_c}{i_b} \frac{i_o}{i_c} = \frac{\beta r_o - R_E}{r_o + R_E + R'_L} \times \frac{R'_L}{R_L}$$

$$\cong \beta \quad \text{if } \beta, \quad r_o \gg R_E + R'_L \quad \text{and} \quad R_L \ll R_C \tag{8.2.36}$$

From KVL around the LHS mesh in Figure 8.2.14 and i_c/i_b from Equation 8.2.36,

$$r_{in} = \frac{v_i}{i_b} = r_\pi + R_E(\beta + 1)\frac{r_o + R'_L/(\beta + 1)}{r_o + R_E + R'_L} \tag{8.2.37}$$

$$\cong r_\pi + (\beta + 1)R_E, \quad \text{for large } r_o$$

$$\cong r_\pi(1 + g_m R_E), \quad \text{for large } r_o \text{ and } \beta$$

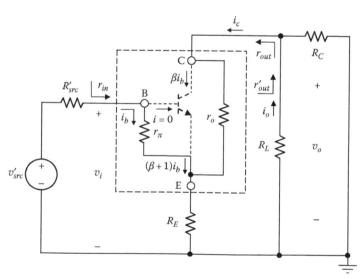

FIGURE 8.2.13
The CE amplifier with unbypassed emitter resistor under ac conditions.

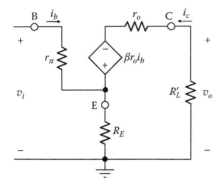

FIGURE 8.2.14
The small-signal equivalent circuit for determining the input resistance.

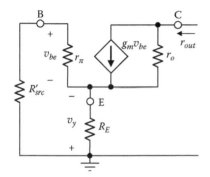

FIGURE 8.2.15
The small-signal equivalent circuit for determining the output resistance.

To apply Equation 8.2.3 and Figure 8.2.2 to determine r_{out}, we note that with $v'_{src} = 0$, the circuit becomes as shown in Figure 8.2.15. It is seen that $R_{E/S(\text{eff})} = R_E \| (r_\pi + R'_{src})$ and the current source in terms of v_y is $g_m v_y r_\pi / (r_\pi + R'_{src})$ so that $g_{m(\text{eff})} = g_m r_\pi / (r_\pi + R'_{src})$. Substituting in Equation 8.2.3:

$$r_{out} = r_o + R_{E/S(\text{eff})} + g_m r_o \frac{R_{E/S(\text{eff})} r_\pi}{r_\pi + R'_{src}}$$

$$= r_o + \frac{R_E}{R_E + r_\pi + R'_{src}} [R'_{src} + r_\pi (1 + g_m r_o)] = r_o + \frac{R_E}{r_\pi + R'_{src} + R_E} [r_\pi + R'_{src} + \beta r_o] \quad (8.2.38)$$

$$\cong r_o(1 + g_m R_E), \quad \text{for large } \beta, \text{ and } r_\pi \text{ large compared to } R_E \text{ and } R'_{src}. \quad (8.2.39)$$

If $R'_{src} = 0$, Equation 8.2.38 reduces to $r_{out} = r_o + (1 + g_m r_o)(R_E \| r_\pi)$, as expected. It follows from Figure 8.2.13 that $r'_{out} = r_{out} \| R_C$.

When the collector terminal is open circuited in Figure 8.2.14, $v_o = -\beta r_o i_b + R_E i_b$, where $i_b = \dfrac{v'_{src}}{r_\pi + R'_{src} + R_E}$. This gives:

$$A_{vo} = -\frac{\beta r_o - R_E}{r_\pi + R'_{src} + R_E} \quad (8.2.40)$$

When the collector terminal is open circuited in Figure 8.2.1, $r'_{out} = r_{out}$. With the polarity of the $G_m v'_{src}$ reversed because the amplifier is inverting, $G_m = -A_{vo}/r_{out}$, so that,

$$G_m = \frac{\beta r_0 - R_E}{r_0(r_\pi + R'_{src} + R_E) + R_E(r_\pi + R'_{src} + \beta r_0)} \tag{8.2.41}$$

$$\cong \frac{g_m}{1 + g_m R_E} \quad \text{for large } \beta, \text{ and } r_\pi \text{ large compared with } R_E \text{ and } R'_{src} \tag{8.2.42}$$

It is seen from Equations 8.2.37, 8.2.39, and 8.2.42 that the factor $(1 + g_m R_E)$ multiplies r_{in} and r_{out} and divides G_m, because of the series–series feedback introduced by R_E.

From Figure 8.2.1, with the polarity of $G_m v'_{src}$ reversed,

$$A'_v = \frac{v_o}{v_{src}} = -G_m(r'_{out} \| R_L) = -G_m(r_{out} \| R'_L) = -G_m R_{out} \tag{8.2.43}$$

A_v can be determined from the relation $A_v = -G_m(r_{out} \| R'_L)$, with $R'_{src} = 0$, or from conventional circuit analysis (Problem P8.2.2), as:

$$A_v = -\frac{\beta r_0 - R_E}{r_\pi(r_0 + R'_L + R_E) + R_E[r_0(1 + \beta) + R'_L]} R'_L \tag{8.2.44}$$

If $r_0 \to \infty$, the input is isolated from the load by the current source, with the following consequences: (1) $r_{out} \to \infty$, (2) R_E can be reflected to the input side to give $r_{in} = r_\pi + R_E(\beta + 1)$, as for the CC amplifier (Equation 8.2.23). A finite r_0 therefore reduces r_{in}, as given by Equation 8.2.37, by diverting current away from R_E. (3) From Equation 8.2.36, $A_i = \beta$, in accordance with the definition of β. (4) From Equation 8.2.41,

$$G_m = \frac{\beta}{r_\pi + R'_{src} + R_E(\beta + 1)} \tag{8.2.45}$$

(5) From Equation 8.2.43,

$$A'_v = -G_m R'_L = -\frac{\beta R'_L}{r_\pi + R'_{src} + R_E(\beta + 1)} \tag{8.2.46}$$

(6) From Equation 8.2.2, $A_v = \dfrac{r_{in} + R'_{src}}{r_{in}} A'_v = -\dfrac{\beta R'_L}{r_\pi + R_E(\beta + 1)} \cong -\dfrac{g_m R'_L}{1 + g_m R_E}$. If β is large, which implies that r_π is large,

$$A'_v \cong A_v \cong -\frac{R'_L}{R_E} \tag{8.2.47}$$

The interpretation is that if β is large, $i_b \cong 0$ and $\alpha \cong 1$. The voltage drop in r_π is negligible so that the input voltage is nearly equal to the voltage across R_E. Equation 8.2.47 follows since the same current flows in R'_L and R_E.

Common-Gate Amplifier

A CG amplifier is illustrated in Figure 8.2.16a, its ac circuit in Figure 8.2.16b, and its small-signal equivalent circuit in Figure 8.2.16c, where $v'_{src} = v_{src}R_S/(R_{src} + R_S)$, $R'_{src} = R_S\|R_{src}$, $R'_L = R_D\|R_L$, and the dependent current source is reversed in terms of v_{sg}. Note that since the gate is grounded and the body terminal is at ac small-signal ground, the effective transconductance is $g_m(1 + \chi)$, as indicated in Figure 8.2.16c, which means that the factor $(1 + \chi)$ multiplies all occurrences of g_m. In the CD amplifier and in the CS amplifier with unbypassed R_S, $v_{gs} \neq v_{bs}$, so that $(1 + \chi)$ does not multiply all occurrences of g_m.

To determine r_{in}, we remove the source v'_{src} and R'_{src}, and apply a test source v_x between S and G. The source current is $i_x = -i_d$ in Figure 8.2.16c. The current in R'_L is i_x and that in r_o is $i_x - g_m(1 + \chi)v_x$. From KVL, $v_x = R'_L i_x + r_o[i_x - g_m(1 + \chi)v_x]$, which gives:

$$r_{in} = \frac{v_x}{i_x} = \frac{r_o + R'_L}{1 + g_m(1 + \chi)r_o} \tag{8.2.48}$$

$$\cong \frac{1}{g_m(1 + \chi)} \quad \text{for } R'_L \ll r_o \text{ and } g_m(1 + \chi)r_o \gg 1 \tag{8.2.49}$$

The term $r_o/[1 + g_m(1 + \chi)r_o]$ in Equation 8.2.48 is $r_o\|[1/g_m(1 + \chi)]$ so that the current source contributes a resistance $1/[g_m(1 + \chi)]$ in parallel with r_o. The term $R'_L/[1 + g_m(1 + \chi)r_o]$ is R'_L *reflected* to the input (Exercise 8.2.10).

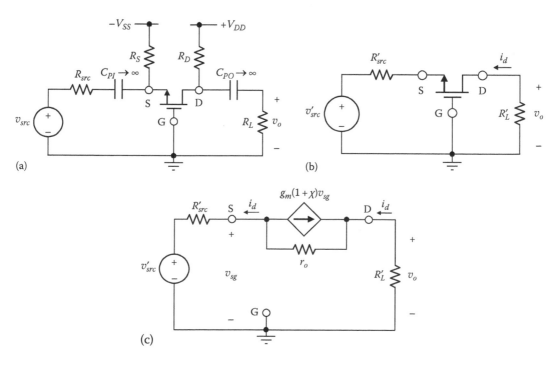

FIGURE 8.2.16
(a) CG amplifier; (b) ac circuit; (c) small-signal equivalent circuit.

When a source v_x is applied between S and G, $i_x = v_x/r_{in}$ and $v_o = R'_L i_x$, so that,

$$A_v = \frac{R'_L}{r_{in}} = \frac{R'_L[1 + g_m(1 + \chi)r_o]}{r_o + R'_L} \cong g_m(1 + \chi)(R'_L \| r_o) \quad \text{for } g_m(1 + \chi)r_o \gg 1 \qquad (8.2.50)$$

A_v is very similar to that of the CS amplifier (Equation 8.2.13), with g_m multiplied by $(1 + \chi)$, and no phase inversion. Referring to Figure 8.2.16b, if the source voltage increases, i_d decreases, which increases v_o, without phase inversion.

The gain $A'_v = v_o/v'_{src}$ is:

$$A'_v = \frac{r_{in}}{r_{in} + R'_{src}} A_v = \frac{R'_L}{r_{in} + R'_{src}} = \frac{R'_L[1 + g_m(1 + \chi)r_o]}{R'_{src} + R'_L + r_o + g_m(1 + \chi)r_o R'_{src}} \qquad (8.2.51)$$

$$\cong \frac{R'_L}{R'_{src}} \quad \text{for } R'_{src} \gg r_{in} \qquad (8.2.52)$$

The approximate result of Equation 8.2.52 is to be expected on the basis that the input and output currents are approximately v'_{src}/R'_{src} so that the output voltage is nearly $(R'_L/R'_{src})v'_{src}$.

To determine the output resistance, we apply Equation 8.2.3 to obtain:

$$r_{out} = r_o + R'_{src} + g_m(1 + \chi)r_o R'_{src} \cong r_o[1 + g_m(1 + \chi)R'_{src}] \qquad (8.2.53)$$

This result is significant in that r_{out} *increases with* R'_{src} *virtually without limit*. Again, it is useful to think of the source resistance as being *reflected* at the output by multiplying it by $1 + g_m(1 + \chi)r_o$. It follows that $R_{out} = r_{out} \| R'_L$.

If the drain in Figure 8.2.16c is short circuited to ground, $i_{sc} = v'_{src}/(r_{in} + R'_{src})$, where r_{in} is given by Equation 8.2.48 with $R'_L = 0$. It follows that:

$$G_m = \frac{1}{r_{in} + R'_{src}} = \frac{1 + g_m(1 + \chi)r_o}{R'_{src} + r_o + g_m(1 + \chi)r_o R'_{src}} = \frac{1}{R'_{src} + r_o \| [1/g_m(1 + \chi)]} \qquad (8.2.54)$$

$$\cong \frac{1}{R'_{src}}, \quad \text{for } 1/g_m(1 + \chi) \ll r_o \text{ and } R'_{src} \qquad (8.2.55)$$

It can be readily verified that $v_o = G_m R_{out} v'_{src} = G_m(R_{out} = r_{out} \| R'_L)v'_{src} = A'_{src} v'_{src}$.

From Figure 8.2.16c the current gain is $i_d/(-i_d)$, which is -1.

Exercise 8.2.10

Transform the current source in Figure 8.2.16c to a voltage source in series with r_o, then show that the resistance seen at the source is given by Equation 8.2.48.   ∎

Exercise 8.2.11

Deduce from Figure 8.2.16c that $A_{vo} = 1 + g_m(1 + \chi)r_o$ ∎

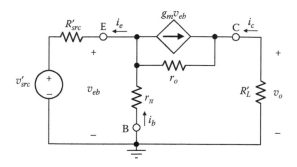

FIGURE 8.2.17
Small-signal equivalent circuit of CB amplifier.

Exercise 8.2.12

Show that Equations 8.5.53, 8.5.54, and $A'_v = G_m(R'_L \| r_{out})$ are consistent with $A'_v = R'_L/(r_{in} + R'_{src})$ from Equations 8.2.2. ∎

Common-Base Amplifier

The analysis of the CB amplifier closely parallels that of the CG amplifier. The small-signal equivalent circuit is shown in Figure 8.2.17, from which it follows that r_{in} is that of the MOSFET (Equation 8.2.48) in parallel with r_π, ignoring χ. Thus,

$$r_{in} = r_\pi \left\| \left(\frac{r_o + R'_L}{1 + g_m r_o} \right) = \frac{r_\pi(r_o + R'_L)}{r_o + R'_L + r_\pi(1 + g_m r_o)} = \frac{r_o + R'_L}{1 + (r_o/r_e) + R'_L/r_\pi} \right. \tag{8.2.56}$$

$$\cong r_e \cong \frac{1}{g_m}, \quad \text{for } R'_L \ll r_o \tag{8.2.57}$$

If $R'_L \ll r_o$, the collector is effectively short-circuited to the base. r_{in} is then the parallel combination of r_o, r_π, and $1/g_m$ due to the source, where $r_\pi \|(1/g_m) = r_e$. Note that $r_{in} = r_e$ was derived in Section 6.1, Chapter 6, for the CB amplifier in the absence of base-width modulation.

If a voltage source v_{eb} is applied in Figure 8.2.17, r_π is redundant, and A_v is the same as for the CG amplifier. From Equation 8.2.50, ignoring χ,

$$A_v = \frac{v_o}{v_e} = \frac{R'_L(1 + g_m r_o)}{r_o + R'_L} \cong g_m(r_o \| R'_L), \quad \text{for } g_m r_o \gg 1 \tag{8.2.58}$$

This is the same as $A_v = -g_m(r_o \| R'_L)$ for the CE configuration (Equation 8.2.8) but without phase inversion. The gain $A'_v = v_o/v'_{src}$ is:

$$A'_v = \frac{r_{in}}{r_{in} + R'_{src}} A_v = \frac{R'_L r_\pi(1 + g_m r_o)}{R'_{src}[r_o + R'_L + r_\pi(1 + g_m r_o)] + r_\pi(r_o + R'_L)} \tag{8.2.59}$$

$$\cong \frac{R'_L}{R'_{src}}, \quad \text{for large } r_o \gg R'_{src} \text{ and } g_m r_o \gg 1 \tag{8.2.60}$$

The approximate result of Equation 8.2.60 is to be expected on the basis that the input and output currents are approximately v'_{src}/R'_{src} so that the output voltage is nearly $(R'_L/R'_{src})v'_{src}$.

The output resistance r_{out} can be determined from Equation 8.2.3 noting that $R_{E/S(eff)} = R'_{src}\|r_\pi$ and $g_{m(eff)} = g_m$. Thus,

$$r_{out} = r_o + (R'_{src}\|r_\pi) + g_m r_o(R'_{src}\|r_\pi) \cong r_o[1 + g_m(R'_{src}\|r_\pi)]$$
$$\cong \beta r_o \quad \text{for large } \beta, \text{ and } r_\pi \ll R'_{src} \text{ and } r_o \tag{8.2.61}$$

Concept: *Whereas r_{out} of the CG amplifier increases virtually without limit with R'_{src}, r_{out} of the CB amplifier is limited for large R'_{src} to βr_o because of r_π (Equation 8.2.61).*

G_m can be determined from the short-circuit current in Figure 8.2.17. With $R'_L = 0$, $i_{src} = -i_e \dfrac{v'_{src}}{R'_{src} + r_\pi\|r_o\|(1/g_m)}$ and $i_{sc} = \dfrac{r_\pi}{r_\pi + r_o\|(1/g_m)} i_{src}$. This gives:

$$G_m = \frac{1}{R'_{src} + r_o\|(1/g_m)[1 + R'_{src}/r_\pi]} \tag{8.2.62}$$

$$\cong \frac{1}{R'_{src}}, \quad \text{for large } \beta, \text{ and } 1/g_m \text{ small compared to } r_o \text{ and to } R'_{src} \tag{8.2.63}$$

From Equations 8.2.2, $A_i = -A_v r_{in}/R_L$, which gives:

$$A_i = -\frac{r_\pi(1 + g_m r_o)}{R'_L + r_o + r_\pi(1 + g_m r_o)} \tag{8.2.64}$$

If $R'_L = 0$ and $r_o \to \infty$, $A_i = -\alpha$, as expected.

The CG/CB amplifier has low input resistance, high output resistance, and, ideally, unity current gain. This is in contrast to the CD/CC amplifier, which has high input resistance, low output resistance and, ideally, unity voltage gain. If the input to a CG/CB amplifier is a current source of relatively low source resistance, the output is a current source of almost the same value but of a high source resistance. In this sense, the CG/CB amplifier can act as a **current buffer**, just as the CD/CC amplifier can act as a **voltage buffer** that transforms a voltage source of relatively high source resistance to a voltage source of almost the same value but of low source resistance.

Table 8.2.1 shows the relations for the three amplifier configurations, the way they are derived, and the approximations that apply under specified conditions. It is an important exercise to derive the approximate relations directly from the relevant ac or small-signal equivalent circuit (Exercise 8.2.13). The properties of the three transistor amplifier are summarized qualitatively in Table 8.2.2 and are examined quantitatively in Exercises 8.2.14 and 8.2.15, as a straightforward application of the equations derived in this section.

Exercise 8.2.13

Derive the approximate relations in Table 8.2.1 directly from the relevant ac or small-signal equivalent circuit. ∎

TABLE 8.2.1

Small Signal Relations for Amplifier Configurations

Configuration	r_{in}	r_{out}	A_i	A_v	G_m
CE	r_π (Equation 8.2.5)	r_o (Equation 8.2.7)	$\beta \dfrac{r_o \parallel R_C \parallel R_L}{R_L}$ (Equation 8.2.6), from r_{in} and current division $\cong \beta,\ R_L \ll R_C$ and r_o	$-g_m(r_o \parallel R_L')$ (Equation 8.2.8)	$G_m = \dfrac{g_m r_\pi}{r_\pi + R_{src}}$ (Equation 8.2.9) $\cong g_m,\ r_\pi \gg R_{src}'$
CS	∞	r_o (Equation 8.2.12)	∞	$-g_m(r_o \parallel R_L')$ (Equation 8.2.13)	$G_m = g_m$
CE + R_E	$r_\pi + R_E(\beta+1)\dfrac{r_o + R_L'/(\beta+1)}{r_o + R_E + R_L'}$ (Equation 8.2.37), from Figure 8.2.14 $\cong r_\pi(1+g_m R_e)$, large β and r_o	$r_o + R_E \dfrac{R_{src}' + r_\pi(1+g_m r_o)}{R_E + r_\pi + R_{src}'}$ (Equation 8.2.38), from (Equation 8.2.3) $\cong r_o(1+g_m R_e)$, large $\beta,\ r_\pi \gg R_E$ and R_{src}'	$\dfrac{(\beta r_o - R_E)R_L'}{R_L(r_o + R_E + R_L')}$ (Equation 8.2.36), from Figure 8.2.14 $\cong \beta,\ r_o \gg R_E + R_L'$ and $R_L \ll R_C$	$\dfrac{(\beta r_o - R_E)R_L'}{r_\pi(R_L'' + R_E) + R_E(R_L'' + \beta r_o)}$ (Equation 8.2.44), from $G_m R_{out}$ with $R_L'' = r_o + R_L'$ $\cong -\dfrac{g_m R_L'}{1+g_m R_E}$, large β and r_o	$\dfrac{\beta r_o - R_E}{r_o(R_{src}'' + R_E) + R_E(R_{src}'' + \beta r_o)}$ (Equation 8.2.41), from A_{vo} with $R_{src}'' = r_\pi + R_{src}'$ $\cong \dfrac{g_m}{1+g_m R_E}$, large $\beta,\ r_\pi \gg R_E$ and R_{src}'
CS + R_S	∞	$r_o + R_S + g_m(1+\chi)r_o R_S$ (Equation 8.2.33), from (Equation 8.2.3) $\cong r_o[1 + g_m(1+\chi)R_S]\ r_o \gg R_S$	∞	$-G_m(r_{out} \parallel R_L')$ (Equation 8.2.35) $\cong -\dfrac{g_m R_L'}{1+g_m(1+\chi)R_S},\ r_o \gg R_L + R_S$	$g_m \dfrac{r_o}{r_{out}}$ (Equation 8.2.34), from A_{vo} $\cong \dfrac{g_m}{1+g_m(1+\chi)R_S},\ r_o \gg R_S$
CC	$r_\pi + (\beta+1)R_E'$ (Equation 8.2.23), resistance in emitter reflected to base in Figure 8.2.8	$r_o \parallel \left(r_e + \dfrac{R_{src}'}{\beta+1} \right)$ (Equation 8.2.31), from the reflection of resistance in base to emitter $\cong 1/g_m$, large β and r_o	$-(\beta+1)\dfrac{R_E'}{R_L}$ (Equation 8.2.26), from current division in Figure 8.2.8 $\cong -(\beta+1),\ R_E' \gg R_L$	$\dfrac{(\beta+1)R_E'}{r_\pi + (\beta+1)R_E'}$ (Equation 8.2.24), from voltage division in Figure 8.2.8 $\cong 1,\ (\beta+1)R_E' \gg r_\pi$	$G_m = \dfrac{\beta+1}{r_\pi + R_{src}}$ (Equation 8.2.28), from Figure 8.2.8

(continued)

TABLE 8.2.1 (continued)

Small Signal Relations for Amplifier Configurations

Configuration	r_{in}	r_{out}	A_i	A_v	G_m
CD	∞	$\dfrac{r_o}{1+g_m(1+\chi)r_o}$ (Equation 8.2.17), from Figure 8.2.6a and b $\cong \dfrac{1}{g_m(1+\chi)}$, $g_m(1+\chi)r_o \gg 1$	∞	$\dfrac{g_m R_S'}{1+g_m(1+\chi)R_S'}$ (Equation 8.2.22), from Figure 8.2.6a and b, with $R_S'=r_o\|R_S\|R_L$ $\cong \dfrac{1}{(1+\chi)}$, $g_m(1+\chi)R_S' \gg 1$	$G_m=g_m$ (Equation 8.2.19), from Figure 8.2.6b
CB	$\dfrac{r_o+R_L'}{1+(r_o/r_e)+R_L'/r_\pi}$ (Equation 8.2.56), from figure 8.2.17 $\cong 1/g_m$, $R_L' \ll r_o$	$r_o+(R_{src}'\|r_\pi)+g_m r_o(R_{src}'\|r_\pi)$ (Equation 8.2.61), from Equation 8.2.3 $\cong \beta r_o$, large β, $r_\pi \ll R_{src}'$ and r_o	$-\dfrac{r_\pi(1+g_m r_o)}{R_L'+r_o+r_\pi(1+g_m r_o)}$ (Equation 8.2.64), from Figure 8.2.17 $\cong -\alpha$, $R_L'=0$ and $r_o \to \infty$	$\dfrac{R_L'(1+g_m r_o)}{r_o+R_L'}$ (Equation 8.2.58), from figure 8.2.17 $\cong g_m(r_o\|R_L)$, $g_m r_o \gg 1$	$\dfrac{1}{R_{src}'+r_o\|(1/g_m)[1+R_{src}'/r_\pi]}$ (Equation 8.2.62), from Figure 8.2.17 $G_m \cong \dfrac{1}{R_{src}'}$, large β, $1/g_m \ll r_o$ and R_{src}'
CG	$\dfrac{r_o+R_L'}{1+g_m(1+\chi)r_o}$ (Equation 8.2.48), from Figure 8.2.16c $\cong \dfrac{1}{g_m(1+\chi)}$, $R_L' \ll r_o$ and $g_m(1+\chi)r_o \gg 1$	$r_o+R_{src}'+g_m(1+\chi)r_o R_{src}'$ (Equation 8.2.53), from Equation 8.2.3 $\cong r_o[1+g_m(1+\chi)R_{src}']r_o \gg R_{src}'$	-1	$\dfrac{R_L'[1+g_m(1+\chi)r_o]}{r_o+R_L'}$ (Equation 8.2.50), from Figure 8.2.16c $\cong g_m(1+\chi)R_L'$, $R_L' \ll r_o$ and $g_m(1+\chi)r_o \gg 1$	$\dfrac{1}{R_{src}'+r_o\|[1/g_m(1+\chi)]}$ (Equation 8.2.54), from Figure 8.2.16c $\cong \dfrac{1}{R_{src}'}$, $1/g_m(1+\chi) \ll r_o$ and R_{src}'

TABLE 8.2.2

Comparison of Amplifier Configurations

Property	CE/CS	CC/CD	CB/CG
Voltage gain	High	Virtually 1	Moderate to high, depending on source resistance
Current gain	High for CE, virtually infinite for CS	High for CC, virtually infinite for CD	Virtually −1
Input resistance	Moderate for CE, virtually infinite for CS	Very high for CC, virtually infinite for CD	Low
Output resistance	Moderate	Low	High

Exercise 8.2.14

Given a BJT having the following parameters: $\beta = 100$, $V_A = 100$ V, and $I_C = 2$ mA. It is required to compare r_{in}, r_{out}, A'_v, and A_i for the (a) CE, (b) CE with $R_E = 0.5$ kΩ, $r_o \rightarrow \infty$; (c) CC, and (d) CB amplifiers, assuming in all cases $R_{src} = 0.5$ kΩ, $R_C = 2$ kΩ, and R_L infinite.

Answers: (a) 1.3 kΩ, 50 kΩ, −107, 96; (b) 51.8 kΩ, ∞, −3.8, 100; (c) 203 kΩ, 18 Ω; 0.99, 101; (d) 13 Ω, 5 MΩ, 3.9, −0.99. ∎

Exercise 8.2.15

Given an E-NMOS transistor having $k_n = 0.5$ mA/V^2, $V_{tn} = 2$ V, $r_o = 100$ kΩ, $\chi = 0.2$, and $I_D = 1$ mA. It is required to compare A_v and r_{out} for the (a) CS, (b) CS with $R_S = 0.5$ kΩ, (c) CD, and (d) CG amplifiers, assuming an active load equivalent to $R_D = 100$ kΩ.

Answers: (a) −50, 100 kΩ; (b) 38.5, 160 kΩ; (c) 0.82, 0.82 kΩ; (d) 61, 50 kΩ. ∎

Before ending this section, it is of interest to highlight some expressions that have been repeatedly encountered in several cases. Consider the BJT small-signal equivalent circuit of Figure 8.2.18, in which the base and collector are connected together to ground and a test source v_x is applied to determine the resistance seen looking into the emitter. From KCL,

$i_x = -\dfrac{v_{be}}{r_\pi} - \dfrac{v_{be}}{r_o} - g_m v_{be} = v_x \left(\dfrac{1}{r_\pi} + \dfrac{1}{r_o} + g_m \right)$. It follows that the effective resistance r_{eff} seen by the source is:

$$r_{eff} = r_\pi \| r_o \| (1/g_m) = \left(\frac{r_\pi}{\beta + 1} \right) \| r_o = r_e \| r_o \cong r_e \cong 1/g_m, \quad \text{for } r_o \gg r_e \qquad (8.2.65)$$

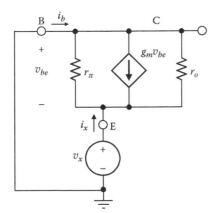

FIGURE 8.2.18
Small-signal equivalent circuit for determining resistance seen by the source.

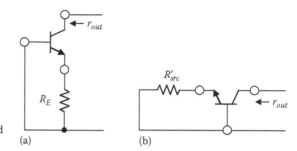

FIGURE 8.2.19
AC circuit for (a) CE amplifier with unbypassed
emitter resistor, and (b) CB amplifier.

For a MOSFET, $r_\pi \rightarrow \infty$, so that,

$$r_{eff} = r_o\|(1/g_m) \cong 1/g_m, \quad \text{for } r_o \gg 1/g_m \tag{8.2.66}$$

But what does r_{eff} represent? It can, in fact, represent one of three resistances:

1. The resistance between the two terminals of a diode-connected BJT transistor (Figure 6.3.8, Chapter 6) or a diode-connected enhancement MOSFET (Figure 5.4.5, Chapter 5). Clearly, the small-signal equivalent circuit is of the form of Figure 8.2.18 in both cases.

2. The output resistance r_{out} of a CC/CD amplifier when $R_{src} \rightarrow 0$, as can be seen from Figure 8.2.4b for a MOSFET and Figure 8.2.8 for a BJT. If $R_{src} = 0$ and r_{out} is to be determined with $v_{src} = 0$, the base and collector will be connected together to ground.

3. The input resistance r_{in} of a CB/CG amplifier when $R_L' \rightarrow 0$, that is, when the output is short-circuited to the ac signal, as can be seen from Figure 8.2.16b for a MOSFET or a BJT.

Another set of common relations is r_{out} as determined by Equation 8.2.2. We have seen that $r_{out} \cong r_o(1 + g_m R)$ for an unbypassed emitter/source resistor, where $R = R_E$ for a BJT and $R = R_S$ for a MOSFET. This same expression for r_{out} applies to the CB amplifier, as can be seen from comparing the ac circuits for the two BJT cases (Figure 8.2.19). Similar considerations apply to the CG amplifier. Note that if $R_E = R_{src}' = 0$, $r_{out} = r_o$ in both cases.

8.3 High-Frequency Response

The frequency response of dc amplifiers, including IC operational amplifiers, extends down to zero frequency. At high frequencies, the response decreases with frequency because of the effects of stored charges and the practical limits on physical processes involved, such as carrier transit time through the base a BJT or through the channel of a MOSFET. The high-frequency effects are accounted for in small-signal equivalent circuits by the addition of internal capacitances, as discussed in Section 5.3, Chapter 5, and in Section 6.2, Chapter 6.

Miller's Theorem

This is a useful theorem that allows rearranging an impedance or admittance so as to simplify the analysis of an amplifier circuit. In Figure 8.3.1a, a voltage v_1 between nodes a

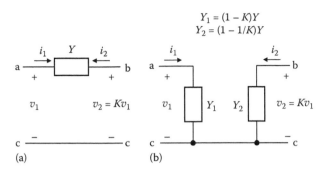

$$Y_1 = (1 - K)Y$$
$$Y_2 = (1 - 1/K)Y$$

FIGURE 8.3.1
(a) Admittance between two nodes related by a voltage amplification factor K; (b) equivalent admittances Y_1 and Y_2 in accordance with Miller's theorem.

and c appears amplified by a factor K between nodes b and c, as when port ac is the input port of an amplifier and port bc its output port. It follows that $i_1 = (v_1 - v_2)Y = v_1(1 - K)Y$ and $i_2 = (v_2 - v_1)Y = v_1(K - 1)Y = v_2(1 - 1/K)Y$. The admittance Y can therefore be replaced by two admittances Y_1 and Y_2, as in Figure 8.3.1b, where $Y_1 = Y(1 - K)$ and $Y_2 = Y(1 - 1/K)$, without altering i_1 and i_2. If K is negative and $|K| \gg 1$, then,

$$Y_1 \cong |K|Y \quad \text{and} \quad Y_2 \cong Y \tag{8.3.1}$$

The effective admittance between nodes a and c is Y magnified by nearly $|K|$. This is known as **Miller's theorem**, or the **Miller effect**, whose importance is in the following:

Concept: *The magnification of internal capacitances between input and output by the Miller effect limits the high-frequency response of some simple amplifier stages.*

It should be pointed out that when applying the Miller effect to amplifiers, then, strictly speaking, K is the voltage amplification in the presence of the capacitance between input and output. This makes K, in general, a function of the capacitance, and hence of frequency, and defeats the purpose of applying the Miller effect to simplify analysis. Hence the **Miller approximation** is usually applied, whereby K is considered to be the amplifier gain with no capacitances in the circuit. This inevitably introduces some error in the calculation.

Poles and Zeros of Transfer Function

Before deriving expressions for frequency response, it is important to clarify the nature of poles and zeros of the transfer functions of amplifiers. The transfer function of a single amplifier stage can be represented, simply but without loss of generality, as having a zero and two poles:

$$H(s) = K \frac{s + \omega_z}{(s + \omega_{p1})(s + \omega_{p2})} \tag{8.3.2}$$

where the zero is at $-\omega_z$ and the poles are at $-\omega_{p1}$ and $-\omega_{p2}$. The following should be noted:

1. In a stable amplifier, ω_{p1} and ω_{p2} are positive real numbers, and the poles are on the negative real axis of the s plane. ω_z is a real number, which could be positive, negative, or zero.

2. The poles $-\omega_{p1}$ and $-\omega_{p2}$ are formal poles and not true poles. A true pole arises from a term, such as $(s^2 + \omega_p^2)$ so that if $s = \pm j\omega_p$, the term is zero and $H(s) \to \infty$. The poles $\pm j\omega_p$ are on the imaginary axis of the s plane and represent a physically realizable frequency ω_p that can be physically generated or measured. Poles that are not on the imaginary axis do not represent physically realizable frequencies (Sabah, 2008, p. 608). Thus, when $s = -\omega_{p1}$ in Equation 8.3.2, for example, $H(s) \to \infty$, but $-\omega_{p1}$ is not a physically realizable frequency. If, at the physical frequency ω_{p1}, we substitute $s = j\omega_{p1}$ in Equation 8.3.2 we obtain:

$$|H(s)| = K \frac{\sqrt{\omega_{p1}^2 + \omega_z^2}}{\sqrt{2\omega_{p1}^2(\omega_{p1}^2 + \omega_{p2}^2)}} \tag{8.3.3}$$

and

$$\angle H(s) = \tan^{-1}\left(\frac{\omega_{p1}}{\omega_z}\right) - \tan^{-1}(1) - \tan^{-1}\left(\frac{\omega_{p1}}{\omega_{p2}}\right) \tag{8.3.4}$$

3. Similar comments apply to the zero $-\omega_z$. When $\omega_z = 0$, the s term in the numerator of Equation is a true zero at the origin of the s plane and is on the imaginary axis. It makes $H(s) = 0$ at $s = 0$, corresponding to dc conditions.

4. It is customary in frequency response analysis of amplifiers to refer to the poles and zeros in Equation 8.3.2 as positive quantities. Thus, a term such as $(s + 3000)$ in the denominator is said to correspond to a pole frequency of $+3000$ rad/s. Equation 8.3.3 is independent of the signs of ω_p and ω_z. In Equation 8.3.4, the argument of the arctan function is the ratio of the imaginary part to the real part so that the correct signs of ω_p and ω_z are automatically included.

Common Emitter/Source Amplifier

The basic, small-signal CE circuit to be analyzed, shown in Figure 8.3.2a, is that of Figure 6.2.10, Chapter 6, but ignoring r_μ, which is normally very large, and r_E and r_C, which are normally small in modern transistors. r_B can be included, in a simplified manner, with the source resistance. $R_L' = R_L \| R_C$, and C_L is the sum of C_{cs}, which is small in modern IC transistors, and load capacitance, which includes the wiring capacitance and the input capacitance of any succeeding amplifier stage. Note that because C_μ appears between base and collector, the replacement of the $g_m v_{be}$ current source by a βi_b current source (Figure 6.2.7, Chapter 6) is not valid.

The circuit of Figure 8.3.2a can be simplified to that in Figure 8.3.2b, where $V_{src}' = V_{src} r_\pi / (r_\pi + r_B + R_{src})$, $R_{src}' = r_\pi \| (r_B + R_{src})$, $R_{out} = r_o \| R_C \| R_L$, and $Z_L = R_{out} \| (1/sC_L)$.

The circuit of Figure 8.3.2b is analyzed rigorously in Section ST8.7. It can also be analyzed using the method of open-circuit and short-circuit time constants (OCSCTC) (Section ST8.9). This is an exact method for determining the poles of the response without deriving the transfer function. Where one of the poles is dominant, an approximate procedure can be applied based on the open-circuit time constants only (OCTC). We will pursue here an approximate analysis using Miller's theorem.

To apply Miller's theorem using the Miller approximation, the following assumptions are made: (1) $1/sC_L \gg R_{out}$ over the frequency range of interest, so that Z_L is approximated by

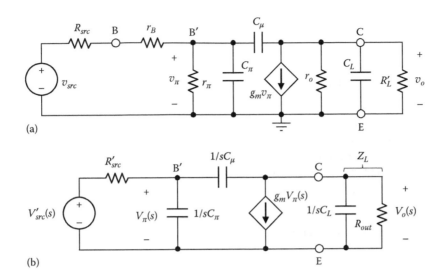

(a)

(b)

FIGURE 8.3.2
(a) High-frequency, small-signal equivalent circuit of CE amplifier; (b) simplified equivalent circuit (a).

R_{out}. (2) $R_{out} \ll 1/sC_\mu$ so that the current in R_{out} is mainly due to the source $g_m V_\pi(s)$, and the current in C_μ is negligible in comparison. Under these conditions, $V_o(s) = -g_m R_{out} V_\pi(s)$, and Miller's theorem can be applied to nodes B' and C, with $Y = sC_\mu$ and $K = -g_m R_{out}$. It follows that C_μ is effectively multiplied by $(1 + g_m R_{out})$ so that the capacitance between B' and E is:

$$C_T = C_\pi + C_\mu(1 + g_m R_{out}) \tag{8.3.5}$$

Moreover, a capacitance $C_\mu[1 + 1/(1 + g_m R_{out})] \cong C_\mu$ appears in parallel with C_L. The circuit becomes as shown in Figure 8.3.3, where $C'_L = C_L + C_\mu$. By multiplying the transfer functions $V_\pi(s)/V'_{src}(s)$ and $V_o(s)/V_\pi(s)$, it follows that:

$$A'_v(s) = \frac{V_o(s)}{V'_{src}(s)} = -\frac{g_m R_{out}}{(1 + sC_T R'_{src})(1 + sC'_L R_{out})} = -\frac{g_m R_{out}}{(1 + s/\omega_{p1})(1 + s/\omega_{p2})} \tag{8.3.6}$$

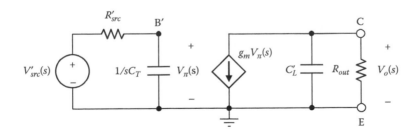

FIGURE 8.3.3
Small-signal equivalent circuit for determining frequency response.

The following should be noted:

1. The poles are at $\omega_{p1} = 1/C_T R'_{src}$ and $\omega_{p2} = 1/C'_L R_{out}$. If ω_{p1} is dominant, the 3-dB cutoff frequency of $A'_v(s)$ is nearly ω_{p1}. If ω_{p1} and ω_{p2} are close together, the 3-dB cutoff frequency is determined as the frequency that makes the denominator of Equation 8.3.6 equal to $\sqrt{2}$.

2. The Miller approximation neglected C_μ in calculating the gain K. In fact, C_μ couples the input and output, with the following consequences: (a) ω_{p1} is no longer dependent on C_T only and ω_{p2} on C'_L only. When ω_{p1} is dominant, a rigorous analysis (Section ST8.7) gives ω_{p1} as the reciprocal of the coefficient of s in the denominator of Equation 8.3.6. This is because the coefficient of s is $\left(\dfrac{1}{\omega_{p1}} + \dfrac{1}{\omega_{p2}}\right)$ and that of s^2 is $\dfrac{1}{\omega_{p1}\omega_{p2}}$. If $\omega_{p1} \gg \omega_{p2}$, the coefficient of s^2 can be neglected in the vicinity of ω_{p1}, and $1/\omega_{p2}$ can be neglected in the coefficient of s, so that,

$$\omega_{p1} = \frac{1}{C_T R'_{src} + (C_\mu + C_L)R_{out}} \tag{8.3.7}$$

(b) The response has a zero that is absent from Figure 8.3.3 based on the Miller approximation, but which can be determined from the circuit of Figure 8.3.2b. If $V_o(s)$ is zero and the output terminals are short-circuited, the current in the short circuit is zero. This means that $g_m V_\pi(s) - sC_\mu V_\pi(s) = 0$, which gives a zero at $s_z = g_m/C_\mu$, or at a frequency:

$$\omega_z = \frac{g_m}{C_\mu} \tag{8.3.8}$$

3. Because the circuit of Figure 8.3.2b has three capacitors, it may be thought that the response has three poles. However, the three capacitors form a loop so that the voltages across them are not independent, and the response in fact has only two poles.

4. ω_{p1} varies inversely with R'_{src}. If $R'_{src} \to 0$, C_T appears across an ideal voltage source and will not affect the frequency response. A rigorous analysis of the circuit of Figure 8.3.2b under these conditions indicates that the response has a pole at $1/R_{out}(C_\mu + C_L)$ and a zero at a higher frequency of g_m/C_μ (Exercise 8.3.2). This shifting of the dominant pole to high frequencies when the source resistance is small should be kept in mind when comparing the response of the CE/CS amplifier with other amplifiers. Note that minimum R'_{src} is $r_B \| r_\pi$. When $R'_{src} \to \infty$, the worst-case 3-dB cutoff frequency is $1/(2\pi\beta r_o C_\mu)$ (Problem P8.3.4).

Exercise 8.3.1
How does Equation 8.3.6 compare with Equation 8.2.10 at very low frequencies? ∎

Exercise 8.3.2
Show that if $R'_{src} = 0$ in Figure 8.3.2b, $\dfrac{V_o(s)}{V_\pi(s)} = -\dfrac{g_m R_{out}(1 - sC_\mu/g_m)}{1 + s(C_\mu + C_L)R_{out}}$, where a pole occurs at $\omega'_{p1} = 1/R_{out}(C_\mu + C_L)$. ∎

Example 8.3.1 Frequency Response of CE Amplifier

It is required to investigate the high-frequency response of the CE amplifier of Exercise 8.2.14, having $g_m = 77$ mA/V, $C_\mu = 1.5$ pF, $C_\pi = 12$ pF, $C_L = 20$ pF, $R_{out} = 2$ kΩ, and $R'_{src} = 1$ kΩ.

SOLUTION

From Equation 8.3.5, $C_T = C_\pi + C_\mu (1 + g_m R_{out}) = 12 + 1.5(1 + 154) \cong 245$ pF. Assuming one pole is dominant, it follows from Equation 8.3.7 that $f_{p1} = \dfrac{1}{2\pi} \dfrac{1}{[C_T R'_{src} + (C_\mu + C_L)R_{out}]} =$

$\dfrac{1}{2\pi(245 \times 1 + 21.5 \times 2)} = \dfrac{1}{2\pi \times 288 \times 10^{-9}} \cong 533$ kHz.   From   Equation   ST8.7.4,

$f_{p2} = \dfrac{1}{2\pi} \dfrac{C_T R'_{src} + (C_L + C_\pi)R_{out}}{[(C_L + C_\mu)C_\pi + C_L C_\mu]R'_{src}R_{out}} = \dfrac{245 + 32 \times 2}{2\pi \times (21.5 \times 12 + 20 \times 1.5) \times 1 \times 2} \cong 85.4$ MHz.

$f_z = \dfrac{g_m}{2\pi C_\mu} = \dfrac{77}{2\pi \times 1.5} \cong 8.17$ GHz. It is seen that f_{p1} is much smaller than f_{p2}, so it is indeed a dominant pole. Since f_{p1} is small compared to f_{p2} and f_z, the 3-dB cutoff frequency is very nearly f_{p1}. A more accurate, but still approximate, expression of the 3-dB frequency is

$\dfrac{1}{\sqrt{1/f_{p1}^2 + 1/f_{p2}^2 - 2/f_z^2}}$ (Equation ST8.8.8, Section ST8.8).

Let us examine the approximations made. In deriving the capacitance due to the Miller effect it was assumed that $R_{out} \ll 1/sC_\mu$. At ω_{p1}, $\dfrac{1}{\omega_{p1} C_\mu} = \dfrac{288 \times 10^{-9}}{1.5 \times 10^{-12}} = 192 \gg 2$. It was

also assumed that $1/sC_L \gg R_{out}$, or, $\dfrac{288 \times 10^{-9}}{20 \times 10^{-12}} = 14.4 \gg 2$. Finally, let us estimate the

frequency $f_T = \dfrac{g_m}{2\pi(C_\pi + C_\mu)}$, the frequency at which the short-circuit current gain becomes unity (Equation 7.2.35, Chapter 7). Substituting, $f_T = 77/(2\pi \times 13.5) \cong 908$ MHz.

The CS amplifier is analyzed in the same manner. The circuits of Figures 8.3.2 and 8.3.3, and the results derived earlier, apply, with $R'_{src} = R_G \| R_{src}$ ($r_\pi \to \infty$, $r_B = 0$), C_π replaced by C_{gs}, and C_μ by C_{gd}. In summary,

Concept: *In a CE/CS amplifier stage, C_μ or C_{gd} is magnified by the Miller effect and introduces a dominant pole that reduces the high-frequency response of the amplifier stage to an extent that depends on the source resistance.*

Common Collector/Drain Amplifier

The small-signal equivalent circuit for the CE amplifier (Figure 8.3.2a) can be readily modified as shown in Figure 8.3.4a for the CC amplifier by grounding the collector and adding R'_E in series with the emitter, where $R'_E = R_E \| R_L \| r_o$. The circuit can be redrawn as in Figure 8.3.4b, where $R'_{src} = R_{src} + r_B$. C_μ now appears between B' and ground. As in the case of the CE amplifier, the three capacitors form a loop so that only two poles are present. The circuit of Figure 8.3.4b is analyzed in Section ST8.7. As done previously, we will pursue here an approximate analysis based on Miller's theorem.

Since $V_\pi(s) = V'_b(s) - V_o(s)$ in Figure 8.3.4b, the current source $g_m V_\pi(s)$ can be expressed as two components $g_m V'_b(s)$ directed upward and a component $g_m V_o(s)$ directed downward. Since the latter component is in the direction of a voltage drop $V_o(s)$, it can be replaced by a

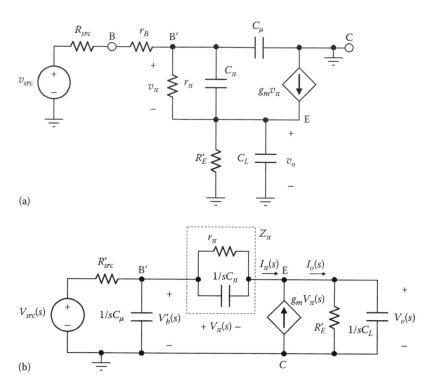

(a)

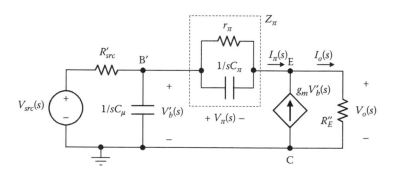

(b)

FIGURE 8.3.4
(a) High-frequency, small-signal equivalent circuit of CC amplifier; (b) the circuit of (a) redrawn.

FIGURE 8.3.5
Modified equivalent circuit for application of Miller's theorem.

resistance $1/g_m$. The circuit becomes as in Figure 8.3.5, where $R_E'' = R_E'||(1/g_m)$ is the small output resistance of the CC amplifier as found previously (Equation 8.2.30). The smallness of this resistance justifies neglecting capacitance at the output because of the small time constant.

The circuit of Figure 8.3.5 is of the form to which Miller's theorem can be applied, considering that the gain from B' to E is $V_o(s)/V_b'(s) = g_m R_E'' = g_m R_E'/(1 + g_m R_E')$. It follows that $Y_\pi = 1/Z_\pi$ appears at the input multiplied by $1 - g_m R_E'/(1 + g_m R_E') = 1/(1 + g_m R_E')$. This means that C_π will be divided by $(1 + g_m R_E')$ and r_π will be multiplied by this factor, as was in fact found previously on the basis of negative feedback considerations (Equation 8.2.23). The total input capacitance is thus $C_\mu + C_\pi/(1 + g_m R_E') \cong C_\mu$ and the resistance at the input, between B' and ground, with $V_{src}(s) = 0$ is $R'_{src}||r_\pi(1 + g_m R'_E) \cong R'_{src}$, assuming $r_\pi(1 + g_m R'_E) \gg R'_{src}$. A pole of the input circuit is at:

$$\omega_{p1} \cong \frac{1}{C_\mu R'_{src}} \tag{8.3.9}$$

This is in sharp contrast to the CE amplifier, where the dominant pole occurs at a much lower frequency because of the magnification of C_μ due to the Miller effect. For the same transistor in Example 8.3.1, $f_{p1} \cong 100$ MHz, compared to about 550 kHz for the CE configuration. As to be expected, gain has been traded for bandwidth through negative feedback.

Following the same argument as for the CE amplifier, a zero exists at the frequency at which the current in Z_π plus that due to $g_m V'_b(s)$ is zero in Figure 8.3.5. This gives $s_z = -(g_m + 1/r_\pi)/C_\pi = -1/C_\pi r_e$, the corresponding frequency being:

$$\omega_z = \frac{1}{C_\pi r_e} \tag{8.3.10}$$

which is close to the unity-gain frequency ω_T.

The same conclusions apply to the CD amplifier, with $r_\pi \to \infty$, and replacing R'_{src} by R_{src}, R'_E by R'_S, C_μ by C_{gd}, and C_π by C_{gs}. The effect of the negative feedback is to divide C_{gs} at the input by the factor $(1 + g_m R'_S)$. The body effect adds a resistance $1/g_m(1 + \chi)$ in parallel with R'_S, as in Figure 8.2.6b, instead of $1/g_m$, to give R''_S.

Concept: *In a CC/CD amplifier stage, $C_\mu(C_{gd})$ appears between input and ground, with no Miller effect. The impedance between base and emitter, or between gate and source, is magnified by approximately $g_m R_{E/S}$ because of the negative feedback.*

Common-Emitter/Source Amplifier with Feedback Resistor

The small-signal equivalent BJT circuit is shown in Figure 8.3.6, where $R'_L = R_L||R_C$. The poles are obtained in Example ST8.9.1, Section ST8.9, by the method of OCSCTC. An instructive approximation is to apply Miller's approximation to derive an expression for the frequency of the dominant pole at the input due to both C_μ and C_π.

r_{in} at dc is given by Equation 8.2.37 and is approximately $r_\pi (1 + g_m R_E)$ for large r_o and β. The effective resistance at the input, between B' and ground with $v_{src} = 0$, is $R'_{src}||r_\pi (1 + g_m R_E)$ as for the CC amplifier, where $R'_{src} = R_{src} + r_B$, bearing in mind that R_E is generally smaller in this case. Ignoring C_L, the gain from base to collector is given by Equation 8.2.44 and is

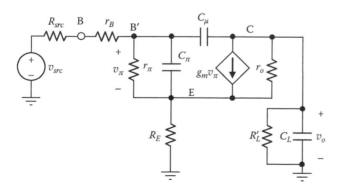

FIGURE 8.3.6
High-frequency, small-signal equivalent circuit of CE amplifier with unbypassed emitter resistor.

nearly $-g_m R'_L/(1 + g_m R_E)$ for large r_o, as explained in connection with Equation 8.2.47. According to Miller's theorem, C_μ is magnified to $C'_\mu = C_\mu[1 + g_m R'_L/(1 + g_m R_E)]$. The gain from base to emitter is very nearly $\dfrac{g_m R_E}{1 + g_m R_E}$ (Equation 8.2.24), so that C_π is reduced, because there is no phase inversion, to $C'_\pi = C_\pi[1 - g_m R_E/(1 + g_m R_E)] = \dfrac{C_\pi}{1 + g_m R_E}$. The frequency of the dominant pole is, therefore:

$$\omega_{p1} = \frac{1}{[R'_{src}\|r_\pi(1 + g_m R_E)](C'_\mu + C'_\pi)} \tag{8.3.11}$$

As in the case of the CE amplifier (Equation 8.3.7), a better approximation is obtained if $(C_\mu + C_L)R'_L$ is added in the denominator.

The CS amplifier with a source resistor can be analyzed in the same manner. The results derived earlier, apply, with R'_{src} replaced by R_{src} ($r_\pi \to \infty$, $r_B = 0$), C_π by C_{gs}, and C_μ by C_{gd}.

Common-Base/Gate Amplifier

The small-signal equivalent circuit of a CB amplifier is shown in Figure 8.3.7a, neglecting r_B. The circuit is redrawn in Figure 8.3.7b, using $V_e(s) = -V_\pi(s)$ and reversing the polarity of the current source. From KCL at node C, $\dfrac{V_o(s)}{Z_L} = g_m V_e(s) + \dfrac{V_e(s) - V_o(s)}{r_o}$, or:

$$\frac{V_o(s)}{V_e(s)} = \frac{g_m + 1/r_o}{1/Z_L + 1/r_o} \cong g_m Z_L \tag{8.3.12}$$

because normally, r_o is much larger than $1/g_m$ and Z_L, where $Z_L = R_L/[1 + s(C_\mu + C_L)R_L]$. This approximation is tantamount to neglecting the current through r_o compared to $g_m V_e(s)$. With r_o removed, the source $g_m V_e(s)$ can be rearranged as two sources, as in Figure 8.3.7c (Sabah 2008, p. 124). The source between collector and base gives Equation 8.3.12 with $r_o \to \infty$, whereas the source between emitter and base is equivalent to a conductance g_m. Hence,

$$\frac{1}{Z_{in}} = \frac{I_e(s)}{V_e(s)} = g_m + \frac{1}{Z_\pi} = g_m + \frac{1}{r_\pi} + sC_\pi = \frac{1}{r_e} + sC_\pi \tag{8.3.13}$$

The input impedance is r_e in parallel with C_π. No Miller effect is present.

To derive $A'_v(s)$, we note that $V_e(s)/V_{src}(s) = Z_{in}/(Z_{in} + R_{src})$. Substituting $V_o(s)/V_e(s) = g_m Z_L$ from Equation 8.3.12 and for Z_{in} from Equation 8.3.13 gives:

$$A'_v(s) = \frac{V_o(s)}{V_{src}(s)} = \frac{\beta}{\beta + 1} \frac{R_L}{r_e + R_{src}} \frac{1}{(1 + s/\omega_{p1})(1 + s/\omega_{p2})} \tag{8.3.14}$$

where,

$$\omega_{p1} = \frac{1}{(C_\mu + C_L)R_L} \quad \text{and} \quad \omega_{p2} = \frac{1}{C_\pi(r_e\|R_{src})} \tag{8.3.15}$$

where the dominant pole is generally ω_{p1}.

The CG amplifier can be analyzed in the same manner, with $r_\pi \to \infty$, $\beta \to \infty$, and replacing C_μ by C_{gd}, and C_π by C_{gs}. The body effect multiplies g_m by $(1 + \chi)$.

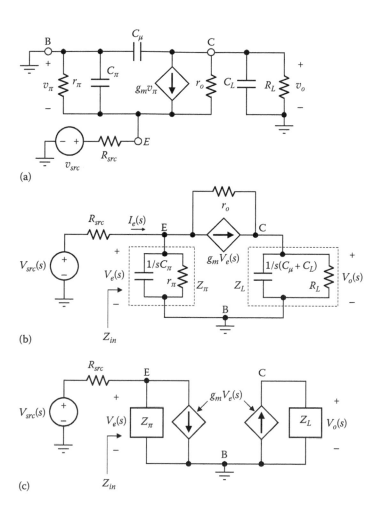

FIGURE 8.3.7
(a) High-frequency, small-signal equivalent circuit of CB amplifier; (b) the circuit of (a) redrawn; (c) $g_m V_e(s)$ source rearranged as two sources.

Concept: *In a CB/CG amplifier, $C_\mu(C_{gd})$ appears between output and ground. The low input impedance gives a pole at much higher frequencies than the input pole of CE/CS amplifier.*

Simulation Example 8.3.2 Frequency Response of BJT Amplifiers

It is required to simulate the high-frequency responses of the CE, CC, and CB amplifiers for the same dc emitter current, load, and source, using a 2N3904 BJT.

SIMULATION

The schematics of the CE, CC, and CB circuits are shown in Figure 8.3.8a through c, respectively. In all cases, a VAC source applies a 1 V ac signal and its dc voltage is adjusted to give an emitter current of 1 mA. The frequency responses are shown in Figure 8.3.9.

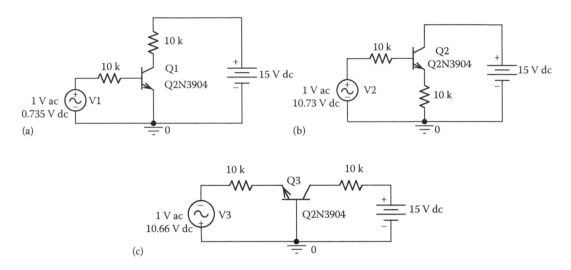

FIGURE 8.3.8
Figure for Simulation Example 8.3.2. Schematics of CE (a), CC (b), and CB (c) configurations.

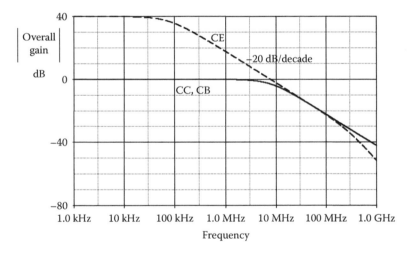

FIGURE 8.3.9
Figure for Simulation Example 8.3.2.

The low-frequency gain of the CE configuration is nearly 40-dB and its 3-dB cutoff frequency is approximately 76 kHz. The low-frequency gains of the CC and CB configurations are practically unity, because in the CB configuration the source and load resistances are equal, and the 3-dB cutoff frequency is 7.73 MHz for both configurations. The GB product is practically the same for all three configurations, since the CE configuration has nearly 100 times the gain but its 3-dB cutoff frequency is 1/100 as much. Alternatively, it can be seen that the unity gain frequency of the CE configuration is the same as the 3-dB cutoff frequency of the CC and CB configurations having a gain of unity. Note how for the CE configuration the slope increases at about 100 MHZ because of the contribution of the higher-frequency pole.

Conclusion: *The CE/CS amplifier has a high gain but a relatively narrow bandwidth because of the Miller effect. The CC/CD and CB/CG amplifiers have a lower gain but a larger bandwidth. However, the gain-bandwidth product is practically the same for the three cases because it fundamentally depends on the unity-gain bandwidth of the transistor used.*

8.4 Composite Transistor Connections

Composite transistors result when two transistors, not necessarily of the same type, are directly connected together so as to obtain a performance that is superior in some respects to that of a single transistor. Several types of composite transistor connections are possible.

Darlington Pair

The Darlington pair, or CC-CC cascade (Figure 8.4.1a), is commonly used to increase the current gain of IC BJTs. It is also available in discrete form as a packaged unit with the three terminals of the composite transistor brought out. The composite transistor has a β_c given by:

$$\beta_c = \beta_1 + \beta_2 + \beta_1\beta_2 \cong \beta_1\beta_2 \tag{8.4.1}$$

(Exercise 8.4.1) and almost twice the V_{BE} of a single transistor having the same emitter current. When Q_1 saturates, its saturation voltage is the collector-base voltage of Q_2, so the BCJ of this transistor cannot become forward biased. A resistor may be connected between Q_2 base and emitter to decrease the turn-off time by providing a path for reverse base current. However, the overall current gain is reduced because the resistor diverts some current from the base of Q_2. Moreover, β_1 may be low because of the low current of Q_1, in which case a bias current source may connected to the emitter of Q_1 to increase the current in this transistor and hence β_1. The input and output resistances of a Darlington pair source follower are derived in Problem P8.2.28.

An alternative compound configuration using an *npn* and a *pnp* transistor is shown in Figure 8.4.1b. Current flows into the upper terminal, whose voltage is one V_{BE} higher than that of the base terminal. The upper terminal therefore acts as the emitter of a composite *pnp* transistor and the lower terminal as its collector, although the output transistor Q_2

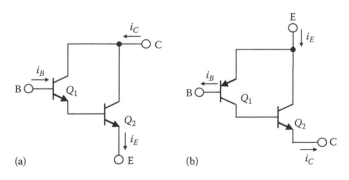

(a)

(b)

FIGURE 8.4.1
Darlington pair: (a) composite *npn* transistor; (b) composite *pnp* transistor.

is *npn*. This feature is sometimes an advantage because of the inferior performance of a single, equivalent *pnp* transistor. The composite transistor has a V_{BE} of a single transistor and a β_c given by:

$$\beta_c = \beta_1(1 + \beta_2) \cong \beta_1\beta_2 \qquad (8.4.2)$$

(Exercise 8.4.1). Again, the BCJ of Q_2 cannot become forward biased.

Exercise 8.4.1

Starting with i_B in Figure 8.4.1a and Figure 8.4.1b, verify Equations 8.4.1 and 8.4.2 by deriving i_C in terms of i_B, β_1, and β_2, neglecting.   ∎

Exercise 8.4.2

Draw the circuit diagrams of a *pnp* Darlington pair and an *npn* composite transistor that is complementary to that of Figure 8.4.1b.   ∎

Application Window 8.4.1 Optoisolators

An **optoisolator** or **optocoupler** consists of an LED and a **phototransistor** mounted in an opaque case (Figure 8.4.2). The base is biased by R_1 and R_2 so as to control the dark current, that is, the transistor current in the absence of light. When the LED is energized by its controlling circuit, the incident light generates electron–hole pairs mainly in the depletion region on the collector side of the reverse-biased BCJ. The photocurrent that crosses the BCJ into the base region is amplified by transistor action by a factor $(1 + \beta_F)$, like I_{CBO}, as explained in connection with Equation 6.1.9, Chapter 6. To increase the current gain further, the phototransistor may be followed by another transistor in a Darlington pair connection. The two transistors are packaged together as a **photodarlington** pair. As its name implies, the optoisolator provides a high degree of electrical isolation between the LED and transistor circuits.

When a control signal is applied to the LED so as to turn the transistor load on and off, the optoisolator becomes a form of **solid-state relay**. The optoisolator can also be used as an amplifier to amplify signals that modulate the intensity of light of the LED. Compared to a two-winding transformer, the optoisolator not only provides a higher degree of isolation but can be used to transmit dc and low-frequency signals without the need to

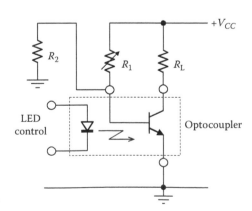

FIGURE 8.4.2
Figure for Application Window 8.4.1.

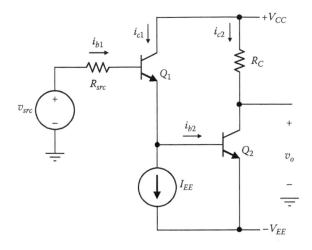

FIGURE 8.4.3
Common-collector common-emitter cascade.

have these signals modulate a carrier of relatively high frequency. The LED and phototransistor can be separate and optically coupled by means of an optical fiber. ∎

Composite transistors are important for improving high-frequency performance. It was demonstrated in the preceding section that the CE/CS configuration has high current and voltage gains but a poor high-frequency performance because of the Miller effect. The CC/CD and CB/CG configurations, on the other hand, have low voltage gain but do not suffer from the Miller effect. It stands to reason, therefore, that much better overall performance can be obtained if the CE/CS configuration is combined with either the CC/CD configuration or the CB/CG configuration.

Common Collector-Common Emitter Cascade

Figure 8.4.3 illustrates a CC-CE cascade commonly used in ICs, using a current source for biasing purposes. Although, in principle, I_{EE} could be dispensed with, because the emitter current of Q_1 can directly flow into the base of Q_2, this would generally result in a low current level in Q_1, and hence a low β, which reduces the overall voltage gain. Neglecting r_{o1}, and considering $\beta \gg 1$, the voltage gain from base to emitter of Q_1 becomes $\dfrac{r_{\pi 2}}{r_{\pi 2} + r_{\pi 1}/\beta}$. This is close to unity as long as $r_{\pi 1}$ is not much different from $r_{\pi 2}$, which is the case when a bias source for Q_1 is included. If not, then $i_{c1} \cong i_{c2}/\beta$. This means that $g_{m1} \cong g_{m2}/\beta$, or $r_{\pi 1}/\beta \cong r_{\pi 2}$. *It follows that if Q_1 is not biased to the same current level as Q_2, the voltage gain from base to emitter of Q_1 is nearly 1/2 instead of being close to unity.*

The superior high-frequency performance of the CC-CE cascade is due mainly to the small input capacitance of Q_1 and its low output resistance, which means a low R'_{src} associated with C_T. This moves the pole of Q_2 to higher frequencies (Exercise 8.3.2).

Example 8.4.1 Comparative Performance of CC-CE Cascade

It is required to compare the performance of the CC-CE cascade with the transistor amplifiers considered earlier, assuming $R_{src} = 1$ kΩ and $R_C = 2$ kΩ (Figure 8.4.1), although these values are likely to be much higher in ICs. As in previous examples, let $\beta = 100$, $r_e = 12.9$ Ω, $r_\pi = 1.3$ kΩ, $r_o = 50$ kΩ, $g_m = 77$ mA/V, $C_\mu = 1.5$ pF, and $C_\pi = 12$ pF for both transistors.

SOLUTION

The voltage gain from source to Q_2 collector can be readily determined from Figure 8.4.3. The effective load at the base of Q_2 is $r_{\pi2}||r_{o1}$. The voltage gain from source to Q_2 base is, from voltage division (Equation 8.2.17), $\dfrac{(\beta+1)(r_{\pi2}||r_{o1})}{R_{src}+r_{\pi1}+(\beta+1)(r_{\pi2}||r_{o1})}=0.982$. The voltage gain from Q_2 base to Q_2 collector is $-g_m(r_{o2}||R_C)=-148$. The overall gain is thus $-0.982\times148=-145$.

The high-frequency equivalent circuit is shown in Figure 8.4.4a. Since $C_{\mu2}$ is a small capacitance, we can assume that the current through it is small compared with $g_{m2}V_{\pi2}$, so that the voltage between C_2 and ground is $-KV_{\pi2}$, where $K=g_{m2}(r_{o2}||R_C)=77\times1.92=148\gg1$. In accordance with Miller's theorem, $C_{\mu2}$ can be split into two capacitances, $(1+K)C_{\mu2}$ between B_2 and ground, and $(1+1/K)C_{\mu2}\cong C_{\mu2}$ between C_2 and ground. $(1+K)C_{\mu2}$ can be combined with $C_{\pi2}$ to give $C_T=C_{\pi2}+(1+K)C_{\mu2}$, as shown in Figure 8.4.4b. We next apply the OCTC method (Section ST8.9). With $C_{\pi1}$, C_T, and $C_{\mu2}$ replaced by open circuits and v_{src} by a short circuit, the resistance appearing across $C_{\mu1}$ is $R_{src}||r_{in1}=1||129=0.99$ kΩ. The open-circuit time constant associated with $C_{\mu1}$ is therefore $\tau_{\mu1}=0.99\times1.5\cong1.5$ ns. The resistance appearing across $C_{\pi1}$ is $r_{\pi1}$ in parallel with the resistance seen by a source $V_{\pi1}$ applied in place of $r_{\pi1}$. By transforming the current source to voltage source, it readily follows that the resistance seen by the $V_{\pi1}$ source is $\dfrac{R_{src}+(r_{o1}||r_{\pi2})}{1+g_{m1}(r_{o1}||r_{\pi2})}\cong23\,\Omega$, which remains practically unaltered when paralleled with $r_{\pi1}$. The open-circuit time constant associated with $C_{\pi1}$ is $\tau_{\pi1}=0.023\times12\cong0.28$ ns. The capacitance $C_T=C_{\pi2}+(1+K)C_{\mu2}=234$ pF. The resistance appearing across C_T is $(r_{o1}||r_{o2})$ in parallel with the output resistance of Q_1, which is $(r_{\pi1}+R_{src})/(\beta+1)=22.8\,\Omega$. The total resistance is about 22 Ω so that the open-circuit time constant associated with C_T is $0.022\times234=5.1$ ns. Finally, the resistance appearing

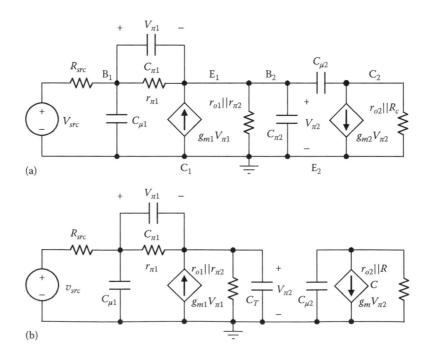

(a)

(b)

FIGURE 8.4.4
Figure for Example 8.4.1. (a) Small-signal equivalent circuit; (b) circuit simplified for application of Miller's theorem.

across $C_{\mu2}$, with $v_{src}=0$, is $r_{o2}||R_C=1.92$ kΩ, which gives an open-circuit time constant of $1.92\times1.5=2.9$ ns. The total time constant is: $1.5+0.28+5.1+2.9=9.78$ ns. Considering one pole to be dominant, the 3-dB cutoff frequency is approximately $1/(2\pi \times 9.78) \cong 16$ MHz, compared to a CE amplifier of 3-dB cutoff frequency of hundreds of kHz. If a pole is not dominant, the more rigorous OCSCTC method should be used (Section ST8.9).

We can examine the current through $C_{\mu2}$ compared to $g_mV_{\pi2}$ at a frequency of about 20 MHz. Taking the voltage across $C_{\mu2}$ to be approximately $150V_{\pi2}$, the current through $C_{\mu2}$ is $150 \times 2\pi \times 20 \times 10^6 \times 1.5 \times 10^{-12}V_{\pi2}=0.028V_{\pi2}$, which is much smaller than $g_mV_{\pi2}=77V_{\pi2}$.

The CD-CS cascade can be analyzed in the same manner (Problem P8.4.2).

Simulation Example 8.4.2 Frequency Response of CC-CE Cascade

It is required to simulate the high-frequency response of the CC-CE Cascade using the same dc emitter current, load, and source as in Simulation Example 8.3.2.

SIMULATION

The schematic is shown in Figure 8.4.5. The VAC source applies a 1 V ac signal and its dc voltage is adjusted to give an emitter current of 1 mA in Q2. The frequency response is shown in Figure 8.4.6. The low-frequency gain is nearly 50.3 dB and the 3-dB cutoff frequency is 2.33 MHz. The slope is -40 dB/decade, which indicates that two poles are very close together.

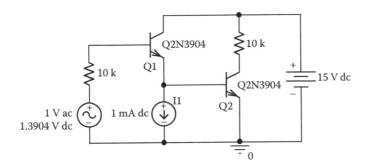

FIGURE 8.4.5
Figure for Simulation Example 8.4.2.

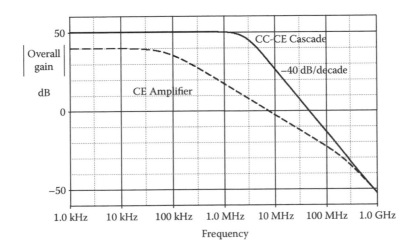

FIGURE 8.4.6
Figure for Simulation Example 8.4.2.

The frequency response for a CE amplifier having the same source and load resistances and dc bias is shown dashed. The low-frequency gain is 40 dB and the 3-dB cutoff frequency is about 76 kHz. The GB product of the CC-CE cascode is nearly a 100 times that of the CE amplifier. Note that part of the improvement in overall gain, from source to output, is due to the larger input resistance of the CC amplifier, relative to R_{src}. Moreover, the location of the dominant pole of the CE amplifier depends on R_{src}.

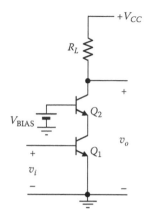

FIGURE 8.4.7
Basic cascode amplifier.

Cascode Amplifier

The cascode amplifier consists of a transistor in the CE/CS configuration coupled to a transistor in the CB/CG configuration. The basic circuit is illustrated in Figure 8.4.7, where it is assumed that the transistors are properly biased. In discrete amplifiers, Q_1 is biased as for CE amplifiers (Section 8.1), and V_{BIAS} is normally derived from V_{CC}, with the base decoupled to ground through a large capacitor. In ICs, a current mirror could be used for biasing, as illustrated later.

As in the CC-CE cascade, the superior high-frequency response of the cascode amplifier is due to the reduction of the Miller effect. This is achieved by having the CE transistor Q_1 loaded by the low input resistance of the CB transistor Q_2, thereby greatly reducing the voltage gain and hence the Miller effect. However, voltage gain at the output is restored because the collector and emitter currents of Q_2 are very nearly equal so that the collector current of Q_1 essentially flows through R_L. In effect, the load R_L is isolated from the collector of Q_1 by the low input resistance of Q_2, thus mitigating the Miller effect. This not only improves the frequency response but reduces the reverse transmission from output to input through C_μ, which is an important consideration in high-frequency amplifiers, such as tuned amplifiers (Section 9.5, Chapter 9).

There are two other important advantages of the cascode amplifier. First, the common-base output resistance is relatively high. This is desirable for high voltage gain with active loads (Section ST8.4) and for improved current mirrors (Section ST8.3). Second, the fact that the input transistor is loaded by the low input resistance of the CB/CG transistor limits the voltage amplitude at the output of the input transistor. Effects of base-width modulation/channel-length modulation, or excessive output voltages, are much less severe. This is particularly important, for example, in the case of transistors having short channels or very thin bases.

MOSFET Cascode

The MOSFET cascode is conveniently analyzed in terms of the open-circuit voltage and the output resistance. When the output is open circuited (Figure 8.4.8), $v_{o1} = -g_{m1}r_{o1}v_{gs1}$. Recall that for the CG amplifier (Figure 8.2.16c), the body effect is accounted for by replacing g_m by $g_m(1+\chi)$. Hence, $v_{o2} = -g_{m2}(1+\chi)r_{o2}v_{gs2}$, where $v_{gs2} = -v_{o1}$. Substituting and adding v_{o1} and v_{o2},

$$v_{oc} = -g_{m1}r_{o1}[1 + g_{m2}r_{o2}(1+\chi)]v_{gs1} \qquad (8.4.3)$$

From Equation 8.2.3, with $v_{gs1} = 0$, $R_{E/S(eff)} = r_{o1}$, and $g_{m(eff)} = g_{m2}(1+\chi)$, it follows that:

$$r_{out} = r_{o2} + r_{o1}[1 + g_{m2}(1+\chi)r_{o2}] \cong g_{m2}(1+\chi)r_{o1}r_{o2} \qquad (8.4.4)$$

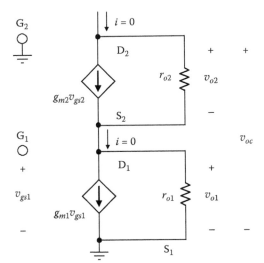

FIGURE 8.4.8
Small-signal equivalent circuit of the MOSFET cascode.

Knowing v_{oc} and $R_{Th} = r_{out}$, the voltage gain can be determined for any R_L. The transconductance G_m can be readily found as:

$$G_m = -\frac{v_{oc}}{r_{out}v_{gs1}} = \frac{g_{m1}r_{o1}[1 + g_{m2}r_{o2}(1+\chi)]}{r_{o1} + r_{o2} + g_{m2}(1+\chi)r_{o1}r_{o2}} = g_{m1}\frac{1 + g_{m2}r_{o2}(1+\chi)}{r_{o2}/r_{o1} + 1 + g_{m2}(1+\chi)r_{o2}} \quad (8.4.5)$$

It is seen that $G_m < g_{m1}$ but the difference is small as long as $g_{m2}(1+\chi)r_{o2} \gg 1 + r_{o2}/r_{o1}$.

It is important to note that if $g_{m2}(1+\chi)r_{o2} \gg 1 + r_{o2}/r_{o1}$, $G_m \cong g_{m1}$ and if $R_L \ll r_o$, the voltage gain is $G_m R_L$, as for a CS amplifier having $R_L \ll r_o$. On the other hand, if R_L is very large, the voltage gain is v_{oc}/v_{gs1}, which is nearly $-(g_{m1}r_{o1})(g_{m2}r_{o2})(1+\chi)$, and is much larger in magnitude than the gain $-g_{m1}r_{o1}$ of a CS amplifier. This emphasizes that:

Concept: *In order to achieve the high voltage gain possible with a cascode amplifier, the load resistance must be large.*

However, a large R_L means a large input resistance of the CG transistor (Equation 8.2.48), and hence a larger gain of the CS transistor and an enhanced Miller effect.

The high-frequency, small-signal equivalent circuit is shown in Figure 8.4.9, where v'_{src} and R'_{src} represent the small-signal TEC at the input, and the internal capacitances of the transistor are indicated together with C_L. Note that although four capacitors are shown, there are only three poles, because three capacitors form a closed loop around $G_1D_1S_2S_1$. A rigorous analysis of this circuit can be made based on the method of OCSCTC.

It is important to note that the increase in the GB product of a cascode amplifier compared to a CS amplifier depends on R'_{src}, because the dominant pole due to the Miller effect in a CS amplifier varies inversely with R'_{src} (Equation 8.3.6). Thus if $R'_{src} = 0$, and v'_{src} appears across C_{gs1}, this capacitance has no effect on the frequency response. Moreover, if $R_L \to \infty$ and $C'_L = C_{db2} + C_{gd2} + C_L$ is large compared to $(C_{db1} + C_{gs2} + C_{sb2})$ and C_{gd1}, these capacitances can be neglected, so that the circuit reduces to that shown in Figure 8.4.10. The dc voltage gain is that given by Equation 8.4.3 and is approximately $-(g_m r_o)^2$,

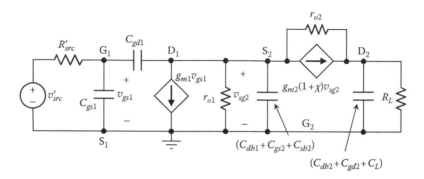

FIGURE 8.4.9
High-frequency small-signal equivalent circuit of the MOSFET cascode.

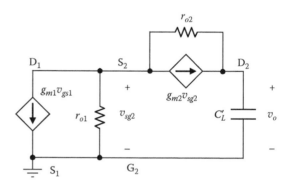

FIGURE 8.4.10
Simplified circuit of Figure 8.4.9.

assuming identical transistors. The resistance seen by the capacitor is r_{out} of Equation 8.4.4 and is approximately $g_m r_o^2$. The 3-dB cutoff frequency is thus $1/C_L' g_m r_o^2$. The dc voltage gain of a CS amplifier of large R_L is $-g_m r_o$ and when $R_{src}' = 0$, the pole of the high frequency response is $1/C_L' r_o$ (Exercise 8.3.2). It is seen that the cascode connection has increased the dc gain by $g_m r_o$ but has also reduced the 3-dB cutoff frequency by the same factor. *Under these conditions, the GB product is practically the same for the cascode as for the CS amplifier.* The advantage of the cascode becomes manifest in the case of a substantial source resistance and high gain, as will also be demonstrated for the BJT cascode.

BJT Cascode

The analysis of the BJT cascode is complicated by the presence of r_π. Referring to Figure 8.4.11, $v_{o1} = -r_{o1}(g_{m1}v_{be1} - i_{b2})$. Substituting $i_{b2} = -v_{o1}/r_{\pi2}$ gives:

$$v_{o1} = -g_{m1}(r_{o1} \| r_{\pi2})v_{be1} \tag{8.4.6}$$

Equation 8.4.6 implies that the resistance between looking into E_2 is $r_{\pi2}$ (Exercise 8.4.3). Moreover,

$$v_{o2} = g_{m2}r_{o2}v_{o1} \tag{8.4.7}$$

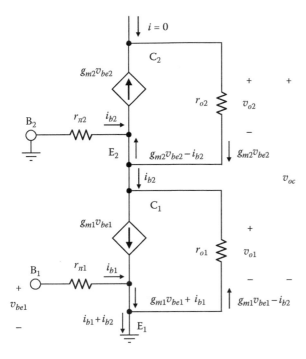

FIGURE 8.4.11
High-frequency small-signal equivalent circuit of the BJT cascode.

and, $v_{oc} = v_{o1} + v_{o2} = (1 + g_{m2}r_{o2})v_{o1} = -g_{m1}(1 + g_{m2}r_{o2})(r_{o1}||r_{\pi2})v_{be1}$. This gives:

$$A_{v0} = \frac{v_{oc}}{v_{be1}} = -g_{m_1}(1 + g_{m_2}r_{o2})(r_{o1}||r_{\pi2}) \cong -g_{m1}\beta_2 r_{o2} \text{ for } g_{m_2}r_{o2} \gg 1 \text{ and } r_{\pi2} \ll r_{o1} \quad (8.4.8)$$

The open-circuit voltage gain from source to output is:

$$A'_{v0} = \frac{v_{oc}}{v_{be1}} \frac{r_{\pi1}}{R_{src} + r_{\pi1}} = -\beta_1 \frac{(1 + g_{m2}r_{o2})(r_{o1}||r_{\pi2})}{R_{src} + r_{\pi1}} \quad (8.4.9)$$

From Equation 8.2.3, with $R_{E/S(eff)} = (r_{o1}||r_{\pi2})$ and $g_{m(eff)} = g_{m2}$

$$r_{out} = r_{o2} + (r_{o1}||r_{\pi2})[1 + g_{m2}r_{o2}] \quad (8.4.10)$$

If the transistors are identical, with $r_o \gg r_\pi$, and $g_m r_o \gg 1$,

$$A_{v0} \cong \frac{\beta^2 r_o}{R_{src} + r_{\pi1}}, \quad r_{out} = \beta r_o, \quad \text{and} \quad G_m \cong \frac{\beta}{R_{src} + r_{\pi1}} \quad (8.4.11)$$

As expected, the maximum output resistance is βr_o, as for the CB amplifier. An approximate result, $v_o \cong -g_{m1}R_L v_{be1}$, can be simply derived as indicated in Exercise 8.4.4.

Exercise 8.4.3

(a) By applying a test source to Q_2 emitter in Figure 8.4.11, show that the resistance looking into E_2, with $v_{be1} = 0$, is $r_{\pi2}$. Note that the $g_{m2}v_{be2}/r_{o2}$ mesh does not contribute to this resistance, since no source current flows through it; (b) Derive Equation 8.4.8 directly from Figure 8.4.11. ∎

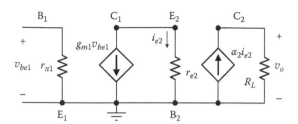

FIGURE 8.4.12
Figure for Exercise 8.4.2.

Exercise 8.4.4

(a) Show that if $g_{m2}r_{o2} \gg 1$, then $v_{o2} \gg v_{o1}$ and if $r_{o1} \gg r_{\pi2}$, then $v_{oc} \cong -g_{m1}\beta_2 r_{o2}v_{be1}$ and $r_{out} \cong \beta_2 r_{o2}$ as for a CB amplifier. (b) By applying these approximations, with $\beta_2 r_{o2} \gg R_L$, show that $v_o = -g_{m1}R_L v_{be1}$. Deduce that this result can be derived from the small-signal equivalent circuit of Figure 8.4.12, assuming $\alpha_2 \cong 1$. ∎

A high-frequency analysis of the cascode amplifier can be made based on the method of OCSCTC, as in Example SE8.2. We will undertake here an approximate analysis based on the simplified equivalent circuit of Figure 8.4.12 with the addition of the internal capacitances and representing the source $\alpha_2 i_{e2}$ as $g_{m2}v_{be2}$ (Figure 8.4.13a). If the small current through $C_{\mu1}$ is neglected compared to $g_{m1}V_{\pi1}$, and with $r_{e2} \ll 1/sC_{\pi2}$ over the frequency range of interest, the voltage $V_{eb2}(s)$ at the emitter of Q_2 is approximately $-g_{m1}r_{e2}V_{\pi1}(s) \cong -V_{\pi1}(s)$, assuming that $r_{e2} \cong r_{e1}$. We can therefore apply Miller's theorem between nodes B'_1 and C_1, replacing $C_{\mu1}$ by a capacitance $2C_{\mu1}$ between B'_1 and E_1 and

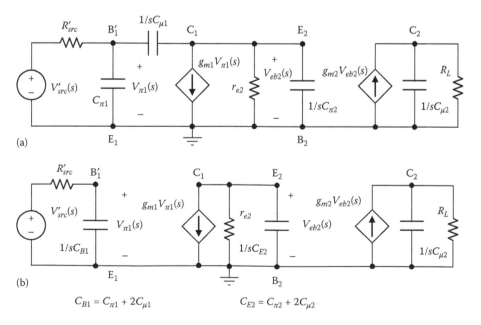

FIGURE 8.4.13
(a) Approximate high-frequency small-signal equivalent of the BJT cascode; (b) the circuit of (a) after application of Miller's theorem.

another capacitance of $2C_{\mu 1}$ between E_2 and B_2 (Figure 8.4.13b). A pole ω_{p1} of frequency $f_{p1} = \dfrac{1}{2\pi R'_{src} C_{B1}} = \dfrac{1}{2\pi R'_{src}(C_{\pi 1} + 2C_{\mu 1})}$ can therefore be separated from the other poles whose frequencies are $f_{p2} = \dfrac{1}{2\pi r_{e2} C_{E2}} = \dfrac{1}{2\pi r_{e2}(C_{\pi 2} + 2C_{\mu 1})}$ and $f_{p3} = \dfrac{1}{2\pi R_2 C_{\mu 2}}$. The pole at f_{p1} is usually dominant, so the 3-dB cutoff frequency is approximately that of the dominant pole. Note that the ratio $f_{p2}/f_{p1} = R'_{src}/r_{e2}$ and depends on the source resistance. Similarly f_{p3} depends on the load resistance. Moreover, because $C_{\mu 1}$ is small, ω_{p2} is practically equal to ω_{T2}, the unity-gain bandwidth of Q_2.

Simulation Example 8.4.3 Frequency Response of Cascode Amplifier

It is required to simulate the frequency response of a cascode amplifier and to compare it with that of a CE amplifier, the source and load impedances being 10 kΩ in both cases.

SIMULATION

The schematic of the cascode is shown in Figure 8.4.14 using Q2N3904 transistors from the PSpice EVAL library. A 1 mA dc source provides the biasing current and is bypassed by a very large capacitor. A 3 V dc bias is applied to Q_2 base in order to have an adequate V_{CE} for Q_1. The frequency response of the cascode over the range 10 kHz to 10 MHz is shown in Figure 8.4.15. Again, it appears from the −40 dB/decade slope that two poles are close together.

With the relatively low load resistance used, the low-frequency gain is approximately 40 dB, the same as the CE amplifier, and the 3-dB cutoff frequency is 2.14 MHz. The 3-dB cutoff frequency, and the GB product, is about 2.14 MHz/76.6 kHZ $\cong$ 28 times that of the CE amplifier. The GB is not as high as that of the CC-CE cascade, and the effect of a higher frequency pole becomes evident at about 100 MHz, as in the CE amplifier.

BiCMOS Cascode

A MOSFET and a BJT can be combined in a cascode configuration, as in Figure 8.4.16, where an active current source is used as a load. The input transistor is the MOSFET, so as

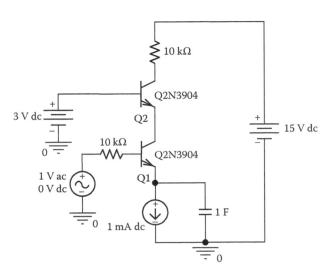

FIGURE 8.4.14
Figure for Simulation Example 8.4.3.

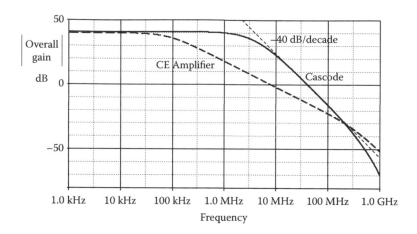

FIGURE 8.4.15
Figure for Simulation Example 8.4.3.

to take advantage of its practically infinite input resistance. Using a BJT for the second transistor, rather than a CG MOSFET, gives higher gain, wider bandwidth, and a large output resistance. Q_1 sees the low input resistance of the CB transistor Q_2, which is smaller than that of a MOSFET Q_2. The Miller effect is therefore reduced and the bandwidth improved.

Considering the output of Q_1 to be effectively short circuited for small signals, the small-signal drain current is $g_{m1}v_i$. This is practically the same as the collector current of Q_2 and is applied to an output resistance r_{ob} that is nearly $\beta r_{o(BJT)}$. Hence, the open-circuit output voltage v_o equals $-g_{m1}r_{ob}v_i$. Not only is $r_{o(BJT)} \gg r_{o(MOS)}$, but is multiplied by β in the CB configuration.

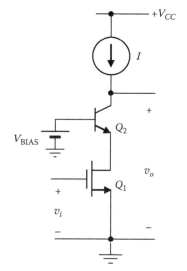

FIGURE 8.4.16
Basic circuit of BiCMOS cascode.

As for the frequency response, it was shown in Example 8.4.2 that the pole ω_{p2} is equal to the unity-gain bandwidth of Q_2. It occurs at a very high frequency and does not affect the bandwidth. If Q_2 is a MOSFET, the corresponding pole will be at a lower frequency and will therefore reduce the 3-dB cutoff frequency.

The cascode combining a MOSFET and a BJT clearly has some advantages over an all-BJT or all-MOSFET amplifiers. This is true of many other BiCMOS circuits. The BiCMOS technology allows combining the two types of transistor in the same IC (Section 4.7, Chapter 4).

Exercise 8.4.5

(a) Draw the small-signal equivalent circuit of BiCMOS cascode of Figure 8.4.16, deduce that the open-circuit gain is $-g_{m1}\beta r_{o(BJT)}$, and show that it can be expressed as $-2\beta V_A / V_{ov}$.

(b) Consider a BJT cascode, as in Figure 8.4.7 but with the addition of a CG CMOS transistor and a current source in place of R_L. Deduce that $G_m = -g_{m1}$, that r_{out} of the modified cascode is nearly $g_{m3}r_{o3}\beta_2 r_{o2}$, and that the open-circuit gain is $-g_{m1}g_{m3}r_{o3}\beta_2 r_{o2}$. ∎

8.5 Cascode Current Sources and Mirrors

The output resistance of an IC current mirror or a transistor current source, which is normally r_o, can be increased by adding a CB/CG transistor in a cascode configuration, as illustrated in Figure 8.5.1a. V_{BIAS1} and V_{BIAS2} for a cascode current source could be derived using the diode-connected transistors Q_3 and Q_4, as in the cascode current mirror shown, or by using more elaborate biasing arrangements (Section 10.4, Chapter 10). Q_3 and Q_4 present a resistance of nearly $1/g_m$ between drain/gate and source, which is low enough to consider the gates of Q_1 and Q_2 to be grounded for ac signals. r_{out1} of Q_1 is r_{o1} and is the source resistance for the CG transistor Q_2 so that r_{out2} of this transistor is approximately $g_{m2}r_{o2}r_{o1}$ (Equation 8.2.3), neglecting the body effect. r_{o1} is therefore multiplied by $g_{m2}r_{o2}$. The high r_{out2} will of course cause much less variation of I_M with the output voltage, V_{D2}. Similarly, adding another CG transistor, say Q_5, on top of Q_2 will result in a double cascode and will make $r_{out5} = g_{m5}r_{o5}(g_{m2}r_{o2}r_{o1})$ so that the output resistance is further multiplied by $g_{m5}r_{o5}$. If the transistors are identical, then each level of cascoding multiplies the output resistance by $g_m r_o$. BJTs generally have larger r_o compared to MOSFETs, because of a larger V_A. If BJTs are used in Figure 8.5.1a, $r_{out2} \cong g_{m2}r_{o2}r_{\pi2}$, that is, r_{o2} is multiplied by β_2, so that it is rarely necessary to have more than one cascode transistor.

A disadvantage of the cascode arrangement of Figure 8.5.1a is that, in the absence of Q_2, the output of Q_1 can swing in the negative direction to $V_{DS(sat)1} = V_{GS1} - V_{tn1} = V_{ov1}$ above ground, with Q_1 remaining in saturation. Adding Q_2 reduces the swing in the negative direction. When $I_M = I_{REF}$, $V_{DS1} = V_{DS3} = V_{GS1}$. Moreover, $V_{G2} = V_{GS4} + V_{GS1}$. Q_2 will be at the edge of saturation when $V_{D2} = V_{G2} - V_{tn2} = V_{GS4} + V_{GS1} - V_{tn2}$. If all the transistors are identical, then Q_2 is at the edge of saturation when $V_{D2} = 2V_{ov} + V_{tn}$ and the swing is reduced by $(V_{ov} + V_{tn})$. Each additional level of cascoding further reduces the swing by $(V_{ov} + V_{tn})$. This can be a serious limitation for low V_{DD}, as is the trend at present.

The **wide-swing current mirror** of Figure 8.5.1b eliminates the threshold term from the most negative swing of the cascode mirror of Figure 8.5.1a. If the gate of Q_2 is biased to $2V_{ov} + V_{tn}$, Q_2 will be at the edge of saturation when its drain voltage

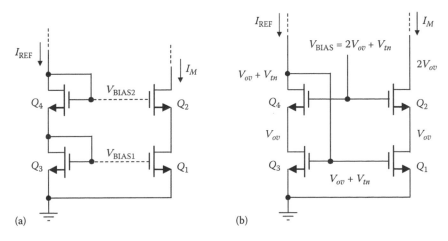

FIGURE 8.5.1
(a) Cascode current mirror; (b) wide-swing current mirror.

is $2V_{ov} + V_{tn} - V_{tn} = 2V_{ov}$. Under these conditions, Q_3 drain is at $2V_{ov} + V_{tn} - V_{GS} = V_{ov}$. With $I_M = I_{REF}$, Q_1 drain is at the same voltage. Hence Q_1 will also be at the edge of saturation. Note that the gate of Q_3 cannot be connected to its drain because of the differences in voltage. The gate of Q_3 is connected instead to the drain of Q_4. V_{GD} of Q_4 is V_{ov} so that Q_4 will be in saturation as long as $V_{ov} < V_{tn}$, which is normally the case. Cascode mirrors are discussed in more detail in Section ST8.3.

Summary of Main Concepts and Results

- The purpose of transistor biasing is to select a quiescent point and stabilize it despite variations in temperature and transistor parameters. There are two general techniques for stabilizing the quiescent point. The technique used in discrete amplifiers is based on negative feedback, whereas in ICs an essentially constant current source is used.

- The three basic amplifier configurations are CE/CS, CC/CD, and CB/CG. The CC configuration has nearly unity voltage gain, high current gain, high input resistance and low output resistance. The CB configuration has high voltage gain, low current gain, low input resistance and very high output resistance. The CE configuration has high voltage gain and high current gain. Its input and output resistances are intermediate between those of the CC and CB configurations. The voltage gains and output resistances of the CS, CD, and CG configurations follow the same pattern as their BJT counterparts. The effect of an emitter/source resistor is to reduce the voltage gain and increase the output resistance as well as the input resistance of the BJT amplifier. The negative feedback introduced by an emitter/source resistor in the CE/CS and CC/CD amplifiers reduces distortion and improves the frequency response.

- The high-frequency response of amplifiers is limited by the effects associated with stored charges and by the speed of response of physical processes. These effects are manifested by internal capacitances.

- The magnification of internal capacitances between input and output by the Miller effect limits the high-frequency response of simple amplifier stages.

- The bandwidth is narrowest for the CE/CS configuration, widest for the CC/CD configuration, and intermediate for the CB/CG configuration with voltage gain larger than unity. However, the gain-bandwidth product is comparable for the three configurations.

- The Darlington connection allows a much higher current gain than is possible with a single, conventional BJT.

- The CC-CE cascade, and its MOSFET counterpart, achieve a wide bandwidth by associating the Miller capacitance of the CE/CS stage with the low output resistance of the CC/CD stage.

- In the cascode amplifier, the Miller effect is mitigated by having a low gain of the CE/CS transistor, which also limits the voltage amplitude at the output of this transistor. In addition, the cascode has the advantages of a high output resistance and a reduced reverse transmission.

- The BiCMOS technology allows combining the advantages of BJTs and MOSFETs in many types of amplifier circuits.
- A cascode current mirror, particularly of the wide-swing type, is commonly used in ICs for improved performance compared to a simple current mirror.

Learning Outcomes

- Design simple amplifying stages based on CE/CS, CC/CD, or CB/CG configurations.
- Articulate the salient features, advantages, and limitations of the three basic transistor configurations and those of composite transistor connections.

Supplementary Examples and Topics on Website

SE8.1 Translinear Gain Cell and Multiplier. Describes the operation of a translinear gain cell that uses current control differentially in order to overcome the dependence of the BJT current on V_{BE}, which introduces nonlinearity and variation with temperature. It is then shown how the translinear gain cell can be used as a four-quadrant multiplier.

SE8.2 High-Frequency Response of Cascode Amplifier. Analyzes the response of a cascode amplifier using the OCSCTC method.

ST8.1 E-NMOS Biasing. Discusses the design of an E-NMOS discrete amplifier using a voltage divider and a source resistance, including the body effect.

ST8.2 MOS Current Mirror with Source Resistors. Analyzes the MOS current mirror when source resistors are included.

ST8.3 Wilson and Cascode Current Mirrors. Analyzes the Wilson current mirror and discusses some aspects of cascode current mirrors.

ST8.4 Common-Drain Amplifier with Current Mirror Load. Analyzes a common-drain amplifier having a current mirror load, including the body effect.

ST8.5 Super Source Follower. Analyzes a source follower circuit having a very low output resistance.

ST8.6 Common-Gate amplifier with Current Mirror load. Analyzes a common-gate amplifier having a current mirror load, including the body effect.

ST8.7 High-Frequency Response of Amplifiers. Analyzes in some detail the high-frequency response of the three amplifier configurations.

ST8.8 3-dB Cutoff Frequency in Terms of Poles and Zeros. Derives approximate expressions for the low-frequency and high-frequency 3-dB cutoff frequencies when the poles and zeros at either frequency extreme are not far apart.

ST8.9 Method of Open-Circuit and Short-Circuit Time Constants for Determining the Frequency Response. Shows how the method can be used for determining the poles of a circuit without having to derive and solve the node-voltage or mesh-current equations of the circuit.

Problems and Exercises

P8.1 Transistor Biasing

P8.1.1   Determine R_B, R_C, and R_E in Figure 8.1.4 so as to give the same quiescent point as in Example 8.1.1, assuming ± 12 V supplies and $R_E(\beta_F + 1) = 10R_B$. Compare the effects of a 50°C rise in temperature and a reduction in β_F to 100.

P8.1.2   Derive Equations 8.1.3 and 8.1.4 including I_{CBO}.

P8.1.3   Consider the common-base circuit of Figure 8.2.16a under quiescent conditions, with the E-NMOS transistor replaced by a BJT, R_S replaced by R_E, and R_D replaced by R_C. Show that $I_C = \dfrac{\beta_F}{\beta_F + 1} \dfrac{V_{EE} - V_{BE}}{R_E} + I_{CBO}$, which is largely independent of β_F as long as β_F is large. How do you explain this result? Consider a circuit having $V_{EE} = 5$ V, $\beta_F = 50$, $R_E = 2$ kΩ, and $I_{CBO} = 0.1$ μA at 25°C. Determine I_C at this temperature and at 125°C.

P8.1.4   Assume that a Darlington pair is used in Figure 8.1.2. Derive an expression for I_C in terms of V_{BE}, β_F, and I_{CBO} of the two transistors. How does this compare with the result of Exercise 8.1.1? What are the relative contributions of I_{CBO1} and I_{CBO2}?

P8.1.5   Consider that a forward-conducting diode represented by a battery V_{DD} to be connected in series with a resistance R_{DD} between base and ground in Figure 8.1.2a. Derive an expression for I_C, neglecting I_{CBO}. Show that: (1) I_C is independent of V_{BE}, if $R_{DD} \ll R_B$ and $V_{DD} = V_{BE}$; (2) I_C is independent of β_F, if $R_E \gg R_{DD}/\beta_F$ and $\beta_F \gg 1$.

P8.1.6   Assume that in Problem P8.1.5 two identical diodes are connected in series with R_{DD}. Show that $dI_C/dT = 0$ if $R_B = R_{DD}$, and $dV_{BE}/dT = dV_{DD}/dT$, neglecting the variation of β_F with temperature.

DP8.1.7 Consider the bias circuit of Figure 8.1.2a with $V_{CC} = 12$ V, $R_C = 10$ kΩ, $R_E = 1$ kΩ, and $\beta_F = 100$. Assume $V_{BE} = 0.7$ V and that for the required bias stability, $(\beta_F + 1) R_E = 10R_B$. Determine R_1 and R_2 so that the transistor operates in the middle of the active region, halfway between saturation and cutoff. Simulate with PSpice.

DP8.1.8 Consider the transistor of Problem P8.1.7 connected in Figure 8.1.4 with ± 12 V supplies and $R_C = 10$ kΩ. Choose $R_B = 100$ kΩ so as to give a high input resistance, the transistor currents being the same. Determine R_E and V_{CE}. How does the ratio $(\beta_F + 1)R_E/R_B$ compare with that of Problem P8.1.7? Simulate with PSpice.

P8.1.9   Consider the transistor of Problem P8.1.7 connected as in Figure 8.1.5 with $V_{CC} = 12$ V and $R_C = 10$ kΩ. If $V_{CE} = 4$ V, what will be the ratio $(\beta_F + 1)R_E/R_B$? Determine I_C and simulate with PSpice.

DP8.1.10 Consider the same transistor of Problem P8.1.7 connected as an emitter follower, with $V_{CC} = 12$ V, $R_E = 10$ kΩ, and R_B connected between V_{CC} and the base. Derive the expression for I_E corresponding to Equation 8.1.1, neglecting I_{CBO}. How does it compare with Equation 8.1.1? Select R_B so that $V_{CE} = 4$ V. How does the ratio $(\beta_F + 1)R_E/R_B$ compare with that of Problem P8.1.9? Simulate with PSpice.

P8.1.11 Let the transistor of Example 8.1.3 be connected in the biasing circuit of Figure 8.1.5. Determine: (a) R_D that makes $I_D = 1$ mA; (b) I_D when V_{tn} changes to 1.8 V.

P8.1.12 An E-NMOS transistor Q_1 having $k_n = 1$ mA/V^2, $V_{tn} = 2.5$ V is used in the circuit analogous to that of Figure 8.1.2a with $V_{DD} = 12$ V, $R_1 = R_2 = 1$ MΩ, $R_D = 1$ kΩ, and $R_S = 1$ kΩ. Determine the quiescent operating of Q_1 and for a replacement transistor Q_2 that has $k_n = 0.8$ mA/V^2 and $V_{tn} = 3.0$ V, neglecting channel-width modulation. Rework this example assuming $R_S = 1.5$ kΩ and R_1 and R_2 are adjusted to give the same drain current in Q_1 as before. How does the change in drain current compare in

P8.1.13 the two cases when Q_1 is replaced by Q_2? Interpret these results graphically by drawing a load line of slope $1/R_S$ on the transfer characteristics.

A D-NMOS transistor Q_1 having $I_{DSS}=5$ mA/V^2 and $\overline{V}_{tnD}=3$ V is used in the circuit of Figure 8.1.6b with $V_{DD}=18$ V, $R_D=2$ kΩ, $R_G=5$ MΩ, and $R_S=1$ kΩ connected to ground instead of a negative supply. Determine the quiescent operating of Q_1 and of a replacement transistor Q_2 that has $I_{DSS}=8$ mA/V^2, $\overline{V}_{tnD}=4$ V, neglecting channel-width modulation. Rework this example with $R_S=1.5$ kΩ and compare the change in I_D between Q_1 and Q_2 when: (1) $R_S=0$, (2) $R_S=1$ kΩ, or (3) $R_S=1.5$ kΩ.

P8.1.14 Consider Q_1 and Q_2 of Problem P8.1.13 to be connected in the circuit of Figure 8.1.2a with $V_{DD}=12$ V, $R_1=R_2=1$ MΩ, $R_D=1$ kΩ, and $R_S=1$ kΩ. Determine the quiescent operating points of Q_1 and Q_2. Rework this example assuming $R_S=1.5$ kΩ, and the same R_1 and R_2. How does the percentage change compare in the two cases? Simulate with PSpice.

P8.1.15 Consider the BJT current mirror of Figure 8.1.7a with $V_{CC}=12$ V and the emitters returned to -12 V; determine: (a) V_{BE} that results in an output current of 100 μA, assuming $I_S=6.7$ fA, $V_A=75$ V, and $V_O=-0.6$ V; (b) R, assuming $\beta=100$. Simulate with PSpice.

P8.1.16 In the MOSFET current mirror of Figure 8.1.11, $V_{DD}=5$ V, Q_1 and Q_2 have $k_n=1$ mA/V^2, $V_{tn}=1$ V, and $\lambda=0.02$ V^{-1}. Determine: (a) R for $I_{REF}=100$ μA; (b) I_M for $V_O=3$ V.

P8.1.17 Consider the Widlar circuit of Figure 8.1.7b with emitter resistors R_{E1} and R_{E2} included for both transistors. Show that $\dfrac{I_{REF}}{I_M}=\dfrac{R_{E2}}{R_{E1}}\left(1-\dfrac{V_T}{R_{E2}I_M}\ln\dfrac{I_{REF}}{I_M}\right)\cong\dfrac{R_{E2}}{R_{E1}}$, as would be the case if V_{BE} of the transistors is negligible compared to the voltage drop across the emitter resistors. How would the output resistance change from that of Figure 8.1.6b? Simulate with PSpice.

P8.1.18 Determine the effective resistance between B and ground, excluding Q_2, in the Widlar circuit of Figure 8.1.7b.

DP8.1.19 Using BJTs instead of MOSFETs in Figure 8.1.12, design a current mirror that provides current sinks of 100 μA, 500 μA, and 1 mA, using *npn* transistors, and current sources of 50, 100, and 500 μA using *pnp* transistors. Assume voltage supplies of ±5 V, a large β_F, negligible base-width modulation, and $V_{BE}=0.7$ V at 1 mA for all transistors. Specify R for $I_{REF}=100$ μA and the ratios of the BEJ areas required of all the transistors.

P8.1.20 The variation of transistor output current with variation in a particular transistor parameter can be expressed in terms of sensitivity. Thus, the sensitivity of I_D with respect to the device transconductance parameter k in a MOSFET is defined as $S_k^{I_D}=\dfrac{dI_D}{dk}\dfrac{k}{I_D}$. Show that for the biasing scheme of Figure 8.1.5 $S_k^{I_D}=\dfrac{1}{1+R_D\sqrt{2kI_D}}$, assuming constant threshold. Determine $S_k^{I_D}$ if $k=0.2$ mA/V^2, with a percentage variability $(\Delta k/k)100=10\%$ and $I_D=0.2$ mA, with a desired percentage variability of $(\Delta I_D/I_D)100=2\%$, and calculate the required value of R_D.

P8.2 Basic Amplifier Configurations

P8.2.1 A voltage amplifier having $r_{in}=10$ kΩ, $r_{out}=1$ kΩ, and $A_{vo}=100$ is loaded by a 5 kΩ resistor and supplied from a source of 5 kΩ resistance. Determine: (a) voltage gain from source to output, (b) current gain, (c) power gain, and (d) NEC at the output.

P8.2.2 Verify that for the CE configuration with emitter resistor R_E and finite r_o, A_v is given by $-\dfrac{\beta r_o-R_E}{r_\pi(r_o+R_L'+R_E)+R_E[r_o(1+\beta)+R_L']}R_L'$, (a) using conventional circuit analysis, and, (b) from the relation, $A_v=-G_m(r_{out}\|R_L')$, with $R_{src}'=0$.

P8.2.3 By applying a test source v_x, verify Equation 8.2.38 for r_{out} of the CE configuration with emitter resistor R_E, considering r_o to be finite.

P8.2.4 Verify that for the CS configuration with emitter resistor R_S and finite r_o, A_v is given by

$$-\frac{g_m r_o R'_L}{r_o(1 + g_m R_S) + R_S + R'_L},$$ ignoring the body effect. Show that this can be obtained from the result of Problem P8.2.2 by letting $r_\pi \to \infty$.

P8.2.5 Rework Exercise 8.2.14 for the CE configuration with emitter resistor and finite r_o, using the results of Problems P8.2.2 and P8.2.3.

P8.2.6 Using the T-model, show that for a CD amplifier $r_{out} = 1/g_m$.

P8.2.7 Consider the CE amplifier of Figure 8.1.4 with $V_{CC} = 12$ V, $R_B = 10$ kΩ, $R_C = 5$ kΩ, $\beta = 100$, and $V_A = 100$ V. R_E is replaced by a current source of 1 mA that is bypassed to ground by a very large capacitor. The signal source connected to the base has a source resistance of 10 kΩ and the load resistance is large. Determine, r_{in}, r_{out}, R_{out}, A_v, and A'_v. Simulate with PSpice.

P8.2.8 For the circuit of Figure P8.2.8, determine: (a) r_{in}, (b) the small-signal gain v_o/v_i, and (c) the current gain i_o/i_i. Assume $\beta = 100$, $V_{BE} = 0.7$ V, and neglect base-width modulation.

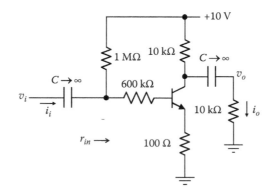

FIGURE P8.2.8

DP8.2.9 It is desired to design a CE amplifier, as in Figure 8.2.3, that gives A'_v of approximately -10 between a 10 kΩ source and a load $R_L = 2$ kΩ using a 12 V supply and a transistor having $\beta = 100$ and $V_A = 100$. It is assumed that R_E is bypassed by a large capacitor. The design guidelines are: (i) the current in R_1 and R_2 is not to exceed one-tenth of the transistor current; (ii) R_E is to be selected for good bias stability. Choose the transistor current to give an appropriate g_m and R_C to have the V_{CE} of the transistor at least 2 V away from saturation or cutoff. How much is A_v? What is the reason for the difference between A_v and A'_v? Simulate with PSpice.

P8.2.10 If the emitter bypass capacitor in Problem DP8.2.9 is omitted, what are the new values of r_{in}, A_v, and A'_v? Simulate with PSpice.

P8.2.11 Determine r_{out} of the transistor in Problem P8.2.10. How does r_{out} change if the biasing is varied so that the collector current is: (a) doubled, (b) halved? Assume that β remains constant.

P8.2.12 The CE amplifier of Example 8.1.1 has $R_E = R_C = 2$ k, $R_{src} = 1$ kΩ, and $V_A = 80$ V. Determine r_{in}, r_{out}, and A_v before and after the 50°C rise in temperature. Simulate with PSpice.

P8.2.13 Replace the BJT in Problem P8.2.7 with an E-NMOS transistor having $k_n = 1$ mA/V^2, $V_{tn} = 3$ V, $r_o = 50$ kΩ, and R_S adjusted to give $I_D = 1$ mA. Determine R_S, A_v and compare with A'_v in Problem P8.2.7. Neglect channel-width modulation and simulate with PSpice.

P8.2.14 Derive the expressions for r_{in}, r_{out}, A_i, and A_v for the amplifier configuration of Figure 8.1.5. Deduce the corresponding expressions for a MOSFET amplifier.

P8.2.15 Given the circuit of Figure 8.1.6b. If an input voltage v_i is applied between gate and ground, and output voltages at the drain and the source are denoted by v_{od} and v_{os}, respectively, derive expressions for the voltage gains v_{od}/v_i and v_{os}/v_i. Note that the circuit allows the generation of two signals of equal magnitudes but opposite phases.

P8.2.16 Given an E-NMOS CS amplifier having an E-PMOS current mirror load supplied from a current source $I_{REF} = 50$ μA and operating between a supply voltage of $+3$ V and ground. The MOSFET parameters are $W/L = 20$ for all transistors, $k'_n = 20$ μA/V², $k'_p = 10$ μA/V², $\lambda_n = 0.04$ V⁻¹, $\lambda_p = 0.08$ V⁻¹, $V_{tn} = 0.7$ V, and $V_{tp} = -0.8$ V. Determine: (a) V_{SG} of the E-PMOS transistors, (b) the limits of v_O that will keep the transistors in saturation, and the corresponding values of v_I, (c) the output resistance r_{out} assuming the output current is I_{REF}, and (d) the small-signal gain v_o/v_i at $v_O = 1.5$ V.

P8.2.17 From a PSpice simulation of the circuit of Problem P8.2.16, (a) derive the v_O–v_I transfer characteristic; how does the linear range compare with the range in (b) of Problem P8.2.16, and how does the slope compare with the small-signal gain in (c)? (b) Bias the input to the middle of the output characteristic, and apply a small signal input of 1 kHz and peak of (i) 20 mV, (ii) 50 mv and observe the resulting waveforms.

P8.2.18 Given an E-NMOS CS amplifier with an E-PMOS current mirror load having $I_{REF} = 100$ μA and operating between a supply voltage of $+12$ V and ground. Assume that $k_n = 2k_p = 0.2$ mA/V², $V_{tn} = \overline{V}_{tp} = 15$ V, and $\lambda_n = \lambda_p = 0.01$ V⁻¹. (a) Determine V_{SGp} of the current-mirror transistors; (b) sketch the load curve on the output characteristics of the E-NMOS transistor, similar to that of Figure 5.5.1b, Chapter 5; (c) neglecting channel-length modulation, determine the output voltage corresponding to point B in Figure 5.5.1b, Chapter 5; (d) derive the relation between input and output voltages over the region BC of the input–output characteristic (Figure 5.5.1c, Chapter 5), by equating the drain currents of the two transistors at the output; (e) from this relation, determine the input voltage corresponding to point B, and the input and output voltages corresponding to the point C; (f) calculate the large-signal voltage gain and compare it to the small-signal voltage gain.

P8.2.19 Consider a CC amplifier that is the BJT counterpart of Figure 8.2.4a with bias provided by a single resistor R_1 adjusted to give $I_C = 1$ mA. Assume $V_{CC} = 5$ V, $R_E = 2$ kΩ, $\beta = 100$, $V_A = 100$ V, $R_{src} = 10$ kΩ, $V_{BE} = 0.7$ V, and very large R_L. Determine r_{in}, R_{out}, A_v, and A'_v. What is the ratio of $\beta R_E/R_1$? Simulate with PSpice.

P8.2.20 Repeat Problem P8.2.19 assuming a voltage divider $R_1 R_2$ designed to give $I_C = 1$ mA, the current through R_1 being 100 μA. Compare with the results of Problem P8.2.19.

P8.2.21 Compare A_v, A_i, r_{in}, and r_{out} of the CB and CE amplifiers if β changes at constant I_C.

P8.2.22 It is desired to use a single voltage supply to bias an ac CG amplifier using an E-NMOS transistor. How would you modify the circuit of Figure 8.2.16a for this purpose?

P8.2.23 Assume that in the CB small-signal equivalent circuit of the CB amplifier r_{ob} can be omitted but a base-spreading resistance r_b appears between base terminal and ground. Determine the expressions for A_v, A_i, r_{in}, r_{out}, and R_{out}.

P8.2.24 The CB amplifier of Figure P8.2.24 is used to match the 50 Ω source and provide a small-signal voltage gain of 30. Determine R_E and R_C assuming $\beta = 100$, $V_{EB} = 0.7$ V, and neglecting base-width modulation.

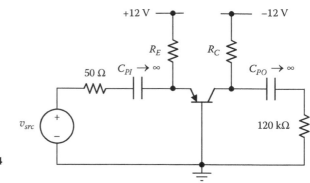

FIGURE P8.2.24

P8.2.25 Determine: (a) v_o/v_i, and (b) v_i/i_i in Figure P8.2.25, assuming $\beta = 100$ and $V_{BE} = 0.7$ V.

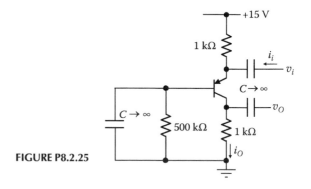

FIGURE P8.2.25

P8.2.26 A source follower has an E-NMOS current-mirror load, voltage supplies of ± 6 V, $I_{REF} = 80$ μA, with all transistors having $V_{tn} = 1$ V and $k_n = 1$ mA/V. Determine: (a) v_{Omax}, v_{Omin}, v_{Imax} v_{Imin}; and the large-signal voltage gain, neglecting the body effect and channel-length modulation; (b) the small-signal v_o/v_i, if $\lambda = 0.05$ V^{-1} and $\chi = 0.15$. Simulate with PSpice.

P8.2.27 The transistor parameters in Figure P8.2.27 are $\beta = 50$, $g_m = 60$ mA/V, and $V_A = 78$ V. Determine: (a) I_C, assuming $V_{BE} = 0.7$ V, and (b) v_o/v_{src}.

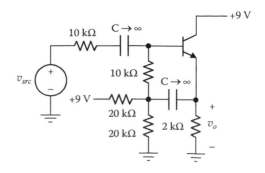

FIGURE P8.2.27

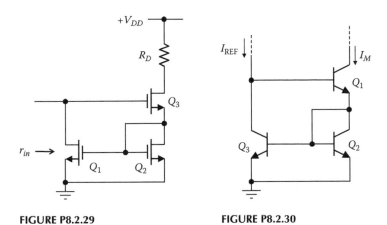

FIGURE P8.2.29   FIGURE P8.2.30

P8.2.28 Show that if a Darlington pair is used in an source follower, $r_{in} \cong \beta_1\beta_2R'_E$ and $r_{out} \cong r_{e2} + r_{e1}/\beta_2 + R'_{src}/\beta_1\beta_2$.

P8.2.29 Show that r_{in} in Figure P8.2.29 is $2/g_m$, assuming all transistors are identical and neglecting channel-length modulation and the body effect.

P8.2.30 Figure P8.2.30 shows a BJT Wilson current mirror (Section ST8.3), where all transistors are assumed identical. Show that: (a) $\dfrac{I_M}{I_{REF}} \cong \left(1 - \dfrac{2}{\beta_F^2}\right)$, as in Exercise 8.1.5; (b) the small-signal input resistance seen by I_{REF} is approximately $2V_T/I_{REF}$, neglecting base-width modulation; (c) the output resistance at Q_1 collector is approximately $\beta r_o/2$.

P8.3 Frequency Response

P8.3.1 Determine r_{in} in Figure P8.3.1 using Miller's theorem, if V_2/V_1 is -5.

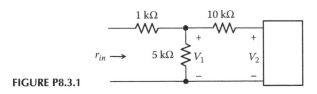

FIGURE P8.3.1

P8.3.2 Consider an ideal amplifier of gain $+2$ and a 100 pF capacitor connected between input and output. What capacitance is reflected at the input of the amplifier? Note that such a connection is sometimes used to neutralize a shunt capacitance at the input of the amplifier.

P8.3.3 The transfer function of an amplifier is $H(s) = \dfrac{10^7}{(s + 1000)(s + 2000)}$. Determine: (a) the poles of the response; (b) the exact 3-dB cutoff frequency; (c) the GB product, and (d) the 3-dB cutoff frequency when two such amplifiers are cascaded without loading effects.

P8.3.4 Consider a CE amplifier driven from an ideal current source and having an ideal current source as a load. Show that the 3-dB frequency is very nearly $1/(2\pi\beta r_o C_\mu)$.

P8.3.5 A CE transistor has $I_C = 2.5$ mA, $R_C = 8$ kΩ, $\beta = 100$, $V_A = 40$ V, $C_\mu = 1$ pF, and $f_T = 300$ MHz. Apply Miller's theorem to determine the effective capacitance at the base of the transistor.

P8.3.6 In the diode-connected transistor of Figure P8.3.6, r_o can be neglected because it appears in parallel with the much smaller r_π and C_μ is short-circuited. Let $Z_\pi = r_\pi \| (1/sC_\pi)$ be the impedance between base and emitter in the absence of transistor action. Deduce that due to transistor action $Z_{in} = Z_\pi/(1 + g_m Z_\pi)$. Show that Z_{in} can be expressed as $Z_{in} = r_e/(1 + sC_\pi r_e)$.

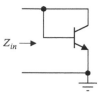

FIGURE P8.3.6

P8.3.7 Consider the CE amplifier of Figure 8.2.3a, with $V_{CC} = 12$ V, R_E replaced by a 1 mA current source, $R_{src} = 4.7$ kΩ, $R_C = 8.2$ kΩ, $R_L = 10$ kΩ, $C_\pi = 52$ pF, $C_\mu = 3.9$ pF, $\beta = 170$, and $V_A = 74$ V. Determine: (a) the low-frequency gain, (b) the 3-dB cutoff frequency, using (i) Miller's theorem, (ii) the open-circuit time constant method, (iii) rigorous circuit analysis.

P8.3.8 Given the CE amplifier of Figure 8.2.3a with $R_{src} = 5$ kΩ, $R_1 = 56$ kΩ, $R_2 = 22$ kΩ, $R_E = 2$ kΩ, $R_C = 4.7$ kΩ, $R_L = 10$ kΩ, $I_C = 1$ mA, $\beta = 150$, $f_T = 800$ MHz, and $C_\mu = 1$ pF. Neglecting r_o, determine the low-frequency gain, the frequencies of the two poles and zero, and the 3-dB cutoff frequency. Simulate with PSpice.

P8.3.9 A CS amplifier has $R_G = 1$ MΩ, $g_m = 1$ mA/V, $r_o = 50$ kΩ, $R_D = 20$ kΩ, $C_{gs} = 2$ pF, and $C_{gd} = 1$ pF, $R_{src} = 200$ kΩ, and $R_L = 10$ kΩ. Determine the low-frequency gain, the frequencies of the two poles and zero, and the 3-dB cutoff frequency. Simulate with PSpice.

P8.3.10 Determine the frequency response of a source follower having $R_{src} = 50$ kΩ, $R_S = 50$ kΩ, $g_m = 1$ mA/V, $\chi = 0.15$, $r_o = 100$ kΩ, $C_{gs} = 0.5$ pF, and $C_{gd} = 0.1$ pF.

P8.3.11 An emitter follower has $R_{src} = 10$ kΩ, $R_E = 1$ kΩ, $f_T = 1$ GHz, $C_\mu = 0.8$ pF, $I_C = 1$ mA, $V_A = 100$ V, and $\beta = 120$. Determine: (1) f_z, (2) f_{p1} and f_{p2} using the method of OCSCTC, (3) the 3-dB cutoff frequency. Compare with the results of rigorous analysis. Simulate with PSpice.

P8.3.12 A CD amplifier has a current mirror load, as in Problem 8.2.26, with $R_{src} = 80$ kΩ and $C_L = 10$ pF. Determine: (1) f_z and (2) the 3-dB cutoff frequency assuming a dominant pole and using the OCTC method. It is given that $g_m = 1$ mA/V, $\chi = 0.15$, and $r_o = 100$ kΩ for all transistors, with $C_{gs} = 0.5$ pF and $C_{gd} = 0.1$ pF.

P8.3.13 Consider the CS amplifier with $R_G = 5$ MΩ connected between gate and drain, as in Figure 8.1.5, $R_{src} = 50$ kΩ, $R_D = 20$ kΩ, $g_m = 1$ mA/V, $r_o = 80$ kΩ, $C_{gs} = 0.2$ pF, and $C_{gd} = 0.05$ pF. Determine: (a) f_z, the poles, and the 3-dB cutoff frequency from a rigorous analysis; (b) the poles using (i) the OCSCTC method and (ii) the OCTC method.

P8.3.14 Apply Miller's approximation to Problem P8.3.13 considering the Miller admittance to be $Y_{gd} = sC_{gd} + 1/R_G$. Determine the pole at the amplifier input and compare with the dominant pole found in Problem P8.3.13.

P8.3.15 A CG amplifier has $C_{gs} = 2.2$ pF, $C_{gd} = 0.12$ pF, $C_L = 3$ pF, $g_m = 4$ mA/V, $\chi = 0.15$, $R_{src} = 1$ kΩ, and $R_L = 20$ kΩ. Neglect r_o and use Figure 8.3.7c to determine: (a) Z_{in}; (b) V_o/V_s; (c) V_o/V_{src} and the poles of the response; and (d) the 3-dB cutoff frequency.

P8.3.16 Write the node-voltage equations for the circuit of Figure ST8.9.1, Section ST8.9, for the three nodes G, D, and S, assuming a current source excitation. Show that the coefficient of s^3 is zero in the determinant of the equations.

P8.4 Composite Transistors

P8.4.1 Show that for the CC/CE cascade of Figure 8.4.3, $G_m = \dfrac{g_{m2}}{1 + (r_{\pi1} + R_{src})/(\beta_1 + 1)r_{\pi2}}$, defined as the ratio of the short-circuit output current to v_{src}. Ignore r_o.

P8.4.2 Derive the voltage gain of a CD-CS amplifier that is the MOSFET counterpart of Figure 8.4.3. Assume an ideal biasing current source but account for finite r_o and the body effect.

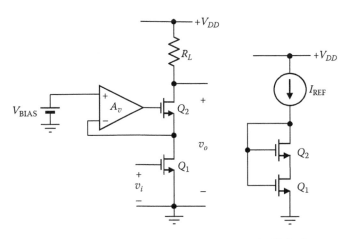

FIGURE P8.4.3   FIGURE P8.4.4

P8.4.3 Figure P8.4.3 shows an active cascode (Säckinger and Guggenbühl, 1990) that realizes a high r_{out} without increasing the number of cascading levels, and hence reducing the output swing. What is r_{out} if the op amp were ideal? Derive r_{out} for finite A_v.

P8.4.4 Assume that Q_1 and Q_2 in Figure P8.4.4 are identical and have $k_n = 1$ mA/V^2 and $V_{tn} = 1$ V, with $I_{REF} = 100$ μA. Because Q_2 is diode connected, it must operate in the active region. Deduce that Q_1 operates in the triode region and determine V_{GS2}, V_{DS1}, and V_{GS1}, neglecting channel-length modulation. Simulate with PSpice.

P8.4.5 Show that if Q_1 and Q_2 in Figure P8.4.4 are identical but the aspect ratio of Q_1 is one-third that of Q_2, $V_{DS1} = V_{ov2} = V_{GS2} - V_{tn}$.

P8.4.6 Figure P8.4.6 shows a BiCMOS Darlington. Determine: (a) V_{BIAS} so that the dc output voltage is 2 V, assuming β_F to be very large, $V_{BE} = 0.7$ V, $k_n = 1$ mA/V^2, $V_{tn0} = 1$ V, $\chi = 0.1$, and neglecting channel-length modulation and base-width modulation; (b) the small-signal gain, assuming, $r_\pi = 2$ kΩ, $\beta = 200$, $\lambda = 0.05$/V, and $V_A = 50$ V. Simulate with PSpice.

P8.4.7 Consider a MOSFET cascode consisting of two identical transistors having $g_m = 2$ mA/V, $\chi = 0.2$, $r_o = 50$ kΩ, $C_\mu = 0.1$ pF, $C_{gs} = 1$ pF, and $C_L = 10$ pF. R'_{src} is small, R_L is infinite, and $C_L = 5$ pF. Determine: (a) r_{out}; (b) low-frequency gain; (c) the approximate 3-dB cutoff frequency.

P8.4.8 Figure P8.4.8 shows a cascode amplifier with an E-PMOS load. (a) Determine: the maximum and minimum values of v_o for which the transistors remain in saturation.

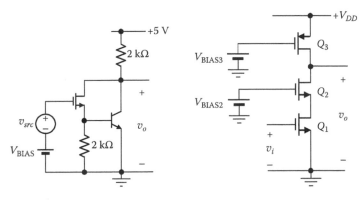

FIGURE P8.4.6   FIGURE P8.4.8

Derive the small-signal, low-frequency equivalent circuit neglecting the body effect of Q_2 and show that $\dfrac{v_o}{v_i} = -g_{m1}\dfrac{r_{o1}r_{o3}[1 + g_{m2}(1 + \chi)r_{o2}]}{r_{o3} + r_{o2} + r_{o1}[1 + g_{m2}(1 + \chi)r_{o2}]}$. Deduce that $v_o/v_i \cong -g_m r_o$ if all transistors are identical, $\chi = 0$, and $r_o \gg 1/g_m$. How do you interpret this result?

P8.4.9 Investigate the high-frequency response of the cascode amplifier of Problem P8.4.8 by adding to the small-signal equivalent circuit capacitors C_1 between G_1 and D_1, C_2 between D_1 and ground, and C_3 between D_2 and ground. (a) What do these capacitances represent? (b) How many poles do you expect the response v_o/v_i to have? (c) Assuming identical transistors, $\chi = 0$, $g_m = 1$ mA/V, $r_o = 50$ kΩ, $C_1 = 0.5$ pF, $C_2 = 2$ pf, and $C_3 = 10$ pF, use the OCSCTC method to determine the poles and the 3-dB cutoff frequency. Which of the poles is dominant?

P8.4.10 Given the BiCMOS cascode of Figure P8.4.10 with $R_{src} = 20$ kΩ. For the MOSFET, $g_m = 4$ mA/V, $C_{gs} = 2$ pF, and $C_{gd} = 0.1$ pF. For the BJT, $\beta = 100$, $g_m = 40$ mA/V, $C_\pi = 2.5$ pF, and $C_\mu = 0.5$ pF, and $C_L = 2$ pF. Neglect channel-length and base-width modulations. Determine: (a) the low- frequency gain v_o/v_{src}; (b) the zero and poles of the CS stage using the OCSCTC method; (c) the pole at the amplifier output; and (d) the 3-dB cutoff frequency.

P8.5 Cascode Current Sources and Mirrors

P8.5.1 In the cascode mirror of Figure 8.5.1a, all transistors have $k_n = 0.5$ mA/V^2, $V_{tn} = 0.8$ V, $V_A = 20$ V, $I_{REF} = 80$ μA, and $V_O = 4$ V. Determine: (a) I_M and compare with that of a simple mirror; (b) output resistance of the mirror; (c) lowest allowable output voltage. Neglect the body effect.

P8.5.2 Consider double-cascode current mirror having a transistor added on top of Q_2 in Figure 8.5.1a and another diode connected transistor added on top of Q_4. Assume $I_{REF} = 0.2$ mA, $V_{D3} = 6$ V, and $k_n = 0.4$ mA/V, $V_{tn} = 1$ V, and $\lambda = 0.02$ V^{-1} for all transistors. Neglecting the body effect, determine: (a) I_M, V_{D1}, and V_{D2}, neglecting channel-length modulation, (b) minimum value of V_{D3}, and (c) the output resistance.

P8.5.3 The technique of Problem P8.4.3 is often used to increase the output resistance of a current mirror, as illustrated in Figure P8.5.3. Show that the output resistance becomes: $r_{out} = r_{o1} + r_{o2} + g_{m2}r_{o1}r_{o2}(A_v + 1)$.

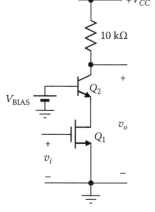

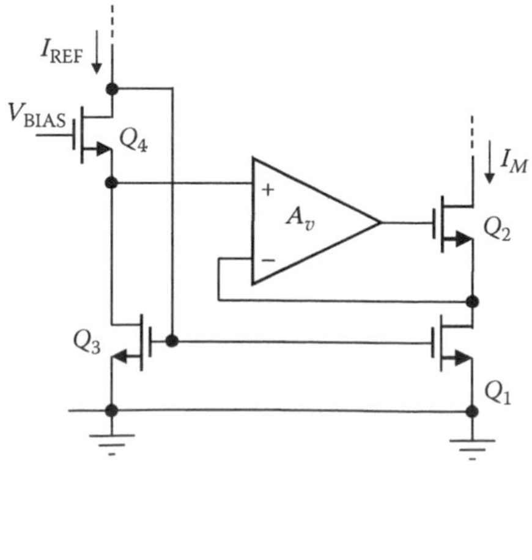

FIGURE P8.4.10

FIGURE P8.5.3

P8.6 Miscellaneous

P8.6.1 The transistors in Figure P8.6.1 are identical and have large β. Show that if a current I is forced into A and a voltage V is applied at B, then the current I flows into B and the voltage V appears at A. Such a circuit is known as a **current conveyor**. If B is grounded, then A will be a virtual ground. If A is connected to a supply voltage V_{CC} through a resistance R, $I = V_{CC}/R$ and this will also be the collector current of an identical transistor whose base and emitter are connected to the respective terminals of Q_1 and Q_2.

P8.6.2 Figure P8.6.2 shows a **peaking current**

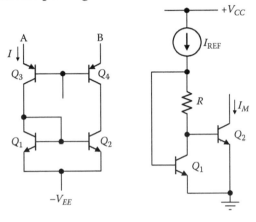

FIGURE P.8.6.1   FIGURE P8.6.2

source that can be used to produce an extremely small source current I_M using reasonable values of I_{REF} and R (Kwok, 1985), even smaller than those attainable with the Widlar current source. Assuming that the two transistors are identical, of very high β, and that for each transistor $i_C = I_s e^{v_{BE}/V_T}$, show that $I_M = I_{REF} e^{-RI_{REF}/V_T}$. Simulate with PSpice and plot I_M vs. I_{REF} in the range 0–10 μA, with $R = 10$ kΩ, and determine the value of I_{REF} for which I_M is a maximum.

P8.6.3 Figure P8.6.3 illustrates a simple offset-bias transconductor circuit that produces a current difference that is proportional to a voltage signal (Bult and Wallinga, 1987). Assuming that the transistors are identical and are in the saturation mode, show that $i_2 - i_1 = k_n/2(V_B - 2V_t)(2v_I - V_B)$. Neglect channel-length modulation.

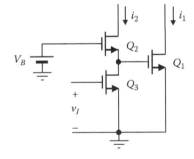

FIGURE P8.6.3

P8.6.4 Consider a CE amplifier having a load of admittance $Y_L = G_L + jB_L \gg sC_\mu$ so that the amplifier gain may be assumed to be $-g_m/Y_L$. Show that the admittance reflected on the input side is $\dfrac{\omega C_\mu g_m B_L}{G_L^2 + B_L^2} + j\omega C_\mu$ $\left(1 + \dfrac{g_m G_L}{G_L^2 + B_L^2}\right)$. Note that if the load is inductive, $B_L < 0$, and the real part is a negative conductance. If the magnitude of this conductance is larger than the positive conductance at the input, the net conductance is negative and will result in oscillations.

9

Multistage and Feedback Amplifiers

The basic design of an amplifier stage was considered in Chapter 8 using a single-transistor or a composite-transistor connection. In many cases, more voltage gain may be required than can be provided by a single amplifier stage, which necessitates cascading amplifier stages. Even if only a modest overall gain is required, it is advantageous to cascade stages for high gain and apply negative feedback to trade some of the gain for better overall performance, such as modified input or output resistance, reduced distortion, and wider bandwidth.

Amplifying stages can be cascaded in a number of ways, depending on the application and type of amplifier. In IC amplifiers, cascaded stages are dc coupled. In discrete amplifiers, stages can be ac coupled using capacitors, magnetic coupling, or opto-coupling. Each of these methods of coupling has its advantages but poses problems that must be addressed.

In calculating the input resistance, output resistance, and overall gain of transistor feedback amplifiers, it is necessary, in general, to account for the loading effect of the feedback circuit on the input and output of the amplifier proper. To accomplish this, a generalized systematic procedure can be applied to the four types of feedback configurations.

When discussing negative feedback in amplifiers, the effect of the inevitable phase shifts due to parasitic capacitances must be considered. These can turn negative feedback into positive feedback at some frequency, thereby causing oscillations. While this is to be avoided in amplifiers, it can be utilized to implement feedback oscillators.

A particular type of ac coupling between cascaded stages is encountered in tuned amplifiers, which are widely used to provide the required gain and frequency selectivity in communication receivers. Introducing positive feedback in what are essentially tuned amplifiers results in another class of widely used oscillators, namely *LC* oscillators.

Learning Objectives

❖ To be familiar with:

• Terminology commonly used with cascaded amplifiers, feedback amplifiers, and oscillators.

❖ To understand:

• How amplifiers are cascaded and dc levels are shifted in the case of dc coupling.
• The shortcut intuitive method of analyzing cascaded amplifiers.
• The effect of coupling capacitors and emitter/source bypass capacitors on the low-frequency response of amplifiers.
• The methods of analyzing transistor feedback amplifiers in the four basic feedback configurations.
• Stability considerations in feedback amplifiers.

- The principle of operation of feedback oscillators.
- The basic features of tuned amplifiers.
- The principle of operation of *LC* oscillators.

9.1 Cascaded Amplifiers

The following coupling methods are used to cascade amplifier stages:

1. Direct coupling, in which the output of one stage is directly coupled to the input of the next stage so that dc signals are transmitted between stages. This can raise two problems: (1) The dc level at the output of a given stage is generally incompatible with the dc level at the input of the next stage, which necessitates some form of dc level shifting. (2) The dc level at the input may be subject to drift, that is, a slow, undesirable variation with time, as may be due, for example, to changes in temperature. The drift, if allowed through a high-gain amplifier, may overload later stages, thereby seriously interfering with the amplification of desired signals.

2. Capacitive coupling, in which series capacitors are used between successive stages. Not only are undesirable dc voltages blocked, but the dc level at the output of a given stage is isolated from the input of the next stage. However, capacitive coupling does have its own disadvantages. Evidently, genuine dc signals cannot be amplified. The presence of capacitors can have undesirable effects on the performance of the amplifier, as will be discussed later. Moreover, because the values of these capacitances are relatively large, they are impractical in ICs. This limits capacitive coupling to discrete amplifier stages. In electrocardiogram (ECG) recording, for example, the ECG signal amplitude is less than a few mVs, and the frequency components of interest are in the range of a fraction of a Hz to about 150 Hz. However, the electrodes that are applied to the body to pick up the ECG signals have a dc offset potential that is subject to drift. Hence ac coupling is used, at least beyond the first stage of amplification.

3. Some form of transformer coupling, also limited to discrete amplifiers. Mutual inductance coupling at high frequencies is used in tuned amplifiers, as discussed in Section 9.4.

4. Optical coupling using an LED as a transmitter and a phototransistor as a receiver (Application Window 8.4.1, Chapter 8).

Cascaded stages can, in general, be analyzed by the method illustrated in Figure 9.1.1 for a two-stage amplifier and outlined as follows:

1. Working forward from the first stage, the output impedance Z_{out1} is determined, taking Z_{src} into account, and Z_{out2} is determined, taking Z_{out1} into account.

2. Working backward from the last stage, the input impedance Z_{in2} of stage 2 is determined, taking Z_L into account, then Z_{in1} is determined, taking Z_{in2} into account.

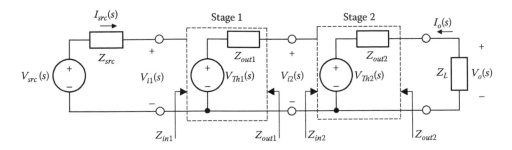

FIGURE 9.1.1
Cascaded amplifier stages.

3. $V_{Th1}(s)$ and $V_{Th2}(s)$ are determined, where $V_{Th1}(s)$ is Thevenin's voltage at the open-circuited output of stage 1, with the source connected, and $V_{Th2}(s)$ is Thevenin's voltage at the open-circuited output of stage 2 with stage 1 connected.

Let $A_{v1o}(s) = V_{Th1}(s)/V_{i1}(s)$ and $A_{v2o}(s) = V_{Th2}(s)/V_{i2}(s)$, be the open-circuit voltage gains of stages 1 and 2, respectively, from the input of the stage to its open-circuited output. It follows that:

$$V_{i1}(s) = \frac{Z_{in1}}{Z_{in1} + Z_{src}} V_{src}(s) \tag{9.1.1}$$

$$V_{i2}(s) = \frac{Z_{in2}}{Z_{in2} + Z_{out1}} V_{Th1}(s) \tag{9.1.2}$$

$$V_o(s) = \frac{Z_L}{Z_L + Z_{out2}} V_{Th2}(s) \tag{9.1.3}$$

Eliminating, through substitution, the intermediate voltages between the source and output:

$$A_v(s) = \frac{V_o(s)}{V_{src}(s)} = \frac{Z_{in1}}{Z_{in1} + Z_{src}} \frac{Z_{in2}}{Z_{in2} + Z_{out1}} \frac{Z_L}{Z_L + Z_{out2}} A_{v1o}(s) A_{v2o}(s) \tag{9.1.4}$$

where
$A_{v1o}(s) = V_{Th1}(s)/V_{i1}(s)$
$A_{v2o}(s) = V_{Th2}(s)/V_{i2}(s)$

If $Z_{in1} \gg Z_{src}, Z_{in2} \gg Z_{out1}, Z_L \gg Z_{out2}$, then $A_v(s) \cong A_{v1o}(s)A_{v2o}(s)$. Hence,

Concept: *When the load impedance at every amplifier node, such Z_{in1}, Z_{in2}, or Z_L is large compared to the source impedance at the given node, the overall transfer function is the product of the open-circuit transfer functions of the individual stages.*

The current gain is:

$$A_i(s) = -\frac{I_o(s)}{I_{src}(s)} = -\frac{V_o(s)}{Z_L} \bigg/ \frac{V_{src}(s)}{Z_{src} + Z_{in1}} = -\frac{Z_{src} + Z_{in1}}{Z_L} A_v(s) \tag{9.1.5}$$

Cascaded stages can also be analyzed by the powerful matrix methods based on two-port circuit theory. In practice, however, short-cut intuitive methods are used for quick, first-order results, as illustrated by Example 9.1.1.

Example 9.1.1 Analysis of Cascaded Amplifier Stages

It is required to determine quickly the small-signal gain v_o/v_i in the circuit of Figure 9.1.2, assuming all transistors have $\beta = 100$ and $r_\pi = 1$ kΩ.

SOLUTION

The method is based on working with small-signal input resistances of stages, and current ratios. The latter are determined by current division at the inputs of transistors and by current gains in the transistors, neglecting r_o of the transistors.

In Figure 9.1.2, Q_3 is connected as an emitter follower. Its input resistance $r_{in3} \cong (\beta + 1)R_{E3} \gg R_{C2}$. Hence, $v_o = v_{c2} \cong R_{C2}i_{c2}$, or,

$$\frac{v_o}{i_{c2}} = R_{C2} = 10 \text{ k}\Omega \tag{9.1.6}$$

For the *pnp* transistor Q_2, i_{b2} and i_{c2} flow out of the base and collector, respectively, and,

$$\frac{i_{c2}}{i_{b2}} = \beta = 100 \tag{9.1.7}$$

The input resistance of Q_2 is $r_{in2} = r_\pi + (\beta + 1)R_{E2} \cong 52$ kΩ. Hence,

$$\frac{i_{b2}}{i_{c1}} = \frac{R_{C1}}{R_{C1} + r_{in2}} = \frac{10}{52} \tag{9.1.8}$$

For Q_1,

$$\frac{i_{c1}}{i_{b1}} = \beta = 100 \tag{9.1.9}$$

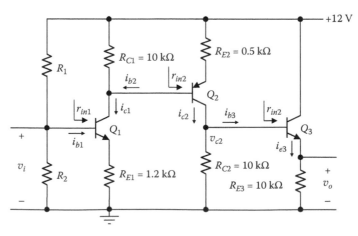

FIGURE 9.1.2
Direct-coupled BJT amplifier stages.

The input resistance of Q_1 is $r_{in1} = r_\pi + (\beta + 1)R_{E1} \cong 122$ kΩ. Hence,

$$\frac{i_{b1}}{v_i} = \frac{1}{r_{in1}} = \frac{1}{122} \text{ mS} \tag{9.1.10}$$

Multiplying Equations 9.1.6 through 9.1.10 gives $v_o/v_i = 10 \times 100 \times (10/52)100 \times (1/122) \cong 160$.

dc Level Shifting

Complementary transistors allow simple dc level shifting. In Figure 9.1.3, Q_1 base is effectively at zero dc voltage. Its emitter is approximately at -0.7 V, which means Q_1 emitter current is 1 mA. Neglecting base currents, Q_1 collector is at $+3.8$ V dc, and Q_2 emitter is at 4.5 V. The emitter current of Q_2 is also 1 mA, so that the dc output voltage is essentially zero.

Complementary MOSFETs in a **folded cascode** (Figure 9.1.4) are often used in ICs to increase the common-mode input range of differential amplifiers (Section 10.3, Chapter 10). The folded cascode also allows shifting the dc level of the output signal with respect to that of the input signal. In Figure 9.1.4, I_{D1} is determined by the dc level of v_I. The dc level of v_O is determined by I_{S2}, which is determined, in turn, by V_{BIAS} and the drain voltage of Q_1. The dc level of the output can therefore be set, mainly by proper choice of V_{BIAS} (Exercise 9.1.1). Compared to the **telescopic cascode** of Figure 8.4.7, Chapter 8, that utilizes two E-NMOS transistors, the cascode of Figure 9.1.4 is "folded" so that both I_{D1} and I_{S2} flow away from node a.

Exercise 9.1.1

Assume that in Figure 9.1.4, the supply voltages are ± 5 V, and the transistors have $k_{n,p} = 100$ μA/V^2 and $|V_t| = 1$ V, $R_1 = 50$ kΩ, and $R_2 = 100$ kΩ. If the dc level of v_I is such that the quiescent current of Q_1 is 10 μA, determine V_{BIAS} so that Q_2 drain is at a dc voltage of zero.
    Answer: 0 V. ∎

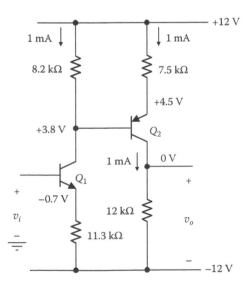

FIGURE 9.1.3
Complementary BJT, direct-coupled amplifier stages.

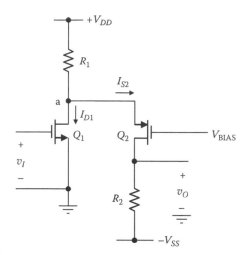

FIGURE 9.1.4
Folded cascode amplifier stage.

Exercise 9.1.2

Draw the low-frequency small-signal equivalent circuit of Figure 9.1.4 with R_1 replaced by an ideal current source I_{BIAS} and show that it is identical to that of a telescopic cascode. What is the effect of including a finite resistance of I_{BIAS}? ∎

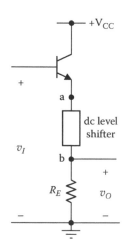

    Emitter followers can be used to shift dc levels. In an *npn* emitter follower, the voltage of the emitter is approximately 0.7 V less than that of the base. Larger voltage differences can be obtained by including, between the emitter and output, devices or subcircuits that can drop substantial dc voltages but have small incremental resistance, so that the small-signal gain is not significantly reduced. For example, several forward-conducting diodes can be connected in series between a and b in Figure 9.1.5, each diode dropping about 0.7 V. Or, a zener diode can be used to drop a larger voltage V_Z (Section 1.4, Chapter 1).

FIGURE 9.1.5
Emitter follower with dc level shifting.

    A useful circuit that can drop a substantial dc voltage but has a relatively small incremental resistance is the V_{BE} **multiplier** illustrated in Figure 9.1.6. Neglecting the base current, $V_{BE} = VR_2/(R_1 + R_2)$, or $V = (1 + R_1/R_2)V_{BE}$. Hence, V is the voltage V_{BE} multiplied by the factor $(1 + R_1/R_2)$. Neglecting r_o, and assuming a large β, the incremental input resistance, is:

$$\frac{v}{i} = \frac{R_1 + R_2}{1 + g_m R_2} \qquad (9.1.11)$$

(Exercise 9.1.3). An attractive feature of this circuit is that the multiplying factor depends on the ratio of two resistances, which can be accurately controlled in an IC. The disadvantage is that the temperature dependence of V is that of V_{BE} multiplied by $(1 + R_1/R_2)$.

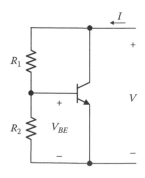

FIGURE 9.1.6
V_{BE} multiplier.

Exercise 9.1.3

Neglecting r_o, show that the incremental input resistance of the V_{BE} multiplier is $\dfrac{v}{i} = \dfrac{R_1 + (R_2\|r_\pi)}{1 + g_m(R_2\|r_\pi)}$ and that it reduces to Equation 9.1.11 for very large β. ∎

9.2 *RC*-Coupled Amplifiers

Coupling and bypass capacitors have been assumed so far to have infinitely large capacitance, which is clearly unrealistic. We will investigate in this section the effect of coupling and emitter/source bypass capacitors on the frequency response of CE/CS amplifiers.

Common-Source Amplifier

Assuming a biasing circuit as in Figure 8.2.3a, Chapter 8, but with the BJT replaced by an E-NMOS transistor, and the resistors labeled accordingly, the small-signal equivalent circuit is shown in Figure 9.2.1. The circuit can be analyzed in a number of ways:

1. "Brute force" circuit analysis using the node–voltage method to derive the overall transfer function $V_o(s)/V_{src}(s)$. The algebra is laborious and the resulting expressions are too complex to give good insight into circuit behavior. A more "tractable" approach is to derive $A_{vo}(s)$ and Z_{out} in a manner similar to that of Section 8.2, Chapter 8.
2. The exact OCSCTC method for determining the poles, as described in Section ST8.9, Chapter 8. In practice, one pole is made dominant by design, in which case the simpler, approximate procedure of short-circuit time constants (SCTCs) can be used. This method is applied later to several cases.
3. An approximate procedure that provides considerable insight into circuit behavior, based on considering each of the three capacitor subcircuits in Figure 9.2.1 in isolation. The input capacitive subcircuit is naturally isolated by the infinite low-frequency

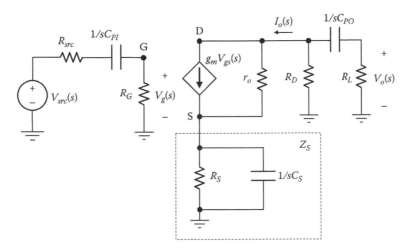

FIGURE 9.2.1
Low-frequency, small-signal equivalent circuit of common-source amplifier.

impedance at the gate of a MOSFET, which allows a straightforward derivation of the transfer function $V_g(s)/V_{src}(s)$. If r_o is not present across the current source, then this source will effectively isolate the two capacitive subcircuits at the output, which allows separate derivations of the transfer functions $I_o(s)/V_g(s)$ and $V_o(s)/I_o(s)$. The overall transfer function is then the product of these three individual transfer functions. r_o can be neglected altogether if the source current $g_m V_{gs}(s)$ is much larger than the current through r_o. A better approximation is to consider r_o to be in parallel with R_D. This is tantamount to assuming that $g_m v_{gs}$ is large compared to the *difference* in the current in r_o between the two cases, when r_o is connected between drain and source or between drain and ground.

The transfer function of the capacitive subcircuit at the input is:

$$\frac{V_g(s)}{V_{src}(s)} = \frac{sC_{PI}R_G}{1 + sC_{PI}(R_{src} + R_G)} \tag{9.2.1}$$

C_{PI} introduces a zero at $s=0$, because it blocks dc voltages, and a pole at $\omega_{pI} = 1/C_{PI}(R_{src} + R_G)$.

When r_o is paralleled with R_D, $I_o(s) = g_m V_{gs}(s) = g_m[V_g(s) - Z_S I_o(s)]$. This gives:

$$\frac{I_o(s)}{V_g(s)} = \frac{1}{Z_S + 1/g_m} = \frac{g_m}{1 + g_m R_S} \frac{1 + sC_S R_S}{1 + sC_S R_S/(1 + g_m R_S)} \tag{9.2.2}$$

The interpretation of Equation 9.2.2 is that ω_{zE}, the zero of $I_o(s)$, occurs when the sum of the currents in R_S and C_S is zero, that is, when $sC_S V_S(s) + V_S(s)/R_S = 0$, which gives $\omega_{zE} = 1/C_S R_S$. The output resistance at the transistor source is $1/g_m$, so that the resistance across C_S is $(R_S \| 1/g_m) = R_S/(1 + g_m R_S)$, which gives a pole at $\omega_{pE} = 1/C_S(R_S \| 1/g_m)$, with $\omega_{zE} < \omega_{pE}$ (Figure 9.2.2). Moreover, $|I_o(s)/V_g(s)| = g_m/(1 + g_m R_S)$, as $s \rightarrow 0$, and $|I_o(s)/V_g(s)| = g_m$, as $s \rightarrow \infty$.

The transfer function from $I_o(s)$ to $V_o(s)$ is:

$$\frac{V_o(s)}{I_o(s)} = -\frac{sC_{PO}R'_D R_L}{1 + sC_{PO}(R'_D + R_L)} \tag{9.2.3}$$

where $R'_D = (R_D \| r_o)$. The overall transfer function is the product of three transfer functions:

$$\frac{V_o(s)}{V_{src}(s)} = -\frac{g_m}{1 + g_m R_S} \frac{s^2 C_{PI} C_{PO} R'_D R_L R_G (1 + sC_S R_S)}{[1 + sC_{PI}(R_{src} + R_G)][1 + sC_S R_S/(1 + g_m R_S)][1 + sC_{PO}(R'_D + R_L)]} \tag{9.2.4}$$

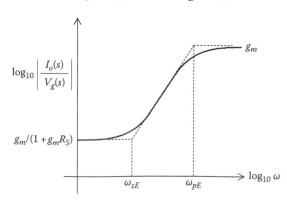

FIGURE 9.2.2
Frequency response due to source-resistor bypass capacitor.

The three independent capacitances give rise to three low-frequency poles. At high frequencies, $s \to \infty$, and the reactances of the capacitors become zero. Since the highest power of s is s^3 in both the numerator and the denominator of Equation 9.2.4, the voltage gain as $s \to \infty$, is $[-g_m R_G / (R_G + R_{src})](R'_D \| R_L)$, as expected for a CS amplifier. The magnitude of the low-frequency voltage gain as $s \to \infty$ is the **midband gain**, because at high enough frequencies the response is dominated by the internal capacitances of the transistor, as discussed in Section 8.3, Chapter 8.

In practice, the lower 3-dB cutoff frequency should be less than a certain value. Since the resistance associated with C_E is generally much smaller than that associated with C_{PI} or C_{PO}, ω_{PE} is made the dominant pole, of highest frequency, in order to keep C_E as small as possible. The 3-dB cutoff frequency will then be nearly that of the dominant pole (Section ST8.8). The other two poles can be chosen to be at the frequency of the zero, which cancels one pole and minimizes the capacitance associated with the other pole (Example 9.2.1).

Design Example 9.2.1 Low-Frequency Response of D-NMOS Amplifier

It is required to design an *RC*-coupled amplifier stage using a D-NMOS transistor in the circuit of Figure 8.2.3a, assuming $V_{DD} = 15$ V, $R_{src} = 50$ kΩ, $R_L = 10$ kΩ, $r_o = 100$ kΩ, $I_{DSS} = 8$ mA, and $V_{tnD} = -2$ V. It is desired to operate the MOSFET at a quiescent point having $I_D = 2$ mA, and $V_{DS} = 5$ V, with R_S at least 2 kΩ, to achieve good insensitivity to variations in transistor parameters. The low 3-dB cutoff frequency should not exceed 50 Hz.

SOLUTION

Neglecting channel-length modulation, $I_D = I_{DSS}(1 - V_{GS}/V_{tnD})^2$. Substituting numerical values, $V_{GS} = -1$ V. If $R_S = 2$ kΩ, the voltage across R_D is $V_{CC} - V_{DS} - R_S I_D = 15 - 5 - 4 = 6$ V, so $R_D = 2$ kΩ. It remains to determine R_{G1} and R_{G2}. Since $V_S = 4$ V, and $V_{GS} = -1$ V, $V_G = 3$ V. R_{G1} and R_{G2} are chosen at a conveniently high value so as to maximize $R_G = R_{G1} \| R_{G2}$. Let $R_{G1} = 1.5$ MΩ. Then $R_{G2} = 375$ kΩ, so that $V_G = 3$ V and $R_G = 300$ kΩ.

From Equation 6.5.5, $g_m = \dfrac{2I_{DSS}}{|V_t|}\sqrt{\dfrac{I_D}{I_{DSS}}} = \dfrac{16}{2}\sqrt{\dfrac{2}{8}} = 4$ mA/V. The midband gain from the source to collector, when the reactances of all capacitors are negligibly small, is

$A'_{vm} = -\dfrac{R_G}{R_{src} + R_G}g_m(R_D \| r_o \| R_L) = -\dfrac{300}{350} \times 4 \times (2\|100\|10) = 5.6$.

As explained previously, the pole due to C_S is dominant at 50 Hz. It follows from Equation 9.2.2 that $C_S = 1/[2\pi f_{pS}(R_S \| 1/g_m)] = 1/[100\pi(2\|0.25)] \cong 14.3$ μF, so a standard value 16 μF may be used. This moves the pole to $50 \times (14.3/16) = 44.7$ kHz. From Equation 9.2.2, the zero due to C_S is at $f_z = 1/2\pi R_S C_S = 1/(2\pi \times 2 \times 10^3 \times 16 \times 10^{-6}) \cong 5$ Hz. We can make this zero cancel one of the poles due to C_{PI} or C_{PO}. The other pole can also be chosen at this frequency so as to minimize the value of the capacitance. It follows from Equation 9.2.1 that $C_{PI} = 1/[2\pi f_z(R_{src} + R_G)] = 91$ nF. From Equation 9.2.3,

$C_{PO} = \dfrac{1}{2\pi f_z[R_L + (R_D \| r_o)]} = \dfrac{1}{10\pi \times 11.96 \times 10^3} \cong 2.7$ μF. The 3-dB frequency can be approximated as $f_L = \sqrt{(47.7)^2 + (5)^2} \cong 48$ Hz (Section ST8.7, Chapter 8).

Common-Emitter Amplifier

The small-signal equivalent circuit of the CE amplifier of Figure 8.2.3a is shown in Figure 9.2.3. The input and the output circuits are no longer isolated from one another

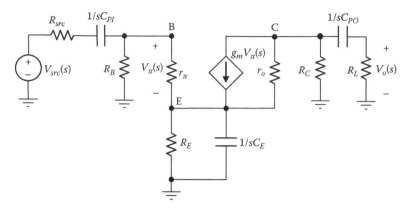

FIGURE 9.2.3
Low-frequency, small-signal equivalent circuit of common-emitter amplifier.

because of r_π. The poles can be determined using the exact OCSCTC method (Section ST8.9, Chapter 8), but since one pole is intentionally made dominant, the simpler SCTC method can be applied. According to this method, the resistance across one capacitor at a time is determined, with the other capacitors short circuited and $V_{src}(s)$ set to zero.

When C_E, is short-circuited, the input circuit is isolated from the output. The resistance appearing across C_{PI}, with $V_{src} = O$, is $R_{PI} = R_{src} + R_B||r_\pi$. With CPI and C_E short circuited, the resistance appearing across C_{PO} is $R_{PO} = R_L + R_C||r_o$. With C_{PI} and C_{PO} short-circuited, the resistance R_{CE} across C_E can be determined from Figure 9.2.4, where $i'_b = -i_b$ and the polarity of the current source is reversed accordingly. If a test source v_x is applied in place of C_E, $i_x = \dfrac{v_x}{R_E} + \dfrac{(\beta+1)v_x}{r_\pi + R_B||R_{src}} + i$, and $v_x = r_o i + (\beta i'_b + i)(R_C||R_L)$. Solving for v_x and simplifying,

$$R_{CE} = R_E||(r_o + R_C||R_L)||\left(\frac{(r_\pi + R_B||R_{src})(r_o + R_C||R_L)}{(\beta+1)r_o + R_C||R_L}\right) \qquad (9.2.5)$$

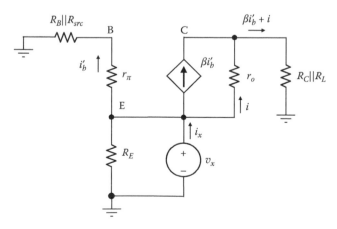

FIGURE 9.2.4
Resistance appearing across the emitter-resistor bypass capacitor.

The 3-dB cutoff frequency ω_{cl} is then:

$$\omega_{cl} \cong \frac{1}{C_{PI}R_{PI}} + \frac{1}{C_{PO}R_{PO}} + \frac{1}{C_E R_{CE}} \tag{9.2.6}$$

Each of the terms in Equation 9.2.6 is the contribution of the respective time constant. The pole due to C_E can be made dominant, that is slightly less than ω_{cl}, and the other two poles placed at the zero due to C_E, as illustrated by the following Example.

Design and Simulation Example 9.2.2 Frequency Response of *RC*-Coupled CE Amplifier

It is required to simulate the frequency response of a CE amplifier in the circuit of Figure 8.2.3a using a 2N2222 transistor with $V_{CC} = 12$ V, $R_{src} = 1$ kΩ, $R_1 = 24$ kΩ, $R_2 = 12$ kΩ, $R_C = 2.7$ kΩ, $R_E = 2$ kΩ, and $R_L = 10$ kΩ.

SIMULATION

From the dc bias point of the simulation, the quiescent values are $I_B = 10$ µA, $I_C = 1.62$ mA, $I_E = 1.63$ mA, $V_B = 3.96$ V, $V_C = 7.62$ V, $V_E = 3.26$ V, which gives $g_m = 1.62/0.026 = 64.8$ mA/V and $\beta_F = 162$. The measured small-signal β is 177. Using this value, $r_\pi = 2.73$ kΩ. Figure 9.2.5 shows the output characteristics with the dc and ac load lines. The quiescent operating point is the intersection of the $I_B = 10$ µA characteristic with the dc load line having a slope of $1/(R_C + R_E) = 1/4.7$ kΩ. The slope of the 10 µA characteristic gives $r_o = 61.5$ kΩ.

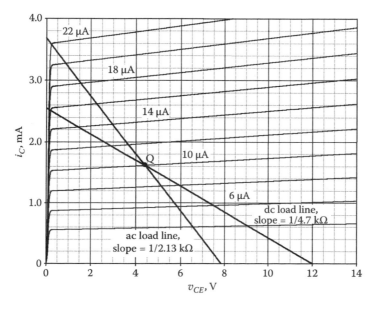

FIGURE 9.2.5
Figure for Design and Simulation Example 9.2.2.

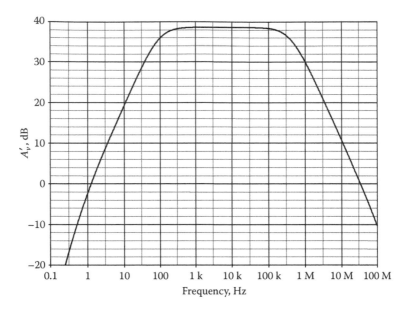

FIGURE 9.2.6
Figure for Design and Simulation Example 9.2.2.

In the midband region, all reactances can be neglected, and the ac load line has a slope of $1/(R_C \| R_L) = 1/2.13$ kΩ. The maximum value of v_{CE} is reduced to about 7.84 V because of the smaller resistance of the ac load line. From Equation 8.2.10, Chapter 8, and using TEC to allow for R_B gives $A'_v = -\dfrac{177(61.5 \| 2.7 \| 10)}{2.73 + 1 \| 8} \times \dfrac{8}{9} = 89.3$. The measured gain at 10 kHz is 85.5.

Suppose a low 3-dB cutoff frequency of less than 100 Hz is desired. From Equation 9.2.5, $R_{CE} \cong R_E \| \lfloor (r_\pi + R_{Bg})/(\beta + 1) \rfloor = 20.8$ Ω. Hence, $C_E = 1/(2\pi \times 100 \times 20.8) = 76.5$ μF, say 80 μF. The zero is at $1/2\pi C_E R_E = 1/(2\pi \times 80 \times 10^{-6} \times 2 \times 10^3) \cong 1$ Hz. Placing the poles due to C_{pI} and C_{pO} at this frequency gives $C_{pI} = 1/[2\pi(1 + 8 \| 2.73) \times 10^3] \cong 50$ μF, and $C_{po} = 1/[2\pi(10 + 2.7 \| 61.5)] \cong 13$ μF. The resulting f_{cl}, from Equation 9.2.6, is about 98 Hz. Note that for the CE amplifier, the zero due to C_E is much less than the pole. A smaller C_{PI} is obtained if the pole due to this capacitor is placed at say 10 Hz.

The amplifier frequency response over the range 0.1 Hz to 100 MHz is shown in Figure 9.2.6. The corner frequencies are approximately at 93 Hz to 400 kHz. The reduced response at high frequencies is of course due to the internal capacitances of the transistor.

The CB/CG and CC/CD amplifiers are easier to analyze as they generally have only a coupling capacitor at the input and another at the output. They are left to problems at the end of this chapter.

9.3 Feedback Amplifiers

In cascaded amplifier stages, it is advantageous to use negative feedback to trade gain for some desirable characteristics of the amplifier, such as modified input or output resistances, reduced distortion, and greater bandwidth, as discussed in Sections 7.3 and 7.4, Chapter 7.

The four basic, negative feedback configurations were introduced in Section 7.3 as two-port circuit implementations of voltage, current, transresistance, and transconductance amplifiers under idealized conditions. In considering practical transistor feedback amplifiers, account should be taken of finite gain, finite load, and the loading effects of the feedback circuit at the input and output of the amplifier. We will still assume, however, negligible reverse transmission in the amplifier and negligible forward transmission in the feedback circuit.

Series–Shunt Feedback

Series–shunt feedback is most conveniently discussed in terms of the h parameters, because the h-parameter equivalent circuit (Figure 7.1.2c, Chapter 7) has series-connected elements at the input and parallel-connected elements at the output. Figure 9.3.1 illustrates a general series–shunt feedback configuration in terms of the h-parameter equivalent circuit of Figure 7.1.2c, where the input ports of both the amplifier and feedback circuits are on the LHS and the output ports are on the RHS. h_{21f}, representing forward transmission through the feedback circuit, is neglected compared to the much larger forward transmission through the amplifier. h_{12a}, representing reverse transmission through the amplifier is also neglected. Some reverse transmission can occur in the CE amplifier, particularly at high frequencies, through r_μ and C_μ, and in the CS amplifier through C_{gd}. However, such transmission is practically absent in the cascode amplifier.

It should be noted that with the assigned positive directions of I_i, $h_{21a}I_i$, and V_o in Figure 9.3.1, the feedback is in fact positive. Thus, if V_o decreases, the feedback signal $h_{12f}V_o$ decreases, which increases I_i and $h_{21a}I_i$. But the increase in $h_{21a}I_i$ further decreases V_o. That the feedback is positive in Figure 9.3.1 should not be surprising since the assigned positive directions are those of two-port circuits and are not intended to give negative feedback. To maintain consistency with the polarity of the feedback signal at the input and with the polarity of the current source in the standard case (Figure 7.3.1, Chapter 7), we

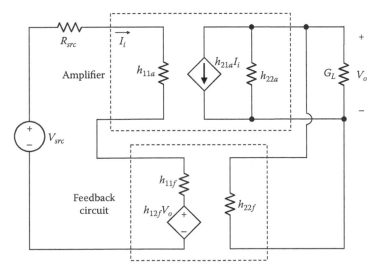

FIGURE 9.3.1
Two-port circuit representation of series–shunt feedback configuration.

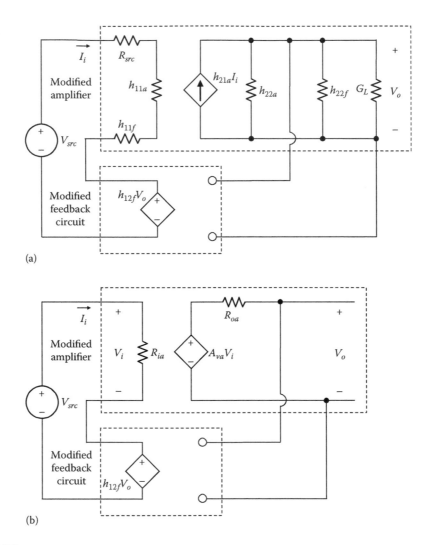

FIGURE 9.3.2
Two-port circuit representation of series–shunt negative feedback configuration: (a) the input impedance and the output admittance of feedback circuit moved to amplifier; and (b) standard form.

will reverse the polarity of the source $h_{21a}I_i$, as in Figure 9.3.2a. h_{11f} and h_{22f} are moved to the amplifier circuit so that (Figure 9.3.2b):

$$R_{ia} = R_{src} + h_{11a} + h_{11f}, \quad R_{oa} = 1/(h_{22a} + h_{22f} + G_L) \qquad (9.3.1)$$

The dependent current source at the output can then be transformed to a voltage source, as in Figure 9.3.2b, where $A_{va}V_i = h_{21a}R_{oa}I_i = h_{21a}(R_{oa}/R_{ia})V_i$, which gives:

$$A_{va} = h_{21a}\frac{R_{oa}}{R_{ia}} \qquad (9.3.2)$$

With the amplifier proper modified as shown, the series–shunt feedback amplifier is now in the standard form of Figure 7.3.1, Chapter 7. The feedback circuit does not load the output, and the feedback signal at the input is applied through a voltage-controlled dependent voltage source. Moreover, because the source and load resistances have been included within the modified amplifier, the feedback amplifier is effectively open-circuited at the output and is supplied at the input from an ideal source of zero source impedance. The objective of the aforementioned procedure is:

Concept: *When the feedback amplifier is represented in standard form, the gain, input resistance, and output resistance of the feedback amplifier can be determined quite simply from the corresponding expressions for the standard feedback amplifier.*

The signal-flow diagram of the series–shunt feedback amplifier of Figure 9.3.2b is illustrated in Figure 9.3.3. From Equations 7.3.1, 7.3.3, and 7.3.4, Chapter 7,

$$A_{vfa} = \frac{V_o}{V_{src}} = \frac{A_{va}}{1 + \beta_f A_{va}}, \quad R_{ifa} = R_{ia}(1 + \beta_f A_{va}), \quad R_{ofa} = \frac{R_{oa}}{1 + \beta_f A_{va}} \tag{9.3.3}$$

where the a, f, and fa subscripts refer to the modified amplifier, feedback circuit, and feedback amplifier, respectively.

The aforementioned procedure will be illustrated by the noninverting op-amp circuit (Figure 9.3.4), taking into account the differential input resistance R_{id}, the output resistance R_o of the amplifier, and the finite gain A_v. Note that the direction of I_o is reversed in this case, as is usual in op amps. The analysis proceeds in a number of steps based on the basic circuit of Figure 9.3.1.

1. To find h_{11f}, as defined in Figure 7.1.2c, Chapter 7, we set the source $h_{12f}V_o$ to zero by short-circuiting the output for the ac signal and determining the resistance of the feedback circuit at the input side. From Figure 9.3.4, this is $(R_r \| R_f)$ and

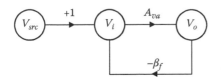

FIGURE 9.3.3
Signal-flow diagram of series–shunt feedback amplifier.

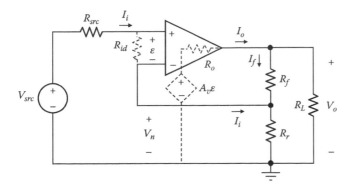

FIGURE 9.3.4
Nonideal, noninverting op-amp circuit.

represents the loading effect of the feedback circuit at the amplifier input. Because of the series connection at the input,

$$R_{ia} = R_{src} + R_{id} + (R_r || R_f) \tag{9.3.4}$$

2. To find h_{22f}, as defined in Figure 7.1.2c, Chapter 7, we set the source $h_{21a}I_i$ to zero by open-circuiting the input for the ac signal ($I_i = 0$) and determining the resistance of the feedback circuit at the output side. From Figure 9.3.4, this is $(R_r + R_f)$ and represents the loading effect of the feedback circuit at the amplifier output. Because of the shunt connection at the output,

$$R_{oa} = R_o || R_L || (R_r + R_f) \tag{9.3.5}$$

3. To find h_{12f}, as defined in Figure 7.1.2c, Chapter 7, we apply a voltage V_o at the output of the feedback circuit, open-circuit the input, and determine the open-circuit voltage at the input of the feedback circuit. From Figure 9.3.4, h_{12f} is the feedback factor β_f given by:

$$\beta_f = \frac{R_r}{R_r + R_f} \tag{9.3.6}$$

4. The feedback amplifier becomes as in Figure 9.3.5. The modified amplifier is the circuit of Figure 9.3.5 with the feedback source $\beta_f V_o$ set to zero (Figure 9.3.2b). The forward transmission A_{va} of the modified amplifier is V_o/V_{src} in Figure 9.3.5 with $\beta_f V_o = 0$. It follows from this figure that $\varepsilon = V_{src}(R_{id}/R_{ia})$ and

$$V_o = A_v \varepsilon \frac{R_L || (R_f + R_r)}{R_o + R_L || (R_f + R_r)} = A_v \varepsilon \frac{R_{oa}}{R_o}. \text{ Substituting for } \varepsilon,$$

$$\frac{V_o}{V_{src}} = A_{va} = A_v \frac{R_{id} R_{oa}}{R_o R_{ia}} \tag{9.3.7}$$

5. Relations 9.3.3 are used to determine A_{vfa}, R_{ifa}, and R_{ofa}.

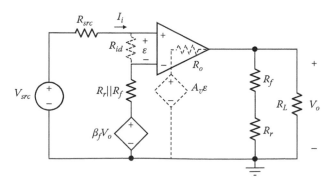

FIGURE 9.3.5
The circuit of Figure 9.3.4 in standard form of a series–shunt feedback amplifier.

Note that Relations 9.3.3 apply to R_{src} and R_L included as part of the modified amplifier (Figure 9.3.2b). The input resistance R'_{ifa} of the feedback amplifier, not including R_{src}, is:

$$R'_{ifa} = R_{ifa} - R_{src} \tag{9.3.8}$$

Similarly, the output resistance R'_{ofa} of the feedback amplifier seen by the load R_L is such that $R_{ofa} = R'_{ofa} \| R_L$. Hence,

$$\frac{1}{R'_{ofa}} = \frac{1}{R_{ofa}} - \frac{1}{R_L} \tag{9.3.9}$$

Exercise 9.3.1

An op amp in the noninverting configuration has $A_v = 5{,}000$, $R_{src} = 10$ kΩ, $R_{id} = 100$ kΩ, $R_o = 1$ kΩ, $R_L = 2$ kΩ, $R_r = 5$ kΩ, and $R_f = 20$ kΩ. Determine A_{vfa}, R'_{ifa}, and R'_{ofa}.
   Answers: $A_{fa} = 5$; $R'_{ifa} = 64.9$ MΩ; and $R'_{ofa} = 1.14$ Ω. ∎

The procedure just outlined can be generalized to apply to the other three feedback configurations, as summarized in Table 9.3.1. The second column identifies the circuit variable at the input or output that is common to both the amplifier and feedback circuit. Evidently, current is the common variable in a series connection, and voltage is the common variable in a shunt connection. The third column emphasizes that when the feedback connection at the input (or output) is series type, the resistances of the source (or load), amplifier, and feedback circuit are added. Similarly, when the feedback connection at the input (or output) is shunt type, the conductances of the source (or load), amplifier, and feedback circuit are added. The fourth column specifies the procedure for determining the loading effect of the feedback circuit at the input. In the case of series–shunt feedback, the resistance at the input of the feedback circuit is determined with the common circuit variable at the output (which is voltage) set to zero, that is, with

TABLE 9.3.1

Determining the Loading Effect of the Feedback Circuit and the Feedback Factor

Connection at Input or Output	Common Circuit Variable at Input or Output	Input or Output Resistance at Amplifier Input or Output	To Determine Loading Effect of Feedback Circuit at Input	To Determine Loading Effect of Feedback Circuit at Output	To Derive the Feedback Factor
Series	Current (factual output)	Resistances add	Determine resistance at input of feedback circuit with common circuit variable at output set to zero	Determine resistance at output of feedback circuit with common circuit variable at input set to zero	With the common circuit variable at input set to zero, determine the other circuit variable at input of feedback circuit when the common circuit variable at output is applied
Shunt	Voltage (factual output)	Conductances add			

the output short circuited. The fifth column specifies the procedure for determining the loading effect of the feedback circuit at the output. In the case of series–shunt feedback, the resistance at the output of the feedback circuit is determined with the common circuit variable at the input (which is current) set to zero, that is, with the input open circuited. The last column specifies the procedure for determining the feedback factor. In the case of series–shunt feedback, the input is open-circuited and the voltage at the input of the feedback circuit is determined when a voltage is applied at the output of this circuit. The following should be noted:

1. Open-circuiting or short-circuiting the input or output refers to the *ac signal only, without disturbing the dc bias conditions*. A large capacitor can be used for an ac short circuit that does not affect dc voltages on either side of the capacitor, whereas a large inductor can be used for an ac open circuit that does not affect the dc current through the inductor.
2. If the feedback at the output is series, the *factual* output is the current that is common to both the amplifier and the feedback circuit and is the current used in feedback analysis. The *designated* output, on the other hand, could be another current or voltage, depending on the circuit. In a CE/CS amplifier with an emitter/source resistor, for example, the feedback is series at the output and the factual output is current, but the designated output is the voltage at the collector/drain. After the feedback analysis is performed using the factual output variable, the voltage gain or output resistance at the designated output is determined, as illustrated in Example 9.3.2. Similarly, if the feedback at the output is shunt, the factual output is voltage.
3. When determining the feedback factor, it must be kept in mind that the assigned positive direction of current is inward at both the input and output terminals of the feedback circuit, in accordance with the convention for two-port circuits.

It should be noted that the preceding analysis is based on the assumptions of negligible reverse transmission in the amplifier and negligible forward transmission in the feedback network. These can be ascertained by comparing the relevant two-port parameters, which in all cases are the "12" parameter of the amplifier compared to that of the feedback circuit, and the "21" parameter of the feedback circuit compared to that of the amplifier. If these assumptions are not justified, the more general method based on two-port circuit theory should be used.

Example 9.3.1 Series–Shunt Feedback of Transistor Amplifier

Figure 9.3.6a is a variation on the "series–series triple" (Example 9.3.2) in which two transistor stages Q_1 and Q_2 are followed by an emitter follower Q_3. Part of the output signal v_o is fed back via R_f to the emitter of Q_1. That the feedback is negative can be readily ascertained. An increase in v_{be1} through an increase in v_{src}, increases the collector current of Q_1, decreases that of Q_2, and increases v_o. The feedback increases v_{e1}, which opposes the increase in v_{be1}. Note that the feedback circuit is not connected directly to the input terminal, that is, Q_1 base. However, it is connected to the emitter of Q_1 and is therefore still in the base-emitter input circuit of Q_1 and subtracts from the applied input. This is similar to applying the source input to the noninverting input of an op amp and the feedback signal to the inverting input.

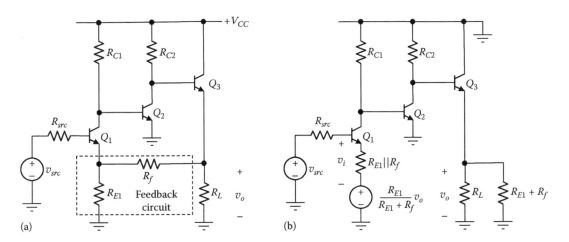

FIGURE 9.3.6
Figure for Example 9.3.1. (a) Feedback amplifier circuit; (b) modified amplifier and feedback source.

It is required to determine A_{fa}, R_{ifa}, and R_{ofa}, assuming that $I_{C1} = 0.6$ mA, $I_{C2} = 1$ mA, $I_{C3} = 0.8$ mA and that $\beta_f = 100$, $R_{E1} = 100\,\Omega$, $R_{C1} = 9$ kΩ, $R_{C2} = 5$ kΩ, $R_L = 3$ kΩ, $R_f = 640\,\Omega$, and $R_{src} = 0$. Circuit variables are lowercase with small subscripts to emphasize small-signal operation.

Solution

The procedure outlined in Table 9.3.1 is applied to Figure 9.3.6a. To determine the loading effect of the feedback circuit at the input of the amplifier, we short circuit the output for ac signals. The input resistance of the feedback circuit, at Q_1 emitter, is $R_{E1}\|R_f$. To determine the loading effect at the output of the amplifier, we open circuit the input for ac signals. The resistance looking into the output of the feedback circuit, at Q_3 emitter, is $R_{E1} + R_f$. To determine the feedback factor, we open circuit the input as before and determine the voltage at the input of the feedback circuit when v_o is applied. This voltage is $\beta_f v_o = v_o R_{E1} / (R_{E1} + R_f)$, which gives $\beta_f = R_{E1}/(R_{E1} + R_f)$. Figure 9.3.6b shows the modified amplifier circuit and the feedback voltage source $v_o R_{E1}/(R_{E1} + R_f)$. As ac operation is assumed, the V_{CC} line is shorted to ground.

The small-signal equivalent circuit of the modified amplifier is shown in Figure 9.3.7, neglecting r_o, which is large compared to the resistance in series with the collectors of Q_1 and Q_2.

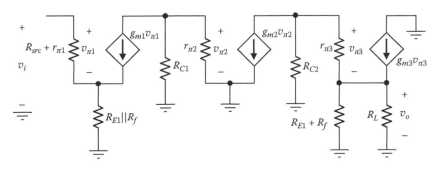

FIGURE 9.3.7
Figure for Example 9.3.1.

From the values given, $g_{m1} = 0.6/0.026 = 23$ mA/V, $r_{\pi 1} = 4.33$ kΩ, $g_{m2} = 1/0.026 = 38.5$ mA/V, $r_{\pi 2} = 2.6$ kΩ, $g_{m3} = 0.8/0.026 = 31$ mA/V, $r_{\pi 3} = 3.25$ kΩ, $R_{E1}||R_f = 86.5$ Ω, $r_{\pi 2}||R_{C1} = 2$ kΩ, and $(R_E + R_f)||R_L = 594$ Ω. The forward transmission through the amplifier $A = v_o/v_i$ is calculated from the voltage gains of the three stages. For the first stage, $v_{\pi 2}/v_i = -100 \times 2/[4.33 + 101 \times 86.5 \times 10^{-3}] = -15.3$ (Equation 8.2.46, Chapter 8). For the second stage we consider the voltage across R_{C2} with Q_3 disconnected. This gain is simply $-g_{m2}R_{C2} = 38.5 \times 5 = -192.5$. The voltage at this collector, with Q_3 disconnected, is $192.5 \times 15.3v_i$. As far as Q_3 is concerned, this is the open-circuit voltage of a source having a source resistance of 5 kΩ. From Equation 8.2.24, Chapter 8, the voltage gain from the source to emitter of the CC configuration is $101 \times 0.594/(5 + 3.25 + 101 \times 0.594) = 0.88$. Alternatively, we can calculate the voltage at Q_2 collector to be $v_{c2} = -g_{m2}v_{\pi 2}R_{C2}||[r_{\pi 3} + (\beta + 1)R_E']$, where $R_E' = R_L||$
$(R_{E1} + R_f)$, and $v_o = \dfrac{(\beta + 1)R_E'}{r_{\pi 3} + (\beta + 1)R_E'}v_{c2}$, which gives $v_o = -g_{m2}\dfrac{R_{C2}(\beta + 1)R_E'}{R_{C2} + r_{\pi 3} + (\beta + 1)R_E'}v_{\pi 2}$, as before.

The forward transmission is therefore $192.5 \times 15.3 \times 0.88 = 2590$. $\beta_f = 100/740 = 0.135$, so that $1 + \beta_f A_{gfa} = 350$. It follows $A_{gfa} = 2590/350 = 7.4$. The input resistance of the modified amplifier is $4.33 + 101 \times 0.0865 = 13.1$ kΩ. Hence, $R_{ifa} = 13.1 \times 350 = 4.6$ MΩ. Since $R_{src} = 0$, $R_{ifa}' = R_{ifa}$. The output resistance of the CC stage, with a source resistance of 5 kΩ is $0.594||[(3.25 + 5)/101] \equiv 72$ Ω. This gives $R_{ofa} = 72/350 = 0.21$ Ω. From Equation 9.3.9, $R_{outfa}' \cong R_{outfa}$.

Series–Series Feedback

Series–series feedback is most conveniently discussed in terms of the z parameters. Figure 9.3.8 illustrates a general series–series feedback configuration based on the z-parameter equivalent circuit of Figure 7.1.2a, Chapter 7, ignoring z_{12a} and z_{21f}. As in the case of series–shunt feedback, the feedback is positive. Thus, if I_o decreases, the feedback signal $z_{12f}I_o$ decreases, which increases I_i and $z_{21a}I_i$. But the increase in $z_{21a}I_i$ further decreases I_o. To maintain consistency with the polarity of the feedback signal at the input

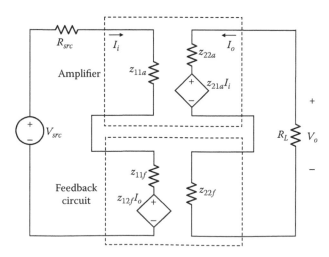

FIGURE 9.3.8
Two-port circuit representation of series–series feedback configuration.

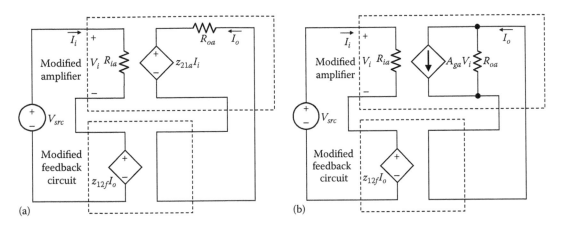

FIGURE 9.3.9
Two-port circuit representation of series–series negative feedback configuration: (a) input and output impedances of feedback circuit moved to amplifier; (b) standard form.

and with the polarity of the current source in the ideal case (Figure 7.3.5, Chapter 7), we will reverse the polarity of the source $z_{21a}I_i$.

The procedure of Table 9.3.1 applies to the derivation of z_{11f} and z_{22f} from Figure 9.3.8. Thus, because of the series connection, current is the common variable at the input and output. If the output is open-circuited, $z_{12f}I_o = 0$, and the resistance seen at the input of the feedback circuit is z_{11f}. If the input is open-circuited, $z_{21a}I_i = 0$, and the resistance seen at the output of the feedback circuit is z_{22f}. With $I_i = 0$, and if a current I_x is applied in the output circuit, the voltage measured at the input of the feedback circuit is $z_{12f}I_x$. The feedback factor is $z_{12f} = \beta_f$.

Figure 9.3.9a depicts the feedback amplifier with the polarity of the source $z_{21a}I_i$ reversed and with z_{11f} and z_{22f} moved to the amplifier circuit to give:

$$R_{ia} = R_{src} + z_{11a} + z_{11f}, \quad R_{oa} = z_{22a} + z_{22f} + R_L \tag{9.3.10}$$

The voltage source $z_{21a}I_i$ is transformed to a current source in Figure 9.3.39b and expressed as $A_{ga}V_i$, where and $A_{ga}V_i = (z_{21a}/R_{oa})I_i = (z_{21a}/R_{oa})(V_i/R_{ia})$, so that,

$$A_{ga} = \frac{I_o}{V_i} = \frac{z_{21a}}{R_{ia}R_{oa}} \tag{9.3.11}$$

The amplifier of Figure 9.3.9 is now in the standard form of Figure 7.3.5, Chapter 7, the signal-flow diagram being as in Figure 9.3.10, Equations 7.3.6, 7.3.8, and 7.3.9, Chapter 7, of the standard series–series feedback amplifier can then be applied to give:

$$A_{gfa} = \frac{I_o}{V_{src}} = \frac{A_{ga}}{1 + \beta_f A_{ga}}, \quad R_{ifa} = R_{ia}(1 + \beta_f A_{ga}), \quad R_{ofa} = R_{oa}(1 + \beta_f A_{ga}) \tag{9.3.12}$$

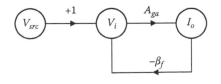

FIGURE 9.3.10
Signal-flow diagram of series–series feedback amplifier.

The input resistance R'_{ifa} of the feedback amplifier, not including R_{src}, is given by Equation 9.3.8. The output resistance R'_{ofa} of the feedback amplifier, not including R_L, is:

$$R'_{ofa} = R_{ofa} - R_L \tag{9.3.13}$$

Example 9.3.2 Series–Series Feedback of Transistor Amplifier

Figure 9.3.11 shows the "series–series triple" with the output taken from Q_3 collector. This is the circuit used in the MC1553 wideband amplifier having a bandwidth of 50 MHz and a gain of 50. In the MC1553, the actual output is taken from an emitter follower stage that follows Q_3. It is required to determine A_{gfa}, R_{ifa}, and R_{ofa}. Assume that the transistors are biased so that $I_{C1} = 0.6$ mA, $I_{C2} = 1$ mA, and $I_{C3} = 4$ mA and that $\beta = 100$, $R_{E1} = 100\ \Omega$, $R_{C1} = 9\ k\Omega$, $R_{C2} = 5\ k\Omega$, $R_{C3} = 600\ \Omega$, $R_{E3} = 100\ \Omega$, $R_f = 640\ \Omega$, and $R_{src} = 0$.

SOLUTION

As in Example 9.3.1, the feedback signal is applied to the emitter of Q_1. It is derived from the voltage across R_{E3}, which is directly proportional to i_o. This makes the feedback series-type at the output, R_{E3} being part of the feedback circuit. However, the designated output is v_o taken from Q_3 collector. The analysis as a feedback system is therefore based on the factual output i_o. After this analysis is made, the designated output is considered.

To determine the loading effect of the feedback circuit at the input of the amplifier, we open-circuit the output for ac signals at, for example, XX'. The resistance appearing at the input of the feedback circuit, at the emitter of Q_1, due to the feedback circuit is $R_{E1}||(R_f + R_{E3})$. To determine the loading effect at the output of the amplifier, we open-circuit the input for ac signals. The resistance appearing at the output of the feedback circuit, at the emitter of Q_3, due to the feedback circuit is $R_{E3}||(R_f + R_{E1})$. To determine the feedback factor, we open circuit the input for ac signals and determine the voltage at the input when i_o is applied at the output of the feedback circuit. The voltage at the input of the feedback circuit is $\beta_f i_o = i_o R_{E1} R_{E3}/(R_{E1} + R_f + R_{E3})$, so the feedback factor is $\beta_f = R_{E1} R_{E3}/(R_{E1} + R_f + R_{E3})$.

The small-signal equivalent circuit of the modified amplifier is shown in Figure 9.3.12 neglecting r_o and assuming that $R_{src} = 0$. From the values given, $g_{m1} = 0.6/0.026 = 23$ mA/V,

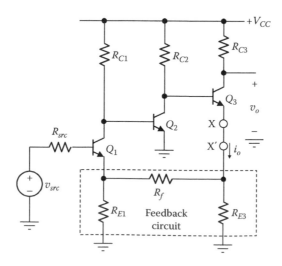

FIGURE 9.3.11
Figure for Example 9.3.2.

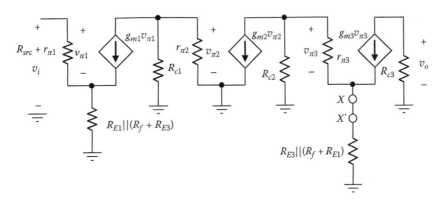

FIGURE 9.3.12
Figure for Example 9.3.2.

$r_{\pi 1} = 4.33$ kΩ, $g_{m2} = 1/0.026 = 38.5$ mA/V, $r_{\pi 2} = 2.6$ kΩ, $g_{m3} = 4/0.026 = 154$ mA/V, $r_{e3} = 6.44\ \Omega$, $R_{E1}||(R_f + R_{E3}) = R_{E3}||(R_f + R_{E1}) = 88\ \Omega$, and $r_{\pi 2}||R_{C1} = 2$ kΩ. For the first stage, $v_{\pi 2}/v_i = -100 \times 2/(4.33 + 101 \times 88 \times 10^{-3}) = -15.1$ (Equation 8.2.46, Chapter 8). The resistance between Q_3 base and ground is $(6.44 + 88) \times 101 \equiv 9.54$ kΩ. The collector load of Q_2 is $9.54||5 = 3.28$ kΩ, and the gain of this stage is $-38.5 \times 3.28 = -126$. Hence, the voltage at Q_3 base $v_{b3} = 15.1 \times 126 v_i$. For the last stage, $i_o = v_{b3}/[r_{e3} + R_{E3}||(R_f + R_{E1})] = v_{b3}/(6.44 + 88) = v_{b3}/94.4$ A. It follows that the forward transmission is $15.1 \times 126/94.4 = 20.2$ S.

The feedback factor $\beta_f = 100 \times 100/840 = 11.9\ \Omega$. The gain with feedback is $A_{fa} = 20.2/(1 + 11.9 \times 20.2) = 20.2/241.4 = 83.7$ mS. The voltage gain, assuming $\beta = 100$, is $-\alpha_3 R_{C3} A_{fa} = -49.7$.

The input resistance of the modified amplifier is $R_i = 4.33 + 101 \times 88 \times 10^{-3} = 13.22$ kΩ. Hence, $R_{ifa} = 13.22 \times 241.4 = 3.19$ MΩ. Since $R_{src} = 0$, $R'_{ifa} = R_{ifa}$. To determine the output resistance of the modified amplifier, we break the connection at XX' and determine the resistance R_{out} between these points, with zero input, so that $g_{m2}v_{\pi 2} = 0$. Because r_o is neglected, R_{C3} does not contribute to R_{out}. It is seen from the equivalent circuit that $R_{oa} = R_{E3}||(R_f + R_{E1}) + r_{e3} + R_{C2}/(\beta + 1) = 88 + 6.44 + 5 \times 10^3/101 = 144\ \Omega$, where the resistance in the base circuit has been referred to the emitter. Hence, $R_{ofa} = 144 \times 241.4 = 34.7$ kΩ. To determine the resistance looking into Q_3 collector, we have to assume a certain value for r_o, say 25 kΩ. Equation 8.2.38, Chapter 8, can then be applied to the modified amplifier to determine the resistance looking into Q_3 collector with zero input. Substituting $R'_{src} = 5||25 = 4.2$ kΩ, $r_{\pi 3} = 6.44 \times 101 \equiv 0.65$ kΩ, and $R_E = 0.088$ kΩ gives $25 + \dfrac{0.088}{0.65 + 4.2 + 0.088}[0.65 + 4.2 + (100 \times 25)] = 69.64$ kΩ. Multiplying this by (1 − loop gain), because the feedback tends to keep the collector current constant, gives $69.64 \times 241.4 = 16.6$ MΩ. The total resistance between Q_3 collector and ground is this resistance in parallel with 600 Ω, which is practically 600 Ω. It may be noted that this is the source resistance for the emitter follower at the output of the MC1553. This stage has an emitter resistor of 3 kΩ and a collector current of 3.2 mA, so that $r_{e4} = 8\ \Omega$. From Equation 8.2.31, Chapter 8 the resistance looking into the emitter of the emitter follower is $8 + \dfrac{600}{101} = 13.9\ \Omega$. When paralleled with the 3 kΩ resistor, the output resistance of the MC1553 is nearly 13.9 Ω.

Having determined i_o, $v_o = -R_{C3}\alpha_3 i_o$.

Exercise 9.3.2

The basic amplifier in a series–series feedback configuration has a short-circuit current gain $(-z_{21a}/z_{22a})$ of -4000, an input resistance of 2 kΩ, and an output resistance of 1 kΩ. The feedback circuit has an input resistance of 1 kΩ, with port 2 open-circuited, an output resistance of 1 kΩ, and a feedback factor of 0.1 V/mA. The source has a resistance of 2 kΩ and the load is a 4 kΩ resistor. Determine A_{ga}, A_{gfa}, R_{ifa}, and R_{ofa}.

Answers: $A_{ga} = 133$ mA/V; $A_{gfa} = 9.3$ mA/V; $R_{ifa} = 71.5$ kΩ; $R_{ofa} = 85.8$ kΩ. ∎

Shunt–Series Feedback

Shunt–series feedback is most conveniently discussed in terms of the g parameters. Figure 9.3.13 illustrates a general shunt–series feedback configuration based on the g-parameter equivalent circuit of Figure 7.1.2d, ignoring g_{12a} and g_{21f}. As discussed in connection with series–series feedback, polarities in Figure 9.3.13 are such that the feedback is in fact positive. Thus, if I_o decreases, the feedback current $g_{12f}I_o$ decreases. More current is diverted to g_{11a}, which increases V_i and $g_{21a}V_i$. But the increase in $g_{21a}V_i$ further decreases I_o. To maintain consistency with the polarity of the feedback signal at the input and with the polarity of the current source in the standard case (Figure 7.3.7, Chapter 7), we will reverse the sign of the source $g_{21a}V_i$.

It can be readily verified that the procedure of Table 9.3.1 applies to the derivation of g_{11f} and g_{22f} from Figure 9.3.13. Thus, because of the series connection at the output, current is the common variable at the output. If the output is open-circuited, $g_{12f}I_o = 0$, and the conductance seen at the input of the feedback circuit is g_{11f}. Voltage is the common variable at the input. If the input is short-circuited, $g_{21a}V_i = 0$, and the resistance seen at the output of the feedback circuit is g_{22f}. With $V_i = 0$, and if a current I_x is applied in the output circuit, the input current of the feedback circuit is $g_{21f}I_x$. The feedback factor is $g_{12f} = \beta_f$.

Figure 9.3.14a depicts the feedback amplifier with the polarity of the source $g_{21a}V_i$ reversed and with g_{11f} and g_{22f} moved to the amplifier circuit to give:

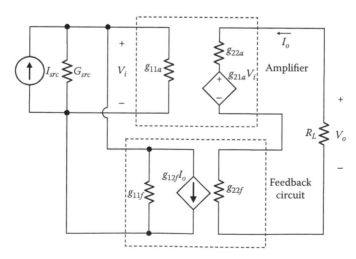

FIGURE 9.3.13
Two-port circuit representation of shunt–series feedback configuration.

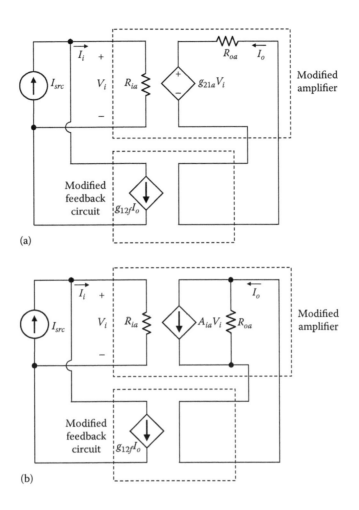

FIGURE 9.3.14
Two-port circuit representation of shunt–series negative feedback configuration: (a) input and output impedances of feedback circuit moved to amplifier; (b) standard form.

$$\frac{1}{R_{ia}} = G_{src} + g_{11f} + g_{11a}, \quad R_{oa} = R_L + g_{22a} + g_{22f} \tag{9.3.14}$$

The voltage source $g_{21a}V_i$ is transformed in Figure 9.3.14b to a current source and expressed as $A_{ia}V_i$, where and $A_{ia}I_i = (g_{21a}/R_{oa})V_i = (g_{21a}/R_{oa})R_{ia}I_i$, so that,

$$A_{ia} = \frac{I_o}{I_i} = \frac{g_{21a}R_{ia}}{R_{oa}} \tag{9.3.15}$$

The amplifier of Figure 9.3.14b is now in the standard form of Figure 7.3.7, Chapter 7, the signal-flow diagram being as in Figure 9.3.15. Equations 7.3.10, 7.3.12, and 7.3.13, Chapter 7, of the standard shunt–series feedback amplifier can then be applied to give:

$$A_{ifa} = \frac{I_o}{I_{srci}} = \frac{A_{ia}}{1 + \beta_f A_{ia}}, \quad R_{ifa} = R_{ia}/(1 + \beta_f A_{ia}), \quad R_{ofa} = R_{oa}(1 + \beta_f A_{ia}) \tag{9.3.16}$$

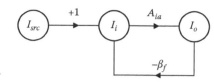

FIGURE 9.3.15
Signal-flow diagram of shunt–series feedback amplifier.

The input resistance R_{if} of the feedback amplifier, not including R_{src}, is:

$$\frac{1}{R'_{ifa}} = \frac{1}{R_{ifa}} - \frac{1}{R_{src}}$$

(9.3.17)

The output resistance R'_{ofa}, not including R_L, is given by Equation 9.3.13.

Simulation Example 9.3.3 Shunt–Series Feedback of Transistor Amplifier

Figure 9.3.16 shows a "current feedback pair" that is used in ICs, the feedback circuit consisting of R_E and R_f. It is required to determine A_{ifa}, R_{ifa}, and R_{ofa}. Assume that the transistors are properly biased so that $I_{C1} = 1$ mA, $I_{C2} = 1.85$ mA, with $\beta = 140$, $r_{o1} = 100$ kΩ, $r_{o2} = 50$ kΩ, $R_{src} = 10$ kΩ, $R_C = 9$ kΩ, $R_f = 10$ kΩ, $R_L = 5$ kΩ, and $R_E = 3.3$ kΩ.

SOLUTION

As in Example 9.3.2, the factual output is i_o, whereas the designated output is v_o. To determine the loading effect of the feedback circuit at the input of the amplifier, we open-circuit the output for the ac signal. The input resistance of the feedback circuit, at Q_1 base, is $R_f + R_E$. To determine the loading effect at the output of the amplifier, we short-circuit the input for the ac signal at Q_1 base. The resistance looking into the output of the feedback circuit, at Q_2 emitter, is $R_E \| R_f$. To determine the feedback factor, we short-circuit the input for the ac signal and determine the current at the input when i_o is applied at the output of the feedback circuit. According to the convention for two-port circuits, as shown in Figure 9.3.14, the assigned positive direction of current at the input of the feedback circuit is inward. This current is $\beta_f i_o = -i_o R_E / (R_E + R_f)$, so the feedback factor is $\beta_f = -R_E / (R_E + R_f) = -3.3/13.3 = -0.248$.

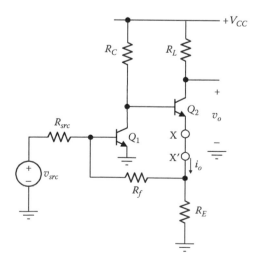

FIGURE 9.3.16
Figure for Example 9.3.3.

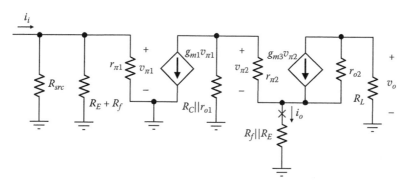

FIGURE 9.3.17
Figure for Example 9.3.3.

The feedback current due to i_o is directed outward at the base of Q_1 so it adds to the source current. That is, the actual sign in the feedback path is positive. This is in accordance with β_f having a negative value, so that $-\beta_f$ in the feedback path is positive. To have a negative loop gain, the forward gain of the modified amplifier should have a negative value.

The small-signal equivalent circuit is shown in Figure 9.3.17. From the values given, $g_{m1} = 1/0.026 = 38.5$ mA/V, $g_{m2} = 1.85/0.026 = 71.2$ mA/V, $r_{\pi1} = 3.64$ kΩ, $r_{o1} = 100$ kΩ, $r_{\pi2} = 1.97$ kΩ, $R_f + R_E = 13.3$ kΩ, and $R_E || R_f = 2.48$ kΩ. It is seen that $v_{\pi1} = [R_{src} || (R_f + R_E) || r_{\pi1}] \times i_i = (10 || 13.3 || 3.64)i_i = 2.22i_i$. Neglecting r_{o2}, which is relatively large, introduces little error in calculating the forward transmission, but is very convenient. Hence, $v_{b2} = -g_{m1}v_{\pi1}R_C || r_{o1} || [r_{\pi2} + (\beta + 1)(R_E || R_f)] = -38.5(9 || 100 || 351.6)v_{\pi1} = -310.6v_{\pi1}$. In addition, $i_o = v_{b2}/[r_{e2} + R_E || R_f] = v_{b2}/2.49$. Multiplying out gives $A_{ia} = i_o/i_i = -276.9$, which is negative, as it should be. The loop gain $-\beta_f A_{ia}$ is $0.248 \times (-276.9) = -68.7$. The gain with feedback is $A_{ifa} = -276.9/(1 + 68.7) = -3.97$. The input resistance of the modified amplifier is $R_{ia} = R_{src} || (R_f + R_E) || r_{\pi1} = 2.22$ kΩ. Hence, $R_{ifa} = 2.22/69.7 \equiv 31.9$ Ω. From Equation 9.3.17, $R'_{ifa} \cong R_{ifa} = 31.9$ Ω. The voltage gain from Q_1 base to Q_2 collector is $-\alpha i_o R_L/i_i R_{ifa} = 0.99 \times 3.97 \times 5 \times 10^3/31.9 = 616$.

To determine the output resistance of the modified amplifier, we open-circuit the output for small signals at Q_2 emitter (X in Figure 9.3.17). The resistance at the output of the feedback circuit, is $R_E || R_f = 2.48$ kΩ. Neglecting the large r_{o2}, the output resistance of the amplifier proper, looking into the emitter, is $[r_{\pi2} + R_C || r_{o1}]/(\beta + 1) = 10.23/141 = 0.07$ kΩ. Hence, the resistance that appears in series with Q_2 emitter in the presence of feedback is $(2.48 + 0.07) \times 69.7 = 178$ kΩ. This is the resistance seen by a voltage source inserted between X and X′ in Figure 9.3.16. To determine the resistance looking into Q_2 collector, we use Equation 8.2.38, with $R'_{scr} = 9 || 100 = 8.26$ kΩ, $r_{\pi2} = 1.97$ kΩ, and $R_E = 0.07$ kΩ. This gives $50 + \dfrac{2.48}{1.97 + 8.26 + 2.48}[1.97 + 8.26 + (140 \times 50)] = 1.42$ MΩ. Multiplied by (1−loop gain), as in Example 9.3.2, the resistance becomes nearly 100 MΩ. The total resistance between Q_2 collector and ground is this resistance in parallel with 5 kΩ, which is practically 5 kΩ.

SIMULATION

When the feedback loop is opened for the purpose of determining the open-loop transfer function, *the loop must be properly terminated on either side of the break*, that is, by the effective source resistance, where the input voltage is applied, and by the effective load impedance, where the voltage is measured. Otherwise, the system will be disturbed by the break and the correct loop gain will not be measured. Where it is not easy to determine these

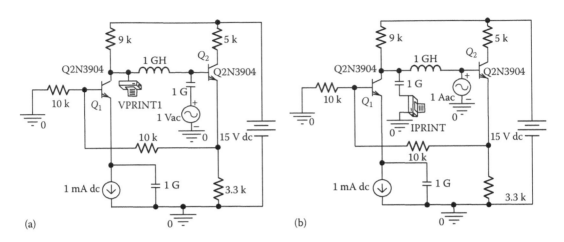

FIGURE 9.3.18
Figure for Example 9.3.3. Schematic for determining the open-circuit voltage transfer function (a), and the short-circuit current transfer function (b).

termination impedances when the loop is opened, the method of determining the open-circuit voltage transfer function and the short-circuit current transfer function of the loop can be used. This method is explained in Section ST9.1 and is particularly convenient for simulations.

The schematic of the circuit used for determining the open-circuit voltage transfer function is shown in Figure 9.3.18a. The transistor currents and parameters of Example 9.3.3 are close to those of the simulated circuit using the Q2N3904 transistors from the PSpice EVAL library. The loop is broken between Q_1 collector and Q_2 base and a large inductor is inserted at this location. In this manner, dc conditions are preserved but the inductor presents an open circuit to the ac signal. A voltage signal of 1 kHz is applied at Q_2 base through a large capacitor and the open circuit voltage is measured at Q_1 collector. The simulation gives −68.59 as the ratio of the output voltage to the input voltage, which is the open-circuit voltage gain of the loop. The schematic of the circuit used for determining the short-circuit current transfer function is shown in Figure 9.3.18b. A current input is applied at the base of Q_2 and the ac short-circuit current is measured at Q_2 collector. The simulation gives −3085 as the ratio of the output current to the input current, which is the short-circuit current gain of the loop. It follows from Equation ST9.1.5 that the magnitude of the loop gain is given by $\dfrac{1}{|A\beta|} = \left(\dfrac{1}{68.59} + \dfrac{1}{3085}\right)$, or $|A\beta| = 67.1$, in good agreement with the value of 68.7 calculated earlier.

It should be emphasized that in Example 9.3.3 the feedback factor has a negative value, so that $-\beta_f$ has a positive value, which means that the feedback current adds to the source current. However, since the forward gain through the amplifier has a negative value, the loop gain is negative, as it should be in a negative feedback system.

Exercise 9.3.3

The basic amplifier in a shunt–series feedback configuration has a short-circuit gain $(-g_{21a}/g_{22a})$ of −400 mA/V, an input resistance of 1 kΩ, and an output resistance of 2 kΩ. The feedback circuit has an input resistance of 10 kΩ, with port 2 open-circuited, an output resistance of 10 kΩ, and a feedback factor of −0.1. The source has a resistance of 10 kΩ and the load is a 10 kΩ resistor. Determine A, A_{fa}, R_{ifa}, and R_{ofa}.

Answers: $A_{ia} = -15.1$; $A_{ifa} = -6.0$; $R_{ifa} = 327$ Ω; and $R_{ofa} = 55.2$ kΩ. ∎

Shunt–Shunt Feedback

Figure 9.3.19 illustrates a general shunt–shunt feedback configuration in terms of the y-parameter equivalent circuit of Figure 7.2.1b, Chapter 7, ignoring y_{12a} and y_{21f}. As discussed in connection with shunt–series feedback, polarities in Figure 9.3.19 are such that the feedback is positive. Thus, if V_o decreases, the feedback current $y_{12f}I_o$ decreases. More current is diverted to y_{11a}, which increases V_i and $y_{21a}V_i$. But the increase in $y_{21a}V_i$ further decreases V_o. To maintain consistency with the polarity of the feedback signal at the input and with the polarity of the voltage source in the standard case (Figure 7.3.8, Chapter 7), we will reverse the sign of the source $y_{21a}V_i$.

It can be readily verified that the procedure of Table 9.3.1 applies to the derivation of y_{11f} and y_{22f} from Figure 9.3.19. Because of the shunt connection, voltage is the common variable at the input and output. If the output is short circuited, $y_{12f}V_o=0$, and the conductance seen at the input of the feedback circuit is y_{11f}. If the input is short-circuited, $y_{21a}V_i=0$, and the conductance seen at the output of the feedback circuit is y_{22f}. With $V_i=0$, and if a voltage V_x is applied at the output, the current at the input of the feedback circuit is $y_{12f}V_o$, so that the feedback factor is $y_{12f}=\beta_f$.

Figure 9.3.20a depicts the feedback amplifier with the polarity of the source $y_{21a}V_i$ reversed and with y_{11f} and y_{22f} moved to the amplifier circuit to give:

$$\frac{1}{R_{ia}} = G_{src} + y_{11f} + y_{11a}, \qquad \frac{1}{R_{oa}} = G_L + y_{22f} + y_{22a} \qquad (9.3.18)$$

The current source $y_{21a}V_i$ is transformed in Figure 9.3.20b to a voltage source and expressed as $A_{ra}I_i$, where $A_{ra}I_i=y_{21a}R_{oa}V_i=y_{21a}R_{oai}R_{ia}I_i$, so that,

$$A_{ra} = \frac{V_o}{I_i} = y_{21a}R_{oa}R_{ia} \qquad (9.3.19)$$

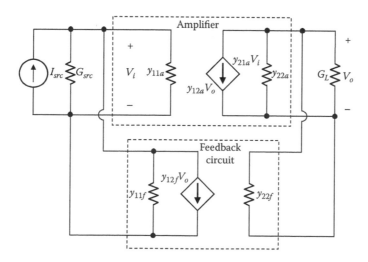

FIGURE 9.3.19
Two-port circuit representation of shunt–shunt feedback configuration.

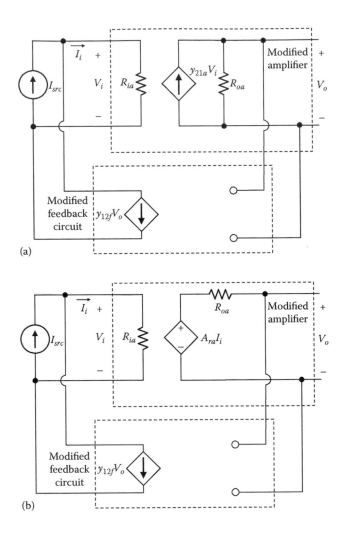

FIGURE 9.3.20
Two-port circuit representation of shunt–shunt negative feedback configuration: (a) input and output impedances of feedback circuit moved to amplifier; (b) standard form.

The amplifier of Figure 9.3.20b is now in the standard form of Figure 7.3.8, Chapter 7, the signal-flow diagram being as in Figure 9.3.21. Equations 7.3.14, 7.3.16, and 7.3.17, Chapter 7, of the standard shunt–shunt feedback amplifier can then be applied to give:

$$A_{raf} = \frac{V_o}{I_{src}} = \frac{A_{ra}}{1 + \beta_f A_{ra}}, \quad R_{iaf} = \frac{R_{ia}}{1 + \beta_f A_{ia}}, \quad R_{oaf} = \frac{R_{oa}}{1 + \beta_f A_{ia}} \qquad (9.3.20)$$

R'_{ifa} of the feedback amplifier, not including R_{src}, is given by Equation 9.3.17, and R'_{ofa} of the feedback amplifier, not including R_L, is given by Equation 9.3.9.

Example 9.3.4 Shunt–Shunt Feedback of Transistor Amplifier

As an example of shunt–shunt feedback, consider the single transistor biasing scheme in which the base is connected to the collector through a resistor. The circuit is shown in

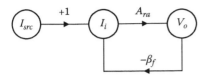

FIGURE 9.3.21
Signal-flow diagram of series–shunt feedback amplifier.

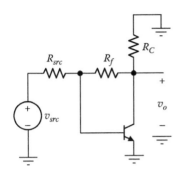

FIGURE 9.3.22

Figure 9.3.22 for the ac signal, with V_{CC} grounded. It is required to determine A_{ra}, R_{ifa}, and R_{ofa}. Assume that $I_{C1} = 2$ mA, with $\beta = 100$, $R_{src} = 10$ kΩ, $R_C = 5$ kΩ, $R_f = 50$ kΩ, and r_o large.

SOLUTION

To determine the loading effect of the feedback circuit at the base of the transistor, we short circuit the collector to ground for the ac signal. The resistance appearing at the base due R_f is simply R_f. To determine the loading effect at the collector, we short circuit the base to the ground for the ac signal. The resistance appearing at the collector due to R_f is again R_f. To determine the feedback factor, we short circuit the base and determine the current at the input when v_o is applied at the collector. As in Example 9.3.2, the assigned positive direction of current at the input is inward. This current is $\beta_f v_o = -v_o/R_f$, so the feedback factor is $\beta_f = -1/R_f = -0.02$ mS.

The small-signal equivalent circuit is shown in Figure 9.3.23, neglecting r_o. From the values given, $g_m = 2/0.026 = 77$ mA/V and $r_\pi = 1.3$ kΩ. It follows that $v_\pi = (R_g||R_f||r_\pi)$ $i_i = (10||50||1.3)i_i = 1.12i_i$, and $v_o = -g_m v_\pi (R_C||R_f) = -77(5||50)v_\pi = -350v_\pi$. This gives $A_{ra} = v_o/i_i = -392$ kΩ. The loop gain $-A_{ra}\beta_f$ is $-(-392)(-1/50)$, and $(1 - $ loop gain) is $1 + 392/50 = 8.84$, and $A_{rfa} = -392/8.84 = -44.3$kΩ.

The input resistance of the modified amplifier is $R_{ia} = (R_{src}||R_f||r_\pi) = 1.12$ kΩ. Hence, $R_{ifa} = 1.12/8.84 \equiv 126.7$ Ω. From Equation 9.3.17, $R'_{ifa} \cong 126.7$ Ω.

The output resistance R_{oa} of the modified amplifier is $R_{oa} = R_C||R_f = 5||50 = 4.55$ kΩ. Hence, $R_{ofa} = 4.55/8.84 = 515$ Ω.

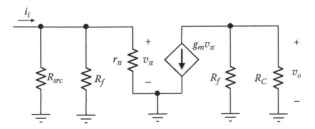

FIGURE 9.3.23
Figure for Example 9.3.4.

Exercise 9.3.4

The basic amplifier in a shunt–shunt feedback configuration has an open-circuit voltage gain $(-y_{21a}/y_{22a})$ of -200, an input resistance of 5 kΩ, and an output resistance of 1 kΩ. The feedback circuit has an input resistance of 10 kΩ, with port 2 short-circuited, an output resistance of 10 kΩ, and a feedback factor of -0.1 mA/V. The source has a resistance of 10 kΩ and the load is a 10 kΩ resistor. Determine A_{ra}, A_{rfa}, R_{ifa}, and R_{ofa}.

Answers: $A_{ra} = -417$ V/mA; $A_{rfa} = -9.77$ V/mA; $R_{ifa} = 58.5$ Ω; and $R_{ofa} = 19.5$ Ω.   ∎

9.4 Closed-Loop Stability

Concept: *A feedback amplifier is susceptible to instability because of unavoidable phase lags due to parasitic capacitances. The feedback becomes positive when the additional phase lag is 180°. If at this frequency the loop gain is at least unity, the amplifier breaks into oscillations.*

To clarify this concept, consider the signal-flow diagram of an amplifier with feedback (Figure 9.4.1a), where A_v represents the voltage gain of an amplifier having a net input voltage ε, and an output voltage v_O. It follows that $v_O/v_I = A_v/(1 - \beta_f A_v)$. Assuming that β_f and A_v are positive, real constants, this means that the feedback is positive. Suppose that while v_I remains constant, v_O increases by a positive increment Δv_O due to a transient disturbance to the circuit, such as interference pick up or a switch-on transient. To sustain the change Δv_O in the output requires a change in ε of $\Delta \varepsilon = \Delta v_O/A_v$. The change Δv_O causes $\Delta \varepsilon = \beta_f \Delta v_O$. Hence, to sustain the change Δv_O, we should have $\Delta v_O(\beta_f A_v) = \Delta v_O$, or $\beta_f A_v = 1$. If $\beta_f A_v < 1$, the fed-back signal is not sufficient to sustain the transient, which therefore dies out.

If $\beta_f A_v = 1$, the fed-back signal is just sufficient to sustain the transient, which will theoretically persist. However, this is a situation of marginal stability, or metastable equilibrium. In practice, $\beta_f A_v$ cannot exactly equal unity for any appreciable time; it is bound to become either slightly less, or slightly more, than unity. If $\beta_f A_v > 1$, the fed-back signal is more than sufficient to sustain the transient, which will therefore continue to grow. Eventually, some nonlinearity limits the magnitude of v_O, as when the amplifier saturates. It is seen that positive feedback causes instability if the loop gain exceeds unity.

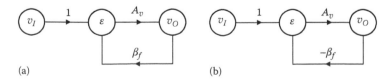

(a)   (b)

FIGURE 9.4.1
Signal-flow diagrams of positive feedback amplifier (a) and of a negative feedback amplifier (b).

Suppose that the sign of β_f is negative (Figure 9.4.1b), so that the feedback is negative. We may assume then that the system is stable. That may well be true under some, but not all, conditions. Consider a more realistic situation where A_v is frequency dependent, because of shunt capacitances associated with transistor amplifier stages. Let us assume, for the sake of argument, that A_v has three poles at frequencies ω_{p1}, ω_{p2}, and ω_{p3}, and that the dc gain is A_{v0}, so that:

$$A_v = \frac{A_{v0}}{(1 + s/\omega_{p1})(1 + s/\omega_{p2})(1 + s/\omega_{p3})} \tag{9.4.1}$$

Figure 9.4.2 shows the asymptotic magnitude plot and phase plot of A_v, where,

$$|A_v(j\omega)| = \frac{A_{v0}}{\sqrt{(1 + (\omega/\omega_{p1})^2)(1 + (\omega/\omega_{p2})^2)(1 + (\omega/\omega_{p3})^2)}} \tag{9.4.2}$$

and

$$\phi = -[\tan^{-1}(\omega/\omega_{p1}) + \tan^{-1}(\omega/\omega_{p2}) + \tan^{-1}(\omega/\omega_{p3})] \tag{9.4.3}$$

There exists a frequency ω_- at which $\phi = 180°$ so that $A_v(j\omega_-) = -|A_v(j\omega_-)|$, and is a negative real number. Since the sign of β_f is negative, the loop gain, and hence the feedback, becomes positive at the frequency ω_-. As discussed earlier for positive feedback, the amplifier is stable if $\beta_f |A_v(j\omega_-)| < 1$, but if $\beta_f |A_v(j\omega_-)| > 1$, the output signal grows in amplitude until limited by some nonlinearity that is inevitably present in the circuit.

The same argument applies if β_f is frequency dependent. ω_- is now the frequency at which the phase angle of $|\beta_f(j\omega_-)A_v(j\omega_-)|$ is 180°. Since the sign of β_f in Figure 9.4.1b is negative, then when $\angle \beta_f(j\omega_-)A_v(j\omega_-) = 180°$, the total phase shift around the loop is zero. Hence:

Criterion of Stability: *A negative feedback system is stable if*
$|\beta_f(j\omega_-)Av(j\omega_-)| < 1$, *at the frequency* ω_- *at which* $\angle\beta_f(j\omega_-)A_v(j\omega_-) = 180°$,
or the total phase shift around the loop is zero. It is unstable if
$|\beta_f(j\omega_-)A_v(j\omega_-)| > 1$, *and is marginally stable if* $|\beta_f(j\omega_-)A_v(j\omega_-)| = 1$. $\qquad$ (9.4.4)

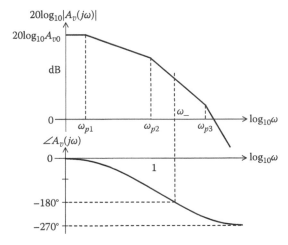

FIGURE 9.4.2
Asymptotic Bode magnitude and phase plots of amplifier gain.

The following should be noted:

1. If the open-loop transfer function is two-pole, rather than three-pole as in Equation 9.4.1, the frequency ω_- would occur at infinity, where the magnitude of the gain is effectively zero. The amplifier will be stable. However, its performance may not be satisfactory because of excessive, underdamped oscillations, as will be discussed later (see Exercise 9.3.1).

2. Since the total phase shift around the loop is zero at a single frequency ω_-, and the only periodic signal that has a single frequency is a sinusoid, the resulting oscillations are essentially sinusoidal. There will be some distortion, however, because of the nonlinearity that limits the amplitude of oscillation.

Stability from Bode Magnitude Plots

The transfer function of a general negative feedback system, as in Figure 9.4.1b, is:

$$G(s) = \frac{A_v(s)}{1 + \beta_f(s)A_v(s)} \tag{9.4.5}$$

The stability of any system is related to the location of the poles of the transfer function in the s-plane. The poles of $G(s)$ are the roots of the denominator, that is the values of s that make $1 + \beta_f(s)A_v(s) = 0$. The system is stable if the poles are located in the left half of the s plane, is unstable if at least one pole is located in the right half of the s-plane, and is marginally stable if the poles are located on the imaginary axis. Stability is examined in Section ST9.2 in terms of the Nyquist plot and the locations of the poles of the closed-loop transfer function of a three-pole system. We will, in what follows, discuss stability in terms of Bode plots.

The magnitude of the loop gain $|\beta_f(j\omega)A_v(j\omega)|$ can be drawn on an asymptotic magnitude Bode plot as the sum of the plots of $20\log_{10}|\beta_f(j\omega)|$ and $20\log_{10}|A_v(j\omega)|$. However, this does not lend itself to a convenient assessment of stability, as demonstrated later. It is more convenient to plot $20\log_{10}|1/\beta_f(j\omega)|$ on the same graph as $20\log_{10}|A_v(j\omega)|$. Where the two curves intersect, $20\log_{10}|A_v(j\omega)| = 20\log_{10}|1/\beta_f(j\omega)|$, so that $|\beta_f(j\omega)A_v(j\omega)| = 1$. This is illustrated in Figure 9.4.3, where it is assumed, for simplicity at this stage, that β_f is independent of frequency, so that the phase angle variation is due to $A_v(j\omega)$ alone. If $\beta_f = 1$, the line $20\log_{10}|1/\beta_f(j\omega)|$ coincides with the 0-dB line, that is the x-axis. The line $20\log_{10}|1/\beta_f(j\omega)|$ moves upward as β_f decreases. An added advantage of drawing $20\log_{10}|1/\beta_f(j\omega)|$ is that when the loop gain is reasonably high, the gain of the feedback amplifier is very nearly $1/\beta_f$.

With reference to Figure 9.4.3, and to the Stability Criterion 9.4.4, Case 1 represents a stable system, because the phase angle is less than 180° at the point of intersection, when $|\beta_f(j\omega)A_v(j\omega)| = 1$, which means that $|\beta_f(j\omega)A_v(j\omega)| < 1$ when $\angle\beta_f(j\omega)A_v(j\omega) = 180°$. Case 2 represents the boundary of instability since the phase angle equals 180° at the point of intersection, whereas Case 3 represents an unstable system, because the phase angle corresponding to the intersection point exceeds 180°, which means that $|\beta_f(j\omega)A_v(j\omega)| > 1$ when $\angle\beta_f(j\omega)A_v(j\omega) = 180°$. *It is seen that the circuit becomes more stable as β_f, and hence the loop gain, is decreased.*

Stability of the amplifier circuit does not, by itself, guarantee an acceptable performance. If, for example, the phase angle at unity magnitude loop gain is only slightly less than 180°, the poles of the closed-loop transfer function (Equation 9.4.5) are in the left half

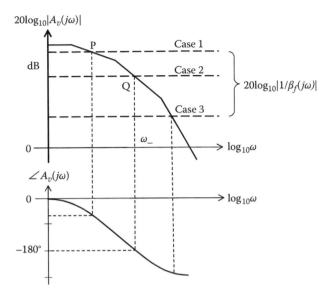

FIGURE 9.4.3
Asymptotic Bode magnitude and phase plots
of amplifier gain and $1/\beta_f$.

of the s plane but are very close to the imaginary axis, Q may be unacceptably large (Exercise 9.4.1), so that the response to a sudden disturbance can be markedly underdamped, having objectionable oscillations. To guarantee an acceptable performance, the phase angle should not exceed $(180° - \theta)$ at the frequency at which the magnitude of the loop gain is unity, where θ is referred to as the **phase margin**. Alternatively, an acceptable performance can be expressed in terms of the **gain margin,** or the difference between unity and the magnitude of the gain $|\beta_f(j\omega_-)A_v(j\omega_-)|$ at the frequency ω_-. The phase and gain margins are illustrated in Figure 9.4.4 based on plots of the magnitude and the phase angle of the loop gain $\beta_f(j\omega_-)A_v(j\omega_-)$. Acceptable gain or phase margins depend on the particular application. Generally speaking, a phase margin not less than $45°$ is considered acceptable, so that the phase shift at the frequency at which the magnitude of the loop gain is unity should not exceed $135°$.

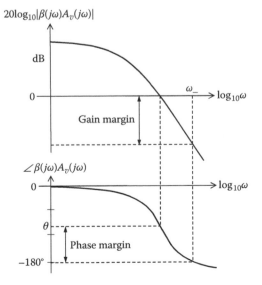

FIGURE 9.4.4
Gain and phase margins.

Exercise 9.4.1

Determine Q in a two-pole system having poles at $-1 + j10^3$ and $-1 - j10^3$ rad/s.
  Answer: 500.   ∎

Working with both magnitude and phase plots is awkward. It would be very convenient, if stability could be assessed from the magnitude plot alone. This is possible, because of the relationship between phase magnitude and phase for the type of system we are dealing with.

Consider the three-pole system presented earlier, where the phase angle at any frequency ω is given by Equation 9.4.3. At $\omega = \omega_{p2}$, the contribution of the pole at ω_{p2} to the phase shift is $-\tan^{-1}(\omega_{p2}/\omega_{p2}) = -45°$. The contribution of the pole at ω_{p1} to the phase shift at $\omega = \omega_{p2}$ is $-\tan^{-1}(\omega_{p2}/\omega_{p1})$ and increases in magnitude as ω_{p1} is located at lower frequencies, further away from ω_{p2}. It approaches a limit of $-90°$ for $\omega_{p1} \ll \omega_{p2}$. If $\omega_{p1} = 0.1\omega_{p2}$, the contribution of the pole ω_{p1} to the phase shift at $\omega = \omega_{p2}$ is $-\tan^{-1}(10) \cong -84.3°$. The contribution of the pole at ω_{p3} to the phase shift at $\omega = \omega_{p2}$ is $-\tan^{-1}(\omega_{p2}/\omega_{p3})$ and decreases in magnitude as ω_{p3} is located at higher frequencies, further away from ω_{p2}. It becomes zero in the limit for $\omega_{p3} \gg \omega_{p2}$. The contribution is $-5.7°$ if $\omega_{p3} = 10\omega_{p2}$. In fact, if $\omega_{p3} = k\omega_{p2}$ and $\omega_{p1} = \omega_{p2}/k$, where k is a positive constant, the phase shift at ω_{p2} due to ω_{p3} and ω_{p1} is $-\tan^{-1}(1/k) - \tan^{-1}(k) = -90°$, and the total phase shift at ω_{p2} is $-45° - 90° = -135°$. This is illustrated by line L_2 in Figure 9.4.5. The phase margin is increased if the intersection point is at a lower frequency on the line of slope -20 dB/decade (line L_1) and is reduced if the intersection point is on the line having a slope of -40 dB/decade (line L_3).

Based on the preceding, an often applied "rule of thumb" is that closed-loop stability is adequate if the frequency-independent $20\log_{10}|1/\beta_f|$ plot intersects the $20\log_{10}|A_v(j\omega)|$ plot at a point on the -20 dB/decade segment, assuming that poles of higher frequency are not close to the frequency at which the intersection occurs.

When β_f is frequency dependent, its phase angle adds algebraically to that of $A_v(f)$. For simplicity, let $A_v(j\omega) = \dfrac{A_{v0}}{(1 + j\omega/\omega_{p1A})(1 + j\omega/\omega_{p2A})}$ and $\beta_f(j\omega) = \dfrac{1}{1 + j\omega/\omega_{p\beta}}$. Figure 9.4.6

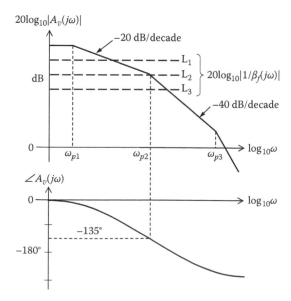

FIGURE 9.4.5
Stability from Bode magnitude plot.

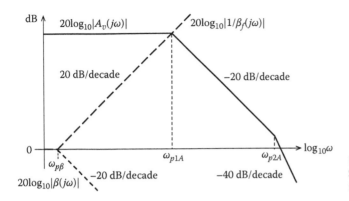

FIGURE 9.4.6
Stability when both amplifier gain and feedback factor are frequency dependent.

illustrates the asymptotic Bode plots for $20\log_{10}|A_v(j\omega)|$, $20\log_{10}|\beta_f(j\omega)|$, and $20\log_{10}|1/\beta_f(j\omega)|$. To simplify the argument further, let the point of intersection of the $20\log_{10}|A_v(j\omega)|$ and the $20\log_{10}|1/\beta_f(j\omega)|$ plots be at ω_{p1A}. Applying the same preceding reasoning to the phase shift of $\beta_f(j\omega)A_v(j\omega)$, the phase shift at ω_{p1A} due to the pole itself is $-45°$, whereas the total phase shift due to ω_{p2A} and $\omega_{p\beta}$ is $-90°$ if these poles are, respectively, at $k\omega_{p1A}$ and ω_{p1A}/k, and will be less than $-90°$ if either pole is further away from ω_{p1A}. It is seen that to achieve the phase margin of $45°$, the slope of the $20\log_{10}|A_v(j\omega)|$ plot *minus* that of $20\log_{10}|1/\beta_f(j\omega)|$ plot, at a frequency just below the point of intersection should not be more negative than -20 dB/decade. In Figure 9.4.6, the slope of the $20\log_{10}|A_v(j\omega)|$ plot at a frequency just less than ω_{p1A} is 0, whereas the slope of the $20\log_{10}|1/\beta_f(j\omega)|$ plot at this frequency is $+20$ dB/decade. Hence $0 - 20 = -20$ gives a phase margin of at least $45°$, assuming that ω_{p2A} and $\omega_{p\beta}$ are adequately spaced from ω_{p1A}. The criterion of adequate stability can be generalized as follows:

> **Criterion of Adequate Stability:** *A circuit has adequate stability if the slope of the $20log_{10}|A_v(j\omega)|$ plot minus that of the $20log_{10}|1/\beta_f(j\omega)|$ plot at their point of intersection, is not more negative than -20 dB/decade, the slopes considered being the those of the asymptotes of the two plots at frequencies just below the point of intersection* (9.4.6)

It should be kept in mind that at the corner frequency where two asymptotes of the *same* function intersect, the gain is actually 3 dB less than that at the corner frequency, and the slope is the average of the slopes of the two asymptotes.

Frequency Compensation
The technique of shaping the frequency variation of $|A_v(j\omega)|$ and/or $|\beta_f(j\omega)|$ to achieve desired gain or phase margins is **frequency compensation**. In principle, it is possible to introduce zeros that cancel poles in the open-loop transfer function, as illustrated by problems at the end of the chapter. Simpler and more common techniques are discussed in this section.

In **dominant-pole compensation**, a new, low-frequency pole is introduced at a frequency ω_{pd} so that the slope of the $20\log_{10}|A_v(j\omega)|$ plot is -20 dB/decade all the way down to 0 dB at ω_{p1}, the lowest pole frequency of the uncompensated amplifier (Figure 9.4.7). The amplifier will be unconditionally stable, even if $\beta_f = 1$. Although stability is assured, the bandwidth of

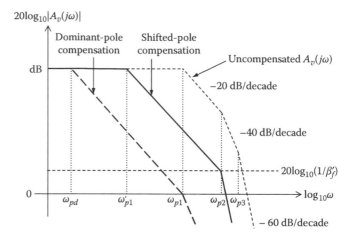

FIGURE 9.4.7
Dominant-pole and shifted-pole frequency compensation.

the amplifier is severely reduced, which compromises the speed of response and decreases the loop gain available for improving performance at higher frequencies.

A generally better approach is to shift the lowest frequency pole ω_{p1} of the amplifier to a lower frequency ω'_{p1}, as illustrated in Figure 9.4.7, resulting in increased bandwidth compared to dominant pole compensation. To determine ω'_{p1}, the line representing $20\log_{10}(1/\beta'_f)$ is drawn, where β'_f is the desired β_f. At the frequency ω_{p2} on this line, a line having a slope of -20 dB/decade is drawn. Where this line intersects the low-frequency asymptote of the amplifier gain is the frequency ω'_{p1}. Example 9.4.1 illustrates these methods of frequency compensation.

In general, the lowest frequency pole can be shifted to a lower frequency by adding capacitance. In amplifiers, a pole usually arises because of series resistance and shunt capacitance. The transfer function of a simple RC voltage divider is $(1/sC)/(R + 1/sC) = 1/(1 + sCR)$, which gives a pole at $\omega = 1/CR$. Adding a capacitance C_x in parallel with C shifts the pole to a lower frequency $1/(C + C_x)R$. Another possibility is to add capacitance to enhance the Miller effect. This can shift one pole to a lower frequency and another pole to a higher frequency (Section ST8.7, Chapter 8, and Section 10.2, Chapter 10).

In practice, IC op amps may be internally compensated or may have terminals brought out for connection of suitable values of R and C so as to achieve a degree of compensation that is appropriate for a particular application.

Example 9.4.1 Amplifier Stability

Assume that in Equation 9.4.1, $A_{v0} = 10^5$, $\omega_{p1} = 1$, $\omega_{p2} = 10$, and $\omega_{p3} = 100$, all in rad/s. It is required to investigate the stability of this amplifier when used in a noninverting configuration.

SOLUTION

It follows from Equation 9.4.2 that:

$$|A_V(j\omega)| = \frac{10^5}{\sqrt{[1 + \omega^2][1 + (\omega/10)^2][1 + (\omega/100)^2]}} \tag{9.4.7}$$

and the phase angle ϕ of A_v is given by:

$$\phi = -[\tan^{-1}\omega + \tan^{-1}(\omega/10) + \tan^{-1}(\omega/100)] \tag{9.4.8}$$

The magnitude and phase plots are illustrated diagrammatically in Figure 9.4.8. If used as a unity-gain amplifier, the $20\log_{10}(1/\beta)$ line coincides with the $\log_{10}\omega$ axis, and Criterion 9.4.6 gives a slope of -60 dB/decade, so that the amplifier is unstable. We can confirm this by determining the phase shift at ω_T, where the $20\log_{10}|A_v(j\omega)|$ plot intersects the x-axis. ω_T is obtained from Equation 9.4.7 by setting $|A_v(j\omega_T)| = 1$ and solving for ω_T. Using the roots(p) command of MATLAB gives $\omega_T = 461$ rad/s. From Equation 9.4.8, the phase shift at ω_T is $-256°$. This means that when the phase shift is $-180°$, the loop gain exceeds unity, which confirms instability.

The poles of the closed-loop transfer function, with $\beta_f = 1$, are the roots of the equation: $1 + A_v(s) = 0$, where $A_v(s) = \dfrac{10^5}{(1 + s)(1 + s/10)(1 + s/100)}$. Using the roots(p) command of MATLAB gives the roots of this equation as -503.4, $(196.2 + j4.002)$, and $(196.2 - j4.002)$ rad/s. A complex pair of roots has positive real parts, which again confirms instability.

Let us choose β_f such that the $20\log_{10}|A_v(j\omega)|$ line intersects the $20\log_{10}(1/\beta_f)$ at the second corner frequency $\omega_{p2} = 10$ rad/s, which, on the asymptotic plots, is the limit of the -20 dB/decade slope. The ordinate is therefore 20 dB below that of $20\log_{10}|A_{vo}|$, that is, at 80 dB (Figure 9.4.8). Setting $20\log_{10}(1/\beta_f) = 80$, gives $\beta_f = 10^{-4}$. The low-frequency gain of the feedback amplifier will be $10^5/(1 + 10^{-4} \times 10^5) \cong 10^4$, or 80 dB. The phase angle at $\omega = 10$ rad/s is, from Equation 9.4.8, $135°$. The feedback amplifier is stable under these conditions, the phase margin being $180° - 135° = 45°$.

It should be noted that the preceding calculations are based on nominal values using asymptotic plots. Thus, from Equation 9.4.7, $|A_v(j10)| = 76.9$ dB, and $\beta_f = 10^{-4}$ gives

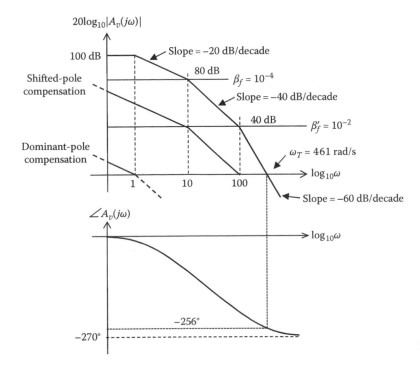

FIGURE 9.4.8

Figure for Example 9.4.1.

a low-frequency gain of 79.2 dB for the feedback amplifier, whereas the phase shift at $\omega_{p2} = 10$ rad/s is, from Equation 9.4.8, exactly $-135°$.

To find the gain margin, we have to determine the frequency at which the phase angle of the loop gain becomes 180° and the gain at this frequency. From Equation 9.4.8, or by determining the frequency at which the imaginary part of the polynomial $P(s)$ of the denominator of $A_V(s)$ vanishes, we find $\omega_- = \sqrt{1110} = 33.32$ rad/s. From Equation 9.4.7, the amplifier gain, $20\log_{10}|A_V(j\omega_-)|$, at this frequency is 58.3 dB. The magnitude of the loop gain, $20\log_{10}|\beta_f A_V(j\omega_-)|$, is therefore, $58.3 - 80 = -21.7$ dB. The gain margin is therefore 21.7 dB.

We can determine the location of the poles of the closed-loop response by finding the roots of the equation $1 + 10^{-4}A_V(s) = 0$. These are $(-4.95 + j9.18)$, $(-4.95 - j9.18)$, and -101.1. Since all the poles have negative real parts, the stability of the feedback amplifier is confirmed under these conditions.

To stabilize the amplifier unconditionally using dominant pole compensation, we have to ensure that the slope of the $20\log_{10}|A_V(j\omega)|$ plot is 20 dB/decade at 0 dB, which means locating the 0-dB crossing frequency at 1 rad/s. Since $20\log_{10}|A_V(j\omega)| = 100$ dB, the corner frequency of the dominant pole ω_{pd} must occur at a frequency that is $100/20 = 5$ decades below ω_{p1}. Thus, $\omega_{pd} = 1 \times 10^{-5} = 10^{-5}$ rad/s. With dominant pole compensation, a pole at ω_{pd} must be added. Equation 9.4.7 becomes:

$$|A_V(j\omega)| = \frac{10^5}{\sqrt{[1 + (\omega/10^{-5})^2][1 + \omega^2][1 + (\omega/10)^2][1 + (\omega/100)^2]}} \tag{9.4.9}$$

If the frequency scaling factor is $10^5/2\pi$, so that $f_{p1} = 10^5$ Hz, then $f_{pd} = 1$ Hz.

Better performance is obtained by shifting the pole at 1 rad/s. Suppose β_f' is to be 10^{-2}, corresponding to a closed-loop gain of $10^5/(1 + 10^{-2} \times 10^5) \cong 10^2$. We draw the line $20\log_{10}(1/\beta_f') = 40$ dB. At 10 rad/s on this line, we draw a line having a slope of -20 dB/decade. Since this line is 60 dB below the low-frequency asymptote of the amplifier, it intersects this asymptote at $10 \times 10^{-3} = 10^{-2}$ rad/s. Hence, the pole at 1 rad/s should be shifted to this frequency.

Application Window Oscillations in Amplifiers

A high-gain amplifier that should be stable may break into oscillations as soon as it is switched on. This is usually caused by some inadvertent feedback from output to input, as through a small but finite impedance of the dc supplies. The feedback becomes positive at some frequency, resulting in oscillations at this frequency. The usual remedy is to "decouple" the dc power supplies by connecting a capacitor of appropriate size between the dc supplies and ground as close as possible to where these supplies feed into the amplifier. In some cases it may be necessary to connect additional decoupling capacitors at appropriate locations along dc supply lines. Decoupling capacitors should be of a type that has very low impedance at the frequency of oscillation. It must be kept in mind that a seemingly innocuous length of wire can have sufficient inductance at high frequencies to act as a coupling impedance between output and input stages. That is, the voltage developed across this impedance due to current from a later amplifier stage can provide a feedback signal to an earlier stage that is sufficient to cause instability. Hence, the physical layout of a high-gain amplifier is important in practice.

It must not be assumed that instability only occurs at high frequencies. In multistage *RC*-coupled amplifiers, dc blocking and emitter/source bypass capacitors introduce poles at low frequencies, and if these poles are not properly located, there may be sufficient gain and phase shift at some low frequency to cause oscillation.

Feedback Oscillators

A sinusoidal oscillator can be implemented by connecting an amplifier and a frequency selective circuit in cascade in a positive feedback loop. The criterion for oscillation is derived from Criterion 9.4.4 as the **Barkhausen criterion**:

$$A(j\omega_0)\beta_f(j\omega_0) = 1, \quad \text{where } \angle\beta_f(j\omega_0)A_v(j\omega_0)) = 0 \qquad (9.4.10)$$

In words, the circuit oscillates at a frequency ω_0 at which the phase shift around the loop is zero, and the magnitude of the loop gain is unity.

In practice, the gain is made greater than unity to ensure oscillation. The output increases until limited by some nonlinearity that effectively reduces the loop gain to unity. The nonlinearity introduces distortion, however, so that some measures must be taken to minimize this distortion, as illustrated later for the Wien-bridge oscillator and the high-Q oscillator.

It is evident from the Barkhausen criterion that the stability of the frequency of oscillation depends on $\left.\dfrac{d\phi}{d\omega}\right|_{\omega=\omega_0}$, where ϕ is the phase shift around the loop. Thus, if ϕ changes by $\Delta\phi$ for any reason, such as a change in the value of some circuit component, the resulting change in frequency is $\Delta\phi \left/ \left(\left.\dfrac{d\phi}{d\omega}\right|_{\omega=\omega_0}\right)\right.$, which is small if $d\phi/d\omega$ is large. Hence, for good frequency stability it is desirable to have a large $d\phi/d\omega$ at the frequency of oscillation.

Wien-Bridge Oscillator

An oscillator can be built using a Wien-bridge and a noninverting op-amp circuit (Figure 9.4.9a). The signal-flow diagram is shown in Figure 9.4.9b. $\beta_f(s) = V_p/V_o = H(s)$ is the transfer function of the bridge, which can be easily shown to be $\dfrac{s/CR}{s^2 + (3s/CR) + \omega_0^2}$, where $\omega_0 = 1/RC$. The loop gain can be expressed as: $\dfrac{1}{3}\dfrac{(3s/RC)}{s^2 + (3s/RC) + \omega_0^2}\left(1 + \dfrac{R_f}{R_r}\right)$ and is $\dfrac{1}{3}\left(1 + \dfrac{R_f}{R_r}\right)$ at $s = j\omega_0$. Setting this equal to unity, gives $R_f = 2R_r$ as the criterion for oscillation at the frequency ω_0.

In order to stabilize the amplitude, **thermistors** (temperature-sensitive resistors whose resistance changes markedly with temperature and hence power dissipated) can be used. Thus, a resistor having a positive temperature coefficient (PTC) of resistance can be used for R_r, or a resistor having a negative temperature coefficient (NTC) of resistance used for R_f. In either case, if the output voltage increases, because the loop gain exceeds unity, the power dissipated in R_r and R_f will increase. As a result either R_r increases, if it is PTC type, or R_f decreases, if it is NTC type. In either case, the negative feedback voltage to the inverting input increases and the overall gain is reduced back to unity.

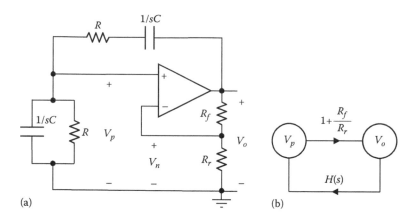

FIGURE 9.4.9
Wien-bridge oscillator. (a) circuit; (b) signal-flow diagram.

The output voltage would thus be maintained nearly constant, with relatively low distortion.

The amplitude of oscillation can also be stabilized by means of diode-limiting circuits. Temperature-sensitive resistors, as well as diodes, are affected by ambient temperature, so that the amplitude of oscillation would change with temperature.

Exercise 9.4.2

Consider three, identical first-order lag circuits as in Figure 9.4.10, where in each circuit, $R_f = R$, $R_r = R/2$, and $C_f = 2C$. Show that if these circuits are connected in a ring, a sinusoidal oscillator results, in which the three outputs form a balanced three-phase system. Determine the frequency of oscillation and verify that at this frequency the loop gain is unity.

   Answer: $\sqrt{3}/2CR$ rad/s   ∎

High-Q Oscillator

An alternative form of feedback oscillator consists of a high-Q bandpass filter of center frequency ω_0, connected to a limiter circuit in a positive feedback loop whose gain exceeds

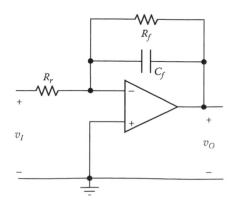

FIGURE 9.4.10
Figure for Exercise 9.4.2.

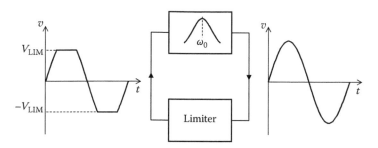

FIGURE 9.4.11
High-Q oscillator.

unity at the frequency ω_0 (Figure 9.4.11). The limiter circuit (Section 1.5, Chapter 1) "clips" the sinusoidal output voltage to a certain level V_{LIM}, as shown in Figure 9.4.11. Since the clipped waveform is periodic of frequency ω_0, it consists of a fundamental component at this frequency and higher harmonics. The high frequency selectivity of the bandpass filter heavily attenuates the harmonics, so that the output of the filter is nearly a pure sinusoid. The high-Q oscillator is simulated in Example 9.4.2. Additional examples of these oscillators are given in the problems at the end of the chapter.

Simulation Example 9.4.2 High-Q Oscillator

It is required to simulate the High-Q oscillator of Figure 9.4.11.

SIMULATION

This example illustrates the use of the Analog Behavioral Modeling (ABM) feature of PSpice, which allows the use of functional blocks to perform some mathematical operations. The ABM parts are located in the ABM library. Three of these parts are used in the present simulation: LAPLACE (for implementation of a transfer function), GLIMIT (for gain with output clipped at specified limits), and SUM (for summing two signals).

When LAPLACE is placed in the schematic, the default function is of the form $1/(s+1)$. To change the numerator, double click on it and enter 50*s in the value field of the Display Properties window. To change the denominator, double click on it and enter s*s + 50*s + PWR(10,8). The transfer function is now $50s/(s^2 + 50s + 10^8)$, which is a unity gain, bandpass response centered at 10^4 rad/s and having a Q of $10^4/50 = 200$. The gain of GLIMIT is set to 1.5 and the voltage limits to ± 1 V.

In a real circuit, oscillation is started by the inevitable disturbances due to switching-on or noise. To ensure oscillations in a simulation, the circuit can be given an initial "kick," by applying a short pulse the SUM part using the VPULSE source, as shown in Figure 9.4.12. The parameters of this source were set as follows: V1 = 0, V2 = 5, TD = 0, TR = 1u, TF = 1u, PW = 100u, and PER = 1 s. A Time Domain (transient) analysis is performed with a "Run to time" of 1 s, because the oscillations take some time to build up to a steady value.

Figure 9.4.13 shows v_O over the interval 0.6–0.602 s. The frequency is 1.59 kHz corresponding to $\omega_0 = 10^4$. A Fourier analysis of the segment from 0.6 to 1 s gives an amplitude of 0.946 V and a total harmonic distortion of 1.18% up to and including the ninth harmonic.

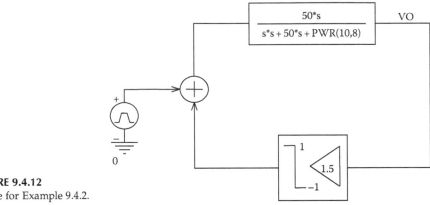

FIGURE 9.4.12
Figure for Example 9.4.2.

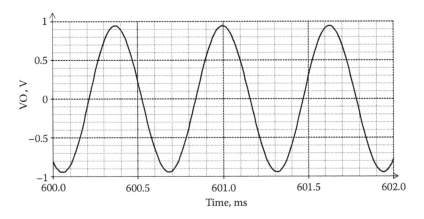

FIGURE 9.4.13
Figure for Example 9.4.2.

9.5 Tuned Amplifiers

When a radio, television, or communication receiver is tuned to a particular rf (radio frequency) input signal, the frequency of the local oscillator in the receiver is set so that when the oscillator output is mixed with the received signal, a fixed frequency, called the **intermediate frequency** (i-f) is obtained. The bulk of amplification of the received signal and discrimination against unwanted frequencies is then conveniently done at this single, fixed frequency. The i-f amplifier has a bandpass response of center frequency in the range of a few hundred kHz to a few hundred MHz, depending on the system, and a bandwidth that depends on the type of modulation used. Tuned amplifiers are usually used to obtain the required bandpass response.

Tuned amplifiers utilize the bandpass properties of parallel *RLC* circuits, which allow high gain for frequencies within the passband, while rejecting frequencies outside the passband. Figure 9.5.1 illustrates a basic tuned-circuit amplifier, where it is assumed that the transistor is properly biased and that the parallel *RLC* circuit is of high *Q*, so that inductor losses are represented by a parallel resistance R_p, in accordance with the high-*Q*

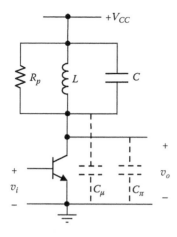

FIGURE 9.5.1
Tuned amplifier.

approximation (Example SE9.2). It follows that at the center frequency $\omega_0 = 1/\sqrt{LC}$ the gain from base to collector is:

$$\frac{v_o}{v_i} = -g_m R_L \tag{9.5.1}$$

where R_L is the effective load resistance that is the parallel combination of R_p, r_o of the transistor, and any resistance that may be added to obtain the desired bandwidth.

The maximum theoretical gain is of interest. Evidently, the maximum R_L is R_p. Substituting $R_p = Q/\omega_0 C$ for the parallel tuned circuit, and $BW = \omega_0/Q$,

$$\frac{v_o}{v_i} = -\frac{g_m}{BW \times C} \tag{9.5.2}$$

For given BW and g_m, the highest gain is obtained by having the smallest C, which in the limit is that due solely to the internal capacitances of the transistor. Considering the Miller effect, recall that the capacitance C_μ between the base and collector is equivalent to a capacitor $C_\mu(1 - K)$ at the base of the transistor, and to a capacitor $C_\mu(1 - 1/K)$ at the collector, where $K = -g_m R_p$ is the gain. If $|K|$ is large, C_μ appears in parallel with the tuned circuit (Figure 9.5.1). To this capacitance must be added that due to the load. Assuming that the transistor is connected to another amplifying stage having an identical transistor, as is usually the case, and that the Miller effect of this stage is negligible, as in the case of a cascode amplifier, for example, the smallest additional capacitance can be considered to be C_π. The minimum total capacitance in Equation 9.5.2 is therefore $C_\mu + C_\pi$, so that,

$$|G| = \frac{g_m}{BW \times (C_\mu + C_\pi)} \tag{9.5.3}$$

where G is the gain. Equation 9.5.3 is significant in not only giving the maximum gain for a given transistor and bandwidth but also in showing that the maximum gain–bandwidth product is $g_m/(C_\mu + C_\pi)$ as given by Equation 6.2.38, Chapter 6. This is a limit on the maximum attainable gain for a given bandwidth and was encountered earlier in connection with various amplifier configurations. Equation 9.5.3 is yet another expression of the following fundamental concept.

Concept: *Gain and bandwidth can be traded for one another, but the limit on their product is set by the transistor.*

Exercise 9.5.1

A tuned amplifier such as that of Figure 9.5.1 is to have a gain of −50 at a center frequency of 5 MHz and a bandwidth of 50 kHz using a transistor of $g_m = 50$ mA/V. Determine L and C, and R that must be added, assuming the Q of the coil is 150 at 50 MHz and neglecting r_o and the capacitances of the transistor.

Answers: 3.18 nF; 79.6 nH; 1.36 kΩ. ∎

The gain and selectivity required of tuned amplifiers cannot in practice be obtained from a single tuned circuit or from a single stage. This means that there will, in general, be tuned circuits at the inputs as well as the outputs of transistors, which causes two problems:

1. In CE/CS stages, the Miller effect adds appreciable capacitance across the input tuned circuit, which will change the frequency and bandwidth of this circuit and can even cause instability (Problem P9.3.6).
2. If all the circuits are tuned to the same frequency, the bandwidth is reduced (Section ST9.4).

The problem of the Miller effect is avoided by using amplifier stages that do not suffer from this effect, such as a cascode. In the CC–CE cascade the Miller effect is mitigated by the low output impedance of the CC transistor. Another useful configuration that does not suffer from the Miller effect is the CC–CB cascade, illustrated in Figure 9.5.2, where the voltage source at the input has been transformed to the equivalent current source. C_μ of the first transistor appears between base and ground and C_μ of the second transistor appears between collector and ground. Each adds to the capacitance of the respective tuned circuit at the input and the output, but neither capacitance is multiplied by the Miller effect. The input and output tuned circuits are effectively isolated from one another. Note that the configuration can equally be viewed as a differential pair with zero collector load for one transistor and with the base of the other transistor grounded (Section 10.1, Chapter 10). Since C_μ is not a precisely known capacitance and is subject to variations with operating conditions and with different transistors of the same type, C_μ is kept small compared to the capacitance of the tuned circuit.

The problem of selectivity can be solved in two ways: (1) magnetic coupling between the tuned circuit at the output of a given stage and the tuned circuit at the input of the following stage (Section ST9.3); (2) stagger tuning the various tuned circuits (Section ST9.4).

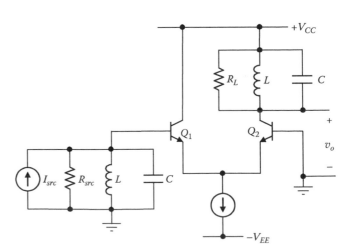

FIGURE 9.5.2
CC–CB cascade tuned amplifier.

9.6 *LC* Oscillators

If part of the voltage developed across the *RLC* circuit of a tuned amplifier is fed back to the transistor input, sustained sinusoidal oscillations will result. In the **Colpitts oscillator**, *C* is replaced by two capacitors in series, the feedback voltage being the voltage across one of these capacitors. In the **Hartley oscillator**, *L* is replaced by two inductors in series, with or without coupling between them, the feedback voltage being the voltage across one of these inductors. Both types of oscillators are widely used in practice, particularly in broadcast receivers.

Figure 9.6.1 illustrates one form of biasing for the Colpitts oscillator. A radio-frequency choke (RFC) presents very high impedance at the oscillation frequency, thereby effectively isolating the circuit from the power supply. Without the RFC, C_2 is effectively grounded at both ends. C_B and C_E have low impedance at the oscillation frequency and therefore appear as short circuits across the corresponding resistors. The conditions for oscillation can be derived from the equivalent circuit of Figure 9.6.2, in which several simplifications have been made to highlight the basic principles involved. R_2 and r_π, both of which appear in parallel with C_2 are neglected on the assumption that they are large compared with the reactance of C_2 at the oscillation frequency. This frequency is assumed not to be high

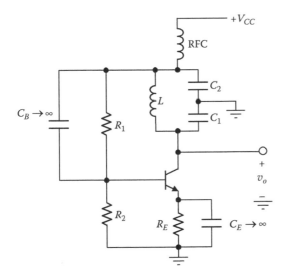

FIGURE 9.6.1
Colpitts oscillator with biasing.

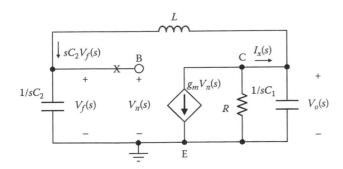

FIGURE 9.6.2
Simplified, small-signal equivalent circuit of Colpitts oscillator.

enough for C_μ to have a significant effect. C_π, which is normally much larger than C_μ, can be included with C_2. R accounts for r_o of the transistor, R_p of the coil, and load resistance.

Assume that the feedback loop is interrupted at X, a voltage source $V_\pi(s)$ is applied between base and emitter (Figure 9.6.2), and the voltage $V_f(s)$ determined. Note that breaking the loop at an open circuit obviates the need for any terminations at the breakpoint. Thus, the voltage $V_f(s)$ is measured at an open circuit, as when the loop is closed, and any source impedance that is associated with the source $V_\pi(s)$ is immaterial because the source is applied at an open circuit. The condition for sustained oscillation is that the voltage $V_f(s)$ that is fed back equals $V_\pi(s)$. Applying KCL to node C: $g_m V_\pi(s) + (sC_1 + 1/R)V_o(s) + sC_2V_f(s) = 0$. From voltage division, $V_f(s) = V_o(s)/(1 + s^2LC_2)$. Eliminating $V_o(s)$ gives:

$$V_\pi(s) = -\frac{1}{g_m}\left[sC_2 + \left(\frac{1}{R} + sC_1\right)(1 + s^2LC_2)\right]V_f(s) \tag{9.6.1}$$

To have sustained oscillations, the quantity in square brackets must equal -1. Replacing s by $j\omega$ and setting the imaginary part to zero and the real part in square brackets to -1,

$$\omega_o = 1\left/\sqrt{L\frac{C_1C_2}{C_1 + C_2}}\right. \tag{9.6.2}$$

$$g_m R = C_2/C_1 \tag{9.6.3}$$

Equation 9.6.2 gives the frequency of oscillation ω_o as that of L in series with C_1 and C_2.

To get a better insight into the operation of the circuit, let us calculate $I_x(s)$ in Figure 9.6.2 with the loop closed. It is seen that $I_x(s) = sC_1V_o(s) + sC_2V_\pi(x)$, where $V_\pi(s) = V_o(s)/(1 + s^2LC_2)$. Or, $I_x(s) = V_\pi(s)[sC_2 + sC_1(1 + s^2LC_2)]$. The bracketed quantity on the RHS is the imaginary part of the quantity in the square brackets of Equation 9.6.1 and is zero. This means that $I_x(s) = 0$, which explains why ω_o (Equation 9.6.2) depends only on L, C_1, and C_2. But if $I_x(s) = 0$, then the current source must present a resistance $-R$, so that R in parallel with $-R$ gives an infinite resistance. This can be verified, since the resistance presented by the current source is $V_o(s)/g_m V_\pi(s)$, which, from voltage division, equals $(1 + s^2LC_2/g_m)$. The real part that was set to -1 in Equation 9.6.1 is: $\frac{1 + s^2LC_2}{g_m R} = -1$. Hence, $\frac{V_o(s)}{g_m V_\pi(s)} = -R$. A negative resistance delivers power. Hence,

Concept: *The positive feedback through the transistor – an active device – supplies power that compensates for the power loss in the oscillating circuit.*

With R and the current source neutralizing one another, the circuit becomes a lossless, LC circuit. The power delivered by the source is $-g_m V_\pi(s)V_o(s)$, and that dissipated in R is $V_o^2(s)/R$. Equating these powers and substituting $V_\pi(s)/V_o(s) = -C_1/C_2$ at the frequency of oscillation, gives Equation 9.6.3.

It should be kept in mind that accounting for the losses in an inductor by means of a parallel resistance involves an approximation that, strictly speaking, makes the equivalent parallel inductance and resistance functions of frequency. Hence, the parallel circuit representation is an idealization that becomes more valid the higher the Q of the circuit.

The conditions for oscillation can also be determined from the circuit equations without opening the loop. This is convenient when it is awkward to determine the proper terminations on either side of the loop break point. The argument is that KVL and KCL must be satisfied even in the presence of oscillations. Referring to Figure 9.6.2, KCL at node C can be written as:

$$sC_1 V_o(s) + \frac{V_o(s)}{R} + \frac{V_o(s)}{sL + 1/sC_2} + \frac{g_m V_o(s)/sC_2}{sL + 1/sC_2} = 0$$

Multiplying out and collecting terms,

$$V_o(s)\left[(1 + sC_1 R)(1 + s^2 L C_2) + R(g_m + sC_2)\right] = 0$$

In the presence of oscillations $V_o(s) \neq 0$, so that the quantity in brackets is zero. Replacing s by $j\omega$ and equating the real and imaginary parts to zero yields Equations 9.6.2 and 9.6.3.

In practice, in order for the oscillations to start and continue reliably, it is necessary to have $g_m > C_2/(C_1 R)$. The oscillations grow in amplitude until transistor nonlinearities reduce g_m to the level required by Equation 9.6.3. Nonlinearities imply distortion. However, the fundamental at the oscillation frequency is amplified by $|g_m R|$, whereas harmonics are relatively attenuated because of the frequency selectivity of the oscillating circuit. The harmonic distortion is therefore low, typically less than 1%. Since the oscillations are limited by transistor nonlinearities rather than special amplitude-limiting circuitry, as in the case of op-amp oscillators, the Colpitts oscillator is an example of a **self-limiting oscillator**.

The Hartley oscillator can be similarly analyzed, as illustrated by Example 9.6.1. Note that when the oscillation frequency exceeds about a quarter of the unity-gain bandwidth of the transistor, a more accurate transistor model that includes internal capacitances should be used.

Example 9.6.1 Hartley Oscillator

Consider the Hartley oscillator in the circuit of Figure 9.6.3, in which it is assumed that the transistor is properly biased. The two coils are usually wound in the same sense on a common former, so that they are mutually coupled as shown. It is required to determine the oscillation frequency as well as the condition for oscillation.

SOLUTION

Method 1: The small-signal equivalent circuit is shown in Figure 9.6.4. Because of mutual coupling, it is more convenient to represent the transistor in terms of $I_b(s)$ and $\beta I_b(s)$. r_π includes the resistance reflected from the L_1 side, which accounts for r_o, load resistance, and losses in the coils. The dot markings are those required for positive feedback. Thus, an increase in $I_b(s)$ increases the

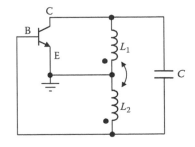

FIGURE 9.6.3
Figure for Example 9.6.1.

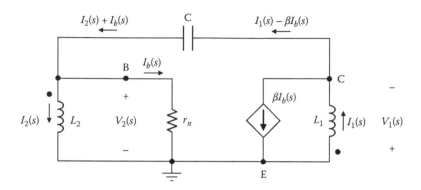

FIGURE 9.6.4
Figure for Example 9.6.1.

collector current $\beta I_b(s)$, which increases $V_1(s)$. According to the dot markings, this increases $V_2(s)$, which in turn increases $I_b(s)$. From KCL at nodes B and C, and equating the currents on both sides of the capacitor, gives:

$$(\beta + 1)I_b(s) = I_1(s) - I_2(s) \tag{9.6.4}$$

From KVL around the outer loop,

$$V_1(s) + V_2(s) + [I_2(s) + I_b(s)]/j\omega C = 0 \tag{9.6.5}$$

where,

$$V_1(s) = j\omega L_1 I_1(s) + j\omega L_2 I_2(s) \tag{9.6.6}$$

and

$$V_2(s) = j\omega L_2 I_2(s) + j\omega M I_1(s) \tag{9.6.7}$$

Substituting $V_1(s) + V_2(s)$ from Equations 9.6.6 and 9.6.7 in Equation 9.6.5, eliminating $I_b(s)$ using Equation 9.6.4, and collecting terms, gives the following equation involving $I_1(s)$ and $I_2(s)$:

$$\left[\omega^2 C(L_1 + M) - \frac{1}{\beta + 1}\right] I_1(s) + \left[\omega^2 C(L_2 + M) - \frac{\beta}{\beta + 1}\right] I_2(s) = 0 \tag{9.6.8}$$

From KVL around the loop containing L_2 and r_π:

$$V_2(s) - r_\pi I_b(s) = 0 \tag{9.6.9}$$

Substituting for $V_2(s)$ from Equation 9.6.7, and for $I_b(s)$ from Equation 9.6.4 gives a second equation involving $I_1(s)$ and $I_2(s)$:

$$\left[j\omega M - \frac{r_\pi}{\beta + 1} \right] I_1(s) + \left[j\omega L_2 + \frac{r_\pi}{\beta + 1} \right] I_2(s) = 0 \qquad (9.6.10)$$

In the presence of oscillations, $I_1(s)$ and $I_2(s)$ are not zero. But since the RHS of Equations 9.6.8 and 9.6.10 is zero, it means that the determinant of the coefficients of $I_1(s)$ and $I_2(s)$ in Equations 9.6.8 and 9.6.10 must be zero. Equating the real part of the determinant to zero, gives the frequency of oscillation:

$$\omega_o = 1/\sqrt{C(L_1 + L_2 + 2M)} \qquad (9.6.11)$$

Equating the imaginary part to zero, gives the condition for oscillation:

$$\beta = \frac{L_2 + M}{L_1 + M} \qquad (9.6.12)$$

To ensure oscillations, β should be larger than this value.

Equation 9.6.11 is that of an *LC* circuit having a total inductance $L_1 + L_2 + 2M$, which is that of two coupled coils in series. If there is no coupling between the coils, $M = 0$ in Equations 9.6.11 and 9.6.12. The condition for oscillation is interpreted in connection with Method 2.

Method 2: The simplicity of Equations 9.6.11 and 9.6.12 suggests a simpler derivation based on the argument that under conditions of oscillation, the amplifier delivers power that compensates for the power dissipated in the circuit. The power delivered by the current source is $\beta I_b(s)V_1(s)$, and that dissipated in r_π is $V_2(s)I_b(s)$. Equating these powers, $\beta = V_2(s)/V_1(s)$. With the current source and r_π neutralizing one another, they can both be removed from the circuit, leaving a lossless *LC* circuit having two coupled coils L_1 and L_2 in series. As the effective inductance is $(L_1 + L_2 + 2M)$, the resonant frequency is given by Equation 9.6.11. The ratio $V_2(s)/V_1(s)$ is $(L_2 + M)/(L_1 + M)$, which gives Equation 9.6.12.

Figure 9.6.5 shows a circuit for a variable frequency Hartley oscillator. A three-coil transformer is used, with a step-up turns ratio from the collector to the tuned circuit. The advantage of such an arrangement is that the base–collector capacitance of the transistor, which varies with V_{CC}, appears as a reduced capacitance across the tuned circuit, and will therefore cause less variation in the frequency of oscillation.

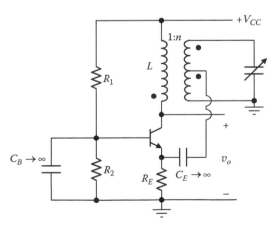

FIGURE 9.6.5
Figure for Example 9.6.1.

9.7 Crystal Oscillators

Some crystals, such as naturally occurring quartz crystals, or artificial ceramic crystals, notably barium titanate, exhibit a **piezoelectric effect**. When a slab cut from the parent crystal is subjected to mechanical deformation in the appropriate direction, a voltage is generated between opposite faces of the slab due to movement of charges. The effect is reversible, so that an applied voltage will cause a change in the dimensions of the slab, the strength of the piezoelectric effect depending on the crystallographic axes along which the crystal is cut. Piezoelectric transducers are extensively used to generate and receive ultrasound waves. The deformable piezoelectric slab exhibits mechanical resonance due to its inertance, which is the mechanical equivalent of inductance, and to its elastance, which is the mechanical equivalent of capacitance. The mechanical resonance can be reinforced, through the piezoelectric effect, to produce sustained oscillations. Thus, a voltage across the crystal at or near the mechanical resonant frequency will produce a magnified mechanical deformation that will in turn, through the piezoelectric effect, reinforce the voltage across the crystal. The resulting oscillations are very stable and have a small temperature coefficient of the order of 1 cycle per million per °C. The frequency of oscillation depends on the dimensions of the crystal and the type of mounting and can range between a few kHz and hundreds of MHz. The waveform of oscillations is rich in harmonics, so that multiples of the fundamental frequency can be selected by appropriate bandpass filters.

The electrical equivalent circuit of the crystal is shown in Figure 9.7.1 together with the crystal symbol. L_m, C_m, and R_m are the electrical equivalents of the mechanical inertance, elastance, and damping, respectively. C_p and R_p are the capacitance and the leakage resistance, respectively, between the faces of the crystal. Typically, L_m can be as large as hundreds of H, R_p in the range of GΩ, C_m as small as a fraction of a fF, and C_p of the order of a few pF, being much larger than C_m. The series resonant circuit has $Q = \omega_0 L_m / R_m$, which is typically several hundred thousand, so that R_m is very small compared to $\omega_0 L_m$. Ignoring R_m and R_p, the reactance between nodes a and b of the equivalent circuit is:

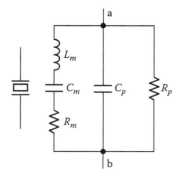

FIGURE 9.7.1
Symbol and electrical equivalent circuit of piezoelectric crystal.

$$X_{ab} = \left(\frac{1}{sC_p}\right) \middle\| \left(sL_m + \frac{1}{sC_m}\right)$$

$$= \frac{1}{sC_p} \frac{s^2 + 1/L_m C_m}{s^2 + 1/L_m(C_m \| C_p)} \qquad (9.7.1)$$

The reactance X_{ab}, illustrated in Figure 9.7.2, has a zero at a frequency:

$$\omega_s = \frac{1}{\sqrt{L_m C_m}} \qquad (9.7.2)$$

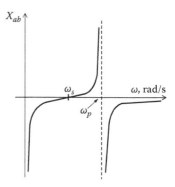

FIGURE 9.7.2
Reactance of piezoelectric crystal.

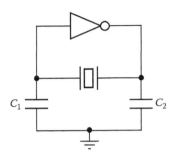

FIGURE 9.7.3
Crystal oscillator.

corresponding to series resonance of L_m and C_m. At a frequency $\omega_p > \omega_s$, the series reactance of L_m and C_m is inductive and will be in parallel resonance with C_p giving a pole of X_{ab} at:

$$\omega_p = \frac{1}{\sqrt{L_m(C_m <> C_p)}} \qquad (9.7.3)$$

where $C_s <> C_p$ is the series capacitance of C_s and C_p. Since $C_p \gg C_m$, ω_s and ω_p are very close together (Exercise 9.7.1). When the crystal is substituted for the inductance in a Colpitts oscillator, the frequency of oscillation will be between ω_s and ω_p. The basic circuit, represented as in Figure 9.7.3, is referred to as a **Pierce oscillator**. Inversion could be obtained between the base and emitter of a BJT, as in Figure 9.6.1, or by using a CMOS inverter that is biased for high-gain by connecting a resistor between its input and output terminals (Section 13.1, Chapter 13). As in Figure 9.6.2, the current that flows in LC_1C_2 is such that the voltages across C_1 and C_2 are in antiphase, so that the connection of the inverter results in positive feedback that reinforces these voltages and compensates for power dissipation, as discussed for the Colpitts oscillator.

Exercise 9.7.1

(a) Express Equation 9.7.1 in terms of the reactance X_{ab}, where $Z_{ab} = jX_{ab}$, and sketch X_{ab} as a function of ω.

(b) Determine ω_s and ω_p if $L_m = 100$ H, $C_m = 1$ fF, and $C_p = 1$ pF.

  Answers: (a) 503.292 kHz; and (b) 503.544 kHz. ∎

Summary of Main Concepts and Results

- Transistor amplifier stages can be directly coupled or coupled using capacitors, transformers, or optical coupling. In dc coupling, the dc level at the output of a given stage may have to be shifted so as to be compatible with the dc level at the input of the following stage. This can be done using diodes, an emitter follower, or a V_{BE} multiplier.

- The low-frequency response of amplifiers is affected by coupling capacitors and emitter/source bypass capacitors. Coupling capacitors introduce a zero at zero frequency and a pole at low frequencies. Emitter/source bypass capacitors introduce a zero at low frequencies and a pole at a higher frequency.

- Transistor feedback amplifiers are conveniently analyzed by neglecting the reverse transmission through the amplifier proper and the forward transmission through the feedback circuit. The feedback configuration can then be reduced to one of four standard configurations, taking into account the loading effects of the feedback circuit at the input and output of the amplifier. These effects, together with the feedback factor, can be determined following some simple, generalized rules. The feedback factor is the "12" two-port parameter of the feedback circuit. The gain, input and output resistances in the presence of feedback follow from the corresponding quantities without feedback, as modified by the loop gain.

- Feedback amplifiers should have an adequate degree of stability. This can be conveniently achieved by having the slope of the asymptote of the $20\log_{10}|A_v(j\omega)|$ plot, minus that of the $20\log_{10}|1/\beta_f(j\omega)|$ plot at their point of intersection, not more negative than -20 dB/decade.

- In feedback oscillators, the phase shift around the loop at the oscillating frequency is zero and the gain is nominally unity.

- In a tuned amplifier, tuned circuits are connected at the input and output of transistor amplifier stages. To increase the bandwidth, the circuits can be stagger tuned, or magnetically coupled. The maximum gain–bandwidth product in a tuned amplifier is that of the transistor.

- In an *LC* oscillator, the transistor provides, through positive feedback, the energy required to compensate for the dissipation in the tuned circuit.

- A piezoelectric crystal can replace the inductor in an *LC* oscillator, resulting in oscillations of high stability and low temperature coefficient.

Learning Outcomes

- Analyze and design basic, dc-coupled or ac-coupled, cascaded stages.
- Analyze feedback amplifiers.
- Articulate the principles of operation of feedback oscillators, tuned amplifiers, and *LC* oscillators.

Supplementary Examples and Topics on Website

SE9.1 Triangular-Waveform Generator. Describes a two-op-amp square-wave/triangular-wave generator.

SE9.2 High-*Q* approximation. Expresses a series *RL* circuit of high *Q* as a parallel *RL* circuit (Sabah 2008, Example SE10.2, p. 393).

SE9.3 High-Performance Source Follower. Analyzes a source follower using an op amp in a series–shunt configuration to enhance the performance of a source follower. A method that is simply based on signal-flow diagrams is presented, as well as a more conventional method.

ST9.1 Determination of Loop Gain from Open-Circuit and Short-Circuit Transfer Functions. Describes a method for deriving the loop gain from the measurement of the open-circuit voltage transfer function and the short-circuit current transfer function.

ST9.2 Stability in Terms of Poles of Closed-Loop Transfer Function. Examines stability in terms of the Nyquist plot and the locations of the poles of the closed-loop transfer function of a three-pole system.

ST9.3 Magnetically Coupled Tuned Circuits. Analyzes magnetically coupled tuned circuits and applies the analysis to the design of an i-f stage in an AM radio receiver (Sabah 2008, Section ST10.7, p. 393).

ST9.4 Stagger-Tuned Circuits. Discusses lowpass-to-narrowband bandpass transformation and applies it to the design of a stagger-tuned i-f amplifier (Sabah 2008, Section ST10.6, p. 393).

Problems and Exercises

P9.1 Cascaded and Feedback Amplifiers

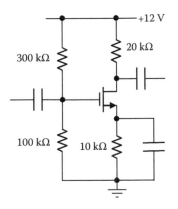

FIGURE P9.1.1

P9.1.1 In the discrete amplifier stage of Figure P9.1.1 the E-NMOS transistor has $k_n = 2$ mA/V^2, $V_{tn} = 2$ V, and $\lambda = 0.05$ V^{-1}, the capacitors being very large. (a) Determine I_D and g_m, neglecting channel-length modulation. (b) The amplifier stage is cascaded with an identical stage and connected between a small-signal voltage source of 50 kΩ source resistance and a 10 kΩ load. Determine the overall small-signal voltage gain from the source to load.

P9.1.2 The transistors in Figure P9.1.2 have $\beta = 100$, $V_{BE} = 0.7$ V, the capacitor and r_o being very large. Determine v_o/v_i, the resistance looking into the bases of Q_2 and Q_1, and i_o/i_i.

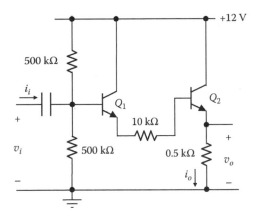

FIGURE P9.1.2

P9.1.3 Determine: (a) the dc currents in each transistor in Figure P9.1.3 assuming $|V_{BE}| = 0.7$ V and very large β; (b) the small-signal input resistance, current gain, and voltage gain, assuming $\beta = 100$ for all transistors, and neglecting r_o.

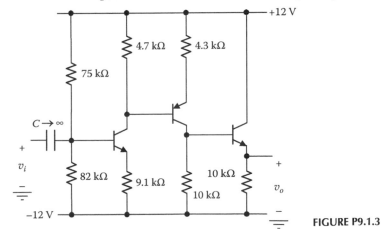

FIGURE P9.1.3

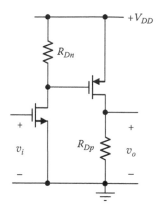

FIGURE P9.1.4

P9.1.4 Derive the expression for the voltage gain v_o/v_i in Figure P9.1.4, neglecting r_o.

P9.1.5 Figure P9.1.5 shows a two-stage dc coupled emitter follower. Calculate V_O, the dc output voltage for zero dc input assuming V_{EB} of Q_2 is 0.7 V and V_{BE} of Q_1 is 0.7 V at 1 mA. Determine r_{in}, r'_{out}, A_v and A'_v, neglecting r_o and assuming $\beta_1 = 30$ and $\beta_2 = 100$. Simulate with PSpice.

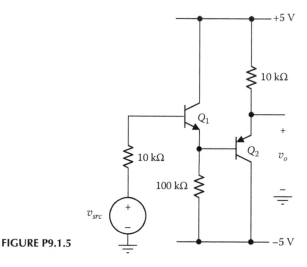

FIGURE P9.1.5

P9.1.6 A CS amplifier with a gate–drain feedback resistor, as in the BJT circuit of Figure 8.1.5, Chapter 8, has $R_G = 5$ MΩ, $R_D = 20$ kΩ, $g_m = 1$ mA/V, $r_o = 80$ kΩ, and coupling capacitors C_{PI} and C_{PO} at the input and output, respectively. If the load resistance is 50 kΩ, determine C_{PI} and C_{PO} using the SCTC method to give a pole at 15 Hz, the second pole being at least a decade lower.

P9.1.7 Consider the two-cascaded stages of Problem P9.1.1, but with the source bypass capacitors removed, the capacitor connected to the source being 5 μF, and the remaining coupling capacitors being 1 μF. Use the SCTC method to find the 3-dB cutoff frequency.

P9.1.8 Consider the circuit of Figure 8.1.4, Chapter 8, with the base connected to a source of 2 kΩ resistance, the output coupled by a capacitor C_{PO} to a 10 kΩ load, and the emitter bypassed to ground by a capacitor C_E. $R_C = R_E = 5$ kΩ, $I_C = 1$ mA, and $\beta = 100$. Neglecting r_o, determine C_E and C_{PO} so that the lower 3-dB frequency is at 50 Hz, the contribution of the nondominant pole to this frequency not exceeding 5%, and the total capacitance is minimized.

P9.1.9 Given a CE amplifier as in Figure 8.2.1a, Chapter 8, having $R_{src} = 2$ kΩ, $R_C = 5.1$ kΩ, $R_E = 3.9$ kΩ, $R_L = 8.2$ kΩ, $R_1 \| R_2 = 20$ kΩ, $I_C = 0.5$ mA, $\beta = 100$, $r_o = 100$ kΩ, $C_E = 10$ μF, $C_{PI} = 5$ μF, and $C_{PO} = 1$ μF. Evaluate the SCTCs and estimate the lower 3-dB cutoff frequency. Simulate with PSpice.

P9.1.10 Repeat Problem P9.1.9 for an E-NMOS transistor having $g_m = 1$ mA/V, $r_o = 100$ kΩ, and the same circuit components as in Problem P9.1.9 in but with $R_G = 1$ MΩ. Determine the SCTCs, and the lower 3-dB cutoff frequency. Simulate with PSpice.

P9.1.11 In the PSpice simulation of the CE amplifier in Problem P9.1.9, observe the waveforms of i_B, i_C, and v_{CE} at maximum swing, noting the considerable distortion of the sinusoidal waveform. Observe the same waveforms without the emitter bypass capacitor and note the reduction of distortion. As the transistor enters saturation, the i_B waveform has a narrow, peaked excursion superposed on the positive maximum of the sinusoid, whose amplitude and width increase with the degree of saturation. What do you ascribe this to? Compare the midband, small-signal voltage gain, and bandwidth with and without the emitter bypass capacitor.

P9.1.12 An amplifier of midband gain of -25 has a phase shift of $135°$ at 50 Hz and $225°$ at 500 kHz. Determine: (a) the bandwidth; and (b) the frequency range over which the gain exceeds unity.

P9.1.13 An amplifier has the transfer function $H(s) = \dfrac{10^{11}s^2}{(s+1)(s+5)(s+10^4)(s+5\times10^4)}$. Determine: (a) the midband gain in dB; (b) the lower 3-dB cutoff frequency; (c) the upper 3-dB cutoff frequency. $\left(\text{Hint: express } H(s) \text{ as: } \dfrac{200}{(1+1/s)(1+5/s)(1+s/10^4)(1+s/(5\times10^4))}\right)$.

P9.1.14 A wideband amplifier can be realized by cascading low-gain stages, as in Figure P9.1.14 (Steininger, 1990), where Q_1 and Q_2 are identical, except for the channel width W, and are biased to the same V_{GS}. Neglect r_o and assume that $C_{gd} \gg C_{gs}$, and the gain of the stage v_o/v_i is small so that the Miller effect is negligible. The small-signal equivalent circuit of the MOSFET can be then represented by a current source $g_m v_{gs}$ at the output and a capacitance g_m/ω_T at the input, where ω_T is the unity-gain bandwidth and is independent of W. Show that if the stage of Figure P9.1.14 is followed by an identical stage, $\dfrac{v_o}{v_i} = -\dfrac{A_0}{1+\dfrac{s}{\omega_T/(A_0+1)}}$, where $A_0 = \dfrac{g_{m1}}{g_{m2}} = \dfrac{W_1}{W_2}$.

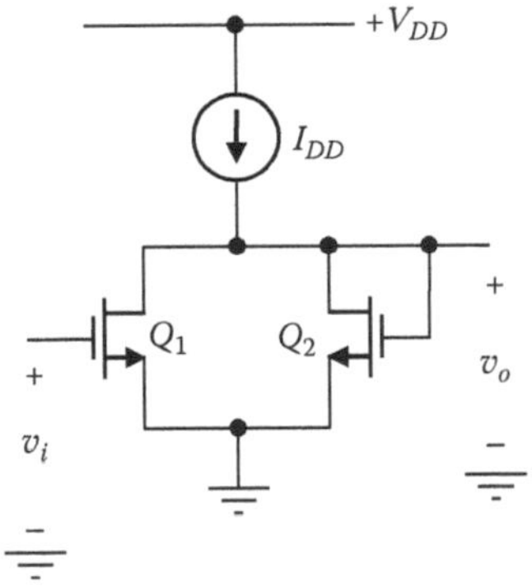

FIGURE P9.1.14

P9.1.15 Consider a source follower as a series–shunt feedback circuit. Derive A_{va}, R_{oa}, A_{vfa}, and R_{ofa} neglecting the body effect. Compare with Equations 8.2.16 and 8.2.20.

P9.1.16 In the circuit of Figure P9.1.16, the series–shunt feedback of the D-NMOS source follower is augmented by Q_2. Assume that Q_1 has $I_{DSS}=5$ mA and $V_{tD}=-2$ V, whereas Q_2 has $V_{EB}=0.7$ V, a large β, and $g_{m2}=30$ mA/V. (a) Determine the quiescent currents

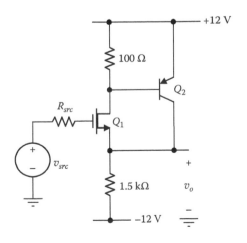

FIGURE P9.1.16

of Q_1 and Q_2 for zero dc input. (b) Derive A_{va}, R_{oa}, A_{vfa}, and R_{ofa} neglecting the body effect, channel-length width modulation, and base-width modulation. What is the effective g_m of Q_1?

P9.1.17 Figure P9.1.17 shows the ac circuit of a shunt–shunt feedback amplifier. Assume that the drain/collector currents are 1 mA. Q_1 has $k_n = 2$ mA/V^2 and $\lambda = 0.02$ V^{-1}, whereas Q_2 and Q_3 have $\beta = 100$ and $V_A = 50$ V. Determine β_f, v_o/v_{src}, R_{ifa}, and R_{ofa}.

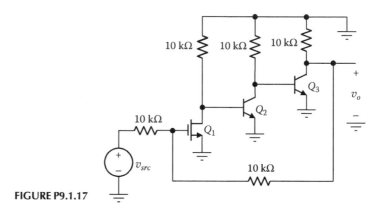

FIGURE P9.1.17

P9.1.18 (a) Specify the feedback configuration in Figure P9.1.18. (b) Determine i_o/v_{src} from feedback analysis assuming that the current mirror is ideal, $k_n = 1$ mA/V^2, $V_{tn} = 1$ V, very large r_o for all transistors, and neglecting the body effect. (c) Verify the result in (b) by direct analysis.

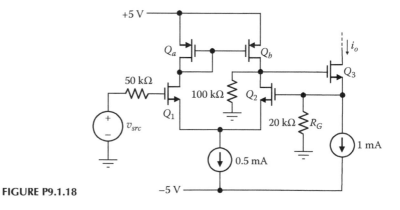

FIGURE P9.1.18

P9.1.19 (a) Specify the feedback configuration in Figure P9.1.19. (b) Apply feedback analysis to determine v_o/v_{src} and the output resistance looking into Q_2 collector. Assume $g_{m1} = g_{m2} = 2$ mA/V, $r_{o1} \rightarrow \infty$, $r_{o2} = 40$ kΩ and neglect the body effect.

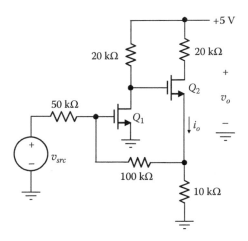

FIGURE P9.1.19

P9.1.20 The h-parameter two-port circuit equations are: $V_1 = h_{11}I_1 + h_{12}V_2$ and $I_2 = h_{21}I_1 + h_{22}V_2$, as follows from Figure 7.1.2c. Apply these equations to the series–shunt circuit of Figure 9.3.1 to show that the equations for the feedback amplifier are $V_{src} = h_{11fa}I_i + h_{12fa}V_o$ and $I_2 = h_{21fa}I_i + h_{22fa}V_o$, where $h_{11fa} = R_{src} + h_{11a} + h_{11f}$, $h_{12fa} = h_{12a} + h_{12f}$, $h_{21fa} = h_{21a} + h_{21f}$, and $h_{22fa} = h_{22a} + h_{22f}$. Setting h_{12a} and h_{21f} to zero, show that A_{vfa}, R_{ifa}, and R_{ofa} derived from the h-parameter equations are the same as those obtained from feedback analysis. Note that the same procedure can be applied to the other feedback configurations using the z-, y-, and g-parameters.

P9.2 Closed-Loop Stability and Feedback Oscillators

P9.2.1 Show that a capacitive load in conjunction with the op-amp output resistance makes the amplifier less stable, in both the inverting and noninverting configurations. Assume that $A_v(f) = \dfrac{A_{v0}}{(1 + j\omega/\omega_{c1})(1 + j\omega/\omega_2)}$, an otherwise ideal op amp, and a purely resistive β_o.

P9.2.2 The op amp shown in Figure P9.2.2 has infinite input resistance, an output resistance of 500 Ω, $A_v(f) = 10^4/(s + 50)$, and $C_L = 1$ µF. Determine the phase margin.

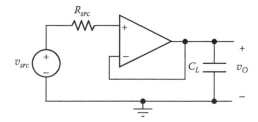

FIGURE P9.2.2

P9.2.3 An op amp has $A_{v0} = 10^5$, a one-pole transfer function, and a unity-gain bandwidth of 1 MHz. Determine: (a) $|A_v(f)|$ at 1 kHz; and (b) the low-frequency closed-loop gain for a 3-dB cutoff frequency of 50 kHz.

P9.2.4 An op amp having the transfer function $A_{v0} \big/ \left(1 + j\frac{f}{100}\right)^3$, where f is in Hz, is to be used as a noninverting amplifier with a feedback factor of 0.01. Determine A_{v0} for: (a) a phase margin of 45°; and (b) a gain margin of 20 dB.

P9.2.5 An amplifier having $A_v(f) = \dfrac{10^4}{(1+jf/10)(1+jf/1000)}$, where f is in Hz, is used in a negative feedback configuration with $\beta_f = 0.1$. Determine: (a) the low-frequency gain of the feedback amplifier; (b) the phase margin; and (c) the gain margin.

P9.2.6 An op amp having $A_v(f) = 10^4/(1+jf/f_{p1})^4$, where $f_{p1} = 10$ kHz, is used in a noninverting configuration of $\beta_o = 0.01$. Working with asymptotes, (a) show that the amplifier is unstable, (b) determine the frequency of a dominant pole that will stabilize the amplifier even for $\beta_o = 1$, and (c) find the low-frequency gain of the feedback amplifier.

P9.2.7 Let $A_v(f) = \dfrac{A_{v0}}{(1+jf/f_{p1})(1+jf/f_{p2})(1+jf/f_{p3})}$. (a) Show that the imaginary part of $A_v(f)$ vanishes at $f = \sqrt{f_{p1}f_{p2}+f_{p2}f_{p3}+f_{p3}f_{p1}}$ and that the real part at this frequency is $-\dfrac{A_{v0}}{2+\frac{f_{p2}+f_{p3}}{f_{p1}}+\frac{f_{p1}+f_{p3}}{f_{p2}}+\frac{f_{p1}+f_{p2}}{f_{p3}}}$. (b) Assuming β_f is frequency independent, deduce that the stability criterion can be expressed as $\beta_f < \dfrac{1}{A_{v0}}\left(2+\dfrac{f_{p2}+f_{p3}}{f_{p1}}+\dfrac{f_{p1}+f_{p3}}{f_{p2}}+\dfrac{f_{p1}+f_{p2}}{f_{p3}}\right)$.

P9.2.8 An op amp having $A_v(f) = 10^4/(1+jf/f_{p1})^3$, where $f_{p1} = 10$ kHz, is to be used in a noninverting configuration of $\beta_f = 0.01$. (a) Use the result of Problem P9.2.7 to show that the amplifier is unstable. (b) A lead-lag circuit having a transfer function $(1+jf/f_{p1})/(1+jf/f_{p0})$ is used to cancel one of the f_{p1} poles and add a pole at f_{p0}. Show that the criterion of stability of Problem P9.2.7 becomes $\beta_f < \dfrac{1}{A_{v0}} \times \left[4+2\left(\dfrac{f_{p0}}{f_{p1}}+\dfrac{f_{p1}}{f_{p0}}\right)\right]$. (c) Determine that largest possible value of f_{p0} that assures stability and calculate the GB product.

P9.2.9 An op amp having $A_v(f) = \dfrac{10^4}{(1+jf/f_{p1})(1+jf/f_{p2})(1+jf/f_{p3})}$, where $f_{p1} = 1$ kHz, $f_{p2} = 5$ kHz, and $f_{p3} = 40$ kHz, is used in an inverting configuration of $\beta_o = 0.01$. Apply the criterion of Problem P9.2.7 and confirm instability by finding the phase shift for $|\beta_o A_v(f)| = 1$.

P9.2.10 Lead-lag compensation is applied to the op amp of Problem P9.2.9, the transfer function multiplying $A_v(f)$ being $\dfrac{1+jf/f_{p1}}{1+jf/f_{p0}}\dfrac{1+jf/f_{p2}}{1+jf/f_{p3}}$. Determine the largest possible value of f_{p0} that assures stability and calculate the GB product.

P9.2.11 Assume that the amplifier in Problem P9.2.9 is compensated by introducing a dominant pole for $\beta_o = 0.01$. Determine the frequency of this pole and the resulting GB product.

P9.2.12 Assume that the amplifier in Problem P9.2.10 is compensated by shifting f_{p1} to a lower frequency. Determine this frequency and the resulting GB product. How does the GB product compare with those of Problems P9.2.10 and P9.2.11?

P9.2.13 Determine the low-frequency closed-loop gain of the amplifier of Problem P9.2.9 for a phase margin of 45°. What is the resulting gain margin?

P9.2.14 In Problem P9.2.9, assume that the low-frequency open-loop gain is increased from a low value. Determine the value of this gain at which poles become located on the imaginary axis.

P9.2.15 The amplifier of Problem P9.2.9 is to be used as an integrator having a time constant $C_f R_r = 2.5$ μs. What is the smallest low-frequency open-loop gain that ensures stability? Is the integrator more stable than the amplifier?

P9.2.16 Repeat Problem P9.2.15 for a differentiator having the same time constant. Is the differentiator more stable than the amplifier?

P9.2.17 An oscillator consists of an RC lag circuit, a phase shifter that introduces a phase lag of $\theta°$, without affecting the gain, and an amplifier of gain $-A$, as shown in Figure P9.2.17. If a frequency of oscillation of 16 kHz is required, determine: (a) θ; and (b) the minimum value of A.

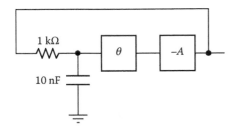

FIGURE P9.2.17

P9.2.18 Show that if the output is connected to the input in Figure P9.2.18, the Barkhausen criterion is satisfied. Determine the frequency of oscillation.

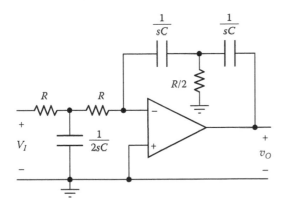

FIGURE P9.2.18

P9.2.19 Figure P9.2.19 illustrates a **phase-shift oscillator**. Derive the expression for the open-loop gain and deduce the frequency and condition for oscillation.

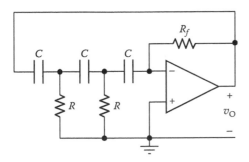

FIGURE P9.2.19

P9.2.20 For the phase-shift oscillator of Figure P9.2.20, $R = 10$ kΩ and $C = 0.5$ nF. Determine the frequency of oscillation and the value of R_f required to sustain oscillation.

P9.2.21 Figure P9.2.21 illustrates a **quadrature oscillator** based on an integrator and the current source circuit of Figure P7.2.7, Chapter 7. Determine the frequency of oscillation and show that v_O and v'_O are in phase quadrature.

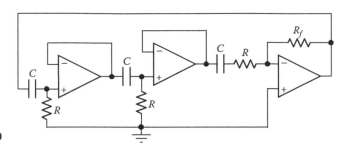

FIGURE P9.2.20

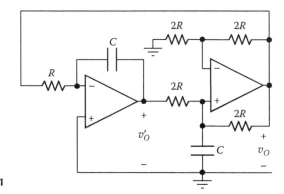

FIGURE P9.2.21

P9.2.22 A Colpitts oscillator has $L = 0.3$ μH, $C_1 = 18$ pF, and uses for C_2 a varactor diode of capacitance $30/\sqrt{1 + V_D/0.7}$ pF, where V_D is the reverse diode voltage. Determine V_D for oscillation in the FM radio band 88–108 MHz.

P9.3 Tuned and Magnetically Coupled Circuits

P9.3.1 Show that if n identical tuned circuits of 3-dB bandwidth BW are cascaded with isolation, the overall 3-dB bandwidth is reduced as $BW\sqrt{2^{1/n} - 1}$, assuming Q is large.

P9.3.2 Consider three identical, isolated, cascaded tuned circuits. Select C and R_p for each circuit so as to have a center frequency of 480 kHz and an overall 3-dB bandwidth of 8 kHz, using inductors of 10 μH. Obtain the Bode plots using PSpice.

P9.3.3 Show that for critical coupling ($kQ = 1$), the bandwidth is given by Equation ST9.3.15.

P9.3.4 A pair of coupled tuned circuits having $Q = 40$ and $k = 0.025$ (critical coupling) is cascaded with an identical pair having $k = (\sqrt{2} + 1)/40$ (3-dB sag). Estimate the maximum variation of gain in the passband using PSpice simulation.

P9.3.5 Obtain the response of Example ST9.3.1 using three stagger-tuned circuits having a center frequency of 470 kHz and a nominally maximally flat response. Derive the 3-dB bandwidth, simulate with PSpice, and compare with Example ST9.3.1.

P9.3.6 A CE amplifier has a load of admittance $Y_L = G_L + jB_L$ that is large compared with that of C_μ so that the amplifier gain may be assumed to be $-g_m/Y_L$. Show that the admittance reflected at the input is $\dfrac{\omega C_\mu g_m B_L}{G_L^2 + B_L^2} + j\omega C_\mu \left(1 + \dfrac{g_m G_L}{G_L^2 + B_L^2}\right)$. Note that if the load is inductive, $B_L < 0$, and the real part is a negative conductance. If the magnitude of this conductance is larger than the positive conductance at the input, the net conductance is negative and will result in oscillations.

10

Differential and Operational Amplifiers

Differential amplifiers are a special and important class of amplifiers, particularly in the form of IC operational amplifiers, because they inherently respond to the difference between two signals applied to their two inputs. This not only lends itself to many useful applications, but it also means that, ideally, an unwanted signal, or interference, that is superimposed on the two inputs will be subtracted out and will not contribute to the output. The basic building block of differential amplifiers is the differential pair, which is a special connection of two transistors.

The chapter therefore starts with a basic analysis of the differential pair, using BJTs or MOSFETs, and examines its essential features, both in terms of the transfer function between input and output and in small-signal operation. Some of the op-amp imperfections, or departures from the ideal, are due to the differential pair and are introduced at this stage. The next logical step is to show how differential pairs are used in IC op amps. This is done for some basic types of CMOS op amps. The essential features of BJT op-amp circuits are considered next, the detailed analysis being left to Supplementary Examples and Topics on the website.

Having introduced nonideal op amps, the behavior of these op amps is examined in some important op-amp applications, namely, inverting or noninverting amplifiers, adders, difference amplifiers, and integrators. The chapter ends by considering two types of circuits in which op amps play an essential role, namely, switched-capacitor circuits and digital/analog converters (DACs).

Learning Objectives

❖ To be familiar with:

- Terminology commonly used with differential and operational amplifiers.
- Principle of operation of switched-capacitor circuits, DAC and analog-to-digital converter (ADC).

❖ To understand:

- The basic features of differential amplifiers, using BJTs or MOSFETs, how they respond to differential and common-mode inputs, and the imperfections they exhibit.
- The essential features and performance limitations of representative CMOS and BJT op amps.
- The effect of op-amp imperfections on the performance of inverting and non-inverting amplifiers, adders, difference amplifiers, integrators, and instrumentation amplifiers (IAs).

10.1 Differential Pair

Basic Operation

The **BJT differential pair**, or **emitter-coupled pair**, is an important building block in both analog ICs, where it is extensively used in operational amplifiers, and in digital ICs in the form of **emitter-coupled logic** (ECL, Section 13.7, Chapter 13). To appreciate the basics of the BJT differential pair, consider Figure 10.1.1, in which Q_1 and Q_2 are matched transistors connected to a current source I_{EE} and have a matched load R_C. Assume that the same dc voltage $V_B = v_{B1} = v_{B2}$ is applied to the bases of Q_1 and Q_2. The common emitter voltage V_E with respect to ground is $V_E = V_B - V_{BE}$, where $V_{BE} \cong 0.7$ V. Since the circuit is symmetrical, the current I_{EE} divides equally between Q_1 and Q_2. The two transistors have a base current, $I_B = I_{EE}/2(\beta_F + 1)$, the same collector current, $\alpha_F I_{EE}/2$, and the same collector voltage, $V_{C1} = V_{C2} = V_{CC} - \alpha_F I_{EE} R_C/2$. It follows that if V_B is replaced by a time-varying v_{cm}, applied in parallel to the bases of both transistors, as indicated by the dotted line in Figure 10.1.1, $v_E = v_{cm} - 0.7$. The voltage v_E changes the voltage across the ideal current source but does not change I_{EE}. The collector currents remain the same, and the collector voltages with respect to ground remain at $V_{CC} - \alpha_F I_{EE} R_C/2$. No signal voltage appears between the collectors and ground. In other words, the ideal differential pair completely rejects a **common-mode voltage** v_{cm} that is applied to both base terminals.

Since collector voltages do not depend on a common-mode input, suppose that $V_B = 0$ and a small-signal voltage v_i is applied in antiphase to the base terminals (Figure 10.1.2). Assuming linearity and operation in the active region, if v_i increases the base current of Q_1 by i_b then $-v_i$ decreases the base current of Q_2 by i_b. The emitter currents of Q_1 and Q_2 are $(I_{EE}/2) + (\beta + 1)i_b$ and $(I_{EE}/2) - (\beta + 1)i_b$, respectively, their sum being I_{EE}. Neglecting r_o, there will now be a small-signal $v_{c1} = -\beta R_C i_b$ and $v_{c2} = +\beta R_C i_b$. It can be readily argued that small-signal $v_e = 0$. Let v_i be applied to Q_1 base with $v_{B2} = 0$ and let the resulting v_e be denoted as v_{e1}. If $-v_i$ is applied to Q_2 base with $v_{B1} = 0$, the resulting v_{e2} equals $-v_{e1}$, from

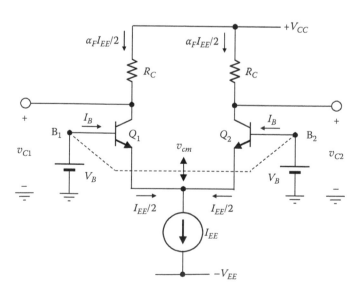

FIGURE 10.1.1
Basic BJT differential pair.

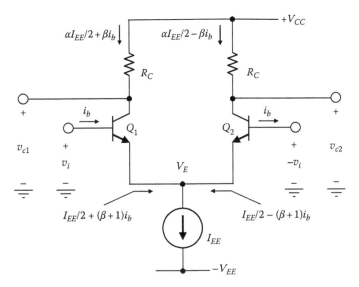

FIGURE 10.1.2
Basic BJT differential pair with equal input signals applied in antiphase.

symmetry. If both inputs are applied together, $v_e = 0$, assuming linearity. Thus, the emitters are effectively grounded for small, antiphase signals.

Concept: *An ideal differential pair completely rejects a common-mode input, which will appear at the common emitters. Equal, antiphase input signals result in equal antiphase output signals and no signal voltage appears at the common emitters.*

If two arbitrary inputs v_{B1} and v_{B2} are applied, we define two inputs v_{cm} and v_i as:

$$v_{cm} = \frac{1}{2}(v_{B1} + v_{B2}) \quad \text{and} \quad v_i = \frac{1}{2}(v_{B1} - v_{B2}) \tag{10.1.1}$$

Solving for v_{B1} and v_{B2}:

$$v_{B1} = v_{cm} + v_i \quad \text{and} \quad v_{B2} = v_{cm} - v_i \tag{10.1.2}$$

Since v_i in Figure 10.1.2 is balanced with respect to ground, it could just as well be applied as a *differential input* $v_{id} = v_i - (-v_i) = 2v_i = v_{B1} - v_{B2}$ between the base of Q_1 and that of Q_2 (Figure 10.1.3a). The currents are the same as before. The output voltage could be taken, as a *single-ended output*, between either collector and ground, or as a *differential output* $v_{od} = v_{c1} - v_{c2} = 2\beta R_C i_b$ that is balanced with respect to ground.

Concept: *Arbitrary inputs applied to the base terminals of a differential pair can be resolved into a common-mode input that is the average of the two inputs, and a differential input that is the difference of the two inputs.*

Thus, a differential input v_{id} can be replaced by two inputs $v_{id}/2$ applied in antiphase as in Figure 10.1.2. Moreover, small-signal currents and voltages of Q_1 and Q_2 in Figure 10.1.3a are equal in magnitude but opposite in sign, the common emitter being a signal ground.

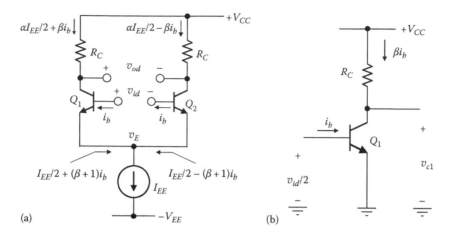

FIGURE 10.1.3
(a) Basic BJT differential pair with differential input and output; (b) equivalent, differential half circuit.

Hence, for the differential small-signal input, we need only to consider a **differential half circuit** consisting of a single transistor (Figure 10.1.3b). If the source that is applied differentially has a source resistance R_{src}, the voltage between the bases B_1 and B_2 is $v_{id} - R_{src}i_b$. Dividing this voltage by 2, it follows that in the differential half circuit the source $v_{id}/2$ has a source resistance $R_{src}/2$.

Example 10.1.1 Differential Pair

It is required to analyze a differential pair having $v_{B1} = v_i$, with $v_{B2} = 0$.

SOLUTION

Let i_b denote Q_1 base current. Q_1 emitter current is $I_{EE}/2 + (\beta + 1)i_b$, as in Figure 10.1.2. Since the total emitter current remains at I_{EE}, Q_2 emitter current must be $I_{EE}/2 - (\beta + 1)i_b$. This implies that $v_e = v_i/2$. In effect, v_i divides as $v_i/2$ between the base and emitter of Q_1 and $v_i/2$ between the emitter and base of Q_2 so that Q_1 emitter current increases by $(\beta + 1)i_b$, while Q_2 emitter current decreases by $(\beta + 1)i_b$. The result is the same as having $+v_i/2$ applied to the base of Q_1 in Figure 10.1.2 and $-v_i/2$ applied to the base of Q_2.

Alternatively, we can apply Equation 10.1.1 to deduce that v_i is equivalent to $v_{cm} = v_i/2$, with $v_i/2$ applied at B_1, and $-v_i/2$ applied at B_2. The signals add to v_i at B_1 and zero at B_2. The differential signal has no effect on v_E, and v_{cm} gives $v_e = v_i/2$.

Transfer Characteristic

The variation of output current with differential input voltage can be derived using the Ebers–Moll equations (Section 6.3, Chapter 6). From Equation 6.3.2, Chapter 6 with $e^{v_{BC}/V_T} \ll 1$ and $e^{v_{BE}/V_T} \gg 1$ in the active region, $i_{E1} = I_{SE}(e^{v_{BE1}/V_T})$, and $i_{E2} = I_{SE}(e^{v_{BE2}/V_T})$. Hence,

$$\frac{i_{E1}}{i_{E2}} = e^{(v_{BE1} - v_{BE2})/V_T} = e^{v_{id}/V_T} \tag{10.1.3}$$

where $v_{BE1} = v_{B1} - v_E$ and $v_{BE2} = v_{B2} - v_E$ so that $v_{BE1} - v_{BE2} = v_{B1} - v_{B2} = v_{id}$ (Equation 10.1.1).

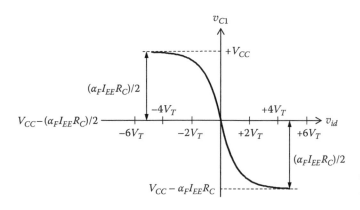

FIGURE 10.1.4
Transfer characteristic of BJT differential pair.

By setting $i_{E1} + i_{E2} = I_{EE}$, and solving for i_{E1} and i_{E2} using Equation 10.1.3:

$$i_{C1} = \alpha_F i_{E1} = \frac{\alpha_F I_{EE}}{1 + e^{-v_{id}/V_T}} \quad \text{and} \quad i_{C2} = \alpha_F i_{E2} = \frac{\alpha_F I_{EE}}{1 + e^{v_{id}/V_T}} \tag{10.1.4}$$

The exponential in the denominators makes i_{C1} and i_{C2} change rapidly with v_{id}. At $v_{id} = 0$, $i_{C1} = \alpha_F I_{EE}/2$, and $v_{C1} = V_{CC} - \alpha_F I_{EE} R_C/2$. As v_{id} varies from $-4V_T$ to $+4V_T$, i_{C1} swings between approximately $0.02\alpha_F I_{EE}$ and $0.98\alpha_F I_{EE}$, whereas v_{C1} swings between $V_{CC} - 0.02 R_{C1}\alpha_F I_{EE} R_C \cong V_{CC}$ and $V_{CC} - 0.98\alpha_F I_{EE} R_C \cong V_{CC} - \alpha_F I_{EE} R_C$ (Figure 10.1.4). The changes in i_{C2} and v_{C2} are the mirror images about the v_{id} axis of the changes in i_{C1} and v_{C1}, respectively. This large excursion of output for a small change in differential input is utilized for switching in digital ECL circuits (Section 13.7, Chapter 13). The switching is fast since the transistors are in the active region and do not suffer from delays due to saturation. The transfer characteristic has a nearly linear region for small v_{id} of either polarity, which makes the differential pair suitable for amplification as well and validates using the small-signal equivalent circuit for small changes (Exercise 10.1.1).

Exercise 10.1.1

Using the relationship $v_{C1} = V_{CC} - \dfrac{\alpha_F R_C I_{EE}}{1 + e^{-v_{id}/V_T}}$, derive the slope dv_{C1}/dv_{id} and deduce that the departure from linearity at $v_{id} = V_T$ is $1 - 4/(1 + e) \cong 7.5\%$. ∎

Small-Signal Differential Operation

We will now introduce the first deviation from the ideal conditions assumed until now, namely, the finite resistance R_{EE} of the I_{EE} source. Note that R_{EE} is a small-signal resistance that does not affect dc conditions so that the dc emitter current of each transistor is still $I_{EE}/2$.

By invoking symmetry and superposition, it can be readily argued that, when v_{id} is applied, $v_e = 0$ in exactly the same manner as when $R_{EE} \to \infty$. The same differential half circuit of Figure 10.1.3b applies, so that $v_{c1} = -g_m(R_C \| r_o)v_{id}/2$. For the other half circuit, $v_{c2} = +g_m(R_C \| r_o)v_{id}/2$. The differential gain A_{dd0}, assuming a differential output, is:

$$A_{dd0} = \frac{v_{c1} - v_{c2}}{v_d} = -g_m(R_C \| r_o) = -\frac{\beta(R_C \| r_o)}{r_\pi} \tag{10.1.5}$$

Although A_{dd0} increases with R_C, the voltage drop $V_{R_C} = R_C I_{EE}/2$ sets a limit to the maximum V_{cm} that can be applied before Q_1 and Q_2 saturate. For these transistors to remain in the active region, we must have $V_{CB} = (V_{CC} - V_{R_C}) - V_{cm} > 0$, or $V_{cm} < V_{CC} - V_{R_C}$. Note that if R_C is an active load, V_{R_C} can be kept low, while the small-signal differential load resistance is high.

If the input is applied differentially, but the output is single ended, then the gain A_{ds0} is:

$$A_{ds0} = \frac{v_{c1}}{v_{id}} = \frac{A_{dd0}}{2} = -\frac{g_m(R_C \| r_o)}{2} \tag{10.1.6}$$

Exercise 10.1.2

Neglecting r_o, Equation 10.1.6 gives $A_{ds0} = -g_m R_C/2$. Use the results of Exercise 10.1.1 to show that this is the same as the slope of the transfer characteristic at $v_{id} = 0$. Consider that $g_m = \alpha_F I_{EE}/2V_T$, which identifies α with α_F.   ■

Since $v_e = 0$, the differential input resistance seen between B$_1$ and B$_2$ is the sum of the two r_π's (Equation 8.2.5, Chapter 8):

$$r_{ind} = 2r_\pi = 2\frac{\beta}{g_m} \cong \beta\frac{V_T}{I_{EE}} \tag{10.1.7}$$

For a large r_{ind}, I_{EE} should be small, which reduces g_m. Note that, Equation 10.1.7 ignores r_μ. Although r_μ may be large, the Miller effect divides r_μ by $1 + |A_v|$ so that $r_\mu/(1 + |A_v|)$ may not be negligible compared to r_π (Example 10.1.2).

The output resistance between either collector and ground is $r_o \| R_C$ and the differential output resistance is $2(r_o \| R_C)$.

High-Frequency Response

The high-frequency, differential response of the differential pair also follows from that of the differential half circuit. By analogy to Equation 8.3.7, Chapter 8, the gain as a function of frequency is determined by a dominant pole ω_{pd}, so that,

$$A_{dd} = \frac{A_{dd0}}{1 + s/\omega_{pd}} \tag{10.1.8}$$

with,

$$\omega_{pd} = \frac{1}{(C_T R'_{src} + (C_\mu + C_L)R'_C)} \tag{10.1.9}$$

where $R'_{src} = [(R_{src}/2) \| r_\pi]$ and $R'_C = r_o \| R_C$. The zero and the other pole occur at much higher frequencies, as for the CE amplifier.

Small-Signal Common-Mode Response

Concept: *Although the resistance R_{EE} does not affect the small-signal differential response, it has a significant effect on the small-signal, single-ended, common-mode response.*

With v_{cm} applied to the differential pair, it follows from symmetry that, for small signals, the circuit can be split into two half circuits (Figure 10.1.5), without changing

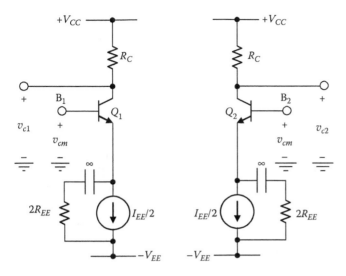

any conditions in the circuit. Note the large capacitor included in series with $2R_{EE}$ to account for the fact that this resistance affects small-signal operation but not dc conditions. The small-signal gain from base to collector is given by Equation 8.2.44, Chapter 8. When βr_o is large compared to R_C and $2R_{EE}$, it follows from Equation 8.2.47, Chapter 8, that:

$$v_{c1} = v_{c2} \cong -\frac{\alpha R_C}{r_e + 2R_{EE}} \cong -\frac{R_C}{2R_{EE}} v_{cm} \qquad (10.1.10)$$

The differential output $v_{c1} - v_{c2}$ is zero. Thus, *although there is a single-ended common-mode response at each collector, the differential common-mode response is zero*, assuming that the two common-mode half circuits are identical. The common-mode gain of the half circuit is:

$$A_{cm0} = \frac{v_{c1}}{v_{cm}} \cong -\frac{\alpha R_C}{r_e + 2R_{EE}} \cong -\frac{R_C}{2R_{EE}} \qquad (10.1.11)$$

It is important to appreciate that dissymmetry between the two halves of the differential pair will result in a common-mode differential output. Thus, if $R_{C1} = R_C + \Delta R_C/2$, and $R_{C2} = R_C - \Delta R_C/2$, the dc collector voltages will not be equal. For small signals, Equation 10.1.10 gives:

$$v_{c1} - v_{c2} = -\frac{R_C + \Delta R_C/2}{2R_{EE}} + \frac{R_C - \Delta R_C/2}{2R_{EE}} = -\frac{\Delta R_C}{2R_{EE}} v_{cm} \qquad (10.1.12)$$

The **common-mode rejection ratio (CMRR)** is the magnitude of the ratio of the differential gain to the common-mode gain. The differential gain could be A_{dd0} or A_{ds0} depending on the circuit being considered. In the case of the differential half-circuit of Figure 10.1.5, it is appropriate to use A_{ds0} since A_{cm0} is taken with respect to a single-ended output, the single-ended differential gain should be used. Assuming $r_o \gg R_C$ in Equation 10.1.6:

$$\text{CMRR}_0 = \left| \frac{A_{ds0}}{A_{cm0}} \right| \cong g_m R_{EE} \qquad (10.1.13)$$

A large CMRR is served by having large g_m and R_{EE}.

A useful relation involving the CMRR is the differential input v_{dcm} that is equivalent to a given v_{cm}. v_{Ocm} due to v_{cm} is $A_{cm0}v_{cm}$. Hence, v_{dcm} is this voltage divided by the differential gain, which is generally A_{dd0}. Thus, $v_{dcm} = A_{cm0}v_{cm}/A_{dd0}$. Using Equation 10.1.13, with A_{dd0} replacing A_{ds0},

$$v_{dcm} = \frac{v_{cm}}{\text{CMRR}_0} \qquad (10.1.14)$$

If the common-mode input resistance of the differential pair is $r_{in(cm)}$, the input resistance of the common-mode half circuit is $2r_{in(cm)}$, because when a common-mode input is applied to B_1 and B_2 joined together, the two half circuits are paralleled. Since $r_{in(cm)}$ is usually very high, r_o and r_μ may have to be taken into account, as shown in the equivalent circuit of Figure 10.1.6a. A useful approximation is to assume that the common-mode gain is low enough so that the collector is essentially at signal ground, as indicated by the dotted ground at the collector. r_μ then appears across the input, R_C is shorted out, and r_o is in parallel with $2R_{EE}$ (Figure 10.1.6b). Transforming the current source to a voltage source, which is then replaced by an equivalent resistance, $r'_{in} = r_\pi + (\beta+1)(2R_{EE}\|r_o)$. Neglecting r_π, $2r_{in(cm)} = r_\mu\|[(\beta+1)(2R_{EE}\|r_o)]$, and,

$$r_{in(cm)} = \left(\frac{r_\mu}{2}\right)\|[(\beta+1)(R_{EE}\|r_o/2)] \qquad (10.1.15)$$

High-Frequency Response

Because of the usually large value of R_{EE}, even small values of output capacitance C_{EE} of the I_{EE} source can have a significant effect on the frequency response of the amplifier. In the common-mode half circuit, C_{EE} appears as $C_{EE}/2$, and the small-signal equivalent circuit becomes as in Figure 10.1.7. Neglecting r_μ and r_o in order to gain better insight into circuit behavior, the gain from the base to collector is simply that given by Equation 10.1.11 with $2R_{EE}$ replaced by $2Z_{EE}$, where $Z_{EE} = (R_{EE}\|1/sC_{EE}) = R_{EE}/(1+sC_{EE}R_{EE})$. Substituting,

$$A_{cm} = \frac{V_{c1}(s)}{V_{cm}(s)} = -\frac{\alpha R_C}{r_e + 2R_{EE}}\frac{1 + sC_{EE}R_{EE}}{1 + (sC_{EE}/2)(r_e\|2R_{EE})} \qquad (10.1.16)$$

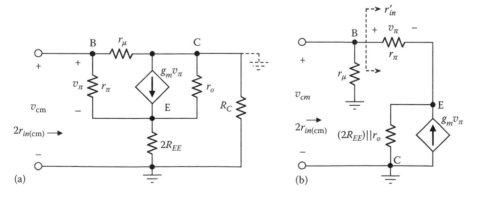

FIGURE 10.1.6
(a) Small-signal equivalent circuit of the common-mode half circuit of the BJT differential pair, including r_o and r_μ; (b) approximate small-signal equivalent circuit, assuming small common-mode gain.

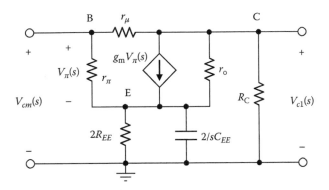

FIGURE 10.1.7
Small-signal equivalent circuit of the common-mode half circuit of the BJT differential pair, taking C_{EE} into account.

At very low frequencies $A_{cm} = A_{cm0}$, whereas at high frequencies, the gain approaches $-\alpha R_C/r_e = -g_m R_C$, as for a CE amplifier without an emitter resistor. The response has a zero at:

$$\omega_{z(cm)} = \frac{1}{C_{EE}R_{EE}} \tag{10.1.17}$$

and a pole at:

$$\omega_{p(cm)} = \frac{2}{C_{EE}(r_e\|2R_{EE})} \cong \frac{2}{C_{EE}r_e} \tag{10.1.18}$$

where $\omega_{p(cm)} \gg \omega_{z(cm)}$ (Figure 10.1.8). At higher frequencies, the effects of C_π and C_μ become significant. The response can then be determined using the methods of Section 9.2, Chapter 9.

Considering that $\omega_{p(cm)}$ is much higher than the frequency range of interest, Equation 10.1.16 can be expressed as $A_{cm} = A_{cm0}(1 + s/\omega_{z(cm)})$. From Equation 10.1.8, $A_{dd} = A_{dd0}/(1 + s/\omega_{pd})$. If we denote A_{dd0}/A_{cm0} by $CMRR_0$, the CMRR as a function of frequency becomes:

$$CMRR = CMRR_0 \frac{1}{(1 + s/\omega_{z(cm)})(1 + s/\omega_{pd})} \tag{10.1.19}$$

The CMRR therefore decreases with frequency over the frequency range of interest so that *common-mode high-frequency signals are not rejected as well as low-frequency signals.*

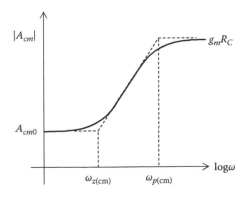

FIGURE 10.1.8
Frequency response of common-mode gain.

Exercise 10.1.3

Assume that in the circuit of Figure 10.1.1, $I_{EE}=0.5$ mA, $V_{CC}=12$ V, $V_{EE}=-12$ V, $R_C=10$ kΩ, and $\beta=100$. Determine: (a) differential gain; (b) differential input resistance; (c) largest common-mode input. Assume $V_T=0.025$ V, $V_{BE}=0.7$ V, $V_{CE(sat)}=0.2$ V, and neglect r_o.
  Answers: (a) 100; (b) 20 kΩ; (c) $\cong 8$ V. ∎

Input Bias Currents

When the transistors are perfectly matched,

$$I_{B1} = I_{B2} = \frac{I_{EE}}{2(\beta_F + 1)} \tag{10.1.20}$$

If the β_F's of Q_1 and Q_2 are unequal so that $\beta_{F1}=\beta_F+\Delta\beta_F/2$ and $\beta_{F2}=\beta_F-\Delta\beta_F/2$, then substituting in Equation 10.1.20, and assuming that $\beta_F \gg 1$ and $\Delta\beta_F/\beta_F$ is small:

$$I_{B1} = \frac{I_{EE}}{2(\beta_F + 1 + \Delta\beta_F/2)} \cong \frac{I_{EE}}{2(\beta_F + 1)}\left(1 - \frac{\Delta\beta_F}{2\beta_F}\right) \tag{10.1.21}$$

and

$$I_{B2} = \frac{I_{EE}}{2(\beta_F + 1 - \Delta\beta_F/2)} \cong \frac{I_{EE}}{2(\beta_F + 1)}\left(1 + \frac{\Delta\beta_F}{2\beta_F}\right) \tag{10.1.22}$$

The **input bias current** I_{ib} is the average of the two input bias currents I_{B1} and I_{B2}. Thus,

$$I_{ib} = \frac{I_{B1} + I_{B2}}{2} \cong \frac{I_{EE}}{2(\beta_F + 1)} \tag{10.1.23}$$

which is the same as Equation 10.1.20 considering β_F to be the average for the two transistors.

The **input offset** current I_{io} is the magnitude of the difference between I_{B1} and I_{B2}. Hence,

$$I_{io} = |I_{B1} - I_{B2}| \cong I_{ib}\left(\frac{\Delta\beta_F}{\beta_F}\right) \tag{10.1.24}$$

Input Offset Voltage

When the inputs are grounded, $V_{C1} - V_{C2} = V_{Od(offset)} \neq 0$ because of dissymmetry. If divided by A_{dd0}, $V_{Od(offset)}$ is referred to the input as an **input offset voltage** V_{io}. Clearly, if $-V_{io}$ is applied between the two inputs, $V_{Od}=0$. As $V_{Od(offset)}$ could be negative or positive, the sign is usually ignored and V_{io} is considered as a positive quantity. It can be readily shown that:

1. If the two R_C's differ by ΔR_C, their average being R_C, V_{io} is (Problem P10.1.18):

$$V_{io} = \left(\frac{\Delta R_C}{R_C}\right)V_T \tag{10.1.25}$$

2. If the BEJ areas of Q_1 and Q_2 are unequal so that the saturation currents differ by ΔI_S, their average being I_S, V_{io} is (Problem P10.1.19):

$$V_{io} \cong \left(\frac{\Delta I_S}{I_S}\right) V_T \qquad (10.1.26)$$

3. If the β_F's of Q_1 and Q_2 are β_{F1} and β_{F2}, respectively, V_{io} is (Problem P10.1.20):

$$V_{io} \cong \left(\frac{1}{\beta_{F1}} - \frac{1}{\beta_{F2}}\right) V_T \qquad (10.1.27)$$

In precision applications, it is the drift in V_{io}, or its variation with temperature, that is of more concern than the magnitude of V_{io} per se (Section 10.6). Assuming that the main variation due to temperature in Equations 10.1.25 through 10.1.27 is due to V_T, it follows that:

$$\frac{dV_{io}}{dT} = \frac{V_{io}}{T} \qquad (10.1.28)$$

The drift in V_{io} will therefore be small, if V_{io} is itself small. When the components V_{io1}, V_{io2}, etc. of V_{io} are uncorrelated, that is, they occur independently of one another, V_{io} is given by:

$$V_{io} = \sqrt{V_{io1}^2 + V_{io2}^2 + \cdots} \qquad (10.1.29)$$

Example 10.1.2 Performance of a BJT Differential Pair

Given the differential amplifier of Figure 10.1.9, with $\beta_F = \beta = 100$, $r_o = 100$ kΩ, and $r_\mu = 2\beta r_o$, it is required to determine: (a) the differential gain and output; (b) the common-mode gain, output, and CMRR$_0$; (c) the common mode gain, output, and CMRR$_0$, assuming a differential output, with a mismatch of 2% in R_C's; (d) the input resistance seen by the source, taking r_μ into account; (e) the common-mode input resistance; (f) the input offset voltage, assuming a mismatch of 2% in R_C and a mismatch of 5% in the emitter–base junction areas

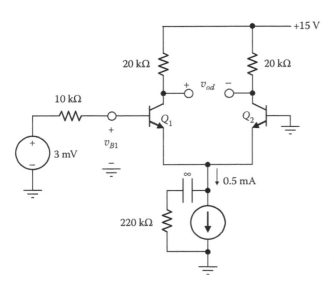

FIGURE 10.1.9
Figure for Example 10.1.2.

of the two transistors; (g) the input bias and offset currents, assuming a mismatch in β of 4%; (h) f_{dp}, f_{zcm}, and ω_{pcm}, assuming $C_\mu = 1.5$ pF and $C_\pi = 12$ pF, and $C_{EE} = 20$ pF.

SOLUTION

$g_m = I_C/V_T = (0.5/2)/0.026 = 9.6$ mA/V, $r_\pi = 100/9.6 \cong 10$ kΩ, $r_e \cong 100$ Ω, and $r_\mu = 2 \times 100 \times 100 \cong 20$ MΩ. The resistance looking into Q_2 emitter is the input resistance of a CB amplifier and is very nearly r_e, or 100 Ω (Equation 8.2.18, Chapter 8). The effective resistance between Q_1 emitter and ground is therefore $r_e\|R_{EE} \cong r_e$. The input resistance at the base of Q_1 is that of a CE amplifier with an emitter resistance r_e. From Equation 8.2.37, Chapter 8, this is equal to $r_\pi + (\beta+1)r_e = 2r_\pi = 20$ kΩ, if r_o, R_C, and r_μ, are neglected. If r_o and R_C, are taken into account (Equation 8.2.37), the input resistance becomes 18.4 kΩ. Since the source sees this resistance, it follows that $v_{B1} = \dfrac{18.4}{10 + 18.4} \times 3 = 1.94 \cong 2$ mV. As explained in Example 10.1.1, this is equivalent to a differential input of 2 mV and a common-mode input of 1 mV.

(a) From Equation 10.1.5, $A_{dd0} = -9.6(20\|100) = 160$. The differential output $v_{c1} - v_{c2}$ is -320 mV. The common-mode output cancels out, assuming that the two halves are symmetrical.

(b) From Equation 10.1.11, the common-mode gain $A_{cm0} = v_{c1}/v_{cm} = -\alpha R_C/(2R_{EE} + r_e) \cong -0.045$. The common-mode output is 0.045×1 mV $= -45$ μV; $A_{ds0} = A_{dd0}/2 = -80$; $CMRR_0 = 80/0.045 \cong 1800$.

(c) From Equation 10.1.12, the common-mode differential output is $\alpha\Delta R_C/(2R_{EE} + r_e)v_{cm} = 0.9$ μV, the common-mode gain being approximately 9×10^{-4}; $CMRR_0 = \dfrac{160}{9 \times 10^{-4}} = 1.78 \times 10^5 \equiv 105$ dB.

(d) $v_{B1} \cong 2$ mV, as determined previously, and $v_{C1} \cong -160$ mV. r_μ is reduced by the Miller effect by a factor of $1/(1 + 160/2) = 1/81$, which gives a resistance of 20 MΩ/81 $\equiv$ 247 kΩ in parallel with the 18.4 kΩ determined above, or 17 kΩ. This makes the input voltage about 1.9 mV, which may still be taken as 2 mV. It should be noted that taking $r_\mu = 2\beta r_o$ is generally on the low side.

(e) From Equation 10.1.15, $r_{in(cm)} = 10\|[101 \times (0.22\|0.050] = 2.9$ MΩ.

(f) V_{io} due to mismatch in R_C is $(\Delta R_C/R_C)V_T = 0.02 \times 26 = 0.5$ mV (Problem P10.1.18). V_{io} due to mismatch in emitter–base junctions is $(\Delta A/A)V_T = 0.05 \times 26 \cong 1.3$ mV (Problem P10.1.19). Since the two components V_{io} are uncorrelated, the total V_{io} is $\sqrt{(0.5)^2 + (1.3)^2} = 1.4$ mV.

(g) From Equations 10.1.23 and 10.1.24, $I_{ib} = 0.5/(2 \times 101) \cong 2.5$ μA and $I_{io} = 2.5 \times 0.04 = 0.1$ μA.

(h) $R'_{src} = (R_{src}/2)\|r_\pi = 5\|10 = 3.33$ kΩ and $R'_C = R_C\|r_o = 10\|100 = 9.1$ kΩ. From Equation 10.1.9, neglecting the term $C_\mu R'_C$, $f_{pd} = 1/2\pi C_T R'_{src} = 1/2\pi[C_\pi + (1 + g_m R'_C)C_\mu]R'_{src} = 1/2\pi[12 + (1 + 9.6 \times 9.1)1.5]3.33 \cong 330$ kHz. From Equation 10.1.17, $f_{z(cm)} = 1/2\pi C_{EE}R_{EE} = 1/2\pi \times 20 \times 220 \cong 36$ kHz. From Equation 10.1.18, $f_{p(cm)} = 2/2\pi C_{EE}(r_e\|2R_{EE}) \cong 1/\pi \times 20 \times 100 \equiv 160$ MHz.

Current-Mirror Load

Concept: *A current mirror load doubles the single-ended voltage gain of a differential pair, and significantly increases common-mode rejection.*

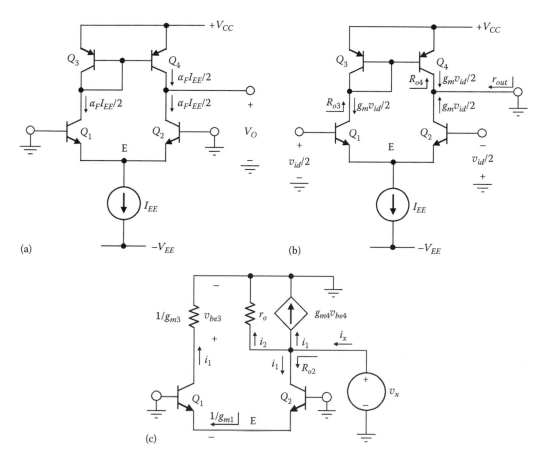

FIGURE 10.1.10
Differential pair with current-mirror load. (a) inputs grounded; (b) differential input applied; (c) test source applied.

The current mirror load therefore has some important advantages. In order to highlight these important features a somewhat simplified analysis will be made.

With the bases of Q_1 and Q_2 grounded in Figure 10.1.10a, I_{EE} divides equally between Q_1 and Q_2, because, ideally, the collector current of Q_4 mirrors that of Q_1. With perfect matching, the collector voltages of Q_3 and Q_4 are equal so that on open circuit, $V_O = V_{CC} - V_{EB3}$.

When a small-signal differential input is applied (Figure 10.1.10b), the resistance R_{o3} seen by Q_1 collector is nearly $1/g_{m3} = r_{e3}$ (Equation 10.1.30). This node is therefore almost a signal ground, so only a single-ended output from Q_2/Q_4 collector is feasible. Note that because of the differences between R_{o3} and R_{o4}, the circuit is no longer symmetrical. We can, however, assume that if the output is short-circuited to ground, the circuit will be nearly symmetrical because the voltage at Q_1/Q_3 collectors is small, and the voltage at Q_2/Q_4 collectors is zero. The common emitters will then be at nearly signal ground, as are the emitters of the current mirror. Under these conditions, the currents flowing in the r_o's of the transistors can be neglected. Assuming that the parameters of all transistors are the same, the small-signal currents i_{c1} and i_{c2} will have a magnitude of $g_m v_{id}/2$ and direction as shown in Figure 10.1.10b. Assuming large β's, i_{c4} mirrors i_{c1} and adds to i_{c2} resulting in a short circuit current of $g_m v_{id}$. The effective transconductance G_m for the differential input v_{id} is therefore g_m.

Knowing G_m, then in order to calculate v_O for a given load we need r_{out}, which is the resistance looking into the output node with the inputs grounded. The resistance looking into Q_1 emitter is nearly $1/g_{m1}$. Hence, both the common emitter node of Q_1 and Q_2, and the common base node of Q_3 and Q_4, can be considered a signal ground. This gives $r_{out} = r_{o2} \| r_{o4}$, to a first approximation. Nevertheless, it is instructive to determine r_{out} by applying a test source v_x, with both bases grounded. Figure 10.1.10c shows the ac circuit, with Q_3 replaced by its equivalent resistance R_{o3}. By inspection in Figure 10.1.10a, R_{o3} consists of the following parallel components: (1) r_π's of Q_3 and Q_4, (2) r_o of Q_3, and (3) controlled current source $g_{m3}v_{eb3}$ of Q_3, equivalent to a resistance $1/g_{m3}$ (Equation 8.2.65, Chapter 8). Hence,

$$R_{o3} = r_{\pi3} \left\| r_{\pi4} \right\| \left(\frac{1}{g_{m3}} \right) \right\| r_{o3} \cong \frac{1}{g_{m3}} \tag{10.1.30}$$

We next determine R_{o2} in Figure 10.1.10c. Q_1 is a CB transistor having a load resistance $1/g_{m3}$. Its input resistance is very nearly $1/g_{m1}$ (Equation 8.2.57, Chapter 8). Q_2 is a CB transistor whose source resistance is $1/g_{m1}$ in parallel with a much larger R_{EE} of I_{EE}. From Equation 8.2.61, Chapter 8, with $1/g_{m1} = r_{e1} \ll r_{\pi2}$ and $g_{m2} = g_{m1}$,

$$R_{o2} \cong r_{o2} + g_{m2}r_{o2} \left(\frac{1}{g_{m1}} \right) = 2r_{o2} \tag{10.1.31}$$

In Figure 10.1.10c, $i_{c3} = i_1 = g_{m3}v_{be3} = g_{m4}v_{be4}$, since $v_{be3} = v_{be4}$ and g_{m3} is assumed equal to g_{m4}. It follows that $i_x = 2i_1 + v_x/r_{o4}$. Substituting $i_1 = v_x/2r_{o2}$ (Equation 10.1.31), we obtain:

$$r_{out} = r_{o2} \| r_{o4} = \frac{r_o}{2} \tag{10.1.32}$$

Since $v_o = g_m v_{id} r_{out}$, the voltage gain is:

$$\frac{v_o}{v_{id}} = g_m r_{out} = \frac{g_m r_o}{2} \tag{10.1.33}$$

If Q_3 and Q_4 were separate transistor loads, $v_o = g_m(v_{id}/2)(r_o/2) = (g_m r_o/4)v_{id}$. Thus, the coupling between the two currents of a current mirror load doubles the single-ended voltage gain.

This coupling has a more dramatic effect on the common-mode response. In Figure 10.1.11, a positive v_{cm} increases the collector currents of Q_1 and Q_2 by $i_{cm1} = g_m v_{cm} = i_{cm2}$, neglecting r_o. For an ideal mirror, $i_{cm4} = i_{cm1} = i_{cm2}$. No current flows into any load at the output, which means that the common-mode output voltage is zero. The CMRR is thus infinite, although R_{EE} is finite. In practice, the mirror is not ideal, so that the CMRR is finite but large.

For a more realistic estimate of the CMRR, we first consider the gain i_{cm4}/i_{cm1} of the current mirror. In Figure 10.1.11 $v_{eb3} = R_{o3}i_{cm1} = v_{eb4}$, with $i_{cm4} = g_{m4}v_{eb4}$, neglecting r_o. From Equation 10.1.30, with r_o neglected, $R_{o3} = r_{\pi3} \| r_{\pi4} \| (1/g_{m3})$. It follows that:

$$\frac{i_{cm4}}{i_{cm1}} = g_{m4}R_{o3} = \frac{\beta}{\beta+2}, \quad \text{for } g_{m3} = g_{m4} \quad \text{and} \quad r_{\pi3} = r_{\pi4} \tag{10.1.34}$$

(as in Equation 8.1.8, Chapter 8). v_E essentially follows v_{cm} so that the small-signal current in R_{EE} is v_{cm}/R_{EE}. Because the emitter voltage follows v_{cm}, Q_1 and Q_2 effectively behave as

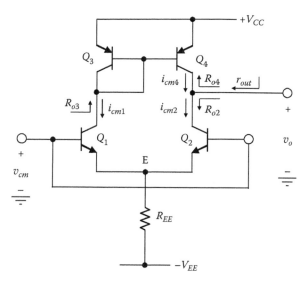

FIGURE 10.1.11
Common-mode input applied to a differential pair with current-mirror load.

CE amplifiers, each with a source resistance $2R_{EE}$. The output resistance of Q_1 and Q_2 (Equation 8.2.38, Chapter 8) is much larger than the resistances presented by Q_3 and Q_4, which implies that the small-signal current in R_{EE} almost divides equally between Q_1 and Q_2. Hence,

$$i_{cm1} \cong i_{cm2} = \frac{v_{cm}}{2R_{EE}} \tag{10.1.35}$$

The effective resistance at the output is r_{o4}, and the common-mode output voltage is:

$$v_{ocm} = (i_{cm4} - i_{cm2})r_{o4} \tag{10.1.36}$$

Substituting from Equations 10.1.34 and 10.1.35 in Equation 10.1.36, with $r_{o4} = r_o$:

$$A_{cm} = \frac{v_{ocm}}{v_{cm}} = \frac{r_o}{2R_{EE}}\left(\frac{\beta}{\beta+2} - 1\right) \cong -\frac{r_o}{\beta R_{EE}} \tag{10.1.37}$$

From Equations 10.1.13, 10.1.33, and 10.1.37

$$\text{CMRR}_0 = \frac{\beta g_m R_{EE}}{2} \tag{10.1.38}$$

Compared to Equation 10.1.13, CMRR_0 is multiplied by $\beta/2$.

The BJT differential pair having a current-mirror load demonstrates that an input offset voltage can result from inherent circuit features, even though the components in the two half circuits are perfectly matched. This type of input offset voltage is a **systematic input offset voltage**, in contrast to the **random input offset** voltage due to a mismatch between corresponding components of the two half circuits. Thus, if both base terminals in Figure 10.1.11 are grounded, and assuming $I_{C1} = I_{C2}$, there will be a current difference

$\Delta I_C = I_{C4} - I_{C2} = \left(\dfrac{\beta_F}{\beta_F + 2} - 1\right)I_{C2} \cong -\dfrac{2}{\beta_F}I_{C2}$ (Equation 8.1.8, Chapter 8). ΔI_C can flow through an externally connected load so that $V_O \neq 0$. To make $V_O = 0$, we should apply a voltage $g_m\Delta I_C = -(2g_m/\beta_F)I_{C2}$ to Q_2 base so as to reduce I_{C2} to make $\Delta I_C = 0$. The magnitude of this voltage is V_{io}. Substituting, $g_m = V_T/I_{C2}$,

$$V_{io} = \frac{2V_T}{\beta_F} \tag{10.1.39}$$

Systematic V_{io} is thus due to finite β_F. If $\beta_F = 100$, V_{io} is about 0.5 mV.

More elaborate current mirrors (Section ST8.3, Chapter 8) can be used to increase the load resistance of the differential pair, or give a very large R_{EE}. Other improvements to the performance of the differential pair are discussed in Section ST10.1.

Simulation Example 10.1.3 Differential Amplifier with Resistors Added to Each Emitter

The circuit is that of the differential amplifier of Figure 10.1.3, with $V_{CC} = 5$ V, $I_{EE} = 0.25$ mA, $R_C = 20$ kΩ, and R_E included in series with each emitter. The transistors are Q2N3904 from PSpice EVAL library. A dc source is connected differentially between the bases of the transistors and is swept between -250 and $+250$ mV. The output is taken differentially as $v_{C2} - v_{C1}$. The transfer characteristics for $R_E = 0$, 0.5 and 1 kΩ are shown in Figure 10.1.12.

It is seen that R_E extends the linear range of operation because of the negative feedback. Thus, if v_{id} is such that Q_2 is cut off, so its emitter current is zero, then v_{id} must increase by $R_E I_{EE}$, which makes the characteristic more linear by shifting it in the direction of positive v_{id} by $R_E I_{EE}$. However, this shift also reduces the slope of the characteristic and hence the gain.

The differential half circuit of Figure 10.1.3b applies with R_E included and becomes a CE amplifier with unbypassed emitter resistor (Section 8.2, Chapter 8). The differential amplifier with emitter/source resistors is left to problems at the end of the chapter.

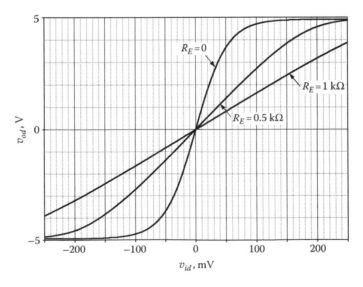

FIGURE 10.1.12
Basic MOSFET differential pair.

MOSFET Differential Pair

A MOSFET differential pair is illustrated in Figure 10.1.13, where it is assumed that Q_1 and Q_2 are perfectly matched and in saturation. In the quiescent state, with both gates grounded, I_{SS} divides equally between Q_1 and Q_2. If a common-mode voltage v_{cm} is applied, the two drain currents must be equal and sum to I_{SS}. Hence, each drain current must remain equal to $I_{SS}/2$. If the drain currents do not change with v_{cm}, then neglecting channel-length modulation, the gate-source voltages must not change so that the common source voltage v_S must follow v_{cm}. If a small-signal differential input is applied, $v_{g1} = v_{id}/2$ and $v_{g2} = -v_{id}/2$. Neglecting r_o, $i_{D1} = (I_{SS}/2) + (g_m v_{id}/2)$ and $i_{D2} = (I_{SS}/2) - (g_m v_{id}/2)$.

To derive the transfer characteristic, the drain currents are expressed as:

$$i_{D1} = \frac{k_n}{2}(v_{GS1} - V_{tn})^2 \quad \text{and} \quad i_{d2} = \frac{k_n}{2}(v_{GS2} - V_{tn})^2 \tag{10.1.40}$$

or,

$$\sqrt{i_{D1}} = \sqrt{k_n/2}(v_{GS1} - V_{tn}) \quad \text{and} \quad \sqrt{i_{D2}} = \sqrt{k_n/2}(v_{GS2} - V_{tn}) \tag{10.1.41}$$

Since $v_{GS1} - v_{GS2} = v_{id}$, Equations 10.1.41 give:

$$\sqrt{i_{D1}} - \sqrt{i_{D2}} = \left(\sqrt{k_n/2}\right)v_d \tag{10.1.42}$$

If we denote $v_{GS1} = v_{GS2} = V_{GS}$ when $v_{id} = 0$, then $i_{D1} = i_{D2} = I_{SS}/2$, and,

$$\frac{I_{SS}}{2} = \frac{k_n}{2}(V_{GS} - V_{tn})^2 = \frac{1}{2}k_n V_{ov}^2 \tag{10.1.43}$$

Equation 10.1.42 can be solved with $i_{D1} + i_{D2} = I_{SS}$ for i_{D1} and i_{D2}. $k_n/2$ is then eliminated using Equation 10.1.43 to give:

$$i_{D1} = \frac{I_{SS}}{2}\left[1 + \frac{v_{id}}{V_{ov}}\sqrt{1 - \left(\frac{v_{id}}{2V_{ov}}\right)^2}\right] \quad \text{and} \quad i_{D2} = \frac{I_{SS}}{2}\left[1 - \frac{v_{id}}{V_{ov}}\sqrt{1 - \left(\frac{v_{id}}{2V_{ov}}\right)^2}\right] \tag{10.1.44}$$

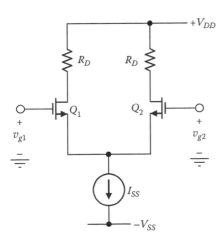

FIGURE 10.1.13
Transfer characteristic of BJT differential pair.

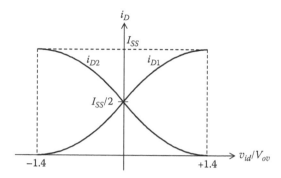

FIGURE 10.1.14
Variation of drain currents with differential input for an MOSFET differential pair.

A plot of i_{D1} and i_{D2} vs. v_{id}/V_{ov} is illustrated in Figure 10.1.14. For small v_{id}, the square root terms in Equation 10.1.44 are very nearly unity so that i_{D1} and i_{D2} vary linearly with v_{id}, the magnitude of the slope being $I_{SS}/2V_{ov}$. To change i_{D1} or i_{D2} between zero and I_{SS}, the term $(v_{id}/V_{ov})\sqrt{1-(v_{id}/V_{ov})^2}$ must change between -1 and $+1$. The term equals 1 at $v_{id}=|v_{id}|_{max}$ given by:

$$|v_{id}|_{max}= \sqrt{2}V_{ov} = \sqrt{2}\frac{I_{SS}}{g_m} \qquad (10.1.45)$$

where $g_m = 2I_D/V_{ov} = I_{SS}/V_{ov}$ (Equation 5.3.11, Chapter 5). Whereas full switching of a BJT depends only on V_T, full switching of a MOSFET depends on I_{SS} and g_m.

The small-signal analysis of the MOSFET differential pair closely follows that of the BJT differential pair. For a differential input v_{id}, the input voltage to each transistor is $v_{id}/2$, the change in drain current is $g_m v_{id}/2$, assuming a large r_o, and the drain voltage v_{d1} is:

$$v_{d1} = \frac{-g_m R_D v_{id}}{2} \qquad (10.1.46)$$

$v_{d2}=-v_{d1}$ and $v_{od}=v_{d1}-v_{d2}=2v_{d1}$. Note that v_S varies with the common-mode signal but not with the differential signal, since the common source is ideally at signal ground for the differential input. Hence, the body effect is manifest for common-mode, but not for differential signals. The frequency response of the MOS differential pair closely parallels that of the BJT and is left to problems at the end of the chapter.

Exercise 10.1.4
Show that the magnitude of the slope di_{D1}/dv_{id} of the transfer characteristic at $v_{id}=0$ is $g_m/2$. ∎

As in the case of the BJT, an input offset voltage V_{io} arises from a mismatch between the two halves of the circuit. However, V_{io} is considerably larger for the MOSFET case, compared to its BJT counterpart, in the ratio of $V_{ov}/2$ to V_T. It can be readily shown that:

1. If the two R_D resistors differ by ΔR_D, the input offset voltage V_{io} is (Problem P10.1.21):

$$V_{io} = \left(\frac{\Delta R_D}{R_D}\right)\frac{V_{ov}}{2} \qquad (10.1.47)$$

2. If W/L of the two transistors differ by $\Delta(W/L)$, W/L being the average aspect ratio, the input offset voltage V_{io}, neglecting r_o is (Problem P10.1.22):

$$V_{io} \cong \left(\frac{\Delta(W/L)}{W/L}\right)\frac{V_{ov}}{2} \tag{10.1.48}$$

3. If the thresholds of Q_1 and Q_2 differ by $\Delta V_{tn} \ll V_{ov}$, V_{io} is (Problem P10.1.23):

$$V_{io} \cong \Delta V_{tn} \tag{10.1.49}$$

The MOSFET differential pair with a current-mirror load is analogous to its BJT counterpart. Assuming all MOSFETs have the same parameters (Section ST10.2):

$$\text{CMRR}_0 = g_m^2 r_o R_{SS} \tag{10.1.50}$$

Note that from Equations 8.2.11 and 8.2.15, Chapter 8, the maximum gain $g_m r_o$ is considerably higher for a BJT differential pair compared to a MOSFET differential pair.

Simulation Example 10.1.4 MOSFET Differential Amplifier with Current Mirror Load

Given the MOSFET differential amplifier with a current-mirror load (Figure 10.1.15), $V_{tn} = \overline{V}_{tp} = 0.7$ V, $k_n = k_p = 1$ mA/V^2, and $\lambda = 0.01$ V^{-1} for all transistors. Neglecting the body effect, it is required to simulate the circuit and determine: (a) r_{out}, (b) v_o/v_{id}, (c) A_{cm} and CMRR, assuming $R_{SS} = 200$ kΩ, and (d) $v_{cm(max)}$.

ANALYSIS AND SIMULATION

Q_1 to Q_4 have a drain current of $I_{SS}/2 = 0.5$ mA and $r_o = 1/\lambda I_D = 200$ kΩ. The schematic is that of Figure 10.1.15; the MOSFETs are from the BREAKOUT library, and their parameters are entered in the PSpice model and the Property Editor spreadsheet.

(a) From Equation 10.1.32, $r_{out} = r_{o2} \| r_{o4} = 100$ kΩ.

(b) Neglecting channel-length modulation, $g_m = \sqrt{2k_{n,p}I_D} = \sqrt{2 \times 1 \times 0.5} = 1$ mA/V, $v_o/v_{id} = g_{mn}r_{out}/2 = 100$. The simulation gives 105.6, taking channel-length modulation into account.

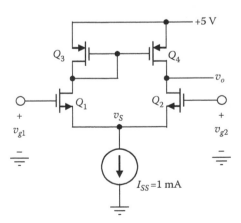

FIGURE 10.1.15
Figure for Simulation Example 10.1.4.

(c) From Equation 10.1.33, the differential gain is $g_m r_o/2$. Substituting in Equation 10.1.50, $|A_{cm}| = 1/(2g_m R_{SS}) = 1/(2 \times 200) = 0.025$; CMRR $= 100/0.025 = 4000$. The measured A_{cm} is 0.0245, so that the CMRR $= 105.6/0.0245 = 4310$.

(d) $v_{cm(max)}$ is determined by Q_1 or Q_2 going out of saturation. To decide which, we note that for Q_1 and Q_3, $V_{DD} = v_{SG3} + v_{DG1} + v_{cm}$. At the edge of the saturation region, $v_{DG1} = -V_{tn}$. This gives $v_{cm(max)} = V_{DD} - v_{SG3} + V_{tn}$. Considering Q_2 and Q_4, $V_{DD} = v_{SD4(sat)} + v_{DG2} + v_{cm}$, where $v_{SD4(sat)}$ is the source-drain voltage of Q_4 at the edge of saturation and equals V_{ov4}. This gives $v_{cm(max)} = V_{DD} - V_{ov4} + V_{tn}$. Since $V_{ov4} < v_{SG3}$, it follows that Q_1 will go out of saturation before Q_2. For Q_3, $v_{SG3} = v_{SG3} = \sqrt{2i_D/k_n} + \overline{V}_{tp} = \sqrt{2 \times 0.5/1} + 0.7 = 1.7$ V. This gives $v_{cm(max)} = 5 - 1.7 + 0.7 = 4$ V.

10.2 Two-Stage CMOS Op Amp

We will illustrate in this section and the next how differential pairs, active loads, and other circuits previously discussed can be used in two examples of basic CMOS op amps.

Figure 10.2.1 illustrates a basic two-stage CMOS op amp consisting of a PMOS differential input stage (Q_1 and Q_2) having an NMOS current mirror (Q_3 and Q_4), followed by an output stage. Q_5, Q_6, and Q_7 are current-steering circuits (Figure 8.1.10, Chapter 8), whereby Q_6 biases the differential pair, and Q_7 acts as a current-source load for the NMOS CS output transistor Q_8. I_{REF} can be of a form described in Section 10.4. C_C is used for frequency compensation. A complementary configuration is also possible as discussed at the end of this section.

The output resistance, $r_{o7} \| r_{o8}$ is relatively high, and is therefore appropriate for high-impedance loads only, such as capacitive loads in CMOS ICs. A CD output stage of relatively low output resistance can be added for increased drive capability.

Input-Offset Voltage

To have $V_{io} = 0$, the output voltage should be zero when both inputs are grounded. Apart from having matched transistors, certain relations between transistor sizes must be satisfied.

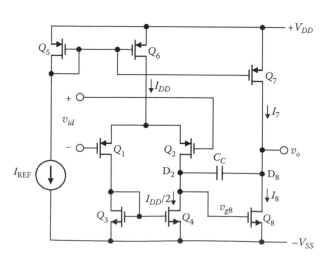

FIGURE 10.2.1
Basic two-stage CMOS op amp.

When the gates of Q_1 and Q_2 are grounded, the drain currents of Q_3 and Q_4 are $I_{DD}/2$. Q_3, Q_4, and Q_8 have the same k'_n and V_{tn}. It follows that $V_{DS4} = V_{DS3} = V_{GS3,4}$, the common gate-source voltages of Q_3 and Q_4. Hence, $V_{GS8} = V_{GS3,4}$, so that the drain current I_8 of Q_8 is related to the drain current $I_{DD}/2$ of Q_4 by their aspect ratios, neglecting channel-width modulation:

$$\frac{I_8}{I_{DD}/2} = \frac{(W/L)_8}{(W/L)_4} \tag{10.2.1}$$

Q_5, Q_6, and Q_7 also have the same k'_p, V_{tp}, and V_{SG}. The drain currents of Q_6 and Q_7 are therefore related by their aspect ratios, neglecting channel-width modulation:

$$\frac{I_7}{I_{DD}} = \frac{(W/L)_7}{(W/L)_6} \tag{10.2.2}$$

On open circuit, $I_7 = I_8$. Dividing Equation 10.2.1 by Equation 10.2.2 and rearranging,

$$\frac{(W/L)_8}{(W/L)_4} = 2\frac{(W/L)_7}{(W/L)_6} \tag{10.2.3}$$

In order to have zero output voltage, assuming $V_{DD} = V_{SS}$, then $V_{SD7} = V_{DS8}$. Equating the drain currents of Q_7 and Q_8,

$$\frac{1}{2}k'_n\left(\frac{W}{L}\right)_8 (V_{ovn})^2(1 + \lambda_{n8}V_{DS8}) = \frac{1}{2}k'_p\left(\frac{W}{L}\right)_7 (V_{ovp})^2(1 + \lambda_{p7}V_{SD7}) \tag{10.2.4}$$

where
$V_{ovn} = V_{GSn} - V_{tn}$
$V_{ovp} = V_{SGp} - \overline{V}_{tp}$

Equation 10.2.4 can be satisfied by having $\lambda_{n7} = \lambda_{p8}$, and $\mu_e(W/L)_8(V_{ovn})^2 = \mu_h(W/L)_7(V_{ovp})^2$. Combining with Equation 10.2.3, the condition $V_{io} = 0$ is:

$$2\mu_e\left(\frac{W}{L}\right)_4 (V_{ovn})^2 = \mu_h\left(\frac{W}{L}\right)_6 (V_{ovp})^2 \quad \text{and} \quad \lambda_{n7} = \lambda_{p8} \tag{10.2.5}$$

Voltage Gain and Output Swing

Analogous to Equation 10.1.33, the low-frequency voltage gain from the differential input V_{id} to the single-ended output of the differential stage is:

$$\frac{v_{g8}}{v_{id}} = -g_{mp}(r_{o2}\|r_{o4}) \tag{10.2.6}$$

where $g_{mp} = 2I_D/V_{ov1,2}$ (Equation 5.3.11, Chapter 5) and $V_{ov1,2}$ is the overdrive voltage od Q_1 and Q_2, assumed to have equal thresholds. Substituting $r_o = V_A/I_D$,

$$\frac{v_{g8}}{v_{id}} = -\frac{v_{A2,4}}{v_{ov1,2}} \tag{10.2.7}$$

where $V_{A2,4}$ is the Early voltage of Q_2 and Q_4, assumed to be equal.

The gain of the output stage is:

$$\frac{v_o}{v_{g8}} = -g_{mn}(r_{o7}\|r_{o8}) \tag{10.2.8}$$

where g_{mn} is g_m of Q_8. Analogous to Equation 10.2.7, Equation 10.2.8 may be written as:

$$\frac{v_o}{v_{g8}} = -\frac{v_{A7,8}}{v_{ov8}} \tag{10.2.9}$$

where V_{A7} and V_{A8} are assumed to be equal, in accordance with Equation 10.2.5. It follows that the overall gain is the product of the gains of the two stages:

$$A_{v0} = g_{mp}g_{mn}(r_{o2}\|r_{o4})(r_{o7}\|r_{o8}) = \frac{V_{A2,4}}{V_{ov1,2}} \frac{V_{A7,8}}{V_{ov8}} \tag{10.2.10}$$

To have a large A_{v0}, V_{ov}'s should be small and V_A's large. V_A is increased by increasing L, and V_{ov} is reduced by increasing W/L (Equations 5.3.7 and 5.3.15, Chapter 5). Typically, A_{v0} is in the range of 1000 to about 5000.

Q_8 is at the edge of the saturation region when $v_{DS8} = v_{GS8} - V_{tn8} = V_{ov8}$, or $v_o^- = -V_{SS} + V_{ov8}$. Similarly, Q_7 is at the edge of the saturation region when $v_{SD7} = v_{SG7} - \overline{V}_{tp7} = V_{ov7}$, or $v_O^+ = V_{DD} - V_{ov7}$. Thus,

$$-V_{SS} + V_{ov8} \leq v_O \leq V_{DD} - V_{ov7} \tag{10.2.11}$$

The output can swing to the magnitude of each supply voltage less V_{ov} of an output transistor. Beyond these limits, one or both transistors will enter the triode region, which will significantly reduce the overall gain.

Common-Mode Response

From Equation 10.1.50, with $R_{SS} = r_o$, the CMRR is approximately $(g_m r_o)^2$.

The maximum positive swing, v_{cm}^+, of v_{cm} is that which brings Q_6 to the edge of the saturation region, that is, when $v_{DS6} = V_{ov6}$. Under these conditions, $v_{cm}^+ + V_{SG1,2} + V_{ov6} = V_{DD}$, or $v_{cm}^+ = V_{DD} - V_{ov1} - \overline{V}_{tp1} - V_{ov6}$. The most negative swing, v_{cm}^-, of v_{cm} is that which brings Q_1 to the edge of the saturation region, as explained in Simulation Example 10.1.4, which happens when $v_{GD1} = \overline{V}_{tp1}$, that is, when $v_{cm}^- + \overline{V}_{tp1} - V_{GS3} = -V_{SS}$, or $v_{cm}^- = V_{ov3} + V_{tn3} - \overline{V}_{tp1} - V_{SS}$. Thus,

$$V_{ov3} + V_{tn3} - \overline{V}_{tp1} - V_{SS} \leq v_{cm} \leq V_{DD} - V_{ov1} - \overline{V}_{tp1} - V_{ov6} \tag{10.2.12}$$

The V_{ov}'s reduce the swing of both v_O and v_{cm}. The swing in v_O and the swing in v_{cm} should ideally be equal because in a unity-gain op amp the output is connected to the inverting input, which under dc conditions makes $V_{cm} = V_O$.

Frequency Response

The small-signal equivalent circuit is shown in Figure 10.2.2, where $g_{mi} = g_{m1} = g_{m2}$, $g_{mo} = g_{m8}$, $R_1 = r_{o2}\|r_{o4}$, $R_2 = r_{o7}\|r_{o8}$, C_1 is the total capacitance to ground at node D_2, and C_2 is the total capacitance to ground at node D_8 (Figure 10.2.1):

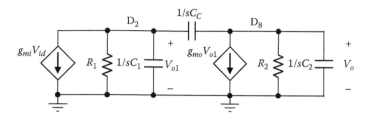

FIGURE 10.2.2
High-frequency, small-signal equivalent circuit of the two-stage CMOS op amp.

$$C_1 = C_{gd4} + C_{db4} + C_{gd2} + C_{db2} + C_{gs8} \tag{10.2.13}$$
$$C_2 = C_{db8} + C_{gd7} + C_{db7} + C_L \tag{10.2.14}$$

where C_L is the load capacitance. The following should be noted concerning these capacitances: (1) The substrates of the NMOS and PMOS transistors are normally connected to the $-V_{SS}$ and $+V_{DD}$ supplies, respectively, which are at ac ground. The well bodies are connected to the sources, which are also at ac ground. Hence, drain-body capacitances are with respect to ground. (2) The gates of the diode-connected transistors Q_3 and Q_5 have a low resitance of $1/g_m$ between the gates and the dc supplies. Hence these gates may be considered grounded so that the drain-gate capacitances of Q_4 and Q_7 are with respect to ground. The gate of Q_2 is assumed to be grounded through the applied source. (3) C_{gd} of Q_8 is in parallel with C_C and is normally small compared to C_C. (4) C_2 includes the load capacitance, which is normally larger than the internal capacitances of the transistors so that C_2 is considerably larger than C_1.

The node-voltage equations for the circuit of Figure 10.2.2 are:

$$\left(\frac{1}{R_1} + sC_1 + sC_C\right)V_{o1} - sC_C V_o = -g_{mi}V_{id}$$
$$-sC_C V_{o1} + \left(\frac{1}{R_2} + sC_2 + sC_c\right)V_o = -g_{mo}V_{o1} \tag{10.2.15}$$

The transfer function can be derived from these equations as:

$$\frac{V_o}{V_{id}} = \frac{g_{mi}(g_{mo} - sC_C)R_1 R_2}{1 + s[C_1 R_1 + C_2 R_2 + C_C(g_{mo}R_1 R_2 + R_1 + R_2)] + s^2[C_1 C_2 + C_C(C_1 + C_2)]R_1 R_2} \tag{10.2.16}$$

The transfer function has a zero and two poles. The zero occurs when $g_{mo} = sC_C$. The interpretation is that when $V_o = 0$, R_2 and C_2 can be replaced by a short circuit through which the current is zero. But the current through the short circuit can be considered as the sum of two components that flow in opposite directions: one component being equal to $g_{mo}V_{o1}$, the other equal to $sC_C V_{o1}$. Equating these currents gives $g_{mo} = sC_C$. Note that the zero does not occur at a real frequency, since $g_{mo}V_{o1}$ has zero phase angle whereas $j\omega C_C V_{o1}$ has a phase angle of 90°.

The denominator $D(s)$ of the transfer function can be expressed as:

$$D(s) = \left(1 + \frac{s}{\omega_{p1}}\right)\left(1 + \frac{s}{\omega_{p2}}\right) = 1 + s\left(\frac{1}{\omega_{p1}} + \frac{1}{\omega_{p2}}\right) + \frac{s^2}{\omega_{p1}\omega_{p2}} \tag{10.2.17}$$

If the amplifier has adequate stability, one pole, say ω_{p1}, is dominant ($\omega_{p1} \ll \omega_{p2}$) so that ω_{p1} is approximately the reciprocal of the coefficient of s in $D(s)$. Moreover, $g_{mo}R_1R_2$ is much larger than either R_1 and R_2, and $g_{mo}C_CR_1R_2$ is normally much larger than C_1R_1 and C_2R_2. Hence,

$$\omega_{p1} \cong \frac{1}{(g_{mo}R_2C_C)R_1} \tag{10.2.18}$$

The interpretation of Equation 10.2.18 is that if C_2R_2 is negligible, C_2 can be ignored. The gain of the second stage, from V_{o1} to V_o is $-g_{mo}R_2$. The capacitance appearing in parallel with R_1 due to the Miller effect is $C_C(1 + g_{mo}R_2) \cong C_Cg_{mo}R_2$, which is assumed to be much larger than C_1.

The second pole is $\omega_{p2} = 1/(\omega_{p1} \times$ coefficient of s^2 in the transfer function). It follows that:

$$\omega_{p2} \cong \frac{g_{mo}C_C}{C_1C_2 + C_C(C_1 + C_2)} \tag{10.2.19}$$

C_C causes **pole splitting** by decreasing ω_{p1} and increasing ω_{p2}. The increase in ω_{p2} depends on the relative magnitudes of the terms in the denominator. If $C_2 \gg C_1$, the second term becomes C_CC_2, and if $C_C \gg C_1$, the denominator reduces to C_CC_2. C_C then cancels out, giving:

$$\omega_{p2} \cong \frac{g_{mo}}{C_2} \tag{10.2.20}$$

For adequate stability of the amplifier, ω_{p1} should be such that if a line of slope -20 dB/ decade is drawn from the point (ω_{p1}, A_{v0}) on the $\log A_v$ vs. $\log \omega$ plot, it intersects the unity-gain line at a frequency ω_T that is sufficiently removed from ω_z and ω_{p2}, so that ω_z and ω_{p2} do not add more than 45° of phase lag, to give a phase margin of at least 45° (Figure 10.2.3). Note that the zero is of the form $(1 - j\omega/\omega_z)$ and adds a phase lag $-\tan^{-1}(\omega/\omega_z)$ like a pole. A slope of magnitude 20 dB/decade is of unity slope on a log–log plot, because moving a decade along the abscissa moves a decade along the ordinate, since $20\log_{10}10 = 20$ dB. It follows from Figure 10.2.3 that $\dfrac{\log(A_{v0}) - \log 1}{\log \omega_{p1} - \log \omega_T} = 1$, or $\dfrac{A_{v0}}{1} = \dfrac{\omega_T}{\omega_{p1}}$.

From Equation 10.2.10,

$$\omega_T = (g_{mi}g_{mo}R_1R_2)\omega_{p1} \tag{10.2.21}$$

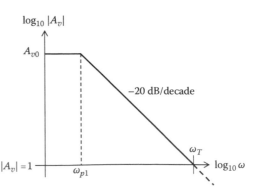

FIGURE 10.2.3
Asymptotic Bode magnitude plot of the gain of the two-stage CMOS op amp.

Substituting for ω_{p1} from Equation 10.2.18,

$$\omega_T \cong \frac{g_{mi}}{C_C} \tag{10.2.22}$$

C_C, and hence ω_{p1}, should be chosen so that ω_z and ω_{p2} are sufficiently larger than ω_T. ω_z is moved by adding a resistance R_Z in series with C_C. This replaces sC_C in Equation 10.2.16 by $1/(R_Z + 1/sC_C)$. The $(g_{mo} - sC_C)$ term becomes $\dfrac{g_{mo} - sC_c(1 - g_{mo}R_Z)}{1 + sC_cR_z}$. The new zero is moved to

$$\omega_Z \cong 1 \left/ \left(C_C \left(\frac{1}{g_{mo}} - R_Z \right) \right) \right. \tag{10.2.23}$$

If $R_Z = 1/g_{mo}$, $\omega_z \rightarrow \infty$. R_Z is thus a **zero-nulling resistor**. If $R_Z > 1/g_{mo}$, ω_Z will be on the negative real axis and will add to the phase margin. Adding R_Z also adds another pole ω_{p3} at $1/C_CR_Z$, which is normally at a frequency considerably higher than ω_{p2}.

Slew Rate*

The slew rate (SR) is a large-signal frequency response limitation of the op amp. It is usually measured by configuring the op amp as a unity-gain amplifier, connecting it to a specified load, and applying to the noninverting input a square wave of relatively large, specified amplitude (Figure 10.2.4). Ideally, the amplifier output should follow the input. In practice, the output is trapezoidal in shape, rising and falling at rates v_o/t_1 and v_o/t_2 (Figure 10.2.4b). The smaller of these two rates is the slew rate of the amplifier and is usually expressed in V/µs. The slew rate arises because a rapid change in output requires an equally rapid charging or discharging of capacitors associated with the amplifier.

If a large, positive step is applied to the base of Q_2 in Figure 10.2.1, Q_2 will turn off and I_{DD} flows through Q_1 and Q_3. Q_4 will mirror this current, which will flow through C_C.

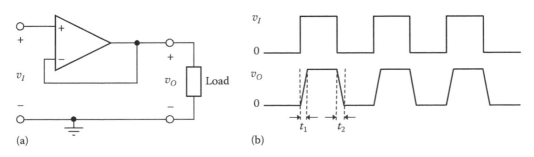

FIGURE 10.2.4
(a) Circuit for measuring the slew rate of an op amp; (b) response to a square-wave input because of slew rate.

---

* Sabah 2008, p. 731.

The voltage at v_O will therefore initially increase at a rate I_{DD}/C_C. Roughly speaking, this may be considered to be the SR. Thus,

$$SR \cong \frac{I_{DD}}{C_C} \qquad (10.2.24)$$

Substituting $I_{DD} = g_{m1,2}V_{ov1,2}$ and $g_{m1,2} = k'_p(W/L)V_{ov1,2}$

$$SR \cong \frac{(V_{ov1,2})^2 k'_p(W/L)}{C_C} \qquad (10.2.25)$$

A large V_{ov} increases SR as well as f_T (Equation 5.2.33, Chapter 5) but reduces the swings of v_O and v_{cm} as well as the gain.

Compared to the complementary configuration in which the input transistors are NMOS and the CS amplifier is PMOS, the circuit of Figure 10.2.1 has the following advantages: (1) For the same maximum W, the smaller k'_p of PMOS transistors means a larger V_{ov} (Equation 5.2.15, Chapter 5) and hence a larger $(V_{ov1,2})^2 k'$ and higher SR, (2) PMOS transistors have less flicker $(1/f)$ noise than NMOS transistors (Section 6.5, Chapter 6), and (3) the NMOS CS transistor has a higher g_{mo}, which enhances pole splitting (Equations 10.2.18 and 10.2.19).

10.3 Folded Cascode CMOS Op Amp

As another example of a CMOS op amp, we will consider the basic folded-cascode CMOS op amp illustrated in Figure 10.3.1. Q_1 and Q_2 are the input differential pair, with Q_3 and Q_4 connected in folded cascode to Q_1 and Q_2, respectively (Section 9.1, Chapter 9). The bias current of Q_1 and Q_2 is $I_{SS}/2$, whereas the bias current of Q_3 and Q_4 is $I_{DD} - I_{SS}/2$, the current sources being derived from single transistors Q_a to Q_c in CS configuration. Q_d to Q_g is a wide-swing current mirror (Figure 8.4.12b, Chapter 8). C_L includes the internal capacitances of the output transistors, load capacitance, and any capacitance that may be added to obtain a desired frequency response. The bias voltages are generally derived as described in Section 10.4.

Common-Mode Input Range

An important advantage of the folded cascode is a wider common-mode input range, which is relevant, for example, to a unity-gain amplifier. When v_{cm} is applied to both inputs connected together, the maximum positive swing, v_{cm}^+, of v_{cm} is determined by the requirement that Q_1 and Q_2 remain in saturation. At the edge of saturation, v_{GD} of these transistors is V_{tn} so that their drain voltage with respect to ground is $v_{cm}^+ - V_{tn}$. If Q_b or Q_c is at the edge of saturation with v_{cm}^+ applied, v_{SD} of these transistors is equal to their overdrive voltage $V_{ovb,c} = v_{SGb,c} - \overline{V}_{tp1}$. Hence $v_{cm}^+ - V_{tn} + V_{ovb,c} = V_{DD}$, or

$$v_{cm}^+ = V_{DD} + V_{tn} - V_{ovb,c} \qquad (10.3.1)$$

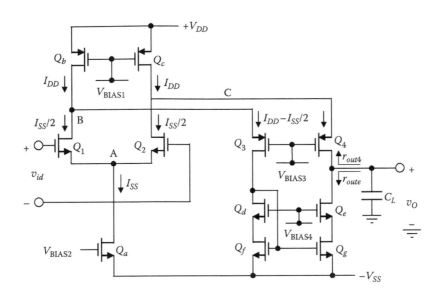

FIGURE 10.3.1
Basic folded-cascode CMOS op amp.

V_{BIAS1} is chosen to give the required I_{DD} while keeping $V_{ovb,c}$ small, say about 0.2 V. If $V_{ovb,c} < V_{tn}$, then $v_{cm}^+ > V_{DD}$. The following may be noted concerning Equation 10.3.1:

1. If Q_b and Q_c are realized using a current mirror, with the gate of Q_b connected to its drain, then $v_{cm}^+ - V_{tn} + V_{SGb} = V_{DD}$, or $v_{cm}^+ = V_{DD} + V_{tn} - V_{SGb}$, which is less than in Equation 10.3.1.

2. If a normal cascode is used instead of the folded cascode, then Q_3 and Q_4 would be NMOS transistors interposed between Q_1 and Q_b, and between Q_2 and Q_c, respectively. Assuming that Q_3 and Q_4 will be at the edge of saturation, then Equation 10.3.1 becomes $v_{cm}^+ - V_{tn} + V_{ovb,c} + V_{ov3,4} = V_{DD}$. This would reduce v_{cm}^+ in Equation 10.3.1 by $V_{ov3,4}$.

The most negative swing, v_{cm}^-, of v_{cm} is determined by the requirement that Q_a remains in saturation. The voltage to ground of the common sources of Q_1 and Q_2 is: $v_{cm}^- - v_{GS1,2} = v_{cm}^- - V_{ov1,2} - V_{tn}$. Hence, $v_{cm}^- - V_{ov1,2} - V_{tn} - V_{ova} = -V_{SS}$, or,

$$v_{cm}^- = -V_{SS} + V_{ov1,2} + V_{ova} + V_{tn} \qquad (10.3.2)$$

Again, V_{BIAS2} is chosen to give the required I_{SS} while keeping V_{ova} small. It is seen that V_{tn} reduces the magnitude of v_{cm}^-, but this is due to the differential pair and not the folded cascode. In fact, the folded cascode offers a solution to this problem. For if a complementary circuit is used, where NMOS and PMOS transistors are interchanged, v_{cm}^- is given by an expression of the form of Equation 10.3.1 and v_{cm}^- is given by an expression of the form of Equation 10.3.2. If two complementary input circuits are paralleled together, then both v_{cm}^+ and v_{cm}^- are given by expressions of the form of Equation 10.3.1 so that v_{cm}^+ can exceed V_{DD} and v_{cm}^- can exceed V_{SS}. This **rail-to-rail folded cascode** structure is discussed in Section ST10.3.

Output Voltage Swing

The most positive swing, v_O^+, of v_O is determined by the requirement that Q_4 and Q_c remain in saturation. This means that $v_O^+ + V_{ov4} + V_{ovc} = V_{DD}$, or,

$$v_O^+ = V_{DD} - V_{ov4} - V_{ovc} \tag{10.3.3}$$

v_O^+ is thus two overdrive voltages less than V_{DD}.

The most negative swing, v_O^-, of v_O is also two overdrive voltages more than $-V_{SS}$, as described in connection with the wide-swing cascode mirror of Figure 8.4.12b. Thus,

$$-V_{SS} + 2V_{ov} \leq v_O \leq V_{DD} - 2V_{ov} \tag{10.3.4}$$

where the same V_{ov} is assumed. If $V_{ov} = 0.2$ V and a peak-to-peak output swing of 2 V is required, then the output swing in each direction is 1 V, and the magnitude of the positive and supply voltages should be at least 1.4 V.

Voltage Gain

Assuming, for simplicity, that $I_{DD} = I_{SS} = I_{DS}$, the four folded cascode transistors have the same drain current of $I_{DS}/2$ and hence the same g_m. Assuming they have the same V_{ov},

$$g_m = g_{m1} = g_{m2} = g_{m3} = g_{m4} = \frac{2(I_{DS}/2)}{V_{ov}} = \frac{I_{DS}}{V_{ov}} \tag{10.3.5}$$

Neglecting the output resistances of Q_b and Q_c and the body effect, the low-frequency voltage gain A_{v0} with a cascode current mirror load is approximately (Equation 10.1.33):

$$A_{v0} = \frac{v_O}{v_{id}} = g_m(r_{out4} \| r_{oute}) \tag{10.3.6}$$

where r_{out4} and r_{oute} are the resistances looking into Q_4 drain and Q_e drain, respectively, with zero input (Figure 10.3.1). To determine r_{out4}, we note that Q_4 is a CG transistor whose source resistance is $(r_{o2} \| r_{oc})$. From Equation 8.2.53, Chapter 8, $r_{out4} \cong g_{m4} r_{o4}(r_{o2} \| r_{oc})$. From Equation 8.4.4, Chapter 8, the output resistance of the cascode, is $r_{oute} \cong g_m r_{oe} r_{og}$. It follows that r_{out}, the output resistance of the amplifier, is:

$$r_{out} = r_{out4} \| r_{oute} = [g_m r_{o4}(r_{o2} \| r_{oc})] \| (g_{me} r_{oe} r_{og}) \tag{10.3.7}$$

and

$$A_{v0} = g_m r_{out} = g_m \{[g_m r_{o4}(r_{o2} \| r_{oc})] \| (g_{me} r_{oe} r_{og})\} \tag{10.3.8}$$

The following should be noted concerning this gain expression:

1. A_{v0} is of the order of $(g_m r_o)^2$. Because of the relatively large r_{out}, the gain can be quite high. Thus, if $g_m = 1$ mA/V and $r_o = 100$ kΩ for all the transistors, $A_{v0} \cong 3300$. A_{v0}'s of more than 10,000 are feasible. If used as a unity-gain follower, with $A_{v0} = 3300$, the closed-loop gain is 3300/3301, which is within 0.03% from unity.

Hence, a single amplifying stage as in Figure 10.3.1 is quite sufficient for most purposes. Where additional gain is needed, another amplifying stage may be added, although this would complicate somewhat frequency compensation.

2. Using the same figures for g_m and r_o, the output resistance r_{out} is 3.33 MΩ. Although this is high, the output resistance with feedback is much lower. For a unity-gain follower, it is reduced to $R_{ofa} \cong R_{oa}/A_{vo} = 1/g_m = 1$ kΩ, using Equation 10.3.6. Such an output resistance is acceptable where the load is the relatively small capacitance of other CMOS inputs. If a low output resistance is required, an output CD stage could be added.

The folded-cascode CMOS op amp is an example of an **operational transconductance amplifier (OTA)**, defined as an op amp having high-impedance nodes only at the input and output, all other nodes having a much lower impedance level. In Figure 10.3.1, the impedance level at the three internal signal nodes A, B, and C is nearly $1/g_m$, as they involve inputs of CG transistors. The advantage of low-impedance internal nodes is that internal capacitances will introduce only high-frequency poles. The frequency response is then determined by a dominant pole due to load capacitance, which could be augmented by additional capacitance, if necessary, to obtain a desired frequency response.

Frequency Response

Since the frequency response is determined by the dominant pole due to C_L,

$$\frac{V_o}{V_{id}} = \frac{g_m r_{out}}{1 + sC_L r_{out}} \tag{10.3.9}$$

as illustrated in Figure 10.3.2, where the frequency f_p of the dominant pole is given by:

$$f_p = \frac{1}{2\pi C_L r_{out}} \tag{10.3.10}$$

At the unity-gain frequency f_T: $f_T \times 1 = f_p A_{vo}$, so that,

$$f_T = g_m r_{out} f_p = \frac{g_m}{2\pi C_L} \tag{10.3.11}$$

If $C_L = 5$ pF, $r_{out} = 3.33$ MΩ, and $g_m = 1$ mA/V, then $f_p \cong 9.5$ kHz and $f_T \cong 32$ MHz.

At f_T, the phase lag due to the dominant pole is $-90°$. If the phase lag due to higher frequency poles is small, the phase margin is nearly $90°$.

Slew Rate

When a large differential signal is applied, Q_2 is cut off and I_{SS} is diverted to Q_1. The current through Q_3 is $I_{DD} - I_{SS}$ and that through Q_4 is I_{DD}. The current through Q_d is $I_{DD} - I_{SS}$, which

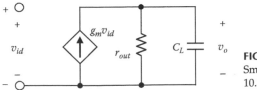

FIGURE 10.3.2
Small-signal equivalent circuit of the amplifier of Figure 10.3.1.

is the mirrored current through Q_e and Q_g. The current into C_L is $I_{DD} - (I_{DD} - I_{SS}) = I_{SS}$. The output voltage increases linearly as $I_{SS}t/C_L$, so that SR is roughly given by:

$$SR = \frac{I_{SS}}{C_L} \qquad (10.3.12)$$

If $I_{SS} = 0.2$ mA and $C_L = 5$ pF, then SR $= 0.4$ V/μs.

As mentioned earlier, the bias current of Q_1 and Q_2 is $I_{SS}/2$, and that in Q_3 and Q_4 is $I_{DD} - I_{SS}/2$. If $I_{DD} = I_{SS}$, the bias current in all the transistors is nominally $I_{SS}/2$. However, if $I_{DD} = I_{SS}$ the mirror will turn off when a large differential signal is applied, which is undesirable, because of the distortion that is introduced. Hence, I_{DD} is made about 10%–20% higher than I_{SS}.

As in the case of the two-stage CMOS op amp, the CMRR is of the order of $(g_m r_o)^2$. Using the figures quoted previously, gives a CMRR of approximately 80 dB.

Simulation Example 10.3.1 Basic Operational Transconductance Amplifier

A basic OTA amplifier is shown in Figure 10.3.3, where it is seen that the internal signal nodes have either a gate-drain or a source connection, which makes them low-impedance nodes.

ANALYSIS

It is assumed that g_m is the same for all MOSFETs, and that Q_6 and Q_8 have widths that are K times those of the other PMOS and NMOS transistors, respectively. When a differential input v_{id} is applied, and since r_o is much larger than the load resistance of nearly $1/g_m$ of Q_1 and Q_2,

$$i_d = \frac{g_m}{2} v_{id} \qquad (10.3.13)$$

Because of coupling through the current mirrors, $i_o = 2Ki_d$. At low frequencies, and assuming Q_6 and Q_8 have the same r_o,

$$v_o = \frac{r_o}{2} i_o = Kr_o i_d \qquad (10.3.14)$$

The low-frequency voltage gain is:

$$A_{v0} = \frac{v_o}{v_{id}} = \frac{v_o}{i_o} \frac{i_o}{i_d} \frac{i_d}{v_{id}} = \left(\frac{r_o}{2}\right)(2K)\left(\frac{g_m}{2}\right) = \frac{Kg_m r_o}{2} \qquad (10.3.15)$$

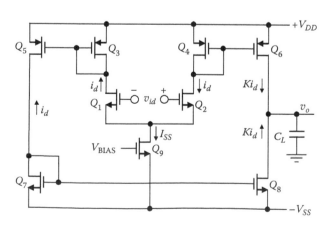

FIGURE 10.3.3
Figure for Simulation Example 10.3.1.

The transconductance of the op amp is:

$$G_m = \frac{i_o}{v_{id}} = \frac{i_o}{i_d}\frac{i_d}{v_{id}} = (2K)\left(\frac{g_m}{2}\right) = Kg_m \qquad (10.3.16)$$

In an OTA, the pole due to C_L is dominant. It is given for the op amp of Figure 10.3.3 by

$$\omega_p = \frac{2}{r_o C_L} \qquad (10.3.17)$$

It follows that the gain at any frequency is:

$$A_v = \frac{A_{v0}}{1 + s/\omega_p} \qquad (10.3.18)$$

This is the same expression that results from replacing $r_o/2$ in Equation 10.3.15 by $(r_o/2)\|$ $(1/sC_L)$. If a large differential step input is applied, Q_1 is turned off and the drain current of Q_2 is I_{SS}. The output current is KI_{SS} and the maximum slew rate is:

$$SR = \frac{v_O}{t} = \frac{i_O}{C_L} = \frac{KI_{SS}}{C_L} \qquad (10.3.19)$$

SIMULATION

The schematic corresponds to Figure 10.3.3, with $V_{DD} = V_{SS} = 5$ V and $C_L = 0.1$ pF. The transistor parameters used correspond to $k_n = k_p = 1$ mA/V^2, $V_{tn} = \overline{V}_{tp} = 1$V, $\lambda = 0.01$ V^{-1}, and $K = 10$. $V_{BIAS} = -3.38$ V to have $I_{SS} \cong 200$ μA. Thus, $g_{m1,2} = \sqrt{2 \times 0.1} = 0.45$ mA/V, neglecting channel-width modulation, and $r_o = 1/(0.01 \times 1) \equiv 100$ kΩ for Q_6 and Q_8. From Equation 10.3.15, $A_{v0} = (10 \times 0.45 \times 100)/2 = 225$, and from Equation 10.3.17, $f_p = 31.8$ MHz. The simulation values are 239 and 32.2 MHz, respectively. From Equation 10.3.19, SR = 2000 V/μs. When a differential step of 5V is applied, v_O changes at an initial rate of about 2000 V/μs as shown in Figure 10.3.4. Note that the quiescent output voltage is about −3.54 V, which is the gate voltage of Q_7 and Q_8 and equals the drain voltages of these transistors.

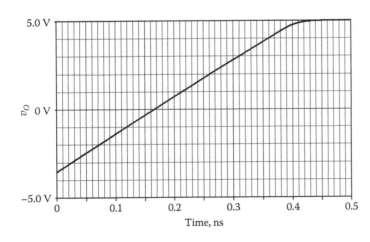

FIGURE 10.3.4
Figure for Simulation Example 10.3.1.

10.4 CMOS Current and Voltage Biasing

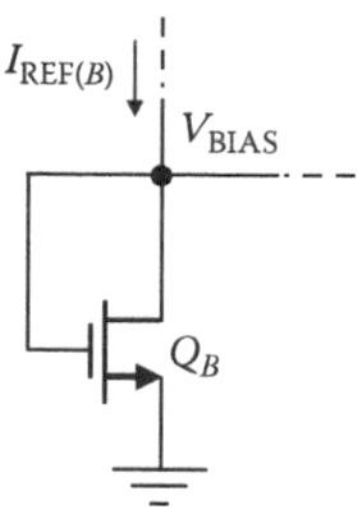

We have encountered many circuits that require current biasing
or voltage biasing. As a matter of terminology, when a biasing
current or voltage is stabilized with respect to variations in supply
voltages and temperature, it is usually referred to as a current
reference or a voltage reference, respectively. In ICs, current biasing is generally of basic importance, since a voltage bias can be
derived from a specified current bias applied to an appropriately
sized, diode-connected MOSFET. This is illustrated in Figure
8.5.1a, Chapter 8, where two bias voltages are derived from the
two diode-connected transistors of a cascode current mirror.

In Figure 10.4.1, $I_{\text{REF}(B)} = (k'_n/2)(W/L)_B(V_{\text{BIAS}} - V_{tn})^2$, so that
$V_{\text{BIAS}} = V_{tn} + \sqrt{2I_{\text{REF}(B)}/k'_n(W/L)_B}$. If a transistor Q_X has the same
k'_n, a bias current I_{REF}, and an overdrive voltage V_{ov}, then
$I_{\text{REF}} = k'_n/2(W/L)_X(V_{ov})^2$. Substituting for $2/k'_n$,

FIGURE 10.4.1
A simple method of obtaining a voltage bias from a
reference current.

$$V_{\text{BIAS}} = V_{tn} + V_{ov}\sqrt{\frac{I_{\text{REF}(B)}(W/L)_X}{I_{\text{REF}}(W/L)_B}} \qquad (10.4.1)$$

If $I_{\text{REF}(B)} = I_{\text{REF}}$, and $\left(\dfrac{W}{L}\right)_B = \dfrac{1}{4}\left(\dfrac{W}{L}\right)_X$, $V_{\text{BIAS}} = V_{tn} + 2V_{ov}$. This is the bias for the wide-
swing cascode in Figure 8.5.1b and could be derived from a diode-connected MOSFET
having the same I_{REF} in Figure 8.5.1b and a (W/L) that is a quarter of the (W/L) of
transistors Q_1 to Q_4.

Self-Biasing

The simple current mirror that relies on a resistor to set the reference current suffers from
an appreciable sensitivity to variations in supply voltages and temperature. To reduce
sensitivity to supply voltage variations, a *self-biasing* scheme is used in which the reference
current is made to depend on the output current, and not directly on the supply voltages.
The basic circuit, illustrated in Figure 10.4.2, consists of transistors Q_1 and Q_2 forming a
Widlar current source that is connected to a PMOS current mirror. Transistors Q_a, Q_b,

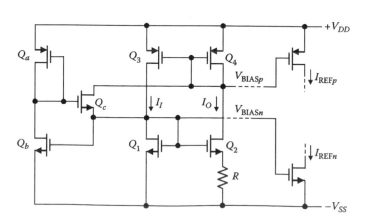

FIGURE 10.4.2
Basic self-biasing circuit.

and Q_c are not involved in normal circuit operation but are required for starting the circuit, as will be discussed later.

It is convenient to denote I_I and I_O as the input and output currents, respectively, of the Widlar current source. The loop formed by Q_1 to Q_4 is a positive feedback loop. Thus, if I_I increases, V_{GS1} increases, which in turn increases I_O. But I_O is the reference current for the PMOS mirror so that the increase in I_O further increases I_I. Because of the positive feedback, the circuit will be unstable if the loop gain exceeds unity (Section 9.4, Chapter 9). The gain of the PMOS current mirror is ideally unity. The small-signal gain of the Widlar current source, from V_{GS1} to I_O, depends on $\dfrac{g_{m2}}{1 + g_{m2}(1 + \chi)R}$, ignoring r_{o2} (Exercise 8.2.8, Chapter 8). If $g_{m2}(1 + \chi)R \gg 1$, which increases the negative feedback due to R, the gain becomes $1/(1 + \chi)R$; g_{m2} is increased by increasing $(W/L)_2$. Typically, $(W/L)_2 = 4(W/L)_1$.

The circuit can be readily analyzed neglecting channel-length modulation. Thus,

$$I_O = \frac{k'_n}{2} \left(\frac{W}{L}\right)_2 (V_{GS2} - V_{tn})^2 \tag{10.4.2}$$

$$I_I = \frac{k'_n}{2} \left(\frac{W}{L}\right)_1 (V_{GS1} - V_{tn})^2 \tag{10.4.3}$$

and

$$V_{GS1} = V_{GS2} + RI_O \tag{10.4.4}$$

If V_{tn} is subtracted from both sides of Equation 10.4.4, $(V_{GS2} - V_{tn})$ and $(V_{GS1} - V_{tn})$ can be substituted from Equations 10.4.2 and 10.4.3, respectively. Setting $I_O = I_I = I_{\text{REF}}$, $K = (W/L)_2/(W/L)_1$, and rearranging gives:

$$I_{\text{REF}} = \frac{2}{k'_n(W/L)_2 R^2} \left(\sqrt{K} - 1\right)^2 \tag{10.4.5}$$

For a given k'_n, I_{REF} is seen to depend on R, $(W/L)_2$ and K. It is independent of V_{tn} and supply voltages in the absence of channel-length modulation. Since R should be a stable and accurate resistor, it is often an off-chip resistor. However, this can add considerable capacitance at Q_2 source, which bypasses R and may increase the gain sufficiently at some frequency to cause oscillation. The circuit of Figure 10.4.2 is also referred to as **a beta multiplier** reference circuit, because $k'_n(W/L)$ is sometimes denoted by β, and β of Q_2 is that of Q_1 multiplied by K.

g_{m2} of Q_2 equals $\sqrt{2k'_n(W/L)_2 I_{\text{REF}}}$. It follows from Equation 10.4.5 that:

$$g_{m2} = \frac{2}{R} \left(\sqrt{K} - 1\right) \tag{10.4.6}$$

Thus, g_{m2} depends only on R and K. Since g_{mx} of any MOSFET Q_x is $\sqrt{2k'_{n,p}(W/L)_x I_{Dx}}$, then any transistor whose bias current is derived from I_{REF} will have g_{mx} given by:

$$g_{mx} = g_{m2} \sqrt{\frac{k'_{n,p} I_{Dx}(W/L)_x}{k'_n I_{\text{REF}}(W/L)_2}} \tag{10.4.7}$$

where $k'_{n,p}/k'_n = 1$ for an NMOS transistor and $k'_{n,p}/k'_n = \mu_e/\mu_h$ for a PMOS transistor. g_{mx} will also be independent of threshold and supply voltages

From Equations 10.4.3 and 10.4.5, with $I_I = I_{REF}$:

$$V_{BIASn} = V_{GS1} = V_{tn} + \frac{2}{k'_n(W/L)_1 R}\left(1 - \frac{1}{\sqrt{K}}\right)^2 \qquad (10.4.8)$$

A self-biasing circuit is not, in general, self-starting, because the state in which $I_I = I_O = 0$ is a stable state, in addition to the normal operating state. Transistors Q_a to Q_c are included in the self-biasing circuit of Figure 10.4.2 for starting purposes. At the start, with $I_I = I_O = 0$, Q_3, Q_1, and hence Q_b are cut off. Q_a will be turned on but does not conduct, because Q_b is off, which makes the drain voltage of Q_a equal to V_{DD}. This turns on Q_c, which pulls up the gates of Q_1 and Q_2, and establishes normal operation. In this state, $V_{GSc} \simeq 0$, Q_c is essentially cut off and does not affect circuit operation.

Let us consider temperature dependence next. Differentiating Equation 10.4.5 with respect to T, where R and k'_n are the only temperature-dependent quantities, and dividing by I_{REF} to obtain the temperature coefficient:

$$\frac{1}{I_{REF}}\frac{dI_{REF}}{dT} = -2\left(\frac{1}{R}\frac{dR}{dT}\right) - \frac{1}{k'_n}\frac{dk'_n}{dT} \qquad (10.4.9)$$

where $\dfrac{1}{R}\dfrac{dR}{dT}$ is the temperature coefficient of R and $\dfrac{1}{k'_n}\dfrac{dk'_n}{dT} = \dfrac{1}{\mu_e}\dfrac{d\mu_e}{dT} = -\dfrac{2.5}{T}$ (Equation 2.6.3, Chapter 2). Hence, if $\dfrac{1}{R}\dfrac{dR}{dT} = \dfrac{1.25}{T}$, the temperature coefficient of I_{REF} will be zero.

Differentiating Equation 10.4.8 with respect to T:

$$\frac{dV_{BIASn}}{dT} = \frac{dV_{tn}}{dT} - \frac{2}{Rk'_n(W/L)_1}\left(1 - \frac{1}{\sqrt{K}}\right)^2\left(\frac{1}{R}\frac{dR}{dT} + \frac{1}{k'_n}\frac{dk'_n}{dT}\right) \qquad (10.4.10)$$

To have $dV_{BIAS_n}/dT = 0$, we choose R to be:

$$R = \frac{2}{(dV_{tn}/dT)k'_n(W/L)_1}\left(1 - \frac{1}{\sqrt{K}}\right)^2\left(\frac{1}{R}\frac{dR}{dT} + \frac{1}{k'_n}\frac{dk'_n}{dT}\right) \qquad (10.4.11)$$

where dV_{tn}/dT is in the range -0.5 to -1.5 mV/°C and K is generally equal to 4, which means that $\left|\dfrac{1}{R}\dfrac{dR}{dT}\right| < \left|\dfrac{1}{k'_n}\dfrac{dk'_n}{dT}\right| = \dfrac{1.5}{T}$, so $R > 0$. This inequality is satisfied if $\dfrac{1}{R}\dfrac{dR}{dT} = \dfrac{1.25}{T}$ in order to have a zero temperature coefficient of I_{REF}.

Although the self-biasing circuit of Figure 10.4.2 considerably reduces the effect of supply voltage variations, it does not eliminate it entirely because of channel-length modulation. Changes in supply voltages will affect the drain-source voltages of the transistors and hence the currents in the circuit. These variations are more pronounced with short-channel transistors because of their increased susceptibility to channel-length modulation. The effect of changes in the supply voltages on the drain-source voltages of Q_1 and Q_2 can be reduced by adding a cascode layer to Q_1 and Q_2 and by making I_I and I_O more nearly equal despite supply voltage variations. There are two approaches to enforcing the equality I_I and I_O: (1) Adding a cascode layer to Q_3 and Q_4 leading to four transistor stacks across the voltage supplies when the Widlar current source is also cascoded. Although four bias supplies become available, from each of the four diode-connected transistors, the transistor stacks set a limit to the smallest supply voltages that can be used. (2) Using an op amp to force equality of the drain voltages of Q_3 and Q_4. The op amp need not be more elaborate than two CS transistors sharing a current-mirror load. Improvements to the basic self-biasing circuit of Figure 10.4.2 are discussed in Section ST10.4 as well as bandgap voltage references having very low temperature coefficients.

Exercise 10.4.1

Determine the temperature coefficient of R in Equation 10.4.9 that will make the temperature coefficient of I_{REF} zero at $T = 300K$.
  Answer: 4167 ppm/K. ■

10.5 BJT Op Amps

A BJT op amp can be constructed along almost exactly the same lines as the two-stage CMOS op amp described in Section 10.3. Another possibility is the four-stage op amp illustrated in Figure 10.5.1. The first two stages are differential amplifiers formed by Q_1–Q_2 and Q_4–Q_5. These stages are differentially coupled and biased by the current mirror formed by Q_9, Q_3, and Q_6. A single-ended output is taken from the collector of Q_5 to the third stage formed by Q_7. The *pnp* transistor shifts the dc level, as discussed in Section 9.1, Chapter 9. The fourth stage is the emitter follower transistor Q_8, which provides a low output resistance and allows the output to swing both positively and negatively. This op amp is analyzed in Section ST10.5 together with the popular, general-purpose μA741 op amp. Some of the characteristics of the μA741 are explored in Simulation Example 10.5.1, which also serves to illustrate some op-amp imperfections and their simulations. In the remainder of this section, we will discuss some circuit techniques that are of special importance in BJT op amps, particularly the input and output stages.

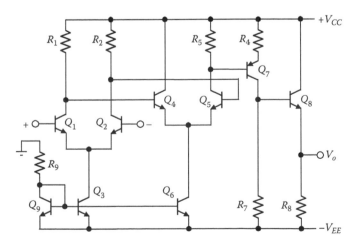

FIGURE 10.5.1
Basic BJT four-stage op amp.

Input Resistance and Bias Current

Evidently, the input stage in a BJT op amp determines the input resistance and the input bias current. Figure 10.5.2 illustrates an input stage based on that of the µA741 op amp (Section ST10.5). Q_1 and Q_2 are emitter followers that increase the input resistance. It can be argued, in exactly the same manner as in Figure 10.1.2, that when a differential input is applied, node A is at signal ground. Hence, Q_3 and Q_4 constitute a differentially connected common-base pair whose load is the current mirror formed by Q_5 and Q_6. As discussed in connection with Figures 10.1.10 and 10.1.11, the current mirror load results in high common-mode rejection and no loss of gain in converting the differential input to a single-ended output. For reasons of lowering manufacturing cost by not adding process steps to those involved in the manufacture of *npn* transistors, Q_3 and Q_4 are of the lateral type (Section 4.7, Chapter 4) having a lightly doped, relatively wide base, with consequent reduction of current gain and unity-gain frequency. However, since the CB configuration has a better high-frequency response than CE stages used in the rest of the amplifier, the overall frequency response is not seriously impaired. Q_3 and Q_4 shift the dc level of the input, but more importantly, having their BEJs in series with the BEJs of Q_1 and Q_2 protects the BEJs of Q_1 and Q_2 from breakdown if a relatively high voltage is accidentally connected between the input terminals. The reverse-voltage breakdown of BEJs of *npn* transistors is normally less than 10 V, whereas that of the lateral *npn* transistors is in the range of 50 V.

An obvious measure to reduce the input bias currents to practically zero is to have a differential MOSFET first stage. The MOSFET transistors, however, will have a lower transconductance than the BJTs they replace. A better approach is to connect the MOSFET transistors as source followers feeding into a BJT differential stage, as illustrated in Figure 10.5.3. Although input bias currents are practically eliminated, the input offset voltage will be increased because of some inevitable mismatch of the MOS transistors and their bias currents.

Another approach to reducing the input bias currents is to use input transistors of very high β_F. HBTs (Section 7.4, Chapter 7) are a possibility, although this would considerably complicate the manufacturing process and add to the cost. Another possibility is to use **superbeta transistors** of very narrow base width, of about 0.1 µm, which increases both the base transport factor and the emitter injection efficiency (Example 6.1.1, Chapter 6) leading to β_F's in the range of a few thousands. At this base width, however, the transistor is

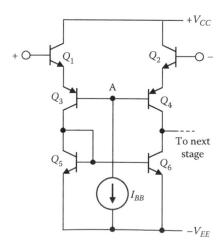

FIGURE 10.5.2
An input stage of a BJT op amp.

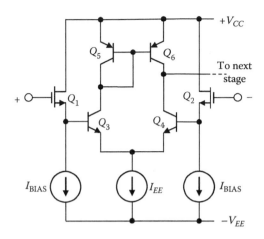

FIGURE 10.5.3
A MOSFET source follower input stage of a BJT op amp.

susceptible to punchthrough (Section 6.2, Chapter 6) and to pronounced base-width modulation. Hence, they should be operated at a base-collector voltage that is near zero. An appropriate cascode arrangement is illustrated in Figure 10.5.4, where Q_1 and Q_2 are superbeta transistors and I_{BIAS} biases the diode-connected transistors D_1 and D_2, which in turn bias the common-base transistors Q_3 and Q_4. D_1 drops almost the same voltage as V_{BE3} and V_{BE4}, whereas D_2 limits V_{CE1} and V_{CE2} to one diode voltage drop. Hence, the base-collector voltage of Q_1 and Q_2 is near zero. Considering that I_{BIAS} flows mainly through D_1 and D_2, the current through Q_1 and Q_2 is $I_{EE} - I_{BIAS}$. If this current is in the range of μAs, the input bias current is a few nAs, and the input offset current is an order of magnitude smaller.

Input Offset Voltage

Op amps such as the μA741 have connections for an external potentiometer for counteracting the input offset voltage and zeroing the output (Section ST10.5). However, in

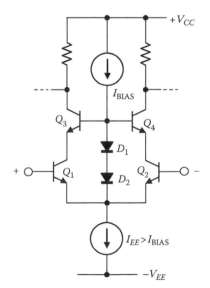

FIGURE 10.5.4
An input stage of a BJT op amp using superbeta transistors Q_1 and Q_2.

precision applications, it is the drift in input offset voltage and input offset current that sets a limit to the stability of the output of a given op amp (Example 10.6.1). These drift components are essentially unpredictable and cannot be compensated. We have seen earlier how to minimize the input bias currents and hence the input offset currents. We will consider next how to minimize the input offset voltage, which will also minimize the drift in this voltage (Equation 10.1.28).

When the gain of the input stage is relatively high, this stage determines the input offset voltage, because the contribution of the stages that follows will be relatively small. Thus, if the gain of the first stage is 20, the contribution of the input offset voltage of the second stage is divided by 20 when referred to the input. The input offset voltage in the first stage arises mainly from mismatch in the components of this stage. Hence, it is desirable to have a simple first stage with the fewest of components, such as the differential pair of Figure 10.5.1. In this case, the mismatch can occur only from the input transistors and their collector resistors. A basic limit is set by the BEJ area of the input transistors and of the dimensions of the resistors relative to the photolithographic resolution of the fabrication process (Section 4.2, Chapter 4). In general, the smaller are these dimensions relative to process resolution, the larger is the proportional effect of a fixed uncertainty in these dimensions, and the greater is the degree of mismatch.

One approach to reducing the input offset voltage is to trim the collector resistors of the input stage on-chip so as to counteract the input offset voltage. This can be done by having one or both collector resistors made up of a fixed component R_{cf} and a number of increments ΔR_{c1}, ΔR_{c2}, etc. that are shorted by fusible metallic links, as illustrated in Figure 10.5.5. The increments could be equal in value or could be binary-weighted, that is progress in multiples of two. Trimming takes place by testing the die and melting the required number of fusible links using a laser or current pulses. Another measure is to use a layout that compensates for process-related gradients along the die, as illustrated in Figure 10.5.6. The input transistors are divided into two parts each, the parts being diagonally distributed as shown. The horizontal and vertical variations of the properties of the die are evened out, leading to better matching.

FIGURE 10.5.5
Fusible links across increments of resistance used to trim the value of the total resistance.

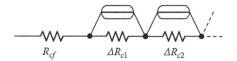

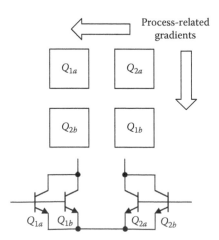

FIGURE 10.5.6
Method of compensation for process-related manufacturing gradients.

Simulation Example 10.5.1 The μA741 in PSpice

This example illustrates simulation techniques for investigating op-amp characteristics and imperfections.

SIMULATION

The μA741 is listed in PSpice EVAL library. Figure 10.5.7 shows the schematic of the op amp with two dc bias supplies of +15 V and −15 V shown connected to pins 7 and 4, respectively. Note the use of VCC_ARROW power connectors from the CAPSYM library so as to avoid cluttering the figure with wiring to the dc supplies (Appendix SB). Also, the output voltage is designated by an off-page connector. The inverting terminal is grounded and a 1 V ac source is connected to the noninverting terminal to obtain $A_v(f)$ of the op amp. In the simulation settings, AC Sweep/Noise type of analysis is selected with logarithmic sweep from 0.1 Hz to 1 MHz, using 100 points per decade. A temperature of 25°C is entered in the Temperature(Sweep) under Options in the Simulation settings window. In Figure 10.5.8

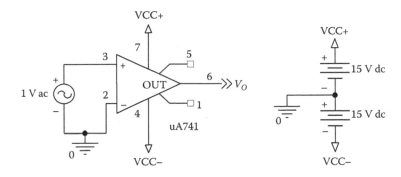

FIGURE 10.5.7
Figure for Simulation Example 10.5.1.

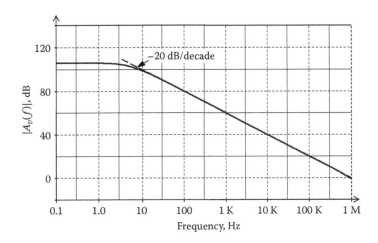

FIGURE 10.5.8
Figure for Simulation Example 10.5.1.

showing $A_v(f)$, the vertical scale extends to tens of megavolts. This does not mean of course that the op amp is capable of such an output. A linear model is assumed in AC Sweep, so the output voltage is not limited by the dc supplies or the physical limitations of the device.

Using Probe cursor, the low-frequency gain A_{vo} is found to be almost 2×10^5. The cursor search command sle(max-3) gives very nearly 5 Hz for the 3-dB corner frequency. The cursor search command sle(0) gives 888.2 kHz for f_T, at which $|A_v(f)| = 1$. This is somewhat less than the 1 MHz expected from the product of A_{vo} and the corner frequency, because the magnitude of the slope starts to increase slightly before f_T due to the influence of higher frequency poles. The μA741 is internally compensated so that the slope of the frequency response is only slightly more negative than -20 dB/decade at f_T.

The schematic of Figure 10.5.9 is used for measuring input bias currents, input offset voltage, and the CMRR. With V1 $=0=$ V2, a bias point simulation is run at 25°C and then at 100°C. The input bias currents and the output voltages are found to be:

$$T = 25°C: \quad I_{ibP} = I_{bN} = 710.54 \, \text{nA}, \quad V_o = 3.861 \, \text{V}$$

$$T = 100°C: \quad I_{ibP} = I_{bN} = 710.59 \, \text{nA}, \quad V_o = 2.911 \text{V}$$

The bias currents are well matched in the simulation and do not vary much with T. Their polarity can be determined from the polarity of the source currents of V1 and V2 in the output file. Since these currents are negative, they flow from the negative to the positive terminal of the source. Hence, the bias currents flow inward at the op-amp inputs. With V1 $=0=$ V2, the output voltage is due to the input offset voltage and decreases by about 25% between 25°C and 100°C.

With V1 $=0$, a few trials with V2 indicate that a value of around 20 μV reduces V_O to zero at 25°C. A DC sweep is therefore performed that varies V2 from 0 to 20 μV at both temperatures. The resulting plots are shown in Figure 10.5.10. Using the cursor command sle(0), it is found that the values of V2 that reduce V_O to zero are 19.29 and 16.91 μV, at 25°C and 100°C respectively. These are the input offset voltages. A_{vo} may be determined from the ratio of V_O, when V2 $=0$, to the value of V2 that reduces V_O to zero. The figures obtained give an A_{vo} that is almost 2×10^5 at 25°C, in agreement with that derived from the frequency response. At 100°C, A_{vo} is reduced to about 1.72×10^5.

To estimate the CMRR, we set V1 to 5 V and determine the value of V2 that makes V_O zero, using DC sweep at 25°C, as was done above. This value is found to be -136.2 μV. The difference between this and the input offset voltage of $+19.29$ μV is 155.5 μV and is the differential equivalent of the common-mode voltage. This gives a CMRR of 5 V/155.5 μV $= 3.2 \times 10^4$.

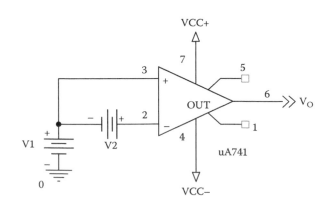

FIGURE 10.5.9
Figure for Simulation Example 10.5.1.

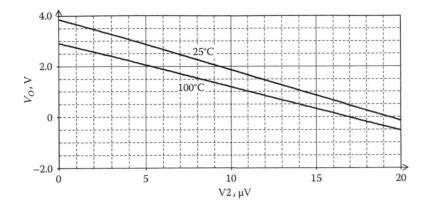

FIGURE 10.5.10
Figure for Simulation Example 10.5.1.

10.6 Some Basic Practical Op-Amp Circuits

We will consider in this section some basic op-amp circuits taking into account op-amp imperfections such as input bias currents, input offset voltage, and finite gain.

Concept: *It is possible, at least in principle, to compensate for the effects of steady input bias currents and offset voltage of an op amp. However, it is impractical to compensate for drift of these quantities, which therefore sets a limit on the output stability of an op-amp circuit.*

In precision op-amp circuits, the drift of input bias current and offset voltage is usually more significant than the absolute values of these quantities. Noise in op amps, which is mostly attributable to the input stage (Section 6.5, Chapter 6), is accounted for by adding a noise voltage source in series with the input offset voltage and noise current sources in parallel with the input bias currents I_{bP} and I_{bN}. An example of noise calculations is given in Problem P10.3.10.

Inverting and Noninverting Op-amp Circuits

In Example 10.6.1 illustrating the design of an inverting amplifier, it is shown that:

Concept: *In order to minimize the effect of input bias currents, the dc resistances in the external circuit, between either op amp input and ground, should be equal. This is true in both the inverting and noninverting configurations.*

Design Example 10.6.1 Practical Inverting Amplifier*

In a given op amp, $A_v = 10,000$, $V_{io} = 2$ mV, $I_{ib} = 200$ nA, $I_{io} = 30$ nA, temperature dependence is 5 μV/°C for V_{io}, 10 nA/°C for I_{ib}, and 0.5 nA/°C for I_{io}. It is required to design an

---

* Sabah 2008, Section ST18.2, p. 755.

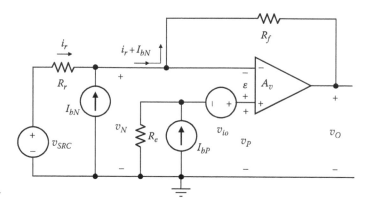

FIGURE 10.6.1
Figure for Design Example 10.6.1.

amplifier that inverts a 100 mV dc signal at unity gain, the amplifier input impedance to be at least 10 kΩ, and to determine the effect on the output of a 50°C change in temperature.

SOLUTION

Figure 10.6.1 shows the inverting configuration taking into account the input offset voltage V_{io} and the input bias currents I_{bN} and I_{bP}. The reason for adding R_e will soon become apparent. It can be readily shown (Exercise 10.6.1) that:

$$v_O = \left(-\frac{R_f}{R_r} v_{SRC} + \frac{R_r + R_f}{R_r} V_{io} - R_f I_{bN} + \frac{R_e(R_r + R_f)}{R_r} I_{bP} \right) \frac{1}{1 + 1/\beta_o A_v} \qquad (10.6.1)$$

If $I_{bP} = I_{bN}$, their effects cancel out when $R_e = (R_r \| R_f)$. If $I_{bP} \neq I_{bN}$, as in fact is the case, the best that can be done is to still satisfy this relation, which makes the contributions of the bias currents equal $(R_r \| R_f)I_{io}$, where $I_{io} = |I_{bP} - I_{bN}|$ is the input offset current and is generally an order of magnitude smaller than I_{bP} or I_{bN}.

Differentiating v_O with respect to T, neglecting the temperature effects on the resistances and on A_v, $\dfrac{dv_O}{dT} = \left(\dfrac{R_r + R_f}{R_r} \dfrac{dV_{io}}{dT} + R_f \dfrac{dI_{io}}{dT} \right) \dfrac{1}{1 + 1/\beta_o A_v} = (2 \times 5 + 10 \times 0.5)$

$\dfrac{1}{1 + 1/5000}$ μV/°C ≅ 15 μV/°C. Over a 50°C change in temperature, this amounts to about 0.75 mV, or about 0.75% change in V_O. A worst case has been assumed in which V_{io} drift and I_{io} drift are directly additive.

It is seen that in order to minimize the effects of drift in V_{io} and I_{io}, R_f should be kept as small as possible. Since an input impedance of 10 kΩ is acceptable with unity gain, we may choose $R_r = R_f = 10$ kΩ. The effective dc resistance seen at the inverting input is 5 kΩ, so a 5 kΩ resistance is inserted in series with the noninverting input.

If finite A_{vo} is taken into account, then from Equation 7.4.5, Chapter 7, with $\beta_o = 0.5$, $V_O = -100/(1 + 1/5000) = 99.98$ mV. If we include 0.1% resistance tolerances, the worst-case output is $99.98 \times 9.99/10.01 = 99.78$ mV. The effect of 0.1% resistance tolerances is 0.22 mV, whereas the effect of finite A_v is 0.02 mV, even though the assumed A_v is on the low side for a good IC op amp.

Exercise 10.6.1

Verify that the governing equations in Figure 10.6.1 are: $v_P = R_e I_{bP} + V_{io}$, $v_N = v_{SRC} - R_r i_r$, $v_O = -R_f(i_r + I_{bN}) + v_N$, and $v_O = A_v \varepsilon = A_v(v_P - v_N)$. Derive Equation 10.6.1 by eliminating

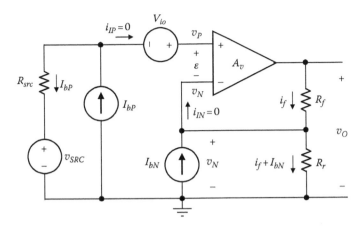

FIGURE 10.6.2
Figure for Exercise 10.6.2.

v_P, v_N, and i_r between these four equations. Note that in Equation 10.6.1, V_{io} is accentuated relative to v_{SRC} by $(R_r + R_f)/R_f$, which can be large if $R_r \gg R_f$, as when amplifier gain is much less than 1. ∎

Exercise 10.6.2

Verify that the governing equations for the noninverting amplifier of Figure 10.6.2 are $v_P = v_{SRC} + R_{src}I_{bP} + V_{io}$, $v_N = R_r(i_f + I_{bN})$, $v_O = R_f i_f + v_N$, $v_O = A_v \varepsilon = A_v(v_P - v_N)$. Eliminate the three unknowns v_P, v_N, and i_f to show that:

$$v_O = \frac{1}{\beta}\left(v_{SRC} + V_{io} + R_{src}I_{bP} - \frac{R_r R_f}{R_r + R_f}I_{bN}\right)\frac{1}{1 + 1/\beta_o A_v} \tag{10.6.2}$$

Note that V_{io} is amplified by the same factor as v_{SRC} and that in order to minimize the effect of input bias currents, $R_{src} = R_r \| R_f$. ∎

The inverting and noninverting adders are analyzed in Section ST10.6 taking into account V_{io}, I_{ib}, and finite A_v. Similar conclusions apply concerning the relative effects of V_{io} and the minimization of the effect of I_{ib}.

Integrator

Input offset voltage and bias currents have a drastic effect on integrators, as they saturate the op-amp output, even when no input is applied. Moreover, to amplify down to very low frequencies, or over long periods, the op-amp gain must be very high (Section 7.4, Chapter 7). The practical limitations of integrators are illustrated by Example 10.6.2.

Design Example 10.6.2 Practical integrator*

It is required to investigate the effects of V_{io}, input bias currents, and finite A_v on the performance of an integrator, and deduce some design guidelines and constraints.

---

* Sabah 2008, Section ST18.3, p. 755.

SOLUTION

In the integrator circuit of Figure 10.6.3, a resistance R is connected in series with the noninverting input to minimize the effect of input bias currents (Example 10.6.1), which makes $v_P = RI_{bP} + V_{io}$. Considering $v_{SRC} = 0$, then $v_N = R(I_{bN} + i_f)$, where i_f is the current through C. Assuming that $A_v \rightarrow \infty$, $v_P = v_N$ and,

$$i_f = \frac{1}{R}(V_{io} + RI_{io}) = \frac{V_{ib}}{R} \tag{10.6.3}$$

It follows that:

$$v_O = \frac{1}{C}\int_0^i i_f dt + v_{O0} = \frac{t}{RC}V_{ib} + v_{O0} \tag{10.6.4}$$

where v_{O0} is v_O at $t = 0$. As t increases, v_O eventually reaches saturation, even though $v_{SRC} = 0$.

V_{io} and I_{io} thus have a drastic effect on the integrator. In practice a switch, such as a MOSFET, is connected across the capacitor so as to keep v_O at a small initial voltage $v_{O0} = v_P = v_N$, when no input is applied ($v_{SRC} = 0$). The switch is opened at the start of integration ($t = 0$).

Consider next the effect of finite A_{v0}. The governing equations are:

$$v_N = v_{SRC} + R(I_{bN} + i_f) \tag{10.6.5}$$

$$v_N = v_P - \varepsilon = v_P - \frac{v_O}{A_{v0}} \tag{10.6.6}$$

$$v_P = RI_{bP} + V_{io} \tag{10.6.7}$$

$$v_O = v_N + \frac{1}{C}\int i_f dt = v_P - \frac{v_O}{A_{v0}} + \frac{1}{C}\int i_f dt \tag{10.6.8}$$

Eliminating v_N and v_P between Equations 10.6.5 through 10.6.7:

$$V_{ib} - v_{SRC} = Ri_f + \frac{v_O}{A_{v0}} \tag{10.6.9}$$

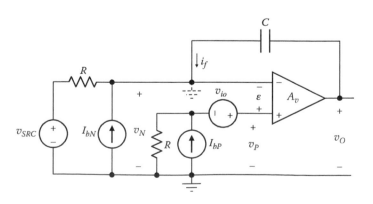

FIGURE 10.6.3
Figure for Design Example 10.6.2.

where $V_{ib} = V_{io} + R(I_{bP} - I_{bN})$. Differentiating Equation 10.6.8 bearing in mind that v_P is constant,

$$\frac{dv_O}{dt}\left(1 + \frac{1}{A_{v0}}\right) = \frac{i_f}{C}$$

(10.6.10)

Eliminating i_f between Equations 10.6.9 and 10.6.10,

$$\frac{dv_O}{dt} + \frac{v_O}{RC(1 + A_{v0})} = \frac{1}{RC}\left(\frac{A_{v0}}{1 + A_{v0}}\right)(V_{ib} - v_{SRC})$$

(10.6.11)

Assuming $v_{O0} = 0$ and v_{SRC} is a step function V_{SRC}, the solution to this equation is:

$$v_O = A_{v0}(V_{ib} - V_{SRC})\left(1 - e^{-t/RC(1+A_{v0})}\right)$$

(10.6.12)

If A_{v0} is very large, $1 - e^{-t/RC(1+A_{v0})} \cong t/RCA_{v0}$ and if, in addition, $V_{ib} = 0$ Equation 10.6.12 reduces to $v_O = -V_{SRC}t/RC$ for a perfect op-amp integrator. It is seen from Equation 10.6.12 that V_{ib} adds algebraically to V_{SRC}. If $V_{io} = 1$ mV, $I_{io} = 20$ nA, and $R = 100$ kΩ, then $V_{ib} = 3$ mV, assuming that the effects of V_{io} and I_{io} are additive. V_{ib} sets a lower limit on the magnitude of the input signal.

Let $V'_{SRC} = V_{SRC} - V_{ib}$ and let $t' = 1/RC$ so as to normalize time. From Equation 10.6.12,

$$v_O = -A_{v0}V'_{SRC}(1 - e^{-t'/(1+A_{v0})})$$

(10.6.13)

The deviation Δv_O of v_O from $-V'_{SRC}t'$ is

$$\Delta v_O = -V'_{SRC}\left[t' - A_{v0}\left(1 - e^{-t'/(1+A_{v0})}\right)\right]$$

(10.6.14)

Expanding the exponential as a power series and neglecting powers of the exponent greater than 2, Equation 10.6.14 becomes:

$$\Delta v_O = -V'_{SRC}\left(\frac{t'}{1 + A_{v0}} + \frac{A_{v0}}{(1 + A_{v0})^2}\frac{(t')^2}{2}\right) \cong -\frac{V'_{SRC}t'}{A_{v0}}\left(1 + \frac{t'}{2A_{v0}}\right)$$

(10.6.15)

According to Equation 10.6.15, no matter how large is a finite A_{v0}, there will come a time when the deviation from linearity becomes significant. For example, if $|\Delta v_O/V'_{SRC}| = 0.005$, which is a 0.5% deviation from linearity, and $A_{v0} = 10^5$, the deviation occurs at $t' = 500$s. If $R = 100$ kΩ and $C = 0.1$ μF, $t = 5$ s. This emphasizes that in order to perform an accurate integration over a long period, A_{v0} must be appropriately large or RC large.

The integrator is considered in Section ST10.7 in the frequency domain, taking into consideration the variation of A_v with frequency. It is shown that the effective frequency range for integration is from $f_1 = 1/2\pi RCA_{v0}$ to $f_T = A_{v0}f_c$, where f_c is the corner frequency of A_v and f_T is the frequency at which $|A_v(f)| = 1$ (Exercise 7.4.6, Chapter 7).

Difference Amplifier*

A basic difference amplifier is shown in Figure 10.6.4. Assuming an ideal op amp to begin with, v_P at the noninverting input is simply the output of a voltage divider:

---

* Sabah 2008, p. 732.

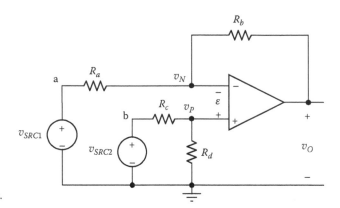

FIGURE 10.6.4
Basic op-amp difference amplifier.

$v_P = v_{SRC2}R_d/(R_c + R_d)$. Applying KCL at the inverting input, $(v_{SRC1} - v_N)/R_a = (v_N - v_O)/R_b$. Setting $v_P = v_N$, and eliminating v_N,

$$v_O = -\frac{R_b}{R_a}v_{SRC1} + \frac{1 + R_b/R_a}{1 + R_c/R_d}v_{SRC2} \qquad (10.6.16)$$

Equation 10.6.16 follows from superposition. When $v_{SRC2} = 0$, $v_{O1} = -v_{SRC1}R_b/R_a$. When $v_{SRC1} = 0$, the circuit reduces to a noninverting op amp with v_{SRC2} applied through a voltage divider. It follows that $v_{O2} = \dfrac{R_d}{R_c + R_d}\left(1 + \dfrac{R_b}{R_a}\right)v_{SRC2}$. Adding v_{O1} and v_{O2} gives Equation 10.6.16.

If $R_b/R_a = R_d/R_c = k$, Equation 10.6.16 reduces to:

$$v_O = k(v_{SRC2} - v_{SRC1}) \qquad (10.6.17)$$

The circuit can therefore be used for directly subtracting one signal from another. Resistance ratios can be closely matched in an IC (Section 4.5, Chapter 4). Signals can also be subtracted by inverting one signal and adding it to the other signal.

It is instructive to derive Equation 10.6.16 from a modification of the relation for an inverting amplifier having $v_{SRC2} = 0$ in Figure 10.6.4, so that $v'_O = -v_{SRC1} R_b/R_a$. With v_{SRC2} applied, $v_P = v_{SRC2} R_d/(R_c + R_d)$. Since $v_N = v_P$ in the ideal case, v_N must change from zero, under virtual ground conditions, to the new v_P. But because of the voltage-divider effect with shunt feedback as illustrated in Figure 7.4.8, Chapter 7, then in order for v_N to change by $v_{SRC2} R_d/(R_c + R_d)$, v_O must change by an amount v''_O such that $v''_O R_a/(R_a + R_b) = v_{SRC2} R_d/(R_c + R_d)$. Hence, $v''_O = [R_d/(R_c + R_d)][(R_a + R_b)/R_a]v_{SRC1}$. Adding v'_O and v''_O gives Equation 10.6.16.

Exercise 10.6.3

Using the same reasoning as above, show that if an input v_{SRC3} is added in series with R_d in Figure 10.6.4, the output is given by:

$$v_O = -\frac{R_b}{R_a}v_{SRC1} + \frac{R_d}{R_c + R_d}\frac{R_a + R_b}{R_a}v_{SRC2} + \frac{R_c}{R_c + R_d}\frac{R_a + R_b}{R_a}v_{SRC3} \qquad \blacksquare$$

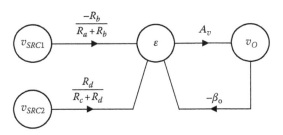

FIGURE 10.6.5
Signal-flow diagram of difference amplifier.

When A_v is finite, we can apply superposition to derive the signal-flow diagram of Figure 10.6.5. If $v_{SRC2} = 0$ in Figure 10.6.4, the signal-flow diagram is that of the inverting configuration shown in Figure 7.4.9b, Chapter 7, with $R_r = R_a$, $R_f = R_b$, and $v_{SRC} = v_{SRC1}$. If $v_{SRC1} = 0$, the circuit and the signal-flow diagram is that of the noninverting configuration (Figure 7.4.3, Chapter 7), with v_P omitted and the multiplier $R_d/(R_c + R_d)$ included between v_{SRC2} and v_P because of voltage division. The two signal-flow diagrams can now be combined as in Figure 10.6.5 where $\beta_o = R_a/(R_a + R_b)$.

With both v_{SRC1} and v_{SRC2} applied, it follows from Figure 10.6.5 that:

$$v_O = -\frac{R_b}{R_a + R_b} \frac{A_v}{1 + \beta_o A_v} v_{SRC1} + \frac{R_d}{R_c + R_d} \frac{A_v}{1 + \beta_o A_v} v_{SRC2} \qquad (10.6.18)$$

This can be simplified to give:

$$v_O = \left(-\frac{R_b}{R_a} v_{SRC1} + \frac{1 + R_b/R_a}{1 + R_c/R_d} v_{SRC2} \right) \frac{1}{1 + 1/\beta_o A_v} \qquad (10.6.19)$$

As $A_v \to \infty$, Equation 10.6.19 reduces to Equation 10.6.16, as expected.

The effect of input bias currents is minimized by equalizing the dc resistances seen by the two inputs, that is, $R_a \| R_b = R_c \| R_d$. This is satisfied if $R_b/R_a = R_d/R_c = k$.

Because significant voltages may be applied to both inputs of a difference amplifier, the common-mode response should be taken into consideration. Note that a common-mode *response* can arise not only due to inherent limitations in the amplifier, as discussed for the differential amplifier in Section 10.1, but also from a common-mode input that causes a *differential* input because of impedance unbalance at the inputs of the op amp. This is illustrated in Figure 10.6.6, where the input impedances Z_1, Z_2, and Z_3 account for the

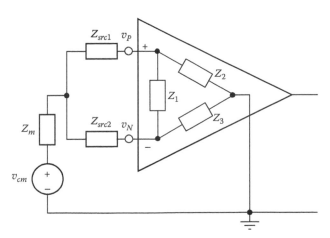

FIGURE 10.6.6
A common-mode input can cause a differential input because of impedance imbalance at the input of an op amp.

differential input impedance Z_{id} between the two inputs, and for the impedances Z_{PG} and Z_{NG} between the noninverting and inverting inputs, respectively, and the common reference. When the two inputs are joined together, $Z_{PG}\|Z_{NG}$ is the common-mode impedance. As discussed for the differential pair, Z_{id} is normally much smaller than Z_{PG} or Z_{NG}. All these impedances are largely resistive but could have small capacitive components due to the internal capacitances of the transistors.

Z_{src1} and Z_{src2} are due to differential sources connected to the op amp, with the differential input signals in series with these impedences set to zero. If $Z_{src1} = Z_{src2}$ and $Z_2 = Z_3$, it follows from symmetry that $v_P = v_N$, so that the differential input due to v_{cm} is zero. More generally, $v_P = v_N$ if a bridge balance condition is satisfied, that is, $Z_{src1}/Z_{src2} = Z_2/Z_3$. Otherwise, a differential input arises due to v_{cm} that gives rise to a corresponding output. Example SE10.2 illustrates a practical example from ECG recording.

Instrumentation Amplifier*

A useful amplifier is the IC instrumentation amplifier (IA) shown in Figure 10.6.7. The input stage consists of two noninverting op amps connected so as to provide a differential input, with their feedback resistors combined to provide a differential output. The second stage is the difference amplifier considered previously. Since the inputs are applied to the noninverting terminals, the input impedance is high. Moreover, as shown later, the overall gain is determined by R_1 only. To minimize the common-mode response, the resistors having the same subscript number should be well matched, as when all components are manufactured on the same chip.

If ideal op amps are assumed, the differential gain of the first stage can be very simply derived by noting that, because of the virtual short circuit at the inputs of an ideal op amp, $v_{P1} = v_{N1}$ and $v_{P2} = v_{N2}$. This means that the voltage across R_1 equals the differential input voltage $v_{SRC1} - v_{SRC2}$, so the current in R_1 is $(v_{SRC1} - v_{SRC2})/R_1$. This same current flows in the two resistances R_1 and R_2. Hence, $v_{O1} - v_{O2} = [(v_{SRC1} - v_{SRC2})/R_1](R_1 + 2R_2)$. The differential gain of the first stage is therefore:

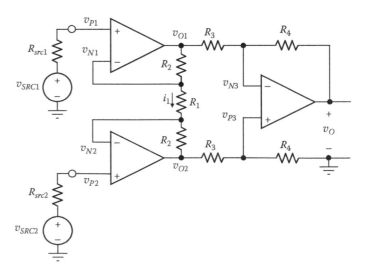

FIGURE 10.6.7
Instrumentation amplifier.

---

* Sabah 2008, p. 734.

$$\frac{v_{O1} - v_{O2}}{v_{SRC1} - v_{SRC2}} = 1 + 2\frac{R_2}{R_1} \tag{10.6.20}$$

In order to obtain the overall gain, this differential gain has to be multiplied by that of the difference amplifier, as given by Equation 10.6.17 with $k = R_4/R_3$:

$$\frac{v_O}{v_{SRC1} - v_{SRC2}} = \left(1 + 2\frac{R_2}{R_1}\right)\frac{R_4}{R_3} \tag{10.6.21}$$

It is seen that the gain can be varied by varying only R_1, which does not have be to be matched to any other resistance.

To determine the effects of amplifier imperfections, consider the first stage, with both op amps having a finite A_v, as shown in Figure 10.6.8. The governing equations are: $v_{SRC1} - v_{SRC2} = \varepsilon_1 + R_1 i_1 - \varepsilon_2$, $v_{O1} - v_{O2} = (R_1 + 2R_2)i_1$, $\varepsilon_1 = v_{O1}/A_v$, and $\varepsilon_2 = v_{O2}/A_v$. Eliminating i_1, ε_1, and ε_2 gives:

$$\frac{v_{O1} - v_{O2}}{v_{SRC1} - v_{SRC2}} = \frac{A_v}{1 + \beta_o A_v} = \frac{1}{\beta_o}\frac{1}{1 + 1/\beta_o A_v} \tag{10.6.22}$$

where $\beta_o = R_1/(R_1 + 2R_2)$. If $A_v \to \infty$, Equation 10.6.22 reduces to Equation 10.6.20.

If we express β_o as $\dfrac{R_1/2}{(R_1 + 2) + R_2}$, the first stage can be considered as consisting of two half circuits, as indicated in Figure 10.6.8 by the dotted line. Thus, $v_{O1} = \dfrac{1}{\beta_o}\dfrac{1}{1 + 1/\beta_o A_v}v_{SRC1}$ and $v_{O2} = \dfrac{1}{\beta_o}\dfrac{1}{1 + 1/\beta_o A_v}v_{SRC2}$. Subtracting these equations gives Equation 10.6.22.

It is clear from symmetry in Figure 10.6.8 that the effects of V_{io} in the two amplifiers cancel out at the outputs, as long as the two halves are identical. Similarly, the common-mode

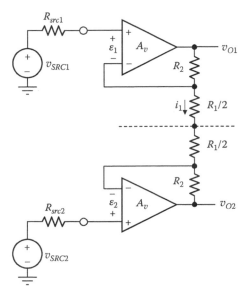

FIGURE 10.6.8
The input stage of the IA divided into two half circuits.

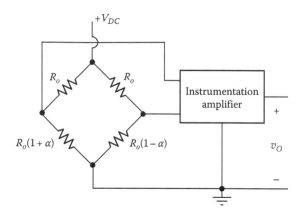

FIGURE 10.6.9
Figure for Application Window 10.6.1.

response at the outputs cancels out so that the common-mode response at the output of the IA depends on the common-mode response of the second stage. In order to minimize the effects of input bias currents, the source resistances R_{src1} and R_{src2} must be equal. It is seen that, as long as the two halves of the IA are properly matched, the IA is a high-performance amplifier.

Application Window 10.6.1 Bridge Amplifier*

An IA can be used to amplify the output of a transducer bridge, as illustrated in Figure 10.6.9 for a strain-gauge bridge. When the resistance of one strain-gauge element increases by αR_o, as a result of strain in a given direction, the resistance of the other element decreases by αR_o, where R_o is the nominal resistance of the elements in the rest position. The upper two resistors are dummy strain gauges that are identical to the active strain gauges but are not subjected to any strain. Such a bridge configuration has two advantages: (1) since the resistances of the lower, active strain-gauge elements change in opposite directions, the bridge output for a given strain is essentially doubled, and (2) because the four elements have the same resistance at rest and are made of the same material, they are equally affected by changes in ambient temperature so that the bridge output is independent of these changes.

10.7 Switched-Capacitor Circuits[†]

Analog IC filters can be implemented as switched-capacitor circuits, the objective being to replace resistors by capacitors and MOSFET switches (Chapter 5). Not only does this enable simulation of large resistors but it also results in improved accuracy of the frequency response, good linearity, and extended dynamic range. The improved accuracy of frequency response is due to filter parameters being determined by capacitance ratios which can be set in ICs to a precision as high as 0.1% compared to a variation as large as 20% for RC time constants.

To appreciate the underlying principle, consider a capacitor C connected to a changeover switch S and to two dc voltage sources V_1 and V_2 (Figure 10.7.1). Assume that the switch is

---

* Sabah 2008, p. 735.
[†] Sabah 2008, Section ST18.6, p. 756.

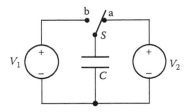

FIGURE 10.7.1

A capacitor used to transfer charge between two voltage sources.

moved from position a to position b and back to a at a frequency f_S, or period $T_S = 1/f_S$, and that for the sake of argument, $V_1 > V_2$. When the switch moves to position b, it acquires a charge CV_1, and when it moves to position a, the charge on the capacitor is CV_2. Since $V_1 > V_2$, charge is transferred from V_1 to V_2. The rate at which this charge is transferred constitutes a "steady" current given by $I = C(V_1 - V_2)/T_S = C(V_1 - V_2)f_S$. If this current were due to a resistance R, then $I = (V_1 - V_2)/R$, where R is an equivalent resistance of the capacitor–switch combination and is given by:

$$R = \frac{1}{Cf_S} \tag{10.7.1}$$

As an example of a basic switched-capacitor circuit, consider the integrator of Figure 10.7.2a. Switches are implemented in practice using MOSFETs, which are well-suited for this purpose because they can have a high off resistance, a relatively low on resistance, and zero offset voltage. Although a single MOSFET can be used as a single-pole, single-throw switch, a transmission gate is more advantageous (Section 5.4, Chapter 5). The changeover switch of Figure 10.7.1 is implemented by two transmission gates

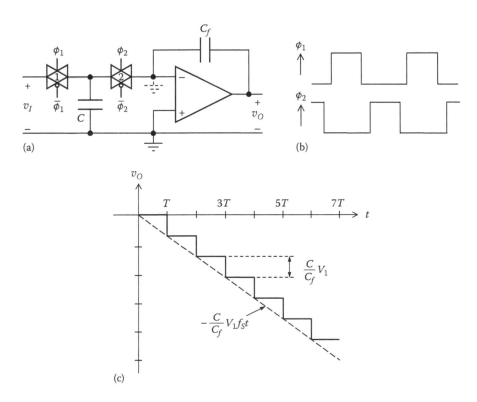

FIGURE 10.7.2

(a) Basic switched-capacitor integrator; (b) control voltage waveforms applied to the circuit in (a); (c) variation of output voltage with respect to time.

controlled by clock signals ϕ_1 and its negation $\overline{\phi}_1$ for one transmission gate, and by the combination ϕ_2–$\overline{\phi}_2$ for the other gate. The clock signals are nonoverlapping (Figure 10.7.2b), which means that at no time are the two clocks simultaneously high, which prevents the two switches from being closed simultaneously.

C_f establishes a feedback path at signal frequency, so that the inverting terminal of the op amp is a virtual ground, and V_2 in Figure 10.7.1 is zero. A charge CV_1 is transferred to C_f every T_S, resulting in a stepwise variation in the output voltage v_O of $-V_1C/C_f$ per step. The average rate of change of v_O per unit time is $-(V_1/T_S)(C/C_f) = -V_1f_SC/C_f$, equivalent to a continuous variation with time of $-V_1f_S(C/C_f)$ (Figure 10.7.2c), or to an integrator gain of $-1/(RC_f) = -f_S(C/C_f)$ (Equation 10.7.1) that depends only on f_S and capacitance ratios.

It should be noted that the above analysis, based on charge transfer, is valid for an input signal v_I that varies at a frequency that is much smaller than f_S, say one-tenth of f_S. For signal frequencies that are comparable to f_S, the circuit is analyzed using the z-transform. This is essentially the Laplace transform in discrete time and is beyond the scope of our discussion.

A limitation of the circuit of Figure 10.7.2a is that it is sensitive to stray capacitances, which are mainly due to: (1) internal capacitances associated with the inputs and outputs of the transmission gates and op amp, and (2) capacitance of interconnections. Some of these capacitances are voltage dependent, so it would not be possible to account for them by adjusting the values of C and C_f. In practice, the circuit is modified as in Figure 10.7.3 to make it insensitive to the dotted stray capacitances. C is now connected in series rather than in shunt, as in Figure 10.7.2, which necessitates the use of four transistor switches instead of two.

Let us first investigate the operation of this circuit ignoring the stray capacitances to begin with. When ϕ_2 goes high, switches 2 and 3 are closed, while switches 1 and 4 are open (Figure 10.7.2b). C is completely discharged, since both of its terminals are grounded. C_f retains the charge it had previously acquired because switch 4 is open. When ϕ_1 is high, one end of C is connected to v_I whereas the other end is connected to virtual ground. C is therefore charged to v_I, but the charging current flows through C_f, effectively adding positive charge to the LHS plate of C_f and negative charge to its RHS plate. The integrator is therefore inverting; the transfer of charge is the same as that in Figure 10.7.2a and the effective gain is $-f_S(C/C_f)$.

Let us verify that the parasitic capacitances do not affect operation, other than slightly prolonging settling times. C_c is permanently connected between true ground and virtual ground. The upper plate of C_b is connected to true ground when ϕ_2 is high and to virtual ground when ϕ_1 is high. C_a is charged and discharged with C, but its charging and discharging currents do not flow through C_f. Thus, parasitic insensitivity is achieved by having C series-connected, whereas the parasitic capacitances are shunt-connected.

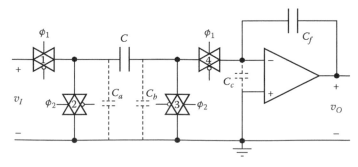

FIGURE 10.7.3
Modified switched-capacitor, inverting integrator that is insensitive to stray capacitances.

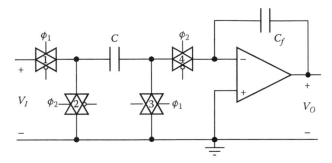

FIGURE 10.7.4
The circuit of Figure 10.7.3 modified to give a noninverting integrator.

An interesting feature of Figure 10.7.3 is that it can be converted to a noninverting integrator by having switch 3 controlled by ϕ_1 and switch 4 by ϕ_2, as illustrated in Figure 10.7.4. C is charged when ϕ_1 goes high and is discharged when ϕ_2 goes high, but the discharge current flows away from virtual ground making the RHS plate of C_f positive and its LHS plate negative.

Exercise 10.7.1

If $f_S = 10^5$ kHz, $C_f = 1\ \mu\text{F}$, $C = 0.01\ \mu\text{F}$, and $V_1 = 5$ V in Figure 10.7.2a, determine: (a) the step size of v_O; (b) the steady time variation of v_O; (c) the equivalent charging resistance.
Answers: (a) -50 mV; (b) $-5000t$ V; (c) 1 kΩ. ∎

10.8 Digital-Analog Conversion*

It is assumed in what follows that the decimal equivalent of an n-bit word is given by:

$$D = b_{n-1}2^{n-1} + b_{n-2}2^{n-2} + \cdots + b_2 2^2 + b_1 2^1 + b_0 2^0 \tag{10.8.1}$$

Thus, if n-bit word is 1110, the decimal equivalent is $1 \times 2^3 + 1 \times 2^2 + 1 \times 2^1 + 0 \times 2^0 = 14$.

A digital-to-analog (DAC) converter converts an n-bit word, where n depends on the particular system, to a voltage in the range of 0 and a limit $V_{\max}$. Since $V_{\max}$ would correspond to the maximum value $(2^n - 1)$ of the binary word, it follows that a change in the binary word of one least significant bit would correspond to $\Delta V = V_{\max}/(2^n - 1)$, where ΔV is the **resolution** of the DAC. Thus, if $n = 8$ and $V_{\max} = 12$V, $\Delta V = 10/255 = 39.2$ mV.

A fairly obvious way of implementing a DAC is to use an inverting adder with the bits controlling resistances weighted in powers of 2. However, when n is larger than 4, it becomes awkward to have exact values for all the resistances. It becomes preferable to use an R–$2R$ ladder for the resistances, as shown in Figure 10.8.1. Starting from node b_0, it can be readily verified that the resistance to the right of each node is $2R$ and the resistance connecting the node directly to ground, or virtual ground, is also $2R$. Thus, current divides equally at each node, and the currents flowing to ground or virtual ground at each node increase in powers of 2, from right to left.

Considering node b_{n-1}, $V_{\text{REF}} = 2R \times 2^{n-1} \times I_O$, or,

$$I_O = \frac{V_{\text{REF}}}{2^n R} \tag{10.8.2}$$

---

* Sabah 2008, p. 705 and Example SE18.8, p. 756.

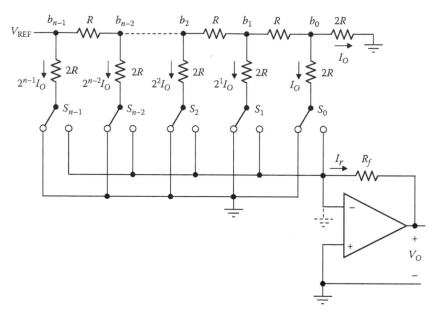

FIGURE 10.8.1
DAC using an inverting op amp adder and an R-2R ladder.

A given switch is placed in the right position, or in the left position, depending on whether the value of the corresponding bit is 1 or 0, respectively. It follows that:

$$I_r = \frac{V_{REF}}{2^n R} \left[b_{n-1}2^{n-1} + b_{n-2}2^{n-2} + \cdots + b_2 2^2 + b_1 2^1 + b_0 2^0 \right] \quad (10.8.3)$$

and,

$$V_O = -\frac{V_{REF}R_f}{2^n R} \left[b_{n-1}2^{n-1} + b_{n-2}2^{n-2} + \cdots + b_2 2^2 + b_1 2^1 + b_0 2^0 \right] \quad (10.8.4)$$

where b_m is 0 or 1, if the mth bit is 0 or 1, respectively, and $m = 0, 1, 2, \ldots, n-1$ for an n-bit word. Equation 10.8.4 confirms that V_O is proportional to the decimal value of the binary word. If all the bits are 1, it follows from Equation 10.8.4 that:

$$V_O = -\frac{R_f}{R} \frac{2^n - 1}{2^n} V_{REF} = -V_{max} \quad (10.8.5)$$

As a simple example, consider the case of $n = 4$, $V_{max} = 15$ V, and $V_{REF} = 5$ V. If $R = 10$ kΩ, then $R_f = (15/5) \times (16/15) \times 10 = 32$ kΩ. The resolution is $\Delta V = V_{max}/(2^n - 1) = 15/15 = 1$ V. In other words, if the binary word increases from zero in steps of 1 least significant bit, the analog output increases in steps of 1 V from zero to 15 V. The resolution in terms of currents is I_O, which, from Equation 10.8.2, is $5/(16 \times 20) = 1/32$ mA. It follows that $\Delta V = R_f I_O = 32/32 = 1$ V, as found previously.

Figure 10.8.2 illustrates a DAC based on a current steering circuit using MOSFETs of widths that increase in powers of 2. Q_r is the diode transistor carrying a current $I_{REF} = V_{REF}/R$. The gate and drain of Q_r are at virtual ground, and the output of op amp A_1 is such that V_{GS} of Q_r, and all the other MOSFETs, is that required for the current I_{REF}. Normally, $I_{REF} = I_O$ so that

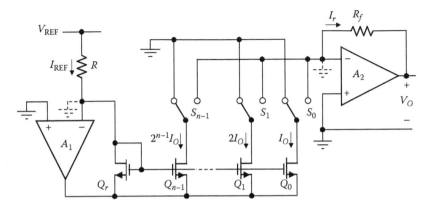

FIGURE 10.8.2
DAC based on a current steering circuit using MOSFETs of widths that increase in powers of 2.

transistors Q_r and Q_0 have the same width, whereas the widths of transistors $Q_0 \ldots Q_{n-1}$ increase in powers of 2. The drains of these transistors are connected to single-pole, double-throw switches so that the currents either flow to ground or to the virtual ground of op amp A_2, as in Figure 10.8.1. A given switch is placed in the right position, or in the left position, if the value of the corresponding bit is 1 or 0, respectively. It follows that:

$$I_r = -I_0 \lfloor b_{n-1}2^{n-1} + b_{n-2}2^{n-2} + \cdots + b_2 2^2 + b_1 2^1 + b_0 2^0 \rfloor \qquad (10.8.6)$$

and

$$V_O = \frac{V_{\text{REF}} R_f}{R} [b_{n-1}2^{n-1} + b_{n-2}2^{n-2} + \cdots + b_2 2^2 + b_1 2^1 + b_0 2^0] \qquad (10.8.7)$$

which is proportional to the decimal equivalent of the binary number. Note that the drains of $Q_0 \ldots Q_{n-1}$ are nominally at ground voltage, like the drain of Q_r, which ensures good current matching.

A single-pole, double-throw switch can be implemented by a differential pair in which the current is switched between two transistors. Figure 10.8.3 illustrates S_m for the mth bit. V_{BIAS} is applied to the gate of Q_{mr} whose drain is permanently grounded. Q_{ms} drain is connected to the inverting terminal of A_2, and a voltage $V_{\text{BIAS}} + v_{bm}$ is applied to Q_{ms} gate, where v_{bm} represents the value of b_m. If $b_m = 0$, v_{bm} is sufficiently negative so that Q_{ms} is turned off and I_m, the drain current of Q_{mr}, is diverted to ground, corresponding to S_m being in the left-hand position. On the other hand, if $b_m = 1$, v_{bm} is sufficiently positive so that Q_{ms} is turned on and Q_{mr} is turned off. I_m is diverted to the inverting input of A_2, corresponding to S_m being in the right-hand position.

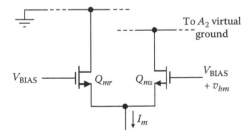

FIGURE 10.8.3
MOSFET differential-pair implementation of a single-pole, double-throw switch.

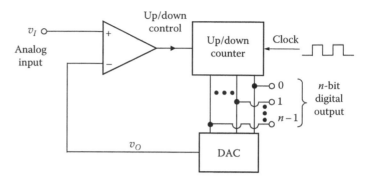

FIGURE 10.8.4
An implementation of an ADC.

The inverse operation, that is, analog-digital conversion can be implemented as in Figure 10.8.4. The analog input v_I is applied to one input of a comparator, the output of which controls an up/down counter fed from a clock generator. The counter output is applied to a DAC, whose output v_O is connected to the other input of the comparator, thus completing the negative feedback loop. If $v_O < v_I$, the comparator output is of one polarity and causes the counter to count clock pulses upward, that is, as an increasing number. The input to the DAC increases and v_O increases until it just exceeds v_I. The output of the comparator then changes polarity, causing the counter to count downward, that is, in a decreasing number of counts. v_I is generally time varying, so the feedback keeps v_O close to v_I, and v_O is said to "track" v_I. Since v_O is the output of a DAC, the input to the DAC, which is the counter output, must be the digital equivalent of v_I.

This implementation of the analog-to-digital converter (ADC) illustrates a useful technique, namely, the insertion of a functional block in a negative feedback loop so as to generate the inverse function of that block.

Summary of Main Concepts and Results

- Arbitrary inputs applied to the base terminals of a differential pair can be resolved into a common-mode input that is the average of the two inputs, and a differential input that is the difference between the two inputs.

- Differential amplifiers discriminate between differential and common-mode inputs, the latter inputs generally being undesirable. The CMRR is the magnitude of the ratio of the differential gain to the common-mode gain.

- Analysis of the differential pair is facilitated by using a differential half circuit that is a CE/CS amplifier, and a common-mode half circuit that is a CE/CS amplifier with a feedback resistor.

- The input bias current I_{ib} is the average of the two input bias currents I_{B1} and I_{B2}, whereas the input offset current I_{io} is the magnitude of the difference between I_{B1} and I_{B2}.

- The input offset voltage is the output offset voltage, measured at the output for zero input, referred to the input. A random input offset voltage arises from dissymmetry in the two halves of the differential amplifier because of random mismatch between components. A systematic input offset voltage can arise from inherent features of the circuit, as in a current-mirror load.

- The performance of the basic differential amplifier can be improved by using active loads and composite transistor connections. In particular, a current mirror load doubles the single-ended voltage gain of a differential pair, and significantly increases common-mode rejection.

- In comparing BJT and MOSFET differential pairs, $v_{id(\text{max})}$ for full switching is nearly $4V_T$ for the BJT and $\sqrt{2}V_{ov}$ for the MOSFET. In the corresponding expressions for input offset voltage, V_T in the case of the BJT is replaced by $V_{ov}/2$ for the MOSFET.

- The folded cascode op amp has the advantage of a relatively wide common-mode input range.

- Operational transconductance amplifiers have low-impedance internal nodes so that the frequency response is determined by a dominant pole due to the high-impedance output node.

- Self-biasing schemes are much less sensitive to variations in supply voltage and temperature.

- Whereas it is possible, at least in principle, to compensate for the effects of steady input bias currents and offset voltage of an op amp, it is impractical to compensate for drift of these quantities, which therefore sets a limit on the output stability of an op-amp circuit.

- In order to minimize the effect of input bias currents in op-amp circuits, the dc resistances in the external circuit, between either op-amp input and ground, should be equal. A common-mode response may be not only due to dissymmetry in the amplifier, but could also arise from a common-mode input that causes a differential input because of impedance unbalance at the inputs of the op amp.

- Input offset voltage and bias currents have a drastic effect on integrators, as they saturate the op-amp output, even when no input is applied. Moreover, to amplify down to very low frequencies, or over long periods, the op-amp gain must be very high.

- The IA is a high-performance IC op amp that has a noninverting differential input stage followed by a difference amplifier as a second stage.

- In switched-capacitor circuits, resistors are replaced by capacitors and MOSFET switches, which allows simulation of large resistors and results in improved accuracy of the frequency response, good linearity, and extended dynamic range.

- A DAC can be based on an inverting adder and an R-$2R$ resistive ladder or MOSFETs of widths that increase in powers of 2. An ADC can be implemented by connecting a DAC converter in a negative feedback loop.

Learning Outcomes

- Articulate the basic properties of the differential pair.
- Articulate the effects of op-amp imperfections on the performance of op-amp amplifiers, adders, and integrators.
- Articulate the basic principles of operation of switched-capacitor circuits, DAC and ADC.

Supplementary Examples and Topics on Website

SE10.1 Simulation of Large Feedback Resistor. A T-circuit is used to simulate a large R_f in an inverting configuration. The performance of the circuit is compared with that using a single resistor for R_f when finite A_v, input bias current, and input offset voltage are taken into account (Sabah 2008, Example SE18.1, p. 756).

SE10.2 ECG Recording. Gives an example of ECG recording that emphasizes the common-mode response of the amplifier due to pick-up from installation wiring (Sabah 2008, Example SE18.3, p. 756).

ST10.1 Improvement of Performance of Differential Pair. Discusses measures for improving the performance of differential amplifiers, including active loads and composite transistors.

ST10.2 CMRR of CMOS Differential Pair. Derives the CMRR for a CMOS differential pair having a current-mirror load.

ST10.3 Improvement of Performance of Folded Cascode Op Amp. Discusses various techniques for improving the performance of the basic folded-cascode op amp.

ST10.4 Improved CMOS Biasing Circuits. Discusses various techniques for improving the performance of the self-bias circuit of Figure 10.4.2 as well as bandgap references.

ST10.5 BJT Op Amps. Analyzes examples of BJT op amps, including the μA741.

ST10.6 Noninverting and Inverting Adders. Analyzes the two types of adders and compares them taking into account input offset voltage, input bias currents, and finite A_v (Sabah 2008, Section ST18.2, p. 755).

ST10.7 Integrator in Frequency Domain. Analyzes an integrator in the frequency domain, including the frequency variation of the op amp gain, and derives the frequency range over which integration is nominally ideal (Sabah 2008, Section ST18.3, p. 755).

Problems and Exercises

P10.1 Differential Pair

P10.1.1 In the differential amplifier of Figure 10.1.1, $V_{CC} = V_{EE} = 10$ V, $I_{EE} = 2$ mA with $R_{EE} = 50$ kΩ, $R_C = 5$ kΩ, $\beta = 150$, $V_A = 100$ V, and $r_\mu = 5$ MΩ. Determine: (a) the dc base currents when the two bases are grounded; (b) the differential input signal currents, A_{dd0}, and r_{ind} when $v_{id} = 10 \sin \omega t$ mV is applied; (c) the common-mode input signal currents, A_{cm0}, and $r_{in(cm)}$ when a common-mode signal $v_{cm} = 100\sin\omega t$ mV is applied. Simulate with PSpice.

P10.1.2 A differential amplifier has $V_{CC} = 15$ V, $V_{EE} = 15$ V, $R_C = 10$ kΩ, with I_{EE} provided by $R_{EE} = 20$ kΩ resistor connected between the emitters and $-V_{EE}$. Determine the single-ended differential gain, common-mode gain, and CMRR, assuming $V_{BE} = 0.7$ V, $\alpha_F \cong 1$, and large r_o.

P10.1.3 A 20 kΩ resistor is connected between the two collectors in Problem P10.1.2. Derive the differential and common-mode half circuits.

P10.1.4 R_{EE} in Problem P10.1.2 is replaced by the π-circuit shown in Figure P10.1.4. Derive the differential and common-mode half circuits.

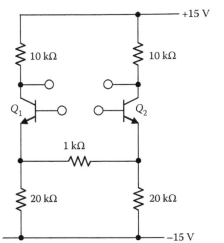

+15 V

10 kΩ 10 kΩ

Q_1 Q_2

1 kΩ

20 kΩ 20 kΩ

−15 V **FIGURE P10.1.4**

P10.1.5 A differential amplifier has a CMRR of 70 dB. The single-ended output voltage is 100 mV when the inputs are $v_1 = 21.5$ mV and $v_2 = 20$ mV. Determine the largest magnitude of the differential gain.

P10.1.6 When the inputs to a differential amplifier are $(\sin t - \cos t)$ mV and $(\sin t + \cos t)$ mV, the single-ended output is $(20{,}000\cos t - \sin t)$ mV. Determine the CMRR.

P10.1.7 In a given differential pair having $I_{EE} = 2$ mA, the BEJ areas are mismatched by 25%. Determine: (a) the emitter currents of the two transistors for zero input; (b) the differential input that equalizes the currents in the two transistors.

P10.1.8 Consider the differential amplifier of Figure 10.1.1 with $V_{CC} = 15$ V, $I_{EE} = 1$ mA, $V_{BE} = 0.7$ V, $\beta = 100$, and large r_o. (a) What v_{id} will cause i_E of Q_1 to become 1% that of Q_2? (b) Under these conditions, what value of R_C will cause Q_2 to reach saturation with $v_{CB} = 0$? (c) For this value of R_C, what is the largest v_{cm} that will not cause the saturation of the transistors with $v_{CB} = 0$? (d) If the negative of v_{cm} found in (c) is applied, what is the common-emitter voltage, and what are the collector-emitter voltages? (e) What is A_{dd0}? Note that the larger R_C, the larger is the differential gain but the lower is the largest common-mode voltage that will not cause the saturation.

P10.1.9 Determine the largest v_{id} that can be applied to a BJT differential amplifier if the departure from linearity is not to exceed 1%, assuming $V_T = 26$ mV. Simulate with PSpice.

P10.1.10 A differential amplifier has $V_{CC} = V_{EE} = 12$ V, $I_{EE} = 2$ mA, $R_C = 10$ kΩ, and $R_E = 0.5$ kΩ in series with each emitter. The transistors have $\beta = 150$, and large r_o. (a) Derive the differential half circuit. (b) Determine A_{dd0} and r_{ind}.

DP10.1.11 A differential amplifier is required to have $A_{dd0} = 25$, $v_{od} = 2$ V, and $r_{ind} = 50$ kΩ using transistors having $\beta = 100$ and large r_o. The signal amplitude across each BEJ is to be limited to 5 mV for good linearity. Design a suitable amplifier using emitter resistors.

P10.1.12 A MOSFET differential amplifier has $V_{DD} = 12$ V, $V_{tn} = 1$ V, $I_{SS} = 9$ mA, $r_o = 0.01$ V^{-1}, $R_D = 2$ kΩ, and $k_n = 4$ mA/V^2. Determine: (a) g_m for zero input; (b) A_{ds0}. Simulate with PSpice.

P10.1.13 Determine the largest common-mode signal that can be applied to the differential amplifier of Problem P10.1.12 without the transistors going into saturation, (a) neglecting the body effect, (b) assuming $\gamma = 0.6$ V$^{1/2}$, $|\varphi_F| = 0.3$ V, and the body connected to −12V.

P10.1.14 A MOSFET differential amplifier has $V_{DD} = V_{SS} = 10$ V, $I_{SS} = 4$ mA with $R_{SS} = 50$ kΩ, $R_D = 5$ kΩ, $k_n = 4$ mA/V^2, and $V_{tn} = 1.2$ V. Determine: (a) v_{od} for $v_{id} = 10\sin\omega t$ mV; (b) v_o for $v_{cm} = 100\sin\omega t$ mV, assuming $\lambda = 0.02$ V^{-1} and no body effect; and (c) CMRR.

P10.1.15 A MOSFET differential amplifier has $I_{SS} = 400$ μA, $k_n = 2$ mA/V^2, $V_{tn} = 1$ V, and large r_o. Determine: (a) V_{GS} and g_m for zero input; (b) v_{id} for full current switching; (c) I_{SS} for full switching to occur for half the v_{id} found in (b); (d) i_{D1}, i_{D2} and v_S for $v_{G1} = 2.5$ V and $v_{G2} = 2$ V, neglecting the body effect; (e) if $R_D = 10$ kΩ, what is $V_{DD(min)}$ that would allow a common-mode input of 2 V with the transistors remaining in saturation? Simulate with PSpice.

P10.1.16 Consider the MOSFET differential amplifier with a current mirror load, as in Figure 10.1.10a, with $V_{DD} = 5$ V, $I_{SS} = 200$ μA, $k_{n,p} = 0.4$ mA/V^2, $|V_t| = 0.8$ V and $\lambda = 0.02$ V^{-1}. Determine: (a) g_m of the transistors, open circuit V_O, and V_S, assuming perfect matching and zero input; (b) if the output is short-circuited, write down the governing equations. Simulate with PSpice.

P10.1.17 For the amplifier of Problem P10.1.16, determine: (a) G_m; (b) the small-signal, open circuit voltage gain; (c) the CMRR. Simulate with PSpice.

P10.1.18 Assume that the two R_C's of a BJT differential amplifier differ by ΔR_C, their average being R_C. Derive V_{C1} and V_{C2} in terms of R_{C1}, R_{C2}, and the collector current $\alpha_F I_{EE}/2$. Determine V_{od} when both inputs are grounded, neglecting r_o, and show that $V_{io} = V_{od}/A_{dd0} = (\Delta R_C/R_C)V_T$.

P10.1.19 Assume that the saturation currents I_{S1} and I_{S2} of a BJT differential pair differ by ΔI_S so that $I_{C1} = (\alpha_F I_{EE}/2)(1 + \Delta I_S/2I_S)$ and $I_{C2} = (\alpha_F I_{EE}/2)(1 - \Delta I_S/2I_S)$. Neglecting r_o, show that $V_{io} = (\Delta I_S/I_S)V_T$.

P10.1.20 Show that if the two transistors of a BJT differential amplifier have large, unequal β_Fs that are denoted by β_{F1} and β_{F2}, (a) $V_{io} \cong V_T(1/\beta_{F1} - 1/\beta_{F2})$, neglecting r_o, and (b) $I_{io} \cong I_{iB}(\Delta\beta_F/\beta_F)$.

P10.1.21 Assume that the two R_D's of a MOSFET differential amplifier differ by ΔR_D, their average being R_D. Divide V_{od} by A_{dd0}, neglecting r_o, and use the relation $g_m = I_{SS}/V_{ov}$ to show that $V_{io} = (\Delta R_D/R_D)(V_{ov}/2)$. Compare with the result of Problem P10.1.18.

P10.1.22 Consider that the two MOSFETs of a differential amplifier differ by $\Delta(W/L)$. Show that
$$I_{D1} = \frac{I_{SS}}{2}\left[1 + \left(\frac{\Delta(W/L)}{2(W/L)}\right)\right] \text{ and } I_{D2} = \frac{I_{SS}}{2}\left[1 - \left(\frac{\Delta(W/L)}{2(W/L)}\right)\right],$$
the average aspect ratio being W/L. Divide V_{od} by A_{dd0}, and use the relation $g_m = I_{SS}/V_{ov}$ to show that
$$V_{io} = \left(\frac{\Delta(W/L)}{2(W/L)}\right)\frac{V_{ov}}{2}.$$

P10.1.23 Consider that the thresholds of a MOSFET differential pair differ by $\Delta V_{tn} \ll 2V_{ov}$, the average being V_{tn}. Divide V_{od} by $g_m R_D$, and use the relation $g_m = I_{SS}/V_{ov}$ to show that $V_{io} = \Delta V_{tn}$.

P10.1.24 A BJT differential amplifier has $I_{EE} = 1$ mA, $R_C = 10$ kΩ, $\beta = 200$, large r_o, $C_\mu = 0.8$ pF, and $f_T = 900$ MHz. A source of 10 kΩ source resistance is applied differentially. Determine the low-frequency gain from v_{src} to v_{od}, the 3-dB cutoff frequency, and the gain-bandwidth product.

P10.1.25 Repeat Problem P10.1.24 assuming that $R_E = 50$ Ω is included in series with each emitter.

P10.1.26 Assume that I_{SS} of Problem P10.1.24 has a source resistance of 200 kΩ and $C_{EE} = 5$ pF. Determine the poles of the CMRR and $f_{p(cm)}$ (Equation 10.1.18).

P10.1.27 A MOSFET differential pair has $I_{SS} = 500$ μA, $R_D = 20$ kΩ, $k_n = 2$ mA/V^2, $V_{tn} = 1$V, $r_o = 500$ kΩ, $C_{gs} = 0.2$ pF, $C_{gd} = 0.04$ pF, and a differential source resistance of 100 kΩ. Determine A_{dd0} and the 3-dB cutoff frequency.

P10.1.28 Assume that I_{SS} of Problem P10.1.27 has $R_{SS} = 2$ MΩ and $C_{SS} = 5$ pF. Determine the poles of the CMRR.

P10.1.29 Show that if r_μ is taken into account in the small-signal hybrid-π model of the differential half circuit, $r_{ind} = 2\left[r_\pi \middle\| \left(\dfrac{r_\mu + (r_o \| R_C)}{1 + g_m(r_o \| R_C)}\right)\right]$.

P10.1.30 Add r_o and r_μ to the small-signal T-model of the common-mode half circuit. Neglecting r_e and assuming $R_C \ll r_\mu$, and $(1-\alpha)R_C \ll r_o$, show that $A_{cm0} = -\dfrac{\alpha R_C}{2R_{EE}}\left[1 - R\left(\dfrac{1}{\beta r_o} + \dfrac{1}{\alpha r_\mu}\right)\right]$, where $R = 2R_{EE} + \alpha R_C$. Note that if the second term inside the square brackets is small compared with unity, A_{cm0} reduces to that given by Equation 10.1.11.

P10.2 IC Op Amps

P10.2.1 Draw the circuit diagram of a two-stage op amp similar to that of Figure 10.2.1 but using complementary transistors. Assume that $V_{DD} = V_{SS} = 2.5$ V, $k_n' = 250$ μA/V^2, $V_{tn} = 0.5$ V, $\lambda_n = 0.05$ V^{-1}, $k_p' = 100$ μA/V^2, $V_{tp} = -0.6$ V, $\lambda_p = 0.1$ V^{-1}, $I_{REF} = 100$ μA, $(W/L) = 20$ for all the transistors except Q_8, and perfect matching. Neglect the body effect. Determine: (a) I_D and V_{ov} of all the transistors and $(W/L)_8$ to make $V_O = 0$ for zero input, neglecting channel-width modulation except in Q_7 and Q_8; (b) the overall gain for a differential input; (c) the common-mode gain and the CMRR; (d) the range of common-mode input; (e) the range of output voltage; and (f) C_c required to give a dominant pole at 100 Hz based on Equation 10.2.18 and the resulting slew rate.

P10.2.2 Add a source follower stage to the op amp of Problem P10.2.1 consisting of two NMOS transistors Q_9 and Q_{10} of $(W/L) = 20$, with Q_9 base connected to Q_7 base. Determine under the same conditions as in Problem P10.2.1: (a) $(W/L)_8$ to make $V_O = 0$ for zero input, (b) the overall gain for a differential input; (c) the range of v_O; and (e) r_{out} compared to Problem P10.2.1.

P10.2.3 In the op amp shown in Figure P10.2.3, $k_n' = 25$ μA/V^2, $k_p' = 10$ μA/V^2, $|V_t| = 0.75$ V, $\lambda = 0.05$ V^{-1}, $I_{REF} = 100$ μA, and (W/L) of the MOSFETs in μm/μm as indicated. Determine: (a) the noninverting input; (b) I_D and V_{ov} of all the transistors and V_O for zero input, neglecting channel-width modulation and the body effect and assuming perfect matching; (c) the overall gain for a differential input; and (d) r_{out}.

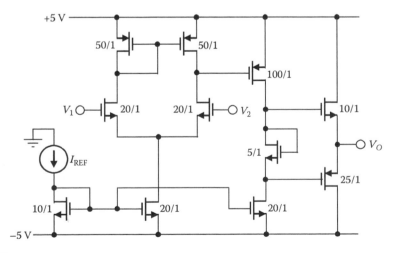

FIGURE P10.2.3

P10.2.4 Consider the folded-cascode op amp of Figure 10.3.1 with ± 2.5 V power supplies, $I_{DD} = 200$ μA, $I_{SS} = 160$ μA, $k_n' = 50$ μA/V^2, $k_p' = 20$ μA/V^2, $V_{tn} = 0.4$ V, $V_{tp} = -0.5$ V,

$V_{ov} = 0.2$ V, $\lambda = 0.04$ V^{-1} for all transistors, and negligible body effect. Determine: (a) V_{BIAS1}, $(W/L)_{b,c}$, $(W/L)_{3,4}$, and V_{BIAS3} assuming $V_{SDc} = V_{SD4} = 1.25$ for $V_O = 0$, and a zero input; (b) $(W/L)_{1,2}$, $(W/L)_a$, and V_{BIAS2}; (c) V_{BIAS4} and $(W/L)_{d-g}$; (d) v_{cm}^+, v_{cm}^-, v_O^+, and v_O^-; and (e) A_{v0}.

P10.2.5 (a) Determine f_p and f_T for the op amp of Problem P10.2.4, assuming $C_L = 50$ pF; (b) show that if R_Z is added in series with C_L, the transfer function is modified to $\dfrac{r_{out}(1 + sC_LR_Z)}{1 + sC_L(r_{out} + R_Z)}$; choose R_z to place the zero at $1.3f_T$; (c) SR.

P10.2.6 Consider a two-stage BJT op amp similar to that of Figure 10.2.1. Assume that $V_{DD} = V_{SS} = 10$ V, $I_{REF} = 200$ μA, and $\beta = 100$, $r_\pi = 2$ kΩ, $|V_{BE}| = 0.7$ V, $V_{CE(sat)} = 0.3$ V, and $V_A = 100$ V for all transistors. Determine: (a) I_C of all the transistors for zero input, neglecting base-width modulation and considering that $V_O = 0$ for zero input; (b) the overall gain for a differential input; (c) A_{cm} and the CMRR; (d) the range of v_{cm}; (e) the range of v_O; and (f) C_c required to give a dominant pole at 100 Hz and the resulting SR.

P10.2.7 For the op amp shown in Figure P10.2.7, assume that $V_D = 0.7$ V for a conducting diode, $V_{BE} = 0.7$ V, and $V_{CE(sat)} = 0.3$ V. The BJTs are identical and have negligible base current. The FETs are identical with $k_n = 1$ mA/V^2 and $V_{tn} = 1$ V. Neglecting base-width and channel-length modulation and the body effect, determine: (a) the noninverting input; (b) V_O when $V_1 = V_2 = 0$; (c) the total circuit power dissipation when $V_1 = V_2 = 0$; and (d) range of v_{cm}.

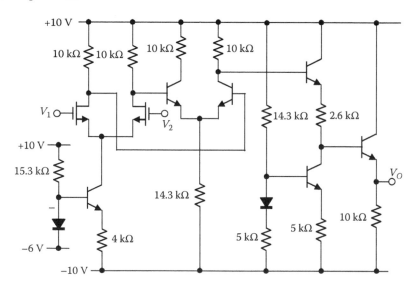

FIGURE P10.2.7

P10.2.8 Assume that in Figure 10.5.2 $I_{C1} = I_{C2} = 10$ μA, β_1 and β_2 have a 5% variation from their average value of 200. Determine: (a) bias and offset currents; (b) r_{id}; and (c) G_m of the input stage, assuming $\alpha = 0.98$ for Q_3 and Q_4.

P10.3 Op Amp Circuits

P10.3.1 An op amp having $V_{io} = 2$ mV is to be used for amplifying a signal of 50 mV, the magnitude of gain being 100. Determine the possible values of output voltage if the amplifier is used in: (a) an inverting configuration; (b) a noninverting configuration. Assume $\beta_o A_v \gg 1$.

P10.3.2   An op amp has input bias currents of 1.2 and 1 μA at the inverting and non-inverting terminals, respectively, and an input offset voltage of 2 mV. It is to be used to amplify a signal of 50 mV at unity magnitude of gain. Determine the possible values of output voltage if the amplifier is used in: (a) an inverting configuration with 10 kΩ resistors; (b) a noninverting configuration with $R_{src} = 10$ kΩ = R_f. Assume $\beta_o A_v \gg 1$.

DP10.3.3 The circuit of Figure P10.3.3 can be used to compensate for the effects of input offset voltage and bias currents in an inverting configuration. The two resistors R_a allow good sensitivity of zero adjustment of V_O by means of the potentiometer. Normally, R_3 is much larger, say 20 times, than R_2 and is also large compared to R_p. Under these conditions, show that suitable guidelines for choosing the resistances in the compensating circuit are: (1) $R_1 + R_2 = R_r \| R_f$ and (2) $|V_p| R_2 / R_3$ is slightly larger than $|V_{io}|$, where $|V_p| \cong |V_{CC}| R_p / (2R_a + R_p)$ is the magnitude of the voltage at either end of the potentiometer with respect to ground. Determine suitable values for the compensating circuit, assuming $R_r = 10$ kΩ, $R_f = 100$ kΩ, $V_{io(max)} = 5$ mV, $I_{io} = 0.2$ μA, and the voltage supplies to be ±10V.

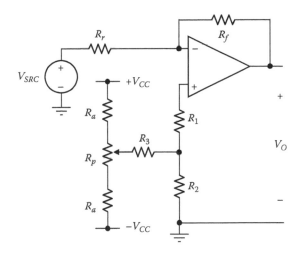

FIGURE P10.3.3

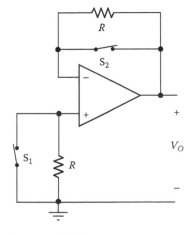

FIGURE P10.3.4

P10.3.4 Consider the circuit of Figure P10.3.4 with the effects of input offset voltage and bias currents included. Determine v_O when: (a) S_1 and S_2 are closed; (b) S_1 is open and S_2 is closed; (c) S_1 is closed and S_2 is open; and (d) S_1 and S_2 are open.

P10.3.5 An op amp having $A_v = 10^5$ and CMRR = 1000 is used in a noninverting configuration with $R_r = R_f = 10$ kΩ. The input is $v_{id} = 100$ mV and $v_{cm} = 1$ V. Determine the percentage error due to finite A_v and due to the common-mode response.

P10.3.6 An op amp has a differential gain of 10^5 and a CMRR of 80 dB. If $v_P = 10$ μV, $v_N = -10$ μV, and $v_{cm} = 0.1$ V, determine A_{cm} and v_O, assuming addition at the output.

P10.3.7 Assume that in Figure 10.6.4 $R_a = R_c = 1$ kΩ, $R_b = R_d = 10$ kΩ, CMRR $= 100$ dB, and large A_{v0}. Determine the worst-case CMRR for $\pm 1\%$ resistor tolerance. Simulate with PSpice.

P10.3.8 The slew rate of an amplifier is 2 V/μs. What is the maximum frequency for an undistorted sinusoidal output voltage of: (a) 1 V peak, (b) 5 V peak.

P10.3.9 A train of 0–1 V rectangular pulses is applied to a noninverting amplifier using an op amp of SR $= 10$ V/μs, the voltage gain being 2. Assuming equal SR on the leading and trailing edges of pulses, determine the minimum pulse width for a rectangular pulse output.

P10.3.10 This problem illustrates noise calculations. Consider an op amp in the inverting configuration having $R_r = 1$ kΩ and $R_f = 10$ kΩ so that the input resistance is R_r and the gain is -10. The total (rms)^2 noise voltage per unit bandwidth (BW) is given by:

$$v_n^2 = 4k_J T R_r + v_{ni}^2 + (R_r i_{ni})^2$$

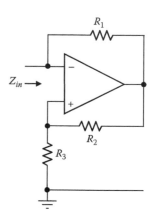

R_1

$Z_{in} \rightarrow$

R_2

R_3

where k_J is Boltzmann's constant and T is the absolute temperature. The first term on the RHS is due to R_r, and the second and third terms are due, respectively, to the noise voltage and current of the op amp. Assuming $4k_J T = 0.3 \times 10^{-20}$ VA/Hz, $v_{ni} = 20$ nV/$\sqrt{\text{Hz}}$, $i_{ni} = 10$ pA/$\sqrt{\text{Hz}}$, and that BW $= 1$ kHz, determine: (a) the input noise power $P_n = (v_n^2/R_r) \times$ (BW); (b) the rms noise voltage at the input, $\sqrt{P_n R_r}$; and (c) the rms noise voltage at the output.

FIGURE P10.4.1

P10.4 Miscellaneous

P10.4.1 (a) Assuming $f \ll f_T$, $A_{v0} \gg 1$, and $A_v(f)$ given by $A_{v0}/(1 + jf/fc)$, show that Z_{in} at the inverting input in Figure P10.4.1 is approximately $(R_1/A_{v0}) + jR_1 f/f_T$, equivalent to an inductance of R_1/ω_T having a Q of $\omega A_{v0}/\omega_T = \omega/\omega_c$.

P10.4.2 Figure P10.4.2 illustrates a basic biasing circuit that makes the g_m of a MOSFET depend only on aspect ratios and the value of a resistor. A high-performance current mirror ensures that $I_{D1} = I_{D2}$. Neglecting channel-length modulation and the body effect, and assuming that Q_1 and Q_2 have the same k_n' and V_{tn}, show that $g_{m1} = (2/R)\left(1 - \sqrt{(W/L)_1/(W/L)_2}\right)$.

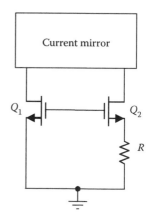

Current mirror

Q_1 Q_2

R

FIGURE P10.4.2

11

Power Amplifiers and Switches

The preceding chapters were chiefly concerned with low-power semiconductor devices. However, in a host of applications in control, communications, and power conversion, semiconductor devices handle substantial power that can exceed tens of kilowatts. In these applications, the semiconductor device may operate in a continuous mode, as was discussed in the preceding chapter, or in a switching mode, that is, being either fully conducting or cutoff. When the device is cutoff, power dissipation is insignificant, whereas in full conduction, the voltage across the device is small and power dissipation is low; however, substantial power can be dissipated in the semiconductor device during switching, when both the current through the device and the voltage across it may be appreciable. Continuous signals, such as audio signals, can still be amplified in the switching mode, using an appropriate modulation scheme.

To enable them to meet the demanding requirements of power amplification and switching, power diodes, MOSFETs and BJTs embody some special structural features that alter their mode of operation and characteristics. The effects of these modifications readily follow from the basic concepts previously discussed. Special precautions have to be taken to ensure that the devices operate within safe limits of current, voltage, and power dissipation. In power amplifiers, signal levels are no longer small, so that issues such as power-conversion efficiency and signal distortion become important. The switching mode entails several additional precautions, particularly concerning maximum allowable rates of change of voltage and current. Because power MOSFETs and BJTs have their advantages and limitations, a device that combines some of the advantages of both, the insulated gate bipolar transistor (IGBT), has gained popularity for medium power applications. In addition, a whole class of semiconductor power latches having some very useful properties can be used in certain power applications.

Learning Objectives

❖ To be familiar with:

- The safe operating area of transistors.
- The power derating curve of semiconductor devices.
- The terminology used in dc–dc and dc–ac power conversion.
- The terminology associated with power latches.

❖ To understand:

- The physical considerations that determine the maximum allowed limits on transistor output current, voltage, and power.
- The dependence of the maximum allowed power dissipation in a semiconductor device on the total thermal resistance with respect to the ambient environment.

- The factors that contribute to thermal instability and how stability may be achieved.
- The distinguishing features of class A, B, AB, and C operation, particularly in terms of power-conversion efficiency.
- The advantages of push–pull operation of power output stages.
- How transistor output stages may be used to augment the current, voltage, and power delivered by a small-signal op amp.
- The advantages of switching in power applications.
- The basic schemes of class D amplification, dc–dc, and dc–ac conversion.
- The nature and effects of structural modifications of power diodes and power transistors and the departure of the behavior of these devices from their small-signal counterparts.
- The principle of operation of IGBTs and several types of power latches.

11.1 General Considerations

Safe Operating Limits

Power transistors have safe operating limits defined in terms of the maximum allowable current, voltage, and power dissipation. On the plot of output current vs. output voltage, these limits define a **safe operating area** (SOA), as in Figure 11.1.1. $v_{out(max)}$ is the maximum instantaneous value that will cause avalanche breakdown if exceeded. $i_{out(max)}$ is the maximum continuous value that, if exceeded, will generate sufficient heat to cause damage (Application Window 11.1.1). Whereas $i_{out(max)}$ is a usable current in a MOSFET, it may not be in a BJT, because at this high current level, β_F may be too low and the base current correspondingly large.

The thermal dissipation hyperbola is defined by $P_{D(max)}$, the maximum power that can be safely dissipated in the transistor. For a MOSFET, $P_{D(max)} = v_{DS}i_{DS} = v_{out}i_{out}$. For a BJT, $P_{D(max)} = v_{BE}i_E + v_{CB}i_C \cong v_{CE}i_C = v_{out}i_{out}$ if $v_{BE}i_B$ is negligible compared to $v_{CB}i_C$. $P_{D(max)}$ is

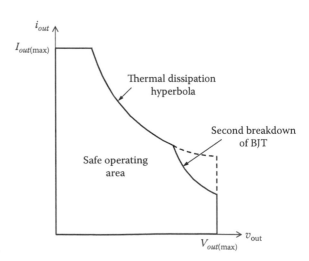

FIGURE 11.1.1
Safe operating limits of a transistor.

limited by the maximum junction temperature $T_{J(max)}$, which for Si transistors is in the range 150°C–200°C. Beyond this temperature, semiconductor devices are irreversibly damaged (Application Window 11.1.1). A lower limit exists in BJTs, referred to as **second breakdown**, due to current crowding at the emitter–base junction (Section 6.1, Chapter 6), which results in localized heating and necessitates reducing $P_{D(max)}$ so that $T_{J(max)}$ is not exceeded at these hot spots. Generally speaking, BJTs have larger $i_{out(max)}$, $v_{out(max)}$, and $P_{D(max)}$, than MOSFETs.

Because of the thermal nature of their effects, $i_{out(max)}$ and $P_{D(max)}$ are dc or rms values for continuous operation. When operation is intermittent, these values may safely be exceeded to an extent that depends on the duty cycle.

Application Window 11.1.1 Effects of Temperature on Semiconductor Devices

Semiconductor devices are affected by high temperatures, low temperatures, and sudden changes in temperature. Military specifications typically specify an ambient temperature range of −55°C to +125°C. High ambient temperatures are encountered in automotive applications, where electronic circuits may be housed close to an internal combustion engine. Space applications expose semiconductors to extreme temperatures in addition to radiation. When a satellite is in earth orbit but in shadow from the sun, temperature can drop to around −230°C, whereas a probe on Venus may be subjected to a temperature of +500°C. The conventional approach has been to use heaters or coolers to keep the temperature of electronic circuits within safe limits, but at the expense of added cost and weight. The recent trend has been to dispense with temperature control in favor of materials and fabrication methods that can cope with extreme temperatures. SiGe and GaAs improve the low-temperature operation of bipolar devices. Silicon carbide shows promise as a semiconductor for very high temperatures.

A rise in temperature generates electron–hole pairs; the additional minority carriers can seriously affect the operation of bipolar devices and can cause thermal runaway at high enough temperatures. MOSFET devices, being unipolar, are much less affected by carrier generation. In SOI technology, the silicon oxide barriers further limit the effect of generated carriers. High temperatures greatly accelerate aging effects in semiconductor devices, leading to rapid deterioration and failure. The high temperatures can melt the solder in contact metallization or the wires that bond the transistor chip to the package. These bond wires can also break due to temperature cycling. The effects of low temperatures are not as clear cut, and may be beneficial under certain conditions, but tests have indicated that bipolar devices fail below about −200°C and CMOS devices below about −230°C.

Thermal Resistance

The relationship between heat flow and temperature can be expressed in terms of thermal resistance, considering that heat flow is analogous to current and temperature difference is analogous to voltage difference (Sabah 2008, p. 783). Hence, thermal resistance θ in °C/W is temperature difference in °C divided by heat flow in watts.

Typically, power transistors are designed and mounted so as to remove heat efficiently from collector or drain, which have a large area and are in direct contact with a metal side or case of the transistor. Heat is generally removed in the following ways that are progressively more efficient thermally: (1) Where power dissipation is small, the transistor is simply mounted in free air. Heat is lost from the exposed plastic case or metal side to the surrounding air through radiation and convection. A small, finned, heat sink may be fitted to the transistor case to increase heat loss. (2) The transistor is bolted to the metal chassis so that the metal side or case

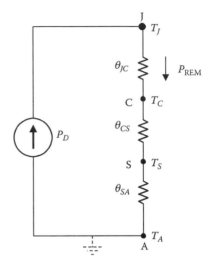

FIGURE 11.1.2
Thermal equivalent circuit of a transistor.

is in intimate contact with the chassis. A thermally conductive paste is usually applied to fill any voids between the two metallic surfaces, which results in better heat conduction from the transistor to chassis. Since the chassis is normally grounded, whereas the collector or drain generally are not, a thin mica wafer is inserted between the transistor and chassis to provide electrical insulation with good thermal conduction. Insulating sleeves or bushings are used to prevent contact between transistor leads and the chassis. (3) The transistor is bolted to a relatively large, finned heat sink that dissipates heat efficiently through radiation and convection. To improve heat loss through radiation, the heat sink is blackened. The same electrical insulation precautions are taken as in the case of direct bolting to the chassis. (4) A fan is used to blow air over the heat sink for more efficient heat removal through convection. (5) A liquid coolant is used to conduct heat away.

In the typical thermal equivalent circuit of Figure 11.1.2, the current source P_D represents the power dissipated in the transistor. Four nodes are identified: junction (J), case (C), heat sink (S), and ambient environment (A). A thermal resistance θ connects adjacent nodes. In the steady state, heat is generated at a rate P_D and removed at an equal rate P_{REM}. It follows that:

$$P_{D(max)} = P_{REM} = \frac{T_{J(max)} - T_A}{\theta_{JA}} = \frac{T_{J(max)} - T_A}{\theta_{JC} + \theta_{CS} + \theta_{SA}} \qquad (11.1.1)$$

where
$T_{J(max)}$ is the maximum allowed junction temperature
$P_{D(max)}$ is the corresponding power dissipation

For a given $(T_{J(max)} - T_A)$, the lower θ_{JA}, the higher is $P_{D(max)}$. If the transistor is mounted in free air, θ_{JC} is unchanged, because it depends on the transistor, but $\theta_{CA} = \theta_{CS} + \theta_{SA}$ is much larger than that with an efficient heat sink, so that $P_{D(max)}$ is much lower.

Exercise 11.1.1

A transistor has $\theta_{JC} = 2.5°C/W$, $\theta_{CS} = 0.4°C/W$, and $\theta_{SA} = 3.6°C/W$. (a) If the transistor dissipates 25 W at $T_A = 25°C$, what is T_J? (b) When mounted without a heat sink, $\theta_{CA} = 27.5°C/W$. What is $P_{D(max)}$ at $T_A = 25°C$, if $T_{J(max)} = 150°C$?
   Answers: (a) 187.5°C; (b) 4.17 W.   ∎

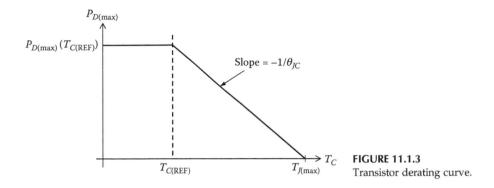

FIGURE 11.1.3
Transistor derating curve.

Manufacturer's data sheets usually specify θ_{JC} and $P_{D(\max)}(T_{C(REF)})$, either explicitly or through a power **derating curve** that relates $P_{D(\max)}$ to T_C (Figure 11.1.3), where $T_{C(REF)}$ is a reference case temperature. As it is generally impractical to maintain T_C at a temperature below $T_{C(REF)}$, $P_{D(\max)}$ is considered constant at $P_{D(\max)}(T_{C(REF)})$ for $T_C \leq T_{C(REF)}$. For $T_C \geq T_{C(REF)}$:

$$P_{D(\max)} = \frac{T_{J(\max)} - T_C}{\theta_{JC}}, \quad T_C \geq T_{C(REF)} \tag{11.1.2}$$

If $T_C = T_{J(\max)}$, then $P_{D(\max)} = 0$, that is, no power dissipation is allowed.

Example 11.1.1 Transistor Derating

A transistor has $T_{J(\max)} = 150°C$, $P_{D(\max)} = 30$ W at $T_{C(REF)} = 25°C$, $\theta_{JC} = 3°C/W$, and $\theta_{CA} = 42°C/W$ when operated in free air, and $T_A = 50°C$. It is required to determine: (a) the maximum power that can be safely dissipated; (b) the thermal resistance of a heat sink that would allow the transistor to safely dissipate 15 W, assuming $\theta_{CS} = 0.5°C/W$.

SOLUTION

(a) In free air, $\theta_{JA} = 3 + 42 = 45°C/W$. Hence, $P_D = \dfrac{150 - 50}{45} = 2.2$ W.

(b) To dissipate 15 W for a temperature rise of junction of 100°C requires a total junction-to-ambient thermal resistance θ_{JA} of $100/15 = 6.7°C/W$. With $\theta_{JC} = 3°C/W$ and $\theta_{CS} = 0.5°C/W$, the thermal resistance of the heat sink should not exceed $6.7 - 3 - 0.5 = 3.2°C/W$.

Exercise 11.1.2

The derating curve of a transistor indicates $P_{D(\max)} = 50$ W at $T_{Cref} = 25°C$, and $T_{J(\max)} = 175°C$. What is the maximum θ_{CA} so that the junction temperature will not exceed 175°C when the transistor dissipates 10 W at $T_A = 50°C$?
    Answer: 9.5°C/W. ∎

Thermal Stability

Concept: *If the output current of a power transistor increases with temperature, then under certain conditions, thermal instability, or thermal runaway, can occur.*

This is because the increase in output current increases power dissipation. If heat is not removed fast enough, the rise in temperature increases the output current, which further increases the temperature, and so on, leading to destruction of the transistor. i_C of a BJT increases with T (Section 6.2, Chapter 6), whereas i_D for a MOSFET increases with T for low i_D and decreases with T for high i_D (Section 5.3, Chapter 5). When the output current increases with temperature, thermal stability is assured by having a low enough thermal resistance from the junction to the ambient environment, or by having a sufficiently high electrical resistance in the output circuit (Section ST11.1). In this case, an increase in the output current leads to a large enough decrease in output voltage, so that power dissipation decreases.

Nonlinear Distortion

Relatively high power outputs imply large-signal operation that departs from linearity and introduces nonlinear distortion. In a BJT, for example, the output characteristics are more crowded at larger values of currents, so that the variation of i_C with i_B is nonlinear and leads to distortion in large-signal operation. This variation can be expressed as a power relation:

$$i_c = b_1 i_b + b_2 i_b^2 + b_3 i_b^3 + \cdots \tag{11.1.3}$$

where b_1, b_2, b_3, ... are constant coefficients. For small-signal variations, higher powers of i_b are neglected, so that $b_1 = \beta$ in this case. If $i_b = i_{bm} \cos\omega t$, then i_c is obtained by substituting for i_b in Equation 11.1.3, leading to the following deductions: (1) Even powers of $\cos\omega t$, such as $\cos^2\omega t$, introduce a dc component, since $\cos^2\omega t = (1 + \cos 2\omega t)/2$. Because the ac signal changes the dc collector current, rectification is said to occur. The nonlinearity is such that rectification decreases the quiescent current. (2) A term in $\cos n\omega t$, $n = 2, 3, \ldots$, introduces an nth harmonic as well as lower-order harmonics that are odd or even, depending on whether n is odd or even, respectively. For example, a term in $\cos^3\omega t$ introduces a third harmonic as well as a fundamental component. A term in $\cos^4\omega t$ introduces a fourth harmonic as well as a second harmonic and a dc component (Appendix SA). (3) If the input signal has more than one frequency, as is usually the case, the nonlinearity also introduces **intermodulation products** having frequencies that are the sums and differences of input frequencies and their harmonics (Example SE1.5, Chapter 1).

Distortion in FET amplifiers is generally greater than in BJT amplifiers, because of the quadratic relation between i_D and v_{GS}. In the presence of distortion, the output current of either a FET or a BJT amplifier can be expressed as a Fourier Series Expansion (FSE):

$$i_{out} = a_0 + a_1 \cos \omega t + a_2 \cos 2\omega t + a_3 \cos 3\omega t + \cdots \tag{11.1.4}$$

The Fourier coefficients can be determined graphically from the i_C waveform, as explained in Section ST11.2. The **harmonic distortion** D_n due to the nth harmonic is defined as:

$$D_n = \left| \frac{a_n}{a_1} \right| \tag{11.1.5}$$

or as a percentage. The **total harmonic distortion** (THD) D is obtained by considering the total power dissipated in a resistance R_L due to the ac component:

$$P = (a_1^2 + a_2^2 + a_3^2 + \cdots) \frac{R_L}{2} = (1 + D_2^2 + D_3^2 + \cdots) P_1 = (1 + D^2) P_1 \tag{11.1.6}$$

where

$$D = \sqrt{D_2^2 + D_3^2 + D_4^2 + \cdots} \qquad (11.1.7)$$

and

$$P_1 = a_1^2 R_L / 2$$

Because of D^2 in Equation 11.1.6, the power due to harmonics is generally negligible compared to that due to the fundamental. Thus, if $D = 0.1$, corresponding to 10% THD, the power due to the harmonics is only 0.01 P_1.

Distortion meters are available for measuring harmonic distortion, which for good fidelity should not exceed a total of 0.1%. Distortion is usually reduced by applying negative feedback from the output to earlier stages of voltage amplification (Problems P.11.2.3 through P11.2.5).

Power-Conversion Efficiency

The **power-conversion efficiency** of an output stage is defined as:

$$\eta = \frac{P_1}{P_S} \times 100 \qquad (11.1.8)$$

where $P_S = V_{CC} a_0$ (Equation 11.1.4) is the power delivered by the dc power supply to the output stage, neglecting the small power due to the base current in the case of a BJT.

Concept: *Not only is high power-conversion efficiency desirable, but also low power dissipation in the absence of the signal.*

Both measures reduce the cost of the power supply, prolong battery life in battery-powered equipment, and reduce the amount of heat that has to be removed through cooling.

Simulation Example 11.1.2 Performance of Power Darlington Pair

Consider a 2N6059 power Darlington pair that is very simply biased as in Figure 11.1.4. It is desired to investigate distortion using PSpice.

SOLUTION

The 2N6059, part number Q2N6059 in PSpice EVAL library, is rated at 100 V, 12 A, 150 W. Its output characteristics, derived from PSpice as in Example 6.2.1, Chapter 6, are shown in Figure 11.1.5, with a 4 Ω load line. Note that $V_{C(sat)} = 1.25$ V at $i_C = 4.2$ A. For illustration purposes, the quiescent point Q is chosen at $V_C = 6.7$ V and $I_{CQ} = 2.82$ A, $I_{BQ} = 0.4$ mA. The corresponding, measured input voltage is 1.56 V, so a base resistor of $(18 - 1.56)/0.4 = 41.1$ kΩ is required.

A sinusoidal current source of 0.25 mA amplitude, 1 kHz, is applied. Because the input resistance is determined to be about 0.6 kΩ, about 98.6% of the source current is the base current of the transistor. The results of a Fourier analysis of the collector current are as follows:

DC component = 2.486682E + 00

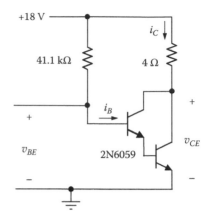

FIGURE 11.1.4
Figure for Example 11.1.2.

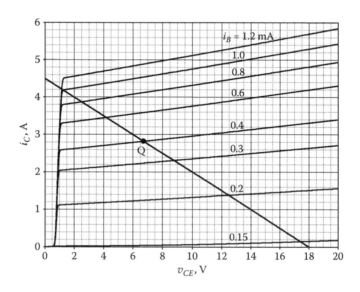

FIGURE 11.1.5
Figure for Example 11.1.2.

Harmonic No	Frequency (Hz)	Fourier Component	Normalized Component	Phase (Deg)	Normalized Phase (Deg)
1	1.000E+03	1.361E+00	1.000E+00	−4.897E+00	0.000E+00
2	2.000E+03	3.827E−01	2.812E−01	7.663E+01	8.643E+01
3	3.000E+03	1.460E−01	1.073E−01	1.566E+02	1.712E+02
4	4.000E+03	6.297E−02	4.627E−02	−1.243E+02	−1.047E+02
5	5.000E+03	2.921E−02	2.146E−02	−4.577E+01	−2.129E+01
6	6.000E+03	1.420E−02	1.043E−02	3.277E+01	6.215E+01
7	7.000E+03	7.152E−03	5.256E−03	1.107E+02	1.449E+02
8	8.000E+03	3.664E−03	2.693E−03	−1.716E+02	−1.324E+02
9	9.000E+03	1.967E−03	1.445E−03	−9.664E+01	−5.257E+01
10	1.000E+04	9.990E−04	7.341E−04	−2.604E+01	2.293E+01

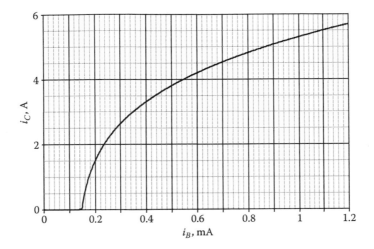

FIGURE 11.1.6
Figure for Example 11.1.2.

Total Harmonic Distortion $= 3.055424E+01\%$
The applied signal decreases the dc collector current from a quiescent value of 2.82 A to 2.49 A. Total harmonic distortion is about 30.5%. The signal power dissipated in the 4 Ω collector resistor is $(1.361)^2 \times 4 = 7.41$ W. The dc supply delivers 50.8 W, which gives a power-conversion efficiency of about 14.6%. The dc power dissipated in the base resistor is only about 6.6 mW and that dissipated in the base input is $0.4 \times 1.56 = 0.6$ mW.

The distortion arises from the nonlinear relation between i_C and i_B (Figure 11.1.6). Since the curvature of this plot is similar to that of the $v_{BE} - i_B$ relation, it would be expected that a voltage drive produces less distortion in i_C than a current drive (Exercise 11.1.3).

Exercise 11.1.3

Compare the i_C waveforms for: (a) a current drive as in Example 11.1.4, and (b) a voltage drive that gives the same peak-to-peak variation in i_C as in (a). What is the total harmonic distortion in the case of voltage drive? ∎

11.2 Class A Operation

Concept: *In class A operation, the output current of the power transistor flows throughout the signal cycle, as in small-signal operation, except that the swings in output current and output voltage are not small.*

In the CE amplifier of Figure 11.2.1, the input circuit is the dc and low-frequency ac Thevenin's equivalent circuit (TEC) of the biasing network and the applied source. From KVL in the output circuit under dc conditions,

$$V_{CEQ} = V_{CC} - R_C I_{CQ} - R_E I_{EQ} = V_{CC} - I_{CQ}\left(R_C + \frac{\beta_F + 1}{\beta_F}R_E\right) = V_{CC} - R_{dc}I_{CQ} \qquad (11.2.1)$$

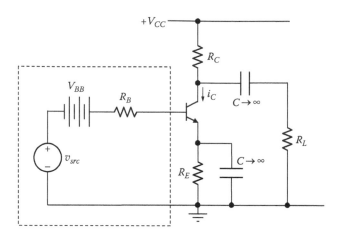

FIGURE 11.2.1
Basic class A amplifier.

where,

$$R_{dc} = R_C + \frac{\beta_F + 1}{\beta_F} R_E \cong R_C + R_E \qquad (11.2.2)$$

The quiescent point Q (Figure 11.2.2a) is the intersection of the output characteristic for the given I_B with the dc load line of slope $-1/R_{dc}$. With large coupling and emitter by-pass capacitors, the ac load is $R_{ac} = R_L \| R_C$, so that when an ac signal is applied, the operating point moves along an **ac load line** of slope $-1/R_{ac}$. Figure 11.2.2a is redrawn in Figure 11.2.2b to show only the ac load line, assuming that $v_{CE(sat)}$ is negligibly small, so that the operating point can move all the way to the vertical axis. Suppose that a sinusoidal base current is applied and that harmonic distortion is negligible; the variations in i_C and v_{CE} are sinusoidal and the operating point moves between the two limits Q′ and Q″, the peak values of i_C and v_{CE} being I_{cm} and V_{cem}, respectively. Maximum signal amplitude occurs when the quiescent point Q is at the midpoint of the ac load line, and Q′ and Q″ lie on the current and voltage axes, respectively, which gives:

$$V_{CEQ} = R_{ac} I_{CQ} \qquad (11.2.3)$$

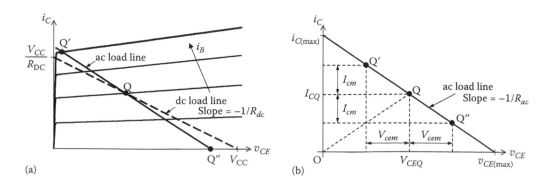

FIGURE 11.2.2
(a) dc and ac load line construction; (b) idealized ac operation.

This is the case of **maximum symmetrical swing**, neglecting distortion and assuming $v_{CE(sat)} = 0$. It can be shown that when Q is chosen for maximum symmetrical swing, the ac load line is tangent to the maximum dissipation hyperbola at Q (Problem P11.3.2).

It follows from Equations 11.2.1 and 11.2.3 that:

$$I_{CQ} = \frac{V_{CC}}{R_{dc} + R_{ac}} \quad \text{and} \quad V_{CEQ} = \frac{R_{ac}}{R_{dc} + R_{ac}} V_{CC} \tag{11.2.4}$$

The ac power delivered to the load is $V_{cem}I_{cm}/2$. The power delivered by the dc power supply is $V_{CC}I_C$, neglecting the power supplied to the base circuit. Applying Equation 11.1.8,

$$\eta = \frac{1}{2} \frac{V_{cem}I_{cm}}{V_{CC}I_S} \times 100 \tag{11.2.5}$$

where I_S is the supply current. η increases with signal amplitude. Under conditions of maximum symmetrical swing, $I_{cm(max)} = I_{CQ} = I_S$ and $V_{cem(max)} = V_{CEQ}$. If, in addition, $R_{ac} = R_{dc}$, then $V_{CEQ} = V_{CC}/2$ (Equation 11.2.4). Substituting in Equation 11.2.5 gives $\eta_{max} = 25\%$.

When $R_E \neq 0$, $\eta_{max} = R_L/(4R_L + 2R_E)100$ (Problem P11.3.3). Moreover, the power delivered by the dc supply in the absence of a signal is $V_{CC}I_C$, the same as in the presence of a signal, neglecting distortion and the power supplied to the base circuit.

Exercise 11.2.1

Verify Equations 11.2.4. Note that in Figure 11.2.2b, $i_{C(max)} = 2I_{CQ}$ and $v_{CE(max)} = 2V_{CEQ}$. ∎

Example 11.2.1 Capacitively Coupled Class A Amplifier

Given a load of 50 Ω that is to be supplied from a CE output stage having $R_E = 5$ Ω for good bias stability. If $\beta_F = 50$ and the power supply is 28 V, it is required to determine the maximum power that can be delivered to the load using capacitive coupling such that the ac power dissipated in R_C does not exceed 25% of the power dissipated in R_L.

SOLUTION

The ac output voltage signal appears across R_C in parallel with R_L. If the ac power dissipated in R_C is not to exceed 25% of the power dissipated in R_L, $R_C = 4R_L = 200$ Ω. Hence, $R_{ac} = 200 \| 50 = 40$ Ω, and $R_{dc} = 200 + 5(51/50) = 205.1$Ω. It follows from Equations 11.2.4 that $I_{CQ} = 28/(40 + 205.1) = 114$ mA and $V_{CEQ} = 4.57$ V. The maximum ac power is $\frac{1}{2}V_{CEQ}I_{CQ} = 0.114 \times 4.57 = 0.52$ W. The power delivered to the 50 Ω load is $0.52 \times 40/50 = 0.42$ W and that dissipated in R_C is 0.11 W. The power delivered by the supply, neglecting that in the base circuit, is $V_{CC} \times I_{CQ} = 3.19$ W. The maximum power-conversion efficiency is $(0.42/3.19) \times 100 \cong 13\%$.

Note that the main reason for this low efficiency is the presence of R_C. If the load R_L is dc coupled, that is, R_L is connected in place of R_C, $P_{L(max)}$ increases to 1.78 W and η_{max} becomes 23.8%. If $R_E = 0$, $P_{L(max)}$ increases to 1.96 W and η_{max} becomes 25% (Exercise 11.2.2).

Exercise 11.2.2

Determine I_{CQ}, V_{CEQ}, and η_{max} in Example 11.2.1 if the 50 Ω load is dc coupled: (a) in the presence of R_E; (b) when $R_E = 0$.

Answers: (a) $I_{CQ} = 267$ mA, $V_{CEQ} = 11.3$ V, $P_{L(max)} = 1.78$ W; $\eta_{max} = 23.8\%$; (a) $I_{CQ} = 280$ mA, $V_{CEQ} = 14$ V, $P_{L(max)} = 1.96$ W; $\eta_{max} = 25\%$.   ∎

Transformer Coupling

Concept: *The maximum theoretical power-conversion efficiency can be doubled by coupling the load through a transformer, basically because of elimination of power dissipation in R_C.*

Transformer coupling of the load is illustrated in Figure 11.2.3a. Applying KVL to the output circuit under dc conditions,

$$V_{CEQ} = V_{CC} - r_p I_{CQ} - R_E I_{EQ} = V_{CC} - I_{CQ}\left(r_p + \frac{\beta_F + 1}{\beta_F} R_E\right) = V_{CC} - R_{dc} I_{CQ} \qquad (11.2.6)$$

where
  r_p is the small dc resistance of the primary winding
  $R_{dc} = r_p + R_E(\beta_F + 1)/\beta_F \cong r_p + R_E$ (Figure 11.2.3b)

If the primary-to-secondary turns ratio is n, the resistance reflected to the primary circuit is $R_P = n^2 R_L$, irrespective of the dot markings of the transformer. If $i_c = I_{cm}\cos\omega t$, the signal voltage across the primary is: $v_p = V_{pm}\cos\omega t$, where $V_{pm} = R_P I_{cm}$, neglecting the small $r_p i_c$ voltage drop. For the ac signal: $v_{ce} = -v_p = -R_P i_c$. The operating point moves along an ac load line passing through Q and having a slope $-1/R_P$. The ac load line is in this case less steep than the dc load line.

The power delivered to the transformer primary is $V_{cem}I_{cm}/2$, where $V_{cem} = V_{pm} = R_P I_{cm}$. This is the same as the power delivered to the load, neglecting losses in the transformer. The power-conversion efficiency is still given by Equation 11.2.5, but with an important difference. During one half-cycle, while i_C increases and the operating point moves toward Q', v_p opposes the increase in i_C and subtracts from V_{CC}. But during the other half-cycle,

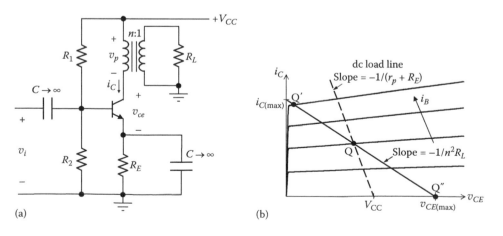

(a)   (b)

FIGURE 11.2.3
(a) Transformer-coupled, class A amplifier; (b) load-line construction.

while i_C decreases and the operating point moves toward Q'', v_p opposes the decrease in i_C and adds to V_{CC}. This means that v_{CE}, the instantaneous collector–emitter voltage now exceeds V_{CC}.

If $R_E = 0 = r_p$ the dc load line is a vertical line at $v_{CE} = V_{CC}$, so that $V_{CEQ} = V_{CC}$. Neglecting the saturation voltage of the transistor, distortion, the power supplied to the base circuit, and transformer losses, the maximum value of I_{cm} is $I_{cm(max)} = I_{CQ} = I_S$, the maximum value of V_{cem} is $V_{cem(max)} = V_{CEQ} = V_{CC}$, and the power supplied is $V_{CC}I_S$. Substituting in Equation 11.2.5, gives a maximum theoretical efficiency of 50% rather than 25%.

R_{ac} for maximum symmetrical swing is given by Equation 11.2.3 and can be determined in a number of ways. If $i_{C(max)}$ is specified, then $I_{CQ} = i_{C(max)}/2$, as in the ac-coupled case. Once I_{CQ} is fixed, V_{CEQ} can be determined from Equation 11.2.6. Or, if $v_{CE(max)}$ is specified, $V_{CEQ} = v_{C(max)}/2$, and I_{CQ} can then be determined from Equation 11.2.6. Alternatively, the quiescent operating point can be determined from the maximum power dissipation hyperbola using the basic geometrical property that the tangent to a hyperbola is bisected by the point of tangency (Problem P11.3.2). The ac load line may be drawn tangential to the hyperbola and passing through Q, $i_{C(max)}$, or $v_{CE(max)}$. Examples of these methods are considered in Problems P11.3.9 through P11.3.14. Once R_{ac} is selected, and R_L is given, n is found from the relation $R_{ac} = n^2 R_L$.

Design Example 11.2.2 Transformer-Coupled Class A Amplifier

A 6.6 Ω load is coupled through a transformer having $r_p = 0.9$ Ω using the same supply and transistor as in Example 11.2.1. It is required to determine the turns ratio, the maximum power delivered to the load, and the power-conversion efficiency, assuming that $i_{C(max)} = 0.5$ A.

SOLUTION

Under conditions of maximum symmetrical swing, $I_{CQ} = i_{C(max)}/2 = 0.25$ A. $R_{dc} = 0.9 + 5 \times 51/50 = 6$ Ω. From Equation 11.2.6, $V_{CEQ} = 28 - 6 \times 0.25 = 26.5$ V. Hence, $R_{ac} = 26.5/0.25 = 106$ Ω. This gives: $n = \sqrt{106/6.6} = 4$. The maximum power delivered to the load, assuming no losses in the transformer is $V_{CEQ}I_{CQ}/2 = 3.31$ W, and $\eta = 100 \times 3.31/(28 \times 0.25) = 47.3\%$. If $R_E = 0$ and r_p is neglected, $V_{CEQ} = 28$ V, the maximum load power delivered to the transformer is $0.5 \times 28 \times 0.25 = 3.5$ W, and $\eta = 100 \times 3.5/(28 \times 0.25) = 50\%$.

Exercise 11.2.3
Verify that Equations 11.2.4 apply to Example 11.2.2. ∎

Emitter Follower

The emitter follower is a useful class A output stage because of its high input resistance and low output resistance, which means that most of the signal voltage appears across the load. Figure 11.2.4a shows an emitter follower biased with what is effectively a dc current source I_{EE}. Figure 11.2.4b corresponds to Figure 11.2.2b and shows only the ac load line drawn with respect to i_E and v_{CE} axes. The quiescent point Q is located on the midpoint of the load line for maximum symmetrical swing. Neglecting the small-signal variation v_{eb}, v_o follows v_i. Assuming sinusoidal operation, the operating point swings between Q' and Q'',

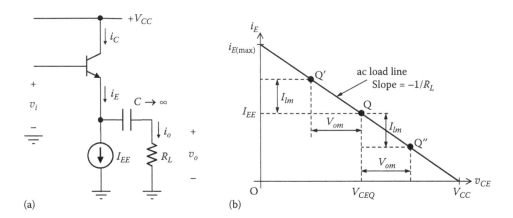

FIGURE 11.2.4
(a) Emitter follower power amplifier; (b) ac load line construction.

where $V_{om} = V_{im}$ and $I_{om} = V_{om}/R_L$. The load power is $P_L = V_{em}I_{om}/2$. The power delivered by the supply is $V_{CC}I_{EE}$. Hence,

$$\eta = \frac{P_L}{P_S} \times 100 = \frac{1}{2}\frac{V_{om}I_{om}}{V_{CC}I_{EE}} \times 100 = \frac{1}{2}\frac{V_{om}^2}{R_L V_{CC}I_{EE}} \times 100 \qquad (11.2.7)$$

Under conditions of maximum symmetrical swing, $V_{om(max)} = V_{CC}/2$ and $I_{om(max)} = I_{EE}$ assuming, as before, that $v_{CE(max)} = 0$. It follows from Equation 11.2.6 that $\eta_{max} = 25\%$. Note that the positive excursion of v_o is limited by transistor saturation, whereas the negative excursion of v_o is limited by the transistor cutting off, when $I_{o(max)} = I_{EE}$.

It was implicitly assumed in all class A operation that the input drive can be adjusted to give maximum swing in the output current and voltage, and hence maximum power output. If the transistor input is fixed, maximum power in the load occurs when the load resistance equals r_{out}, the output resistance of the transistor. This is the condition for maximum power transfer between the source and load as derived in electric circuit analysis.

Example 11.2.3 Emitter Follower Class A Amplifier

Consider Example 11.2.1, with $\beta_F = 50$, $V_{CC} = 12$ V, and the 100 Ω load capacitively coupled to an emitter follower. It is required to determine the maximum power delivered to the load and the maximum efficiency.

SOLUTION

With a +12 V supply, $V_{CEQ} = 6$ V, and $V_{om(max)}$ may be, 5.8 V, at the saturation limit. The power delivered to the load is $(1/2)[(5.8)^2/100] \cong 170$ mW. The current source should be at least $I_{om(max)} = 5.8/R_L = 58$ mA, say 60 mA. The power delivered by the dc power supply is $12 \times 0.060 = 0.72$ W. The power-conversion efficiency is $100 \times 0.17/0.72 \cong 23\%$. Had we taken the maximum amplitude of load voltage to be 6 V, the load power would be $(1/2)[(6)^2/100] \cong 0.18$ W and the power-conversion efficiency would be 25%.

The power delivered by the power supply is 0.72 W, 0.36 W being dissipated in the current source, and 0.36 W in the transistor when no signal is applied. With the maximum signal applied, and neglecting nonlinearities, the power delivered by the supply, and that dissipated in the source, does not change. When 0.17 W of ac power is delivered to the load, then from conservation of energy, the power dissipated by the transistor must decrease by 0.17 W.

Exercise 11.2.4

Consider the emitter-coupled amplifier of Figure 11.2.4a operating with maximum symmetrical swing, negligible saturation voltage, and with the lower limit set by the transistor cutting off. Assuming a sinusoidal input, describe the time variation of v_{CE}, i_C, the voltage across the current source, and the instantaneous power dissipation in the transistor, and in the current source. ■

11.3 Class B Operation

Concept: *In class B operation, two output transistors are used in push–pull so that the current of each transistor flows for only half a cycle.*

Suppose that the current source in the emitter follower circuit of Figure 11.2.4a is replaced by a matched complementary transistor, with the bases of the two transistors connected to the input v_I (Figure 11.3.1a). When $v_I = 0$, both transistors are cut off, and $v_O = 0$. Q_n begins to conduct when v_I equals the cut-in voltage V_γ of about 0.5 to 0.6 V. When Q_n conducts, it acts as an emitter follower, and $v_O = v_I - 0.7$ V (Figure 11.3.1b). While Q_n conducts, the BEJ of Q_p is reverse biased by 0.7 V, so Q_p is cut off. When v_I increases in the negative direction, the sequence of events is reversed. Q_p does not conduct until $v_I \cong -0.6$ V, and when Q_p is fully conducting, Q_n is cut off. This type of operation is described as **push–pull**, since the two transistors are essentially connected to the load in phase opposition. Because the collector current of each transistor flows practically for half a cycle, the operation is class B.

With no signal applied, both transistors are cut off, so $P_S = 0$. When a transistor conducts, the dc supply current equals the load current, so η_{max} is high, theoretically 78.5% (Section ST11.3). If the two transistors are perfectly matched, even harmonics cancel out from the output (Example SE11.1). A serious drawback, however, is **crossover distortion** that occurs at the zero crossings of v_I, because v_O does not change from zero until $|v_I| > V_\gamma$

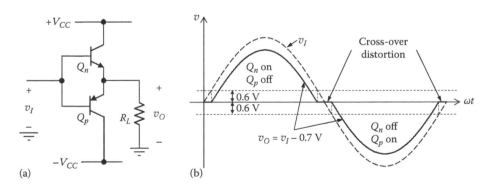

FIGURE 11.3.1
(a) Basic class B amplifier; (b) voltage waveforms.

(Figure 11.3.1b). Crossover distortion can be substantially reduced by negative feedback (Problem P11.4.2). In principle, it can be eliminated by biasing the transistors so that they are just on the verge of conduction when no signal is applied. However, variations in circuit and transistor parameters make this impractical. Instead, Q_n and Q_p are biased slightly into conduction, so as to eliminate crossover distortion. Since operation is now intermediate between class A and class B, it is class AB.

Exercise 11.3.1

Assume that in Figure 11.3.1 $V_{CC} = 12$ V, $V_\gamma = 0.6$ V, $V_{BE} = 0.7$ V when the transistor is conducting, and that the circuit is operating under ideal conditions with maximum input. (a) At what percentage of the input amplitude does conduction begin? (b) What is the width in degrees of the deadband due to crossover distortion?

Answers: (a) 4.72%; (b) 5.42°.   ∎

11.4 Class AB Operation

A basic IC, class AB output stage is illustrated in Figure 11.4.1. In the quiescent state, V_{DD} across the conducting diodes, which are normally diode-connected transistors, biases Q_n and Q_p to conduct a quiescent current I_{EQ} that is related to I_{DD}, the current in the diodes, by:

$$I_{EQ} = kI_{DD} \cong kI_{BB} \qquad (11.4.1)$$

where
 k is the ratio of the BEJ area of Q_n and Q_p to that of the diodes
 I_{BB} is a biasing current source that is normally much larger than I_{Bn}, the quiescent base current of Q_n

In the quiescent state, $v_O = 0$, the emitter current of Q_n and Q_p is I_{EQ}, and $v_I = -V_{EBpQ}$ of Q_p.

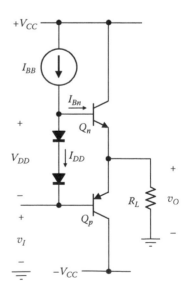

FIGURE 11.4.1
Basic class AB amplifier.

The largest positive excursion of output voltage occurs when Q_n is at the edge of saturation and is supplying load current $I_{om(max)}$, with $I_{Bn} = I_{om(max)}/(\beta_F + 1)$. I_{BB} should then at least equal I_{Bn}. Substituting for I_{BB} from Equation 11.4.1:

$$k < (\beta_F + 1)\frac{I_{EQ}}{I_{om(max)}} \tag{11.4.2}$$

If I_{EQ} is to be a small fraction of $I_{om(max)}$, so as to avoid excessive quiescent power dissipation, then for a given β_F, Equation 11.4.2 sets an upper limit on k.

An advantage of diode biasing in Figure 11.4.1 is that the temperature variation of the voltage drop across the biasing diodes nearly matches that of V_{BE}'s of Q_n and Q_p, so as to practically eliminate the variation of I_{EQ} due to the effect of temperature on the V_{BE}'s.

Exercise 11.4.1

Under quiescent conditions, $v_O = 0$ in Figure 11.4.1. What is the current that the source v_I must sink, taking into account the base currents of the two transistors? Verify this result by applying KCL to a surface that encloses the two diodes and the two transistors. Does the source v_I deliver or absorb power under quiescent conditions?

Answer: I_{BQ}; deliver power.   ∎

Example 11.4.1 Class AB Amplifier

Consider the basic circuit of Figure 11.4.1 with ±12 V supplies, $I_{BB} = 2.2$ mA, and a 100 Ω load. Let Q_n and Q_p have a saturation current $I_S = 0.2$ pA and a constant $\beta_F = 50$. The amplitude of the sinusoidal voltage v_O is to be 10 V, with the quiescent current of Q_n and Q_p not exceeding 5% of the peak current. It is required to analyze circuit operation.

Solution

Since the amplitude of the load current is $10/0.1 = 100$ mA, then $I_{EQ} = 5$ mA, and the maximum base current is $100/51 = 1.96$ mA. If $I_{BB} = 2.2$ mA, then the minimum current through the diodes is $2.2 - 1.96 = 0.24$ mA.

Under quiescent conditions, the base current is $5/51 \cong 0.1$ mA, which is much less than I_{BB}, and the current through the diodes is $2.2 - 0.1 = 2.1$ mA. This means that $k = 5/2.1 = 2.4$, and I_S of the diodes is about $0.2/2.4 = 0.083$ pA. The power delivered by the supplies is: $2 \times 12 \times (2.2 + 5 \times 50/51) \cong 170$ mW. The average signal current of each supply at full load is $(100/\pi) \times (50/51) = 31.2$ mA, and the power delivered by the supplies is $2 \times 12 \times (31.2 + 2.2) \cong 802$ mW. The load power is $(10)^2/200 \cong 500$ mW, so $\eta = (500/802) \times 100 \cong 62\%$. The efficiency could be improved by driving the transistors to near saturation, at the expense of increased distortion.

Under quiescent conditions, $V_{DDQ} = 2V_T\ln(2.1 \text{ mA}/0.083 \text{ pA}) = 1.246$ V. Since $v_O = 0$, $V_{BpQ} = -V_{DDQ}/2 = v_{IQ} = -0.623$ V, and $V_{BnQ} = 0.623$ V. The current that the v_I source must sink is 2.2 mA, and the power delivered by the source is $2.2 \times 0.623 = 1.37$ mW.

When $v_O = +10$ V, Q_n base current is 1.96 mA, and the diode current drops to 0.24 mA, as determined above. $v_{DD}^+ = 2V_T \ln(0.24 \text{ mA}/0.083 \text{ pA}) = 1.133$ V. The emitter currents and base-emitter voltages of Q_n and Q_p can be determined from: $v_{DD}^+ = v_{BEn}^+ + v_{EBp}^+ = V_T\ln\frac{i_{En}^+}{I_S} + V_T\ln\frac{i_{En}^+}{I_S}$. This gives: $V_T\ln\frac{i_{En}^+\left(i_{En}^+ - I_{lm}\right)}{I_S^2} = 2V_T\ln\frac{0.24 \text{ mA}}{0.083 \text{ pA}}$, or $\frac{i_{En}^+\left(i_{En}^+ - 100\right)}{I_S^2} =$

$\left(\dfrac{0.24}{0.083}\right)^2$, or $i_{En}^+\left(i_{En}^+ - 100\right) = 0.33$. Solving this quadratic equation, we obtain i_{En}^+ very close to 100 mA, as assumed above, and $v_{BEn}^+ = V_T \ln(100\ \text{mA}/0.2\ \text{pA}) \cong 0.700\ \text{V}$; $v_{EBp}^+ = 1.133 - 0.7 = 0.433\ \text{V}$, so that $i_{Ep}^+ \cong 0.v_{Bp}^+ = v_I^+ = v_O - v_{EBp}^+ = 9.57\ \text{V}$. The positive excursion of v_I is $9.57 - V_{BpQ} = 9.57 - (-0.623) \cong 10.2\ \text{V}$. The swing in i_{En} is from 5 to 100 mA.

When $v_O = -10\ \text{V}$, Q_n base current can be neglected, so I_{BB} passes through the diodes. $v_{DD}^- = 2V_T \ln(2.2\ \text{mA}/0.083\ \text{pA}) = 1.248\ \text{V}$. Following the same procedure as above, $i_{Ep}^-(i_{Ep}^- - 100)/I_S^2 = (2.2/0.083)^2$, or $i_{Ep}^-(i_{Ep}^- - 100) = 28.1$. This gives $i_{Ep}^- = 100.28\ \text{mA}$ and $v_{EBp}^- = V_T \ln(100.28\ \text{mA}/0.2\ \text{pA}) \cong 0.700\ \text{V}$. Hence, $i_{En}^- = 0.28\ \text{mA}$, and $v_{BEn}^- = 1.248 - 0.7 = 0.548\ \text{V}$. $v_{Bp}^- = v_I^- = v_O - v_{EBp}^- = -10 - 0.7 = -10.7\ \text{V}$. The negative excursion of v_I is $-10.7 - (-0.62) \cong -10.1$, which is slightly different in magnitude from the positive excursion, because v_{DD} is not constant. V_{DD} can be maintained more nearly constant by having I_{BB} large compared to i_{Bn}, which implies that either I_{BB} or β_F should be large. The total swing in v_I is 20.27 V, so that the large-signal gain is $20/20.27 = 0.99$.

If the load is accidentally shorted to ground while the drive signal is applied, then at the peak of the positive excursion, $v_{Bp+} = 9.57\ \text{V}$, as determined above. If $v_O = 0$, $V_{EBp} = -9.57\ \text{V}$. $v_{Bn+} \cong 0.7\ \text{V}$, so the diode-connected transistors are also off. The base drive of Q_n is therefore limited by I_{BB} and the current through the short circuit is approximately $51 \times 2.2 \cong 112\ \text{mA}$. A small value of I_{BB} therefore limits the current through Q_n under these conditions. At the peak of the negative excursion, v_{Bp-} becomes about $-0.7\ \text{V}$, and the base drive of Q_p depends on the current that the circuit providing v_I can sink. Q_n and Q_p can be protected against short circuits, as described later.

It is of interest to determine the small signal output resistance of Q_n and Q_p, assuming the ac resistance between the bases of both transistors and ground is negligibly small. Under quiescent conditions, $r_e = V_T/I_{EQ} = 26/5 = 5.2\ \Omega$, and $r_{out} = 5.2/2 = 2.6\ \Omega$. At a current of 100 mA, $r_e = r_{out} = 26/100 = 0.26\ \Omega$, since only one transistor is conducting.

An adaptation of the circuit of Figure 11.4.1 for discrete transistors and a single power supply is shown in Figure 11.4.2. Under quiescent conditions, the voltage of node A is $V_{CC}/2$, assuming that the transistors and the resistors R_E are matched. These resistors,

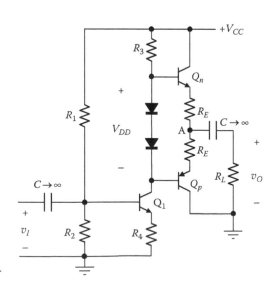

FIGURE 11.4.2
Class AB amplifier having a single power supply.

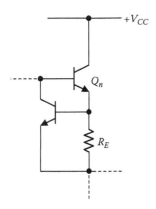

FIGURE 11.4.3
Excess current protection of output transistor.

typically less than 1 Ω each, introduce negative feedback that reduces the effects of mismatches in the transistors, particularly β_F, and also reduces dI_{EQ}/dT_J, which helps protect against thermal runaway.

The output transistors can be protected against excess current by connecting a transistor as in Figure 11.4.3 for Q_n, with a similar arrangement for Q_p. With the rated maximum output current flowing through Q_n, R_E is chosen so that the voltage drop across it is about 0.5 V, and the protection transistor is normally off. An excess output current biases the protection transistor into conduction, so that its collector current draws the current away from the base of Q_n and limits the output current. The voltage drop across R_E reduces the maximum swing of v_O by about 0.5 V.

Exercise 11.4.2

Assume that in Figure 11.4.3, $R_E = 0.5 \ \Omega$, β_F of Q_n is 20 and its normal peak emitter current is 1 A. It is required that the protection transistor draw 50 mA from the base drive of Q_n when the peak emitter current of Q_n reaches 1.2 A. (a) What should be I_S of the protection transistor? (b) What current would this transistor draw from the base drive of Q_n when the peak emitter current of Q_n reaches is 1 A?
Answers: (a) 4.75 pA; (b) 1.1 mA. ∎

In order to reduce the base current drive requirement of the output transistors, a composite transistor is sometimes used. A circuit using matched complementary Darlington pairs is analyzed in Example SE11.4. MOSFETs can also be used in class AB amplifiers, as discussed in Examples SE11.5 and SE11.6. The output MOSFETs can be driven by a complementary BJT emitter follower that rapidly charges and discharges the gate capacitors. This is particularly important in high-speed switching and operation at relatively high frequencies.

11.5 Class C Operation

In a class C amplifier, illustrated in Figure 11.5.1a, the output transistor is biased well into cut off so that, in response to a sinusoidal input voltage, the output current flows only for short intervals near the peaks of the input voltage (Figure 11.5.1b). If this current were to flow in a normal resistive load, the load voltage would be severely distorted compared to the sinusoidal input voltage. Hence, class C amplifiers are used only at radio frequencies with a tuned circuit load whose resonant frequency is equal to that of the input signal. The selectivity of the tuned circuit heavily attenuates the harmonics so that the output voltage is nearly sinusoidal. Because the dc power supply delivers power only when collector current flows, the power-conversion efficiency is very high, typically in excess of 90%.

Class C amplifiers are commonly used for power amplification of carrier-modulated radio frequency (rf) signals in transmitters for radio, TV, and mobile phones. The modulated signal is generally narrow-band around the carrier frequency, so that a tuned-circuit load is acceptable. The design of class C amplifiers is a specialized topic that will not be pursued here.

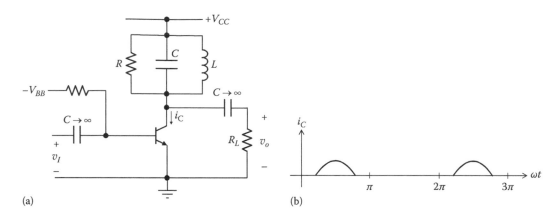

(a)   (b)

FIGURE 11.5.1
(a) Basic class C amplifier; (b) waveform of collector current.

11.6 Power Operational Amplifiers

Small-signal op amps are generally capable of delivering up to about 300 mW at voltages close to maximum supply voltages of ± 20 V. These op amps can be augmented with an appropriate power output stage for delivering substantial power to a low-impedance load. Figure 11.6.1 illustrates an op amp connected to a class AB output stage through an emitter follower transistor Q_3, with negative feedback from the output to the inverting input of the op amp. Zener diodes protect the output of the op amp in case of accidental connection to the higher voltages $\pm V_{CCO}$. R_o is a small resistance that limits the maximum op-amp current.

Current and voltage boosting may be incorporated internally in an IC power op amp, allowing power outputs of tens of watts, with an appropriate heat sink, at currents of few amperes and voltages in excess of a hundred volts. The op amp may be protected by a thermal

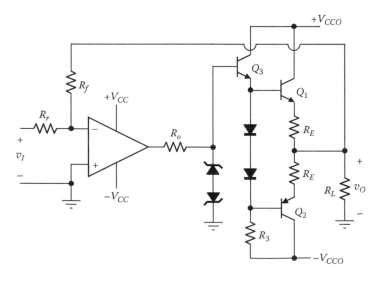

FIGURE 11.6.1
Op amp with class AB power output stage.

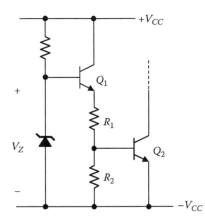

FIGURE 11.6.2
Basic thermal shut-down circuit.

shutdown circuit such as that illustrated in Figure 11.6.2. As the temperature of the chip rises, V_Z increases, because of its positive temperature coefficient, while V_{BE} of Q_1 decreases. This increases the current through Q_1 and hence the voltage drop across R_2. When this becomes sufficiently large, Q_2 conducts and diverts current from some input stage, thereby reducing the drive to the output stages. Limiting the maximum temperature of the chip is important so as not to adversely affect the long-term reliability of the IC power op amp.

Exercise 11.6.1

The circuit of Figure 11.6.2 is required to shut down the amplifier at 125°C. At 25°C, $V_Z = 7.2$ V and Q_1 and Q_2 have $V_{BE} = 0.7$ V at a current of 0.2 mA. (a) Choose R_1 and R_2 so that a current of 0.2 mA flows in Q_1 and Q_2 at 125°C; (b) Determine the current in Q_2 at 25°C, assuming that the transistor β's are high and that the temperature coefficients are $+2$ mV/°C for the zener diode, -2 mV/°C for V_{BE}.
   Answers: (a) $R_1 = 32$ kΩ, $R_2 = 2.5$ kΩ; (b) 0.03 μA. ∎

Exercise 11.6.2

Figure 11.6.3 shows two power op amps having equal magnitude of voltage gain connected to a load in a **bridge configuration**. Show that the load voltage and the slew rate are doubled.

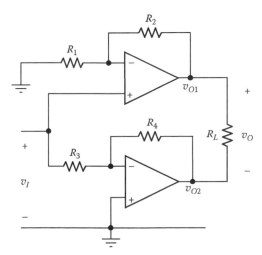

FIGURE 11.6.3
Power op amps connected in bridge configuration.

11.7 Power Switching

In many applications, the power transistor is operated in a switching mode, in which it is repeatedly turned on and off. The main advantage is that power efficiency is high because the transistor is either fully conducting or cut off. In either case, the power dissipated in the device is low, so that devices of relatively small power rating are capable of controlling much larger load power. On the negative side, switching inevitably generates noise in the system and generally slows down the response. Power switching is the domain of power electronics. A detailed discussion is beyond the scope of this book, so only some basic principles will be covered.

Class D Amplifier

In a class D amplifier, the power transistors are connected in push–pull and operated in a switching mode, as illustrated in Figure 11.7.1a. To allow amplification of a continuously varying signal v_S, such an audio signal, the signal is applied to one input of an op amp, and a triangular waveform v_Δ, of frequency much higher than the highest frequency of v_S, is applied to the other input. For the polarities shown, the op-amp output v_G is at its positive limit V^+ when $v_S > v_\Delta$, and is at its negative limit V^- when $v_S < v_\Delta$, where V^+ and V^- have the same magnitude but opposite polarities. The output v_G is pulse-width modulated (PWM), consisting of rectangular pulses whose amplitude alternates between V^+ and V^-, the width of each pulse varying with the instantaneous value of v_S (Figure 11.7.1b).

v_G is applied to the common gate terminal of CS, matched MOSFETs Q_1 and Q_2 connected in push–pull. When $v_G = -V^-$, Q_1 is on and Q_2 is off, v_D is less than V_{SS} by the voltage drop across Q_1. Conversely, when $v_G = V^+$, Q_2 is on, Q_1 is off. When $v_S = 0$, v_D is a square wave of high frequency and zero average value, so that the output of the low-pass filter is ideally zero, and no signal is applied to the load. When v_S is applied, the output of the low-pass filter is the average of the PWM waveform, which ideally reproduces the signal v_S. In practice, more elaborate circuits are used to improve performance.

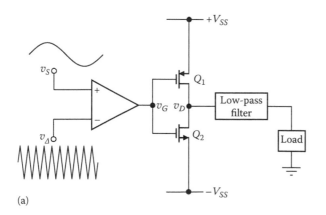

(a)

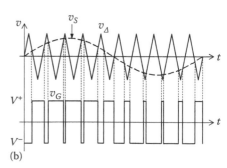

(b)

FIGURE 11.7.1
(a) Basic class D amplifier; (b) voltage waveforms in (a).

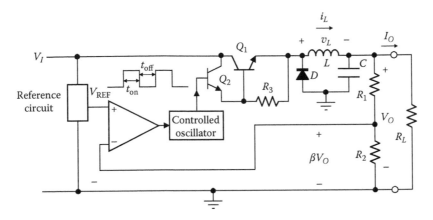

FIGURE 11.7.2
Basic switching voltage regulator.

Switched Regulated Supplies

A basic switching voltage regulator is illustrated in Figure 11.7.2, where the input V_I is an unregulated but well-filtered dc supply. A fraction $\beta_f V_O$ of the dc output voltage, where $\beta_f = R_2/(R_1 + R_2)$, is compared with a reference voltage V_{REF}. The difference between $\beta_f V_O$ and V_{REF}, that is, the error signal, is amplified by a differential amplifier and applied to a controlled oscillator connected to Q_2 and Q_1 in a sense that reduces the error through negative feedback. If the loop gain is large enough, the steady-state error is almost zero, so that $\beta_f V_O = V_{REF}$, or,

$$V_O = \left(1 + \frac{R_1}{R_2}\right) V_{REF} \tag{11.7.1}$$

The controlled oscillator turns Q_2 on and off for predetermined intervals at frequencies in the range of few tens of kilohertz to more than a hundred kilohertz. The emitter current of Q_2 establishes through R_3 the base–emitter voltage of the power transistor Q_1, switching this transistor between cut off and saturation. When Q_1 is in saturation, V_I is applied to the LC filter, and current flows through L to the load. Freewheeling diode D is reverse-biased under these conditions. Assuming Q_1 cuts off instantly, the current in L does not change at the instant of switching and is diverted through D. If the forward voltage drop in D is neglected, the input voltage to the LC filter is now zero. It follows that the input to the filter is a rectangular waveform of amplitude V_I during the interval t_{on} and is zero during the interval t_{off}. The average value of this rectangular waveform is the dc output voltage V_O. Thus:

$$V_O = \frac{t_{on}}{t_{on} + t_{off}} V_I = V_I T \tag{11.7.2}$$

where $T = t_{on}/(t_{on} + t_{off})$ is the duty cycle. The reactance of L at the switching frequency $f_S = 1/(t_{on} + t_{off})$ is much larger than the reactance of C, which means that the fundamental and harmonics are heavily attenuated. The controlled oscillator varies T so as to make $V_O = V_{REF}/\beta$, in accordance with Equation 11.7.1. For example, if V_O tends to decrease because of increased load current, the error increases and the controlled oscillator

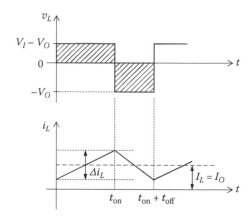

FIGURE 11.7.3

Inductor voltage and current in basic switching voltage regulator.

increases T. As a result, V_O increases (Equation 11.7.2) and maintains $V_O = V_{REF}/\beta$. The output of the controlled oscillator is usually a pulse-modulated signal in which t_{on} is varied while keeping $(t_{on} + t_{off})$ constant. Turn-off and turn-on are examined in more detail in Section ST11.4.

It is seen that the voltage across the inductor v_L is (Figure 11.7.3):

$$v_L = V_I - V_O, \quad 0 \le t \le t_{on} \quad \text{and} \quad v_L = -V_O, \quad t_{on} \le t \le t_{on} + t_{off} \quad (11.7.3)$$

Since v_L is constant during t_{on} and t_{off}, $i_L = (1/L) \int v_L dt$ varies linearly as shown in Figure 11.7.3. In the steady state, Δi_L is of constant magnitude, which implies that the positive and negative areas of v_L have equal magnitudes. Moreover, the average current I_L equals the average load current I_O because the average current through C is zero and R_1 and R_2 are relatively large compared to R_L. Hence,

$$I_L = I_O \quad (11.7.4)$$

As long as I_L is sufficiently large, i_L always flows (Figure 11.7.3). However, if I_L falls below a certain value (Problem P11.6.1), i_L becomes zero before the end of t_{off}, leading to a discontinuous mode of operation, and the above analysis is no longer valid.

Since Q_1 is either in saturation or is cut off, power dissipation is small, and the power efficiency is typically more than 90%, much higher than that of a linear voltage regulator (Example SE11.9). Switching regulators are available as ICs, some including Q_1. An added advantage of the switching regulator is its versatility, as discussed later. Switching regulators are extensively used to provide dc power in electronic equipment and in chargers for rechargeable batteries used in many portables, such as electric shavers and cellular phones. These regulators readily accommodate a range of voltages and frequencies, typically 100–240 V, 50–60 Hz.

Example 11.7.1 Output Ripple of Switching Regulator

It is required to determine the output ripple of the switching regulator of Figure 11.7.2.

SOLUTION

The ac component of i_L in Figure 11.7.3 can be assumed to flow through C (Figure 11.7.4), since the reactance of C is much smaller than R_L at the switching frequency. The variation

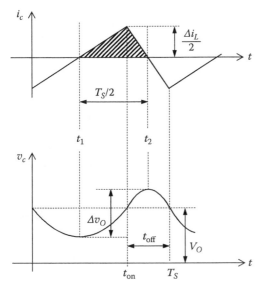

FIGURE 11.7.4
Figure for Example 11.7.1.

in v_O is related to i_C by: $\Delta v_O = (1/C)\int_{t_1}^{t_2} i_C dt$. The integral, which is the shaded area shown is $(1/2)(\Delta i_L/2)(T_S/2)$. Hence, $\Delta v_O = (T_S/8C)\Delta i_L$. From Figure 11.7.3, $\Delta i_L = (t_{off}/L)V_O$. Eliminating Δi_L gives:

$$\frac{\Delta v_O}{V_O} = \frac{T_S}{8LC}t_{off} = \frac{T_S^2}{8LC}\frac{t_{off}}{t_{on}+t_{off}} = \frac{\pi^2}{2}\left(\frac{f_o}{f_S}\right)^2\frac{t_{off}}{t_{on}+t_{off}} = \frac{\pi^2}{2}\left(\frac{f_o}{f_S}\right)^2(1-T) \qquad (11.7.5)$$

where
$f_S = 1/T_S$ is the switching frequency
$f_o = 1/2\pi\sqrt{LC}$

As to be expected, the ripple is small when $f_o \ll f_C$ and $t_{off} \ll t_{on}$, that is, when the duty cycle is almost unity.

Exercise 11.7.1

In the steady state, the increase in i_L during the on period must equal its decrease during the off period, which implies that the two shaded areas in Figure 11.7.3 must be equal. Show that Equation 11.7.2 follows from this equality. ∎

Exercise 11.7.2

Neglecting power losses in the circuit, deduce from conservation of power that: $(I_O/I_I) = (V_I/V_O) = 1/T$, where I_I is the average input current. This relation may be considered as that of a step-down dc transformer having an electronically variable turns ratio a. ∎

The configuration of Figure 11.7.2 is that of a **step-down** regulator, or a **buck converter**, in which $V_O < V_I$ (Equation 11.7.2). If the power switch is connected in shunt, after the inductor, a **step-up** regulator, or **boost converter**, is obtained, in which $V_O > V_I$, as illustrated in Figure 11.7.5, where the regulator is shown controlling the switch. Feedback

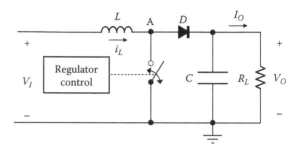

FIGURE 11.7.5
Basic step-up regulator.

is derived from the output voltage as before. During t_{on}, the switch is closed, and V_I causes i_L to flow through the switch. D is reverse biased, and C discharges through R_L. The time constant $R_L C$ is made much larger than t_{on}, so that the variation in V_O is small. When Q_1 turns off, i_L does not change at the instant of switching and flows in D, but tends to decrease because the voltage at node A has changed from zero to nearly V_O. The polarity of the induced voltage in L is now that of a voltage *rise* in the direction of i_L so as to keep it flowing. The induced voltage therefore adds to V_I, so that $V_O > V_I$. The v_L and i_L waveforms are very similar to those of Figure 11.7.3, except that during t_{on}, when the switch is closed, the positive amplitude of v_L is V_I and, during t_{off}, when the switch is open, the negative amplitude of v_L is $-(V_O - V_I)$.

Exercise 11.7.3

Assuming a lossless system, deduce from conservation of power that: $(I_O/I_I) = (V_I/V_O) = 1 - T$, for the step-up regulator, analogous to that of Exercise 13.7.2.   ∎

dc-to-ac Converters

Also known as **inverters**, dc-to-ac converters transform a dc input to an ideally sinusoidal output of desired amplitude and frequency. They are extensively used in diverse applications such as variable speed ac-motor drives, standby and uninterruptible power supplies (UPS), and high voltage dc transmission. If followed by a rectifier, a dc-to-dc converter is obtained, in which the output voltage can be of the same polarity, or of opposite polarity, as the input voltage, and whose magnitude is less than, equal to, or greater than, that of the input voltage. Moreover, the transformer in these converters isolates the input from the output. Although switched regulated supplies are a form of dc-to-dc converter, they are not as versatile as the two-stage dc-ac-dc converters. In addition to regulated power supplies, dc-to-dc converters are used in sophisticated dc-motor drives. The ac inputs and outputs could be single-phase or three-phase, but only single-phase is discussed here in order to illustrate the basic concepts involved.

A simple converter is illustrated in Figure 11.7.6a, in which two semiconductor switches Q_1 and Q_2 are activated by a control circuit. When Q_1 is turned on, with Q_2 off, V_I is impressed across one-half of the transformer primary winding, and $v_{P1} = V_I$ (Figure 11.7.6b). When Q_2 is turned on, with Q_1 off, v_I is impressed across the second half of the transformer primary winding, and $v_{P2} = -V_I$. Because of the configuration of dotted terminals of the transformer, and assuming the transformer is ideal, a square ac waveform appears across the transformer secondary, whose amplitude depends on the transformer turns ratio. The square-wave output can be readily converted to a dc output using rectifier diodes as shown. In practice, of course, the square wave is distorted by the inductances and capacitances of the transformer.

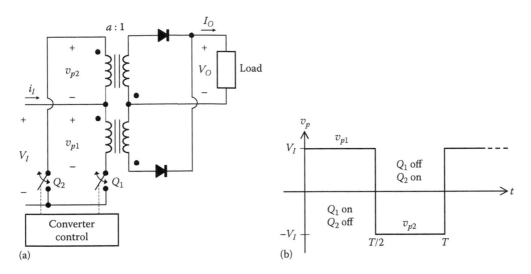

FIGURE 11.7.6
(a) A simple dc-to-dc converter; (b) voltages of transformer primary.

A more complex arrangement is that of the **half-bridge** illustrated in Figure 11.7.7a. A general load is shown, which could be the transformer-rectifier-load of Figure 11.7.6a. Ignoring D_1 and D_2 for the moment, it is seen that $v_O = V_I/2$ when Q_1 is turned on, and $v_O = -V_I/2$ when Q_2 is turned on (Figure 11.7.7b). The control circuitry ensures that Q_1 and Q_2 are not turned on simultaneously, so as not to short-circuit the input.

For a purely inductive load, $v_O = L di_O/dt$. Hence, when $v_O = V_I/2$, i_O increases linearly at a rate of $V_I/2L$ (Figure 11.7.7b). Under steady-state conditions, the change in i_O over half a period is $(V_I/2L)(T/2) = V_I T/4L$, from $-V_I T/8L$ to $+V_I T/8L$. For $0 \leq t \leq T/4$, $i_O < 0$, and $v_O i_O < 0$, which means the inductor is returning stored energy to the supply. Negative i_O cannot flow through Q_1 because the *npn* transistor cannot conduct current from the emitter to collector while its BEJ is forward biased. Hence, D_1 is provided to allow a negative i_O that returns energy to the supply. For $(T/4) \leq t \leq (T/2)$, $i_O > 0$, and passes through Q_1 but not D_1. In an analogous manner, for $(T/2) \leq t \leq (3T/4)$, $i_O > 0$ but $v_O i_O < 0$, so that the inductor returns energy to the supply. Q_2 is on, but cannot pass positive i_O. Hence, D_2 is provided for this purpose. When the load is purely resistive, i_O is a square waveform in phase with v_O, and the diodes do not conduct. As shown in Figure 11.7.7a, a center-tapped supply is needed, but in some cases, a single supply can be used together with two large capacitors of equal value connected as a voltage divider.

Example 11.7.2 Half-Bridge Inverter with Resistive Load

Consider that the half-bridge inverter of Figure 11.7.7a has $V_I = 48$ V and a 3 Ω resistive load. It is required to analyze the circuit, assuming ideal conditions.

SOLUTION

The output v_O is a square wave of amplitude 24 V. The rms value of this waveform is the same as the amplitude, that is, 24 V. The total output power is $(24)^2/3 = 192$ W. From the Fourier series expansion of a square wave, the fundamental has an amplitude of $2V_I/\pi = 30.6$ V and

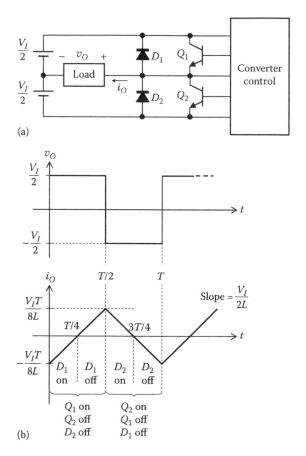

(a)

(b)

FIGURE 11.7.7
(a) Half-bridge configuration; (b) waveforms of output voltage and current.

an rms value of $2V_I/\pi\sqrt{2} = 21.6$ V. The output power due to the fundamental is $(21.6)^2/3 \cong$ 156 W. The total harmonic distortion is $D = \sqrt{(\pi/2\sqrt{2})^2 - 1} = 0.483$. The waveform has odd harmonics only, the lowest-order harmonic being the third, and the amplitude of the nth order harmonic is $1/n$ that of the fundamental. The peak current of each transistor is $24/3 = 8$ A, and the average current is half of this, that is, 4 A. The peak V_{CE} voltage across a transistor while the other transistor is conducting is $V_I = 48$ V.

The extension of the half-bridge to a **full-bridge**, or **H-bridge** is illustrated in Figure 11.7.8, the operation being analogous to that of Figure 11.7.7a. In one half-cycle, Q_1 and Q_3 are on, whereas Q_2 and Q_4 are off, and $v_O = V_I$. In the second half-cycle, Q_2 and Q_4 are on, whereas Q_1 and Q_3 are off, and $v_O = -V_I$. Diodes D_1–D_4 are provided to feed power back to the supply as in the half-bridge. The diagram of Figure 11.7.7b applies with the transistors and diodes having the odd subscripts 1 and 3 working together, and the transistors and diodes having the even subscripts 2 and 4 working together.

Ideally, the voltage across the load is a square wave of amplitude $V_I/2$ for the half-bridge and of amplitude V_I for the full bridge. The full bridge does not need a center-tapped supply and supplies four times as much power to the same load as the half bridge, the peak

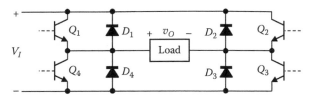

FIGURE 11.7.8
H-bridge configuration.

and average transistor currents being twice as much. Voltage control of outputs of converters is usually based on PWM (Section ST11.4).

Exercise 11.7.4

Consider that the full-bridge rectifier is used with $V_I = 48$ V and $R_L = 3\,\Omega$. As in Example 11.7.2, determine under ideal conditions: (a) the rms value of v_O; (b) the total output power; (c) the rms of the fundamental; (d) the output power associated with the fundamental; (e) peak current per transistor; (f) peak V_{CE} voltage per transistor.

Answers: (a) 48 V; (b) 768 W; (c) 43.2 V; (d) 623 W; (e) 16 A; (f) 48 V. ∎

Application Window 11.7.1 Uninterruptible Power Supplies

In many cases, an emergency power supply is required to supply loads in case of failure of the mains supply. ac generators driven by gasoline or diesel engines are used for this purpose when it is acceptable to interrupt power for the time it takes to start the engine and bring the ac generator online, the whole process being performed automatically on power outage. There are critical loads, however, where the supply should not be interrupted, as in the case of computers performing important functions and medical equipment in operating rooms or for intensive care. In such cases, an uninterruptible power supply (UPS) is used. These supplies may also be used to protect against under-voltage, overvoltage, frequency variations, transients, and noise.

Figure 11.7.9 illustrates the block diagram of a UPS. Basically, a battery supplies an inverter that provides ac power to the load. The battery is kept charged by a rectifier from the ac mains supply, which could be single-phase or three-phase. ac static switches (Application Window 11.10.1) connect the load to the mains supply or to the inverter

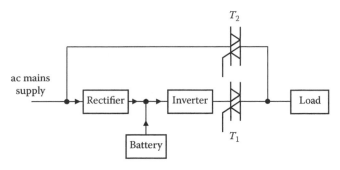

FIGURE 11.7.9
Figure for Application Window 11.7.1.

output. The system can be operated in one of two modes: (1) Standby mode, in which under normal conditions, the inverter and switch T_1 are off, while switch T_2 connects the load to the mains supply. The battery is trickle charged, that is, at a low enough current that maintains it fully charged. When the mains supply is interrupted, T_2 disconnects this supply from the load, and T_1 connects the load to the inverter output. The transfer is automatic and may take a few milliseconds, but the inverter output should be synchronized with the mains supply for minimal supply disturbance to the load. (2) Online mode, in which the load is normally connected to the inverter, and the battery is continuously charged from the mains supply. In case the UPS fails for some reason, the static switches disconnect the load from the inverter and connect it to the mains supply. Not only is the load completely oblivious to the state of the mains supply, but the inverter allows close control of the amplitude, waveform, and frequency of load voltage.

The battery is normally of the lead–acid, nickel–cadmium, or more recently, lithium ion type. Although more expensive, the latter two types have the advantages of no emission of corrosive and flammable gas, and ability to better withstand overheating and excessive discharging. The inverter usually includes an isolation transformer at its output as well as an ac filter to minimize the harmonic content of the output.

11.8 Power Diodes

The forward current of a power diode can be more than several kA and the reverse voltage in excess of 1 kV. A typical structure of a power diode designed to handle such currents and voltages is illustrated in Figure 11.8.1. It consists of a heavily doped n^+ substrate ($N_D \cong 10^{19}/cm^3$) to which the cathode contact is made. A lightly doped n^- epitaxial layer ($N_D \cong 10^{14}/cm^3$, Section 4.3, Chapter 4) is grown on top of the substrate. Acceptor impurities are

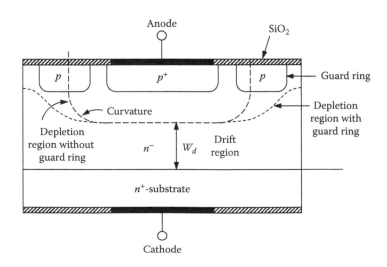

FIGURE 11.8.1
Diagrammatic structure of power diode.

then added to form a heavily doped p^+ region ($N_A \cong 10^{19}/\mathrm{cm}^3$) to which the anode contact is made. p-type guard rings are also formed around the anode.

The large current-carrying capacity is subserved by a large junction area. The depletion region extends much more into the lightly doped n^- side, which is made wide enough (100 μm or more), to support the required reverse voltage (Example 11.8.1). The diffusion of acceptor impurities to form the p^+ region leads to a p^+n^- junction with some curvature (Figure 11.8.1), so that the electric field becomes nonuniform and of larger magnitude in the region of curvature than in a plane p^+n^- junction. Breakdown will then occur at a lower voltage. The guard ring is electrically floating and acquires a reverse voltage with respect to the anode that is less than the cathode–anode voltage. Its depletion layer merges with that of the p^+n^- junction and flattens it, thereby making the electric field more uniform and increasing the reverse breakdown voltage.

The presence of the n^- region significantly affects diode behavior in the steady-state and under transient conditions. It would appear that the low conductivity and the relatively large depth of this region will increase the ohmic drop across the diode and hence the power dissipation. In fact, holes are injected in the forward direction from the p^+ region at a density that far exceeds the low equilibrium concentration n_{no-} of electrons in the n^- region, so high-level injection occurs at modest values of current. The positive charge due to injected holes attracts electrons from the n^+ region so as to maintain the n^- region almost electrically neutral. Assuming the diffusion length $L = \sqrt{D\tau} > W_d$, the concentrations of electrons and holes in the n^- region, also referred to as a **drift region**, are almost equal, uniform, and typically a hundred times n_{no-}. The resulting increase in conductivity, or **conductivity modulation,** reduces the voltage drop and power dissipation significantly. Nevertheless, these remain appreciable, so that power loss in the on-state is what determines the diode's power handling capability. Moreover, the voltage drop v_{drift} across the n^- region varies almost linearly with the current i_D at relatively large currents. Thus,

$$i_D = \frac{\sigma_{n-} A v_{\mathrm{drift}}}{W_d} = \frac{|q|(\mu_e + \mu_h) n_{n-} A v_{\mathrm{drift}}}{W_d}$$

or,

$$v_{\mathrm{drift}} = \frac{W_d}{|q|(\mu_e + \mu_h) n_{n-} A} i_D \qquad (11.8.1)$$

where
   n_{n-} is the concentration of electrons or holes in the n^- region
   A is the cross-sectional area of the diode

At low to moderate i_D, and hence relatively low n_{n-}, the mobilities change little with n_{n-}; v_{drift} does not change much with i_D since n_{n-} increases with i_D in Equation 11.8.1. However, at high n_{n-}, the mobilities become almost inversely proportional to n_{n-} and v_{drift} becomes nearly directly proportional to i_D.

The total diode voltage drop v_D in the forward direction is the sum of v_J – the voltage across the p^+n^- junction – and the near-ohmic voltage drop $R_{\mathrm{on}} i_D$ due to all the resistive paths in the diode. As v_D increases, v_J increases at a slow rate, since it varies as $\ln i_D$, and is eventually swamped by $R_{\mathrm{on}} i_D$, making the forward i–v characteristic more linear as v_D increases.

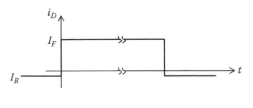

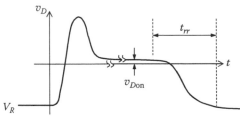

FIGURE 11.8.2

Transcient response of power diode.

Concept: *A low-conductivity "drift" region is used in practically all power semiconductor devices to control the breakdown voltage. In bipolar devices, conductivity modulation reduces the voltage drop across this region in the on state.*

The n^- region affects the transient behavior of the diode, as illustrated in Figure 11.8.2. Assume that a forward current I_F is applied at $t = 0$ from an initially reverse state. After the space charge under reverse bias is neutralized, the p^+n^- junction becomes forward biased. However, conduction modulation has not yet set in, so the resistance of the n^- region is relatively high, and v_D increases rapidly to values that could reach few tens of volts. As charge in the n^- region builds up and the conductivity increases, v_D drops to almost its on-state value v_{Don} in few μs.

At turn-off, current is reversed to I_R. Stored charge is removed by recombination and by I_R. The junction remains forward biased as long as carrier concentrations are above equilibrium values. v_D drops because of the now opposing voltage drop in the n^- region due to I_R. When carrier concentrations at the junction are near zero, the diode becomes reverse biased, and the depletion region rapidly expands into the n^- region. The total time from initiation of diode turn-off to its completion is the **diode reverse recovery time** t_{rr}, which is quoted in manufacturer's data sheets for specified reverse-bias conditions.

Example 11.8.1 Breakdown Voltage in a Power Diode

It is required to derive a relation between the width of the drift region and the diode breakdown voltage.

SOLUTION

Assuming the depletion approximation applies and that $N_A \gg N_D$, the breakdown voltage BV is the area of a triangle whose base is W_d, the width of the drift region and whose height is the breakdown electric field ξ_{brk} (Figure 3.1.3, Chapter 3). Thus,

$$W_d = \frac{2BV}{\xi_{brk}} \qquad (11.8.2)$$

where the sign of ξ is being disregarded for convenience, since only magnitudes are being considered. If $\xi_{brk} = 2 \times 10^5$ V/cm, then $W_d = 10^{-5}$ BV cm. If the breakdown voltage is to be 1 kV, for example, the width of the drift region must be at least 10^{-2} cm, or 100 μm.

In a **punchthrough diode** (Section ST11.5), the width of the drift region is reduced by decreasing its conductivity and allowing the depletion region to extend to the n^- n^+ junction before voltage breakdown occurs.

11.9 Power Transistors

Bipolar Junction Transistors

Power BJTs can have current and voltage ratings in excess of few hundred amps and 1.5 kV, respectively. As in the case of power diodes, such ratings necessitate some special structures.

The typical, vertical-type structure of an *npn* power transistor illustrated in Figure 11.9.1a provides a large cross-sectional area that supports a high current with minimal voltage drop and hence power dissipation. It also minimizes thermal resistance, as required to maintain a low junction temperature for a given power dissipation (Section 11.1). The doping levels of the n^+ emitter and the collector terminal regions are high, typically about 10^{19}/cm³. The base region, as usual, has a much lower doping level ($N_A \cong 10^{16}$ to 10^{17}/cm³) for high emitter efficiency (Section 6.1, Chapter 6). An n^- drift region ($N_D \cong 10^{14}$/cm³) is interposed between the base and n^+ region of the collector, for the same reason as in power diodes, namely, to control the breakdown voltage. Depending on this voltage, the width of the n^- region ranges from several tens to few hundred microns. However, the base cannot be made too thin in order to have a large transport factor. Otherwise, the depletion region in the base expands all the way to the emitter junction, leading to punchthrough (Section 6.2, Chapter 6). The usual base thickness of several microns to few tens of microns is a compromise between β_F and punchthrough. When the base is

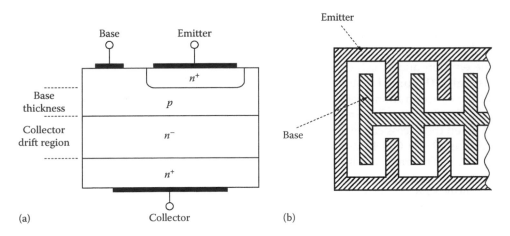

FIGURE 11.9.1
(a) Diagrammatic structure of power transistor; (b) interleaving of base and emitter regions.

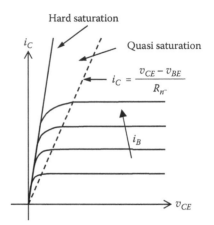

FIGURE 11.9.2
Output characteristics of power BJT.

relatively wide, β_F may be quite low, about 10 or so. In such cases it is not unusual to combine two transistors as a Darlington pair in order to have an acceptably large current gain.

Another structural feature of power BJTs is the interleaving of emitter and base regions (Figure 11.9.1b). The interleaving shortens the path of the base current, which reduces the base spreading-parasitic resistance and the base–emitter voltage drop. More importantly interleaving reduces current crowding, and hence the possibility of second breakdown (Figure 11.1.1).

The n^- region also affects the steady-state and transient behavior of the BJT. The transition from the active region to the saturation region becomes more gradual, leading to a distinction between **quasi saturation** and **hard saturation** (Figure 11.9.2). To appreciate this distinction, suppose that v_{CE} is decreased in the active region at constant v_{BE}. For a given i_C:

$$v_{CE} = v_{n^-} + v_{CB} + v_{BE} \tag{11.9.1}$$

where v_{n^-} and v_{CB} are the voltage drops across the n^- region and CB n^-p junction, respectively. Because of the additional voltage drop v_{n^-}, the CBJ becomes forward biased at a higher v_{CE} than in the absence of the n^- region. When the CBJ just becomes forward biased, the excess minority carrier concentration at the CBJ is zero (Figure 11.9.3a), and i_C is proportional to the slope of the nearly linear concentration profile of excess carriers in the base. As v_{CE} decreases further and the CBJ becomes forward biased, holes are injected from the base to the n^- region. There will be more recombination with electrons injected from the emitter, and which cross the base into the n^- region. In effect, the base spreads into this region (Figure 11.9.3b). i_C is reduced, as indicated by the reduced magnitude of the slope of the concentration profile, and the transistor is now in quasi saturation. The added recombination implies of course a smaller β_F. The limit of quasi saturation is illustrated by line AB when the excess carrier concentration just reaches the n^+ region. Further increase of forward bias of the CBJ leads to additional charge storage, as illustrated by line AC in Figure 11.9.3c. The transistor is now in hard saturation, the n^- region being part of the base, and the ratio of i_C to i_B is β_{forced}, as in a low-power transistor.

Quasi saturation sets in when $v_{CB} = 0$, and the n^- region is part of the collector. From Equation 11.9.1, this occurs when $v_{n^-} = R_{n^-}i_C$, or,

$$i_C = \frac{v_{CE} - v_{BE}}{R_{n^-}} \tag{11.9.2}$$

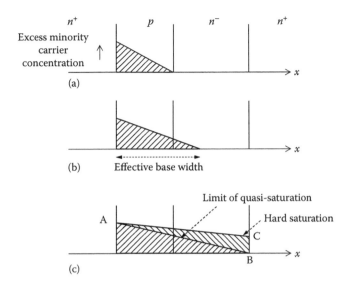

FIGURE 11.9.3
(a) Onset of quasi saturation; (b) quasi saturation; (c) hard saturation.

where R_{n-} is the ohmic resistance of the n^- region (Figure 11.9.2). Since, for a given i_C, v_{CE} is larger in quasi saturation than in hard saturation, more power is dissipated. Hence it is desirable to have the transistor in hard saturation when switched on.

The voltage across the transistor when in hard saturation is given by:

$$v_{CE(sat)} = v_{BE(on)} - v_{BC(sat)} + v_{n-(on)} + (R_{en^+} + R_{cn^+})i_{C(on)} \tag{11.9.3}$$

where
$v_{BC(sat)}$ is the forward voltage drop across the BCJ
R_{en^+} and R_{cn^+} are the ohmic resistances of the n^+ emitter and collector regions, respectively

At large currents, the voltage drops across the n^+ regions may become significant. However, the major contribution to $v_{CE(sat)}$ at large currents is from $v_{n-(on)}$. As noted for the power diode, this increases rather rapidly at large currents because of the decrease in mobilities and lifetime (Equation 11.8.1). Note that the current becomes proportional to $e^{v_{BE}/2V_T}$ under conditions of high-level injection.

The additional charge stored in the n^- region clearly increases the turn-on and turn-off times, essentially because it takes more time to build up this charge during turn-on and to remove it during turn-off. The manner in which the voltages and currents change depends not only on the parasitic elements of the transistor but also on the external circuit. The interactions can be quite complex and are beyond the scope of the present discussion. However, the following points should be noted: (1) the parasitic transistor elements that play a significant role in switching are the resistances and capacitances previously discussed as well as inductances of leads that carry a substantial current, such as emitter and collector leads. (2) When the transistor is conducting in the steady state, the current is large but the voltage across the transistor is low. When the transistor is off, the current is negligible but the voltage may be high. In both cases, the power dissipated is relatively low. However, under transient conditions, there may be short periods when the transistor

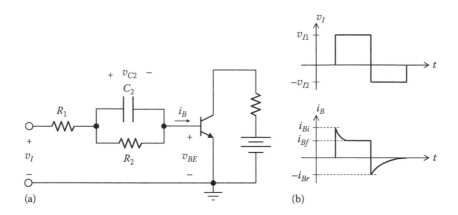

FIGURE 11.9.4
(a) Connection of a speed-up capacitor; (b) voltage and current waveforms.

is subjected to both a large current and a high voltage (Section ST11.11). The instantaneous power dissipation is high and the transistor may be particularly susceptible to second breakdown. (3) The transistor may be subject to a high rate of rise of current during turn-on and a high rate of rise of voltage during turn-off. If these fast rates are more than the transistor can handle, they have to be limited by means of additional circuitry (Section ST11.11).

Turn-off and turn-on times can be reduced by using some form of a speed-up capacitor circuit, as explained in connection with Figure 6.3.4a, Chapter 6, and illustrated by the more versatile circuit in Figure 11.9.4a. It can be readily shown that when v_I suddenly assumes the value v_{I1} (Figure 11.9.4b), the capacitor voltage does not change at that instant, so that,

$$i_{Bi} = \frac{v_{i1}}{R_1} \tag{11.9.4}$$

where i_{Bi} is the initial peak of i_B, assuming that $v_{BE} \cong 0$ when i_{Bi} starts to flow. The capacitor charges with a time constant $C_2(R_1 \| R_2)$ to a final voltage $v_{C_2} = (v_{I1} - v_{BE(sat)})R_2/(R_1 + R_2)$, neglecting source and base-emitter resistance.

When the capacitor is fully charged, the final value of i_B is:

$$i_{Bf} = \frac{v_{i1} - v_{BE(sat)}}{R_1 + R_2} \tag{11.9.5}$$

When v_I changes to $-v_{I2}$, the initial reverse base current is:

$$i_{Br} = \frac{v_{C_2} + v_{I2} + v_{BE(sat)}}{R_1} \tag{11.9.6}$$

i_{Br} removes the charge stored in the BEJ. Once this charge is removed and the BEJ becomes reverse biased, i_B drops to a low value. C_2 discharges through R_2 and the reverse bias across the BEJ is practically $-v_{I2}$. Reverse base current is effective in removing the charge stored in the p-region of the base but not from the n^- drift region. The latter charge is removed mainly through the relatively slow process of recombination, so that the collector current decays with a rather long tail (Figure 11.9.4b).

MOSFETs

A large MOSFET current requires a large W/L ratio. L cannot be made too small, as this would severely limit the breakdown voltage because of punchthrough. A different structure is therefore required if the transistor is to handle large currents and high voltages. Most power MOSFETs in common use are of the vertical, double-diffused type (VDMOS), diagrammed in Figure 11.9.5, which allows a large cross-sectional area that minimizes the electrical and thermal resistances of the drain. An epitaxial (Section 4.3, Chapter 4) n^- drift region is grown on an n^+ substrate, with p and n^+ regions diffused in succession in the n^- region to form the body and source, respectively. The doping concentrations are roughly the same as in the BJT, about $10^{19}/cm^3$ for the n^+ region, $10^{16}/cm^3$ for the p-region, and $10^{14}/cm^3$ for the n^- region. The oxide thickness is usually 0.1 μm. The channel is induced in the p-body underneath the gate, its length being in the range of 1.5 to few μm. The source consists of a large number, which could be many thousands, of small, polygon-shaped areas, or cells, often hexagonal, that are connected in parallel. W becomes roughly equal to the perimeter of a source cell multiplied by the number of cells. This gives a large W so the W/L ratio is correspondingly large, typically 10^5 or more.

The gate extends over the n^- region. When the transistor is turned on, the gate is positive with respect to the source, and electrons are attracted to the n^- region, forming an accumulation layer underneath the gate, which reduces the on-resistance $r_{DS(on)}$. When the transistor is cutoff, with $V_{GS} \cong 0$, a depletion region is formed in the n^- region, underneath the gate, which merges with the pn^- depletion region and flattens it, thereby making the electric field more uniform, as in the case of the power diode. Contrary to a low-power MOSFET, the structure of Figure 11.9.5 is not symmetrical, that is, the drain and source cannot be interchanged.

The MOSFET has two parasitic, active structures (Figure 11.9.6). A parasitic JFET is formed in the current path between the source and drain, with the n^- region acting as the channel of the JFET and the p-body acting as the gate. Pinch-off can occur in the JFET channel, which increases $r_{DS(on)}$. The n^+ source, p-body, and n^- drift region form an npn BJT, which can adversely affect performance. First, the base of this transistor is open-circuited, so that the drain–source breakdown voltage is reduced to BV_{CEO} (Section 6.2, Chapter 6). Second,

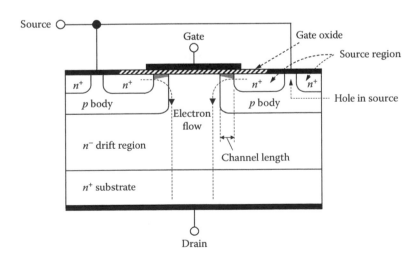

FIGURE 11.9.5
Vertical, double-diffused power MOSFET.

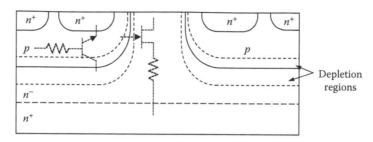

FIGURE 11.9.6
Parasitic structure in power MOSFET.

if this transistor conducts, it provides a current path between source and drain that is not controlled by the gate. Not only is power dissipation increased, but conduction in this path can only be stopped by interrupting the drain current in the external circuit. To prevent these undesirable effects, a "hole" is formed in the middle of each source cell, so that the p-body reaches the source metallization (Figure 11.9.5), which cuts off the *npn* transistor by shorting its emitter and base. A diode structure is also created between the source and drain. In normal operation, this diode is reverse biased, but it may be used as the antiparallel diode in half-bridge and full-bridge circuits (Figures 11.7.7 and 11.7.8).

As in the power diode and BJT, the n^- region increases the breakdown voltage. But there is no conductivity modulation, because the MOSFET is a majority-carrier device. Punch-through occurs when the depletion region in the p-body extends all the way to the source, which creates a current path between the drain and source and results in **soft breakdown**, characterized by a gradual increase of i_D with v_{DS}. **Reachthrough** occurs when the depletion layer in the n^- region extends all the way to the n^-n^+ junction. Any further increase in v_{DS} rapidly increases the electric field to its breakdown value, as in the punchthrough diode (Section ST11.5).

$r_{DS(on)}$ is made up of several components, as illustrated in Figure 11.9.7:

$$r_{DS(on)} = r_S + r_{ch} + r_A + r_J + r_{drf} + r_{sub} + r_{con} \tag{11.9.7}$$

where the resistances are those of the source region (r_S), the channel (r_{ch}), the accumulation layer in the n^- region under the gate (r_A), the parasitic JFET channel between the p-body regions (r_J), the rest of the n^- region (r_{drf}), the substrate (r_{sub}), and the connections and

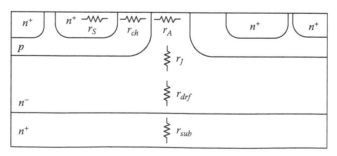

FIGURE 11.9.7
Components of the on resistance of a power MOSFET.

contacts (r_{con}). For power MOSFETs having a low breakdown voltage of about 50 V, $r_{DS(on)}$ is a few tens of mΩ and is dominated by r_{ch} ($\cong$30%), $r_J + r_{drf}$ ($\cong$30%), and r_{con} ($\cong$20%). In power MOSFETS having breakdown voltages of 500–1000 V, the n^- region is wide. $r_{DS(on)}$ may be as high as 10 Ω and is dominated by r_{drf} ($\cong$65%) and r_J ($\cong$20%), with r_{ch} contributing only about 5%. It can be shown that ideally, the resistance of a drift region varies as BV^2 in a majority-carrier device, such as a MOSFET, and as BV in a bipolar device (Section ST11.6). For a given BV, therefore, a BJT has a lower on-resistance, and hence power loss, than a MOSFET.

$r_{DS(on)}$ has a positive temperature coefficient, because mobility decreases with temperature. This is of practical importance, as it automatically ensures safe current sharing between paralleled power MOSFETs of the same type. If the current in a given MOSFET increases, its temperature rises because of increased power loss. The increase in $r_{DS(on)}$ reduces current in the given MOSFET, which tends to equalize currents among the paralleled MOSFETs.

The transfer characteristic of a power MOSFET departs from the square law at large currents, where it becomes almost linear, mainly because of the decrease in mobility at large carrier densities and high electric field in the pinch-off region of the channel.

The comparison between power MOSFETs and BJTs can be summarized as follows: BJTs can handle larger currents and higher voltages, have lower losses in the on-state, especially when voltage ratings are high, but require large base-drive currents, and are subject to second breakdown. Being majority carrier devices, MOSFETs can be turned on and off much faster. Moreover, they require practically zero gate power in the conducting state, but sufficient current must be supplied to rapidly charge and discharge the input capacitance. The maximum switching frequency is a few tens of kHz for power BJTs but more than 1 MHz for power MOSFETs. Complementary *pnp* power BJTs and *p*-channel power MOSFETs are available. The switching behavior of a power MOSFET in a step-down regulator circuit is examined in Section ST11.7.

Insulated Gate Bipolar Transistors (IGBTs)

IGBTs combine the advantages of MOSFETs and BJTs while having intermediate power handling capability and speed of response. The basic IGBT structure and symbol are shown in Figure 11.9.8 together with the polarities of voltage and current in the on state. The structure is similar to that of a power MOSFET (Figure 11.9.5) with the addition of a p^+ hole injecting layer at the collector terminal. A complementary structure to that of Figure 11.9.8 is possible, and its symbol will have an inward pointing arrow at the emitter, as in a *pnp* transistor.

When v_{GE} is less than the threshold, no channel is formed in the *p*-region underneath the gate, and the device is in the off-state. When v_{GE} exceeds the threshold, a channel is formed in the *p*-region. With the collector positive with respect to the emitter, electrons flow from the n^+ emitter, through the channel and the n^- region, toward the collector, as in any MOSFET (Figure 11.9.8). However, since J_1 is forward biased, holes are injected from the p^+ region through the thin n^+ region and are attracted by electrons injected into the n^- region. Conductivity modulation occurs in this region since the equilibrium concentration of electrons is low and is much smaller than that of injected electrons and holes. The total voltage drop in the on-state can be expressed as:

$$v_{CE(on)} = v_{BE(on)} + v_{n^-} + v_{channel} \qquad (11.9.8)$$

$v_{BE(on)}$, the voltage drop across the forward-biased J_1 *pn* junction, is in the range 0.7–1 V. $v_{channel}$ is comparable to that in a power MOSFET, but v_{n^-} is much less, due to conductivity

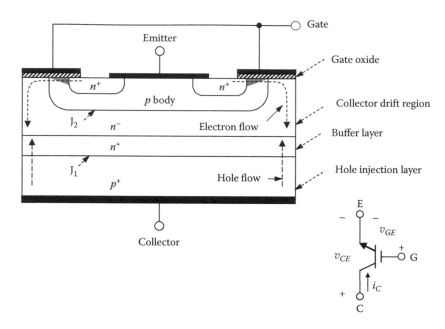

FIGURE 11.9.8
Diagrammatic structure and symbol of IGBT.

modulation. It could be further reduced by allowing reachthrough in the n^+ layer. The low $v_{CE(on)}$, comparable to that of a power BJT, gives the IGBT its principal advantage over the MOSFET.

Consider the forward and reverse voltages that can be blocked in the off state, assuming to begin with that there is no n^+ buffer layer. In the forward blocking state ($v_{CE} > 0$), the voltage is supported by the J_2 junction between the p-body and the n^- region. In the reverse blocking state ($v_{CE} < 0$), the voltage is supported by the junction between the n^- and p^+ regions. In both cases, the depletion layer extends mainly in the n^- region, because it is the more lightly doped. The forward and reverse blocking voltages would be the same. Adding the n^+ layer makes the forward blocking depend on junction J_1 and is only about few tens of volts because both sides of the junction are heavily doped. The purpose of the n^+ buffer layer is to reduce the thickness of the drift region by reducing its conductivity and allowing the depletion region to extend all through the drift region at voltages considerably less than the breakdown voltage, as in the case of the punchthrough diode (Section 11.8). The n^+ buffer layer prevents reachthrough of the depletion region to the p^+ layer.

Although the main path for holes is as shown in Figure 11.9.8, an alternative path exists through the p-body to the emitter metallization. At large currents, or when the IGBT is suddenly turned off, latchup may occur, which can be explained by the action of parasitic *npn* and *pnp* transistors, as discussed in Section ST11.8. Doping levels and profiles would then have to be adjusted for optimum performance.

Simulation Example 11.9.1 IGBT Characteristics

The IXGH40N60 from the PSpice EVAL library is a high-speed IGBT rated at a maximum V_{CE} of 600 V and a maximum I_C of 75 A at 25°C. To simulate the output characteristics, a dc V_{CE} source is swept from 0 to 20 V, while a dc V_{GE} source in secondary sweep is stepped at the values indicated in Figure 11.9.9. It is seen that the output characteristics are similar to those of an *npn* BJT, except that the parameter of the characteristics is v_{GE} rather than the base current. The transfer characteristic is shown in Figure 11.9.10 for $V_{CE} = 20$ V and at two junction temperatures of 25°C and 125°C. The characteristic is similar to that of a MOSFET. It is seen that the threshold is about 4.2 V and that between $V_{GE} = 4.8$ V and 5.8 V I_C varies from about 15 to 95 A, giving an average transconductance of about 80 A/V.

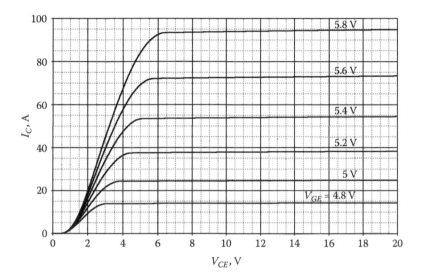

FIGURE 11.9.9
Figure for Example 11.9.1.

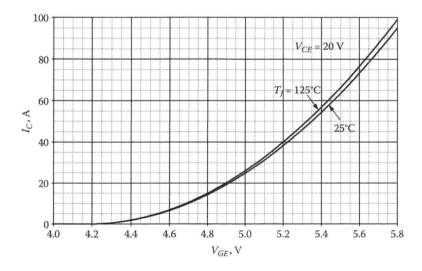

FIGURE 11.9.10
Figure for Example 11.9.1.

11.10 Power Latches

In all the transistor switches discussed so far, the input must be applied in order to maintain the device in the on state. In contrast, after a latching device is turned on, the input may be removed, but the device remains on, that is, it "latches" in the on state, until turned off by some means. This can be an advantage in some applications, particularly since the device can be turned on by a narrow trigger pulse, so little power is expended by the drive circuitry. However, unlike transistors, latching devices are not useful for amplifying continuous signals. The most important power latch is the **thyristor** and its many variations.

Thyristor

Also known as a **silicon controlled rectifier** (SCR), or a **reverse blocking triode thyristor**, the thyristor is a four-layer *pnpn* device having the general structure and symbol illustrated in Figure 11.10.1. The heavily doped p^+ and n^+ regions have doping concentrations of about $10^{19}/cm^3$, the *p*-region about $10^{17}/cm^3$, and the n^- region about 10^{13}–$10^{15}/cm^3$. Although the maximum switching frequency of 1 kHz or so makes thyristors the slowest of all power semiconductor switching devices; they nevertheless, have the greatest current and voltage capabilities, in excess of 3 kA and 5 kV, respectively.

The *i–v* characteristic of a thyristor is illustrated in somewhat exaggerated form in Figure 11.10.2. In the reverse direction ($v_{AK} < 0$), only a small current flows, until a breakdown voltage V_{brk} is reached, when the current rapidly increases. Junctions J_1 and J_3 are reverse-biased, with J_1 supporting the reverse voltage, because of the n^- layer. In the forward direction ($v_{AK} > 0$), and with the gate open-circuited ($i_G = 0$) only a small current flows as long as $v_{AK} < V_{BO}$, the **breakover voltage**, and the thyristor is in the forward blocking state. Junction J_2 supports the forward blocking voltage, again with the depletion region being

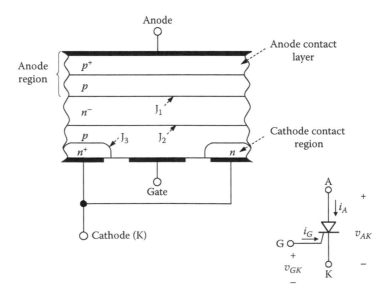

FIGURE 11.10.1
Diagrammatic structure and symbol of thyristor.

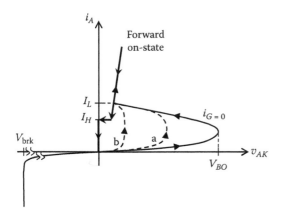

FIGURE 11.10.2
Illustration of i–v characteristic of thyristor.

mostly in the n^- layer. Hence the n^- layer is made thick in order to support high forward and reverse blocking voltages.

Once V_{BO} is exceeded, the characteristic exhibits an unstable, negative resistance region, followed by a positive resistance region of small resistance. This is the forward on-state, in which a large i_A can flow with only a small voltage drop of few volts or less. When a gate trigger pulse of sufficient duration is applied, the thyristor turns on at a lower v_{AK} that depends on the magnitude of i_G, as illustrated in Figure 11.10.2 by the dotted curves a and b. The **latching current** I_L is the smallest i_A that results just when the thyristor changes state from forward blocking to forward conduction, with $i_G = 0$. Once the thyristor is in the on-state, i_A must be reduced to a smaller current I_H, the **holding current**, before the thyristor can revert to the forward blocking state. Once turned on, the thyristor can only be turned off by reducing i_A to less than I_H for a minimum duration. In ac circuits, the thyristor automatically turns off when v_{AK} goes through zero, resulting in **natural commutation**. In dc circuits i_A must be sufficiently reduced or potentially reversed, resulting in **forced commutation** (Section ST11.9).

The operation of the thyristor is conveniently described in terms of an equivalent circuit consisting of a *pnp–npn* regenerative pair (Example 6.3.2, Chapter 6), as illustrated in Figure 11.10.3. The turn-on condition is (Equation 6.3.23, Chapter 6):

$$\alpha_{F1} + \alpha_{F2} = 1 \qquad (11.10.1)$$

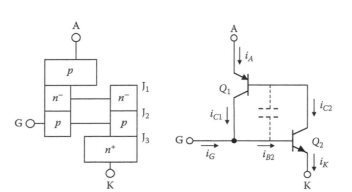

FIGURE 11.10.3
The thyristor considered as a regenerative pair.

considering the avalanche multiplication factors to be unity. In terms of the equivalent circuit, the current through the transistors is small in the forward blocking state. This makes the α's small, so $\alpha_{F1} + \alpha_{F2} < 1$. In normal thyristor operation, a momentary i_G increases i_{B2} and i_{C2}, and hence i_{C1}. The α's increase sufficiently so that their sum reaches unity, regeneration takes place, and both transistors are driven hard into saturation.

In terms of the *pnpn* structure (Figure 11.10.1), assume that in the presence of a given, forward v_{AK}, i_G of sufficient magnitude is applied for an appropriate duration. This causes the flow of an appreciable forward current across the forward-biased J_3 junction, from the *p*-region to the n^+ region. The current is mainly due to electron injection from the n^+ region to the *p* region, because of the large difference in doping levels of the two regions. The injected electrons diffuse through the *p*-region and are swept at the reverse-biased junction J_2 into the n^- region. The negative charge due to these electrons induces a forward hole current to flow across junction J_1 into the n^- region to maintain near space-charge neutrality. The injected holes diffuse across the n^- region and are swept at junction J_2 into the *p* region, where they further induce electron injection across junction J_3. A regenerative process is thus established, which results in a large i_A that is limited only by external circuit resistance. When i_A exceeds the latching current, it injects enough carriers into the base regions to latch the thyristor on, with junctions J_1 and J_3 forward biased. Because of its low equilibrium carrier concentration, the n^- region becomes heavily conductivity modulated, the *p* base region of J_3 less so.

In the on-state, the voltage across the thyristor can be expressed as:

$$v_{AK(\text{on})} = v_{J_1} - v_{J_2} + v_{J_3} + v_{n^-} + v_{\text{par}} \qquad (11.10.2)$$

where the v_J's are forward voltage drops across the respective junctions and v_{par} is a parasitic resistance that accounts for the remaining voltage drops in the body and contacts of the thyristor.

Once the thyristor is turned on, the gate loses all control over the current. A negative gate current cannot turn off the thyristor, because of the relatively large cathode area. A negative gate current can divert some current away from the cathode, but the diverted current is normally not sufficient to remove the forward bias from the whole of J_3. In a **gate turn-off** (GTO) switch, the structure is modified to allow the gate to turn off the thyristor (Section ST11.10).

Other than by a gate pulse, a *pnpn* device can be turned on in a number of ways, resulting in a number of practical devices. In a **silicon unilateral switch** (SUS), also known as a **Schokley diode**, a **current latch**, or a **reverse blocking diode thyristor**, conduction occurs because of avalanche breakdown initiated by v_{AK}, which increases M (Example 6.3.2, Chapter 6). Conduction can be initiated by radiant energy generating electron–hole pairs, as in a **light-activated silicon-controlled rectifier** (LASCR). Two SUS's in inverse parallel, so that conduction is bilateral, constitute a **diac**. A **triac** consists essentially of two thyristors in inverse parallel, with their gates connected together. It can be triggered in either direction by a positive or a negative voltage of appropriate magnitude applied between its gate and the electrode that is closer to the gate in the triac symbol (Figure 11.10.4). Once triggered, the triac conducts in the direction determined by the polarity of the trigger voltage and will continue to conduct until the current falls below the holding current.

A thyristor may be triggered inadvertently by a rise in temperature. The reverse current of J_2 in the forward blocking state approximately doubles for every 10°C rise in temperature, so a sufficiently high temperature may raise the current to a level that turns on the

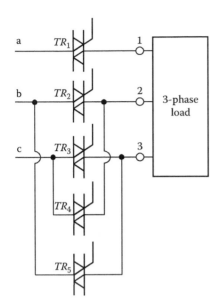

FIGURE 11.10.4
Figure for Application Window 11.10.1.

thyristor. A similar effect may be caused by a large dv_{AK}/dt that increases the device current through capacitances of the junctions. Thyristors are also subject at turn-on to a maximum di_{AK}/dt limitation, because of the finite time it takes to spread the current uniformly throughout the junctions. If the rate of rise is too high, localized hot spots may occur, which may cause excessive heating and lead to device failure.

Application Window 11.10.1 Static Switches

Semiconductor switches are extensively used in electric power systems and in industrial electronics to replace electromechanical switches. They offer the advantages of: (1) very high switching speeds, (2) accurate timing of closure of the circuit, and (3) elimination of moving parts and mechanical contacts, which are generally noisy and subject to wear. Contacts generally produce arcing or sparking, which can be hazardous in flammable surroundings, and which damages the contact with time and increases the contact resistance. Moreover, mechanical contacts usually bounce upon closing, that is, they rapidly open and close the circuit many times before settling to a closed position, which can be very troublesome in digital circuits.

For low-power dc applications, transistors are used. But for high power dc and for ac applications thyristors or triacs are generally the natural choice. Figure 11.10.4 illustrates the use of triacs to reverse the phase sequence to a three-phase motor so as to reverse its direction of rotation. When TR_1, TR_2, and TR_3 are activated by gate pulses, while gate pulses to TR_4 and TR_5 are suppressed, the phase sequence abc is seen at load terminals 123. But when TR_1, TR_4, and TR_5 are activated by gate pulses, while gate pulses to TR_2 and TR_3 are suppressed, the phase sequence acb is seen at load terminals 123. If the load is resistive, gate pulses are applied at the zero voltage crossing on each half cycle. With a switching time of few microseconds, the whole ac voltage is applied to the load as if the switch were absent. In the case of an inductive load, switching should occur on the zero crossings of current so as to eliminate switching transients.

Application Window 11.10.2 Incandescent Lamp Dimmer

The mechanical switch that controls lamps can be replaced by an ac switch having the added feature of continuous dimming of the light with very little expenditure of energy. Figure 11.10.5a illustrates a simple circuit for dimming an incandescent lamp. The triac is triggered by a diac designed to break down when the voltage across it exceeds a certain level, usually about 10 V of either polarity, as illustrated in Figure 11.10.5b. Once it breaks down, the voltage across the diac drops to about 1 V.

The circuit operates as follows: At the beginning of a half-cycle, the triac is off, because the current through it has just passed through zero. Negligible current passes through the lamp, so the full supply voltage is impressed across the gate trigger circuit consisting of the diac, C, and variable resistance R. C is charged through R by the supply voltage. When the voltage across C just exceeds the breakover voltage of the diac, the diac conducts at $\omega t = \theta$, triggering the triac and discharging C. The diac stops conducting when the current through it drops to below the holding current. R_G limits the discharge current through the diac and the triac gate. When the triac conducts, practically the full supply voltage is applied to the lamp until the end of the half cycle, when the triac again turns off, and the whole process is repeated over the next half cycle (Figure 11.10.5c). The supply voltage is thus applied to the lamp at a conduction angle θ after the beginning of a half cycle and until the end of the half cycle. θ can be varied within a range that is nominally 0°–180° by varying R. This method of controlling power in a load by varying the fraction of the half-cycle during which the supply voltage is impressed across the load is an example of **phase control**. It should be noted that the first and last 30° of each half cycle contribute only a total of

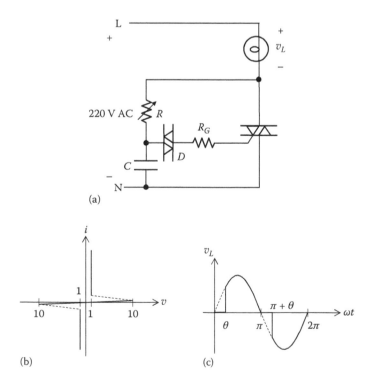

(a)

(b)

(c)

FIGURE 11.10.5
Figure for Application Window 11.10.2. (a) Dimming circuit; (b) diac i–v characteristic; (c) load voltage wave form.

about 6% of the power (Problem P11.7.6). Thus, varying θ from 30° to 150° during each half-cycle in Figure 11.10.5c varies the load power from nearly 97% to 3% of full power.

Summary of Main Concepts and Results

- Transistors have a safe operating area (SOA) delimited by a maximum output current, maximum output voltage, and maximum power dissipation. In addition, BJTs have a second breakdown because of current crowding at the base–emitter junction.

- For a given maximum allowed junction temperature above the ambient temperature, the lower the thermal resistance with respect to the ambient environment, the larger is the maximum allowed power dissipation in the semiconductor device.

- Thermal runaway can occur if the output current increases with temperature. Thermal stability is assured by having a low enough thermal resistance from the junction to ambient environment, or by having a sufficiently high electrical resistance in the output circuit.

- Harmonic distortion occurs in large-signal operation because the output current varies nonlinearly with the input quantity, which is the base current in a BJT and the gate–source voltage in a MOSFET. Distortion in MOSFETs is generally larger than in BJTs.

- Power conversion efficiency is the ratio of the power delivered at signal frequency to the power supplied by the dc power supply. Not only is high power-conversion efficiency desirable, but also low power dissipation in the absence of the signal.

- In class A operation, the output current of the power transistor flows throughout the signal cycle. The maximum power-conversion efficiency is 25% for a capacitively coupled load and 50% for a transformer-coupled load.

- In ideal class B operation, the output transistor current flows for half a signal cycle and no current flows in the absence of a signal. The output voltage has a deadband around the zero crossing, because of the cut-in voltage of the transistors, which results in cross-over distortion.

- In class AB operation, the output transistors are biased so as to have a small quiescent current. The output current flows for slightly less than half a cycle, and the power-conversion efficiency is higher than that of class A operation.

- Class AB amplifiers are normally operated in push–pull so as to combine the half cycles of output from each transistor. Even harmonics are eliminated by push–pull operation.

- In class C operation, the output current flows for only a small fraction of a cycle. The power-conversion efficiency is high, but the load can only be a tuned circuit of high selectivity so as to reduce distortion to an acceptable level.

- Switching enables a transistor of a given power rating to control much larger load power. Switching is used for class D amplification, dc–dc conversion, and dc–ac conversion.

- A lightly doped drift region is introduced in power diodes to increase the breakdown voltage. Conductivity modulation occurs in this drift region so that the voltage drop across it is much less than what would be expected from its dimensions and equilibrium carrier concentrations. The presence of the drift region affects the transient response and causes a prominent peak in the voltage across the diode at turn on.

- Power BJTs also have an n^- collector drift region that allows a large breakdown voltage. In addition to affecting the transient behavior, the n^- region introduces a quasi-saturation region in the output characteristics. The base and emitter regions are interleaved to reduce the base spreading resistance and minimize current crowding.
- The structure of a power MOSFET differs from that of a low-power MOSFET. An n^- region is again used to increase the breakdown voltage. The transfer characteristic departs from the square law at large currents, becoming almost linear.
- Compared to MOSFETs, BJTs can handle larger currents and higher voltages, have lower losses in the on-state, especially at higher blocking voltage ratings, but require large base-drive currents, and are subject to second breakdown. On the other hand, MOSFETs can be turned on and off much faster and require practically zero gate power in the conducting state.
- IGBTs are intermediate between BJTs and MOSFETs in terms of power handling capability and speed of operation. Like MOSFETs they do not require any gate power in the on state. They are liable to latchup at high currents and high current turn-off rates.
- Power latches are based on a *pnpn* structure. Various structural modifications lead to a number of useful devices having particular features.

Learning Outcomes

- Design basic types of class A and class AB power amplifiers.
- Articulate the special features of power amplifiers and power semiconductor devices.

Supplementary Examples and Topics and on Website

SE11.1 Distortion in Push–Pull Amplifier. Shows that if the two output transistors are matched, even harmonics cancel out in the output current.

SE11.2 Transformer-Coupled Push–Pull Amplifiers. Analyzes such an amplifier for class A and ideal class B operation.

SE11.3 Complementary Symmetry BJT Power Amplifier. Analyzes Class AB operation, under conditions of no load and finite load, of such an amplifier having CE driving transistors.

SE11.4 Complementary Symmetry Darlington Pair Power Amplifier with V_{BE} Multiplier. Analyzed class AB operation of such an amplifier.

SE11.5 Complementary Symmetry MOSFET Power Amplifier with Current Mirror. The amplifier that is the MOSFET analog of that of Figure 11.4.1 is analyzed for class AB operation and small-signal operation around the quiescent operating point.

SE11.6 Complementary Symmetry MOSFET Power Amplifier with BJT Drive. Analyzes Class AB operation of such an amplifier.

SE11.7 Heat Sink Selection. Discusses the selection of a heat sink for a power transistor for both natural and forced cooling.

SE11.8 Basic Regulated Supply with Series Transistor. Analyzes a basic regulated power supply consisting of a zener diode and a series pass transistor.

SE11.9 Feedback Regulated Power Supply. Discusses a regulated power supply of the type used in IC regulators.

ST11.1 Thermal Runaway. Derives the condition for thermal runaway.

ST11.2 Graphical Determination of Fourier Coefficients of Output Current. Shows how the Fourier coefficients can be determined graphically from a recorded waveform of the output current.

ST11.3 Theoretical Maximum Efficiency in Class B Operation. Derives the condition for maximum power efficiency in ideal class B operation.

ST11.4 Pulse-Width Modulation Control. Explains how PWM is used in the control of regulated supplies and converters.

ST11.5 Punchthrough Diode. Analyzes the operation of this type of power diode, in which the drift region is reduced by decreasing its conductivity and allowing the depletion region to extend to the $n^- n^+$ junction before voltage breakdown occurs.

ST11.6 Relation between On-resistance and Breakdown Voltage. Derives the relation between the on-resistance of the drift region and the breakdown voltage in a majority-carrier device and in a bipolar device under conditions of conductivity modulation.

ST11.7 Switching Behavior of a Power MOSFET in a Step-down Regulator. Investigates the switching behavior of a power MOSFET in a step-down regulator.

ST11.8 Latchup in IGBT. Investigates latchup in an IGBT and how it can be avoided.

ST11.9 Forced Commutation of Thyristors. Illustrates a simple method of forced commutation, in which the firing of one thyristor turns off the other thyristor.

ST11.10 Modified Thyristors. Discusses modified thyristors, such the reverse-conducting thyristor (RCT), the gate-turn-off thyristor (GTO), and the MOS-controlled thyristor (MCT).

ST11.11 Snubbers. Considers circuits that are used to limit di/dt and dv/dt in power semiconductor devices to safe levels.

Problems and Exercises

P11.1 Thermal Considerations

P11.1.1 Identify from manufacturer data sheets on the Web various power transistor cases.

P11.1.2 A power transistor in a TO-3 case has $\theta_{JC} = 1.8°\text{C/W}$ and $\theta_{CA} = 25°\text{C/W}$ when mounted in free air. Determine the case and junction temperatures when the transistor dissipates 5 W at an ambient temperature of 30°C, with the transistor: (a) mounted in free air; (b) used with a heat sink having $\theta_{CS} = 0.5°\text{C/W}$ and $\theta_{SA} = 2.5°\text{C/W}$.

P11.1.3 A transistor has $T_{J(max)}$ of 120°C and $P_{D(max)}$ of 160 W at $T_{C(REF)} = 40$°C. If $\theta_{CS} = 0.3$°C/W and $\theta_{SA} = 0.2$°C/W, determine the maximum allowed power dissipation at 80°C ambient temperature.

P11.1.4 A transistor has $T_{J(max)} = 120$°C and $P_{D(max)} = 5$ W at $T_{C(REF)} = 25$°C in free air. Determine θ_{JC} and the power dissipation when the case temperature is 80°C.

P11.1.5 For the transistor of Problem P11.1.4, if $\theta_{CS} = 0.5$°C/W and $\theta_{SA} = 5$°C/W, what is the maximum allowed power dissipation at an ambient temperature of 40°C?

P11.1.6 A transistor is rated at 115 W at 25°C ($T_{C(REF)}$) and is derated above this temperature at the rate of 0.8°C/W. Determine: (a) $T_{J(max)}$; (b) the maximum power dissipation at a case temperature of 80°C.

P11.2 Distortion in Amplifiers

P11.2.1 Show that if V_{rms} and V_{1rms} are, respectively, the rms values of a periodic waveform and its fundamental, the total harmonic distortion is given by: $D = \sqrt{(V_{rms}/V_{1rms})^2 - 1}$.

P11.2.2 Show that if two amplifying stages having harmonic distortion D_a and D_b, respectively, are cascaded, the overall harmonic distortion is approximately $\sqrt{D_a^2 + D_b^2}$, provided $D_a D_b \ll 1$.

P11.2.3 From a signal flow diagram showing the distortion signal at the output of a power amplifier, deduce that the harmonic distortion in the presence of negative feedback is reduced by $(1 + \beta A_v)$ where βA_v is the loop gain.

P11.2.4 Given a feedback power amplifier having an open-loop gain of 100 and a feedback of 9% (feedback factor $\beta = 0.09$). (a) If 10% harmonic distortion is produced when the magnitude of the fundamental at the amplifier output is 10 V, determine the harmonic distortion in the presence of feedback. (b) Compare the input signal with and without negative feedback. What is the price paid for the reduction in harmonic distortion?

P11.2.5 Given two amplifying stages, each of gain 100. Assume that the percentage distortion in the output is 1% at an output level of 10 V and is negligible at smaller levels. Consider two cases: (a) the two stages are cascaded with 9% feedback in each stage and a final output level of 10 V; (b) the two stages are cascaded without feedback in each stage but with feedback from the output of the second stage to the input of the first stage so as to give the same percentage distortion as in case (a) at an output level of 10 V. Compare the input signal in both cases.

P11.3 Class A Amplifiers

P11.3.1 Show that the maximum signal power output in class A operation, according to Figure 11.2.2b is $V_{CC}^2/8R_C$, assuming the load is R_C and $R_E = 0$.

P11.3.2 Show that when the quiescent point of a class A amplifier is biased for maximum symmetrical swing as in Figure 11.2.2b, the ac load line is tangent to the maximum dissipation hyperbola at the Q point. Deduce that $I_{CQ} = \sqrt{P_{D(max)}/R_{AC}}$ and $V_{CEQ} = \sqrt{P_{D(max)}R_{AC}}$.

P11.3.3 Consider a class A amplifier having the circuit of Figure 11.2.1a but with the load R_L connected in place of R_C. Neglecting distortion and the small base current, show that: (a) $P_S = V_{CC}^2/(2R_L + R_E)$; (b) $P_{L(max)} = V_{CC}^2 R_L/2(2R_L + R_E)^2$; and (c) $\eta_{max} = 100R_L/(4R_L + 2R_E)$. How is $P_{L(max)}$ related to the quiescent power dissipated in the transistor?

P11.3.4 Given a class A direct-coupled amplifier as in Problem P11.3.3, with $R_E = 1$ Ω, $R_L = 10$ Ω, and $V_{CC} = 24$ V. Determine I_{CQ}, V_{CEQ}, P_S, $P_{L(max)}$, and η_{max}. Neglecting the power dissipated in the bias network supplied from V_{CC} and the power due to the base current, what are the maximum and minimum values of the power dissipated in the transistor?

P11.3.5 Consider the ac coupled amplifier of Figure 11.2.1 with $R_E = 2\ \Omega$, $\beta_F = 50$, $R_C = 150\ \Omega$, $R_L = 220\ \Omega$, and $V_{CC} = 24$ V. Determine under conditions of maximum symmetrical swing, as in Example 11.2.1, I_{CQ}, V_{CEQ}, P_S, $P_{L(max)}$, and η_{max}.

P11.3.6 If R_L in Problem P11.3.5 is variable, under what conditions will η_{max} be maximum?

P11.3.7 η_{max} of the class A amplifier of Figure 11.2.1 can be increased by reducing the power dissipated in R_C, as when R_C is replaced by an inductor of small dc resistance R_{ind} and whose reactance at the lowest signal frequency is much larger than R_L. (a) Sketch the dc and ac load lines; (b) derive the expressions corresponding to Equations 11.2.4; (c) determine η_{max}.

P11.3.8 The amplifier of Problem P11.3.7 has $R_E = 1\ \Omega$, $R_{ind} = 0.8\ \Omega$, $V_{CC} = 18$ V, $\beta_F = 50$, and $P_{D(max)} = 10$ W. Determine R_L for maximum symmetrical swing, P_S, $P_{L(max)}$, and η_{max}.

P11.3.9 Given a transformer-coupled amplifier as in Figure 11.2.3a having $R_E = 0.5\ \Omega$, $R_L = 2\ \Omega$, $\beta_F = 40$, $V_{CC} = 18$ V, $r_p = 1.5\ \Omega$, the secondary resistance is $0.2\ \Omega$, the transformer turns ratio is 4:1, and the transformer leakage reactance is negligible at the frequencies of interest. Determine I_{CQ}, P_S, the minimum values of $i_{C(max)}$, BV_{CEO}, $P_{L(max)}$, and $P_{D(max)}$.

P11.3.10 If the transistor of Problem P11.3.9 is to be operated with an $i_{C(max)}$ of 0.5 A, determine the transformer ratio for maximum symmetrical swing, $P_{L(max)}$, and η_{max}, assuming $r_p = 1.5\ \Omega$ and neglecting the transformer secondary resistance.

P11.3.11 The transistor of Problem P11.3.9 has biasing resistors $R_1 = 220\ \Omega$, $R_2 = 33\ \Omega$, with $V_{BEQ} = 0.73$ V. Determine I_{CQ} for maximum symmetrical swing and the transformer ratio for maximum load power, assuming $r_p = 0.5\ \Omega$, $R_L = 1.5\ \Omega$, and negligible secondary resistance.

P11.3.12 A transformer-coupled amplifier uses a transistor whose $P_{D(max)} = 5$ W, $i_{C(max)} = 1$ A, and $V_{CE(max)} = 50$ V. $R_E = 0$, and the transformer resistances may be neglected. Choose the quiescent operating point for maximum symmetrical swing based on $P_{D(max)} = 5$ W and $i_{C(max)}$. Determine the transformer ratio for a load of 1.25 Ω, V_{CC}, $P_{L(max)}$, and η_{max}.

P11.3.13 Repeat Problem P11.3.12, choosing the quiescent operating point for maximum symmetrical swing based on $P_{D(max)} = 5$ W and $v_{CE(max)} = 20$ V.

P11.3.14 Assume that $V_{CC} = 15$ V in the circuit of Problem P11.3.12. Choose the quiescent operating point for maximum symmetrical swing based on $P_{D(max)} = 5$ W. What are $i_{C(max)}$ and $V_{CE(max)}$ in this case?

P11.3.15 Given a power amplifier in which v_{CE} and i_C have minimum instantaneous values of $v_{CE(min)}$ and $i_{C(min)}$, respectively. Deduce that for maximum symmetrical swing between $v_{CE(min)}$ and $i_{C(min)}$, $R_{ac} = \dfrac{V_{CEQ}}{I_{CQ}} = \dfrac{v_{CE(min)}}{i_{C(min)}}$. If the second relation between V_{CEQ} and I_{CQ} is KVL under dc conditions, show that $I_{CQ} = \dfrac{V_{CC} i_{C(min)}}{v_{CE(min)} + R_{dc} i_{C(min)}}$ and

$$V_{CEQ} = \dfrac{V_{CC} v_{CE(min)}}{v_{CE(min)} + R_{dc} i_{C(min)}}.$$

P11.3.16 Repeat Problem P11.3.15 considering that the second relation between V_{CEQ} and I_{CQ} is

$$V_{CEQ} \times I_{CQ} = P_{D(max)}.\ \text{Show that}\ I_{CQ} = \sqrt{\dfrac{i_{C(min)}}{v_{CE(min)}} P_{D(max)}}\ \text{and}\ V_{CEQ} = \sqrt{\dfrac{v_{CE(min)}}{i_{C(min)}} P_{D(max)}}.$$

P11.3.17 A transistor as in Problem P11.3.16 has $v_{CE(min)} = 0.5$ V, $i_{C(min)} = 5$ mA, and $P_{D(max)} = 1$ W. Determine: (a) R_{ac}, V_{CEQ}, and I_{CQ} for maximum symmetrical swing; (b) V_{CC} if $R_{dc} = 4\ \Omega$.

P11.3.18 Figure P11.3.18 shows a dc coupled emitter follower having a current-mirror connected to the source. Assume $V_{CC} = V_{EE} = 12$ V, $R_L = 2$ kΩ, with $v_{CE(sat)} = 0.4$ V, $v_{BE} = 0.75$ V, and $\beta_F = 40$ for all transistors. Determine R for maximum symmetrical output voltage swing, with the lower limit set by: (a) Q_2 saturating; (b) Q_1 cutting off. Specify the corresponding $v_{I(min)}$.

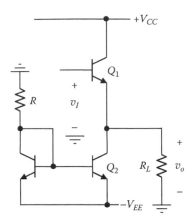

FIGURE P11.3.18

P11.3.19 Consider the emitter follower of Figure P11.3.18 operating under conditions of maximum symmetrical swing, with negligible $v_{CE(sat)}$, and with the negative limit set by Q_1 cutting off. If $V_{CC} = V_{EE} = 12$ V, $R_L = 50$, determine P_L, η_{max}, and the power dissipated in the output transistor. What would these quantities be if the signal amplitude is reduced to one-half?

P11.3.20 Repeat Problem P11.2.19 for maximum swing and $V_{CC} = V_{EE}$: (a) 6 V, and (b) 18 V.

P11.3.21 The BJT in Figure 11.2.4 is replaced by an E-NMOS transistor having $V_{tn} = 2$ V, $k_n = 20$ mA/V^2, with $V_{DD} = 12$ V, $R_L = 100$ Ω. For symmetrical swing, determine: (a) I_{DQ}; (b) $v_{O(max)}$ and $v_{I(max)}$; (c) $v_{O(min)}$ and $v_{I(min)}$; (d) η. Neglect the body effect and channel-length modulation.

P11.3.22 Two transistors having the same parameters as in Problem P11.3.12 are used in a transformer-coupled class A push–pull amplifier (Section SE11.2). Choose the quiescent operating point for maximum symmetrical swing based on $P_{D(max)} = 5$ W and $i_{C(max)}$. Determine the transformer ratio for a load of 1.25 Ω, V_{CC}, $P_{L(max)}$, and η_{max}.

P11.4 Class B Amplifiers

P11.4.1 Analyze the behavior of the circuit of Figure 11.3.1a when a single V_{CC} supply is used, with the collector of Q_p connected to ground and R_L coupled through a large capacitor.

P11.4.2 Consider the input terminals of the amplifier of Figure 11.3.1a to be connected to the output of an op amp of voltage gain A_v. Let the output of the transistor amplifier be connected to the inverting input of the op amp so as to provide negative feedback, the input signal v_I being applied to the noninverting input of the op amp. Show that the width of the deadband due to crossover distortion is reduced to $1/A_v$ of its value without feedback.

P11.4.3 Simulate the circuit of Figure 11.3.1a using the matched Darlington pair QN6502 and QN6509 from the PSpice EVAL library, with $V_{CC} = 12$ V and $R_L = 100$ Ω. Determine the width of the deadband and the effect of negative feedback using an ideal op amp as in Problem P11.4.2.

P11.4.4 The complementary symmetry amplifier of Figure 11.3.1a is biased for ideal class B operation, with $V_{CC} = 12$ V and $R_L = 50$ Ω. Assuming ideal operation, with the largest sinusoidal input signal, negligible cross-over distortion and negligible v_{CEsat}, determine (a) P_L, P_S, P_D for each transistor and η_{max}; (b) $P_{D(max)}$ and the corresponding V_{om}, P_L, P_S, and η.

P11.4.5 Repeat Problem P11.4.4 assuming $v_{CE} = 0.4$ V.

P11.4.6 The complementary symmetry amplifier of Figure 11.3.1a is biased for ideal class B operation, with $V_{CC} = 5$ V and $R_L = 10\ \Omega$. Assuming ideal operation, with a square-wave input signal and negligible $v_{CE(sat)}$, determine P_L, P_S, P_D for each transistor and η when a maximum signal is applied. Repeat for an output signal that is half the maximum.

P11.4.7 The BJTs in Figure 11.3.1a are replaced by matched MOSFETs. (a) What is the quantity corresponding to V_γ that determines the start of conduction? (b) What peak value of v_I will bring Q_n to the edge of saturation? (c) What is the influence of the body effect?

P11.4.8 The MOSFETs in Problem P 11.4.7 have $|V_t| = 1$ V, $k_{n,p} = 1$ mA/V^2, with ±5 V supplies and, $R_L = 10\ \text{k}\Omega$. Assuming the largest sinusoidal output for which the transistors remain in saturation, determine v_i, P_L, P_S, P_D for each transistor. Neglect body effect and channel-length modulation.

P11.5 Class AB Amplifiers

P11.5.1 The amplifier of Figure 11.4.1 is required to have an incremental output resistance about the quiescent operating point of 6.5 Ω. If $I_{BB} = 0.2$ mA, determine k in Equation 11.4.2.

P11.5.2 The amplifier of Figure 11.4.1 has $R_L = 100\ \Omega$. The incremental gain for small signals about the quiescent operating point is to be at least 0.98. Determine: (a) the quiescent current, assuming $\beta = 50$. (b) V_{DD}, assuming the saturation current of the transistors is 10 pA.

P11.5.3 In the circuit of Figure 11.4.1, what limits the largest negative swing in v_O? What limits the largest positive swing in v_O if I_{BB} is large? If I_{BB} is small?

P11.5.4 Repeat Example 11.4.1 assuming that I_{BB} is doubled to 4.4 mA.

P11.5.5 Repeat Example 11.4.1 assuming that β_F is doubled to 100.

P11.5.6 Consider the circuit of Figure 11.4.2. For the output transistors, assume a quiescent current of 5 mA, $v_O = \pm5$ V, $V_{CC} = 15$ V, $R_L = 50\ \Omega$, $R_E = 0.47\ \Omega$, $\beta_F = 50$, and $V_{BE} = 0.7$ V at 10 mA. Assume $V_{BE} = 0.6$ V at 1 mA for the diode-connected transistors, and for Q_1, $V_{BE} = 0.7$ V and $\beta_F = 100$. Determine: (a) R_3 such that the minimum current through it is approximately 0.2 mA larger than needed to drive Q_n; (b) R_4 such that the minimum v_{CE} of Q_1 is not less than 0.4 V; (c) base voltage of Q_1 under quiescent conditions; (d) v_I for $v_O = \pm5$ V?

P11.5.7 Assume that for the output transistors of Figure 11.6.1, the quiescent current is 5 mA, $v_O = \pm5$ V, $V_{CCO} = 15$ V, $R_L = 50\ \Omega$, $R_E = 0.47\ \Omega$, $\beta_F = 50$, and $V_{BE} = 0.7$ V at 10 mA. Assume $V_{BE} = 0.6$ V at 1 mA for the diode-connected transistors and $V_{BE} = 0.7$ and $\beta_F = 100$ for Q_3. (a) Determine R_3 such that the current through it at minimum output is 2.5 mA; (b) calculate the output of the operational amplifier for $v_O = \pm5$ V, neglecting the voltage drop across R_o; (c) what are the corresponding values of v_I if $R_r = 10\ \text{k}\Omega$ and $R_f = 100\ \text{k}\Omega$?

P11.6 Power Switching

P11.6.1 The limit for continuous operation of the switched regulator of Figure 11.7.2 is when i_L in Figure 11.7.3 becomes zero at the end of the off period. Let the average value of i_L under these conditions, which is also the average load current, be denoted by $I_{O(LIM)}$. Show that $I_{O(LIM)} = (t_{on}/2L)(V_I - V_O) = (T/2L)(V_I - V_O)T_S$, where T_S is the switching period and $T = t_{on}/(t_{on} + t_{off})$ is the duty cycle. Deduce that when V_O is constant, $I_{O(LIM)} = (T_S V_O/2L)(1 - T)$.

P11.6.2 Consider the switching regulator of Figure 11.7.2 having an output voltage of 18 V. $f_S = 50$ kHz, $L = 1$ mH, and T in the range 0.1–0.9. (a) Using the result of Problem P11.6.1, what is the maximum load resistance that allows continuous operation? (b) From Equation 11.7.5, what value of C will give a maximum ripple of 1%?

P11.6.3 Sketch the variation of i_L during the on and off periods for the step-up regulator, analogous to Figure 11.7.3. Show that at the limit for continuous operation of the step-up regulator $I_{L(LIM)} = (1/2)(V_I/L)t_{on} = T_S V_O T(1 - T)/2L$ using the result of Exercise 11.7.3. Recognizing that in the step-up regulator, the average inductor current is equal to the average input current, deduce that at the limit of continuous operation, $I_{O(LIM)} = T_S V_O T(I - T)^2/2L$.

P11.6.4 A full bridge rectifier having $V_I = 48$ V supplies an inductive load of impedance $3 + j5$ Ω at 1 kHz. Determine: (a) the amplitudes of the fundamental, third harmonic and fifth harmonic of the load current; (b) the power delivered by each of these components. Which of these is useful power for a heating load? An ac motor?

P11.6.5 Consider the step down regulator of Figure 11.7.2 with a MOSFET used as the switching transistor. When the MOSFET is on, the source voltage is close to the drain voltage V_I. The gate voltage must then be higher than V_I to turn the MOSFET on. The higher voltage is usually derived from V_I using a **charge pump**, an example of which is illustrated in Figure P11.6.5. Describe how the circuit operates, noting that it is essentially a voltage doubler that is supplied from a dc voltage through the dual switch Q_{n1} and Q_{p1}, instead of an ac supply. Assume that the MOSFET gate and circuitry require a rate of charge transfer equivalent to $i_G = 10$ μA, that $V_I = 18$V and that V_{GG} is to be 24 V, neglecting the voltage drops of the diodes and the saturation voltages of the transistors. Deduce that, theoretically, $C_{1(min)}$ is given by $f_S C_{1(min)} V_I = i_G$, where $f_S = 100$ kHz is the frequency of the control voltage. Choose $C_1 = 10 C_{1(min)}$, and choose C_2 so that $C_2 > 10 C_1$ and $C_2 R_L$ is large compared with the rate at which the gate is turned on and off, say $C_2 R_L \geq 1$ ms. Simulate using PSpice and observe the build-up of voltage across C_2.

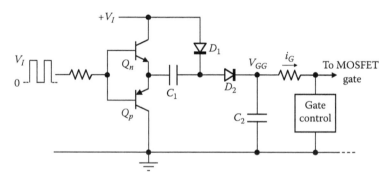

FIGURE P11.6.5

P11.7 Miscellaneous

P11.7.1 A power diode is to have a breakdown voltage of 2 kV. Determine W_d and N_D, assuming $\xi_{brk} = 2 \times 10^5$ V/cm, $N_A = 10^{19}$/cm³, and $N_A \gg N_D$.

P11.7.2 The effect of differences in paralleled transistors can be mitigated by connecting small resistances in series with the sources or emitters. Two E-NMOS power transistors having $k_n = 2$ A/V², $V_{tn1} = 3$ V, and $V_{t2} = 3.5$ V are paralleled to conduct a total current of 20 A. Determine the current in each transistors if: (a) the corresponding terminals are connected together; (b) 0.1 Ω resistors are included in series with each source.

P11.7.3 A power Darlington transistor pair is switched on and off at 10 kHz. The off state lasts for 76 μs, with $I_{CEO} = 5$ mA and $V_{CE} = 200$ V. The on-state lasts for 20 μs, with and $V_{CE(sat)} = 2$ V and $I_{C(sat)} = 100$ A. I_C and V_{CE} can be assumed to change linearly for 1 and 3 μs during turn-on and turn-off, respectively. Determine the average collector power dissipation during each of these four periods.

P11.7.4 Determine the average base power dissipation in Problem P11.7.3 if $v_{BE(sat)} = 2.2$ V, $\beta_{forced} = 20$, bearing in mind that the base current is applied for 21 μs.

P11.7.5 Assuming v_l changes between ± 6 V and $v_{BE(sat)} = 2$ V, choose R_1, R_2, and C in Figure 11.9.4 so that: (1) $i_{Bf} = 5$A, (2) $i_{Bi} = 10$ A, and (3) the time constant is 1 μs. Determine i_{Br}.

P11.7.6 Consider a switched supply where conduction starts at an angle θ after the start of each half-cycle and continues to the end of the half-cycle. Show that if such a voltage waveform is applied across a resistor R, the ratio of the power dissipated in the resistor to that dissipated for full conduction is $1 - (2\theta - \sin 2\theta)/2\pi$. (a) Calculate this ratio for $\theta = 30°$ and $\theta = 150°$. (b) If the diac in Figure 11.10.5a triggers at 10 V and $C = 1$ mF, determine the range of values of R required to vary the conduction angle from 30° to 150° in each half-cycle, assuming that C discharges fully when the diac conducts. Simulate using PSpice.

P11.7.7 Consider the turn-off snubber of Figure ST11.11.2, Section ST11.11 and assume that v_{CE} increases quadratically with time during t_{off} as shown in Figure ST11.11.1b and in accordance with Equation ST11.11.3. Assume that C_s is determined by the need to limit power dissipation during turn off. Show that the energy dissipated in the transistor is $\frac{1}{12}\eta V_I I_O t_{off}$, where ηV_I is the voltage across the capacitor at $t = t_{off}$, when $i_C = 0$. Deduce that $C_s = I_O t_{off}/2\eta V_I$. Calculate C_s if $V_I = 200$ V, $I_O = 100$ A, $t_{off} = 1$ μs, and $\eta = 0.2$. What is the average power dissipated in the transistor during turn-off at a switching frequency of 50 kHz?

P11.7.8 Consider the combined snubber of Figure ST11.11.3a, Section ST11.11 during turn-on and assume that v_{CE} decreases linearly as shown in Figure ST11.11.1b and that the capacitor current can be neglected compared to i_C. If L_s is determined by the need to limit power dissipation during turn on, show that $L_s = V_I t_{on}/2\gamma I_O$, where γI_O is the transistor current at $t = t_{on}$, when $v_{CE} = 0$. Calculate L_s if $V_I = 200$ V, $I_O = 100$ A, $t_{on} = 0.5$ μs, and $\gamma = 0.3$.

12

Basic Elements of Digital Circuits

This chapter and the next deal exclusively with digital circuits. The present chapter begins with a brief review of some basic principles of digital signals and processing in order to emphasize some fundamental features of digital systems. Some basic elements of digital systems are then discussed, namely, gates, flip-flops, and memories.

Logic gates are the essential components of combinational digital circuits and are discussed in terms of the main gate types and the properties of ideal gates, irrespective of how the gates are implemented. However, to make the discussion more concrete, CMOS inverters and basic gates are used for illustration purposes. Special emphasis is placed on properties such as the voltage transfer characteristic, noise margins, propagation delays, and power dissipation.

The essential component of sequential digital circuits is the latch, or its clocked version, the flip-flop. After discussing the basic latch, various types of flip-flops are presented with emphasis on any undesirable features of each type and how they are overcome. The discussion serves to illustrate some important points concerning edge-triggering, timing of signals, and circuit implementation in contrast to logic implementation.

Memory circuits that store program instructions and data are essential elements of practically all digital systems. The most important attributes and features of memory are first discussed in general terms, followed by the main features of static and dynamic read/write memory (R/W memory) and its peripheral circuits. Read-only memories (ROMs) of various types are then presented with special emphasis on flash memory, which is commonly used as an electrically erasable and programmable type of ROM.

Learning Objectives

❖ To be familiar with:

- Terminology used with digital circuits.
- The properties of ideal gates.
- Some general features of digital memory.

❖ To understand:

- Some of the fundamental principles and features of digital signals and processing.
- Gate properties such as the voltage transfer characteristic, noise margins, propagation delays, and power dissipation.
- Why a latch has two stable states.
- The operation of the JK-FF and the D-FF, including the master–slave configuration.
- The essential features of R/W memory and ROM.

12.1 Digital Signals and Processing

Digital Signals

Concept: *The distinguishing characteristic of analog signals is that they assume values over a continuous range.*

This is true, for example, of physical quantities such as displacement, temperature, or pressure. Voltages or currents are described as analog if they are *analogous* to physical quantities in that they are also defined over a continuum of values. Consider, for example, a simple, analog sound reproduction system in which the input sound waves applied to a microphone generate a voltage or current signal that is subsequently processed through amplification and possibly frequency shaping. The air pressure variations of the sound signal at the microphone assume, at every instant, values over a continuous range. The same is true of the voltage and current signals in the analog sound reproduction system.

The conversion of an analog signal to a digital signal involves the following three essential steps:

1. **Sampling:** The analog signal is sampled at a regular rate $f_S = 1/T_S$, where T_S is the sampling period, or interval between successive samples, as illustrated in Figure 12.1.1. According to the sampling theorem, f_S must be at least equal to the Nyquist rate, $2f_B$, where f_B is the highest frequency present in the signal. This means that the highest frequency component in the signal is sampled, on average, at least twice per period, so that no information in the signal is lost because of sampling. In practice, the sampling rate is considerably larger than the Nyquist rate. The sampled signal can then be represented as a sequence of values: $\ldots, f_{n-2}, f_{n-1}, f_n, f_{n+1}, f_{n+2}, \ldots$

2. **Quantization:** The range of values of the samples is divided into a number of discrete levels, and the value of each sample is expressed as a number equal to that of the nearest, or lower, discrete level. The sample values are thus *quantized*.

3. **Coding:** The quantized value of each sample is coded in the form of some digital code, usually a binary code in which a number is expressed as a string of 1's and 0's.

To illustrate steps 2 and 3, suppose that the range of values of a signal is divided into 8 levels, including the 0 level, and consider three sample values, say, 6.2, 7.3, and 3.4, as in Figure 12.1.2. The three samples are assigned value of 6, 7, and 3, respectively. In binary code, these numbers are expressed in the binary number system, in which only the binary digits, or

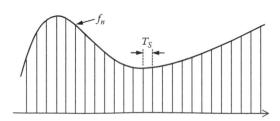

FIGURE 12.1.1
Sampling of an analog waveform.

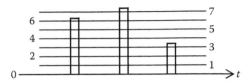

FIGURE 12.1.2
Quantization of analog samples.

bits, 0 and 1 are used. Since we have 8 levels 0 to 7, and $2^3 = 8$, we need three bits to represent a number in the range of 0–7. The number is expressed as the sum of three terms, where each term is a bit multiplied by 2 raised to a power equal to the position of the bit in the number, starting from a power of zero for the least significant position. The number 4, for example, is represented as 100, since $1 \times 2^2 + 0 \times 2^1 + 0 \times 2^0 = 4$. The three values 6, 7, and 3 are represented, respectively, by the sequence of three binary numbers 110 ($1 \times 2^2 + 1 \times 2^1 + 0 \times 2^0 = 6$), 111 ($1 \times 2^2 + 1 \times 2^1 + 1 \times 2^0 = 7$), and 011 ($0 \times 2^2 + 1 \times 2^1 + 1 \times 2^0 = 3$). In practice, many more bits are used, typically 32 or 64 bits. A group of 8 bits is a **byte**. In digital circuits, a binary digit of 0 is represented by a low voltage level that is close to zero, whereas a binary digit of 1, or logic 1, is represented by a higher voltage level that may be close to +5 V, say.

Concept: *The distinguishing characteristic of digital systems is that digital signals are represented in binary code and processed according to the rules of Boolean algebra.*

Processing in analog systems is performed by specialized units, such as adders, integrators, amplifiers, etc., as discussed in preceding chapters. Digital signals, on the other hand are processed according to the rules of Boolean algebra, as will be explained later.

Digital systems possess some highly important advantages over analog systems, which can be summarized as follows:

1. *Accuracy*: The accuracy of a digital system is determined by the number of bits used for digital representation. In a 16-bit system, for example, there are 2^{16}, or 65,536, levels. The corresponding accuracy is $100/2^{16} = 0.0015\%$. In analog systems, on the other hand, elaborate measures are generally necessary for accuracies of 0.1% or better. It should be noted, however, that in digital computations, errors can accumulate, so that the overall accuracy may not be as high as suggested by the number of bits used.

2. *Reliability*: Voltage levels of 1 or 0 correspond to electronic devices such as transistors being either fully turned on, or fully turned off. Noise immunity is high because misreading a 0 level as a 1 level, or conversely, can occur with only relatively large noise inputs. In analog circuits, on the other hand, transistors operate on a continuum of levels over a given range, so that errors and corruption by noise are more likely to occur.

3. *Versatility*: Digital systems are more versatile than analog systems in several respects:

   a. In **combinational digital circuits**, whose outputs depend only on present inputs, all Boolean logic can theoretically be performed by only one type of logic gate (NAND or NOR). In **sequential digital circuits**, whose outputs depend on present as well as past inputs, and which therefore incorporate some memory elements, only a single type of memory element (a bistable) need, in theory, be used.

b. The processing in digital systems can be controlled by software. In other words, an operation, or the parameters of a given operation can be altered, not by changing or adjusting hardware, as in analog systems, but by programming, as in the case of digital computers, digital filters, or computer-controlled communication and control systems.

c. Some operations, such as data storage, data compression, error detection, and correction, are implemented much more conveniently and effectively in digital systems than in analog systems.

4. *Large-Scale Integration*: Digital circuits lend themselves very effectively to "modular design." Gates, gate assemblies, memory blocks, or blocks performing complex functions can be used as "generic" blocks that are readily connected together based on some relatively simple rules or guidelines that the designer adheres to. This has contributed to increasing levels of integration, from small-scale-integration (SSI), of about 10 gates per chip, to medium-scale-integration (MSI), of about 10–100 gates per chip, to large-scale-integration (LSI), of about 100–10^3 gates per chip, to very-large-scale-integration (VLSI), of about 10^3–10^4 gates per chip, and to ultra-large-scale-integration (ULSI), of more than 10^4 gates per chip. 100 million gates per chip are projected for the near future.

Boolean Algebra

Digital signals are processed according to the rules of Boolean algebra, which is the algebra of two-valued logic. This type of logic deals exclusively with statements that are unambiguously either true or false, such as: "It is raining outside." If this statement is denoted by a logic variable, say A, then A can assume one of two values: true (T) or false (F). If a statement is true, its negation is evidently false. Thus, if the statement "It is raining outside" is true, its negation, "It is not raining outside" is false.

Logic variables can be combined to form logic expressions of arbitrary complexity using two fundamental operations: **conjunction** and **disjunction**.

Definition: *The conjunction of two or more logic variables A, B, C, etc. is true only if each of the statements involved is true. The conjunction of two or more logic variables is written as ABC . . . and read as A AND B AND C, etc.*

For example, if A is the statement: "It is raining outside" and B is the statement: "The temperature is less than 15°C," the compound statement: "It is raining outside *and* the temperature is less than 15°C," is true only if both sub-statements are true. Logic relations can be illustrated by means of a **truth table**, as shown in Table 12.1.1 for the conjunction of two variables. The four possible combinations of values of A and B are entered, together with the value of AB for each combination.

Definition: *The disjunction of two or more logic variables A, B, C, etc. is false only if each of the statements involved is false. The disjunction of two or more logic variables is written as $A + B + C$ . . . and read as A OR B OR C, etc.*

TABLE 12.1.1

Conjunction of Two Logic Variables

A	B	AB
F	F	F
T	F	F
F	T	F
T	T	T

For example, if A is the statement: "He did not come because he was busy" and B is the statement: "He did not come because he forgot," the compound statement: "He did not come because he was busy *or* because he forgot" is true if either statement, or both, are true. It is false only if both sub-statements are false. The corresponding truth table is shown in Table 12.1.2.

The three basic operations of negation, conjunction, and disjunction are not independent. Conjunction can be expressed in terms of disjunction and negation, and disjunction can be expressed in terms of conjunction and negation. To illustrate this, let us evaluate the expression $\overline{\overline{A} + \overline{B}}$ by constructing its truth table, as shown in Table 12.1.3, where a bar above a logic variable, or expression, denotes the negation of that variable, or expression. The first two columns show the four combinations of A and B. The next two columns are their negations. The fifth column is the disjunction of $\overline{A}$ and $\overline{B}$, and the last column is the negation of this disjunction. Comparing this column with the rightmost column of Table 12.1.1, it follows that:

$$\overline{\overline{A} + \overline{B}} = AB \tag{12.1.1}$$

Similarly, it can be shown that:

$$\overline{\overline{A}\,\overline{B}} = A + B \tag{12.1.2}$$

(Exercise 12.1.1). Equations 12.1.1 and 12.1.2 are an expression of:

De Morgan's law: *If an expression is negated, conjunction or disjunction immediately beneath the negation symbol is replaced, respectively, by disjunction or conjunction of the negated variables or expressions.*

For example, if the LHS of Equation 12.1.1 is negated, it becomes $\overline{A} + \overline{B}$. The conjunction of the negated variables is AB, as in Equation 12.1.1.

Exercise 12.1.1

Verify Equation 12.1.2 by constructing the corresponding truth table. ∎

TABLE 12.1.2

Disjunction of Two Logic Variables

A	B	AB
F	F	F
T	F	T
F	T	T
T	T	T

TABLE 12.1.3

Conjunction in Terms of Disjunction and Negation

A	B	$\overline{A}$	$\overline{B}$	$\overline{A} + \overline{B}$	$\overline{\overline{A} + \overline{B}}$
F	F	T	T	T	F
T	F	F	T	T	F
F	T	T	F	T	F
T	T	F	F	F	T

Boolean algebra is related to digital signal processing by the following conceptual step:

Concept: *Since the binary number system uses only two bits, 1 and 0, and the logic variables in two-valued logic assume only two values, T and F, then if the 1 and 0 are identified with true and false, digital signals consisting of strings of 1's and 0's can be processed according to the rules of Boolean algebra.*

Conventionally, **positive logic** is used, in which 1 and 0 are identified with T and F, respectively. **Negative logic**, in which 1 and 0 are identified with F and T, respectively, is also possible, but is seldom used. Examples of digital signal processing using the laws of Boolean algebra are considered in the next section. Boolean algebra is expounded in Section ST12.1.

12.2 Logic Gates

Gate Types

Logic gates are electronic circuits that perform some logic, or related, function. The main types of logic gates are shown in Table 12.2.1 together with their symbols, the operations they perform, and the corresponding truth tables. The inverter performs logic negation, which is denoted as usual by a bubble. Conjunction and disjunction are performed by AND gates and OR gates, respectively. A NAND gate is an AND gate with inverted output and is logically equivalent to an AND gate followed by an inverter; similarly for a NOR gate. The Exclusive-OR (XOR) gate differs from an OR gate, which is sometimes referred to as inclusive-OR gate, in that when all of its input are 1, the output is zero. The XOR operation is denoted by a circled plus sign. Although it can be implemented using the aforementioned gates (Exercise 12.2.1), the XOR gate is encountered in many important operations (Example 12.1.1), so that it is convenient to show it as a separate gate. If its output is inverted, the XOR gate becomes an exclusive-NOR (XNOR) gate. The transmission gate was discussed in Section 5.5, Chapter 5. When the control input C is 1, the gate

TABLE 12.2.1

Logic Gates

Gate	Symbol	Operation	Truth Table		
Inverter	A —▷o— O	$O = \overline{A}$			

A	O
0	1
1	0

Gate	Symbol	Operation
AND	A, B —⊃— O	$O = AB$

A	B	O
0	0	0
1	0	0
0	1	0
1	1	1

TABLE 12.2.1 (continued)

Logic Gates

Gate	Symbol	Operation	Truth Table		

NAND

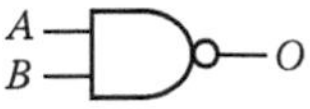

Symbol: A, B — NAND gate — O

Operation: $O = \overline{AB}$

A	B	O
0	0	1
1	0	1
0	1	1
1	1	0

OR

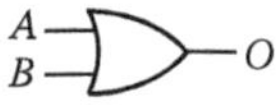

Symbol: A, B — OR gate — O

Operation: $O = A + B$

A	B	O
0	0	0
1	0	1
0	1	1
1	1	1

NOR

Symbol: A, B — NOR gate — O

Operation: $O = \overline{A + B}$

A	B	O
0	0	1
1	0	0
0	1	0
1	1	0

XOR

Symbol: A, B — XOR gate — O

Operation: $O = A \oplus B$

A	B	O
0	0	0
1	0	1
0	1	1
1	1	0

XNOR

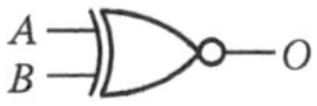

Symbol: A, B — XNOR gate — O

Operation: $O = \overline{A \oplus B}$

A	B	O
0	0	1
1	0	0
0	1	0
1	1	1

Transmission gate

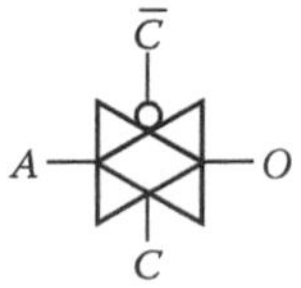

Symbol: $\overline{C}$, A — transmission gate — O, C

Operation: Output is controlled by a control signal

C	A	O
1	0	0
1	1	1
0	0	FL
0	1	FL

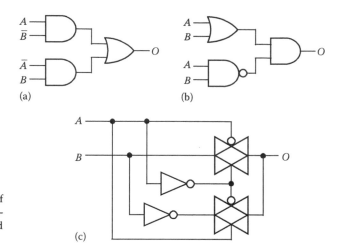

FIGURE 12.2.1
Figure for Exercise 12.2.1. Implementation of
XOR function using complemented variables (a), uncomplemented variables (b), and
transmission gates (c).

input A is transmitted to the output. But when C is 0, both transistors of the gate are off, and the gate output is floating, that is, it has a high impedance to ground and will appear as an open circuit to whatever is connected to the output. The transmission gate is therefore useful for connecting together and disconnecting subcircuits of a system. An output that can have a high-impedance state, in addition to high and low, is referred to as **tristate**.

When increased current drive capability is required, which also speeds up the response because of increased rates of capacitor charging and discharging, **buffer gates** are used whose output transistors can source or sink larger currents. An example is a buffer inverter. A pure **buffer** has a single input and a single output and is implemented as two inverters in cascade, the output inverter having increased drive capability (Example 13.1.5, Chapter 13). Its logic symbol is like that of an inverter but without the bubble at the output.

It was mentioned in Section 12.1 that Boolean expressions can be implemented using inversion and conjunction or inversion and disjunction. If all the inputs of a NAND gate, or a NOR gate, are connected together, an inverter results. If a NAND gate is followed by an inverter, the overall logic function is conjunction. Similarly, if a NOR gate is followed by an inverter, the overall logic function is disjunction. It follows that, in principle, Boolean expressions can be implemented using only NAND gates or only NOR gates.

Exercise 12.2.1

Verify using truth tables that the circuits in Figure 12.2.1 are equivalent to an XOR gate. ∎

Example 12.2.1 Applications of XOR Logic

It is required to use XOR gates to: (a) add two bits; (b) generate an odd parity bit.

SOLUTION

(a) From the truth table for adding two bits A and B (Table 12.2.2), the sum digit $S = A \oplus B$, whereas the carry digit $CD = AB$. These digits can be derived as in Figure 12.2.2a.
(b) Given four bits A, B, C, and D, it is required to generate an output bit only if the number of 1's in ABCD is odd. Three

TABLE 12.2.2

Addition of Two Bits

A	B	S	CD
0	0	0	0
1	0	1	0
0	1	1	0
1	1	0	1

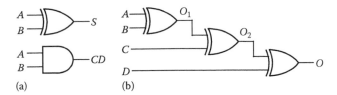

FIGURE 12.2.2
Figure for Example 12.2.1. Using XOR gates to generate a sum bit (a), or an odd-parity bit.

XOR gates can be used for this purpose, as in Figure 12.2.2b. The output O_1 is 1 only if A and B have an odd number of 1's. Similarly, the output O_2 is 1 only if the number of 1's in ABC is odd, and the output O is 1 only if the number of 1's in $ABCD$ is odd.

CMOS Gate Examples

In order to have some concrete idea about gates before we discuss gate characteristics, we will illustrate how some of the logic gates can be implemented in CMOS. The basic CMOS logic inverter, mentioned in Section 5.5, Chapter 5 is illustrated in Figure 12.2.3 and is analyzed in detail in Section 13.1, Chapter 13. When v_I is "high," corresponding to logic 1, the E-NMOS transistor is fully on, whereas the E-PMOS transistor is off. The output is "low" and is connected to ground through the E-NMOS transistor, which *sinks* current at the gate output. Conversely, when v_I is "low," corresponding to logic 0, the E-NMOS transistor is off, whereas the E-PMOS transistor is fully on. The output is high and is connected to V_{DD} through the E-PMOS transistor, which *sources* current at the gate output. In gate terminology, the E-PMOS transistor is an active **pull-up** transistor, whereas the E-NMOS transistor is an active **pull-down** transistor.

Figure 12.2.4 illustrates a CMOS two-input NAND gate, formed from the inverter by adding an E-NMOS transistor in series, and an E-PMOS transistor in parallel, with the transistors of the inverter, and connecting their gates together to provide an additional input. If either input A or B is low, the corresponding E-NMOS transistor is off, whereas the corresponding E-PMOS transistor is on. The output O is high. The output goes low only if both inputs are high, so that the two series E-NMOS transistors will be on. The logic implemented is NAND.

Figure 12.2.5 illustrates a CMOS two-input NOR gate, formed from the inverter by adding an E-NMOS transistor in parallel, and a E-PMOS transistor in series, with the transistors of the inverter, and connecting their gates together to provide an additional input. If either input A or B is high, or both are high, then one or both E-NMOS transistors will be on, whereas one or both E-PMOS transistor will be off. The output O is low, that the logic implemented is NOR.

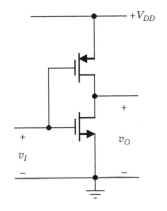

FIGURE 12.2.3
Basic CMOS logic inverter.

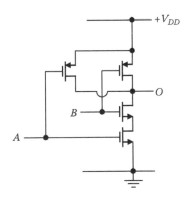

FIGURE 12.2.4
CMOS two-input NAND gate.

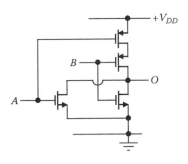

FIGURE 12.2.5
CMOS two-input NOR gate.

CMOS gates are discussed in more detail in Chapter 13. Note that in CMOS, NAND and NOR gates are more "natural" than AND and OR gates. To implement the latter gates, an additional inverting stage is needed.

Gate Performance

An ideal gate has the following properties:

1. It can have an unlimited number of inputs.
2. It can drive any number of gates connected to its output.
3. It performs the required logic function on its inputs with zero delay at the output.
4. The high voltage level, corresponding to logic 1, equals the positive supply voltage, whereas the low voltage level, corresponding to logic 0, equals the most negative supply voltage, which is usually zero. This results in the maximum possible logic swing, referred to as **rail-to-rail**.
5. The voltage transfer characteristic has an abrupt transition between high and low levels in the middle of the voltage swing. This maximizes noise immunity at both high and low levels.
6. It should have good noise performance. This is governed by the following factors:
   a. Relatively large and equal high and low noise margins, as discussed later.
   b. Low impedance levels of signal lines, whether in the high or low states. As illustrated in Figure 12.2.6, a low resistance to ground of a signal line that is at a low voltage level reduces both the magnitude and time constant of the noise voltage induced through capacitive, or inductive, coupling. Similar considerations apply to the high voltage level, because the V_{DD} supply is an ac ground.
   c. When the gate output changes between high and low levels, the current drawn from the voltage supply should remain constant. As will be discussed in Chapter 13, the change in output can cause current pulses, or "spikes" through the supply, which necessitates careful decoupling and layout of the supply lines, so as not to induce noise in other gates.
7. The gate should not dissipate any power.

Practical gates deviate from ideal gates to different extents depending on the logic family involved, that is, the type of circuit that implements the gate. For example, the gate circuit could use MOS transistors or BJTs in different types of circuits, as explained in Chapter 13. For present purposes, we will examine some general aspects of performance of practical gates, irrespective of the logic family to which they belong.

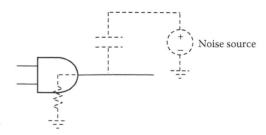

FIGURE 12.2.6
Effect of low gate output resistance on induced noise.

The number of inputs to a gate is the **fan-in**. The **fan-out** is the number of similar gates connected to the output of a given gate, all gates being of the same logic family. The maximum fan-in and fan-out depend on device characteristics and circuit configuration of the gates under consideration. The constraints on maximum fan-out could apply under static, that is, dc conditions, or under dynamic, that is, time-varying, conditions because of capacitive loading, as discussed in Chapter 13. Having a large fan-in and fan-out can simplify chip design and reduce the total number of gates required. A gate is often specified by adding the fan-in to the gate type. Thus, a two-input NAND gate has a fan-in of 2 and is designated as a NAND2 gate.

The voltage transfer characteristic relating the input of a gate to its output under dc, or quasi-static, conditions gives some important information about the behavior of all the gates belonging to a particular logic family. Consider, for example, one inverter driving another (Figure 12.2.7a), the inverter transfer characteristic being as in Figure 12.2.7b. V_{OL} and V_{OH} are the low and high voltage levels, respectively. For the CMOS inverter of Figure 12.2.4, these are nominally zero and V_{DD}, respectively. When $v_{I2} = V_{OL}$ from the output of inverter 1, $v_{O2} = V_{OH}$, and when $v_{I2} = V_{OH}$, $v_{O2} = V_{OL}$. At the **threshold voltage** V_{TH}, $v_{I2} = v_{O2}$, which represents a "boundary point" above or below which the input and output are inversely related.

The logic swing is:

$$LS = V_{OH} - V_{OL} \qquad (12.2.1)$$

As v_{I2} increases above V_{OL}, a level V_{IL} is reached at which the slope of the transfer characteristic is -1. Similarly, as v_{I2} decreases below V_{OH}, a level V_{IH} is reached, at which the slope of the transfer characteristic is -1. For $V_{IL} < v_I < v_{IH}$, the magnitude of the small-signal voltage gain dv_O/dv_I exceeds unity, so that noise is amplified. The range between V_{IL} and V_{IH} is therefore avoided in normal operation. A **low-level noise margin** is defined as:

$$NM_L = V_{IL} - V_{OL} \qquad (12.2.2)$$

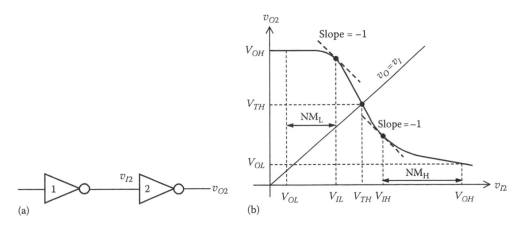

FIGURE 12.2.7
(a) One inverter driving another; (b) inverter transfer characteristic.

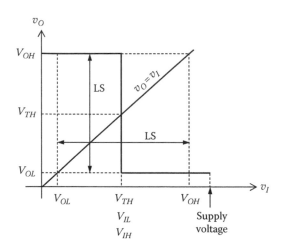

FIGURE 12.2.8
Idealized inverter transfer characteristic.

Similarly, a **high-level noise margin** is defined as:

$$NM_H = V_{OH} - V_{IH} \qquad\qquad (12.2.3)$$

Clearly, it is desirable to have $NM_L = NM_H$ so as to have the same noise immunity in both logic states, assuming that other factors contributing to noise immunity, such as the impedance level to ground, are the same.

The following should be noted concerning the transfer characteristic:

1. An idealized inverter transfer characteristic is illustrated in Figure 12.2.8. The transition between the two levels is sharp and occurs in the middle of the LS, at $V_{TH} = (V_{OH} + V_{OL})/2$. Moreover, $V_{IL} = V_{IH} = V_{TH}$, and $NM_L = NM_H = (V_{OH} - V_{OL})/2$. If $V_{OL} = 0$ and $V_{OH} = 5$ V, NM_L and NM_H will have the maximum value of 2.5 V.

2. In the case of gates of fan-in larger than 1, inactive inputs are tied to logic one or logic zero, as appropriate. For a NAND3 gate, for example, inactive inputs are tied to the positive supply, whereas for a NOR3 gate, inactive inputs are grounded.

3. The transfer characteristic of a noninverting gate is the mirror image of that of Figure 12.2.7b about a vertical line through the middle of the LS (Figure 12.3.1b). The same considerations discussed for the inverting transfer characteristic apply.

The dynamic behavior of a gate is described by the propagation delay from input to output and depends in part on the rise and fall times of the transitions between the two levels. Inverters are usually used for describing dynamic behavior. Consider an inverter driven by a gate, which could be another inverter, as in Figure 12.2.7a. The upper waveform in Figure 12.2.9 represents the input to the inverter and changes from V_{OH} to V_{OL} and back to V_{OH}. The inverter output changes in the opposite direction, as in the lower trace. Rise and fall times, t_r and t_f, respectively, are defined in terms of transitions between the 10% and 90% levels of the LS, as in Figure 12.2.9. There is a propagation delay t_{pLH} for the low-to-high transition of the inverter output, and another propagation delay t_{pHL} for the high-to-low transition. Both t_{pLH} and t_{pHL} are measured from the 50%

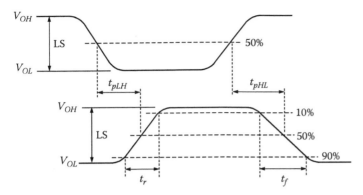

FIGURE 12.2.9
Gate propagation delay.

level in the middle of the LS, as in Figure 12.2.9. In general, $t_{pLH} \neq t_{pHL}$, so that an average propagation delay is defined as:

$$t_p = \frac{1}{2}(t_{pLH} + t_{pHL}) \tag{12.2.4}$$

Concept: *The propagation delay sets a limit to the maximum frequency of operation of the gate.*

At a high enough frequency, the low-to-high and high-to-low transitions become sufficiently close together so that the interpretation of a logic 1 or a logic 0 becomes uncertain. When gates are cascaded, the total propagation delay is the sum of the individual propagation delays plus the delays due to the interconnections between the gates.

The propagation delay of a gate has, in general, two components: (1) a component due to internal capacitances of the gate, and (2) a component due to the output resistance r_{out} of the gate in conjunction with the total capacitance at the output of the gate. If the gate output is an emitter follower, as in the case of emitter-coupled logic (ECL) the second component is relatively unimportant. On the other hand, small MOSFET output transistors have a relatively large r_{out}. The propagation delay in this case may be dominated by the second component.

The power dissipation of a gate belonging to a particular logic family is an important figure that sets a limit to the maximum number of gates that can be included on a chip without having to take special measures for cooling. Power dissipation may have a static, or dc, component P_{dc} proportional to the supply voltage, and a dynamic component P_{ac} proportional to frequency. When P_{dc} is different for the high and low states, an average P_{dc} is taken. In the case of CMOS gates, the static power dissipation is negligible, as there is never a nonconducting path between V_{DD} and ground because of some transistors being cutoff (Figures 12.2.4 through 12.2.6). During switching, however, current flows and power is dissipated that depends on the capacitance appearing at the gate output. In Figure 12.2.10, C charges from a supply of voltage V_{DD} through a nonlinear resistor representing a conducting pull-up transistor. The energy stored in C, when fully charged, is $\frac{1}{2}CV_{DD}^2$. The instantaneous power delivered by the supply during charging is $V_{DD}i_C$, and the total energy delivered during charging is $V_{DD}\int_0^\infty i_C dt = V_{DD}Q_{Cf}$, where Q_{Cf} is the charge delivered to C. Assuming that C is initially uncharged and that when C is fully charged, the voltage drop across R_{pu} is zero, $Q_{Cf} = CV_{DD}$. The energy delivered by the supply is, therefore, CV_{DD}^2, which means that the energy dissipated in R during charging is $\frac{1}{2}CV_{DD}^2$, the same as the final energy stored in C. During discharge, the energy dissipated in the pull-down transistor is $\frac{1}{2}CV_{DD}^2$, the energy stored in C. The total energy dissipated in the output transistors is thus CV_{DD}^2. If the gate operates at a frequency f, the dynamic dissipated power is:

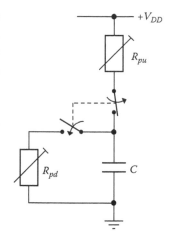

FIGURE 12.2.10
Charging and discharging of capacitor at output of gate.

$$P_{ac} = fCV_{DD}^2 \qquad (12.2.5)$$

According to Equation 12.2.5, P_{ac} increases as V_{DD}^2. This implies that where P_{ac} is dominant compared to P_{dc}, as in CMOS, it is particularly advantageous to reduce V_{DD} as much as possible. Modern CMOS circuit are designed to operate with $V_{DD} = 1.5$ V. Note that dynamic power dissipation can also result from *glitches* that occur in multilevel gates because of propagation delays (Problem P12.2.16).

In general, dc power dissipation can be reduced by increasing resistance values in the circuit. However, for given capacitances, the propagation delay is increased, and conversely. Thus, if the output transistors are resized so as to double their resistance, the dc power dissipation is halved, but the reduced source and sink currents approximately double the propagation delay. A figure of merit for the gate is the power-delay product (PDP), where,

$$\text{PDP} = (P_{dc} + P_{ac})t_p \qquad (12.2.6)$$

PDP can only be significantly improved by using transistors with a higher gain-bandwidth product and by reducing t_p through improved circuit configuration and layout. In adiabatic logic (Section ST12.2) charge is partly conserved when capacitors discharge.

Simulation Example 12.2.1 Ring Oscillator

It is required to explain the behavior of a three-inverter ring oscillator and to simulate it.

EXPLANATION

If the input and output terminals of an inverter are connected together, both terminals will be at a voltage V_{TH}, since this is the voltage at which $v_I = v_O$ (Figure 12.2.7). The circuit is stable, like connecting the output of an op amp to the inverting input. Similarly, when three or more inverters are connected in a ring, the feedback around the ring is negative at dc. However, the output resistances of the inverters, in conjunction with the capacitances at the inverter outputs, introduce a phase shift which turns the feedback positive at some frequency, resulting in sustained oscillations at this frequency. Figure 12.2.11a shows three inverters, assumed identical, connected in a ring. The resistance and capacitance associated with the output of each inverter are illustrated in Figure 12.2.11b. The phase shift of each of the three RC circuits is less than 90°, so that at some frequency ω_0 the total phase shift due to the three RC circuits becomes 180°. The net phase shift around the ring becomes zero, resulting in oscillations at the frequency ω_0. The period of oscillation can be expressed in terms of the total propagation delay, as illustrated in Figure 12.2.11c, where the waveforms involved are idealized for simplicity and clarity. When v_1 at the input of inverter 1 changes

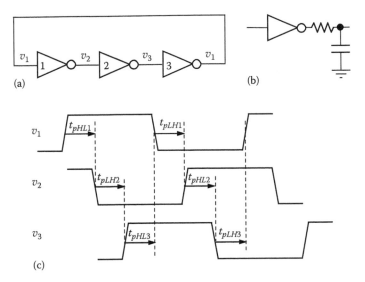

(a)

(b)

(c)

FIGURE 12.2.11
Figure for Example 12.2.1. (a) Three inverters connected in a ring; (b) resistance and capacitance at the output of an inverter; (c) waveforms of inverters.

from low to high, the output of this inverter changes from high to low after a propagation delay t_{pHL1}. This change is propagated though inverters 2 and 3 and back to inverter 1 as a change from high to low of v_1. This again propagates through inverters 2 and 3 and back to inverter 1 as a change from low to high of v_1, thereby completing the cycle. The period T is: $T = t_{pHL1} + t_{pLH1} + t_{pHL2} + t_{pLH2} + t_{pHL3} + t_{pLH3}$. Since the three inverters are assumed identical, $T = 2 \times 3 \times t_p$, where t_p is the average propagation delay of the inverter. When an odd number N of identical inverters are connected in a ring, $f_0 = 1/2Nt_p$.

The ring oscillator is used in practice for the measurement of inverter propagation delay. N is made large enough to bring the frequency down to a value that is convenient to measure.

SIMULATION

The schematic of the three-inverter ring oscillator is shown in Figure 12.2.12, where a 10 pF capacitor is connected at the three inverter outputs. The E-NMOS transistors are MBreakN3 from the BREAKOUT library whose parameters entered in the PSpice Model are: Kp = 50u, Vto = −1, lambda = 0.015, with L = 1u, W = 2u entered in the Property Editor spreadsheet. The corresponding parameters of the E-PMOS transistor are: Kp = 20u, Vto = −1, lambda = 0.015, L = 1u, W = 5u. The waveform of the oscillations is practically sinusoidal because the total phase shift around the loop is zero at a single frequency. Taking the output V_o from a fourth inverter restores the waveform to nearly square (Figure 12.2.13). $f_0 \cong 3$ MHz.

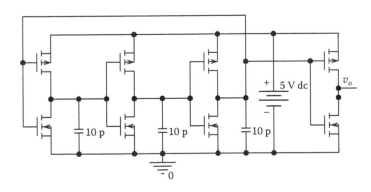

FIGURE 12.2.12
Figure for Example 12.2.1.

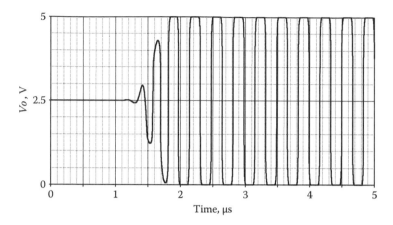

FIGURE 12.2.13
Figure for Example 12.2.1.

12.3 Flip-Flops

Basic Latch

Consider two inverters connected in a loop (Figure 12.3.1a). Figure 12.3.1b illustrates the voltage transfer characteristic v_O vs. v_I when the loop is broken at X. As to be expected, v_O is low when v_I is low and is high when v_O is high. When the loop is closed, $v_O = v_I$, and the operating point can be determined from the intersections of the line $v_O = v_I$ with the transfer characteristic. Point P_1 is a stable operating point, for if v_I increases, for example, due to noise fluctuations, v_O must increase by an equal amount, along the line $v_O = v_I$, in order to at least sustain the change in v_I. In actual fact, v_O increases very little, along the transfer characteristic. The increase in v_O is therefore insufficient to sustain the increase with v_I, because the gain is less than 1, so that v_I reverts back to its value at P_1. Similarly, P_2 is a stable operating point. On the other hand, Q represents an unstable operating point. For if v_I increases, the increase in v_O is more than that required to sustain the increase in v_I,

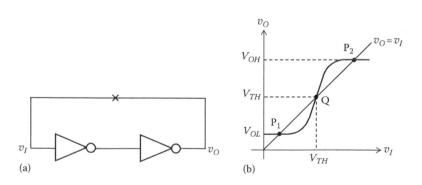

FIGURE 12.3.1
(a) A two-inverter loop; (b) voltage transfer characteristic of two cascaded inverters.

because the transfer characteristic at Q is steeper than the line $v_O = v_I$, that is, the gain is more than 1. The increase in v_I is reinforced, so that both v_I and v_O increase until point P_2 is reached. Similarly, if v_I decreases, the operating point moves to P_1. The circuit is a **latch**, or **bistable**, since it has two stable states, one with v_O high, the other with v_O low. If v_I is set to zero, by temporarily grounding the input of the first inverter, for example, the stable state of the latch corresponds to P_1 in Figure 12.3.1b. Similarly, if the input of the first inverter is temporarily connected to V_{DD}, the state of the latch changes to the stable state corresponding to P_2 in Figure 12.3.1b.

Concept: *The latch is a memory element in that it "remembers" its last state and keeps this state until it is forced to change it.*

The reason the circuit latches to one of the two stable states is due to the positive feedback, which *maintains the given state despite the removal of the cause of the change of state.*

The presence of memory elements is the distinguishing feature of sequential digital circuits. In combinational circuits, consisting of interconnected logic gates without positive feedback, the output at any instant depends on the inputs at that instant. In sequential circuits, the output at any instant depends not only on inputs at that instant but also on the state of the memory elements at that instant.

The bistable is a member of the family of **multivibrator** circuits, which includes, in addition, the **astable**, and **monostable**. The astable does not have a stable state but continuously oscillates between these two states. The monostable has a single stable state. When triggered into the other state, it remains in this state for a preset interval, after which it reverts to the stable state. An output pulse is produced whose duration T is determined by the value of a capacitor C, as illustrated in Figure 12.3.2. Each of the trigger and output pulses can be positive or negative depending on the particular monostable.

The Schmitt trigger is a special type of bistable that exhibits *hysteresis*, so that transitions between states have different thresholds (Figure 12.3.3). When $v_I = 0$, $v_O = V_{LIM}^+$ (point A). If v_I increases to a threshold voltage V_{Th}^+, v_O switches to a negative voltage V_{LIM}^- (point B). If v_I decreases to V_{Th}^-, v_O reverts to V_{LIM}^+. A hysteresis loop results because $V_{Th}^- < V_{Th}^+$. This hysteresis loop is useful in that, once v_I exceeds V_{Th}^+, v_O stays at V_{LIM}^-, despite fluctuations in v_I, as long as v_I does not decrease to less than V_{LIM}^-. All the four voltages V_{LIM}^+, V_{LIM}^-, V_{Th}^+, and V_{Th}^- can be set to suit a particular application, and the hysteresis loop can be centered anywhere along the v_I axis. The characteristic of Figure 12.3.3 is described as inverting, because an increasing v_I decreases v_O from the positive limit to the negative limit. The symbol for the inverting Schmitt trigger is an inverter with a hysteresis loop. Noninverting Schmitt triggers are also possible.

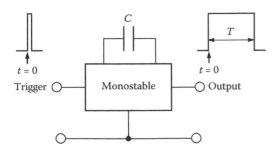

FIGURE 12.3.2
Monostable multivibrator.

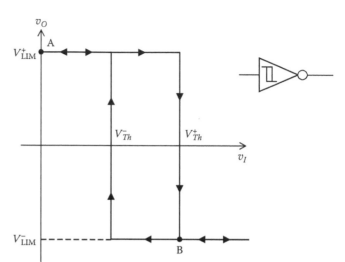

FIGURE 12.3.3
Voltage transfer characteristic of the Schmitt trigger.

The principles of operation of multivibrators and the Schmitt trigger are discussed in Section ST12.2.3. CMOS versions of the monostable, astable, and Schmitt trigger are presented in Problems P12.3.8, P12.3.9, and Example SE12.1, respectively. A versatile IC, the 555 timer, which can be used as a timer, monostable, or astable, is discussed in Section ST12.4.

Exercise 12.3.1

Refer to Simulation Example 12.2.1. Connect two inverters in a loop, as in Figure 12.3.1a. Verify that upon starting the simulation, the inverters will be in the state corresponding to Q in Figure 12.3.1b. This is because there is no "noise" in the simulation to force the state to that of P_1 or P_2. Use a normally closed switch that opens at $t = 0$ to force the inverters to be in either of these states.

SR Latch

The basic latch of Figure 12.3.1 does not allow a convenient means for changing the state. This can be achieved by replacing the inverters of the basic latch with two-input inverting gates, which provides additional inputs to the latch, as illustrated in Figure 12.3.4 using NOR2 gates. When the S and R inputs are both 0, they are inactive, that is, they have no effect on the output, and the circuit reduces essentially to the two inverters of the basic latch (Figure 12.3.1). The complementary outputs Q and $\overline{Q}$ remain in whatever state they were in. This is indicated in the truth table (Table 12.3.1) by having Q_{n+1}, the output after the change of input takes place, the same as Q_n, the output before the change is applied.

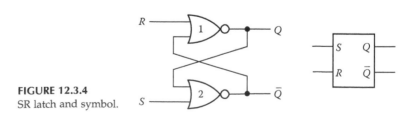

FIGURE 12.3.4
SR latch and symbol.

TABLE 12.3.1

Truth Table for SR Latch

S	R	Q_{n+1}	$\overline{Q}_{n+1}$	Operation
0	0	Q_n	$\overline{Q}_n$	Hold
1	0	1	0	Set
0	1	0	1	Reset
1	1	0	0	Not used

The operation is a *hold*. When the set input S is 1, $\overline{Q}$ is forced to 0 if it is 1 and remains at 0 if it is 0, and if $R=0$, then $Q=1$. The latch is said to be *set*. Similarly, when the reset input $R=1$, Q is forced to 0 if it is 1 and remains at 0 if it is 0, and if $S=0$, then $Q=0$. The latch is said to be *reset*. If both S and R are 1, Q and $\overline{Q}$ become 0 simultaneously and are no longer complementary outputs. If S and R both return to zero at exactly the same time, the state is indeterminate. In practice, however, S and R do not return to zero exactly simultaneously, and the state of the latch is determined by whichever input is 1 when the other input is zero. For these reasons the inputs $S=1$ and $R=1$ are not normally used. The block symbol of the SR latch is shown in Figure 12.3.4. The SR latch can also be implemented using NAND2 gates rather than NOR2 gates (Exercise 12.3.2).

Figure 12.3.5 illustrates how the SR latch can be used in a simple example in which a timer is available that starts timing when its input is connected to logic 1 (high) but stops timing and is reset when its input is logic 0 (low). When PB_1 is pressed, $S=1$, $Q=1$, and the timer starts timing. When PB_1 is released, the latch maintains $Q=1$. To stop timing and reset the timer, PB_2 is pressed, which makes $R=1$ and $Q=0$. Releasing PB_2 does not change this state.

The operation of the SR latch just described is asynchronous, in the sense that outputs can change, after some propagation delay, following a change in S or R, in accordance with the truth table. In **synchronous** digital circuits, the outputs should only change at particular instants defined by a master timing signal, or clock (CK). This can be arranged using additional gates, as in Figure 12.3.6. When CK is low, the outputs of the AND gates are forced to zero, which isolates R and S from the NOR2 gates. Only when CK is high can the levels of R and S affect the NOR2 gates in accordance with the truth table (Table 12.3.1). A clocked latch is referred to as a **flip-flop** (FF), although the terms flip-flop and latch are often used interchangeably.

Exercise 12.3.2

Implement the basic SR latch using NAND2 gates and label S, R, Q, and $\overline{Q}$. ∎

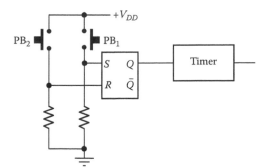

FIGURE 12.3.5

Illustration of the latching property of an SR latch.

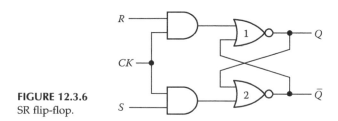

FIGURE 12.3.6
SR flip-flop.

JK Flip-Flop

The clocked SR-FF still suffers from the inconsistency of having both Q and $\overline{Q}$ zero when S and R are 1. This situation can be avoided in one of two ways: (1) if R and S are the negation of one another, they cannot be 1 simultaneously, but neither can they be 0 simultaneously. This method is implemented in the D-FF, as discussed later. (2) A feedback connection is added so that Q and $\overline{Q}$ would block one of the two inputs S and R, resulting in a JK-FF, as illustrated in Figure 12.3.7. If $J=0$ and $K=0$, R and S are forced to zero irrespective of Q and $\overline{Q}$. The flip-flop retains its state, as in the case of the clocked SR-FF discussed previously. When $J=1$ and $K=0$, R is forced to zero. If $Q=0$, which means $\overline{Q}=1$, S becomes 1 one propagation delay after $CK=1$, as shown in the truth table (Table 12.3.2). After a second propagation delay, $\overline{Q}$ becomes 0, with Q still momentarily zero. After a third propagation delay two changes occur simultaneously: (1) $Q=1$, because $R=0$, and (2) $S=0$ because $\overline{Q}=0$. The FF is set, and its state will not change thereafter. If the FF was set to begin with, it remains set.

When $J=0$ and $K=1$, S is forced to zero. If $Q=1$, which means $\overline{Q}=0$, R becomes 1 one propagation delay after $CK=1$, as shown in the truth table. After a second propagation delay, Q becomes zero, with $\overline{Q}$ still momentarily zero. After a third propagation delay two changes occur simultaneously: (1) $\overline{Q}=1$, because $S=0$, and (2) $R=0$, because $Q=0$. The FF is reset, and its state will not change thereafter. If the FF was reset to begin with, it remains reset.

Finally, if $J=1$ and $K=1$, these inputs will be neutral. When $CK=0$, both R and S are 0. The behavior of the FF after CK becomes 1 at $t=0$ can be followed on the timing diagram illustrated in Figure 12.3.8. If $Q=1$, which means, $\overline{Q}=0$, R becomes 1 one propagation delay after $CK=1$. After a second propagation delay, Q becomes zero, with $\overline{Q}$ still momentarily zero. After a third propagation delay, $\overline{Q}=1$, with $Q=0$, so that the FF changes state. If CK becomes zero, $\overline{Q}=1$ is not transmitted back to S, through the lower AND3 gate, and the FF remains in the new state. However, if CK remain 1 then after a fourth propagation delay, $S=1$, and after a fifth propagation delay, $\overline{Q}=0$, with Q still momentarily zero. After a sixth propagation delay, $Q=1$, $\overline{Q}=0$, $R=0$, and $S=0$, so that the FF changes back to the initial state. As long as $CK=1$, the FF toggles between the two states at a rate that depends

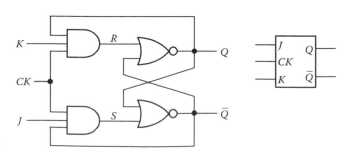

FIGURE 12.3.7
JK flip-flop and symbol.

TABLE 12.3.2

Truth Table for JK Flip-Flop

J	K	Q_n	S	R	Q_{n+1}	Operation
0	0	0	0	0	0	Hold
		1	0	0	1	
1	0	0	1	0	1	Set
		1	0	0	1	
0	1	0	0	0	0	Reset
		1	0	1	0	
1	1	0	1	0	1	Toggle
		1	0	1	0	

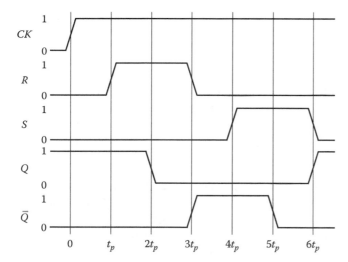

FIGURE 12.3.8
Timing diagram of JK flip-flop.

on the propagation delays. The final state depends on the duration of the 1 state of the clock, which is undesirable. This problem can be solved using a **master–slave** configuration, illustrated in Figure 12.3.9.

Two JK-FFs are cascaded, the CK signal being applied to the first, or master FF, whereas $\overline{CK} = 1$ is applied to the second, or slave FF. The feedback is applied from the output of the slave FF to the input gates of the master FF. Let us illustrate the operation of this FF when $J = 1$, $K = 0$, $Q = 0$, and $\overline{Q} = 1$. When $CK = 1$, the master FF is set, so that $Q_m = 1$ and $\overline{Q}_m = 0$,

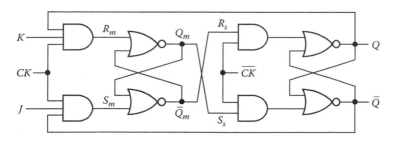

FIGURE 12.3.9
Master–slave JK flip-flop.

as described previously in connection with Figure 12.3.7. When $\overline{CK} = 1$, the J and K inputs to the master FF are disabled. The slave FF is set, as in the SR-FF of Figure 12.3.6. Next, let us check the operation if $J = 1$, $K = 1$, $Q = 1$, and $\overline{Q} = 0$. When $CK = 1$, the operation proceeds as discussed with reference to Figure 12.3.8 until after the third propagation delay. But, because feedback to the lower AND3 gate is taken from $\overline{Q}$ and not $\overline{Q}_m$, the master FF remains in the new state, $Q_m = 0$, and $\overline{Q}_m = 1 = 1$. When $\overline{CK} = 1$, the state of the master FF is transmitted to the slave FF, so that $Q = 0$, and $\overline{Q} = 1$. When $CK = 1$, the master FF toggles to the new state $Q_m = 1$, and $\overline{Q}_m = 0$. When $\overline{CK} = 1$, the state of the master FF is transmitted to the slave FF, so that $Q = 1$, and $\overline{Q} = 0$. The master–slave flip-flop thus toggles between the two states with each clock pulse, irrespective of the duration of the 1 or 0 states of the clock pulse.

The following should be noted concerning the master–slave JK-FF:

1. While $CK = 1$, changes in J and K affect the output of the master FF, after one propagation delay. In this sense, the master FF is said to be *level sensitive*. The changes in J and K may be "glitches" due to noise and may cause unwanted transitions in the output of the master FF.

2. On the other hand, the output of the slave FF changes when CK goes low and is not affected by changes in J and K. It appears, therefore, that the output of the master–slave FF changes on the negative-going transition of the clock pulse. Such operation is described as **edge-triggered**. The operation can be changed from negative-edge-triggered to positive-edge-triggered simply by applying $\overline{CK}$ to the master and CK to the slave.

A FF that toggles with each clock pulse is a **T flip-flop**. As just described, such a FF can be realized by connecting together the J and K terminals of a master–slave JK-FF and labeling them as a T input (Figure 12.3.10). If $T = 0$, the FF retains its state, but if $T = 1$, the FF toggles with each clock pulse. The symbol $>$ at the CK terminal denotes positive-edge triggering. Adding a bubble as shown denotes negative edge triggering.

D Flip-Flop

The data or D-FF is an important type of FF that is used for temporary storage of data or as a delay element. Logically, it can be derived from the JK-FF of Figure 12.3.7 simply by labeling the J terminal as D and connecting it to the K terminal through an inverter (Figure 12.3.11). If $D = 0$, the FF is reset when $CK = 1$, corresponding to $J = 0$, $K = 1$ in Figure 12.3.7. On the other hand, if $D = 1$, the FF is set when $CK = 1$, corresponding to $J = 1$, $K = 0$. When $CK = 0$, the state of the FF is preserved. The cases $J = 0$, $K = 0$, or $J = 1$, $K = 1$ do not arise.

We will use the D-FF to illustrate two important concepts concerning FFs in general.

Concept: *In practice, FFs can be implemented based on a circuit representation that uses a smaller number of transistors than that suggested by the logic representation.*

If the D-FF is to be implemented in CMOS according to Figure 12.3.7, the NOR2 gates would require four transistors each.

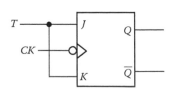

FIGURE 12.3.10
Negative-edge-triggered T flip-flop.

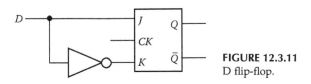

FIGURE 12.3.11
D flip-flop.

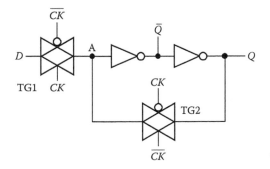

FIGURE 12.3.12
D flip-flop implementation using transmission gates.

An AND3 gate can be implemented using a six-transistor NAND3 gate followed by an inverter. A total number of 25 transistors would be required, including the inverter between the J and K inputs. Now consider the circuit of Figure 12.3.12 consisting of two inverters and two transmission gates, or a total of 10 transistors, including the inverter for $\overline{CK}$. We will show that such a circuit implements a D-FF. When $CK = 1$, TG1 is closed and TG2 is open. The input D is transmitted through TG1 and the two inverters so that $Q = D$. If $D = 0$, the FF is reset and if $D = 1$, the FF is set. When $CK = 0$, TG1 is opened isolating the D input. TG2 is closed, so that the two-inverter loop behaves as a latch and preserves the last state. If when $CK = 1$ again, D is still 1, Q remains 1. If D has changed to 0, Q will change to 0 and will be maintained at zero when $CK = 0$.

Concept: *Because of propagation delays, an active input must be stable for a short time before and a short time after the active clock transition.*

These times are referred to as the **setup time**, t_{st}, and the **hold time**, t_{hd}, respectively. The need for a setup time and a hold time can be appreciated with reference to Figure 12.3.12. Clearly, when CK goes to zero, the input D should have reached the output Q. Hence D should be set up at a certain time t_{st} before CK goes low to account for the delays in TG1 and the two inverters. Similarly, should TG1 close slightly after TG2 opens, D should be held for a certain time t_{hd}. Otherwise, if D changes while TG1 and TG2 are open, the change in D will be transmitted to Q. The setup and hold times are illustrated in Figure 12.3.13. For CMOS, these times are typically about few nanoseconds. IC manufacturers provide these figures for their chips. To avoid uncertain operation of FFs, the setup and hold times should be carefully adhered to.

The circuit of Figure 12.3.12 is level sensitive, since changes in the D input will affect Q as long as $CK = 1$. It can be made into an edge-triggered FF by cascading it with another FF in a master–slave configuration (Figure 12.3.14a). NAND2 gates are used to provide the set and rest inputs, $\overline{S}$ and $\overline{R}$, respectively. These inputs are active low, which means that the corresponding change in the output takes place when $\overline{S}$ or $\overline{R}$ is zero. When $\overline{S} = 1 = \overline{R}$,

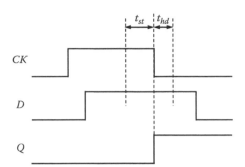

FIGURE 12.3.13
Timing diagram of D flip-flop.

these inputs are neutral, and the NAND2 gates will behave as inverters with respect to the other inputs not connected to $\overline{S}$ or $\overline{R}$. Moreover, these inputs are asynchronous, that is, the change in the output will follow the change in these inputs, after some propagation delay, independently of the state of the clock. When $CK=0$, TG1 and TG4 are closed whereas and TG2 and TG3 are open. Input D appears inverted at B and the previous output Q is held by the positive feedback in the second, slave stage. When $CK=1$, TG1 and TG4 are open whereas TG2 and TG3 are closed. The first, master stage holds the output at B and transmits this to the output Q, which is now the same as D during the preceding $CK=0$ phase. The change in D while $CK=0$ is thus transmitted to Q at the next low-to-high transition when $CK=1$. The FF is therefore **positive-edge triggered**. Inverting CK makes the FF negative-edge triggered.

When $\overline{R}=0$ in Figure 12.3.14a, the outputs of gates 1 and 4 are forced to 1. If TG3 is closed (which means that TG4 is open), $Q=0$ and is held when TG4 closes. If TG4 is closed when $\overline{R}=0$, then $Q=0$ and is held. Similarly, if $\overline{S}=0$, then $Q=1$ and the output of gate 2 is 1. If TG4 is closed the output $Q=1$ is held. If TG4 is open, TG2 and TG3 are closed, so that B is held at 0 and transmitted to the second stage as $Q=1$, which ensure that Q remains 1. If both R and S are 1, then both Q and $\overline{Q}$ will be 1, which may or may not be desirable, depending on the rest of the circuit. The asynchronous set and reset inputs can of course be made into synchronous inputs by gating them with a clock signal (Exercise 12.3.3).

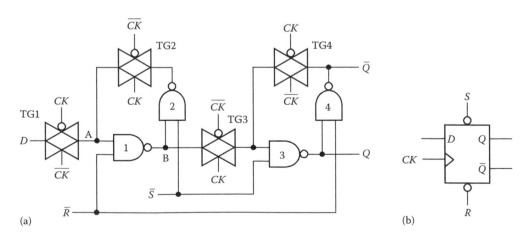

FIGURE 12.3.14
Positive-edge-triggered D flip-flop with active low S and R inputs (a) and its symbol (b).

For proper operation of the D-FF of Figure 12.3.14a, the transmission gates should turn on and off in the proper timing sequence, otherwise there could be an interval when all gates are closed, which would cause malfunctioning. Two factors may contribute to improper timing of clock signals: (1) appreciable rise and fall times of the clock pulses, which necessarily cause an overlap between a clock signal and its complement during their rise and fall times; hence clock signals should have short rise and fall times; (2) the delay in long clock lines causes differences in arrival times of clock signals to their targets, an effect known as **clock skew**. A common remedy is to use separate and *nonoverlapping* clock signals. Thus, the clock signal applied to TG1 and TG4 could be distinct from that applied to TG2 and TG3, with the two clock signals nonoverlapping, that is, their transitions are sufficiently separated in time so that that it would not be possible to have all the transmission gates closed at the same time. Edge-triggering can also be achieved in another way, as described for the D-FF in Section ST12.5.

The symbol for a positive-edge triggered D-FF is shown in Figure 12.3.14b, where a bubble at the S and R inputs indicates that these inputs are active low.

Exercise 12.3.3

Show how the set and reset inputs in Figure 12.3.14 can be made synchronous ∎

12.4 Digital System Memories

Various types of memory are almost universally used in digital systems to store and retrieve digital information. This information is generally of two types: (1) program instructions that direct the operation of the system, and (2) data on which instructions operate. The most important attributes that determine the type of memory to be used for a particular purpose are:

1. **Volatility**, which refers to the ability of memory to retain the stored information in the absence of power to the memory. *Volatile* memory loses stored information when power is removed, whereas *nonvolatile* memory does not. Clearly, memory for long-term storage of information should be nonvolatile, but memory used for temporary storage can be volatile.

2. **Read/write ability; read/write (R/W) memory** allows storage and retrieval of information at comparable speeds. On the other hand, **read-only memory** (ROM) allows retrieval of information at high speeds but restricts writing of information. In some types of ROM, such as CD-ROM, information can be written only once. In erasable ROMs, on the other hand, data can be written repeatedly but at reduced speed. ROM is generally nonvolatile.

3. **Access time** is the time between the initiation of a read operation and the retrieval of the read information. This is a crucial factor that determines the overall speed of operation of the digital system. It depends not only on how the bits of information are actually stored in the various memory locations but also on how the locations are accessed. The fastest access is in **random access memory** (RAM), in which the access time is independent of the physical location of the information in memory. In contrast, in serial memory, as in magnetic tape, information

can be written or read only sequentially, which increases average access time quite considerably. In disc drives, which are used for mass storage in computers, information is stored on circular tracks. The tracks are accessed essentially randomly by R/W heads, but the data in each track is accessed essentially sequentially.

In general, the larger the memory capacity, the longer is the access time. In computers and large digital systems, the highest memory capacity is provided by magnetic or optical disc drives. These are nonvolatile but slow. The working or main memory in computer systems is volatile RAM but should be more properly referred to as R/W memory, because semiconductor ROM memory is also random access. Some memory is provided on the processor chip itself. Although limited in capacity, this memory is optimized for speed of access.

4. **Power consumption** under static and dynamic conditions is important for limiting power dissipation in memory chips and in low-power applications.

5. **Cost per bit**, which depends on the number of stored bits per unit area of the chip. Single-chip semiconductor memory represents the densest packing of transistors per chip, because of the regularity of memory structure and the relatively low percentage area occupied by interconnections. Single-chip memory capacity has been approximately doubling every two years and is expected to reach several tens of gigabits by the year 2015.

Classification of Semiconductor Memories

Figure 12.4.1 shows the classification of semiconductor memories, which are discussed in the following sections. R/W memory is of two types: static RAM (SRAM) and dynamic RAM. SRAM stores information bits in latches. Dynamic RAM stores information as charges on capacitors, which means that these charges must be refreshed periodically to compensate for inevitable charge leakage. **Mask ROM** is factory programmed using a photo mask and cannot subsequently be erased or reprogrammed. **Programmable ROM** (PROM) is similar except that it is programmed by the user, as in fuse ROM. **Erasable PROM** (EPROM) is erased by ultraviolet radiation, which requires removal of the chip from the board. In contrast **electrically erasable PROM** (EEPROM) is erased in place. **Flash memory** is a special type of EEPROM that allows quick erasure of large blocks of stored information. The number of allowable write operations in EPROM and EEPROM is of the order of a million or so.

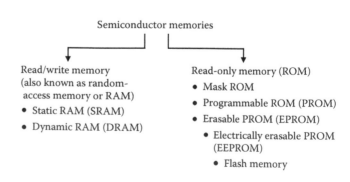

FIGURE 12.4.1
Classification of RAM.

Organization of RAM

RAM, particularly that in the form of a single chip, is generally organized as illustrated in Figure 12.4.2. Memory locations or cells are arranged in a rectangular array of 2^M rows and 2^N columns, which gives a total of 2^{M+N} cells, each cell storing a single bit and located at the intersection of a row and a column. A 64-Mbit chip would have $M+N=26$, since 1 Mb $= 1024 \times 1024 = 1{,}048{,}576 = 2^{20}$ bits and $64 \times 2^{20} = 2^{26}$. Such a memory is designated as 64M $\times$ 1 memory, or simply a 64-Mb memory. If each word in this memory is 16 bits, the memory is designated as 4M $\times$ 16 and would have $2^2 \times 2^{20}$ addresses.

A **row decoder** (described later) receives the M address bits and selects just one of the 2^M row lines. When a row is selected, all the 2^N cells in this row are also selected. Similarly, a **column decoder** decodes the column address. The column decoder also transfers the stored data in and out of the memory chip. For this reason, the 2^N column lines are referred to as **bit lines**, whereas the 2^M row lines are referred to as **word lines**. In addition, writing/reading of information requires some sort of a driver/sense amplifier as discussed in Section 12.5.

A memory chip can have all the bits individually addressable, in which case one of the 2^N lines is selected at a time. The data relevant to L of these one-bit memory chips are combined into L-bit words, where L is typically 8, 16, or 32. In some cases, the memory chip may be organized around L-bit words, in which case L bit lines are selected together. The cell or L cells whose word lines and bit lines are simultaneously selected can then be accessed for reading or writing.

As the number of cells in the memory array increases, the lengths of the word lines and bit lines also increase, which increases the resistance and capacitance of these lines and hence memory access time. The memory on the chip is then split into a number of "banks," and each bank may be subdivided into arrays, as in Figure 12.4.2. A 1-Gb memory, for example, may be divided into eight banks of 128 Mb ($2^3 \times 2^7 \times 2^{20} = 2^{30}$); each 128 Mb bank may be further divided into 512, 256-kb arrays ($2^9 \times 2^8 \times 2^{10} = 2^{27}$), each array having 512, or 2^9, word lines and 512 bit lines.

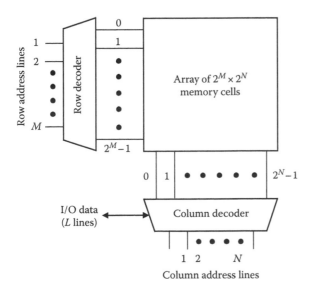

FIGURE 12.4.2
Organization of RAM.

12.5 Read/Write Memory

Static Memory Cell

A six-transistor cell of a CMOS SRAM is illustrated in Figure 12.5.1. Q_1 to Q_4 are two cross-coupled inverters that constitute a latch, as in Figure 12.3.1a. The NMOS and PMOS transistors of the inverters are matched. Q_{T1} and Q_{T2} are access transistors connecting the inverter outputs and the complementary bit lines. Typically Q_{T1} and Q_{T2} have two to three times the width of Q_1 and Q_3. When the word line is not selected, it is low and Q_{T1} and Q_{T2} are off. All capacitors shown in broken lines are parasitic. The sequence of events during a read operation can be summarized as follows:

1. The B and $\bar{B}$ lines are *precharged*, usually to $V_{DD}/2$, for better noise immunity and reduced power consumption. After precharging, the bit lines are left to float.
2. The word line for the given cell goes high, which turns on Q_{T1} and Q_{T2}.
3. Assuming $Q = 1, \bar{Q} = 0$, Q_2 is turned on but is not conducting because its drain is at V_{DD}. Current initially flows from C_Q, through Q_{T1} to the B line, thereby charging C_B and increasing the B line voltage to more than $V_{DD}/2$ (Figure 12.5.2a). As C_Q discharges, current flows through Q_2. Simultaneously, current initially flows from the $\bar{B}$ line through Q_{T2} and to $C_{\bar{Q}}$, at 0 V, discharging $C_{\bar{B}}$ and decreasing the $\bar{B}$ line voltage to less than $V_{DD}/2$ (Figure 12.5.2b). As $C_{\bar{Q}}$ charges, current flows through Q_3. The differential voltage $v_{B\bar{B}}$, of magnitude normally less than 100 mV for $V_{DD} = 5$ V, is positive if $Q = 1, \bar{Q} = 0$, and is negative if $Q = 0, \bar{Q} = 1$.
4. The sense amplifier detects the voltage change $v_{B\bar{B}}$ and amplifies it to a full logic swing, making the B line at V_{DD} and the $\bar{B}$ line at 0, if $Q = 1, \bar{Q} = 0$, and making the B line at 0 and the $\bar{B}$ line at V_{DD}, if $Q = 0, \bar{Q} = 1$. The stored bit is thus read and transmitted the output line.
5. World line goes low.

The read operation is *nondestructive*, that is, it does not change the stored bit. When a 1 is read, for example, with Q_2 and Q_3 on, and Q_1 and Q_4 off, the drop in voltage at Q_2 drain,

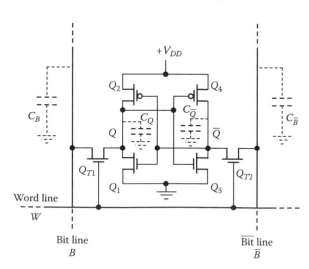

FIGURE 12.5.1
A six-transistor cell of a CMOS SRAM.

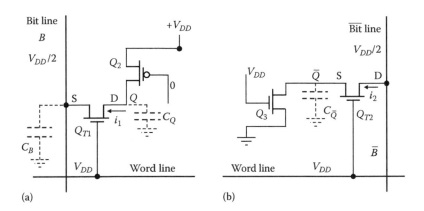

FIGURE 12.5.2
Read operation with $Q=1$ (a) and $\overline{Q}=0$ (b).

should not make Q_4 conduct an appreciable current, while the sense amplifier pulls the voltage of the B line to V_{DD}. The state of the latch will change if the input to the Q_3–Q_4 inverter falls below V_{TH}. Similarly, the rise in voltage at Q_3 drain should not make Q_1 conduct an appreciable current while the sense amplifier pulls the voltage of the $\overline{B}$ line to 0.

To illustrate a write operation assume that $Q=1$, $\overline{Q}=0$, and a 0 is to be written in the cell. The sequence of events during this write operation can be summarized as follows:

1. The B line is pulled to 0 and the $\overline{B}$ line to V_{DD}.

2. The word line goes high, turning on Q_{T1} and Q_{T2}. Both transistors will be in saturation, having $v_{DS} = v_{GS} = V_{DD}$ (Figure 12.5.3). Current initially flows from C_Q through Q_{T1} to the B line. As C_Q discharges, Q_2 conducts and the voltage Q falls toward $V_{DD}/2$, the threshold of the Q_3–Q_4 inverter. Simultaneously, current initially flows from the $\overline{B}$ line through Q_{T2} to $C_{\overline{Q}}$, which charges it and makes Q_3 conduct, and the voltage $\overline{Q}$ rises toward $V_{DD}/2$, the threshold of the Q_1–Q_2

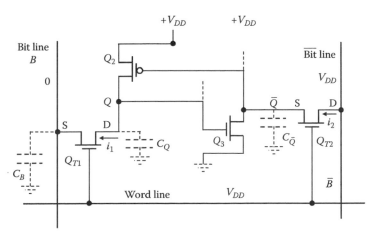

FIGURE 12.5.3
Write operation with $Q=1$ and $\overline{Q}=0$.

inverter. The source of Q_{T1} is held at 0 V by the buffer driving the B line, so there is no body effect, whereas the voltage of Q_{T2} source increases with $\overline{Q}$. Hence, Q_{T2} remains in saturation, but with its threshold increased and its current reduced. This means that Q reaches $V_{DD}/2$ before $\overline{Q}$. The Q_3–Q_4 inverter will switch first and force a change of state of the latch. A 1 is thus written into the cell.

3. The word line goes low.

The write time depends on: (1) the rise time of the voltage on the word line, (2) the time it takes to change the voltage of C_Q or $C_{\overline{Q}}$ by $V_{DD}/2$, and (3) the time it takes the inverters to switch, which is essentially t_d of the inverter. The read time depends on: (1) the rise time of the voltage on the word line, (2) the time it takes to change the voltage of C_B or $C_{\overline{B}}$ by the $v_{B\overline{B}}$ needed to activate the sense amplifier, and (3) the time taken by this amplifier to drive the respective bit lines to V_{DD} and 0. Note that $C_B \gg C_Q$ (Problem 12.4.2).

Dynamic Memory Cell

A dynamic memory (DRAM) cell consists of an E-NMOS access transistor Q_A and a capacitor C_S associated with Q_A, often in the form of a trench capacitor in the IC that adds to the parasitic capacitance of the transistor (Figure 12.5.4). Only a B line is used, with no $\overline{B}$ line. The overall result is a very high density of memory cells on the chip, so that DRAMs have the highest level of integration of all ICs and typically occupy a quarter of the area of an SRAM of the same memory capacity.

If 1 is to be written in the cell, the B line is first raised to V_{DD}, then the W line, which turns Q_A on. If C_S is not fully charged, current flows from B line, charging C_S to $V_{DD} - V_{tn}$, because Q_A stops conducting when the voltage across C_S is $V_{DD} - V_{tn}$. Note that $V_{tn} > V_{tn0}$ due to the body effect. Q_A is in saturation since its drain, connected to the B line, is at the same voltage as its gate. In order to charge C_S more rapidly, and to V_{DD}, the voltage on the W line is often boosted to at least $V_{DD} + V_{tn}$ (Problem 12.4.8).

If 0 is to be written in the cell, the B line is held at zero. When the W line is raised to V_{DD}, Q_A is turned on. If C_S is charged, it is discharged to 0, with the terminal connected to the B line acting as the source. Q_A is in the triode region, because its maximum v_{DS} is $V_{DD} - V_{tn}$, whereas its v_{GS} is V_{DD}, and the threshold is $V_{tn0} < V_{tn}$.

A complication of the DRAM is that the charge on C_S leaks, through the channel of Q_A and its source–substrate and drain–substrate junctions. The charge on C_S must therefore be refreshed every few ms or so, as specified for the particular chip. Refreshing involves rewriting the stored bits and is done by the sense amplifier, as explained later. The periodic refresh is performed one row at a time, with the memory block not available

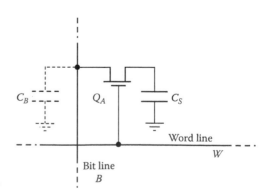

FIGURE 12.5.4

A dynamic memory cell.

for reading and writing during the periodic refresh. This is not a serious limitation, however, because the time required to refresh the entire block is typically less than about 2% of the time between refresh cycles, so that the block is available for reading or writing nearly 98% of the time. Note that when a cell is to be read or written into, and the W line is raised to V_{DD}, all the access transistors in the selected row are turned on, which connects the storage capacitors of all the cells in the selected row to their respective B lines. The resulting change in voltage on each of these bit lines is detected and amplified by the corresponding sense amplifier. The amplifier output signal, whether 0 or V_{DD}, is reapplied to the bit line, thus restoring the signal to its original level and automatically refreshing all the cells in the selected row.

If a cell is to be read, the B line is precharged to $V_{DD}/2$ and left to float. When the W line is pulled to V_{DD} and Q_A is turned on, C_B and C_S are connected together (Figure 12.5.4), with $C_B \gg C_S$. C_B is partly due to the metal-substrate capacitance of the B line but mostly due to the depletion capacitance of the drain or source of each access transistor connected to the B line. C_S is typically about 20 fF, and C_B is about 1 pF. Connecting C_S to C_B changes the voltage on the B line by ΔV_B, where from conservation of charge:

$$C_S V_{CS} + C_B \frac{V_{DD}}{2} = C_S + C_B \left(\frac{V_{DD}}{2} + \Delta V_B \right) \tag{12.5.1}$$

and V_{CS} is the initial charge on C_S. Solving for ΔV_B:

$$\Delta V_B = \frac{C_S}{C_B + C_S} \left(V_{CS} - \frac{V_{DD}}{2} \right) \cong \frac{C_S}{C_B} \left(V_{CS} - \frac{V_{DD}}{2} \right) \tag{12.5.2}$$

If 1 is to be read, and the voltage on the W line is V_{DD}, then $V_{CS} = V_{DD} - V_{tn}$, so that,

$$\Delta V_B(1) = \frac{C_S}{C_B} \left(\frac{V_{DD}}{2} - V_{tn} \right) \tag{12.5.3}$$

If 0 is to be read, $V_{CS} = 0$, so that,

$$\Delta V_B(0) = -\frac{C_S}{C_B} \left(\frac{V_{DD}}{2} \right) \tag{12.5.4}$$

If $C_S/C_B = 1/40$, $V_{DD} = 5$ V, and $V_{tn} = 1.2$ V, ΔV_B (1) $= 32.5$ mV and ΔV_B (0) $= -62.5$ mV. It is seen that ΔV_B is quite small. Moreover, because the voltage on C_S changes to $(V_{DD}/2 + \Delta V_B)$ during read, the read operation of a DRAM is *destructive*.

Sense Amplifier and Precharge Circuit

A precharge and equalization circuit is illustrated in Figure 12.5.5 for an SRAM. When the control signal CS_P goes high at the beginning of a read operation, all three transistors are turned on. Q_{P1} and Q_{P2} apply $V_{DD}/2$ to the B and $\bar{B}$ lines. Q_E equalizes the voltage between the two bit lines, this being a critical condition for correct operation of the sense amplifier.

A sense amplifier for an SRAM, based on the regenerative action of a latch, is illustrated in Figure 12.5.6. When the control signal CS_S is low, Q_5 and Q_6 are off, and the latch is isolated from V_{DD} and ground, so that no power is dissipated. But, when CS_S goes high, Q_5 and Q_6 are turned on, and the sense amplifier is powered. With Q_E in Figure 12.5.5 turned

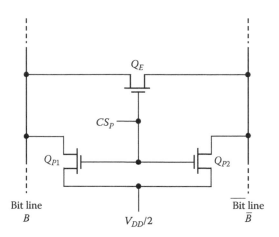

FIGURE 12.5.5
Precharge and equalization circuit.

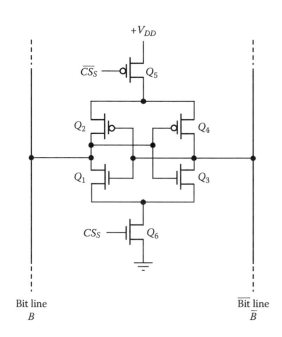

FIGURE 12.5.6
A sense amplifier for SRAM.

on, the B and $\overline{B}$ nodes of the sense amplifier are forced to be at the same voltage, so that when Q_E is turned off, the state of the latch is that of unstable equilibrium, corresponding to point Q in Figure 12.3.1. Whether the state of the latch moves to P_1 or P_2 depends on the differential signal $v_{B\overline{B}}$ applied to the bit lines by the cell being read. If $v_{B\overline{B}} > 0$, the regenerative action brings the B line to V_{DD} and the $\overline{B}$ line to zero (point P_2). Conversely, if $v_{B\overline{B}} < 0$, the regenerative action brings the $\overline{B}$ line to V_{DD} and the B line to zero (point P_1). The following should be noted:

1. The regenerative action is speeded up by the gain of the latch at Q, which is the slope of the voltage transfer characteristic at this point.
2. Since $v_{B\overline{B}}$ is differential, a common-mode input due to coupling from the word line or other parts of the circuit, is rejected as long as the components of the latch are well matched.

3. While Q_E is turned on and the latch is powered, the four transistors of the latch are in saturation, and power is dissipated (Problem P12.4.10). Although this power is small, the number of sense amplifiers that could be active at the same time can be several thousands. Hence, the time during which the amplifier is in precharge must be kept small. More complex clocked sense amplifiers are used in practice to minimize this power dissipation (Section ST12.6).

Since the DRAM cell does not have a $\overline{B}$ line, some means should be provided to have the single-ended output of the cell applied differentially to the sense amplifier. To do so, each B line is split into two halves and a dummy cell consisting of a transistor and a capacitor $C_D = C_S$ is added to each half, as illustrated in Figure 12.5.7. When a cell connected to the right half of a B line, say, is to be read, the two halves of the B line are precharged to $V_{DD}/2$ and their voltages are equalized. Simultaneously, the capacitors of the two dummy cells are precharged to $V_{DD}/2$. The word line of the given cell is selected and the control signal CP_D is made high, so that the dummy cell on the other half of the B line, that is the left half in this case, is connected to the left half of the B line. As a result, a voltage $V_{DD}/2$ appears on the input of the sense amplifier from the dummy cell connected to the left half of the B line, and a voltage $V_{DD}/2 + \Delta V_B$ appears on the input of the sense amplifier from the given cell connected to the right half of the B line. The sense amplifier detects a differential input when enabled and changes the B-line voltage, through regenerative action, to V_{DD} if 1 is read, or to 0 if 0 is read. This also serves to refresh the memory cell and counteract the "destructiveness" of the read operation.

As mentioned previously, the read operation not only refreshes the cell being read, but all the cells in the selected row are refreshed at the same time. This is because when a cell is read, and the voltage of the word line is raised to V_{DD}, all the access transistors in the selected row are turned on and connect the storage capacitors of all the cells in the selected row to their respective bit lines. As just described, the sense amplifiers connected to these bit lines would automatically make the B-line voltage go to V_{DD} if 1 is stored in the respective cell, or leave it at 0 if 0 is stored.

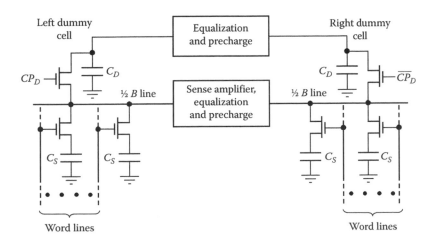

FIGURE 12.5.7
Sense amplifier connections for DRAM using dummy cells.

Simulation Example 12.5.1 Sense Amplifier

It is required to simulate the operation of the sense amplifier.

SIMULATION

The schematic is shown in Figure 12.5.8. The transistors are the same as in Example 12.2.1. The bit line capacitances are assumed to be 1 pF and are precharged to 2.5 V. The current source disturbs the voltage on one B line by applying 1 mA for 1 ps. The resulting V_o is shown in Figure 12.5.9. For the indicated polarity of the current source, the voltage V_o increases from 2.5 to 5 V. For the opposite polarity of the current source V_o decreases from 2.5 to 0 V.

Row Decoder

A row address decoder decodes an M-bit address so that only one of 2^M word lines is selected. Several decoding schemes may be used, one of which is the decoder based on NOR logic and illustrated in Figure 12.5.10 for a three-bit word address. The objective is to select the W_0 line, that is make it high, when the address word $A_2A_1A_0$ is 000, to select the W_1 line when the address word is 001, to select the W_2 line when the address word

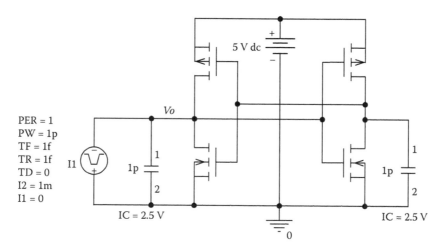

FIGURE 12.5.8
Figure for Simulation Example 12.5.1.

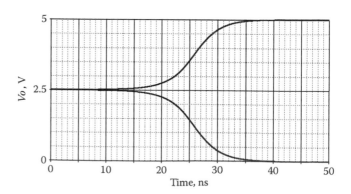

FIGURE 12.5.9
Figure for Simulation Example 12.5.1.

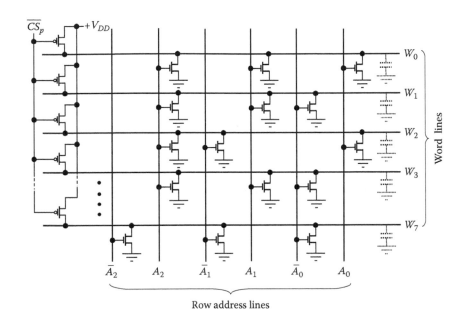

FIGURE 12.5.10
Row-address decoder based on NOR logic.

is 010, etc. The logic is $W_0 = \overline{A_0 + A_1 + A_2}$, $W_1 = \overline{\overline{A_0} + A_1 + A_2}$, $W_2 = \overline{A_0 + \overline{A_1} + A_2}$, etc., which is NOR. Basically, the three transistors connected to a given word line behave as a 3-input NOR gate whose inputs are connected to the address bits or their complements, as appropriate.

Prior to a word line selection, all the row address lines and their complements are low so that all the E-NMOS transistors in the array are off. The precharge control signal $\overline{CS_p}$ is high, which means that the E-PMOS transistors are also off. Hence, no static power is dissipated in the circuit. Prior to initiating a decoding operation $\overline{CS_p}$ goes low, which charges all the word line capacitances to V_{DD}. $\overline{CS_p}$ then goes high again. Following this, the row address bits and their complements are applied to the address lines. All the word lines will be discharged, except the one corresponding to the selected address. Thus, if the address is 001, A_2, A_1, and $\overline{A_0}$ will be low, the three transistors connected to W_1 will remain off and W_1 will be high. All the other word lines will be discharged by at least one conducting transistor and will therefore go low.

After the decoder outputs are stabilized, the word lines are connected to the word lines of the memory cells, usually through clock-controlled buffers. The following should be noted:

1. If there are M uncomplemented address lines, M E-NMOS transistors are required for each of 2^M word lines, or a total of $M2^M$ E-NMOS transistors, plus 2^M E-PMOS transistors.

2. If $\overline{CS_p}$ is permanently low, then when the address bits and their complements are applied, there will be a path to ground from V_{DD} through the conducting E-NMOS transistors, and power is dissipated. But, because the $\overline{CS_p}$ signal is temporary there is no path between V_{DD} and ground when E-NMOS transistors conduct, which significantly reduces power dissipation.

Concept: *Static power dissipation can be eliminated by using dynamic logic.*

The operation of the NOR decoder just described is an example of *dynamic* logic, which relies on charging and discharging of parasitic capacitances (Section 13.4, Chapter 13).

Exercise 12.5.1

Construct a decoder similar to that of Figure 12.5.6 but based on NAND logic.   ∎

Column Decoder

The column decoder is required to transfer data in and out of the memory cells selected. This can be accomplished by a NOR type decoder, as previously described (Figure 12.5.10), together with a multiplexer (Problem 12.2.7), as illustrated in Figure 12.5.11. Having just one I/O data line is in accordance with the assumption that each memory location stores one bit. When a row is selected, all the cells in the row are connected to their respective B lines. However, only the line selected by the column decoder, out of the 2^N output lines of the decoder, will go high. This line will therefore turn on the corresponding E-NMOS transistor of the multiplexer and connect the B line of the selected cell to the I/O data line.

The operation of the E-NMOS transistor of the multiplexer, also referred to as a pass transistor, is illustrated in Figure 12.5.12. If the gate is at V_{DD} and the input is zero (Figure 12.5.12a), the transistor is turned on and the capacitance at the output is discharged, the output terminal acting as the drain and the input terminal as the source. The drain voltage becomes zero when the output capacitance is fully discharged. If the input is at V_{DD}, the transistor is still turned on, but the input now acts as the drain and the output as the

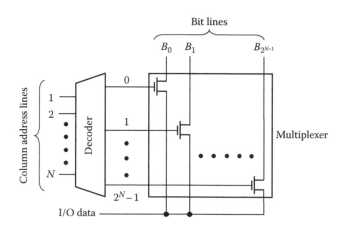

FIGURE 12.5.11
Column decoder and multiplexer.

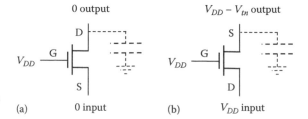

FIGURE 12.5.12
Operation of the pass transistor when the input is 0 (a) or 1 (b).

source. The output voltage is $V_{DD} - V_{tn}$, where $V_{tn} > V_{t0}$ because of the body effect (Figure 12.5.8b). If the gate is grounded, the transistor is off and the output line is in a high-impedance state. When a 1 is being written, the I/O line (corresponding to the input line in Figure 12.5.12b) is at V_{DD} and the B line goes to $V_{DD} - V_{tn}$. The regenerative sense amplifier can be used to pull the data up to V_{DD}. When a 1 is being read, the B line (now corresponding to the input line in Figure 12.5.12b) is at V_{DD}. The data buffer of the I/O line can be made to boost the $V_{DD} - V_{tn}$ output of the pass transistor to V_{DD}.

It may be noted that in large memory chips row and column addresses may be multiplexed, that is the same pins are used to pass the row and column addresses at different times in order to reduce the total number of pins required. A 1-Gb memory, for example, would require a 15-bit row address and another 15-bit column address to access the 2^{30} memory locations. But only 15 pins need be used if the row address is clocked by a signal into a 15-bit latch to hold the row address, and the same 15 pins are used to clock at a different time the column address into another 15-bit latch to hold the column address. Clearly, multiplexing is most efficient when the memory array is square, having the same number of rows and columns.

12.6 Read Only Memory

ROM is *nonvolatile* memory that is extensively used to store program instructions and data in microprocessor and dedicated processor systems as well as in many consumer applications such as computer game cartridges, personal digital assistants (PDAs), digital cameras, media players, mobile phones, and smart cards. A ROM can be considered as a combinational digital circuit whose input is the address of a given memory location and whose output is the information that is stored in that location. The memory locations are randomly accessed, as in RAM. Various types of ROM are discussed in this section.

Mask ROM

A simple diode ROM that stores eight 4-bit words is shown in Figure 12.6.1. The output bit lines B_0–B_3 are pulled down to 0 voltage through resistors. The eight locations are accessed by a 3-bit word address through a decoder, so that for each of the eight addresses one of the eight lines W_0–W_7 goes high. When a given W line goes high, the B lines go high if a diode is connected at the junctions between the W line and the four B lines. If no diode is present the B line remains low. If line W_7, for example, goes high because of a 111 address applied to the decoder, the bit lines $B_3 B_2 B_1 B_0$ will read 1001.

Two points should be noted about this ROM:

1. When a memory location is accessed, current flows to ground through conducting diodes, with consequent power dissipation.
2. The 4-bit words that are stored in the ROM are set during fabrication. However, rather than locate diodes at certain junctions of the W and B lines according to the words to be stored, it is more economical to include during manufacturing a diode at each junction. Then after the final metallization coating of the wafer, a mask is used to selectively etch away portions of the metal, thereby disconnecting some diodes from the lines so as to obtain the desired diode connections. Hence, the ROM is referred to as a **mask-programmable ROM** or a **mask ROM**.

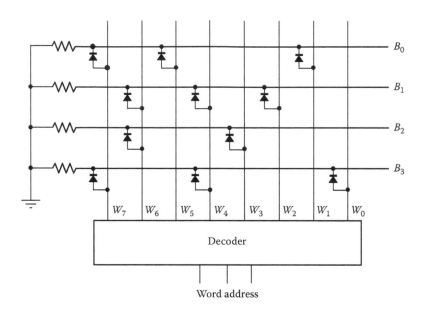

FIGURE 12.6.1
Diode ROM.

A dynamic CMOS ROM that stores eight 4-bit words is shown in Figure 12.6.2. Prior to reading, $\overline{CS_p}$ goes low, thereby charging the B line capacitances, then goes high again. When one of the W lines goes high, the E-NMOS transistors whose gates are connected to this line will conduct and discharge the corresponding B line capacitances. The remaining B lines

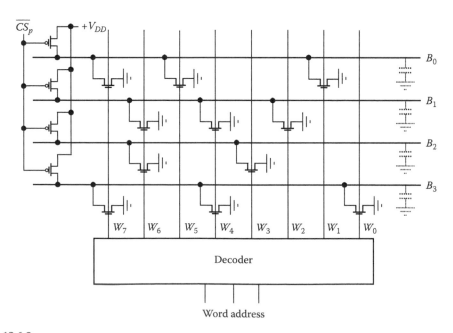

FIGURE 12.6.2
NOR ROM.

remain high as long as their capacitances retain charge. It follows that a given B line, say B_k, is low if any of the n W lines goes high and has a transistor between it and the B_k line. The logic performed is: $\overline{B_k} = W_0 + W_1 + \cdots + W_n$, or $B_k = \overline{W_1 + W_2 + \cdots + W_n}$, which is NOR logic. The ROM is referred to as a **NOR ROM**. The same words can be stored in a **NAND ROM**, as illustrated in Figure 12.6.3. In this case a given B_k is high if any of the transistors connected in series with this line is off, that is if any of the n W lines is low. The logic performed is $B_k = \overline{W_0} + \overline{W_1} + \cdots + \overline{W_n} = \overline{W_0 W_1 \ldots W_n}$, which is NAND logic. In this case, the bit line capacitances are precharged, with all W lines low. During reading, the selected W line is kept low while the remaining W lines go high. If W_0 is low, for example, the transistor at the intersection of the W_0 and B_3 lines is turned off and keeps B_3 high. The other bit lines are discharged low by the series-connected transistors that are turned on by the remaining high W lines. The advantage of the NAND ROM is that it is more compact than the NOR ROM because of elimination of the ground line through the array, the disadvantage being a reduction in the read speed because of the series connection.

Exercise 12.6.1

Compare the 4-bit words stored at the same address locations in the three ROMs of Figures 12.6.1 through 12.6.3. ∎

Programmable ROM

For small volumes, or for prototyping purposes, it is convenient to have the ROM programmed by the user. Figure 12.6.4 illustrates a cell in a BJT ROM. This is similar to the diode ROM of Figure 12.6.1 but with the BEJ replacing the diode and with a microfuse in

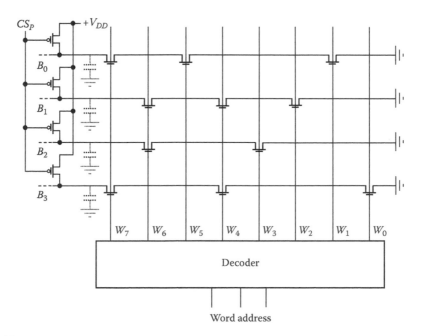

FIGURE 12.6.3
NAND ROM.

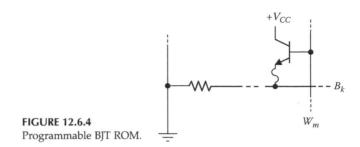

FIGURE 12.6.4
Programmable BJT ROM.

series with the emitter. Where a BJT is connected at the junction of a W line and a B line, and when the W line goes high the transistor conducts and pulls the B line high. If no BJT is present at the junction the B line is low. The ROM is manufactured with a BJT at every junction and a microfuse in series with each emitter. The microfuse could be made from polysilicon or nichrome wire. Where it is desired not to have a BJT the transistor is made to conduct heavily, thereby blowing the fuse. Compared to the diode ROM of Figure 12.6.1, the BJT ROM output is that of an emitter follower, which provides better noise immunity and dynamic performance.

Microfuses can also be used with MOS ROMs. Programming with microfuses is of course irreversible.

Erasable Programmable ROM

An erasable programmable ROM (EPROM) can be erased and reprogrammed up to about a million times. It is based on a specially designed E-NMOS transistor having a second, floating gate, the transistor symbol being as shown (Figure 12.6.5). If the floating gate is uncharged, the threshold is increased only slightly because of the reduced capacitance between the normal gate and substrate. The transistor will turn on as usual when V_{DD} is applied to the gate. If, however, the floating gate is charged negatively, the threshold is increased, because a larger positive voltage is now required to form an n-channel between the source and drain. If the threshold exceeds V_{DD}, the transistor will never turn on in normal operation and will behave as if it has been removed from the circuit. A programmed floating gate transistor (FGT) can be considered to store logic 0 or logic 1 depending on whether it pulls the bit line to ground or to V_{DD}, respectively. Whether this occurs due to a transistor being on or off depends on the type of ROM, that is, NOR or NAND, as explained in connection with Figures 12.6.2 and 12.6.3. An unprogrammed or erased FGT would of course store the complementary bit compared to a programmed FGT. The FGT is included during fabrication at the junction of every W and B line for programming by the user.

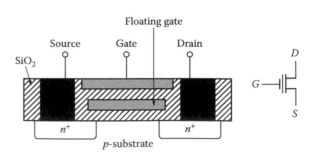

FIGURE 12.6.5
FGT used in erasable, programmable ROM.

To select a given transistor for programming a voltage of about 25 V is applied to the W line (Figure 12.6.2 or Figure 12.6.3). The gates of all the transistors connected to the same line will be at this voltage. A voltage of about 20–25 V is applied to the B line connected to the given transistor, which applies this voltage to the drain of the given transistor, the remaining B lines being at zero voltage. The selected transistor will conduct, the high drain–source voltage accelerating the electrons in the channel to a high velocity in the pinch off region near the drain. The energy of these "hot electrons" can be sufficient to overcome the energy barrier between the channel and the floating gate. Electrons that reach the floating gate get trapped in this gate, charging it negatively. The effect is self-limiting in that the accumulating negative charge on the floating gate increases the threshold voltage, which reduces the current and hence the rate of accumulation of electrons in the floating gate to practically zero. Because of the excellent insulating properties of the silicon dioxide, the trapped charge on the floating gate is retained virtually indefinitely.

To erase the FGT, ultraviolet radiation of appropriate wavelength is applied, typically 253.7 nm. The radiation imparts enough energy to the electrons trapped in the floating gate to enable them to overcome the energy barrier and return to the substrate. To allow application of the ultraviolet radiation, the EPROM package is provided with a quartz window. However, the EPROM has to be removed from the board for erasure.

Flash ROM

An EEPROM can be erased in situ. It is based on a floating gate tunnel oxide (FLOTOX) transistor, which is a modification of the FGT transistor in which the thickness of the oxide between the floating gate and the substrate is reduced to 1 nm or less. This allows programming and erasure of the transistor by electron tunneling (Section 2.1, Chapter 2) between the floating gate and the substrate, although programming can still be done by hot electron injection as in the EPROM. A **flash memory** is the most common type of EEPROM as it allows quick erasure of large blocks of stored information. It is extensively used to store and erase information in cell phones, digital cameras, media players, and as a portable memory that can be plugged to USB ports of computers. The symbol for a FLOTOX transistor is a modification of that of a FGT transistor, as indicated in Figure 12.6.6.

Figure 12.6.6a illustrates the programming of a FLOTOX transistor. With source floating and the drain, p-well, and n-substrate grounded, a voltage of 15–20 V is applied to the gate, which is sufficient to cause tunneling of electrons from the p-well to the floating gate. As this gate acquires negative charge the voltage driving tunneling is reduced, so that the operation is self-limiting, as in the EPROM. To erase the transistor, the gate is grounded, with the drain and source floating, and 20 V is applied between the p-well contact and n-substrate connected together and ground (Figure 12.6.6b). The tunneling voltage is now reversed and electrons tunnel back from the floating gate to the substrate. Again the tunnel current drops as the floating gate becomes less negative. For the erasure times normally used, the floating gate acquires a net positive charge. This makes the threshold of the FLOTOX transistor negative and turns it into a depletion device. Note that when 20 V is applied to the p-well, and if the source and drain are not isolated, the junctions between the well and the source and drain will be forward biased.

Figure 12.6.7 shows a memory cell of a four-word NAND ROM having four FLOTOX transistors connected to four word lines W_0–W_3. Two E-NMOS select transistors are used. The Select 1 transistor reduces the capacitive loading of the B line and provides better isolation between this line and the memory cell, whereas the Select 2 transistor is required in the control of programming and erasing operations, as explained later. As in

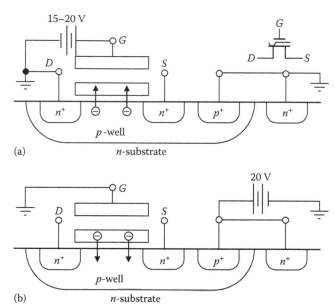

FIGURE 12.6.6
Programming (a) and erasure (b) of a
FLOTOX transistor used in flash memory.

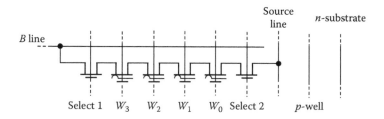

FIGURE 12.6.7
A memory cell in a four-word NAND ROM showing a B line and four FLOTOX transistors connected to four word lines W0–W3.

the NAND ROM of Figure 12.6.3, a read operation begins with precharging the B line to V_{DD}. During read, the source line, p-well line, and n-substrate line are grounded, whereas the Select 1 line and the Select 2 lines are pulled to V_{DD}, as are all the word lines, except the selected word line, say W_0, which is held low. The transistors connected to lines $W_1–W_3$ are therefore turned on. If the transistor connected to line W_0 is programmed, that is stores 1, its threshold is high and the 0 V at its gate would turn it off. The precharge of the B line is held and 1 is read on this line. If the transistor is erased, that is it stores 0, its threshold is negative, and the 0 V at its gate will turn it on. A path will therefore exist to ground through all the transistors, the B line is discharged, and 0 is read on this line.

The programming of the memory cell is illustrated in Figure 12.6.8 and follows the scheme of Figure 12.6.6a. The transistor, shown in gray, at the junction of the B_k line and the W_0 word is to be programmed. Its gate is pulled to 20 V, whereas the gates of the other FLOTOX transistors and the Select 1 transistor are at 5 V, which turns these transistors on. The B_k line is grounded, so that the drain of the given transistor is grounded through the on transistors. The gate of the Select 2 transistor is grounded so that this transistor is off and the source of the given transistor is floating. Electron tunneling occurs in the given transistor, thereby programming it. It will be noted that the gates of the other transistors at

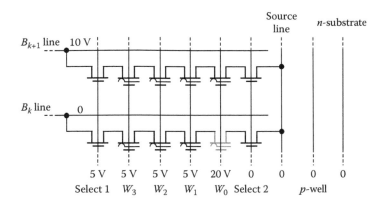

FIGURE 12.6.8
Programming of a FLOTOX cell.

the junction of the W_0 line and the other B lines will also be at 20 V. To prevent undesired tunneling in these transistors, the corresponding B line, shown as B_{k+1} line in Figure 12.6.8 is held at a high enough voltage, say 10 V, so that tunneling does not occur in this transistor, which if erased will remain erased.

Erasure is illustrated in Figure 12.6.9 and is performed as in Figure 12.6.6b. The p-well and the n-substrate are driven to 20 V, the B line is floated, and Select 1 and Select 2 transistors are turned off, which floats the sources and the drains. The gates of the transistors connected to the W lines are grounded. Since the p-well and the n-substrate are common to all the memory cells, and not just to the cell shown, the whole memory can be erased in this way. It is this bulk erasure feature that gives a flash memory its name. Note that the voltage converters required for the higher voltages are included in the chip.

Simulation Example 12.6.1 Charge Pump

It is required to analyze and simulate the voltage multiplier circuit of Figure 12.6.10, referred to as a **Dickson charge pump**, which can be used to generate arbitrarily high voltages, limited by breakdown of the capacitors or the gate oxide of the MOSFETs.

ANALYSIS

Each of the E-NMOS transistors is diode connected. At the beginning, with all capacitors discharged and the CK line at 0 V, C_1 is charged to $V_1 = V_{DD} - V_{t1}$; when $CK = V_{DD}$, $\overline{CK} = 0$,

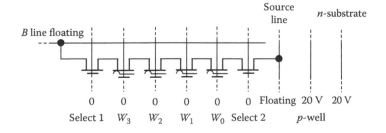

FIGURE 12.6.9
Erasure of a FLOTOX cell.

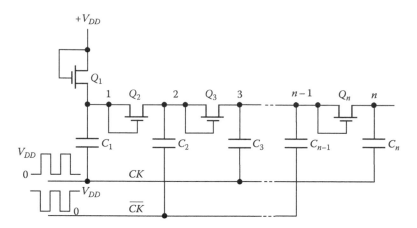

FIGURE 12.6.10
Figure for Simulation Example 10.6.1.

V_1 is boosted to $2V_{DD} - V_{t1}$ and C_2 is charged to $V_2 = 2V_{DD} - (V_{t1} + V_{t2})$; when $CK = 0$, $\overline{CK} = V_{DD}$, V_2 is boosted $3V_{DD} - (V_{t1} + V_{t2})$ and C_3 is charged to $V_3 = 3V_{DD} - (V_{t1} + V_{t2} + V_{t3})$, and so on. Eventually, when $CK = 0$, $\overline{CK} = V_{DD}$, C_n is charged to $V_n = nV_{DD} - (V_{t1} + V_{t2} + \cdots + V_{tn})$ and when $CK = V_{DD}$, $\overline{CK} = 0$, V_n becomes $(n + 1)V_{DD} - (V_{t1} + V_{t2} + \cdots + V_{tn})$. V_n therefore fluctuates by V_{DD} between these two limits. In practice, C_n can be made larger than the other capacitors so as to reduce the ripple. Note that the thresholds of the transistors may be different because of the body effect.

SIMULATION

A three-stage charge pump is simulated having $n = 3$ (Figure 12.6.11). The transistors are MbreakN from the BREAKOUT library. kp = 50u and Vto = 1 are entered in the PSpice model and L = 1u and W = 10u are entered in the Property Editor spreadsheet. The parameters of VSRC1 are: V1 = 0, V2 = 5, TD = 0, TR = 1n, TF = 1n, PW = 0.5u, and PER = 1u. VSRC2 has the same parameters except that V1 = 5 and V2 = 0. The plot of the output V3 is shown in Figure 12.6.12. Since no parameters were entered for the body effect, this effect is not involved in the simulation. The steady-state voltage of V3 is seen to fluctuate between 12 V and 17 V, in accordance with the results of the analysis for $n = 3$ and $V_{t1} = V_{t2} = V_{t3} = 1$ V. Other circuits for generating positive or negative voltages are discussed in Section ST12.7.

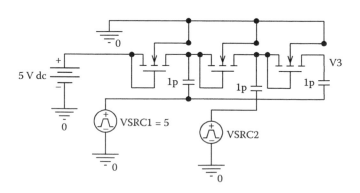

FIGURE 12.6.11
Figure for Simulation Example 10.6.1.

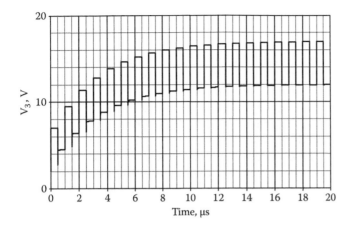

FIGURE 12.6.12
Figure for Simulation Example 10.6.1.

12.7 Ferroelectric RAM

The principal disadvantage of semiconductor flash memories is that the number of erase-program cycles is limited to about a million. Whereas such a number of erase-program cycles is adequate for many consumer applications, where programming frequency is of the order of once per day, it is not suitable for computationally intensive applications, as in digital computers. Hence, other types of nonsemiconductor, nonvolatile, and electrically erasable and programmable memories have been developed. A common example is the ferroelectric RAM (FRAM), capable of at least a billion write–erase cycles.

A ferroelectric material, made from compounds such as lead zirconium titanate, is the electric analog of a ferromagnetic material. It exhibits hysteresis and retains a residual polarization after removal of a bias voltage. Because of hysteresis, this polarization can be of either polarity. Hence, if the ferroelectric material is used as the dielectric of a capacitor, one polarity of the capacitor voltage indicates storage of 1, whereas the opposite polarity indicates storage of 0.

An FRAM cell consists of a ferroelectric capacitor and an access transistor (Figure 12.7.1). The selection of the word line turns on the access transistor for reading or writing. To store a bit, a voltage of appropriate polarity is applied to the bit line and an equal voltage of opposite polarity is applied to the plate line. To read the stored bit, the word line and bit line are driven high. A positive voltage pulse is applied to the plate line and the magnitude of the resulting current pulse indicates whether a 1 or a 0 is stored. Because the read operation is destructive, the cell has to be refreshed after the read operation.

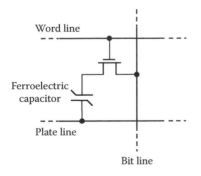

FIGURE 12.7.1
Memory cell of FRAM.

12.8 Metallic Interconnect

Aluminum, copper, and sometimes doped polysilicon, provide the necessary connections between transistors on a chip and to the chip's terminals. These are referred to as the **interconnect**. Increasing attention is being paid to the role that the interconnect plays in the overall performance of digital circuits because, as chips get larger in size and contain an increasing number of transistors of smaller dimensions and faster speeds, the effects of interconnect become more marked, particularly as it becomes necessary to run the inter-connect at different levels in the chip. These effects are due to the parasitic capacitance, resistance, and inductance of the conductors. A detailed treatment of the interconnect is beyond the scope of this chapter; instead, we will briefly outline some of the main considerations involved.

Capacitance

Capacitance of the interconnect takes three main forms:

1. Capacitance to grounded structures such as substrates, which generally are either grounded directly or are at ac ground because they are connected to a dc supply. This capacitance has two components: a parallel-plate capacitance due to the width of the interconnect and a fringing field capacitance due to the electric field at the sides of the interconnect (Section ST12.8).

2. Interlevel capacitance between conductors at different levels in the chip, the most significant being usually between adjacent levels. Chips may contain ten levels or more of the interconnect to allow all the required routing of connections.

3. Interwire capacitance between wires at the same level. As the width and spacing of the wires were decreased, the wire thickness did not decrease in the same propor-tion, so that adjacent wires came to look more like parallel-plate capacitors.

As previously discussed, capacitance to ground at circuit nodes, especially output nodes, increases propagation delay of logic gates and the dynamic power dissipation. Interwire capacitance and interlevel capacitance can increase the *effective* capacitance to ground (Problem 12.5.1) and can distort and delay signals because of **cross talk**, which is interfer-ence due to coupling of signals between neighboring lines. This coupling is not only capacitive but could be magnetic as well. Cross talk is minimized by avoiding long stretches of parallel, adjacent signal lines and by having signal lines on adjacent levels run in orthogonal directions. Parasitic capacitance of interconnect can be reduced by using **low-k dielectrics** of relative permittivity less that of S_iO_2, which is about 3.9. Doping with carbon reduces the relative permittivity of the oxide to about 2.8, and silicon-based polymers have a relative permittivity of 2.2 or less.

Resistance

Based on a bulk resistivity of 1.7×10^{-6} Ω cm for copper, 2.8×10^{-6} Ω cm for aluminum, and 10^{-2} Ωcm for n-type polysilicon, the resistance per unit length of a connection 0.5 μm high and 1 μm wide is 340 Ω/cm for copper, 560 Ω/cm for aluminum, and 2 MΩ/cm for n-type polysilicon, at low frequencies. At high frequencies, the resistance is increased by the **skin effect**. This effect can be understood in a simplified manner by considering the conductor to be made up of a large number of parallel, current-carrying filaments. If the current density were to be uniform in all the filaments, then the magnetic flux surrounding the filaments near the center of the conductor will be greater than that surrounding the filaments near the surface. The central filaments will therefore have an inductance that is larger than that of the filaments near the surface and will consequently carry less current. As a result, the current will be nonuniformly distributed across the cross section of the conductor, being more concentrated toward the surface, which reduces the effective cross-sectional area and increases the resistance. The bulk of the current flows through a **skin depth** δ given at a frequency f by (Section ST12.9):

$$\delta = \sqrt{\frac{\rho}{\pi f \mu}} \tag{12.8.1}$$

where
  ρ is the resistivity
  μ is the permeability

This means that a conductor of rectangular section of width w and thickness h has an effective area at high frequencies of $2\delta(w+h-2\delta)$ instead of wh (Problem P12.5.2). For aluminum δ is 2.7 μm at 1 GHz and 0.27 μm at 100 GHz.

The resistance of interconnect should not cause a voltage drop that exceeds a specified value when the interconnect carries its maximum current. The maximum allowed current density in an interconnect is limited by metal electromigration (Section 5.3, Chapter 5).

Distributed Models

Strictly speaking, voltage and current signals propagate along an interconnect as waves traveling at a speed $u = u_{light}/\sqrt{\mu_r \varepsilon_r}$, where μ_r and ε_r are, respectively, the relative permittivity and relative permeability of the dielectric and u_{light} is the speed of light in vacuum. If these are assumed to be unity and 3.9, respectively, for S_iO_2, then $u \cong u_{light}/2$. Moreover, R, L, and C of the interconnect are distributed along the length of the interconnect and cannot be considered as lumped at nodes or between nodes unless the wave nature of the propagation can be neglected.

The distributed nature of R, L, and C is accounted for by considering the interconnect as a transmission line. A basic property of a transmission line is its **characteristic impedance** Z_0:

$$Z_0 = \sqrt{\frac{r + j\omega l_w}{g + j\omega c}} \tag{12.8.2}$$

where
  r, l_w, and c are, respectively, the resistance, inductance, and capacitance per unit length of the conductor
  g is the conductance per unit length of the dielectric

If the transmission line is lossless, $r=g=0$, and $Z_0 = R_0 = \sqrt{l_w/c}$. In the case of IC interconnect, g is practically zero but r is generally not negligible. When the line is terminated at its receiving end by Z_0, there is no reflection of signals. Otherwise, a fraction of the signal is reflected back to the sending end, where it is partially reflected again if the impedance at the transmitting end is not equal to Z_0, and so on, until the reflected signal becomes negligibly small. In the case of pulse-type signals, reflections cause damped oscillations, or **ringing**, about the final value. Intuitively, if the reflected signal occurs during the rise time of the pulse at the transmitting end, it may only slow the rising edge of the pulse and make it less smooth. However, if the reflection arrives after the rising edge, then it will corrupt the signal level and may cause logic gates to malfunction.

Most of the interconnect on a chip is short and dense, of length less than a millimeter or so, and having minimum line widths and spacing dictated by the technology used. For these lines, the resistance can be neglected and the interconnect represented by a capacitance C_w that is part of the load capacitance C_L considered thus far.

For a longer interconnect, whose resistance is not negligible, a so-called **distributed RC model** is used, according to which the line is represented by one or more *RC* sections (Figure 12.8.1). For a uniform line of length l, represented by N identical sections, the delay is approximated as (Example 12.8.1):

$$t_{pl} = (\ln 2)rcl^2 \frac{N+1}{2N} \tag{12.8.3}$$

Note that the propagation delay t_{Pl} varies as l^2. That is why long lines, such as clock, data, and control lines are usually broken into shorter segments separated by buffer gates called **repeaters**. Equation 12.8.3 is a simple form of the **Elmore delay**. It is interpreted in Example 12.8.1 and considered more generally in Section ST12.9.

The total delay of a gate, plus interconnect, can be considered as:

$$\text{Total delay} = t_{pgate} + (\ln 2)\frac{rcl^2}{2} \tag{12.8.4}$$

where the second term on the RHS is the limit of t_{pl} (Equation 12.8.3) for large N. t_{pgate} is the gate delay due to a total capacitance at the output that includes cl. If t_{pgate} is dominant in Equation 12.8.4, the second term in Equation 12.8.4 is ignored, and the interconnect is modeled as a single lumped capacitance that is included in t_{pgate}. If not, then a distributed *RC* model or a transmission line model is used.

As a rule of thumb, a transmission line model, rather than a distributed RC model, is used if the **time of flight** t_{flight} exceeds t_{pgate}, where $t_{flight} = l/u$ is the time it takes the wave of speed u to travel the length l of the line. For a gate having t_{pgate} of 25 ps, $l \cong 3.75$ mm.

It should be noted that minimizing delay along interconnect is not the only consideration. Cross talk should be minimized as well. These are sometimes conflicting requirements. For example, using wider interconnect, which has less resistance per unit length, and driving it by low output resistance devices reduces the delay but can aggravate crosstalk due to magnetic coupling between lines.

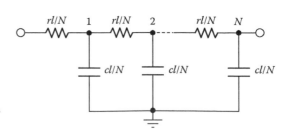

FIGURE 12.8.1
RC representation of distributed parameters of an interconnect.

Simulation Example 12.8.1 Delay of *RC* Distributed Line

It is required to interpret Equation 12.8.3 and to simulate it for a five-section and a ten-section 100 μm distributed line having a resistance of 560 Ω/cm and a capacitance of 1 pf/cm.

ANALYSIS

Consider first a single *RC* section to which a pulse of amplitude V_p is applied. The voltage across *C* is:

$$v(t) = V_p(1 - e^{-t/RC})$$

(12.8.5)

The delay t_p for this section is estimated by setting $v(t) = 0.5V_p$ and solving for t_P:

$$t_p = (\ln 2)RC \cong 0.7RC$$

(12.8.6)

If the line is represented by a single section, then Equation 12.8.6 gives the delay at terminal node 1, with $R = R_T = rl$, the total resistance of the line, and $C = C_T = cl$, the total capacitance. If the line is represented by two sections, each having a resistance $R_{T/2}$ and a capacitance $C_{T/2}$, where $R_{T/2} = R_T/2$ and $C_{T/2} = C_T/2$, the delay at terminal node 2 is considered as:

$$t_{p2} = (\ln 2)\left[R_{T/2}C_{T/2} + 2R_{T/2}C_{T/2}\right]$$

(12.8.7)

Following the same argument, if the line represented by *N* sections, the delay at terminal node *N* is:

$$t_{pN} = (\ln 2)\left[R_{T/N}C_{T/N} + 2R_{T/N}C_{T/N} + \cdots + kR_{T/N}C_{T/N} + \cdots + NR_{T/N}C_{T/N}\right]$$

(12.8.8)

where
$R_{T/N} = R_T/N$
$C_{T/N} = C_T/N$

The *k*th term on the RHS of Equation 12.8.8 is the product of the capacitance $C_{T/N}$ at node *k* and the total resistance $kR_{T/N}$ from the input to node *k*, where $k = 1, 2, \ldots, N$. Substituting for $R_{T/N}$ and $C_{T/N}$, Equation 12.8.8 becomes:

$$t_{pN} = (\ln 2)rc\frac{l^2}{N^2}(1 + 2 + \cdots + N) = (\ln 2)rcl^2\frac{N+1}{2N}$$

(12.8.9)

as in Equation 12.8.3.

For the given line, $R_T = 560 \times 10^{-4} \times 100 = 5.6 \, \Omega$ and $C_T = 1 \times 10^{-4} \times 100 \times 10^3 = 10\,\text{fF}$. It follows that the delay for a single section is $(\log 2) \times 5.6 \times 10 = 38.8$ fs (Equation 12.8.6). For a large number the sections, the delay is $(\ln 2)(R_TC_T/2) = 19.4$ fs (Equation 12.8.4), which is half as much as the delay assuming a lumped parameter representation. For $N = 5$, the delay is $(\ln 2)R_TC_T[(5 + 1)/10] = 23.3$ fs.

SIMULATION

The simulation results are shown in Figure 12.8.2. For the five-section *RC* line, the simulated propagation delay is 25.4 fs, compared to the calculated value of 23.3 fs, which is about 8% less. For the ten-section line the simulated propagation delay is 23.3 fs, compared to the calculated value of 21.3 fs, which is also about 8% less.

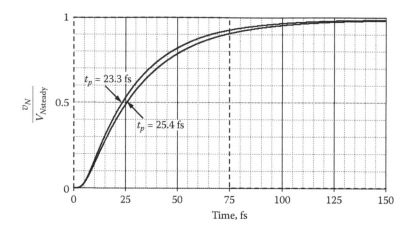

FIGURE 12.8.2
Figure for Example 12.8.1.

Summary of Main Concepts and Results

- The distinguishing characteristic of analog signals is that they assume values over a continuous range.

- The distinguishing characteristic of digital systems is that digital signals are represented in binary code and processed according to the rules of Boolean algebra.

- The major advantages of digital systems are: accuracy, reliability, versatility, and large-scale integration.

- An ideal gate has unlimited fan-in and fan-out, no propagation delay, no static or dynamic power dissipation, rail-to-rail logic swing, a step-shaped voltage transfer characteristic, maximal and equal high-level and low-level noise margins, and low impedance levels of signal lines in both the high and low states.

- Since gate power dissipation and propagation delay are, in general, inversely related, an important figure of merit for a gate is the power-delay product.

- The latch has two stable states that are maintained by positive feedback. It is a memory element in that it "remembers" its last state and keeps this state until it is forced to change it.

- The master–slave clocked JK-FF is a general type of flip-flop that could be made into an SR-FF, a T-FF, or a D-FF. Its output appears to be edge triggered.

- Flip-flops can be implemented based on a circuit representation that uses a smaller number of transistors than that suggested by the logic representation.

- Because of propagation delays, an active input must be stable for a short time before, and a short time after, the active clock transition.

- The most important attributes that determine the type of memory to be used for a particular purpose are: volatility, read/write ability, access time, power consumption, and cost per bit.

- R/W memory can be of the static or dynamic type, the latter representing the highest density of storage but requires periodic refreshing. Static power dissipation can be eliminated by using dynamic logic.

- The most versatile type of ROM is flash memory based on a special type of E-NMOS transistor using an additional floating gate that allows programming and erasure of stored information through electron tunneling between the floating gate and the substrate.

- Interconnect plays an increasingly important role in the overall performance of digital circuits as chips get larger in size and include an increasing number of transistors of smaller dimensions and faster speeds. The interconnect not only affects the signal in or between chips but also crosstalk due to capacitive and inductive coupling between conductors. The interconnect can be represented by a single capacitor, a distributed *RC* line, or a transmission line, depending on the length, resistance, capacitance, and inductance of the interconnect, and on the gate delay.

Learning Outcomes

- Articulate the types of logic gates and the properties of an ideal gate.
- Articulate the basic types of flip-flops, R/W memory, and ROM and their main features.

Supplementary Examples and Topics on Website

SE12.1 CMOS Schmitt Trigger. Analyzes the operation of a CMOS Schmitt trigger.

ST12.1 Boolean Algebra. Discusses some aspects of Boolean algebra, including axioms, theorems, duality, and properties.

ST12.2 Adiabatic Logic. Presents the basic principles of adiabatic logic.

ST12.3 Multivibrators and Schmitt trigger. Discusses the basic principles of operation of multivibrators and Schmitt trigger in terms of op amps (Sabah 2008, p. 752).

ST12.4 The 555 Timer. Describes the basic operation of the 555 timer IC and its use as a monostable or astable.

ST12.5 Edge-Triggered D-FF. Presents an alternative version of this FF.

ST12.6 Clocked sense amplifiers. Discusses various aspects of these amplifiers.

ST12.7 Voltage-Generating circuits. Discusses various types of on-chip voltage-generating circuits.

ST12.8 Fringe Capacitance and Elmore Delay. Gives the expression for the fringe capacitance of a conductor and discusses the general form of the Elmore delay.

ST12.9 Skin Depth. Derives the expression for the skin depth in a conductor.

Problems and Exercises

P12.1 Digital Signals and Boolean Algebra

P12.1.1 Given the function $f(t) = \sin(100\pi t) + \sin^3(100\pi t)$. (a) What is the Nyquist frequency? (b) Obtain samples of the function over half a period at $10/3$ times the Nyquist frequency.

(c) Scale the samples to a maximum value of 15, quantize them to the nearest integer, and express them as 4-bit binary numbers.

P12.1.2 Show that: $\overline{A}\,\overline{B}\,\overline{C}\,\overline{D} + \overline{A}\,\overline{B}\,C D + \overline{A}\,B C \overline{D} + \overline{A}\,B C D = \overline{A + B}$.

P12.1.3 Show that: $(A + B + C)(\overline{AB + AC + BC}) = A\overline{B}\,\overline{C} + \overline{A}B\overline{C} + \overline{A}\,\overline{B}C$.

P12.1.4 A room has three doors with a two-position light switch at each door. It is desired to be able to turn the light in the room on or off by changing the position of any one of the switches. If $f = 1$ denotes the light is on and $f = 0$ denotes the light is off, a value of 1 for each of the switch variables A, B, and C denotes the "up" position, and a value of zero denotes the "down" position, and assuming that the light is initially off with all switches in the down position, show that the required logic is given by: $f = ABC + A\overline{B}\,\overline{C} + \overline{A}B\overline{C} + \overline{A}\,\overline{B}C = A \oplus B \oplus C$ or by its dual form: $f = (A + B + C)(A + \overline{B} + \overline{C})(\overline{A} + B + \overline{C})(\overline{A} + \overline{B} + C)$.

P12.1.5 Show that when three bits A, B, and C are added, the sum digit S is given by: $S = A \oplus B \oplus C = ABC + A\overline{B}\,\overline{C} + \overline{A}B\overline{C} + \overline{A}\,\overline{B}C = ABC + (A + B + C)\overline{C}\,\overline{D}$, and the carry digit is given by: $CD = AB + AC + BC$; this function is also known as a **majority function** because it equals 1 when any of the two inputs are 1.

P12.2 Logic Gates

P12.2.1 Show that with negative logic (low level $\equiv 1$ and high level $\equiv 0$), AND, OR, NAND, NOR, XOR, and XNOR gates perform, respectively, OR, AND, NOR, NAND, XNOR, and XOR operations.

P12.2.2 Verify that the circuit of Figure P12.2.2 implements XOR logic.

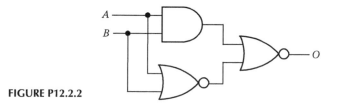

FIGURE P12.2.2

P12.2.3 Derive the output $O_3O_2O_1$ in Figure P12.2.3 when the input bits CBA increase progressively from 0 (000) to 7 (111). The output is a 3-bit **Gray code**, in which consecutive numbers differ by only one bit.

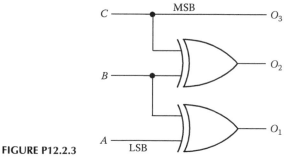

FIGURE P12.2.3

P12.2.4 Modify the circuit of Figure P12.2.3 so that the input is: (a) 4-bit binary; (b) a 3-bit Gray code, as determined in Problem P12.2.3, and the output is a binary number.

P12.2.5 Verify that the expression for f in Problem P12.1.4 can be implemented by a cascade of two XOR2 gates.

P12.2.6 Implement 3-bit addition (Problem P12.1.5) using AND and OR gates.

P12.2.7 A **multiplexer** is a logic circuit whose output equals one of several inputs depending on the binary value of a select input, S. Figure P12.2.7 shows the symbol of a 2-to-1 multiplexer and its truth table. The output equals X_1 or X_2, depending on whether S is 0 or 1, respectively. Implement a 2-to-1 multiplexer using two AND2 gates, one NOR2 gate, and an inverter.

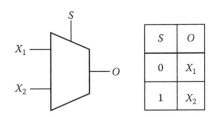

S	O
0	X_1
1	X_2

FIGURE P12.2.7

P12.2.8 Show that the circuit of Figure P12.2.8 implements the majority function $O = X_1X_2 + X_1X_3 + X_2X_3$.

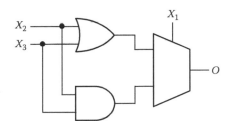

FIGURE P12.2.8

P12.2.9 Show that the circuit of Figure P12.2.9 implements the function $O = X_1 \oplus X_2 \oplus X_3$.

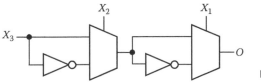

FIGURE P12.2.9

P12.2.10 Figure P12.2.10a shows a 4-to-1 multiplexer and its truth table. Show that the circuit in Figure P12.2.10a implements the function $O = X_1X_2 + X_1X_3 + X_2X_3$.

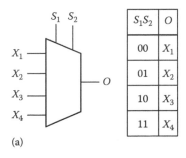

S_1S_2	O
00	X_1
01	X_2
10	X_3
11	X_4

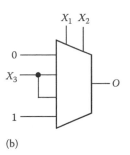

(a)   (b)

FIGURE P12.2.10

P12.2.11 Figure P12.2.11 shows a general connection of two transmission gates as a 2-to-1 multiplexer, where $O=A$ if $S=0$ and $O=B$ if $S=1$. Show how the circuit can be made to implement the following gates: (a) OR; (b) AND; (c) XOR; and (d) XNOR.

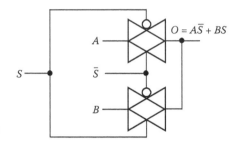

FIGURE P12.2.11

P12.2.12 Sketch the voltage transfer characteristic of a buffer and identify the input and output voltage levels corresponding to Figure 12.2.7b. If the output is connected to the input, will the circuit be stable? Explain.

P12.2.13 An inverting gate operating from a 1.5 V supply has a symmetrical voltage transfer characteristic, a logic swing of 80% of the supply voltage, with $V_{IH} - V_{IL}$ being 20% of the logic swing. Determine V_{TH}, V_{IL}, V_{IH}, V_{OL}, V_{OH} and the noise margins.

P12.2.14 An inverter has $t_p = 2$ ns. (a) Determine t_{pLH} and t_{pHL} if the charging resistance is 1.5 times the discharging resistance. (b) If adding a 1.2 pF capacitor at the output of the inverter increases t_p by 40%, determine the effective capacitance at the inverter output, assuming t_p is directly proportional to capacitance.

P12.2.15 A gate operating from a 3 V supply at 100 MHz has $t_p = 1.5$ ns and PDP $= 0.9$ pJ. The gate draws under static conditions a current of 75 μA in one state and 25 μA in the other state. Determine: (a) the current drawn from the supply; (b) the effective capacitance at the gate output; (c) the frequency, if the supply is increased to 5 V but keeping the same power dissipation.

P12.2.16 The AND2 gates in Figure P12.2.16 have a finite t_p. At $t = t_o$, A and B change from high to low and C changes from low to high. Show that a spurious pulse of width t_p is produced. Note that this "glitch" will not arise if an AND3 gate is used.

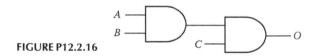

FIGURE P12.2.16

P12.3 Flip-Flops

P12.3.1 Mechanical switches bounce from their contacts when switched from one position to another. The switch rapidly opens and closes a number of times for up to about 10 ms before settling to its final position. If the switch is connected to a digital circuit, a series of high and low levels will be applied that may cause the circuit to malfunction. Figure P12.3.1 illustrates a **switch debouncing circuit** using a simple SR latch. Explain how the circuit operates.

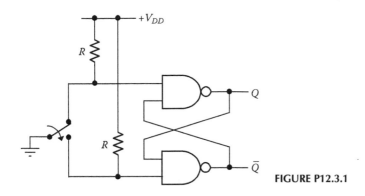

FIGURE P12.3.1

P12.3.2 A **register** is a set of FFs that are used for temporary storage of data. In a **shift register**, the stored bits can be shifted between successive FFs. Figure P12.3.2 illustrates a 3-bit shift register using positive-edge-triggered D-FFs. Assuming an initial state of 000, determine the outputs of each FF for an input sequence of 1, 0, 1, 1, 0. Note that shifting bits to the right is equivalent to multiplication by 2, whereas shifting bits to the left is equivalent to division by 2.

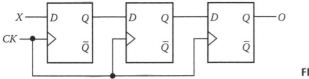

FIGURE P12.3.2

P12.3.3 Figure P12.3.3 illustrates a 3-bit, up, ripple counter using negative-edge-triggered T-FFs. Verify counter operation, starting with an initial 000 count. How would you modify the circuit to use positive-edge-triggered T-FFs?

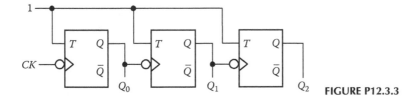

FIGURE P12.3.3

P12.3.4 Modify the connections of the FFs in Figure P12.3.3 to make the counter count downward starting with 111, and verify operation by drawing a timing diagram.

P12.3.5 Based on the connections in Problems P12.3.3 and P12.3.4, suggest a gating scheme that will make the counter count up if a control input $C=1$ and count down if $C=0$.

P12.3.6 Figure P12.3.6 illustrates an **arbiter circuit** in which O_1 goes high if X_1 goes high before X_2, and O_2 goes high if X_2 goes high before X_1. The inverters can be as in Figure 12.2.3, but with the supply voltage of the inverter is taken from one of the NAND gates as shown. This ensures that O_1 and O_2 can never be high at the same time. Verify circuit operation.

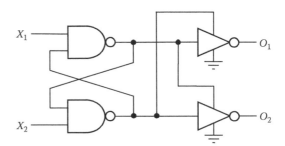

FIGURE P12.3.6

P12.3.7 Implement a type of clocked D-FF in CMOS using two inverters (Figure 12.2.3), whereby the output of one inverter is connected to the input of the other inverter. Add at each output a MOSFET having its gate connected to CK, and its remaining terminal to S or R where $R = \bar{S}$. Label the S and R inputs and describe circuit operation. Note the similarity to the static memory cell (Figure 12.5.1).

P12.3.8 Figure P12.3.8 illustrates a CMOS monostable. Sketch the time course of the voltages in the circuit following a trigger pulse, assuming the inverter threshold is $V_{DD}/2$ and neglecting the output resistance of the inverter and propagation delays. Show that the duration of the output pulse is $T = RC \ln(2)$. If $V_{DD} = 5$ V and $R = 10$ kΩ, determine C for $T = 1$ ms.

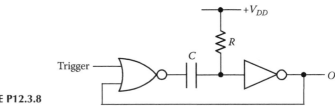

FIGURE P12.3.8

P12.3.9 (a) In the monostable of Figure P12.3.8, what is the effect of a finite inverter output resistance? (b) If the total propagation delay through the inverters is t_p, what should be the minimum width of the trigger pulse? (c) What determines the **recovery time** of the monostable, that is, the time after which a trigger pulse would produce a normal output pulse? Assume that the input of the second inverter is protected against electrostatic discharge (Figure 4.6.4, Chapter 4), draw the circuit that applies during capacitor discharge and show how to analyze it assuming the diode can be modeled by a battery V_{D0}.

P12.3.10 Figure P12.3.10 illustrates a two-inverter, CMOS astable. Assuming that the two transistors of the inverters are matched, show that the frequency of oscillation is $1/2RC \ln 3$.

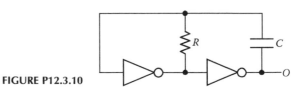

FIGURE P12.3.10

P12.3.11 A D-FF on a chip has a setup time of 3 ns and a hold time of 2 ns. If the change in the data signal and the corresponding active clock transition occur at the same time at the chip terminals, and if change in the data signal is delayed by 1 ns from the chip terminal to the D terminal of the flip-flop, what is the minimum delay of the active clock transition that will not violate the setup time of the flip-flop? What should be the minimum duration of the change in the data signal that will not violate the hold time?

P12.4 Semiconductor Memories

P12.4.1 A 1-Gb memory is divided into 16 banks, each of which is divided into 256 square arrays having the same number of word and bit lines. Determine the number address bits required: (a) within each array, (b) within each bank, and (c) for the whole chip.

P12.4.2 Consider a bit line in an array of Problem P12.4.1. Let the capacitance of the access transistor connected to the bit line be 0.3 fF for each memory cell. Assume the cells are spaced 300 nm apart along the bit line and that this line has a capacitance of 10 aF/μm. (a) Estimate the total capacitance of a bit line. (b) If the average current through the access transistor is 50 μA, estimate the time required to change the voltage on the bit line by 50 mV.

P12.4.3 A six-transistor SRAM cell has $k'_n = 50$ μA/V^2, $(W/L)_n = (5/1)$ for the access transistors, $V_{DD} = 3$ V, $V_{tn0} = 0.7$ V, $\gamma = 0.5$ V$^{1/2}$, and $2\phi_F = 0.6$ V. Determine the initial currents in the access transistors when 1 is read, assuming a step change in the word line voltage.

P12.4.4 Consider the SRAM of Problem P12.4.3 with 0 being written when $Q = 1$ (Figure 12.5.3). Determine: (a) the initial current through C_Q; (b) the current through C_Q when $V_Q = V_{DD}/2$, assuming $k'_p = 20$ μA/V^2, $(W/L)_p = (5/1)$; (c) the time it takes V_Q to become, $V_{DD}/2$, using the average of the currents calculated in (a) and (b), and assuming $C_Q = 60$ fF.

P12.4.5 A DRAM cell has $k_n = 100$ μA/V^2, $V_{DD} = 3.3$ V, $V_{t0} = 0.7$ V, $\gamma = 0.5$V$^{1/2}$, $2\phi_F = 0.6$ V, and $C_S = 20$ fF. If 1 is to be written in a cell that stores 0, (a) determine the final V_{CS}; (b) estimate the time it takes to reach V_{CS} using the average charging current. (c) If 0 is to be written in the cell after 1 is stored as in (a), estimate the time it takes to discharge C_S.

P12.4.6 If in Problem P12.4.5, $C_B = 1$ pF, determine $\Delta V_B(1)$ and $\Delta V_B(0)$ if the B line is pre-charged to: (a) $V_{DD}/2$; (b) to half the maximum V_{CS}.

P12.4.7 If in Problem P12.4.5 a 25% loss of voltage of a stored 1 can be tolerated, and if the average leakage current is 1.2 pA, what should be the maximum refresh cycle?

P12.4.8 Consider the DRAM cell of P12.4.5 with the W line voltage boosted to charge C_S to V_{DD}. (a) Determine the voltage that should be applied to the W line; (b) recalculate the charging time determined in Problem P12.4.5, part (b).

P12.4.9 A square, 1-Mb DRAM memory bank has a 4 ms refresh cycle and is to be available 98% of the time for reading and writing. What is maximum time for refreshing a row?

P12.4.10 The transistors in the latch of the sense amplifier of Figure 12.5.6 are matched and have $k_{n,p} = 100$ μA/V^2 and $V_{t0} = 0.7$ V, with $V_{DD} = 3.3$ V. Determine the current through the supply and the power dissipated during precharge, neglecting the voltage drops in Q_5 and Q_6.

P12.4.11 Extend the row decoder of Figure 12.5.10 to 512 word lines. How many NMOS and how many PMOS transistors are required, including inverters for the address bits?

P12.4.12 Compare the number of transistors required for a 512 column decoder if: (1) one multiplexer is used and each bit is separately inverted, (2) two multiplexers are used, one for the true I/O bit and another for its complement.

P12.4.13 Figure P12.4.13 illustrates a tree decoder for column decoding. How many transistors are required for decoding 512 bit lines? Note that although fewer transistors are required compared to a NOR decoder/multiplexer, the N levels involved reduce the speed of operation.

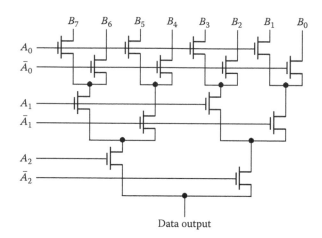

FIGURE P12.4.13

P12.4.14 The E-PMOS transistors of the dynamic ROMs of Figures 12.6.2 and 12.6.3 have $k_p' = 20$ $\mu A/V^2$, $(W/L)_p = (10/1)$, and $V_{tp} = -0.7$ V, with $V_{DD} = 3$ V. If $C_B = 1.2$ pF, estimate the precharge time using an average current that is one-half the initial E-PMOS current.

P12.5 Interconnect

P12.5.1 Consider a line L adjacent to another line L_x, the coupling capacitance between the two wires being C_x and the capacitance to ground of L being C_G (Figure P12.5.1). Determine the effective capacitance C_w to ground of L when L_x: (a) is a ground line or a voltage supply line, (b) carries the same signal as L, and (c) carries a signal that is in antiphase with that carried by L.

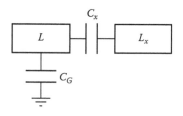

FIGURE P12.5.1

P12.5.2 Compare the resistance per unit length at 100 GHz of aluminum and copper interconnects of 0.8 μm height and 1.2 μm width. Assume that the metal has the permeability of vacuum ($4\pi \times 10^{-9}$ H/cm) and that ρ is 2.8×10^{-6} Ω cm for aluminum and 1.7×10^{-6} Ω cm for copper.

P12.5.3 Determine the capacitance to ground per unit length for a metallic interconnect 0.25 μm high and 0.3 μm wide separated from a grounded substrate by a 0.5 μm thick layer of S_iO_2. Include the fringing field capacitance (refer to Section ST12.8).

P12.5.4 Consider an aluminum interconnect 0.5 mm long of dimensions as in Problem P12.5.3. Compare the delays calculated by considering the line to be: (a) a single RC section; (b) two RC sections using Equation 12.8.7.

P12.5.5 Determine t_{flight} for a copper interconnect 0.5 cm long.

P12.5.6 Consider the interconnect of Example 13.8.1 and let half the line be of reduced width so that it has four times the resistance and 0.3 times the capacitance per unit length. Determine the delay in either direction of propagation by simulation and by calculation using the Elmore delay.

P12.6 Miscellaneous

P12.6.1 Consider a CMOS circuit that is connected to V_{DD} and ground through wires having resistances R_1 and R_2. No current flows in the wires under static conditions. But, under dynamic conditions, the circuit draws 100 μA for 10 ns. What size of decoupling capacitor should placed across the circuit so that the voltage across the circuit does not drop by more than 50 mV?

P12.6.2 Using a signal flow diagram, show that for the differential sensing amplifier (Figure 12.5.6), $\dfrac{V_o(s)}{V_i(s)} = \dfrac{-sC_BG_m}{1 - (sC_BG_m)^2}$, at small t, where $V_i(s)$ and $V_o(s)$ are, respectively, the Laplace transform of the variations in input and output voltages, and $G_m = g_{mn} + g_{mp}$ of the identical inverters. Deduce that if v_i is a negative unit impulse, the output voltage is $\dfrac{V_{DD}}{2} + \dfrac{G_m}{C_B}\cosh\left(\dfrac{G_m}{C_B}t\right)$. Note that at $t = 0$, $dv_O/dt = 0$ and that for small t, v_O is nearly exponential.

13

Digital Logic Circuit Families

The designation "digital logic circuit family" refers to digital ICs that are manufactured using the same technology and have the same circuit structure, such as the CMOS logic family. The circuits of a given logic family share common features that are characteristic of the given family, including the voltage levels that represent logic 1 and logic 0, noise immunity, propagation delay, rise and fall times of pulse signals, power dissipation, operating temperature range, availability of complex functions, and cost. The circuit modules of the same family can be readily connected together, but interconnecting modules from different families require special interface circuits because of differences in the voltages representing logic 1 and logic 0. The basic features of the important logic families at present are considered in this chapter.

Because CMOS is the dominant logic family today, it is discussed in considerable detail, together with some of its variations that are intended for special purposes, namely, pseudo NMOS, pass transistor logic, and dynamic logic. BiCMOS is considered next as a logic family that is basically CMOS based but has a bipolar junction transistor (BJT) output stage that enhances the current-drive capability. For the most part, long-channel behavior is assumed so as to provide some insight into the fundamental concepts involved. Deviations due to short-channel behavior are emphasized where appropriate, with the understanding that in view of the approximations and the empirically derived nonlinear relations that govern short-channel behavior, practically useful results for short-channel transistors are normally obtained through simulation. A popular program for this purpose is SPICE level 4, also known as BSIM (acronym for Berkeley Short-Channel IGFET Model), which has evolved through several versions.

Two BJT families are considered: transistor–transistor logic (TTL) and emitter-coupled logic (ECL). A version of TTL that is still popular to some extent is briefly discussed as an example of a weakly saturating BJT logic. ECL deserves special mention as an example of nonsaturating transistor logic and the fastest logic family in silicon at present.

Learning Objectives

❖ To be familiar with:

• The terminology used in connection with various logic families.
• The approximations made in deriving some important performance parameters of digital ICs.

❖ To understand:

• The static and dynamic behavior of the CMOS inverter, including the voltage transfer characteristic (VTC), noise margins, propagation delay, and power dissipation.
• The basic concepts underlying the design and sizing of CMOS gates.

- The effects of sizing and scaling of CMOS transistor parameters.
- The advantages and limitations of pseudo NMOS logic, pass-transistor logic, dynamic logic, and BiCMOS logic.
- The basic operation and essential features of TTL and ECL gates.

13.1 CMOS

CMOS Inverter

Static Behavior

The CMOS inverter (Figure 13.1.1) embodies most of the salient features of the CMOS logic circuit family. It was briefly considered in Section 5.5, Chapter 5, and in Section 12.2, Chapter 12. It is analyzed in detail in this section, beginning with its transfer characteristic.

When $v_I = V_{DD}$, $v_{SGp} = 0 < \overline{V}_{tp}$. Q_p is cut off, so no current flows through the transistors. $v_{GSn} = V_{DD} > V_{tn}$. Q_n is turned on, but since its drain current is zero, its quiescent point is at the origin, in the triode region of its output characteristics. It follows that $V_{OL} = v_{DSn} = 0$. The output is isolated from V_{DD} and is connected to ground through a low resistance $r_{DSn}|_{V_{DSn}=0}$ (Equation 5.2.3, Chapter 5).

When v_I is low, $v_{GSn} = 0 < V_{tn}$. Q_n is cut off, so again no current flows through the transistors. $v_{SGp} = V_{DD} > \overline{V}_{tp}$. The quiescent point of Q_p on its output characteristics is the origin, so $v_{SDp} = 0$. It follows that $V_{OH} = V_{DD} - v_{SDp} = V_{DD}$. The output is isolated from ground and connected to V_{DD} through a low resistance $r_{DSp}|_{V_{SDp}=0}$ (Equation 5.2.3, Chapter 5).

If v_I is such that the magnitude of the gate-source voltage of both transistors exceeds its threshold, both transistors conduct and current flows from V_{DD} to ground. But this happens only under *dynamic* conditions, when v_I is changing between logic 0 and 1. The VTC under static conditions can be determined using the load-curve construction of Figure 5.5.1, Chapter 5. As v_I increases from zero, Q_n remains off, and $v_O = V_{DD}$, until $v_I = V_{tn}$ (Point B, Figure 13.1.2a). A small further increase in v_I causes i_D to flow, Q_n being in saturation and Q_p in the triode region. As v_I continues to increase, v_{GSn} increases and v_{SGp} decreases. The "active" output characteristic of Q_n moves upward, while that of Q_p moves downward. v_O decreases as the operating point moves from V_{DD} to lower values of v_{DSn}. There will come a point when Q_p just enters the saturation region so that both transistors are in saturation (Point C, Figure 13.1.2a). Because the output characteristics in the saturation region are almost horizontal, a small additional increase in v_I moves the quiescent point all the way to D (Figure 13.1.2b), where Q_p is in saturation, but Q_n is just about to enter the triode region. Further increase in v_I causes Q_n to enter the triode region, and v_O eventually falls to zero (Point E, Figure 13.1.2b). The VTC (Figure 13.1.3) is simulated in the following example.

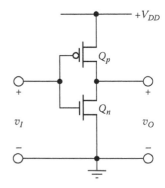

FIGURE 13.1.1
CMOS inverter.

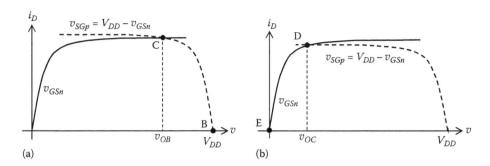

FIGURE 13.1.2
Graphical construction for deriving the VTC of the CMOS inverter. (a) Q_p at the edge of the transition region; (b) Q_n at the edge of the transition region.

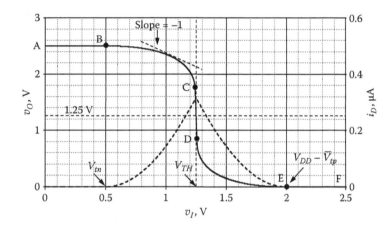

FIGURE 13.1.3
Figure for Example 13.1.1.

Simulation Example 13.1.1 Transfer Characteristic of CMOS Inverter

It is required to simulate the VTC of a CMOS inverter and the variation of i_D with v_I.

SIMULATION

The NMOS and PMOS transistors are based on 0.25 μm technology. The NMOS transistor parameters entered in the PSpice Model are Kp=0.286m, Vto=0.5, and lambda=0.05, with L=0.25u and W=0.9u entered in the Property Editor spreadsheet. The PMOS transistor parameters entered in the PSpice Model are Kp=0.117m, Vto=-0.5, and lambda=0.05, with L=0.25u and W=2.2u entered in the Property Editor spreadsheet. A VDC source connected as v_I is swept from 0 to VDD=2.5 V. The plots of the VTC and of i_D are shown in Figure 13.1.3.

The transistors are matched, because $k'_n(W/L)_n = k'_p(W/L)_p$ and $V_{tn} = \overline{V}_{tp}$, so the VTC is symmetrical about the midpoint at $v_I = v_O = 1.25$ V. i_D is symmetrical and is maximum at the midpoint. This maximum value is determined by applying Equation 5.3.1. For the NMOS transistor, with $v_{GS} = v_{DS} = 1.25$ V, $i_{D\,max} = (0.286/2)(0.9/0.25)(1.25 - 0.5)^2$ $(1 + 0.05 \times 1.25) \cong 0.31$ mA, which agrees with the value from the plot.

It is seen that there are five segments of the VTC:

Segment AB: Q_n is cut off, Q_p is in the triode region
Segment BC: Q_n is in saturation, Q_p is in the triode region
Segment CD: Both transistors are in saturation
Segment DE: Q_n is in the triode region, Q_p is in saturation
Segment EF: Q_n is in the triode region, Q_p is cut off

Basically, there are two unknowns in the inverter circuit under static conditions, namely, i_D and v_O for a given v_I. Assuming long-channel behavior, Equations 5.3.1 and 5.3.2, Chapter 5, apply to the two transistors in their respective regions. In these equations, $v_{GSn} = v_I$, $v_{SGp} = V_{DD} - v_I$, $v_{DSn} = v_O$, $v_{SDp} = V_{DD} - v_O$. Making these substitutions and setting $i_D = i_{Dn} = i_{Dp}$ gives an equation for determining v_O for a given v_I. Once v_O is known, i_D can be determined from either Equation 5.3.1 or 5.3.2, Chapter 5, as applicable.

V_{TH}, the threshold of the inverter, is derived by equating the currents of both transistors in the saturation region, setting $v_I = v_O = V_{TH}$, and solving for V_{TH}. This gives (Exercise 13.1.2):

$$V_{TH} = \frac{V_{DD} - \overline{V}_{tp} + V_{tn}\sqrt{r}}{1 + \sqrt{r}} \tag{13.1.1}$$

where $r = k_n/k_p$ and channel-length modulation is neglected.

Some aspects of the static behavior of the long-channel inverter are explored in Example 13.1.2. For the short-channel inverter, the current in the triode and saturation regions is given by Equations ST5.1.46 and ST5.1.47, respectively. In view of the approximations made and the empirically derived nonlinear relations involved, practically useful results for the short-channel inverter are normally obtained through simulation.

Exercise 13.1.1

Assuming that $k_n = k_p \cong 1$ mA/V^2, $V_{tn} = \overline{V}_{tp} = 0.5$ V, and $V_{DD} = 2.5$ V, as in Simulation Example 13.1.1, determine $r_{DSn}|_{V_{DSn}=0}$ and $r_{DSp}|_{V_{SDp}=0}$ (Equation 5.2.3).
   Answer: 500 Ω.   ∎

Exercise 13.1.2

Derive Equation 13.1.1. Note that if the two transistors are matched, $V_{TH} = V_{DD}/2$.   ∎

Example 13.1.2 Static Behavior of CMOS Inverter

It is required to determine: (a) the coordinates of points C and D on the VTC (Figure 13.1.3), and (b) the v_O–v_I relation over the segments BC and DE, neglecting channel-width modulation and assuming matched transistors.

SOLUTION

   (a) At C, Q_p is at the edge of the triode region (Figure 13.1.2a) so that $v_{SDp} = v_{SGp} - \overline{V}_{tp}$. But $v_{SDp} = V_{DD} - v_{OC}$ and $v_{SGp} = V_{DD} - v_{IC}$, where v_{OC} and v_{IC} are the output and input voltages, respectively, at point C. Substituting for v_{SDp} and v_{SGp} gives:

$$v_{OC} = v_{IC} + V_t \tag{13.1.2}$$

where $V_t = \overline{V}_{tp} = - V_{tp} = V_{tn}$. Similarly, at D, Q_n is at the edge of the triode region, so that $v_{DSn} = v_{GSn} - V_{tn}$. But $v_{DSn} = v_{OD}$ and $v_{GSn} = v_{ID}$. Substituting for v_{DSn} and v_{GSn} gives:

$$v_{OD} = v_{ID} - V_t \tag{13.1.3}$$

If $\lambda = 0$, the output characteristics have zero slope so that C and D lie on the same horizontal line in Figure 13.1.2a and b. The segment CD in Figure 13.1.3 is vertical and occurs at $v_I = V_{TH}$, with $v_{GSn} = V_{TH}$ and $v_{SGp} = V_{DD} - V_{TH}$. Since the two transistor currents are equal, and the transistors are matched, $v_{GSn} = v_{SGp}$, which gives $V_{TH} = V_{DD}/2$. When the segment CD in Figure 13.1.3 is vertical, $v_{IC} = v_{ID} = V_{TH}$. It follows from Equations 13.1.2 and 13.1.3 that $v_{OC} = V_{TH} + V_t$ and $v_{OD} = V_{TH} - V_t$.

(b) Over segment BC, $i_D = \dfrac{1}{2} k_{n,p}(v_I - V_t)^2$ for the NMOS transistor in saturation, where $k_{n,p} = k_n = k_p$. For the PMOS transistor, $i_D = k_{n,p}\left[(v_{SGp} - V_t)v_{SDp} - \dfrac{1}{2}v_{SDp}^2\right]$ in the triode region. Solving for v_{SDp} gives, $v_{SDp} = (v_{SGp} - V_t) \pm \sqrt{(v_{SGp} - V_t)^2 - 2i_D/k_{n,p}}$. Substituting $2i_D/k_{n,p} = (v_I - V_t)^2$ and $v_{SGp} = V_{DD} - v_I$ gives, $v_{SDp} = (V_{DD} - v_I - V_t) \pm \sqrt{(V_{DD} - 2V_t)((V_{DD} - 2v_I)}$. The minus sign should be retained, because when $v_I = V_t$ (point B) $v_{SDp} = 0$. It follows that $v_O = V_{DD} - v_{SDp}$ is:

$$v_O = v_1 + V_t + \sqrt{(V_{DD} - 2V_t)(V_{DD} - 2v_I)} \tag{13.1.4}$$

When $\lambda = 0$, segment CD is vertical, and $v_{IC} = V_{DD}/2$. Equation 13.1.4 gives $v_{OC} = v_I + V_t$, (Equation 13.1.2). With $v_I = 1.25$ V and $V_t = 0.5$ V, $v_{OC} = 1.75$ V.

Over segment DE, a procedure similar to the preceding gives (Problem P13.1.1):

$$v_O = v_I - V_t - \sqrt{(V_{DD} - 2V_t)(2v_I - V_{DD})} \tag{13.1.5}$$

Noise Margins

The noise margins are normally determined by assuming matched transistors and $\lambda = 0$. Equation 13.1.4 applies over segment AB (Figure 13.1.3). Differentiating, gives:

$$\frac{dv_O}{dv_I} = 1 - \frac{V_{DD} - 2V_t}{\sqrt{(V_{DD} - 2V_t)(V_{DD} - 2v_I)}} \tag{13.1.6}$$

Setting $dv_O/dv_I = -1$ and multiplying out,

$$2\sqrt{(V_{DD} - 2V_t)(V_{DD} - 2v_I)} = V_{DD} - 2V_t \tag{13.1.7}$$

Squaring both sides and canceling a $(V_{DD} - 2V_t)$ term,

$$4(V_{DD} - 2v_I) = \pm(V_{DD} - 2V_t) \tag{13.1.8}$$

Solving for v_I gives two values, the larger being V_{IH} and the smaller being V_{IL}. Thus,

$$V_{IH} = \frac{1}{8}(5V_{DD} - 2V_t) \quad \text{and} \quad V_{IL} = \frac{1}{8}(3V_{DD} + 2V_t) \tag{13.1.9}$$

A simple check on these values is that from symmetry, $V_{IH} - V_{DD}/2 = V_{DD}/2 - V_{IL}$, or $V_{IL} + V_{IH} = V_{DD}$, which agrees with Equation 13.1.9. It follows that the noise margins are:

$$\text{NM}_\text{H} = V_{OH} - V_{IH} = V_{DD} - V_{IH} = \frac{1}{8}(3V_{DD} + 2V_t) \tag{13.1.10}$$

$$\text{NM}_\text{L} = V_{IL} - V_{OL} = V_{IL} - 0 = \frac{1}{8}(3V_{DD} + 2V_t) \tag{13.1.11}$$

It is seen that $\text{NM}_\text{L} = \text{NM}_\text{H}$, as expected from symmetry.

Exercise 13.1.3

Derive Equations 13.1.9 in the same manner as Equation 13.1.5.   ■

The ideal, symmetrical VTC of an inverter (Figure 12.2.8, Chapter 12) is reproduced in Figure 13.1.4. It has a logic swing of V_{DD} and $V_{TH} = V_{IH} = V_{IL} = V_{DD}/2$, which makes both noise margins $V_{DD}/2$. The CMOS inverter has a logic swing of V_{DD} and the two transistors can be matched to make $V_{TH} = V_{DD}/2$ and $\text{NM}_\text{H} = \text{NM}_\text{L} = 0.375V_{DD} + 0.25V_t = 0.425V_{DD}$ if $V_t = 0.2V_{DD}$.

Summary: *CMOS has a rail-to-rail logic swing, and can be designed to have a threshold of $V_{DD}/2$ and near-ideal noise margins. On the downside: (1) a relatively high output resistance in the range of a few hundred ohms to a few kΩ, and (2) i_D during switching results in current pulses through the voltage supply lines.*

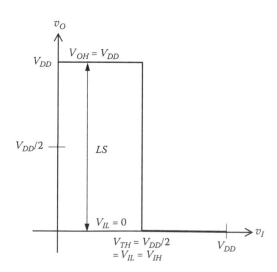

FIGURE 13.1.4
Ideal VTC of an inverter.

Propagation Delay

When input and output voltages alternate between their high and low values, the behavior of the inverter is very much affected by the load capacitance C_L, generally expressed as:

$$C_L = C_o + C_i + C_w \tag{13.1.12}$$

where
  C_o is the output capacitance of the inverter
  C_i is the input capacitance of the gates that are driven by the inverter
  C_w is the capacitance due to wiring between the inverter and these gates

C_o and C_i are expressed in terms of the transistor capacitances in Section ST13.2.

Concept: *The effect of C_L is twofold: (1) the charging and discharging times slow down the rate at which the inverter can be switched between the two states, and (2) power is dissipated in the inverter during charging and discharging.*

The power dissipation due to C_L is generally much greater than that due to the current in the two transistors during switching.

The effect of C_L on the speed of response is illustrated in Figure 13.1.5a. If v_I changes instantaneously between 0 and V_{DD}, v_O takes a finite time for a full swing (Figure 13.1.5a). Since the time for complete charging or discharging of a capacitor is not well defined, it is usual to consider the time t_{PHL} it takes v_O to change from V_{DD} to $V_{DD}/2$, and the time t_{PLH} it takes v_O to change from 0 to $V_{DD}/2$. $t_{PHL} = t_{PLH}$ if the transistors are matched and C_L is constant during charging and discharging. In general, the propagation delay, t_P, is $(t_{PHL} + t_{PLH})/2$ (Equation 12.2.4, Chapter 12). The calculation of t_{PHL} and t_{PLH} is complicated by the following factors:

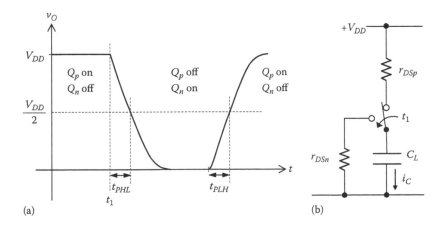

FIGURE 13.1.5
Effect of C_L on the speed of response of the CMOS inverter. (a) Variation of output with time; (b) charging and discharging of C_L.

1. In general, C_L is voltage dependent (Section 5.3, Chapter 5). Only if C_L is dominated by a large, fixed capacitance, as in the case of a long off-chip interconnection, can it be considered constant. Hence, C_L is approximated in various ways, as discussed in Section ST13.2.

2. i_C is not constant during charging and discharging. For example, if v_I increases stepwise from zero to V_{DD}, v_O is initially maintained at V_{DD} by C_L and Q_n will be in saturation, because $V_{DD} > V_{DD} - V_{tn}$. However, as C_L discharges, Q_n will move into the triode region if $v_{DS(sat)} > V_{DD}/2$. This is generally true of long-channel devices, where $v_{DS(sat)} = V_{DD} - V_{tn}$. If, for example, $V_{DD} = 5$ V and $V_{tn} = 1$ V, then $v_{DS(sat)} = 4$ V, and the transistor will go into the triode region before $V_{DD}/2$ is reached. In short-channel devices, however, $v_{DS(sat)}$ is low and the relation $v_{DS(sat)} = V_{DD} - V_{tn}$ no longer holds (Section 5.3, Chapter 5). Applying Equation 5.3.32, Chapter 5, for example, to a 0.13 μm device, considering $V_{DD} = 1.2$ V, $V_{tn} = 0.4$ V, and $\xi_c = 10^5$ V/cm, gives $v_{DS(sat)} = 0.5$ V, which is less than $V_{DD}/2$. This means that Q_n will be in saturation when $V_{DD}/2$ is reached. Even then, i_C cannot be considered constant because of the relatively large variation of i_C with v_{DS} in the saturation region of a short-channel transistor (Section 5.3, Chapter 5). It is usual to consider an average i_C during charging and discharging in both long-channel and short-channel devices.

3. A step input is assumed in calculating t_{PHL} and t_{PLH}, which implies that when the input changes suddenly from low to high, Q_p cuts off instantly and only conduction through Q_n need be considered. Similarly, when the input changes suddenly from high to low, Q_n cuts off instantly and only conduction through Q_p needs to be considered. In actual fact, the input changes with finite rise and fall times (Figure 12.2.9, Chapter 12), so that both transistors conduct during the gradual transition. An empirical approximation is sometimes used to correct t_{PHL} and t_{PLH} to take into account the finite rise and fall times, t_r and t_f, of the input. Thus,

$$t_{PHL}(\text{corrected}) = \sqrt{t_{PHL}^2 + \left(\frac{t_r}{2}\right)^2} \qquad PLH(\text{corrected}) = \sqrt{t_{PLH}^2 + \left(\frac{t_f}{2}\right)^2} \qquad (13.1.13)$$

t_{PHL} and t_{PLH} are calculated in Section ST13.3 assuming constant C_L, a step input, and i_C to be partly a saturation current and partly a current in the triode region. Various approximation methods are also discussed. In all cases, it remains true that:

$$t_{PHL} \propto \frac{C_L}{k_n V_{DD}} f(V_{tn}/V_{DD}) \qquad (13.1.14)$$

where $f(V_{tn}/V_{DD})$ is some function of the ratio V_{tn}/V_{DD}. A similar expression applies for t_{PLH} and to the PMOS transistor. It is seen that t_{PHL} is directly proportional to C_L and inversely proportional to k_n and to V_{DD}. Larger k_n and V_{DD} means a larger current and hence faster discharging of C_L; t_{PHL} also decreases with V_{tn}/V_{DD} over the normal range of values of V_{tn}, again because a smaller V_{tn} for a given V_{DD} means larger transistor currents. However, too small a V_{tn} would mean that when the NMOS transistor is cut off, v_{GS} is close to V_{tn}, so that the subthreshold current and power dissipation will increase (Equation 13.1.18); similarly for the PMOS transistor. Typically, $V_{tn} = 0.2V_{DD}$, so that Equation 13.1.14 becomes:

$$t_P \equiv \frac{1.6C_L}{k_n V_{DD}} \qquad (13.1.15)$$

Concept: *t_P sets an upper limit to the frequency f of operation, because when $1/f$ becomes comparable to $t_{PHL} + t_{PLH}$, the voltage levels are no longer 0 and V_{DD}, so operation becomes unreliable. Moreover, the maximum allowed propagation delay determines the maximum fan-out under dynamic conditions.*

Thus, if the maximum propagation delay is t_{Pmax} for a maximum load capacitance of C_{Lmax}, and if C_i is the input capacitance of a CMOS load gate, then the maximum fan-out of the given gate is the largest integer that is less than C_{Lmax}/C_i.

Exercise 13.1.4

Determine the propagation delay for a long-channel inverter having matched transistors of $k_{n,p} = 0.5$ mA/V^2 and $V_t = 1$ V, with $V_{DD} = 5$ V and $C_L = 20$ pF. Use the average value of i_C for $v_O = 5$ V and $v_O = 2.5$ V.
   Answer: 13.4 ns. ∎

Simulation Example 13.1.3 Propagation Delay of CMOS Inverter

It is required to simulate the propagation delay of a CMOS inverter.

SIMULATION

The same inverter of Simulation Example 13.1.1 is used with $C_L = 2$ pF. The input voltage is VPULSE with V1 = 0, V2 = 2.5, TD = 0, TR = 1p, TF = 1p, PW = 100n, PER = 500n. The output voltage is shown in Figure 13.1.6. The propagation delay from the simulation is 11.45 ns. Equations 13.1.15 and ST13.3.4, Section ST13.3, give 12.8 ns and 11.13 ns, respectively.

Power Dissipation

As discussed in Section 12.2, Chapter 12, the total power dissipation is, in general, the sum of P_{DC}, the static power dissipation, and P_{AC}, the dynamic power dissipation. P_{AC} has two components: (1) a component P_{SW} due to the current during switching (Figure 13.1.3), and

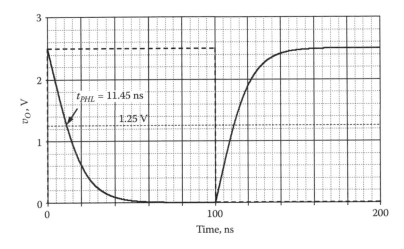

FIGURE 13.1.6
Figure for Example 13.1.3.

(2) a dominant component P_C due to the capacitance at the switching nodes (Equation 12.2.5, Chapter 12), given by:

$$P_C = f_{CK}V_{DD}\sum_{i=1}^{n}\alpha_i C_i V_{SWi} \tag{13.1.16}$$

where
α_i is an activity factor of the ith node that takes into account that in complex CMOS gates, not all the nodes switch at the clock frequency f_{CK}
V_{SWi} is the average voltage swing of the ith node, which could be less than the maximum swing V_{DD}

P_C varies nearly as V_{DD}^2. Whereas V_{DD} can be as low as 1 V, f_{CK} has been steadily increasing in the GHz range. P_C is typically a few tens of μW for an inverter operating at 500 MHz with $V_{DC}=2.5$ V and C of about 10 fF.

The component P_{SW} of P_{AC} can be calculated assuming a symmetrical inverter and a linear variation of v_I between 0 and V_{DD} over an interval τ (Figure 13.1.7). For $V_t < v_I < V_{DD}/2$, Q_n is in saturation, whereas for $V_{DD}/2 < v_I < V_{DD} - V_t$, Q_p is in saturation. Multiplying the saturation currents by V_{DD} and integrating over the corresponding time intervals gives (Problem P13.1.12):

$$P_{SW} = \frac{k\tau f_{CK}}{12}(V_{DD} - 2V_t)^3, \quad V_{DD} > 2V_t \tag{13.1.17}$$

An important feature of CMOS gates is that P_{DC} is negligible compared to P_{AC}, because under static conditions, only leakage currents flow between V_{DD} and ground. The three leakage currents, in decreasing order of importance, are: (1) the subthreshold leakage current in the nonconducting transistor, (2) the leakage current through the reverse-biased drain–substrate or source–substrate junction, as the case may be, and (3) the leakage current through the gate oxide.

The subthreshold power dissipation in a CMOS inverter can be expressed as:

$$P_{subthr} = \frac{W}{L}V_{DD}I_t e^{(v_{GS}-V_t)/\zeta V_T} \tag{13.1.18}$$

where $\zeta = 1 + C_{js}/C_{ox}$, C_{js} is the gate-channel depletion capacitance, and I_t is the drain current when $v_{GS} = V_t$, $W = L$, and $v_{DS} \gg V_T$ (Equation ST5.1.37, Section ST5.1.).

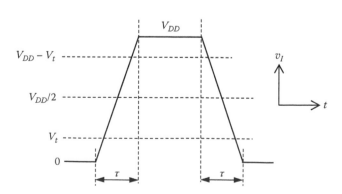

FIGURE 13.1.7
Calculation of power dissipation in an inverter assuming linear rise and fall times of input voltage.

In subthreshold operation $v_{GS} < V_t$, so that the exponent is negative. When the transistor is cut off in a CMOS gate, $v_{GS} = 0$ and P_{subthr} depends on V_t. Note that for SOI MOS transistors $\zeta \cong 1$ (Section 4.6, Chapter 4), which reduces P_{subthr}.

In the CMOS inverter, when Q_n is off, with its substrate grounded, V_{DD} appears between drain and substrate. There will be a leakage current through the reverse-biased drain–substrate junction equal to the saturation current of this junction (Equation 3.2.15, Chapter 3). Similar considerations apply when Q_p is off.

The gate oxide leakage current is negligible when the gate oxide is thicker than about 1.5 nm. As the oxide thickness is reduced, electron tunneling through the oxide can become significant and sets a limit to the minimum oxide thickness.

CMOS Gates

NAND and NOR Gates

The scheme of Figures 12.2.5 and 12.2.6, Chapter 12, can be extended to m input NAND gates and m input NOR gates as illustrated in Figures 13.1.8 and 13.1.9, respectively, where a pull-up network (PUN) of E-PMOS transistors is connected between O and V_{DD}, and a pull-down network (PDN) of E-NMOS transistors is connected between O and ground. The following should be noted concerning these gates:

1. The transistors are sized so that the gate has at least the same current-sourcing or current-sinking capability as the inverter, which means that the output resistance of the gate should not exceed that of the inverter in either state. Since L is nominally the same for all transistors of the gate, sizing means choosing a suitable W. The conductance of an MOS transistor is proportional to W, so that when m identical transistors are paralleled, W is effectively multiplied by m, and when they are connected in series W is effectively divided by m. If one input of the NAND gate is low, the PDN isolates the output from ground, and the output is pulled up high by one of the paralleled transistors in the PUN. To have at least the same current-sourcing capability as the inverter, all the PMOS transistors of the PUN should have the same size as that of an inverter. On the other hand,

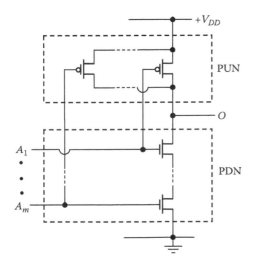

FIGURE 13.1.8
CMOS NAND gate.

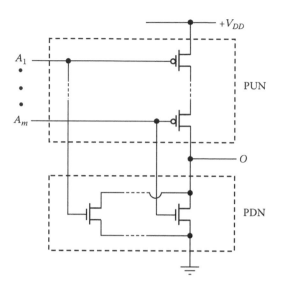

FIGURE 13.1.9
CMOS NOR gate.

when all inputs are high, the PUN isolates the output from V_{DD}, and the output is pulled low by m transistors in series. To have the same current-sinking capability as an inverter, each of the NMOS transistors in the PDN network should have its W sized to m times that of the inverter. Similarly, in the case of the NOR gate of Figure 13.1.9, each paralleled NMOS transistor of the PDN can have the same W as that of the inverter, but each series PMOS transistor of the PUN should have its W sized to m times that of the inverter.

2. With this sizing, a NAND gate occupies less area on the chip than a NOR gate of the same fan-in. This is because W of the PMOS transistors is already 2.5–3 times that of the NMOS transistors to compensate for a smaller μ_h. Multiplying W of the PMOS transistors by m results in a larger area than multiplying the smaller W of the NMOS transistors by the same factor.

3. Adding one input to a CMOS gate adds the input capacitances of two transistors, one NMOS and one PMOS. The increase in the size of transistors to maintain current-drive capability also increases the input capacitance.

3. If we consider inputs A_1 and A_2 to a 2-input gate, the PUN implements a logic function ($O = 1$) on the complemented variables $\overline{A}_1$ and $\overline{A}_2$, because the inputs A and B are essentially inverted by the PMOS transistors, as indicated by the bubble at the gates. Thus, a PUN of two PMOS transistors in parallel implements the function $O = \overline{A}_1 + \overline{A}_2$, since O goes high if either input A_1 or A_2 is low, whereas a PUN of two PMOS transistors in series implements the function $O = \overline{A}_1\overline{A}_2$, since O goes high only if both inputs A_1 and A_2 are low.

4. Similarly, the PDN implements a logic function ($\overline{O} = 1$) on the uncomplemented variables using NMOS transistors whose inputs are not inverted by the transistors. Thus, a PDN of two NMOS transistors in parallel having inputs A_1 and A_2 implements the function $\overline{O} = A_1 + A_2$, since O goes low if either of the inputs A_1 and A_2 is high, whereas, a PDN of two NMOS transistors in series implements the function $\overline{O} = A_1A_2$, since O goes low if both inputs A_1 and A_2 are high.

5. The same PUN and PDN configurations implement complementary functions on the complemented variables. Thus, if $O = \overline{A}_1 + \overline{A}_2 = \overline{A_1A_2}$ for a PUN of two PMOS

transistors in parallel, a PDN of two NMOS transistors in parallel implements the function $O = \overline{A_1 A_2} = \overline{A_1 + A_2}$, which is NOR. The complementary function is OR, and the OR of the complemented variables is $\overline{A}_1 + \overline{A}_2 = \overline{A A_2}$, which is the same NAND function implemented by the PUN.

6. Comparing the PUNs and PDNs of each of the NAND and NOR gates, it is seen that:

Concept: *In a CMOS gate the PUN and PDN are dual circuits.*

This follows from the fact that the dual of a series connection of one type of transistor is a parallel connection of the complementary type of transistor, and the dual of a parallel connection of one type of transistor is a series connection of the complementary type of transistor. Once the PUN or PDN is designed, the other network can be derived from duality (Sabah 2008, p. 403).

CMOS Gate Design

The design of CMOS gates that implement a given logic function is based on the preceding observations. For example, suppose an XOR function $O = A\overline{B} + \overline{A}B$ is required. Starting with the PUN, the term $A\overline{B} = \overline{\overline{A}\,B}$ is implemented as a series connection of two PMOS transistors having the complementary inputs $\overline{A}$ and B. Similarly, the term $\overline{A}B$ is implemented as a series connection of two PMOS transistors having the complementary inputs A and $\overline{B}$. The PUN is then the parallel combination of these two series branches (Figure 13.1.10).

The PDN can be derived as the dual circuit of the PUN just considered (Exercise 13.1.4), or by expressing the XOR function in terms of $\overline{O}$. Thus $\overline{O} = \overline{A\overline{B} + \overline{A}B} = (\overline{A\overline{B}})\,(\overline{\overline{A}B}) = (\overline{A} + B)(A + \overline{B}) = AB + \overline{A}\,\overline{B}$. The term AB is implemented as a series connection of two NMOS transistors having inputs A and B. Similarly, the term $\overline{A}\,\overline{B}$ is implemented as a series connection of two NMOS transistors having inputs $\overline{A}$ and $\overline{B}$. The PDN is then the parallel connection of these series branches (Figure 13.1.10). Note that 12 transistors are required

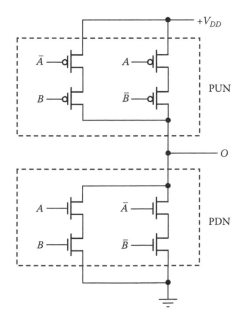

FIGURE 13.1.10
CMOS XOR gate.

---

for this CMOS XOR gate, including 4 transistors for the inverters for $\overline{A}$ and $\overline{B}$. Only eight transistors are required for transmission gate implementation of Figure 12.2.1c, Chapter 12.

Exercise 13.1.5

Derive the PDN that is the dual of the PUN of Figure 13.1.10.
   Answer: The circuit of Figure 13.1.11.   ∎

Design Example 13.1.4 Design of CMOS Gate

It is required to design a CMOS gate that implements the function $O = A(BC + D)$.

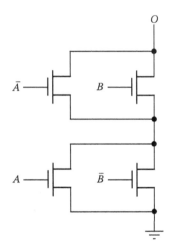

FIGURE 13.1.11
Figure for Exercise 13.1.5.

DESIGN

Suppose we start with the PDN in terms of $\overline{O}$. Thus $\overline{O} = \overline{A(BC + D)} = \overline{A} + \overline{BC + D} = \overline{A} + \overline{(BC)}\overline{D} = \overline{A} + (\overline{B} + \overline{C})\overline{D}$. All the inputs of the PDN transistors will have to be inverted. It is better, therefore, to implement the function $O' = \overline{O}$ and use a single inverter to obtain O. The expression $O' = \overline{A} + (\overline{B} + \overline{C})\overline{D}$ is then that of a PUN having PMOS transistors whose inputs are the complemented variables, without inversion (Figure 13.1.12a).

   The PDN follows from $\overline{O'} = A(BC + D)$. It can also be derived as the dual of the PUN. Q_{p1} is a PMOS transistor in parallel with a subcircuit of other transistors. Its dual is an NMOS transistor in series with a subcircuit of other transistors (Figure 13.1.12b). In the PUN, the subcircuit is Q_{p4} in series with a parallel combination of Q_{p2} and Q_{p3}. The dual subcircuit is Q_{n4} in parallel with a series combination of Q_{n2} and Q_{n3}. The same inputs are applied to transistors of the same number in the PUN and PDN.

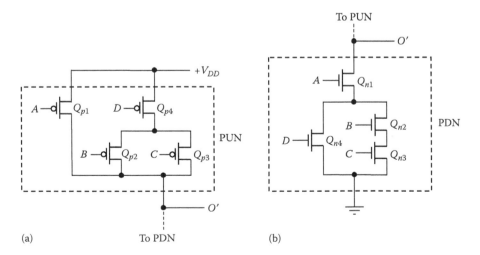

FIGURE 13.1.12
Figure for Example 13.1.4. (a) PUN; (b) PDN.

The sizing of the transistors to have the same current-drive capability as an inverter is based on worst cases. Since Q_{p1} can conduct alone, its W should be the same as that of the PMOS transistor of the inverter. Q_{p4} can conduct in series with Q_{p2} or Q_{p3}. Hence, W of each of these transistors should be twice that of the PMOS transistor of the inverter. In the PDN, the worst case is when Q_{n1}, Q_{n2}, and Q_{n3} conduct in series. Hence, W of each of these transistors must be three times that of NMOS of the inverter. W of Q_{n4} can be one-half that of Q_{n2} or Q_{n3}.

Exercise 13.1.6

Verify that the gate circuit of Figure 13.1.12 followed by an inverter implements the function $O = A(BC + D)$. ∎

Effects of Sizing and Scaling

It is instructive to consider the effects on some CMOS transistor parameters of: (1) sizing, that is, multiplying W by λ, (2) scaling by κ, that is, dividing all dimensions by κ and dividing V_{DD} by the same factor, and (3) scaling by κ while keeping V_{DD} constant. The results are shown in Table 13.1.1. The following should be noted:

1. Only a very small fraction of the gates on the chip have off-chip loads that can be considered to have a fixed C_L that is not affected by sizing or scaling. Almost all the gates on the chip have a C_L that is given by Equation 13.1.12. If C_w is negligible, and if the gate under consideration and the gates connected to it are sized or scaled by the same factor, C_L is affected as C_i in Table 13.1.1, in which the overlap capacitance C_{ol} (Equation 5.3.21, Chapter 5) is neglected.

2. Sizing multiplies $k_{n,p}$ and C_L by λ; t_P is not affected, and P_{AC} of the sized gate is multiplied by λ. The effect on the packing density and the power density, or the power per unit area, depends on the number of transistors being sized.

TABLE 13.1.1

Effect of Sizing and Scaling on CMOS Transistor Parameters

Quantity	Relation	Sizing	Scaling	Scaling (Same V_{DD})
			Multiplying Factor	
L	—	1	$1/\kappa$	$1/\kappa$
W	—	λ	$1/\kappa$	$1/\kappa$
Area	WL	λ	$1/\kappa^2$	$1/\kappa^2$
t_{ox}	—	1	$1/\kappa$	$1/\kappa$
C_{ox}^*	ε_{ox}/t_{ox}	1	κ	κ
$k_{n,p}$	$(\mu\varepsilon_{ox}/t_{ox})(W/L)$	λ	κ	κ
C_i	WLC_{ox}^*	λ	$1/\kappa$	$1/\kappa$
V_{DD}, V_t	—	1	$1/\kappa$	1
t_P	$\propto C_L/(k_{n,p}V_{DD})$	1	$1/\kappa$	$1/\kappa^2$
P_{AC}	$fC_LV_{DD}^2$	λ	$1/\kappa^3$	$1/\kappa$
Packing density	$\propto 1/(WL)$	—	κ^2	κ^2
Power density	$\propto P_{AC}/(WL)$	—	$1/\kappa$	κ
N_A, N_P	$d\xi/dx = -\rho/\varepsilon$	—	κ	κ^2

3. Three highly important parameters are t_P, the packing density, and the power density. Scaling by κ, including V_{DD}, multiplies the packing density by κ^2 while dividing t_P and the power density by κ, which is highly advantageous.

4. Scaling by κ at constant V_{DD} also multiplies the packing density by κ^2 but is more advantageous in that t_p is divided by κ^2 because of larger currents. However, power density is multiplied by κ, which may necessitate special cooling measures such as heat sinks and fans.

5. Scaling dimensions and V_{DD} by the same factor keeps the electric field nearly constant. From Poisson's equation in the form $d\xi/dx = \rho/\varepsilon$, if ξ is kept constant while x is reduced by κ, then the charge density ρ, and hence doping concentrations, must be increased by κ. If V_{DD} is maintained constant, ξ increases by κ, so that ρ must be increased by κ^2 to maintain the same relation.

Example 13.1.5 CMOS Buffer

It is required to analyze the behavior of a CMOS buffer (Figure 13.1.13) consisting of an inverter driving another inverter whose transistors have five times W, with $C_L = 50$ pF.

Solution

Q_{p1} and Q_{p2} parameters are based on the 0.25 μm technology having $L = 0.25$ μm, $W_n = 0.9$ μm, $W_p = 2.2$ μm, $t_{ox} = 7$ nm, and $V_t = \pm 0.5$ V, with $V_{DD} = 1.8$ V.

$$K'_{p1} = \frac{\mu_p \varepsilon_{ox}}{t_{ox}} = \frac{230 \times 3.9 \times 8.85 \times 10^{-14}}{7 \times 10^{-7}} \equiv 0.113 \text{ mA/V}^2;$$

$$k_p == k'_p \left(\frac{2.2}{0.25}\right) = 0.998 \text{ mA/V}^2.$$

$$K'_{n1} = \frac{\mu_n \varepsilon_{ox}}{t_{ox}} = \frac{580 \times 3.9 \times 8.85 \times 10^{-14}}{7 \times 10^{-7}} \equiv 0.286 \text{ mA/V}^2; \quad k_n == k'_n \left(\frac{0.9}{0.25}\right) = 1.03 \text{ mA/V}^2.$$

Hence, we can assume that $k_{n,p} = 1$ mA/V^2 for both input transistors. Neglecting the overlap capacitance C_{ol}, the input capacitance of the second inverter is (Equation ST13.2.2) $C_{i2} = LC^*_{ox}(W_{n2} + W_{p2}) = 0.25 \times 10^{-4} \times 3.9 \times 8.85 \times 10^{-14}(11 + 4.5) \times 10^{-4}/(7 \times 10^{-7}) = 19.1$ fF.

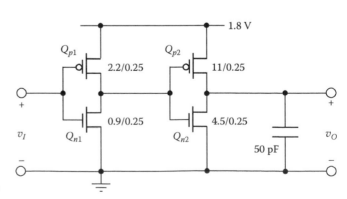

FIGURE 13.1.13
Figure for Example 13.1.5.

C_{o1} of the first inverter is mostly due to the drain–substrate junction capacitance of both transistors (Section ST13.2). Although these can be calculated as explained in Section 5.3, Chapter 5, we will calculate C_{i1} of the first inverter and consider $C_{o1} = C_{i1}$. This is an over-estimate of C_{o1}, since C_{o1} is typically 0.5–0.7 times C_{i1}. In any case, the propagation delay is completely dominated by C_L, as will be demonstrated shortly. By analogy with C_{i2}, $C_{o1} = C_{i1} = LC_{ox}^*(W_{n1} + W_{p1}) = 0.25 \times 10^{-4} \times 3.9 \times 8.85 \times 10^{-14}(2.2 + 0.9) \times 10^{-4}/ (7 \times 10^{-7}) = 3.82$ fF. Neglecting the wiring capacitance between the two inverters, $C_{L1} = 19.1 + 3.82 = 22.9$ fF.

To calculate the propagation delay, we check $v_{DS(sat)}$ for the short-channel transistor at $V_{DD} = 1.8$ V and $\xi_c L = 10^5 \times 0.25 \times 10^{-4} = 2.5$ V. From Equation 5.3.32, Chapter 5, $v_{DS(sat)} = 2.5(1.8 - 0.5)/(1.8 - 0.5 + 2.5) = 0.86$ V. Since this is less than $V_{DD}/2$, we can assume that the output transistor remains in saturation during charging and discharging. $i_D = 0.5 (1.8 - 0.5)^2 = 0.85$ mA. Assuming i_D remains constant, $t_{P1} = 22.9 \times 10^{-15} \times 0.9/ (0.85 \times 10^{-3}) = 24.2$ ps.

For the second inverter, C_{o2} is negligible compared to C_L, $k_{n,p} = 5$ mA/V^2, and $i_D = 2.5(1.8 - 0.5)^2 = 4.23$ mA. Hence, $t_{P2} = 50 \times 10^{-12} \times 0.9/(4.23 \times 10^{-3}) = 10.64$ ns. The total propagation delay $t_{P1} + t_{P2}$ remains at 10.6 ns, to three significant figures. If the sizing of the output stage is 10 instead of 5, t_{P2} is halved. The penalty paid is a larger area on the chip.

Low-Power CMOS

Low-power CMOS is becoming increasingly important for reducing battery drain in portable devices, and for ameliorating problems associated with power dissipation and cooling in large ICs. Basically, the most effective way to reduce power in CMOS circuits is scaling, including V_{DD}, since this reduces P_{AC} by $1/\kappa^3$ (Table 13.1.1). Reducing V_{DD} further, without scaling transistor dimensions, reduces power dissipation but increases t_d. SOI technology is particularly effective in addressing these problems because of the following considerations: (1) Drain–substrate capacitance is substantially reduced, which reduces P_{AC} by at least 20%. (2) In fully depleted SOI MOSFETs (Figure 4.6.5, Chapter 4), a depletion layer exists all the way to the isolating oxide layer. The depletion capacitance C_{jS} is practically zero, which makes ζ in Equation 13.1.18 nearly 1. Since ζ appears in the exponential, the subthreshold power dissipation is typically reduced by an order of magnitude. The reduction in subthreshold leakage allows the reduction of V_{tn}, which improves the dynamic performance, as noted in connection with Equation 13.1.14. Thresholds of 0.1 V, with $V_{DD} = 0.4$ V, are possible with SOI CMOS.

Summary

CMOS has almost ideal characteristics for a logic circuit family. It has the maximum possible logic swing (rail-to-rail), relatively large and equal noise margins, and practically zero static power dissipation. It can have equal t_{PHL} and t_{PLH}, and with continued decrease in transistor sizes, CMOS can have on-chip gate delays in the ps range. It is relatively easy to fabricate, which reduces cost. Its main disadvantages are: (1) two transistors are needed for each additional gate input, so that high fan-in gates would require considerable additional chip area, with consequent increase in total capacitance and attendant increase in propagation delay and dynamic power dissipation, (2) limited current-drive capability, because of relatively large output resistance, and (3) increased susceptibility to noise because of this relatively large output resistance and the current pulses that occur during switching.

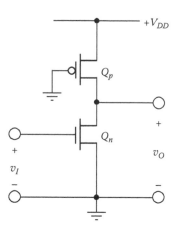

FIGURE 13.2.1
Pseudo NMOS inverter.

13.2 Pseudo NMOS

Pseudo NMOS and pass-transistor logic (Section 13.3) are essentially a form of CMOS that is not intended so much to replace conventional CMOS but rather to supplement it for special purposes by overcoming some of the disadvantages of CMOS noted previously.

The pseudo NMOS inverter (Figure 13.2.1) differs from the CMOS inverter in that the gate of the E-PMOS transistor is grounded instead of being connected to the input. It is called pseudo NMOS to distinguish it from NMOS logic that was popular in the 1970s and 1980s. The earlier form of NMOS had a diode-connected E-NMOS load transistor, whereas the later form had a diode-connected D-NMOS load transistor. Depletion MOSFETs may require an additional ion-implantation processing step, which increases fabrication cost.

Since $v_{SGP} = V_{DD}$, Q_p is always on. When v_I is low, Q_n is off, no current flows, and $v_O = V_{DD}$, as in the CMOS inverter. When v_I is high, Q_n is on, v_O is low, but a conduction path exists between V_{DD} and ground. The resulting static power dissipation is a disadvantage of pseudo NMOS. To see a principal advantage, consider Figure 13.2.2. The output O is low if any of the inputs $A_1 \ldots A_m$ is high, which implements a NOR function. Only $m + 1$ transistors are required, without sizing, compared to $2m$ transistors in CMOS, which also requires sizing of m series-connected transistors. The packing density of pseudo NMOS can be four times that of CMOS.

Pseudo NMOS gates are implemented by designing an appropriate PDN. If the $A_1 \ldots A_m$ transistors of Figure 13.2.2 are connected in series, a NAND results, which, however, requires sizing of the PDN transistors to have the same current-sinking capability as an inverter. Figure 13.2.3 illustrates an XOR2 gate that requires only five transistors. If B is low and A is high, Q_1 conducts, with Q_2 off, and Y is pulled low. If B is high and A is low, Q_2 conducts, with Q_1 off, and Y is again pulled low. If A and B are both high or low, Q_1 and Q_2 are off. The logic performed is XNOR: $\overline{Y} = A\overline{B} + \overline{A}B$. Adding an inverter gives an XOR2 gate.

Since the D-PMOS transistor is not activated by gate inputs, it is considered a passive load.

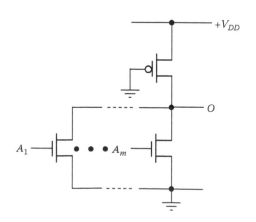

FIGURE 13.2.2
Pseudo NMOS NOR gate.

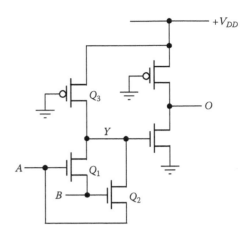

FIGURE 13.2.3
Pseudo NMOS XOR2 gate.

Concept: *An active PDN and a passive load allow wired-AND logic, whereas an active PUN and a passive load allow wired-OR logic.*

Wired logic is useful in that it can save transistors. Figure 13.2.4 shows two NAND2 gates sharing a common passive load. O will be low if both A_1 and A_2 are high, or if both B_1 and B_2 are high, that is $\overline{O} = A_1A_2 + B_1B_2$, or $O = \overline{A_1A_2 + B_1B_2} = \overline{A_1A_2}\,\overline{B_1B_2}$. If each gate has its own passive load, the gate outputs are $O_1 = \overline{A_1A_2}$ and $O_2 = \overline{B_1B_2}$. Hence, $O = O_1O_2$, the AND of O_1 and O_2. Using wired logic has saved on using three NAND2 gates, for O_1, O_2, and their AND combination, followed by an inverter, a total saving of six transistors.

With an active PUN, as in CMOS, wired-AND logic is not feasible, because of the possibility that both the PUN and PDN may conduct for some combination of inputs, which results in a short between V_{DD} and ground. The ECL family having active PUN and passive PDN allows wired-OR logic (Section 13.7).

Exercise 13.2.1

Figure 13.2.5 illustrates a NOR4 gate. Show that it can be regarded as two NOR2 gates that are wire-ANDed together. ∎

Static Operation

The VTC characteristic of the pseudo NMOS inverter can be derived in the same manner as that of the CMOS inverter by drawing the $v_{SGp} = V_{DD}$ characteristic of the PMOS transistor as a load curve on the output characteristics of the NMOS transistor (Figure 13.2.6). An important feature of these curves is that the saturation current of Q_p is made considerably smaller than that of Q_n by having $k_n = rk_p$, where r is typically between 4 and 10. This reduces the current when both transistors are on and keeps V_{OL} low. When $v_I < V_{tn}$, Q_n is off and $v_O = V_{DD}$. As v_I increases beyond V_{tn}, Q_n conducts in the saturation region, and Q_p in the triode region. For larger v_I, there is a small part of the

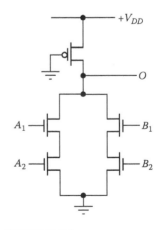

FIGURE 13.2.4
Pseudo NMOS wired-AND logic.

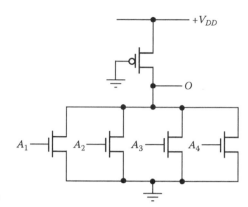

FIGURE 13.2.5
Figure for Exercise 13.2.1.

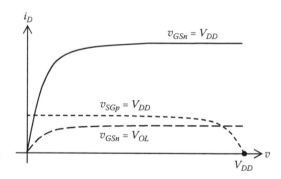

FIGURE 13.2.6
Graphical construction for deriving the VTC of the
pseudo NMOS inverter.

VTC when both transistors are in saturation. For larger v_I still, Q_n is in the triode region
and Q_p in saturation.

The VTC of the pseudo NMOS inverter is derived in Section ST13.4 based on long-
channel behavior. The extension to short-channel behavior is also discussed. The static
current I_{stat} of the long-channel transistor, when $v_I = V_{DD}$, is the saturation current of Q_p:

$$I_{stat} = i_{D\,max} = \frac{1}{2}k_p(V_{DD} - V_t)^2 \tag{13.2.1}$$

where $V_t = V_{tn} = \overline{V}_{tp}$. It is shown in Section ST13.4 that:

$$V_{OL} = (V_{DD} - V_t)\left(1 - \sqrt{1 - 1/r}\right) \tag{13.2.2}$$

where $r = k_n/k_p$ is the ratio of device transconductance parameters,

$$V_{TH} = V_t + \frac{V_{DD} - V_t}{\sqrt{r + 1}} \tag{13.2.3}$$

$$V_{IL} = V_t + \frac{V_{DD} - V_t}{\sqrt{r(r + 1)}} \tag{13.2.4}$$

$$V_{IH} = V_t + \frac{2(V_{DD} - V_t)}{\sqrt{3r}} \tag{13.2.5}$$

$$\text{NM}_L = V_{IL} - V_{OL} = V_t - (V_{DD} - V_t)\left(1 - \frac{1}{\sqrt{r(r+1)}} - \sqrt{1 - \frac{1}{r}}\right) \tag{13.2.6}$$

$$\text{NM}_H = V_{DD} - V_{IH} = (V_{DD} - V_t)\left(1 - \frac{2}{\sqrt{3r}}\right) \tag{13.2.7}$$

Because of the dependence of V_{OL} on r, pseudo NMOS is said to be a **ratioed** logic family. In CMOS, on the other hand, V_{OL} does not depend on the ratio of k_n/k_p, so CMOS is **ratioless**. Since V_{DD}, k'_n, k'_p, and V_t are governed by process technology, r and (W/L) are the main design parameters. A large r reduces V_{OL}, increases the noise margins, and for a given $(W/L)_n$, reduces I_{stat} and the static power dissipation. However, too large an r increases the asymmetry in the dynamic response between t_{PLH} and t_{PHL} (Equations 13.2.9 and 13.2.10) and increases the gate area for a given $(W/L)_p$. Larger gate area means larger C_i and C_o.

For short-channel transistors,

$$V_{OL} = \frac{I_{Dp(sat)}}{k_n(V_{DD} - V_{tn})} \tag{13.2.8}$$

where $I_{Dp(sat)}$ is the saturation current of the PMOS transistor (Equation ST13.4.13).

Exercise 13.2.2

Show that if the two transistors are matched, $V_{OL} = V_{DD} - V_t$. ∎

Dynamic Operation

The dynamic operation of the pseudo NMOS inverter can be analyzed following the same procedure as for the CMOS inverter. If $V_t = 0.2V_{DD}$, it can be shown that (Problem 13.2.1):

$$t_{PLH} \cong \frac{1.7C_L}{k_p V_{DD}} \tag{13.2.9}$$

$$t_{PHL} \cong \frac{1.7C_L}{k_n(1 - 0.46/r)V_{DD}} \tag{13.2.10}$$

$t_{pLH} = t_{PHL}$ for $r = 1.46$. If r is large, then $t_{PLH} \cong r t_{PHL}$, so that a large r increases the asymmetry between t_{PLH} and t_{PHL}.

Simulation Example 13.2.1 Static and Dynamic Operation of Pseudo NMOS Inverter

It is required to simulate a pseudo NMOS inverter and derive its VTC and the variation of i_D with v_I.

SIMULATION

The transistor parameters are $k'_n = 120u$, $k'_p = 40u$, $(W/L)_n = (0.375 \ \mu m/0.25 \ \mu m)$, and $(W/L)_p = (0.125 \ \mu m/0.25 \ \mu m)$, which makes $r = 9$. $\lambda = 0.05 \ V^{-1}$ and $V_{DD} = 2.5$ V. The VTC and the current are shown in Figure 13.2.7, from which $I_{Dstat} = 44.7 \ \mu A$, $V_{OL} = 0.128$ V, $V_{TH} = 1.14$ V, $V_{IL} = 0.69$ V, $V_{IH} = 1.29$ V, $\text{NM}_H = 1.21$ V, $\text{NM}_L = 0.56$ V.

The dynamic behavior of the inverter is simulated as in Simulation Example 13.1.2 using $C_L = 2$ pF. The simulated values of t_{PHL} and t_{PLH} are indicated in Figure 13.2.8.

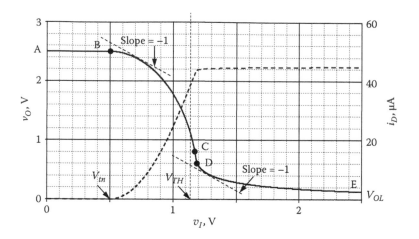

FIGURE 13.2.7
Figure for Example 13.2.1.

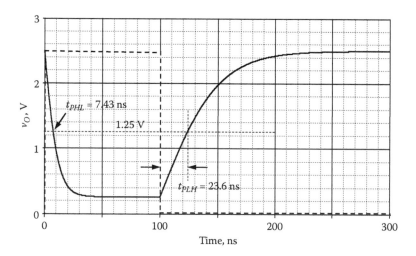

FIGURE 13.2.8
Figure for Example 13.2.1.

13.3 Pass-Transistor Logic

Pass-transistor logic, or **transmission-gate logic**, uses single, pass transistors or CMOS transmission gates for logic operations. A pass transistor was encountered in the column decoder (Section 12.5, Chapter 12) and its operation explained in connection with Figure 12.5.8, Chapter 12. Transmission-gate logic was illustrated by the XOR gate (Figure 12.2.1c, Chapter 12) and the D-FF (Figures 12.3.12 and 12.3.14, Chapter 12). We will consider in this section the operation of these logic elements in more detail, particularly under dynamic conditions.

Figure 13.3.1 illustrates a pass transistor Q_{PT} whose gate is connected to V_{DD} and whose input is suddenly changed from 0 to V_{DD}, v_O being initially zero. The following should be noted:

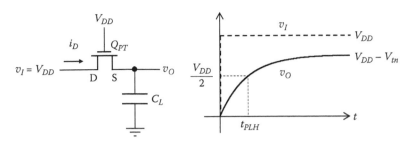

FIGURE 13.3.1
Time variation of output of an NMOS pass transistor when the input changes from 0 to V_{DD}.

1. When v_I changes to V_{DD}, with C_L initially discharged, the transistor source and drain will be as shown, Q_{PT} is initially in saturation and maximum i_D flows. As v_O increases, i_D decreases partly because of reduced v_{GS} and v_{DS} and partly because of increased V_{tn} due to the body effect (Section 5.3, Chapter 5). Hence, v_O increases at a slower rate than if there were no body effect, which increases t_{PLH}.

2. When $i_D = 0$, C_L is charged to a maximum voltage of $V_{DD} - V_{tn}$, with $v_{GS} = V_{tn} = v_{DS}$. Q_{PT} remains in saturation throughout the charging of C since $(v_{DS} = V_{DD} - v_O) > (v_{GS} = V_{DD} - v_O) - V_{tn}$, where V_{tn} is larger than V_{tn0} because of the body effect. The reduced output voltage of $V_{DD} - V_{tn}$ can cause conduction in a CMOS gate connected to the output node resulting in static power dissipation. Thus, if an inverter is connected to the output node, an output of $V_{DD} - V_{tn}$, with $V_{tn} > \overline{V}_{tp}$ of the PMOS transistor of the inverter, will make the inverter operate in the region DE of its VTC (Figure 13.1.3) resulting in current flow through the inverter. It would seem possible to reduce V_{tn} of Q_{PT} to zero and therefore have the voltage of the output node equal to V_{DD}. However, this would make Q_{PT} sensitive to noise and would increase its subthreshold current (Equation 13.1.18). This problem can be avoided in a number of ways, as will be discussed later.

3. If both v_I and the gate voltage go low, which can occur if the gate is driven by a clock signal, two things happen if there is no conducting path between the output node and ground: (a) Q_{PT} is turned off and the output node will go into a high-impedance state, which may or may not be desirable, depending on the circuit, and (b) v_O will be initially maintained by C_L at $V_{DD} - V_{tn}$. It will not follow v_I and will go low only after the charge on C_L has leaked. The current that discharges C_L when both v_I and the gate voltage go low, is due to: (a) leakage between what is now the drain terminal of Q_{PT} and the substrate, and (b) the subthreshold current of Q_{PT} between the new drain and the source. Noise coupled from neighboring interconnect (Section 12.8, Chapter 12) can also contribute to discharging C_L. Radiation generates electron–hole pairs in the silicon substrate, with electrons flowing to the drain and holes to the substrate terminal, which constitutes a current that aids in discharging C_L.

The situation when v_I goes low, with the gate maintained at V_{DD}, and $v_O = V_{DD} - V_{tn}$, is illustrated in Figure 13.3.2. Since what is now the source is at 0 V, there is no body effect. Initially, $v_{DS} = V_{DD} - V_{tn}$, where V_{tn} is the threshold at that end of the charging process and is larger than V_{tn0}. Because $V_{DD} - V_{tn} < V_{DD} - V_{tn0}$, Q_{PT} will initially be in the triode region and will remain in this region as v_{DS} decreases while C_L discharges to 0.

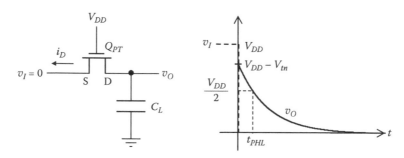

FIGURE 13.3.2
Time variation of output of an NMOS pass transistor when the input changes from V_{DD} to 0.

The output voltage can be boosted to V_{DD} by making $v_I = V_{DD} + V_{tn}$, as is sometimes done in semiconductor memories (Section ST12.5, Chapter 12). Alternatively, positive feedback can be used, as in Figure 13.3.3. If $v_I = 0$, $v_C = 0$ with C_L discharged, and $v_O = V_{DD}$, which turns off Q_F. If v_I is high, then as v_C increases, v_O falls, and Q_F conducts and contributes to the charging of C_L. Eventually, with v_O low, and the current through C_L is zero, $v_C = V_{DD}$. Circuit behavior is analyzed in more detail in Example SE13.2.

t_{PLH} and t_{PHL} can be readily calculated using the approximation based on the average current (Exercise 13.3.1).

The case of a PMOS transistor is similar (Figure 13.3.4). When v_I changes from low to high the transistor is turned on with $v_{SG} = V_{DD}$. The capacitor charges to V_{DD} because

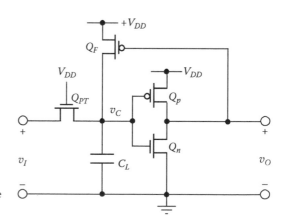

FIGURE 13.3.3
Boosting of capacitor voltage to V_{DD} using positive feedback.

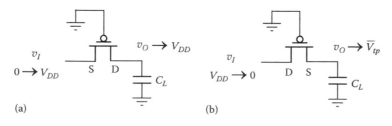

FIGURE 13.3.4
Output of a PMOS pass transistor when input changes from (a) 0 to V_{DD}; (b) V_{DD} to 0.

$v_{SG} > \overline{V}_{tp}$. When v_I returns to zero, the output terminal is the source. As v_O decreases, the transistor cuts off when $v_{SG} = \overline{V}_{tp}$.

Based on Figures 13.3.1, 13.3.2, and 13.3.4, it is said that an NMOS transistor passes a logic 0 well but not a logic 1, whereas a PMOS transistor passes a logic 1 well but not a logic 0. It follows that a transmission gate, which combines the two types of transistor, will pass both logic levels equally well. In fact, the performance of transmission gates as switches is far superior to that of a pass transistor. As discussed in Section 5.5, Chapter 5, the on-resistance remains practically constant for a wide range of input voltage, and C_L charges to V_{DD} and not to a lower voltage. The dynamic operation of the transmission gate is considered in Example 13.3.1.

A versatile pass-transistor gate is analyzed in Example SE13.2.

Example 13.3.1 Dynamic Operation of Transmission Gate

It is required to analyze and simulate the operation of the transmission gate under dynamic conditions, assuming $k'_n = 0.286$ mA/V², $k'_p = 0.117$ mA/V², $(W/L)_n = (W/L)_p = (0.75 \ \mu m/0.25 \ \mu m)$, $V_{tn0} = \overline{V}_{tp0} = 0.5$ V, $\gamma = 0.5$ V$^{1/2}$, and $\phi_f = 0.3$ V, with $V_{DD} = 2.5$ V and $C_L = 1$ pF.

ANALYSIS

At $t = 0$, v_I changes from 0 to V_{DD}, with v_O initially 0 (Figure 13.3.5). Consider Q_p for $t > 0$. v_{SG} is constant at V_{DD}, and the substrate is at V_{DD}, which means that there is no body effect. v_{SD} is initially at V_{DD}, so Q_p is in saturation, and $i_{Dp}(0) = \frac{1}{2}(0.117 \times 3)(2.5 - 0.5)^2 = 0.702$ mA. As v_O increases, v_{SD} decreases and the transistor goes into the triode region. When $v_O = 1.25$ V, $i_{Dp}(t_{PLH}) = (0.117 \times 3)\left[(2.5 - 0.5)1.25 - \frac{1}{2}(1.25)^2\right] = 0.603$ mA. Q_p stops conducting when $v_{SD} = 0$, that is, when C is charged to V_{DD}, because $v_{SG} = V_{DD} > \overline{V}_{tp}$.

Consider Q_n. Initially, $v_{DS} = V_{DD} = v_{GS}$, and $V_{BS} = 0$. $i_{Dn}(0) = i_{Dn}(0) = \frac{1}{2}(0.286 \times 3)(2.5 - 0.5)^2 = 1.716$ mA. As v_O increases, v_{DS} and v_{GS} decrease by the same amount

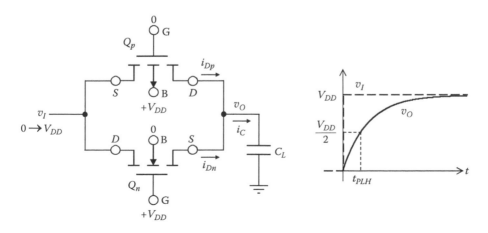

FIGURE 13.3.5
Figure for Example 13.3.1.

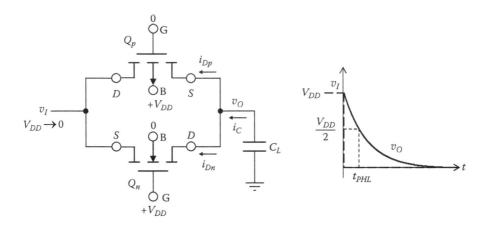

FIGURE 13.3.6
Figure for Example 13.3.1.

while the threshold increases because of the body effect. Q_n remains in saturation and stops conducting when $v_{GS} = V_{tn}$, that is when $v_O = V_{DD} - V_{tn}$. When $v_O = 1.25$ V, $V_{SB} = 1.25$ V and $V_{tn} = 0.5 + 0.5(\sqrt{1.25 + 0.6} - \sqrt{0.6}) = 0.793$ V. Hence, $i_{Dn}(t_{PLH}) = \frac{1}{2}(0.286 \times 3)(1.25 - 0.793)^2 = 0.090$ mA. This gives $i_C(0) = 1.716 + 0.702 = 2.418$ mA, $i_C(t_{PLH}) = 0.603 + 0.09 = 0.693$ mA and $i_{Cavg} = \frac{1}{2}(2.418 + 0.693) = 1.556$ mA. $t_{PLH} = C(V_{DD}/2)/i_{Cavg} = 10^{-12} \times 1.25/(1.556 \times 10^{-3}) = 0.803$ ns.

When v_I changes from V_{DD} to 0, with v_O initially at V_{DD} (Figure 13.3.6), then for $t > 0$, v_{GS} of Q_n remains at V_{DD}, with $V_{SB} = 0$ and V_{DS} initially at V_{DD}. Q_n is initially in saturation but goes into the triode region as v_O and hence v_{DS} decrease. Thus $i_{Dn}(0) = \frac{1}{2}(0.286 \times 3)$ $(2.5 - 0.5)^2 = 1.716$ mA, and $i_{Dn}(t_{PHL}) = (0.286 \times 3)\left[(2.5 - 0.5)1.25 - \frac{1}{2}(1.25)^2\right] = 1.475$ mA.

Q_p is initially in saturation, since $v_{SD} = V_{DD} = v_{SG}$, with $V_{BS} = 0$. Thus, $i_{Dp}(0) = \frac{1}{2}(0.117 \times 3)(2.5 - 0.5)^2 = 0.702$ mA. As v_O decreases, both $v_{SD} = V_{DD} = v_{SG}$ decrease by the same amount while the threshold increases because of the body effect. Q_p remains in saturation. When $v_O = 1.25$ V, $V_{BS} = 1.25$ V, and $\overline{V}_{tp} = 0.5 + 0.5(\sqrt{1.25 + 0.6} - \sqrt{0.6}) = 0.793$ V. Hence, $i_{Dp}(t_{PHL}) = \frac{1}{2}(0.117 \times 3)(1.25 - 0.793)^2 = 0.0367$ mA. This gives $i_C(0) = 0.702 + 1.716 = 2.418$ mA, $i_C(t_{PHL}) = 1.475 + 0.0367 = 1.512$ mA, and $i_{Cavg} = \frac{1}{2}(2.418 + 1.512) = 1.965$ mA. $t_{PHL} = C(V_{DD}/2)/i_{Cavg} = 10^{-12} \times 1.25/(1.965 \times 10^{-3}) = 0.636$ ns. The propagation delay $t_P = \frac{1}{2}(0.803 + 0.636) = 0.72$ ns.

SIMULATION

The schematic for the low-to-high change in the input is shown in Figure 13.3.7. The NMOS transistor parameters entered in the PSpice model are kp = 0.286 m, Vto = 0.5, gamma = 0.5, phi = 0.6, whereas those of the PMOS transistor are kp = 0.117 m, Vto = −0.5, gamma = 0.5, phi = 0.6. L = 0.25u and W = 0.75u are entered for both transistors in the

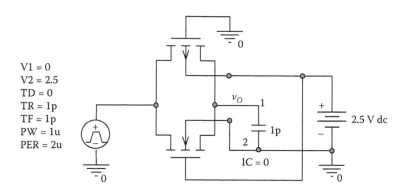

FIGURE 13.3.7
Figure for Example 13.3.1.

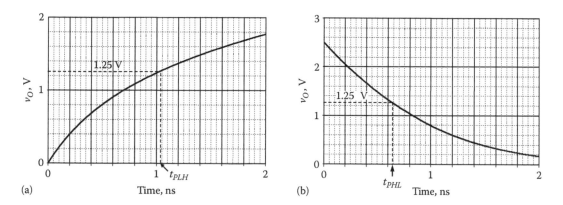

FIGURE 13.3.8
Figure for Example 13.3.1.

Property Editor spreadsheet. The schematic for the high-to-low change in the input is the same except that IC of C is 2.5 V, and V1 and V2 of the source are 2.5 V and 0, respectively. The simulation results are shown in Figure 13.3.8, from which it is seen that $t_{PLH} = 1.015$ ns and $t_{PHL} = 0.648$ ns.

Exercise 13.3.1

Consider a pass transistor having $k_n = 100 \ \mu A/V^2$, $V_{tn0} = 1$ V, with $V_{DD} = 5$ V, $C_L = 30$ fF, $\gamma = 0.5$ $V^{1/2}$ and $\phi_F = 0.3$ V. Determine (a) t_{PLH}, (b) t_{PHL} assuming that v_O is initially equal to V_{DD}. In both cases consider i_D through C_L to be the average of $i_D(0)$ and $i_D(t_{PLH})$ or $i_D(t_{PHL})$, as appropriate, so that the average i_D flowing for a duration t_{PLH} or t_{PHL} changes v_C by $V_{DD}/2$.
   Answers: (a) 0.18 ns; (b) 0.1 ns. ∎

Exercise 13.3.2

Determine v_O at which Q_p cuts off in Figure 13.3.5.
   Answer: 0.861 V ∎

13.4 Dynamic Logic

In dynamic logic, logic 1 is represented as a voltage on what is essentially a parasitic capacitance, which reduces the number of transistors required to implement a given logic function and increases the packing density. In contrast to pseudo NMOS, the static power dissipation is zero. Dynamic logic was encountered in the dynamic decoder and dynamic ROMs discussed in Sections 12.5 and 12.6, Chapter 12.

Basic Configuration

The basic configuration of a dynamic logic gate is illustrated in Figure 13.4.1. When CK is low, the NMOS evaluate transistor Q_{EV} is turned off, whereas the PMOS precharge transistor Q_{PR} is turned on, charging the capacitor C_L at the output node to V_{DD}. When CK goes high, Q_{PR} is turned off and Q_{EV} is turned on. The logic function implemented by the PDN appears at the output O during this phase. For example, if the PDN consists of a single NMOS transistor (Figure 13.4.2a), then the logic implemented is inversion, $\overline{O} = A$. If A is low, Q_A is off and O remains high during the evaluate phase when CK goes high. If A is high, then Q_A is on and C_L is discharged when Q_{EV} conducts during the evaluate phase,

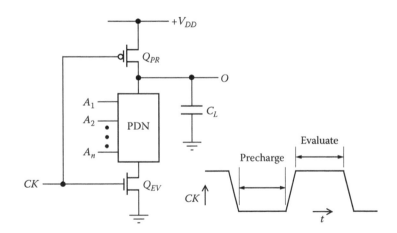

FIGURE 13.4.1
Basic dynamic logic gate.

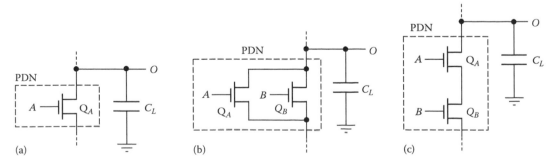

FIGURE 13.4.2
PDNs for dynamic logic gates. (a) Inverter; (b) NOR2 gate; (c) NAND2 gate.

pulling O low. For the PDN of Figure 13.4.2b, $\overline{O} = A + B$, that is, C_L is discharged if A or B is high, so that Q_A, Q_B, or both conduct. The logic implemented is NOR. Similarly, the logic function implemented in Figure 13.4.2c is NAND, $\overline{O} = AB$, because C_L is discharged only if both A and B are high, so that both Q_A and Q_B conduct. The following should be noted concerning dynamic logic:

1. There is no static power dissipation, since Q_{PR} and Q_{EV} are not turned on simultaneously. There is, however, dynamic power dissipation due to charging and discharging of capacitors.
2. The output is available only during the evaluate phase, the precharge phase being "dead time."

Exercise 13.4.1
Verify that the PDN of Figure 13.1.12b, when used in the circuit of Figure 13.4.1, implements the logic function $O = \overline{A(BC + D)}$. ∎

Limitations of Dynamic Logic
Consider the NAND2 gate of Figure 13.4.3. With Q_{PR} and Q_A off, the charge on C_L leaks mainly through the reverse-biased drain-substrate junctions of these transistors. The leakage currents may be of the order of pA, and approximately double for every 10°C rise in temperature. The charge on C_L must be replenished at a minimum frequency so that the output voltage, if high, will not fall below a certain level. Another problem is that if Q_A turns on, while Q_{PR} and Q_B are off, the charge on C_L will be distributed between C_L and C_X, the capacitance appearing at the junction between Q_A and Q_B, causing the voltage at O to fall below V_{DD}. This *charge sharing* can be of critical importance in the design of dynamic logic circuits. It can be eliminated by precharging internal nodes at the expense of added circuit complexity and node capacitances.

In the case of the inverter (Figure 13.4.2a), Q_A begins to conduct during the evaluate phase when $v_A = V_{tn}$, and C_L begins to discharge. Hence, we may consider both V_{IL} and V_{IH} to be approximately equal to V_{tn} (see Figure 13.1.4), which gives,

$$\text{NM}_\text{L} \cong V_{tn} \quad \text{and} \quad \text{NM}_\text{H} \cong V_{DD} - V_{tn} \tag{13.4.1}$$

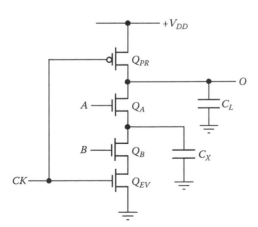

FIGURE 13.4.3
Charge sharing in a dynamic logic gate.

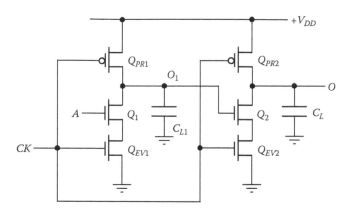

FIGURE 13.4.4
Cascaded dynamic logic inverters.

This makes NM_L low and NM_H high, although a high output may be reduced because of leakage and charge sharing. Moreover, a high output is necessarily a high impedance node except during precharging and discharging, which makes it susceptible to noise.

A serious limitation of dynamic logic circuits is that they cannot be readily cascaded. To appreciate the reason, consider two cascaded inverters (Figure 13.4.4). Assume that A is high, so that O should also be high. At the end of the precharge phase, C_{L1} and C_L will be charged to V_{DD}. Ideally, C_{L1} discharges instantly at the start of the evaluate phase, so that O_1 is pulled low, Q_2 is turned off, and O remains high. But C_{L1} takes time to discharge. As long as the voltage of O_1 is higher than V_{tn} of Q_2 plus the small voltage drop across the conducting Q_{EV2}, Q_2 is on and C_L will discharge. Consequently, the output voltage may no longer be interpreted as high. To allow cascading of dynamic circuits, various schemes may be used, as described next.

Domino Logic

In one form of domino logic, also known as **NP Domino** logic or **NORA CMOS** logic, dynamic logic gates are alternated between those using PDNs and PUNs, with an inverted CK applied to the gates having PUNs. Figure 13.4.5 illustrates two inverters cascaded in this manner. When CK is low, $\overline{CK}$ is high, C_{L1} is precharged to V_{DD}, whereas C_L is "precharged" by Q_{PR2} by discharging it to 0. During the evaluate phase, CK is high and

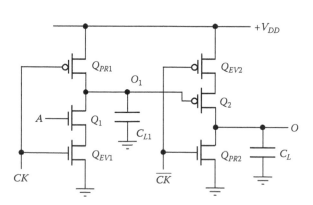

FIGURE 13.4.5
Cascaded inverters in NP domino logic.

$\overline{CK}$ is low. If A is high, C_{L1} discharges, pulling O_1 low. As the voltage of O_1 falls, Q_2 is turned on, and with Q_{EV2} on, C_L charges to V_{DD}, making O high. If A is low, C_{L1} remains high, Q_2 is off, and O remains low.

In the general case, cascaded stages have alternating PDNs and PUNs that implement a desired logic function. During the evaluate phase, the output of each stage is applied to the next stage. In effect, evaluated outputs ripple through the cascade, like a chain of falling dominos.

Other implementations of domino logic are possible (see Problem P13.2.17).

Pipelined Single-Phase Clock Architecture

An interesting variation of domino logic is the **pipelined single-phase clock architecture** of Figure 13.4.6. The "single-phase clock" refers to using only CK, without $\overline{CK}$. This avoids clock timing problems (Section 12.3, Chapter 12) and therefore allows a higher clock frequency. In contrast to domino logic, in which the output of one stage ripples through the other stages during the evaluate phase, the precharge phase and the evaluate phase alternate between successive stages in a pipelined architecture so that consecutive sets of data are processed in alternating stages of the pipelined system during the two phases of the clock cycle.

In Figure 13.4.6, each of the n-stage and the p-stage consists of a precharge-evaluate circuit followed by what is effectively a tristate inverter that maintains the output voltage at its last value, subject to charge leakage. When CK is low, output O_1' is precharged, while O_1 is maintained at its previous value by the cutoff transistors Q_1 and Q_1' at the output. At the same time, O_1 is applied to the p-stage, which is in the evaluate phase. The evaluate output O_2' appears inverted at O_2. When CK goes high, O_2' is precharged, O_2 is maintained, while O_1' is evaluated and appears inverted at O_1. Problem P13.2.19 illustrates a D-FF based on this architecture.

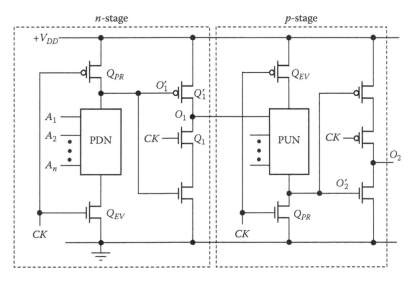

FIGURE 13.4.6
Illustration of pipelined single-phase clock architecture for dynamic logic.

13.5 BiCMOS Logic

Basic Operation

BiCMOS is an important technology that seeks to combine the advantages of CMOS and BJTs. CMOS has advanced to the stage where propagation delays between gates on a chip are in ps. However, when it comes to driving appreciable loads, of capacitance greater than about 1 pF, for example, CMOS is slow because of its limited current-drive capability. Such appreciable loads may be encountered off-chip, that is, at the chip's outputs, or on-chip by gates having a large fan-out. BJTs, on the other hand, can provide the necessary currents to charge and discharge capacitors at high speeds. This is where BiCMOS can be advantageous, although a BJT typically occupies 10 times the area of a MOSFET.

The basic BiCMOS inverter, illustrated in Figure 13.5.1, differs from the CMOS inverter (Figure 13.1.1) by having two *npn* transistors that amplify the currents of Q_p and Q_n. The transistor combinations Q_p–Q_1 and Q_n–Q_2 are reminiscent of the Darlington pair (Section 8.4, Chapter 8). That the circuit behaves as an inverter can be readily verified. If v_I is low, Q_n and Q_2 are turned off, Q_p conducts, supplying base current to Q_1, which in turn conducts and supplies a relatively large current that rapidly charges C_L. When C_L is fully charged, the base current of Q_1 is practically zero, so that v_{SD} of Q_p is negligibly small, v_{BE1} equals the cut-in voltage of the BEJ of Q_1, and $v_O = V_{DD} - v_{BE1\gamma}$. If v_I goes high, Q_p and Q_1 are turned off, while Q_n and Q_2 are turned on, discharging C_L. When C_L is fully discharged, the base current of Q_2 is practically zero, so that v_{DS} of Q_n is negligibly small, and $v_O = v_{BE2\gamma}$ where $v_{BE2\gamma}$ is the cut-in voltage of the BEJ of Q_2. The following should be noted concerning the circuit of Figure 13.5.1:

1. Q_1 and Q_2 do not conduct simultaneously under static conditions. However, a current from V_{DD} to ground flows through them under dynamic conditions, as in a CMOS inverter.

2. Q_1 and Q_2 are not normally driven to saturation, since their BCJs cannot become forward biased because of Q_p and Q_n that are connected across these junctions. However, Q_1 and Q_2 may saturate if their collector resistances (r_c) are high and

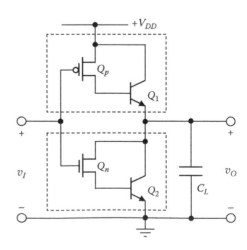

FIGURE 13.5.1
Basic BiCMOS inverter.

their collector currents are large. Under these conditions, and considering Q_p-Q_1 for example, $v_{DS} \cong 0 = r_{c1}i_{c1} + v_{cb1}$, so that Q_1 saturates if v_{cb1} is sufficiently negative. Saturation is undesirable as it reduces β and slows the turn-off of Q_1.

The basic circuit of Figure 13.5.1 suffers from two disadvantages: (1) The logic swing at the output is $V_{DD} - 2V_{BE\gamma}$, where $V_{BE\gamma} = V_{BE1\gamma} = V_{BE2\gamma} \cong 0.6$ V. This precludes small values of V_{DD}, because of too small a logic swing and correspondingly reduced noise margins. (2) Although the turn-on times of Q_1 and Q_2 can be small, their turn-off times are prolonged due to the absence of a circuit path for fast removal of base charge.

The second disadvantage is overcome by adding transistors Q_3 and Q_4 (Figure 13.5.2). If v_I goes high, Q_p and Q_1 turn off and Q_3 turns on, which provides a path for rapid removal of charge stored in Q_1 base. If v_I goes low, Q_3, Q_n, and Q_2 turn off, whereas Q_p, Q_4, and Q_1 turn on, with Q_4 providing a path for rapid removal of charge stored in Q_2 base. When Q_3 is on, Q_4 is off.

The circuit of Figure 13.5.3 is an example of rail-to-rail BiCMOS that overcomes the first disadvantage as well, though not completely satisfactorily. R_1 and R_2 clearly provide a path for the leakage of charge stored in the bases of Q_1 and Q_2. Moreover, when C_L is fully charged, and Q_1 stops conducting, there will not be any voltage drop in Q_p and R_1, which means that C_L will charge to V_{DD}. However, after Q_1 turns off, when $v_O = V_{DD} - v_{BE1\gamma}$, C_L will continue to charge to V_{DD} through the relatively high resistance of R_1 but at a slower rate than when Q_1 is conducting. Similarly, R_2 will allow C_L to discharge all the way to zero, but at a slower rate after Q_2 stops conducting. Moreover, when Q_p or Q_n turns on, R_1 and R_2 divert current from the bases of Q_1 and Q_2, which slows somewhat the turn-on process. The current diversion sets a lower limit to R_1 and R_2, which are typically in the range of few hundred ohms to few kilo-ohms. When V_{DD} is small, the range over which Q_1 and Q_2 conduct is limited, and the speed advantage of BiCMOS is severely compromised. If $V_{DD} = 1.5$ V, for example, Q_1 and Q_2 conduct over the range $0.6 \leq v_O \leq 0.9$ V, and most of the charging and discharging occurs through R_1 and R_2.

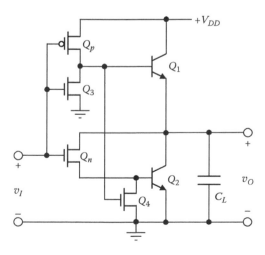

FIGURE 13.5.2
BiCMOS inverter with turnoff transistors Q_3 and Q_4.

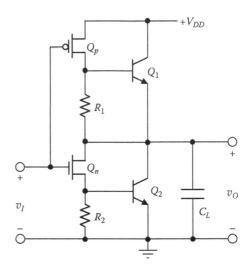

FIGURE 13.5.3
Rail-to-rail BiCMOS inverter.

Propagation Delay

The analysis of the BiCMOS inverter to determine the propagation delay is complicated by operation of the transistors in different modes during the charging and discharging of C_L, so that simulation is usually resorted to (Example 13.5.1). Nevertheless, the following conclusions apply: (1) For C_L larger than about 1 pF, BiCMOS is clearly faster than CMOS, even buffered CMOS in which the output transistors are sized to occupy the same chip area as the BJTs. The larger C_L the more marked is the speed advantage of BiCMOS. (2) For on-chip loads, which are typically less than 100 fF, CMOS is actually faster because the added junction capacitances of the BJTs at the internal node offset the speed advantage of the BJTs. (3) BiCMOS dissipates more power than CMOS, so that the power-delay product is practically the same for both.

BiCMOS is being continually improved. Operation from V_{DD} of 3 V or less is now possible, and SiGe is used (Section 4.8, Chapter 4) for better performance of the BJTs.

Simulation Example 13.5.1 Static and Dynamic Behavior of BiCMOS Inverter

It is required to simulate the static and dynamic behavior of the BiCMOS inverter.

SIMULATION

The inverter simulated is that of Figure 13.5.3, with $R_1 = R_2 = 500$ Ω, $C_L = 2$ pF, and $V_{DD} = 2.5$ V. The MOSFET parameters are the same as those of Example 13.1.1, for comparison. The BJT parameters are IS $= 1.0$f, BF $= 100$, BR $= 1$, TF $= 0.2$n, CJE $= 0.5$p, CJC $= 0.75$p, VAF $= 100$. The VTC, shown in Figure 13.5.4a, confirms that the logic swing is rail to rail, 0 to 2.5 V. Note that the circuit is not symmetrical because of R_1, R_2, and the two *npn* BJTs. The propagation delays are shown in Figure 13.5.4b. $t_{PHL} = 3.27$ ns and $t_{PLH} = 2.22$ ns, compared to 10.54 ns for the CMOS inverter alone. The difference becomes more pronounced for larger C_L.

BiCMOS Gates

The logic performed by BiCMOS gates is based on the same PUNs and PDNs as in CMOS gates, with the addition of the BJTs at the output and the discharge resistors across the BEJs

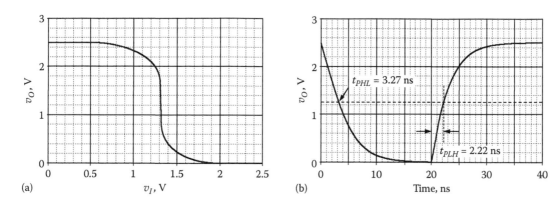

FIGURE 13.5.4
Figure for Example 13.5.1. (a) Voltage transfer characteristic; (b) propagation delays.

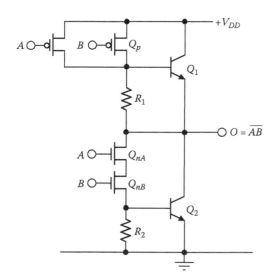

FIGURE 13.5.5
NAND2, rail-to-rail BiCMOS gate.

of these transistors in the case of rail-to-rail BiCMOS. For a NAND2 BiCMOS gate, for example, Q_p is replaced by two PMOS transistors in parallel and two Q_n NMOS transistors in series (Figure 13.5.5). The same remarks concerning sizing apply as for CMOS gates.

Application Window 13.5.1 Low Voltage to High Voltage Level Shifting

It is sometimes required to interface CMOS gates having a low supply voltage to CMOS or BiCMOS gates having a higher supply voltage. The circuit of Figure 13.5.6 can be used for this purpose, where v_{IL} is the input from the circuit having a low-voltage supply V_{DDL} and v_{OH} is the output of the circuit having a high-voltage supply V_{DDH}, which also supplies the two inverters. When $v_{IL} = 0$, Q_{n1} is off, Q_{n2} is on, and $v_{OH} = 0$. Conversely, when $v_{IL} = V_{DDL}$, Q_{n1} is on, Q_{n2} is off, and $v_{OH} = V_{DDH}$, as required. The pseudo NMOS inverter is designed to have a threshold of $V_{DDL}/2$, which requires that (Problem P13.2.27):

$$\frac{k_{n1}}{k_{p1}} = \frac{(V_{DDH} - V_t)(V_{DDH} - V_{DDL}/2) - (V_{DDH} - V_{DDL}/2)^2/2}{(V_{DDL}/2 - V_t)^2/2} \qquad (13.5.1)$$

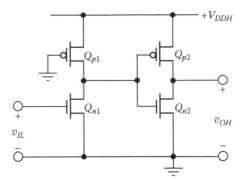

FIGURE 13.5.6
Figure for Application Window 13.5.1.

where $V_t = V_{tn1} = \overline{V}_{tp1}$. The absolute values of k_{n1} and k_{p1} are chosen so as to minimize the static power dissipation due to Q_{p1} while maintaining the drive to the CMOS inverter at the required bit rate. A high voltage level to low voltage level shifter is analyzed in Example SE13.4.

13.6 Transistor–Transistor Logic

TTL was for many years the dominant logic family. But with the advent of VLSI, TTL was no match to CMOS, which had some decisive advantages: (1) high packing density, since a MOSFET typically occupies one-tenth of the chip area as a BJT, (2) low power dissipation of the order of nanowatts per gate, and (3) fewer fabrication steps, and hence lower cost. Moreover, the speed of standard TTL is rather slow because of transistor saturation. On the other hand, the output BJT transistors of TTL have a relatively high current-drive capability. To meet the requirements of some applications, TTL has evolved into several versions, some of which are still the preferred choice in some cases. One of these versions is the advanced low-power Schottky TTL, which will be briefly and qualitatively presented in this section, after discussing the basic TTL inverter.

Basic TTL Inverter

The basic, standard TTL inverter is shown in Figure 13.6.1. It is seen that $V_{OL} = 0.2$ V, that of Q_3 in saturation. If v_I is at this level, then Q_1 is forward biased. The base of Q_1 will therefore be at about 0.9 V with respect to ground. The 0.9 V is applied across the BCJ of Q_1, in the forward direction, in series with the BEJ of Q_2, and the BEJ of Q_3 in parallel with a 1 kΩ resistor. This means that BEJs of Q_2 and Q_3 are biased to less than the cut-in voltage so that these transistors are cut off, and the collector current of Q_1 is practically zero. The base current of Q_1 is $(5 - 0.9)/4 \cong 1$ mA, which is also the emitter current of Q_1 sourced by the low input.

With Q_2 and Q_3 cut off, the steady state output current is very small, as will be argued shortly, which means that the base and collector currents of Q_4 are very small, with negligible voltage drops across the 1.6 kΩ and 130 Ω resistors. The collector-to-base voltage of Q_4 is nearly zero. The BEJ of Q_4 is just above its cut-in voltage, say 0.6 V. The voltage across D_2 is also just above its cut-in voltage, say 0.5 V. The high-level output voltage is $v_{OH} \cong 5 - 0.6 - 0.5 = 3.9$ V.

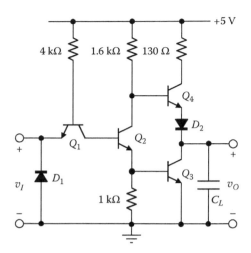

FIGURE 13.6.1
Basic, standard TTL inverter.

If the input suddenly becomes high, the BEJ of Q_1 is reverse biased, whereas its BCJ remains forward biased. Q_1 is therefore in the inverse active mode, with $\beta_R \cong 0.02$. The base voltage of Q_1 is $(0.7 + 2 \times 0.8) = 2.3$ V, where V_{BC} of Q_1 is considered to be 0.7 V and V_{BE} of Q_2 and Q_3 is 0.8 V, assuming Q_2 and Q_3 are in saturation. To check this, $I_{B1} = (5 - 2.3)/4 \cong 0.7$ mA, and $I_{B2} = (1 + \beta_R)I_{B1} \cong I_{B1}$, since the collector of Q_1 is now acting as an emitter. I_{B2} is large enough to saturate Q_2, which in turn applies a voltage of about 0.8 V to the base of Q_3. The emitter current of Q_1 is $\beta_R I_{B1} \cong 0.02 \times 0.7 \equiv 14$ μA directed into the emitter, which is now acting as a collector.

When Q_2 is saturated, with Q_3 base at 0.8 V, Q_2 collector is at about $0.8 + 0.2 = 1$ V, which is also the base voltage of Q_4. The collector of Q_3 is initially maintained at about 3.9 V by C_L, so a reverse bias of about 3 V is applied to D_2 and the BEJ of Q_4 is therefore cut off. The current through the 1.6 kΩ resistor is $(5 - 1/1.6) = 2.5$ mA. Since Q_4 is cut off, its base current can be neglected, so that the collector current of Q_2 is 2.5 mA. With $I_{B2} = 0.7$ mA, as determined above, I_{E2} is 3.2 mA. Thus, $I_{B3} = 3.2 - 0.8/1 = 2.4$ mA. Q_3 is initially in the active mode, and its collector current starts the discharge of C_L, which proceeds rapidly at nearly constant current. Eventually, as C_L discharges and the output voltage falls, Q_3 is saturated.

In the steady state, Q_2 and Q_3 are saturated and Q_4 is cut off. The output voltage drops to the low voltage level of about 0.2 V. The maximum fan-out of a TTL gate is determined by the need to keep Q_3 saturated and prevent the output from rising significantly above 0.2 V while sinking the current that is sourced by all inputs connected to the low output of the gate.

With $V_{C3} = V_{OL} = 0.2$ V and $V_{B4} = 1$ V, this leaves 0.8 V across both D_2 and the BEJ of Q_4, so that D_2 and Q_4 will be cut off. D_2 is included to ensure that Q_4 is cut off in the steady state. If D_2 is omitted, Q_4 will conduct. This would increase power dissipation in Q_3 and reduce the maximum fan-out by increasing the collector current of Q_3, without serving any useful purpose.

When v_I goes low, Q_1 base voltage instantly becomes $0.2 + 0.7 = 0.9$ V, and its collector-to-base voltage becomes $0.8 + 0.8 - 0.9 = 0.7$ V, because the base-emitter voltages of Q_2 and Q_3 remain at 0.8 V while the charge stored in the base is removed. Q_1 is now in the forward active mode and its collector current is a reverse base current of Q_2, which rapidly removes the charge stored in Q_2 base. The 1 kΩ resistor from base of Q_3 to ground helps to reduce the turn-off time of this transistor. After Q_2 turns off, the collector current of Q_1 becomes very small.

When Q_2 and Q_3 are cut off, the output voltage is maintained momentarily at 0.2 V by C_L. It can be readily shown that Q_4 will go into saturation. Assuming that it does, $I_{B4} = (5 - 0.8 - 0.7 - 0.2)/1.6 \cong 2$ mA, whereas $I_{C4} = (5 - 0.2 - 0.7 - 0.2)/0.13 \cong 30$ mA. Hence as long as its β_F exceeds $30/2 = 15$, Q_4 will be in saturation, as assumed. While Q_4 is in saturation, C_L rapidly charges toward V_{CC}. As the output voltage increases, the current in Q_4 decreases, and Q_4 comes out of saturation. In the steady state, the current of Q_4 is small, as assumed earlier, since Q_4 only supplies the 14 μA currents of the TTL gate inputs connected to the high output.

Using the combination of the 130 Ω resistor, Q_4, and D_2 has two advantages: (1) When the output is low, Q_4 and D_2 do not conduct and do not add to the current that Q_3 is sinking. This current equals the current that the input of a TTL gate sources, which is approximately $(5 - 0.8)/4 \cong 1$ mA, multiplied by the number of gates connected to the output. (2) If a resistor is used in place of the D_2-Q_4-130 Ω resistor combination, the value of this resistor must be at least a few kΩ in order to keep the additional current through Q_3 at a reasonable value. With such a resistance, the charging time of C_L is considerably increased. Q_4 acts as an emitter follower and provides active pull-up at the gate output, which decreases the charging time of C_L and hence t_{PLH}. Q_3 provides active pull-down. The rapid charging and discharging of C_L plus the rapid removal of the stored charge in the bases of Q_2 and Q_3 improve the speed of response. Heavy saturation of the transistors is avoided by using Schottky transistors, as discussed later. The output circuit of the TTL gate of Figure 13.6.1 is referred to as a **totem-pole** output.

As discussed earlier, the 130 Ω resistor limits the current in the output circuit when Q_4 turns on or in the event of accidentally short-circuiting the output when high. Even in the absence of C_L, Q_4 turns on slightly before Q_3 turns off, because of the storage delay of Q_3. In the absence of the 130 Ω resistor, there will be a short interval during which a low-resistance path exists between V_{CC} and ground. The 130 Ω resistor limits the current during the low-to-high transition. Even then, the current pulses are undesirable because they generate noise in the system and cause dynamic power dissipation that increases with frequency. Since the V_{CC} supply inevitably has some source impedance, the pulses result in negative-going voltage spikes in the supply line. These are transmitted to other circuits connected to the same supply and may cause false switching elsewhere. The voltage spikes are minimized by connecting by-pass capacitors at several locations between V_{CC} and ground. The capacitors are large enough to be able to supply the current spikes with only a small change in capacitor voltage.

Transistor Q_2 acts as a phase-splitter that drives Q_3 and Q_4 in antiphase. An antiringing diode D_1 is used to damp down oscillations, or ringing, that can occur at high frequencies due to the inductance of the interconnect (Section 12.8, Chapter 12). The oscillations can swing the input voltage positively to more than v_{IH} or negatively to less than zero. The positive swing is harmless because it will only increase the reverse-bias of the BEJ of Q_1, the reverse current being limited by the 4 kΩ resistor. The negative swing, however, can forward bias the *pn* junction isolating Q_1 collector from the substrate, which can cause malfunctioning or even damage to the IC. The diode limits the negative excursion to about -0.7 V.

It may be noted that TTL allows wired-AND logic, as discussed in connection with Figure 13.2.4, by having open collector outputs in which the output consists of Q_3 alone, with its collector terminal brought out for connection to a passive pull-up load resistance R_L between this terminal and V_{DD}, as in Figure 13.6.2. If $O_1, O_2, \ldots, O_n$ are the outputs of the individual gates, O goes low if any of the outputs of the individual gates goes low. The logic performed is $\overline{O} = \overline{O}_1 + \overline{O}_2 + \cdots + \overline{O}_n$, or $O = O_1 O_2 \ldots O_n$.

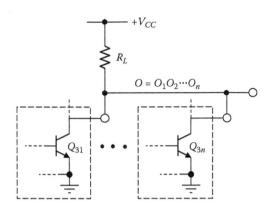

FIGURE 13.6.2
TTL wired-AND logic.

Exercise 13.6.1

Based on the currents and voltages calculated above for the basic TTL inverter, and assuming a fan-out of 8, determine the dc power dissipation when the output is: (a) high; (b) low.

Answers: (a) 5.7 mW; (b) 17.2 mW. ∎

Advanced Low-Power Schottky TTL

In advanced low-power Schottky (ALS) TTL, heavy saturation of transistors is avoided by paralleling the BCJ with a Schottky diode (Section 3.6, Chapter 3). The low forward voltage of about 0.3 V for a silicon-platinum silicide Schottky diode limits the forward voltage of the BCJ. Moreover, since this diode is majority-carrier operated, it has a fast turnoff.

Figure 13.6.3 shows a NAND2 ALS TTL gate that, apart from using Schottky transistors where saturation occurs, has several important features. *pnp* emitter followers at the input reduce the input current for low inputs. Q_2 is an *npn* emitter follower whose V_{BE}

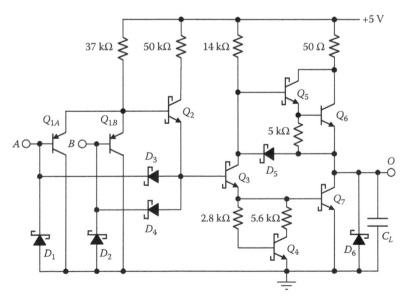

FIGURE 13.6.3
NAND2 ALS TTL gate.

compensates for the V_{EB} of the *pnp* transistors, which increases V_{IL} and V_{IH}. This, in turn, increases NM_L and brings it closer to NM_H. Q_2 also provides a higher current drive to the phase splitter transistor Q_3. Diodes D_3 and D_4 remove base charge from Q_3 when v_I goes low and Q_3 is turned off. Q_5 provides emitter follower drive to Q_6, as does Q_3 to Q_7.

Q_4 together with the 2.8 and 5.6 kΩ resistors improves the static and dynamic performance. Compared to a resistor between the base and emitter of Q_7, the conduction of Q_3 is delayed until V_{BE} of Q_7 exceeds the cut-in voltage, which makes the VTC more square, thereby further increasing V_{IL} and NM_L. Q_4 speeds up the turn-on of Q_7 by initially being off, so that the current of Q_3 flows into the base of Q_7 and is not partly diverted through a resistor between the base and emitter of Q_7. Q_4 also speeds up the turnoff of Q_7 by being initially in the active mode, its collector current rapidly removing the stored charge of Q_7. The 5.6 kΩ resistor prevents v_{BE} of Q_7 from being clamped at v_{CE} of Q_4.

Transistors Q_5 and Q_6 constitute a Darlington pair which increases the current gain, and hence the current sourcing capability of Q_6, thereby reducing t_{PLH}. Q_5 prevents saturation of Q_6, so that this transistor need not be Schottky clamped. Q_5 also obviates the need for a diode in series with Q_6 so as to prevent its conduction when the output is low. Diode D_5 speeds up the discharge of C_L and improves t_{PHL}. Antiringing Schottky diodes, which function as described for the standard TTL gate, are added at the input and output.

The resistors are scaled to reduce power dissipation, and the transistors are reduced in size and are isolated using S_iO_2 rather than a *pn* junction, which decreases parasitic capacitances. Typically, $P_{DC} = 1$ mW for an ALP TTL gate and $t_p = 4$ ns for $C_L = 15$ pF.

13.7 Emitter-Coupled Logic

ECL is the fastest logic family in silicon, surpassed in speed only by GaAs circuits. ECL gates can have an off-gate propagation delay of less than 25 ps, comparable to the on-gate delay of CMOS, and a power-delay product of less than 100 fJ. Even faster rates can be achieved by using heterojunction bipolar transistors in ECL. The main application of ECL at present is in high-speed interchip and intermodule links in massively parallel architectures and in some critical digital communications applications that require high off-chip bit rates. We will discuss in this section the basic operation of ECL before considering in more detail one of the fully compensated versions, the ECL 100k.

Basic Circuit

ECL is based on the BJT differential pair considered in Section 10.1, Chapter 10, where it was shown that current can be diverted from one transistor to another by applying a small differential voltage of about $8V_T$ (Figure 10.1.4, Chapter 10). The input v_I is applied to one transistor, while the input of the other transistor is held at a fixed voltage V_{REF} (Figure 13.7.1). The high and low levels of v_I are centered around V_{REF} and differ by less than 1 V.

When v_I is low, Q_1 is off and I_{EE} flows through Q_R, so that $v_{C1} = V_{CC}$ and $v_{C2} = V_{CC} - R_C I_{EE}$. When v_I is high, Q_1 is on, Q_R is off, and I_{EE} flows through Q_1, so that $v_{C1} = V_{CC} - R_C I_{EE}$ and $v_{C2} = V_{CC}$. The following should be noted about this basic circuit:

1. In the two states, one transistor is off and the other transistor is in the active region. The avoidance of saturation and the relatively small logic swing result in very fast switching.

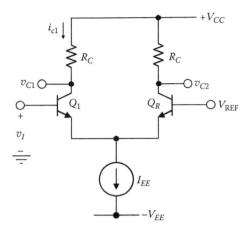

FIGURE 13.7.1
Basic ECL gate.

2. v_{C1} and v_{C2} switch in opposite directions, so that complementary outputs are available simultaneously. This simplifies logic design and obviates the need for an additional inverter.

3. The high and low outputs are V_{CC} and $V_{CC} - R_C I_{EE}$. They can be made independent of variations of V_{CC}, even transiently, by grounding the upper rail, that is making $V_{CC} = 0$.

4. The current through the upper, grounded rail is constant at I_{EE}. This, coupled with the rejection of common signals, significantly improves noise performance.

5. With $V_{CC} = 0$, all voltages are negative with respect to ground. To make the output voltage levels compatible with the input voltage levels, each of the outputs is taken through an emitter follower, which, in addition, provides a high current-drive capability at low output resistance.

6. The ECL gate is affected by temperature, mainly because of the variation of V_{BE} by about -1.5 to -2 mV/°C. Hence, measures have to be taken to ensure that the output voltage levels remain centered around V_{REF} so as not to adversely affect the noise margins.

ECL 100k

The ECL 100k family is an improved ECL version. Its lower supply voltage reduces power consumption, and the speed is improved because of smaller parasitic capacitances due to the isolation of transistors using SiO_2 rather than a *pn* junction. The basic ECL 100k inverter-buffer shown in Figure 13.7.2 has two separate ground lines, to prevent switching transients on the V_{CC1} line from affecting the V_{CC2} line, which does not have switching transients, as noted earlier. Diodes D_1 and D_2 in conjunction with the 500 Ω resistor improve temperature compensation resulting in a temperature variation of V_{OL} and V_{OH} of less than 0.1 mV/°C. V_{REF1} is independent of temperature and supply voltage. V_{REF2} is independent of temperature but varies directly with V_{EE} so that $V_{REF2} - V_{EE}$ is constant, which makes I_{EE} independent of V_{EE}. Thus,

$$I_{EE} = \frac{V_{REF2} - V_{BEA}(Q_E) - V_{EE}}{R_E} = \frac{-3.2 - 0.7 + 4.5}{300} \equiv 2 \text{ mA} \qquad (13.7.1)$$

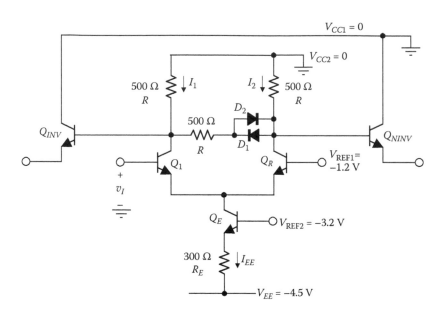

FIGURE 13.7.2
ECL 100k inverter-buffer.

When v_I is low, Q_1 is off and Q_R conducts, its collector voltage going low. The output at the emitter of Q_{NINV} is V_{OL}. From KVL: $0 = RI_2 + V_{BEA}(Q_{NINV}) + V_{OL}$, where V_{BEA} is V_{BE} in the active region. This gives:

$$I_2 = -\frac{V_{BEA}(Q_{NINV}) + V_{OL}}{R} \tag{13.7.2}$$

Since Q_1 is cut off, $0 = 2RI_1 + V_{D2} + V_{BEA}(Q_{NINV}) + V_{OL}$, so that,

$$I_1 = -\frac{V_{D2} + V_{BEA}(Q_{NINV}) + V_{OL}}{2R} \tag{13.7.3}$$

Neglecting the base current of Q_{NINV}, $I_1 + I_2 = I_{EE}$. Solving for V_{OL},

$$V_{OL} = -\frac{2RI_{EE}}{3} - \frac{V_{D2}}{3} - V_{BEA}(Q_{NINV}) \tag{13.7.4}$$

Assuming $V_{D2} = V_{BEA}(Q_{NINV}) = 0.7$ V, V_{OL} is nominally -1.6 V.

When v_I is high, Q_1 conducts and Q_R is off, its collector voltage going high. The output voltage at the emitter of Q_{NINV} is V_{OH}. As before, $0 = RI_2 + V_{BEA}(Q_{NINV}) + V_{OH}$, which gives:

$$I_2 = -\frac{V_{BEA}(Q_{NINV}) + V_{OH}}{R} \tag{13.7.5}$$

With Q_R cut off and the collector of Q_1 low, I_2 flows through D_1, so that $2RI_2 + V_{D1} = RI_1$. Substituting for I_2 from Equation 13.7.4,

$$I_1 = \frac{-2(V_{BEA}(Q_{NINV}) + V_{OH}) + V_{D1}}{R} \tag{13.7.6}$$

Neglecting the base current of Q_{INV}, $I_1 + I_2 = I_{EE}$. Solving for V_{OH},

$$V_{OH} = -\frac{RI_{EE}}{3} + \frac{V_{D1}}{3} - V_{BEA}(Q_{NINV}) \tag{13.7.7}$$

Assuming $V_{D1} = V_{BEA}(Q_{NINV}) = 0.7$ V, V_{OH} is nominally -0.8 V. The logic swing is:

$$LS = V_{OH} - V_{OL} = \frac{RI_{EE}}{3} + \frac{2V_D}{3} \tag{13.7.8}$$

Assuming $V_{D1} = V_{D2} = 0.7$ V, LS is nominally 0.8 V.
  The average of V_{OH} and V_{OL} is:

$$\frac{V_{OH} + V_{OL}}{2} = -\frac{RI_{EE}}{2} - V_{BEA}(Q_{NINV}) \tag{13.7.9}$$

Using the nominal values of V_{OL} and V_{OH}, V_{REF1} is midway between V_{OH} and V_{OL}.
  As for temperature compensation, it is seen from Equation 13.7.1 that:

$$\frac{dI_{EE}}{dT} = -\frac{1}{R_E}\frac{dV_{BEA}(Q_E)}{dT} \tag{13.7.10}$$

Assume that V_{EE} is temperature compensated. Since $\dfrac{dV_{BEA}(Q_E)}{dT} < 0$, I_{EE} has a positive temperature coefficient. Substituting in Equations 13.7.4 and 13.7.7,

$$\frac{dV_{OL}}{dT} = \frac{2R}{3R_E}\frac{dV_{BEA}(Q_E)}{dT} - \frac{1}{3}\frac{dV_{D2}}{dT} - \frac{dV_{BEA}(Q_{NINV})}{dT} \tag{13.7.11}$$

$$\frac{dV_{OH}}{dT} = \frac{R}{3R_E}\frac{dV_{BEA}(Q_E)}{dT} + \frac{1}{3}\frac{dV_{D1}}{dT} - \frac{dV_{BEA}(Q_{NINV})}{dT} \tag{13.7.12}$$

The variation of the diode forward voltage drop with temperature depends on the emission coefficient η and on the current (Section 3.2, Chapter 3). The terms of opposite signs on the RHS of Equations 13.7.11 and 13.7.12 can be made to nearly cancel one another, giving a magnitude of temperature variation of V_{OL} and V_{OH} of less than 1 mV/°C.
  The ECL 100k bias driver is illustrated in Figure 13.7.3. Q_{SH} is a shunt regulator that maintains I_{CS1} through R_2 constant at a value:

$$I_{CS1} = \frac{V_{BEA}(Q_{SH})}{R_2} \tag{13.7.13}$$

Thus, if I_{CS1} increases, $V_{BEA}(Q_{SH})$ increases, which increases Q_{SH} emitter current and the current through R_1, thereby reducing Q_{SH} emitter voltage and restoring I_{CS1} to its original

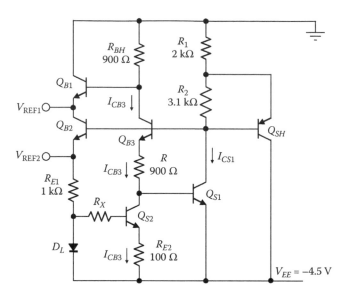

FIGURE 13.7.3
ECL 100k bias driver.

value. With I_{CS1} constant, $V_{BEA}(Q_{S1})$ is maintained constant, so that the currents in Q_{S2} and Q_{B3} are also stabilized against variations in V_{EE}.

As for temperature compensation, the voltage drop across D_L exceeds $V_{BE}(Q_{S2})$, so that,

$$I_{CB3} = \frac{V_{DL} - V_{BEA}(Q_{S2})}{R_{E2}} \tag{13.7.14}$$

neglecting the base current flowing in R_X. This resistor is used to compensate for variations in transistor parameters due to the manufacturing process.

Neglecting base currents,

$$V_{REF1} = -R_{BH}I_{CB3} - V_{BEA}(Q_{B1}) \tag{13.7.15}$$

and

$$V_{REF2} = V_{EE} + V_{BEA}(Q_{S1}) + RI_{CB3} + V_{BEA}(Q_{B3}) - V_{BEA}(Q_{B2}) \tag{13.7.16}$$

I_{CB3} in Equation 13.7.14 is made to have a positive temperature coefficient, so that $R_{BH}I_{CB3} = RI_{CB3}$ compensates for the negative temperature coefficient of $V_{BEA}(Q_{B1})$ and $V_{BEA}(Q_{B2})$. This makes V_{REF1} and V_{REF2} independent of temperature. Moreover, V_{REF2} varies directly with V_{EE}, so that I_{EE} is independent of V_{EE}, as noted earlier (Equation 13.7.1).

For a differential pair, V_{IL} and V_{IH} are considered to be the input voltages when one transistor conducts 1% of I_{EE}, the other 99%. From Equation 10.1.4, Chapter 10:

$$V_{BE}|_{Q_R} - V_{BE}|_{Q_1} = V_T \ln 99 = 0.115 \text{ V} \tag{13.7.17}$$

Since V_{REF1} is in the middle of the transition region, $V_{IL} \cong -1.2 - 0.115 = -1.315$ and $V_{IH} = -1.2 + 0.115 = -1.085$. It follows that NM_L is nominally $-1.315 - (-1.6) = 0.285$ V, and NM_H is nominally $-0.8 - (-1.085) = 0.285$ V.

ECL Gates

With complementary outputs available, the ECL inverter of Figure 13.7.2 is an inverter-buffer. OR-NOR gates can be implemented quite simply by paralleling identical transistors at the input, as illustrated in Figure 13.7.4 for an OR2-NOR2 gate, where Q_1 of Figure 13.7.2 has been replaced by two parallel transistors Q_{1A} and Q_{1B}, the two inputs A and B being applied to the bases of these transistors. Clearly, if either A or B goes high, the corresponding transistor is turned on and Q_R is turned off. Only if both A and B go low with Q_R turned on, its collector will go low. The logic implemented is OR at the NINIV output and NOR at the INV output.

AND-NAND functions can be implemented by having a series connection of input transistors rather than a parallel connection, as illustrated in Figure 13.7.5 for an AND2-NAND2 gate. However, this introduces some complications. If input B is applied directly to the gate of Q_{1B}, then when A and B are high, Q_{1A} and Q_{1B} are turned on, but because their bases are at the same voltage, the collector of Q_{1B} will be at a negative voltage of $V_{BE}(Q_{1A})$ with respect to its base. In other words, Q_{1B} will be in saturation. To prevent this, the emitter follower Q_{IB} is used to shift the base voltage of Q_{1B} negatively by V_{BE}. Moreover, it becomes necessary to have two reference transistors, Q_{RA} for Q_{1A} and Q_{RB} for Q_{1B}, with V_{REF1B} also shifted negatively by V_{BE} so that it remains centered between the high and low voltage levels applied to the base of Q_{1B}. The number of series stages is limited by V_{EE} and the voltage drop across the current source I_{EE} (Problem P13.3.16). Alternatively, since inverted inputs are available, AND-NAND can be implemented as OR-NOR of the inverted inputs. Thus, $\overline{\overline{A}+\overline{B}} = AB$, and $\overline{A}+\overline{B} = \overline{AB}$.

Since conventional ECL has passive pull-down, wired OR logic is possible by simply tying the NINV or INV outputs of gates together (Section 13.2). In Figure 13.7.6,

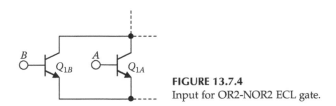

FIGURE 13.7.4
Input for OR2-NOR2 ECL gate.

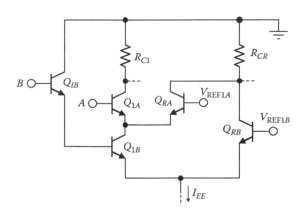

FIGURE 13.7.5
Basic AND2-NAND2 ECL gate.

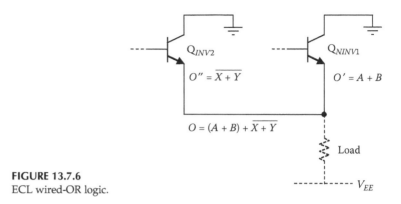

FIGURE 13.7.6
ECL wired-OR logic.

for example, if one gate has an output $O' = A + B$ and the other gate has an output $O'' = \overline{X + Y}$, then tying these outputs together to a common load resistance gives $O = (A + B) + \overline{X + Y}$.

ECL gate inputs are usually connected internally to V_{EE} through a 50 kΩ resistor. This means that an unused input that should be permanently low can be left open. The 50 kΩ resistor will not interfere with high or low active inputs. If the unused input is to be permanently high, it should not be connected to grounded V_{CC1} or V_{CC2}, because this will saturate the associated transistor and can damage internal circuitry. The gate input is connected to a resistive, or diode-resistor, voltage divider that applies -0.7 to -0.9 V to the input. ECL outputs can be connected to standard transmission lines of 50 Ω characteristic impedance (Section 12.8, Chapter 12), in which case the line is terminated at the input of another ECL gate with a 50 Ω resistor tied to a negative supply that is usually -2 V (Figure 13.7.7). Outputs not connected to transmission lines are terminated in the same manner, as shown in the figure. The -2 V limits the current through the output emitter follower and reduces the dc power dissipation.

Several modifications of ECL exist. ECL can operate at voltages as low as 3.3 V and with positive voltages instead of negative. Active pull-down can be used to speed up the high-to-low transition at the output. This is desirable, because when the base of the emitter-follower transistor goes low, the emitter voltage is initially held high by the output capacitance, which turns off the transistor. The voltage across this capacitance decreases passively through the resistance connected to the output, which is usually 50 Ω, until the emitter voltage of the output transistor drops sufficiently for this transistor to conduct (Problem P13.3.17).

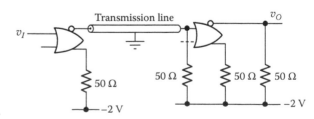

FIGURE 13.7.7
Input and output terminations of ECL gates.

Summary of Main Concepts and Results

- In any logic family, the effect of capacitance at a node is to slow the rate at which the voltage at the node can be switched between the two states, and to dissipate power because of charging and discharging. This capacitance is unavoidable and consists of the output capacitance of the driving gate, the input capacitance of the driven gates, and the capacitance of the wiring involved.

- The propagation delay sets an upper limit to the frequency of operation and determines the maximum fan-out under dynamic conditions.

- CMOS is an almost ideal logic circuit family. It has a rail-to-rail logic swing, practically zero static power dissipation, can be designed to have a threshold of $V_{DD}/2$ and near-ideal noise margins, and can be produced at low cost. Its main limitations are limited drive capability, the need for two transistors for each added gate input, and the current pulses through the voltage supply lines during switching.

- Pseudo NMOS gives a higher packing density compared to conventional CMOS, but has static power dissipation, asymmetrical noise margins, and a relatively large V_{OL}.

- Pass-transistor logic can be used to reduce the number of transistors required to implement a given logic function. Unless special measures are taken, or transmission gates used, the output voltage is reduced by the threshold of the pass transistor.

- Dynamic logic reduces to a minimum the number of transistors required to implement a given logic function, with zero static power dissipation. A limitation of dynamic logic is that periodic refreshing by a clock signal is required. The high output is a high-impedance output, except during capacitor charging and discharging, which makes it susceptible to noise. The output is available only during the evaluate phase. Dynamic logic gates cannot be cascaded except through some special circuit arrangements.

- In BiCMOS logic, *npn* BJTs are used at the output to speed up charging and discharging of relatively large load capacitances. The disadvantage is that the logic swing is reduced by the sum of the cut-in voltages of the BEJs of the output BJTs, unless special measures are taken, which may reduce somewhat the speed advantage.

- In advanced low-power Schottky TTL, heavy saturation of BJTs is avoided by using Schottky diodes. The totem pole output stage allows rapid charging and discharging of load capacitance. Objectionable current spikes occur on switching, which necessitates special precautions in decoupling the power supply lines.

- In ECL, very fast switching is achieved by avoiding saturation and by having a small logic swing. Emitter-followers at the complementary outputs provide high-current drive capability and low output resistance. The noise performance is improved by this low output resistance and by having a constant current through the voltage supply during switching. Measures are taken to ensure that the output voltages do not vary with temperature.

- An active PDN and a passive load allow wired AND logic, whereas an active PUN and a passive load allow wired OR logic.

Learning Outcomes

- Articulate the basic operation and principal features of various logic circuit families.

Supplementary Examples and Topics on Website

SE13.1 Minimum Delay for Large Capacitive Loads. Shows that a large capacitive load can be driven with minimum delay by a string of inverters in which transistor widths are successively multiplied by a factor e.

SE13.2 Analysis of Booster Circuit. Analyzes in more detail the booster circuit of Figure 13.3.3.

SE13.3 A Versatile Pass-Transistor Logic Gate. Analyzes the operation of a pass-transistor gate that can implant AND/NAND, OR/NOR, and XOR/XNOR by interchanging the signals and their complements.

SE13.4 A high Voltage to Low Voltage Level Shifter. Discusses the design of an inverter that can be used for shifting from a high voltage level to a low voltage level.

ST13.1 Small-Signal Equivalent Circuit CMOS Inverter. Derives and discusses the small-signal equivalent circuit of the CMOS inverter.

ST13.2 Input and Output Capacitances of CMOS Inverter. The capacitance is expressed in terms of the capacitances of the individual transistors for the case of an inverter driving another inverter. A method is presented for taking into account the voltage dependence of the depletion capacitance.

ST13.3 Propagation Delay of CMOS Inverter. Presents various methods for determining the propagation delay of a CMOS inverter.

ST13.4 VTC of Pseudo NMOS inverter. Derives the transfer characteristic of the pseudo NMOS inverter and the most important associated quantities for long-channel and short-channel behavior.

Problems and Exercises

P13.1 CMOS Logic Circuits

P13.1.1 Verify Equation 13.1.5

P13.1.2 Show that when the transistors of a CMOS inverter are not matched, segment BC of the VTC is given by: $v_O = v_I + \overline{V}_{tp} + \sqrt{(V_{DD} - \overline{V}_{tp} - v_I)^2 - r(v_I - V_{tn})^2}$, where $r = k_n/k_p$. Hence deduce that V_{IL} satisfies the equation: $V_{DD} - \overline{V}_{tp} - rV_{tn} + (r-1)V_{IL} - 2\sqrt{(V_{DD} - \overline{V}_{tp} - V_{IL})^2 - r(V_{IL} - V_{tn})^2} = 0$.

P13.1.3 Show that when the transistors of a CMOS inverter are not matched, segment DE of the VTC is given by $v_O = v_I - V_{tn} - \sqrt{(v_I - V_{tn})^2 - (V_{DD} - v_I - \overline{V}_{tp})^2/r}$, where $r = k_n/k_p$. Hence show that V_{IH} satisfies the equation: $V_{DD} - \overline{V}_{tp} - rV_{tn} + (r-1)V_{IH} - 2r\sqrt{(V_{IH} - V_{tn})^2 - (V_{DD} - V_{IH} - \overline{V}_{tp})^2/r} = 0$.

P13.1.4 Consider a symmetrical CMOS inverter operating from a 3 V supply. The NMOS transistor has $k_n' = 120$ $\mu A/V^2$, $(W/L)_n = 4$, $V_{tn} = 0.6$ V, and the PMOS transistor has $k_p' = 50$ $\mu A/V^2$, $(W/L)_p = 9.6$, and $V_{tp} = -0.6$ V. Determine: (a) the noise margins; (b) the output resistance when $v_I = 0$; (c) the output resistance when $v_I = V_{DD}$.

P13.1.5 Consider a CMOS inverter operating from a 3 V supply. The NMOS transistor has $k_n' = 120$ $\mu A/V^2$, $(W/L)_n = 4$, $V_{tn} = 0.6$ V, and the PMOS transistor has $k_p' = 50$ $\mu A/V^2$, $(W/L)_p = 6$, and $V_{tp} = -0.7$ V. Calculate: (a) V_{TH}, (b) NM_L, and (c) NM_H. Use the results of Problems P13.1.2 and P13.1.3. Compare them with the results of Problem P13.1.4.

P13.1.6 Consider the symmetrical CMOS inverter of Problem P13.1.4. Using PSpice, plot the VTC for: (a) $\lambda = 0.05$ V^{-1}, (b) $\lambda = 0.1$ V^{-1}, and (c) $\lambda = 0.2$ V^{-1}. Determine the noise margin in each case from the plots using the derivative feature of PSpice Probe.

DP13.1.7 Design a CMOS inverter to operate from a 2.5 V supply, the transistors having $k_n' = 286$ $\mu A/V^2$, $V_{tn} = 0.45$ V, $k_p' = 117$ $\mu A/V^2$, $V_{tp} = -0.55$ V, $L = 0.25$ μm, and $L = 0.25$ μm, if: (a) $W_n = 0.5$ μm, determine W_p so that $V_{TH} = 1.1$ V, (b) V_{tn} varies by $\pm 10\%$ and V_{tp} varies by $\pm 15\%$ due to manufacturing tolerances, determine the resulting variation in V_{TH}.

P13.1.8 Consider a CMOS inverter having $k_n' = 0.286$ mA/V^2, $V_{tn} = 0.5$ V, $k_p' = 0.117$ mA/V^2, $V_{tp} = -0.5$ V, $L = 0.25$ μm, and $W_p = 2.2$ μm, with $V_{DD} = 2.5$ V and $\lambda = 0.05$ V^{-1} for both transistors. Use PSpice to plot the VTC for the cases: (a) $k_n'(W/L)_n = 3$, $k_p'(W/L)_p$; (b) $k_n'(W/L)_n = (W/L)_p$, as in Example 13.1.1; and (c) $k_n'(W/L)_n = k_p'(W/L)_p/3$. Compare the thresholds with those calculated from Equation 13.1.1. What is the effect of varying the ratio k_n/k_p?

P13.1.9 Consider properly sized NAND2 and NOR2 CMOS gates. Using Equation 13.1.1, determine the threshold in each case when the two inputs of each gate are tied together, assuming $V_{DD} = 3$ V and $V_{tn} = \overline{V}_{tp} = 0.6$ V.

P13.1.10 Consider the NAND2 gate of Figure 12.2.4, Chapter 12. (a) Determine the threshold from a PSpice simulation with input B connected to V_{DD}. Use the same transistor parameters of Example 13.1.1, but with proper sizing and $V_{DD} = 2.5$ V. What are the states of the four transistors? (b) Repeat (a) with input A connected to V_{DD}. Note the small difference in threshold.

P13.1.11 Consider the NOR2 gate of Figure 12.2.5, Chapter 12; (a) Determine the threshold from a PSpice simulation with input B grounded. Use the same transistor parameters of Example 13.1.1, but with proper sizing and $V_{DD} = 2.5$ V. What are the states of the four transistors? (b) Repeat (a) with input A grounded. Note the small difference in threshold.

P13.1.12 Verify Equation 13.1.17.

P13.1.13 Given a symmetrical inverter having $k_{n,p} = 1$ mA/V^2, $V_t = 0.5$ V, $V_{DD} = 2.5$ V, $C_L = 1$ pF, $f = 100$ MHz, and rise and fall times $\tau = 1$ ns. Compare P_C and P_{SW}.

P13.1.14 Determine the total area occupied by: (a) a CMOS NAND4, and (b) a CMOS NOR4 gate, assuming: (1) proper sizing so as have the same current-driver capability as an inverter, and (2) that for the inverter, $L = 0.5$ μm, $W_n = 2$ μm, and $W_p = 4.8$ μm.

P13.1.15 Consider two implementations of an AND4 CMOS gate: (a) a NAND4 gate followed by an inverter; (b) two NAND2 gates whose outputs are connected to a NOR2 gate. Compare the areas in the two cases, using the result of P13.1.14, and compare the propagation delays.

P13.1.16 Consider two implementations of an OR4 CMOS gate: (a) a NOR4 gate followed by an inverter; (b) two NOR2 gates whose outputs are connected to a NAND2 gate. Compare the areas in the two cases, using the result of P13.1.14, and compare the propagation delays.

P13.1.17 The transistor Q_n of a CMOS inverter has $k_n = 0.28$ mA/V^2, $V_{tn0} = 0.6$ V, with $V_{DD} = 3$ V, and $C_L = 1$ pf. If the input changes from 0 to 3 V, determine the time it takes the output to fall from the 90% level, that is 2.7 V, to the 10% level.

P13.1.18 (a) Consider a NAND3 gate in which the transistors are matched and properly sized to have the same current-drive capability as the inverter. Let t_{PLH} and t_{PHL} denote the propagation delays of the gate when its output is dominated by a fixed load capacitance. How do t_{pLH} and t_{pHL} compare with those of a gate that is not sized? (b) Repeat (a) for a NOR3 gate.

P13.1.19 Let the symmetrical inverter of Example 13.1.1 have $L_n = L_p = 0.25$ μm, $W_n = 0.9$ μm, $W_p = 2.2$ μm, $C_{ox}^* = 10$ fF/μm^2, $C_{ol} = 0.8$ fF/μm of gate width, and an effective drain-substrate capacitance of 2.0 fF/μm of gate width. Determine for both states of the inverter: (a) the input capacitance, and (b) the output capacitance. Refer to Section ST13.2.

P13.1.20 If one inverter whose parameters are as in Problem P13.1.19 drives another symmetrical inverter having five times the area, determine the propagation delay, assuming $k_n = 0.286$ mA/V^2, $V_{DD} = 2.5$ V, and $V_{tnp} = 0.5$ V.

DP13.1.21 A manufacturing process gives the following transistor parameters: $L = 0.5$ μm, for both NMOS and PMOS transistors, $k_n' = 240$ μA/V^2, $k_n' = 120$ μA/V^2, V_{tn} 0.6 V, and $V_{tp} = -0.8$ V. The inverter is to operate with $V_{DD} = 3$ V. It is required to determine the smallest W_n and W_p so as to satisfy the following constraints: (1) t_{PHL} and t_{PLH} are less than 0.2 ns for $C_L = 0.25$ pF and (2) $V_{TH} = 1.5$ V. (Hint: use Equations ST13.3.3, ST13.3.4, Section ST13.3, and Equation 13.1.1.)

P13.1.22 Figure P13.1.22 illustrates a tristate inverter where the control signal C switches the inverter in or out of the tristate. Describe the operation of the circuit and indicate how the transistors should be sized with respect to a standard inverter.

DP13.1.23 It is required to implement the coincidence function $Y = AB + \overline{A}\,\overline{B}$ in CMOS. (a) Derive the PUN directly and a PDN that is the dual circuit; (b) Derive the PDN directly and a PUN that is the dual circuit;

DP13.1.24 Implement a three-bit odd-parity checker (Figure 12.2.3, Chapter 12) in CMOS.

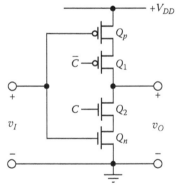

FIGURE P13.1.22

P13.2 Other MOS Logic Circuits

P13.2.1 Verify Equations 13.2.9 and 13.2.10 using average charging and discharging currents and assuming $V_t = 0.2V_{DD}$.

P13.2.2 Plot NM$_H$ and NM$_L$ of the pseudo-NMOS inverter as a function of r and determine r for equal noise margins, assuming $V_{DD} = 2.5$ V and $V_{tpn} = 0.5$ V. Determine r for which NM$_H = 0$ and r for which NM$_L$ is maximum. Note that these values are independent of V_{DD} and V_{tpn}.

P13.2.3 Given the inverter of Figure P13.2.3 consisting of an NMOS transistor Q_n and a resistor R_L. Assume that the current V_{DD}/R_L is in the triode region of Q_n, that $V_{OL} < V_{tn}$, and that long-channel relations apply. Determine V_{OH}, V_{OL}, V_{IL}, V_{IH}, and V_{TH}.

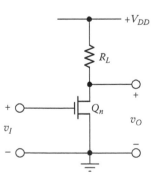

FIGURE P13.2.3

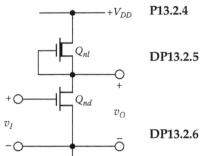

FIGURE P13.2.4

P13.2.4 Sketch the VTC of the inverter shown in Figure P13.2.4 assuming that the saturation current of Q_{nd} is nearly three times that of Q_{nl}.

DP13.2.5 Design a pseudo-NMOS inverter to operate from a 3 V supply with $V_{OL} = 0.15$ V using transistors that have $V_{tpn} = 0.6$ V, $k'_n = 0.286$ mA/V^2, $(W/L)_n = 0.9$ μm/0.25 μm, $k'_p = 0.117$ mA/V^2. Determine W_p, NM$_L$, NM$_H$, t_{PLH} and t_{PHL}, assuming $C_L = 50$ fF.

DP13.2.6 Redesign the pseudo-NMOS inverter of Problem DP13.2.5 to have maximum NM$_L$ ($r = 2.102$). Determine W_p, NM$_L$, NM$_H$, and t_{PHL}.

DP13.2.7 Redesign the pseudo-NMOS inverter of Problem DP13.2.5 to have $t_{PLH} = t_{PHL}$ ($r = 1.46$). Determine W_p, NM$_L$, NM$_H$, and t_P.

P13.2.8 Following the reasoning of Section ST13.2, determine the input and output capacitances of the pseudo-NMOS inverter in terms of the capacitances of the NMOS and PMOS transistors in both states.

DP13.2.9 Implement the coincidence function of Problem P13.1.23 in pseudo NMOS. Show that the two branches of the PDN can be considered to be wire-ANDed together.

P13.2.10 What is the state of the output if the gate is: (a) grounded in Figure 13.3.1 or Figure 13.3.2? (b) connected to V_{DD} in Figure 13.3.4?

P13.2.11 Determine v_C in Figure P13.2.11a and b when $v_I = V_{DD}$.

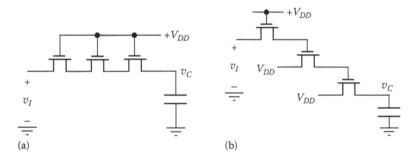

FIGURE P13.2.11

P13.2.12 Redraw the CMOS NOR2 gate as in Figure P13.2.12 and interpret its behavior in terms of pass-transistor logic but with V_{DD} applied to the input of the gate, and the external inputs A and B applied to the transistor gates. What happens if the NMOS transistors are omitted?

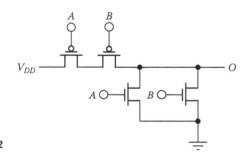

FIGURE P13.2.12

P13.2.13 By interchanging the PMOS and NMOS transistors in Figure P13.2.12, as well as V_{DD} and ground, interpret the CMOS NAND2 gate in terms of pass-transistor logic.

P13.2.14 Assume that a CK signal is applied to the gate of Q_p in the pass-transistor circuit of Figure P13.2.14. Let $V_{tn0} = 0.5$ V, $\gamma = 0.5$ V$^{1/2}$, and $\phi_f = 0.3$ V, with $V_{DD} = 2.5$ V. Determine: (a) the final value of v_O with $CK = 2.5$ V; (b) v_O when CK goes low, assuming that $C_x = 0.08$ fF and $C_O = 0.25$ fF in Figure P13.2.14. This effect is referred to as **clock feedthrough**. What are C_x and C_O due to?

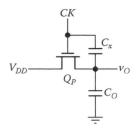

FIGURE P13.2.14

P13.2.15 Bearing in mind that during a low-to-high transition (Figure 13.3.1) the pass transistor is in saturation, and that during the high-to-low transition (Figure 13.3.2) the pass transistor is in the triode region, what are the input and output capacitances in terms of transistor capacitances?

P13.2.16 Consider the dynamic logic NAND2 gate of Figure 13.4.3, with $C_L = 15$ fF and $C_X = 5$ fF and the output voltage to be 3 V at the end of the precharge phase. Assume that when CK goes high, A goes high, B is low, and C_X is fully discharged. Determine the change in the output voltage if Q_A has $V_{tn0} = 0.6$ V, $\gamma = 0.5$ V$^{1/2}$, and $\phi_f = 0.3$ V.

P13.2.17 Show that if a standard CMOS inverter is added at the output O of the dynamic logic gate of Figure 13.4.1, a form of dynamic logic results that is cascadable. Use an inverter for the PDN for verifying cascadability.

P13.2.18 Show that the reduction in capacitor voltage due to charge sharing can be overcome in the following ways: (1) by adding another precharge transistor to precharge an internal node, such as that at which C_x appears in Figure 13.4.3; what can be an added advantage of this scheme? (2) in the manner of Figure 13.3.3, by using a PMOS transistor having its gate connected to the output of the standard CMOS inverter that is added for cascadability (P13.2.17).

P13.2.19 Verify the operation of the dynamic logic, positive-edge triggered D-FF of Figure P13.2.19.

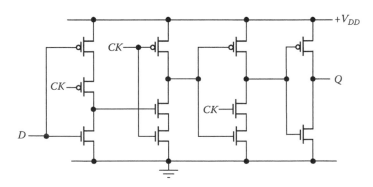

FIGURE P13.2.19

P13.2.20 It is required to estimate t_{PLH} and t_{PHL} of the NAND2 gate of Figure 13.4.3 using the average current approximation. Assume that Q_{PR} has $k_p = 100$ μA/V^2, and $\overline{V}_{tp} = 0.6$ V, with $V_{DD} = 3$ V, such that $C_L = 20$ fF and is initially discharged, and that the three NMOS transistors are identical, each having $k_n = 150$ μA/V^2 and $V_{tn} = 0.5$ V.

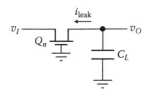

FIGURE P13.2.21

P13.2.21 Consider that $C_L = 50$ fF in Figure P13.2.21 is initially charged to $V_{DD} = 3$ V and that i_{leak} is such that C_L discharges exponentially with a time constant of 15 μs. Determine the time it takes v_O to drop by 0.5 V, assuming that v_I equals: (a) 0; (b) $V_{DD}/2$. Repeat the preceding calculations with v_O initially zero and determine the time it takes v_O to become 0.5 V, assuming v_I equals: (c) V_{DD} and (d) $V_{DD}/2$. If the precharge interval is 20% of the evaluate interval, what is the minimum refresh frequency? What will be the precharge interval? Note that making the final voltage for charging or discharging the load capacitance $v_{DD}/2$ is a common technique for reducing leakage current in DRAMs, where the voltage of the bit lines is made $V_{DD}/2$.

P13.2.22 Show that if the basic BiCMOS inverter of Figure 13.5.1 is unloaded, $i_{Dn}/i_{Dp} = (1 + \beta_1)/(1 + \beta_2)$.

P13.2.23 Consider the basic BiCMOS inverter of Figure 13.5.1 with $V_{DD} = 5$ V, $V_{tn} = \overline{V}_{tp} = 1$ V, $k_{n,p} = 200$ μA/V^2, $V_{BEon} = 0.6$ V, $V_{BEA} = 0.7$ V, $\beta = 100$, $C_L = 2$ pF, and $C_{int} = 10$ fF, where C_{int} is between base and ground of each BJT, and consists of the input capacitance of the BJT and the output capacitance of the corresponding MOSFET (Figure P13.2.23). Determine t_{PLH}, taking into account the initial charging of C_{int} to V_{BEon}, using average charging currents and assuming that C_L is initially fully discharged and that $v_I = V_{BEon}$.

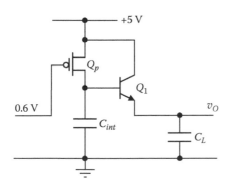

FIGURE P13.2.23

P13.2.24 Determine t_{PHL} for the BiCMOS inverter of Figure P13.2.24, assuming that $v_I = V_{DD} - V_{BEon} = 4.4$ V with v_O initially at 4.4 V.

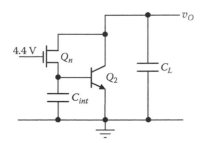

FIGURE P13.2.24

P13.2.25 Repeat P13.2.23 assuming $R_1 = 1$ kΩ in Figure 13.5.3 and neglecting C_{int}.

P13.2.26 Repeat P13.2.23 assuming $R_2 = 1$ kΩ in Figure 13.5.3 and neglecting C_{int}.

P13.2.27 Verify Equation 13.5.1.

P13.3 TTL/ECL Logic Circuits

P13.3.1 Calculate the average power dissipation in the standard TTL gate of Figure 13.6.1 using the figures of Section 13.6, assuming a fan-out of 10.

P13.3.2 Calculate the maximum fan-out of the standard TTL gate of Figure 13.6.1 using the figures of Section 13.6 and assuming: (a) that β_F of Q_3 is 70, and (b) to ensure that Q_3 remains well in saturation, thereby keeping V_{OL} sufficiently low, $I_C(Q_3)$ should not exceed $0.5\beta_F$ times its base current, which allows Q_3 to sink a current equal to $I_C(Q_3)$.

P13.3.3 Figure P13.3.3 shows a standard TTL inverter that is modified to a tristate inverter. When the control input v_C is high, the inverter functions in a normal manner having input v_I and inverted output v_O. When v_C is low, Q_3 and Q_4 are turned off, the output impedance going high. Verify circuit operation, considering that Q_1 is a dual-emitter BJT.

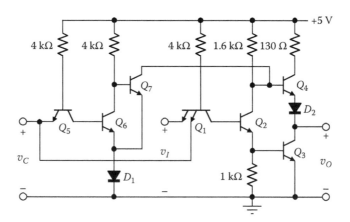

FIGURE P13.3.3

P13.3.4 Consider the circuit of Figure P 13.3.3 with v_I low and Q_7 conducting. At what v_I will Q_7 be cut off, assuming that the voltage across all forward biased junctions of D_5, Q_6, and D_2 is 0.7 V?

P13.3.5 Consider a Schottky transistor whose emitter is grounded and whose base and collector are connected to 5 V supply through 10 and 1 kΩ resistors, respectively. Assume that $V_{BE} = 0.75$ V, the voltage across the Schottky diode is 0.35 V, and β_F of the transistor proper is 40. Determine the currents through the Schottky diode, and in the base and collector of the transistor proper.

P13.3.6 Figure P13.3.6 illustrates a simpler form of Schottky TTL gate, known as STTL. Determine V_{OL}, V_{OH}, V_{IL}, V_{IH}, and the average power dissipation, assuming that transistors begin to conduct at $v_{BE} = 0.7$ V, are fully conducting at $v_{BE} = 0.8$ V, that the forward voltage drop across a Schottky diode is 0.5 V, and that the high output current is negligibly small. Consider V_{IL} to occur when Q_3 and Q_7 begin to conduct and V_{IH} to occur when they are fully conducting.

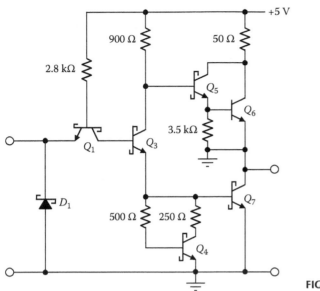

FIGURE P13.3.6

P13.3.7 Figure P13.3.7 illustrates a basic 10k ECL gate. Determine: (a) V_{OL} and V_{OH} assuming V_{BEA} of the differential pair is 0.75 V, β_F of the transistors is 100, and $V_{BE}(Q_{NINV}) = 0.75$ V at $I_{ENINV} = 1$ mA; (b) NM_L and NM_H assuming that at V_{IL} and V_{IH} the corresponding transistor conducts 1% of the current I_{EE} in R_{EE}.

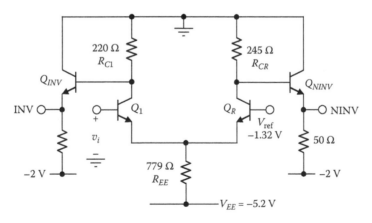

FIGURE P13.3.7

P13.3.8 Determine the average DC power dissipation in the gate of Figure P13.3.7, as the average of the power dissipation when $v_I = V_{OL}$ and $v_I = V_{OH}$.

P13.3.9 Determine the average power dissipation of the ECL 100k gate of Figure 13.7.2, using the values derived in Section 13.7, assuming each output to be connected by a 50 Ω resistor to a −2 V supply.

P13.3.10 Consider the noninverting output of the ECL 100k gate of Figure 13.7.2 to be high at −0.8 V, with a 10 pF capacitance between output and ground, and a 50 Ω resistor connected to a −2 V supply. When v_I goes low, Q_{NINV} cuts off and the capacitor voltage drops to $V_{OL} = -1.6$ V. Q_{NINV} begins to conduct when its base-emitter voltage surpasses the cut-in voltage of the BEJ, which is about 0.5 V compared to the 0.7 V assumed for conduction, when the output is V_{OL}. Determine the time it takes the output voltage to drop to -1.6 V assuming that Q_{NINV} does not conduct till then.

P13.3.11 Determine for the ECL gate of Figure P13.3.7: (a) the input currents for a low or high input, assuming the input is connected to V_{EE} by a 50 kΩ resistor and the transistor β_F is 100; (b) the maximum fan-out, assuming that a change of 0.1 V is allowed in V_{OH} and V_{OL}.

P13.3.12 Perform a PSpice simulation of the ECL gate of Figure P13.3.7 using Q2N3904 transistors and omitting, for simplicity, the emitter followers at the outputs. Interpret the shapes of the inverting and noninverting outputs.

P13.3.13 Figure P13.3.13 shows the circuit used to derive V_{REF} for the ECL gate of Figure P13.3.7. Verify that $V_{REF} = -1.32$ V, neglecting the base current and assuming 0.75 voltage drops for all forward-biased pn junctions.

P13.3.14 Consider the voltage reference circuit of Figure P13.3.13. Neglecting the base current and assuming that all forward-biased pn junctions have a temperature coefficient of -2 mV/°C, show that $dV_{REF}/dT = +1.4$ mV/°C.

P13.3.15 Consider the ECL gate of Figure P13.3.7. Neglecting base currents and assuming that all forward-biased pn junctions have a temperature coefficient of -2 mV/°C,

show that (a) $\dfrac{dV_{OH}}{dT} \cong 2$ mV/°C; (b) $\dfrac{dV_{OL}}{dT} = -\dfrac{R_{C1}}{R_{EE}}$

$\left(\dfrac{dV_{OH}}{dT} - \dfrac{dV_{BEA}(Q_1)}{dT}\right) - \dfrac{dV_{BEA}(Q_{NINV})}{dT} \cong 1$ mV/°C, by

considering $v_I = V_{OH}$ and the output taken from Q_{INV}. (c) Verify that the average of dV_{OH}/dT and dV_{OL}/dT nearly equals dV_{REF}/dT found in Figure P13.3.14.

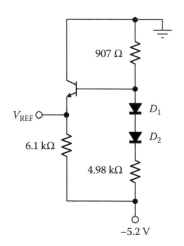

FIGURE P13.3.13

P13.3.16 Show that the maximum number of stages in a series connection of transistors in an

ECL gate, as in Figure 13.7.5, is given by $N_{max} \le \dfrac{-V_{EE} - V_{CS} + V_{OL}}{V_{BE}} + 1$, where V_{CS} is

the voltage across the current source.

P13.3.17 Figure P13.3.17 shows a basic circuit for active pull-down in an ECL gate, where $V_{REG} = V_{OL} - V_{BE}$. Explain circuit behavior when v_I suddenly changes from low to high. Note that a complementary output is no longer available.

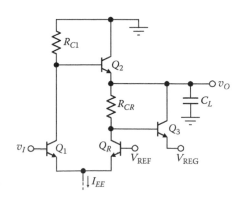

FIGURE P13.3.17

Appendix A: Reference Material

Taylor Series Expansion

A Taylor series expansion of a function $f(x)$ that is infinitely differentiable in the neighborhood of $x = a$ is the power series:

$$f(x) = f(a) + \frac{f^{(1)}(a)}{1!}(x - a) + \frac{f^{(2)}(a)}{2!}(x - a)^2 + \ldots + \frac{f^{(n)}(a)}{n!}(x - a)^n + \ldots \qquad (A1)$$

where $f^{(n)}(a)$ is the nth derivative of $f(x)$ at $x = a$,

If $(x-a)$ is small and is denoted by Δx, then terms involving powers of $(x-a)$ that are greater than one can be neglected, so that the expansion reduces to:

$$\Delta f(x) = \frac{df(x)}{dx} \Delta x \qquad (A2)$$

where $f(x) - f(a) = \Delta f(x)$.

If $f(x, y, z)$ is a function of three variables, Equation A2 becomes:

$$\Delta f(x, y, z) = \frac{\partial f(x, y, z)}{\partial x} \Delta x + \frac{\partial f(x, y, z)}{\partial y} \Delta y + \frac{\partial f(x, y, z)}{\partial z} \Delta z \qquad (A3)$$

Appendix B: Basic PSpice Models

PSpice symbols in Tables B.1 through B.3 and in the equations are written with a capital first letter. However, PSpice entries, including those in the PSpice model editor, are case insensitive.

The default value indicated in the tables is that used by PSpice if no value is specified by the user, or in the PSpice model, or in the Property Editor Spreadsheet.

pn Junction Diode

TABLE B.1

Selected PSpice Diode Parameters

Book Symbol	PSpice Symbol	Description	Default Value	Units
I_S	Is	Saturation Current	10^{-15}	A
R_s	Rs	Series resistance	0	Ω
η	N	Emission coefficient	1	—
τ_T	Tt	Mean transit time	0	s
C_{J0}	Cjo	Zero-bias capacitance	0	pF
V_{npo}	Vj	Equilibrium junction potential	1	V
m	M	Grading coefficient	0.5	F/m
BV, V_{ZK}	Bv	Breakdown voltage	∞	V
I_{ZK}	Ibv	Reverse current at BV	—	A

The diode model is based on the following equations (Figure B.1):

$$v_D = v_D' + (\text{Rs})i_D$$

$$i_D' = (\text{Is})\left[\exp\left(\frac{v_D'}{NV_T}\right) - 1\right], \text{ Equation 1.1.1, Chapter 1}$$

$$C_D = \left(\frac{(\text{Tt})}{NV_T}\right)i_D' + \frac{(\text{Cjo})}{\left(1 - \frac{v_D'}{(\text{Vj})}\right)^M}, \text{ where the first term is the diffusion capacitance}$$

(Equation 3.2.15) and the second term is the depletion capacitance (Equation 3.2.13, Chapter 3)

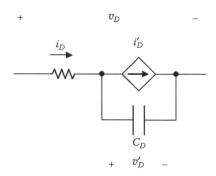

FIGURE B.1
PSpice *pn* junction diode model.

MOSFET

TABLE B.2

Selected PSpice MOSFET Parameters (Level 1)

Book Symbol	PSpice Symbol	Description	Default Value	Units
L	L	Channel length	100	μ
W	W	Channel width	100	μ
k'	Kp	Process transconductance parameter	2×10^{-5}	A/V^2
V_{t0}	Vto	Threshold with $V_{SB} = 0$	0	V
λ	Lambda	Channel-length modulation parameter	0	V^{-1}
γ	Gamma	Body-effect parameter	0	$V^{1/2}$
$2\phi_F$	Phi	Surface inversion potential	0.6	V
C_{gs}	Cgso	Gate–source capacitance per unit gate width	0	F/m
C_{gd}	Cgdo	Gate–drain capacitance per unit gate width	0	F/m
C_{gb}	Cgbo	Gate–body capacitance per unit gate width	0	F/m
C_{bdo}	Cbdo	Body–drain capacitance per unit gate width	0	F/m
C_{bso}	Cbso	Body–source capacitance per unit gate width	0	F/m

Level 1 MOSFET model is based on the following equations:

$$i_D = \frac{(Kp)}{2}\left(\frac{W}{L}\right)(v_{GS} - Vto)^2[1 + (Lambda)v_{DS}], \text{ (saturation region, Equation 5.3.1,}$$

Chapter 5)

$$i_D = (Kp)\left(\frac{W}{L}\right)\left[(v_{GS} - Vto)v_{DS} - \frac{v_{DS}^2}{2}\right][1 + (Lambda)v_{DS}], \text{ (triode region, Equation 5.3.2,}$$

Chapter 5)

$$V_t = (Vto) + (Gamma)\left[\sqrt{(Phi) + |V_{SB}|} - \sqrt{(Phi)}\right], \text{ (Equation 5.3.17, Chapter 5)}$$

BJT

TABLE B.3

Selected PSpice BJT Parameters

Book Symbol	PSpice Name	Description	Default Value	Units
I_S	Is	Saturation current	10^{-16}	A
β_F	Bf	Forward current gain	100	—
β_R	Br	Reverse current gain	1	—
n_e	Ne	Emission coefficient of BEJ	1	—
n_c	Nc	Emission coefficient of BCJ	2	—
V_A	Vaf	Forward early voltage	∞	V
	Var	Reverse early voltage	∞	V
r_B	Rb	Base series resistance	0	Ω
r_C	Rc	Collector series resistance	0	Ω
r_E	Re	Emitter series resistance	0	Ω
τ_F	Tf	Forward transit time	0	s
τ_R	Tr	Reverse transit time	0	s
C_{je0}	Cje	Zero-bias base-emitter capacitance	0	F
V_{je0}	Vje	Equilibrium BEJ potential	0.75	V
m_e	Mje	Grading coefficient of BEJ	0.5	—
C_{jc0}	Cjc	Zero-bias base-collector capacitance	0	F
V_{jc0}	Vjc	Equilibrium BCJ potential	0.75	V
m_c	Mjc	Grading coefficient of BCJ	0.5	—
C_{js0}	Cjs	Zero-bias collector-substrate capacitance	0	F
V_{js0}	Vjs	Equilibrium collector-substrate junction potential	0.75	V
m_s	Mjs	Grading coefficient of collector-substrate junction	0.5	—

The basic augmented model for the BJT is based on the EB transport model (Figure 6.3.2, Chapter 6), with I_S in Equation 6.3.8, Chapter 6 modified as in Equation 6.3.25, Chapter 6 and including the emission coefficients. The equations for the collector and emitter currents of an *npn* transistor become:

$$i_C = (\text{Is})\left[\exp\left(\frac{v_{BE}}{(\text{Ne})V_T}\right) - \exp\left(\frac{v_{BC}}{(\text{Nc})V_T}\right)\right]\left(1 - \frac{v_{BC}}{(\text{Vaf})}\right) - \frac{(\text{Is})}{(\text{Br})}\left[\exp\left(\frac{v_{BC}}{(\text{Nc})V_T}\right) - 1\right]$$

$$i_E = (\text{Is})\left[\exp\left(\frac{v_{BE}}{(\text{Ne})V_T}\right) - \exp\left(\frac{v_{BC}}{(\text{Nc})V_T}\right)\right]\left(1 - \frac{v_{BC}}{(\text{Vaf})}\right) + \frac{(\text{Is})}{(\text{Bf})}\left[\exp\left(\frac{v_{BE}}{(\text{Ne})V_T}\right) - 1\right]$$

As in the case of the diode, the capacitances are given by:

$$C_{BE} = \frac{(\text{Tf})}{(\text{Ne})V_T} \exp\left(\frac{v_{BE}}{(\text{Ne})V_T}\right) + \frac{(\text{Cje})}{\left(1 - \frac{v_{BE}}{(\text{Vje})}\right)^M}$$

$$C_{BC} = \frac{(\text{Tr})}{(\text{Nc})V_T} \exp\left(\frac{v_{BC}}{(\text{Nc})V_T}\right) + \frac{(\text{Cjc})}{\left(1 - \frac{v_{BC}}{(\text{Vjc})}\right)^M}$$

$$C_{CS} = \frac{(\text{Cjs})}{\left(1 - \frac{v_{CS}}{(\text{Vjs})}\right)^M}$$

where
  C_{BE} is the base-emitter capacitance
  C_{BC} is the base-collector capacitance
  C_{CS} is the collector-substrate capacitance

The quantities on the RHS are defined in Table B.3.

References and Bibliography

Allen, P.E. and Holberg, D.R., *CMOS Analog Circuit Design*, 2nd ed., Oxford University Press, New York, 2002.

Arora, N.D., Hauser, G.R., and Roulston, D.J., Electron and hole mobilities in silicon as a function of concentration and temperature, *IEEE Transactions on Electron Devices*, ED-29(2), 292, 1982.

Attia, J.O., *PSPICE and MATLAB for Electronics*, CRC Press, Boca Raton, FL, 2002.

Ayers, J.E., *Digital Integrated Circuits*, CRC Press, Boca Raton, FL, 2004.

Baker, R.J., *CMOS*, 2nd ed., IEEE Press, Wiley-Interscience, Piscataway, NJ, 2005.

Brown, S. and Vranesic, Z., *Fundamentals of Digital Logic with Verilog Design*, McGraw-Hill, Boston, MA, 2003.

Bult, K. and Wallinga, H., A class of analog CMOS circuits based on the square-law characteristics of a MOS transistor in saturation, *IEEE Journal of Solid-State Circuits*, 22, 357, 1987.

Campbell, S.A., *The Science and Engineering of Microelectronic Fabrication*, Oxford University Press, New York, 2001.

Cathley, J.J., *2000 Solved Problems in Electronics*, McGraw-Hill, New York, 1990.

Cochrun, B.L. and Grabel, A., A method for the determination of the transfer function of electronic circuits, *IEEE Transactions on Circuit Theory*, CT-20(1), 16, 1973.

Cui, Z.C., *Micro-Nanofabrication*, Springer, Berlin, Germany, 2005.

De Man, H.J.J., The influence of heavy doping on the emitter efficiency of a bipolar transistor, *IEEE Transactions on Electron Devices*, ED-18(10), 833, 1971.

Dimitriev, S., *Principles of Semiconductor Devices*, Oxford University Press, New York, 2006.

Erickson, R.W. and Maksimović, D., *Fundamentals of Power Electronics*, 2nd ed., Kluwer Academic Publishers, Norwell, MA, 2001.

Feinberg, B.N., *Applied Clinical Engineering*, Prentice Hall, Englewood Cliffs, NJ, 1986.

Franco, S., *Design with Operational Amplifiers and Analog Integrated Circuits*, 3rd ed., McGraw-Hill, Boston, MA, 2002.

Gilbert, B., Current-mode circuits from a translinear viewpoint: A tutorial. In *Analog IC Design: The Current-Mode Approach*, C. Toumazou, F.J. Lidgey, and D.G. Haigh, eds., Peter Peregrinus, London, U.K., 1990.

Gray, P.G., Hurst, P., Lewis, S.H., and Meyer, R.G., *Analysis and Design of Analog Integrated Circuits*, 4th ed., John Wiley & Sons, New York, 2001.

Green, M.A., Intrinsic concentration, effective densities of states, and effective masses in Si, *Journal of Applied Physics*, 67, 2944, 1990.

Haznedar, H., *Digital Microelectronics*, The Benjamin/Cummings Publishing Company, Redwood City, CA, 1991.

Herniter, M.E., *Schematic Capture with Cadence PSpice*, 2nd ed., Prentice Hall, Upper Saddle River, NJ, 2003.

Hodges, D.A., Jackson, H.G., and Saleh, R., *Analysis and Design of Digital Integrated Circuits*, 3rd ed., McGraw-Hill Book Company, Boston, MA, 2004.

Hume-Rothery, W. and Coles, B.R., *Atomic Theory for Students of Metallurgy*, The Institute of Metals, London, U.K., 1968.

Jaeger, R.C. and Blalock, T.N., *Microelectronic Circuit Design*, 3rd ed., McGraw-Hill, Boston, MA, 2008.

Johns, D. and Martin, K., *Analog Integrated Circuit Design*, John Wiley & Sons, New York, 1997.

Kang, S.M. and Leblebici, Y., *CMOS Digital Integrated Circuits*, 3rd ed., McGraw-Hill, Boston, MA, 2003.

Kano, K., *Semiconductor Devices*, Prentice Hall, Upper Saddle River, NJ, 1998.

Kassakian, J.G., Schlecht, M.F., and Verghese, G.C., *Principles of Power Electronics*, Addison-Wesley, Reading, MA, 1991.

Klaassen, D.B.M., A unified model for device simulation—II. Temperature dependence of carrier mobility and lifetime, *Solid-State Electronics*, 35(7), 961, 1992.

Kwok, C.Y., Low-voltage peaking complementary current generator, *IEEE Journal of Solid-State Electronics*, SC-20, 816, 1985.

Levinson, H.L., *Principles of Lithography*, 2nd ed., Spie Press, Bellingham, WA, 2005.

Millman, J. and Grabel, A., *Microelectronics*, 2nd ed., McGraw-Hill, New York, 1987.

Misiakos, and Tsamakis, D., Accurate measurements of the silicon intrinsic carrier density from 78 to 340 K, *Journal of Applied Physics*, 74, 3293, 1993.

Mohan, N., Undeland, T.M., and Robbins, W.P., *Power Electronics*, 3rd ed., John Wiley & Sons, New York, 2003.

Muller, R.S., Kamins, T.I., and Chan, M., *Device Electronics for Integrated Circuits*, 3rd ed., John Wiley & Sons, New York, 2002.

Neame, D.A., *Microelectronics: Circuit Analysis and Design*, 3rd ed., McGraw-Hill, New York, 2007.

Park, C.S. and Schaumann, R., A high-frequency linear transconductance element, *IEEE Transactions on Circuits and Systems*, 33, 1132, 1986.

Paynter, R.T., *Introductory Electronic Devices and Circuits*, 6th ed., Prentice Hall, Upper Saddle River, NJ, 2002.

Pulfrey, D.L. and Tarr, N.G., *Introduction to Microelectronic Devices*, Prentice Hall, Englewood Cliffs, NJ, 1989.

Rashid, M.H., *Power Electronics*, 3rd ed., Prentice Hall, Englewood Cliffs, NJ, 2003.

Research & Education Association, *The Electronics Problem Solver*, Research & Education Association, Piscataway, NJ, 2000.

Rosenstark, S., *Feedback Amplifier Principles*, Macmillan Publishing Company, New York, 1986.

Roulston, D.J., *An Introduction to the Physics of Semiconductor Devices*, Oxford University press, New York, 1999.

Sabah, N.H., *Electric Circuits and Signals*, CRC Press, Boca Raton, FL, 2008.

Säckinger, E. and Guggenbühl, W., A high-swing, high- impedance MOS cascode, *IEEE Journal of Solid-State Electronics*, 25, 289, 1990.

Schroder, D.K., Carrier lifetimes in silicon, *IEEE Transactions on Electron Devices*, 44(1), 160, 1997.

Sedra, A.S. and Smith, K.C., *Microelectronic Circuits*, 5th ed., Oxford University Press, New York, 2004.

Sodagar, A.M., *Analysis of Bipolar and CMOS Amplifiers*, CRC Press, Boca Raton, FL, 2007.

Steininger, J.M., Understanding wideband MOS transistors, *IEEE Transactions on Circuits and Devices*, 6(3), 26, 1990.

Suzuki, K. and Smith, B.W., eds., *Microlithography*, 2nd ed., CRC Press, Boca Raton, FL, 2007.

Sze, S.M., *Semiconductor Devices, Physics, and Technology*, 2nd ed., John Wiley & Sons, New York, 2002.

Sze, S.M. and Ng, K.K., *Physics of Semiconductor Devices*, 3rd ed., John Wiley & Sons, New York, 2007.

Tsividis, Y., *Operation and Modeling of the MOS Transistor*, 3rd ed., Oxford University Press, New York, 2004.

Webster, J.G., ed., *Medical Instrumentation: Application and Design*, 3rd ed., John Wiley & Sons, New York, 1997.

Weste, N.H.E. and Harris, D., *CMOS VLSI Design*, 3rd ed., Addison-Wesley, Boston, MA, 2005.

Index

A

Accelerated testing process, *see* Burn-in process
Acceptor impurity, 78–79
Access time, 611–612
Active mode, 244
Amplifiers
   configurations, 395–398
      common-base amplifier, 363
      common-collector amplifier,
         353–356
      common-drain amplifier, 350–353
      common-emitter amplifier, 348–350
      common-emitter amplifier with emitter
         resistor, 358–359
      common-gate amplifier, 361–363
      common-source amplifier,
         350, 356–358
      comparison, 365, 367
      equivalent transconductance, 347
      n-channel MOSFET transistor, 348
      small-signal equivalent circuit, 349,
         351–352, 361
      small-signal parameters, 346–347
      small signal relations, 364–366
   definition, 180
   feedback amplifier, 301, 302,
      304, 305
   hypothetical amplifying device, 181–182
   ideal amplifiers, 298–300
   inequality, 181
   operational amplifier, 310–325
   unity gain amplifier, 313–314
Analog signals, 588–589
Area defects, 173–174
Aspect ratio, 187
Auger recombination, 96, 97
Avalanche breakdown, 121–122, 201
Avalanche noise, 286

B

Ball grid array (BGA) package, 161
Bandpass filter, 446–447
Base collector junction (BCJ), 242–243
Base emitter junction (BEJ), 242–244
Base metallization, *see* Metallization
Base transport factor, 245
   logic gates
      CMOS Gate, 595–596
      gate performance, 596–602
      types of, 592–595
   metallic interconnect
      capacitance, 632
      distributed models, 633–636
      resistance, 632–633
Basic diode circuits, *see* Diode circuits
Basic op-amp circuits
   difference amplifier
      basic circuit, 511, 512
      common-mode input, 513, 514
      signal-flow diagram, 513
   instrumentation amplifier (IA)
      basic circuit, 514
      bridge amplifier, 516
      input stage, 515
   integrator, 509–511
   inverting and noninverting op-amp,
      507–509
Bias currents, operational amplifiers, 502–503
Biased pn junction
   breakdown
      avalanche breakdown, 121–122
      doping level, 123
      region, 121
      temperature effects, 123–124
      zener breakdown, 122–123
   capacitance, 119–120
   charge–current relation, 118
   charge distributions and currents
      bulk region, 115
      diffusion length, 114
      under forward bias, 114
      injected minority carriers, 113–114
      law of junction, 114
      transition region, 115
   current–voltage relation, 117–118
   rectifier, 112–113
   temperature effects, 120–121
   varactors, 120
Bias point
   calculation, 7–8, 44–46
   definition, 6
   using PSpice, 8–10

BiCMOS, 171–172
    cascode, 389–390
    logic
        basic operation, 678–680
        gates, 680–682
        propagation delay, 680
    technology, 393
Bipolar junction transistors (BJT)
    active mode, 244
    basic structure, 169–170
    BiCMOS, 171–172
    cascode, 386–389
    capacitances, 268–271
    breakdown voltages, determination, 266–268
    h-parameters *vs.* hybrid-π parameters, 265
    hybrid equivalent circuit
        base-width modulation, 260
        CE parameters, 263
        transistors parameters, determination, 264–265
    npn transistor, 268–269
    parasitic resistances, 271, 280
    temperature effects, 265–266
    transit time, 269
    unity-gain bandwidth
        CE current gain, frequency variation, 272
        equivalent circuit, simplified version, 271
        frequency, calculation, 273
        high-frequency response, 274–275
    cutoff mode, 244
    vs. FETs, 287–288
    inverse active mode, 245
    modified structure, 170–171
    npn power transistor, 563
    op-amps
        four-stage op-amp, 501
        input offset voltage, 505–507
        input resistance and bias current, 502–503
        superbeta transistors, 502, 503
    operation
        common-base dc current gains, 245–246
        common-emitter configuration, 247–250
        current gains variations, 252–254
        input and output characteristics, 242–244
        modes, 244–245
        small-signal current gains, 250–254
        small-signal equivalent circuits, 254–256
        structure, IC *npn* transistor, 242, 246–247
        T-model, 255–256
    parasitic, 568
    PSpice parameters, 707–708
    quasi and hard saturation, 564–565
    reverse base current, 566

        saturation mode, 244–245
        secondary effects
            input and output characteristics, 258–259, 261
            base-width modulation, 256–260
            Early effect, 256
        speed-up capacitor circuit, 566
BJT, *see* Bipolar junction transistors (BJT)
Bloch function, 66
Bloch's theorem, 66
Bode magnitude plots, 438–441
Body effect, 202–204
Boltzmann relations, 94
Bonding pads, 159
Bond wire, 160
Boolean algebra, 590–592
Boule, *see* Ingot, definition
Bragg reflection, 67
Breakdown diodes, *see* Zener diodes
Breakdown region, 121
Bridge amplifier, 516
Bridge rectifier, 12–13
Bridging sites, 175
Brillouin zone, 67
Buffer gates, 594
Bulk region, 115
Bulk resistance, 4
Burn-in process, 161
Burst noise, 286

C

Capacitances, 119–120, 205–207, 268–271
Capacitor-input filter
    approximate analysis, 16–18
    definition, 14
    idealized conditions, 15
    simulation, 19–20
    voltage and current waveforms, 14–15
Capacitors, 163
    charging and discharging, 40–41, 600, 653, 678
    dc blocking, 337, 339, 348
    emitter-resistor bypass, 414, 445
    emitter-source bypass, 457
    ferroelectric, 631
    MOS, 128
    source-resistor bypass, 412
    speed-up, 279, 281–282, 566
    storage type, 617, 619
    subcircuits, 411
    switched-capacitor circuits, 516–519

Carrier concentrations
   carrier generation and recombination, 87–88
   density of states, 85–87
   energy bands in a *pin* structure, 92–94
   Fermi level
      extrinsic semiconductor, 91–92
      position, 89
   intrinsic semiconductor, 88
   n-type semiconductor, 89–90
   p-type semiconductor, 90–91
Carrier generation, 87–88
Carrier mobility
   dopant concentration variation, 94–95
   temperature variation, 95
Carrier recombination, 87–88
   Auger recombination, 96
   equilibrium, 95
   imperfections, 96
   minority carrier lifetime, 97
   traps, 96
Cascaded amplifiers, 459–463
   analysis, 408–409
   coupling methods, 406
   dc level shifting, 409–411
   direct-coupled BJT amplifier stages, 408
   direct-coupled complementary BJT amplifier
      stages, 409
   emitter follower, dc level shifting, 410
   folded cascode, 409–410
   stages, 406–407
   telescopic cascode, 409
   Thevenin's voltage, 407
Cascode amplifier, 384–389
Cascode current sources and mirrors, 402–403
   amplifier configurations, 392
   BiCMOS technology, 393
   cascode configuration, 391
   discrete amplifiers, 392
   wide-swing current mirror, 391
CC–CB cascade tuned amplifier, 450
Center-tapped transformer (CTT), 11–14
Channel stop, 164, 165, 170–172
Chanstop, *see* Channel stop
Charge–current relation, 118
Charged-coupled device (CCD), 128
Charge distributions and currents
   bulk region, 115
   diffusion length, 114
   under forward bias, 114
   law of junction, 114
   transition region, 115
Chemical mechanical polishing (CMP),
   150–151, 153

Chemical potential
   definition, 80
   energy, 79
   expression, 79–81
Chemical vapor deposition (CVD), 158
Chip, definition, 150
Clamping circuit, 38–39
Class A operation, 577
   ac and dc load line, 540
   emitter follower, 543–545
   maximum symmetrical swing, 541
   Thevenin's equivalent circuit (TEC), 539
   transformer coupling
      ac and dc load line, 543
      class A amplifier, 543
      power-conversion efficiency, 542–543
Class AB operation, 577
   amplifier, 547–549
   diode biasing, 545–546
   quiescent state, 546
Clean room environment, 159–160
Clipping circuits, 33, 49–50
Clock skew, 611
Closed-loop stability
   asymptotic bode magnitude, 437, 439
   Barkhausen criterion, 445
   bode magnitude plots, 438–441
   feedback oscillators, 445, 463–466
   frequency compensation
      amplifier stability, 442–444
      dominant-pole compensation, 441–442
      high-gain amplifier, 444
      shifted-pole frequency, 441–442
   gain margin, 439
   high-*Q* oscillator, 446–448
   open-loop transfer function, 438
   phase margin, 439
   positive and negative feedback amplifiers,
      436–438
   Wien-Bridge oscillator, 445–446
CMOS fabrication
   silicon-on-insulator (SOI) technology,
      168–169
   transistor formation, 164–168
   transistor isolation, 163–164
Collector metallization, *see* Metallization
Column decoder, 613, 622–623
Common-collector amplifier
   ac conditions, 353–354
   emitter follower, 355
   series-shunt feedback, 356
Common collector-common emitter cascode,
   381–384

Common collector/drain amplifier, 373–375
Common-drain amplifier
    implementation, 351
    source follower, 351–353
Common-emitter amplifier with emitter resistor,
        358–359
Common emitter/source amplifier, 370–373
Common-source amplifier with source resistor,
        358–359
    series-series negative feedback, 357
    unbypassed source resistor, 356
Common-base amplifier, 363
Common-base dc current gains
    base transport factor, 245
    dc CB current gain, 245–246
    emitter injection efficiency, 245
Common-collector amplifier, 353–356
Common-drain amplifier, 350–353
Common-emitter amplifier, 348–350,
        413–416
Common-emitter configuration
    dc CE current gain, 247
    terminal voltages and currents, *npn* transistor,
        247–248
    transistors parameters, α_F and β_F,
        247–250
Common-gate amplifier, 361–363
Common-mode rejection ratio (CMRR),
        473–474
Common-source amplifier, 350, 411–413
Compensated semiconductor, 94
Complementary metal-oxide-semiconductor
        (CMOS), digital
    gates, 595–596
        CMOS buffer, 662–663
        gate designs, 659–661
        NAND and NOR gates, 657–659
        sizing and scaling effects, CMOS,
            661–662
    inverter
        static behavior, 650–651
        transfer characteristic, 649–650
        voltage transfer characteristic (VTC),
            648, 649
    logic circuits, 694–696
    low-power CMOS, 663
    noise margins, 651–652
    power dissipation
        calculation of, 656
        gate oxide leakage current, 657
    propagation delay
        charging and discharging, 653
        general factors, 653, 654

Complementary MOSFET
    amplifier, transfer characteristics
        circuit diagram, 219
        diode-connected D-PMOS transistor,
            221–222
        output characteristics, 219
        small-signal equivalent circuit, 220
Composite transistor connections, 400–402
    BiCMOS cascode, 389–390
    BJT cascode, 386–389
    cascode amplifier, 384–389
    cascode amplifier, frequency response,
        389
    CC-CE cascade
        comparison, 381–383
        frequency response, 383–384
        common collector-common emitter cascade,
            381–384
    MOSFET cascode, 384–386
    npn and *pnp* transistor, 379
    open-circuit voltage gain, 387
Conduction band, 65
Conduction electrons, 66
Conductivity, intrinsic silicon, 84
Contacts, dissimilar materials
    metal–metal contacts, 135–136
    metal–semiconductor contacts,
        136–138
    principles, 134–135
    Schottky diode, 139
Coupling methods, 406
Covalent bond, 75
Cross talk, 632
Crystal defects, 173–174
Crystal growth, 150
Crystal imperfections, 96
Crystal lattice, 74
Crystal momentum, 68
Crystal orientations, 150–151
Crystal oscillators
    pierce oscillator, 457
    piezoelectric effect and reactance, 456
Crystal structure, 74–75
Current mirror
    BJT current mirror
        basic and widlar, comparison, 343
        biasing IC transistors, 341
        current-steering circuit, 344
        small-signal conditions, 343
        Widlar current mirror, 342
    MOSFET current mirror
        E-NMOS and E-PMOS transistors, 345
        E-NMOS current mirror, 344

Current-mirror load, differential pair
    basic MOSFET, 482
        common-mode input, 480, 481
        systematic/random input offset
            voltage, 481
Current-steering circuit, 344
Current–voltage relation, 117–118
Cutoff mode, 244
Czochralski growth technique, 150

D

Damascene process, 164
Darlington pair, 379–381
dc blocking capacitor, 337
dc restorer, 36–37; *see also* Clamping circuit
De Broglie's hypothesis, 56
Degenerate semiconductors, 86
De Morgan's law, 591–592
Density, electron states, 71–73, 85–87
Depletion approximation, 107–110
Depletion capacitance, 118
Depletion-type, *n*-channel MOSFET (D-NMOS)
    output characteristics, 214
    symbols, 215
    transfer characteristics, 214
    voltage polarities, 215–216
Deposition processes
    chemical vapor deposition (CVD), 158
    clean room environment, 159–160
    diffusion, 157–158
    ion implantation, 157
    metallization, 158–159
Device transconductance parameter, 187
D flip-flop
    clock skew, 611
    implementation of, 609
    positive-edge-triggered, 610, 611
    timing diagram, 609, 610
Dice, definition, 150
Dicing, 160
Dickson charge pump, 629–631
Dielectric relaxation time, 97
Differential pair
    basic operation
        BJT differential pair, 468–470
        common-mode voltage, 468
        differential half circuit, 470
    current-mirror load
        basic MOSFET, 482
        common-mode input, 480, 481
        systematic/random input offset voltage,
            481

input bias currents, 476
input offset voltage, 476–478
MOSFET differential pair
    current mirror load, 485–486
    drain currents, 484
    transfer characteristics, 483
small-signal common-mode response
    common-mode rejection ratio (CMRR),
        473–474
    high-frequency response, 474–475
    split circuits, 472, 473
small-signal differential operation,
    471–472
transfer characteristics, 470–471
Diffusion, 157–158; *see also* Electron diffusion;
    Hole diffusion; Outdiffusion;
    Updiffusion
Diffusion capacitance, 119–120
Diffusion constant, 80
Diffusion current, 112
Diffusion length, 114
Digital-analog conversion (DAC)
Digital circuit elements
    flip-flops (FF), 640–643
        basic latch, 602–604
        D flip-flop, 608–611
        JK flip-flop, 606–608
        SR latch, 604–606
        T flip-flop, 608
    ADC implementation, 522
    current steering circuit, 520, 521
    inverting op-amp adder, 519, 520
    MOSFET implementation, 521
Digital logic circuit families
    BiCMOS logic
        basic operation, 678–680
        gates, 680–682
        propagation delay, 680
    complementary metal-oxide-semiconductor
        (CMOS)
        gates, 657–663
        inverter, 648–651
        logic circuits, 694–696
        low-power CMOS, 663
        noise margins, 651–652
        power dissipation, 655–657
        propagation delay, 653–655
    dynamic logic
        basic configuration, 674–675
        domino logic, 676–677
        limitations of, 675–676
        pipelined single-phase clock architecture,
            677

emitter-coupled logic (ECL)
    basic circuit, 686–687
    ECL 100k, 687–690
    gates, 691–692
    logic circuits, 700–702
  MOS logic circuits, 696–700
  pass-transistor logic
    positive feedback, 670
    time variation, 668–670
    transmission gate operation, 671–673
  pseudo NMOS
    dynamic operation, 667–668
    inverter, 664
    static operation, 665–667
  transistor–transistor logic (TTL)
    advanced low-power schottky TTL,
      685–686
    logic circuits, 700–702
    TTL inverter, 682–685
Digital signals and processing
  Boolean algebra
    conjunction, 590–591
    De Morgan's law, 591–592
    disjunction, 590–591
  digital signals
    advantages of, 589–590
    analog-to-digital conversion (ADC),
      588–589
Digital system memories
  attributes of, 611–612
  column decoder, 613
  ferroelectric RAM, 631
  flash memory, 612
  random access memory (RAM)
    classification, 612
    organization, 613
  read only memory (ROM)
    Dickson charge pump, 629–631
    erasable programmable ROM (EPROM),
      626–627
    flash ROM, 627–631
    floating gate transistor (FGT), 626, 627
    mask ROM, 623–625
    programmable ROM (PROM), 625–626
  read/write memory
    column decoder, 622–623
    dynamic memory cell, 616–617
    row decoder, 620–622
    sense amplifier and precharge circuit,
      617–620
    static memory cell, 614–616
  semiconductor memories, 643–644
  row decoder, 613

Diode-capacitor circuits
  capacitor charging and discharging, 40–41
  clamping circuit, 38–39
  dc restorer, 36–37
  voltage multiplication, 41–42
Diode circuits
  bias point
    calculation, 7–8, 44–46
    definition, 6
    using PSpice, 8–10
  characteristics, 44–46
  clamping circuits, 51–52
  clipping circuits, 49–50
  diode-capacitor circuits
    capacitor charging and discharging,
      40–41
    clamping circuit, 38–39
    dc restorer, 36–37
    voltage multiplication, 41–42
  ideal diode
    definition, 2
    Si *pn* junction diode, 2–3
  incremental diode resistance, 4
  offset voltage, 5
  Piecewise linear approximation, 5–6
  practical diodes, 3–4
  rectifier circuits
    capacitor-input filter, 14–20
    definition, 10
    full-wave rectifier, 11–14
    half-wave rectifier, 10–11
    regulated power supply, 20–23
    simulation, 19–20
    smoothing of output, 14
    X-ray tube, 23
  rectifiers and regulated power supplies, 46–49
  small-signal model, 6–7
  voltage limiters
    using back-to-back zener diodes, 34
    circuit diagram, 32
    definition, 32
    overload protection of meter, 35
    piecewise-linear approximation, 33
    protection of transistor switch, 35–36
    simulation, 34
    surge protection, 36
  zener voltage regulator
    analysis, 25–28
    circuit diagram, 25
    design example, 29–30
    load regulation and line regulation, 28
    simulation example, 30–31
    voltage–current characteristics, 24–25

Diode clamp, *see* dc restorer
Diode-connected BJT transistor, 282, 368
Diode-connected MOSFET, 217–218, 368
Direct-bandgap semiconductors, 129
Direct-coupled complementary BJT amplifier
    stages, 408
Direct recombination, 87
Discharge resistors, 680; *see also* Resistors
Discrete transistors, 336–341
    CE amplifiers, 336
    dc blocking capacitor, 337
    E-NMOS biasing, 340–341
    load-line construction, output characteristics,
      336
    using shunt feedback, 340
    source resistance effect, thershold variations,
      340–341
    Thevenin's equivalent circuit, 337
    using voltage divider, 338–339
Dislocation, 174
Distribution function, *see* Fermi–Dirac distribution
Donor impurity, 77–78
Dopant concentration variation, 94–95
Doping, 77
Drain currents, 484
Drift current, 112
Drive-in (diffusion), 157
Dry etching, 156
Dry oxidation, 151–152
Dual-in-line packages (DIPs), 160
Dust control, 159
Dust count, 159
Dynamic logic
    basic configuration, 674–675
    domino logic, 676–677
    limitations of, 675–676
    pipelined single-phase clock architecture, 677
Dynamic memory cell, 616–617

E

Ebers–Moll (EM) model, 275–278
Effective mass, 68
Electric conduction
    density, electron states, 71–73
    effective mass, 68
    electrons in periodic potential, 66–67
    Fermi–Dirac distribution, 73–74
    hole conduction, 69–71
Electrochemical potential
    chemical potential, 79–81
    conductivity, intrinsic silicon, 84
    expression, 81

Fermi level, 84–85
    generalized Ohm's law, 82–83
    nonuniformly doped semiconductor, 84
    state of equilibrium, 83–84
Electrolytic capacitor, 22; *see also* Capacitors
Electron
    definition, 60
    periodic potential, 66–67
Electron diffusion, 106
Electron–hole pair generation and
    recombination, 76
Electronic grade silicon, 150
Electron state, definition, 64
Emitter-coupled logic (ECL)
    basic circuit, 686–687
    ECL 100k, 687–690
    gates, 691–692
    logic circuits, 700–702
Emitter injection efficiency, 245
Emitter metallization, *see* Metallization
Emitter resistor, 358; *see also* Resistors
Energy band
    diagrams, 65, 77–78
    pin structure, 92–94
    structure, 64–66
E-NMOS
    characteristics, 199–200
    and E-PMOS transistors, 345–346
    transistor, 184–185
Epitaxial layer transfer (ELTRAN) process, 168
Epitaxy, 158, 169
Equilibrium diffusion and drift currents,
    110–112
Erasable programmable ROM (EPROM),
    626–627
Etching, 156
Extrinsic defects, 173
Extrinsic semiconductors, 77–79

F

Feedback amplifiers, 459–463
    series–series, 425–428
    series–shunt
      dc bias conditions, 422
      h-parameter equivalent circuit, 417
      loading effect of feedback circuit
        and feedback factor, 421
      modified amplifier circuit, 423
      negative feedback configuration, 418
      nonideal, non inverting op-amp circuit,
        419–420
      signal-flow diagram, 419

transistor amplifier, 422–424
two-port circuit representation, 417
shunt–series
  negative feedback configuration,
    428–429
  open-circuit voltage transfer function,
    determination, 432
  signal-flow diagram, 429–430
  transistor amplifier, 430–431
  two-port circuit representation, 428
shunt–shunt
  negative feedback configuration, 433–434
  signal-flow diagram, 435
  transistor amplifier, 434–436
  y-parameter equivalent circuit, 433
Feedback oscillator, 445, 463–466
Feedback resistor, 514, 522; *see also* Resistors
  common-emitter/source amplifier,
    375–376
Fermi–Dirac distribution, 73–74
Fermi level, 84–85
  extrinsic semiconductor, 91–92
  position, 89
Ferroelectric RAM (FRAM), 631
Fick's law, 80
Field-effect transistors (FET)
  amplifiers
    definition, 180
    hypothetical amplifying device, 181–182
    inequality, 181
  JFET
    channel profile, 227
    current regulator diode, 229–231
    current–voltage relation, 227–229
    operation, 226
    output characteristics, 227
    output resistance, 231–232
    parasitic, 568
    structure, 225–226
    threshold, 226
    transfer characteristics, 227
  MESFET
    GaAs technology, 232–233
    high-mobility devices, 234
    operation, 233–234
    structure, 233
  MOSFET
    body effect, 202–204
    breakdown, 201
    capacitances, 205–207
    channel-length modulation, 196–198
    complementary (CMOS), 218–225
    current–voltage relations, 187–190

depletion-type, 214–218
E-NMOS characteristics, 199–200
improved performance, 212–213
operation, 184–187
overdrive voltage, 201
p-channel MOS transistor, 190–191
short-channel effects, 209–212
small-signal operation, 191–196
structure, 183–184
temperature effects, 201
thin-film transistors, 213–214
transconductance, 199
types, 183
unity-gain bandwidth, 207–209
Field oxide (FOX), 164
Filters, *see* Capacitor-input filter; Inductor-input
  filter
Finite energy barrier and tunneling, 62–64
Flash memory, 612, 627, 628
Flash ROM
  flash memory, 627, 628
  FLOTOX cell programming, 629
  FLOTOX transistor, 627–629
Flicker noise, 286
Flip-chip process, 160
Flip-flops (FF)
  basic latch
    monostable multivibrator, 603
    schmitt trigger, 603, 604
  D flip-flop
    clock skew, 611
    implementation of, 609
    positive-edge-triggered, 610, 611
    timing diagram, 609, 610
  JK flip-flop
    edge-triggered, 608
    master–slave configuration, 607–608
    symbols, 606
    timing diagram, 606, 607
  SR latch
    latching property, 605
    symbols, 604
  T flip-flop, 608
Floating gate transistor (FGT), 626, 627
FLOTOX transistor, 627–629
1/f noise, *see* Flicker noise
Folded cascode CMOS op-amps
  amplifier stage, 409
  common-mode input range, 492–493
  frequency response, 495
  operational transconductance amplifier
    (OTA), 495–497
  output voltage swing, 494

slew rate, 495–497
voltage gain, 494–495
Forward biased junction, 112
Fourier series expansion (FSE), 536
Free atoms, 64
Frenkel defect, 173
Full-wave rectifier
bridge, 12–13
center-tapped transformer (CTT), 11–12
CTT *vs.* bridge, 13–14
definition, 11

G

Gain-bandwidth product (GB) product
asymptotic magnitude bode plot, 322–323
negative feedback trades, 323–324
Gate turn-off (GTO) switch, 574
Generalized Ohm's law, 82–83
Generation–recombination centers, 96
Group velocity, 59
Gummel–Poon model, 284

H

Half-wave rectifier, 10–11
Harmonic distortion, 536, 577
Heisenberg's uncertainty principle, 57
Hermetically sealed packages, 161
Heterojunction, 139–143
Heterojunction bipolar transistor (HBT), 284
High-frequency response, 399–400
BJT amplifiers, frequency response, 377–378
common-base/gate amplifier, 376–377
common collector/drain amplifier, 373–375
common emitter/source amplifier, 370–373
feedback resistor, common-emitter/source
amplifier, 375–376
frequency response determination, 371
Miller approximation, 369, 372, 375
Miller's theorem, 368–370
modified equivalent circuit, 374
poles and zeros, transfer function, 369–370
rigorous analysis, 372
simplified equivalent circuit, 370–371
unbypassed emitter resistor, 375
High-level injection, 117
High-purity silicon, 150
High-*Q* oscillator, 446–448
Hole conduction, 69–71
Hole diffusion, 106
Holes, definition, 71
Hypothetical amplifying device, 181–182

I

IC *npn* transistor, 246
Ideal diode
definition, 2
i–v characteristics, 2
Si *pn* junction diode, 2–3
symbol, 2
Ideal operational amplifier, 325, 327–334
definition, 310
extraneous signals, 321–322
frequency response, 324–325
gain-bandwidth product
asymptotic magnitude bode plot,
322–323
negative feedback trades, 323–324
inverting configuration
differentiator, 318–319
input and output fraction, 315
integrator op-amp, 317–318
op amp shunt–shunt feedback
configuration, 314
signal-flow diagram, 315
virtual ground, 314
voltage source, 316
noninverting configuration, 311–312
op-amp configurations, 319–321
symbol, 310–311
unity-gain amplifier, 313–314
IGBTs, *see* Insulated gate bipolar transistors
(IGBTs)
Imperfections, *see* Crystal imperfections
Implantation of oxygen (SIMOX) process, 168
Impurity activation, 157
Incremental diode resistance, 4
Indirect-bandgap semiconductors, 129
Indirect recombination, 96
Inductor-input filter, 14, 22
Infinite potential well, 60–62
Ingot, definition, 150
Instrumentation amplifier (IA)
basic circuit, 514
bridge amplifier, 516
input stage, 515
Insulated gate bipolar transistors (IGBTs), 578
characteristics, 570–571
conductivity modulation, 569–570
diagrammatic structure, 570
Integrator, 509–511
Interstitial defect, 173
Intrinsic defects, 173
Intrinsic semiconductors, 75–77, 88
Inverse active mode, 245

Inversion layer, 184
Inverter, CMOS, 225
    static behavior, 650–651
    transfer characteristic, 649–650
    voltage transfer characteristic (VTC), 648, 649
Inverting and noninverting op-amps, 507–509
Inverting configuration, ideal operational
        amplifier
    differentiator, 318–319
    input and output fraction, 315
    integrator op-amp, 317–318
    op amp shunt–shunt feedback configuration,
        314
    signal-flow diagram, 315
    virtual ground, 314
    voltage source, 316
Ion implantation, 157, 176
Ionized impurity scattering, 94
Ion milling, 156

J

JK flip-flop
    edge-triggered, 608
    master–slave configuration, 607–608
    symbols, 606
    timing diagram, 606, 607
Johnson noise, *see* Thermal noise
Junction field-effect transistor (JFET)
    channel profile, 227
    current regulator diode, 229–231
    current–voltage relation, 227–229
    operation, 226
    output characteristics, 227
    output resistance, 231–232
    structure, 225–226
    threshold, 226
    transfer characteristics, 227
Junction potential, 106–107

K

Knee current, 24

L

Large-signal modes
    augmented models, 284
    diode connection, 282
    Ebers–Moll (EM) model
        ideal transistor, 276
        npn and *pnp* transistors, 278
        transport version, 276–277
    Gummel–Poon model, 284
    offset voltage, 281
    regenerative pair, 282–283
    saturation mode
        BJT inverter, 278–279
        electron concentration profile, 280–281
        transistor, 280–281
    Schottky transistor, 282
    speed-up capacitor, 281
Latch
    monostable multivibrator, 603
    Schmitt trigger, 603, 604
Lattice scattering, 94
Law of the junction, 114
Law of mass action, 87
LC oscillators
    colpitts oscillator, 451, 453
    Hartley oscillator, 451, 453–455
    radio-frequency choke (RFC), 451
    self-limiting oscillator, 453
Light-emitting diodes
    application, 129–131
    classification, 129
    energy-band diagrams, 129
Lightly doped drain (LDD), 166
Line defects, 173–174
Line regulation, 28
Lithography
    chemical mechanical polishing
        (CMP), 153
    mask pattern, 154–155
    optical stepper, 155
    photolithography, 153
    photoresists, 153–154
    registration error, 155
    resolution, 155–156
    wafer surface, 154
Load line, 6–7
Load regulation, 28
Local oxidation (LOCOS), 164
Logic gates
    CMOS gate, 595–596
    gate performance
        high-level noise margin, 598
        ideal gate properties, 596
        low-level noise margin, 597
        propagation delay, 599–600
        ring oscillator, 600–602
        threshold voltage, 597
        transfer characteristics, 598, 599
    types of, 592–593
        buffer gates, 594
        XOR logic applications, 594–595

M

Majority carriers, 77
Mask read only memory (ROM)
    diode ROM, 623–624
    NOR ROM, 624–625
Mask preparation, 152–153
Master–slave, configuration, 607–608
Mean lifetime, 76
Mean transit time, 118
Metallic interconnect
    capacitance, 632
    distributed models
        characteristic impedance, 633
        RC distributed, delay, 635–636
        RC model, 634
    resistance
        skin depth, 633
        skin effect, 632, 633
Metallization, 158–159, 170–172, 176
Metallurgical grade silicon, 150
Metal–metal contacts, 135–136
Metal–semiconductor contacts
    energy-band diagram, 136–138
    energy barrier, 137–138
    work function, 136–137
Metal–semiconductor field-effect transistor
        (MESFET)
    GaAs technology, 232–233
    high-mobility devices, 234
    operation, 233–234
    structure, 233
Miller effect, 450
Miller's theorem, 368–369
Minimum feature size, 153, 209
Minority carriers, 77
    lifetime, 97
    suppression, 88
Mobility, definition, 80
Modular voltage multiplier, *see* Voltage
        multiplier
MOSFET
    cascode, 384–386
    complementary (CMOS)
        amplifier, 219–222
        inverter, 225
        transmission gate, 222–225
    current mirror, 344–346
    current–voltage relations
        aspect ratio, 187
        large-signal equivalent circuit, 189
        process and device transconductance
            parameters, 187

        symbols, 189–190
        transfer characteristics, 188
    depletion-type, *n*-channel MOSFET
        (D-NMOS)
        diode connection, 217–218
        output characteristics, 214
        symbols, 215
        transfer characteristics, 214
        voltage polarities, 215–216
    differential pair
        current mirror load, 485–486
        drain currents, 484
        transfer characteristics, 483
    operation
        channel-length modulation, 187
        drain current variation, 185
        drain–source voltage, 185
        E-NMOS transistor, 184–185
        hydraulic analogy, 186–187
        inversion layer, 184
        pinch-off region, 186
        saturation and triode regions, 186
    p-channel enhancement-type MOSFET
        (E-PMOS), 190–191
    PSpice parameters, 384–386
    secondary effects
        body effect, 202–204
        breakdown, 201
        capacitances, 205–207
        channel-length modulation, 196–198
        E-NMOS characteristics, 199–200
        improved performance, 212–213
        overdrive voltage, 201
        short-channel effects, 209–212
        temperature effects, 201
        thin-film transistors, 213–214
        transconductance, 199
        unity-gain bandwidth, 207–209
    small-signal operation
        basic circuit, 191
        equivalent circuit, 195
        load line construction, 192
        quiescent operating point,
            192, 194
        small-signal analysis, 193
        transconductance, 194
        transfer characteristics, 192
    structure, 183–184
    types, 183, 216–217
MOS logic circuits, 696–700
Multichip modules (MCMs), 161
Multijunction cells, 128

N

NAND and NOR gates
   CMOS NAND gate, 657
   CMOS NOR gate, 657, 658
n-channel MOSFET transistor, 348
Negative feedback, ideal amplifiers see also
      Feedback Amplifiers
   factor and amplifier, 301
   series–series feedback
      feedback amplifier, 307
      Kirchhoff's current law (KCL), 306
      test voltage source, 306
   series–shunt feedback
      factor and feedback amplifier, 301
      high-gain amplifier, 302–303
      test current source, 303
      Thevenin's impedance, 302
   shunt–series feedback, 307–308
   shunt–shunt feedback
      input and output impedances, 310
      test current source, 309
   signal-flow diagrams
      loop gain, 303–306
      negative feedback system, 304–305
Negative photoresist, 154
Nernst–Einstein relation, 80–81
Noise margins, 651–652
   high-level, 598
   low-level, 597
Nondegenerate semiconductors,
      85, 87, 90, 91
Nonlinear resistor, 599; *see also* Resistors
Nonrepetitive peak surge current, 22
Nonuniformly doped semiconductor, 84
n-type semiconductor, 89–90
Nyquist noise, *see* Thermal noise

O

OCSCTC method, 383, 385
OCTC method, 382
Off-chip resistor, 499; *see also* Resistors
Operational amplifiers
   basic op-amp circuits
      difference amplifier, 511–514
      instrumentation amplifier, 514–516
      integrator, 509–511
      inverting and noninverting op-amp,
         507–509
   BJT op-amps
      four-stage op-amp, 501
      input offset voltage, 503–507

input resistance and bias current, 502–503
      superbeta transistors, 502, 503
   CMOS op-amps
      common-mode response, 488
      current and voltage biasing, 498–501
      frequency response, 488–491
      input-offset voltage, 486–487
      slew rate, 491–492
      voltage gain and output swing, 487–488
   CMOS op-amps folded cascode
      common-mode input range, 492–493
      frequency response, 495
      output voltage swing, 494
      slew rate, 495–497
      voltage gain, 494–495
   IC op-amps, 527–528
   op-amp circuits, 528–530
   switched-capacitor circuits
      control voltage waveforms, 517–518
      inverting integrator, 518–519
      noninverting integrator, 519
      switched-capacitor integrator, 517
Operational transconductance amplifier (OTA),
      495–497
Optical lithography, *see* Photolithography
Optical stepper, *see* Stepper
Optoisolators or optocoupler, 380–381
Outdiffusion, 171, 172
Overdrive voltage, 201
Oxidation, 151–152

P

Packaging, semiconductor, 160–161
Particles and radiant energy, 57
Pass-transistor logic
   positive feedback, 670
   time variation, 668–670
   transmission gate operation, 671–673
Patterning
   definition, 152
   processes
      etching, 156
      lithography, 153–156
      masks preparation, 152–153
Pauli's exclusion principle, 64
Peaking current source, 403
Phase-shift oscillator, 465
Photocell, 126–127
Photoconductive cell, 124
Photoconductive current, 124–125
Photo Darlington pairs, 380
Photodiode, 125–126

Photolithography, 153
Photon, definition, 56
Piecewise linear approximation, 5–6
Pinch-off region, 186
Pin-grid array (PGA) packages, 160
pn junction
  biased type
    breakdown, 121–124
    capacitance, 119–120
    charge–current relation, 118
    charge distributions and currents, 113–117
    current–voltage relation, 117–118
    depletion region width, 113
    rectifier, 112–113
    temperature effects, 120–121
    varactors, 120
  charge flow, under forward bias, 115
  equilibrium
    depletion approximation, 107–110
    equilibrium diffusion and drift currents, 110–112
    junction potential, 106–107
  under forward bias, 114
  under reverse bias, 116
pn junction diode, 4, 145–147
  doping, 122
  fabrication, 162–163
  forward current, 117
  ideal silicon type, 2–3
  light effect, 125
  practical type, 3–4
  PSpice parameters, 705–706
  reverse current, 144
  zener diode, 24
Point defects, 173–174
Pole splitting, 490
Polycide, 166
Popcorn noise, *see* Burst noise
Positive photoresist, 153–154
Potential energy relations and current, 112–113
Power amplifiers
  class A operation
    ac and dc load line, 540
    emitter follower, 543–545
    maximum symmetrical swing, 541
    Thevenin's equivalent circuit (TEC), 539
    transformer coupling, 542–543
  class AB operation, 546–549
  class B operation, 545–546
  class C operation, 549–550
  general considerations
    nonlinear distortion, 536–537
    power-conversion efficiency, 537

    power Darlington pair performance, 537–539
    safe operating limits, 532–533
    semiconductor devices effects of temperature, 533
    thermal resistance, 533–535
    thermal stability, 535–536
    transistor derating, 535
  insulated gate bipolar transistor (IGBT), 531
  power diodes
    breakdown voltage, 562–563
    conductivity modulation, 561
    diagrammatic structure, 560–561
    diode reverse recovery time, 562
    drift region, 561
  power operational amplifiers, 550–551
Power Darlington pair, 537–539
Power latches
  incandescent lamp dimmer, 576–577
  static switches, 575
  thyristor
    breakover voltage, 572
    forward blocking state, 572–573
    gate turn-off (GTO) switch, 574
    i–v characteristic, 572–573
    light-activated silicon-controlled rectifier (LASCR), 574
    pnp–npn regenerative pair, 573
    silicon controlled rectifier (SCR), 572
    silicon unilateral switch (SUS), 574
Power switching, 583–584
  class D amplifier, 552–553
  dc-to-ac Converters, 556–557
  half-bridge inverter, 557–559
  switched regulated supplies
    basic switching voltage regulator, 553
    controlled oscillator, 553–554
    output ripple, 554–556
  uninterruptible power supplies (UPS), 559–560
Power transistors
  bipolar junction transistors (BJT)
    npn power transistor, 563
    parasitic transistor elements, 565
    quasi and hard saturation, 564–565
    reverse base current, 566
    speed-up capacitor circuit, 566
  insulated gate bipolar transistors (IGBTs)
    characteristics, 570–571
    conductivity modulation, 569–570
    diagrammatic structure, 570
  MOSFETs
    bipolar junction transistors (BJT), 569
    components, 568

diode structure, 568
  parasitic JFET, 567–568
  vertical, double-diffused type (VDMOS),
    567
Practical diodes, 3–4
Predeposition, 157
Process transconductance parameter, 187
Programmable ROM (PROM), 625–626
Pseudo NMOS
  dynamic operation, 667–668
  inverter, 664
  static operation, 665–667
PSpice parameters
  BJT, 707–708
  MOSFET, 706
  pn junction diode, 705–706
  symbols, 705
p-type semiconductor, 90–91
Pulse-width modulation (PWM),
    37, 552

Q

Quad-in-line packages (QIPs), 160
Quadrature oscillator, 465
Quantum theory, 56
Quasi-Fermi level, 85

R

Radiant energy, 56–57
Radicals, 156
Random access memory (RAM)
  classification, 612
  organization, 613
RC-coupled amplifiers
  brute force circuit analysis, 411
  common-emitter amplifier, 413–416
  common-source amplifier, 411–413
  D-NMOS amplifier, low frequency response,
    413
  emitter-resistor bypass capacitor, 414
  frequency response, 415–416
  midband gain, 413
  OCSCTC method, 411, 414
  source-resistor bypass capacitor, frequency
    response, 412
Reactive ion etching (RIE), 156
Read only memory (ROM)
  Dickson charge pump, 629–631
  erasable programmable ROM (EPROM),
    626–627

flash ROM
  flash memory, 627, 628
  FLOTOX cell programming, 629
  FLOTOX transistor, 627–629
  floating gate transistor (FGT), 626, 627
mask ROM
  diode ROM, 623–624
  NOR ROM, 624–625
programmable ROM (PROM), 625–626
Read/write memory
  column decoder, 622–623
  dynamic memory cell, 616–617
  row decoder, 620–622
  sense amplifier and precharge circuit
    dynamic random access memory (DRAM),
      619
    precharge and equalization circuit, 617,
      618
    static random access memory (SRAM),
      617, 618
  static memory cell
    read operation, 614–615
    write operation, 615–616
Rectifier circuits
  capacitor-input filter, 14–20
  definition, 10
  full-wave rectifier, 11–14
  half-wave rectifier, 10–11
  regulated power supply, 20–23
  simulation, 19–20
  smoothing of output, 14
  X-ray tube, 23
Registration error, 153, 155, 166
Regulated power supply, 20–23, 46–49
Resistor fabrication, 161–162
Reticle, 155
Retrograde profile, 157
Reverse biased junction, 112
Reverse recovery time, 120
Ring oscillator, 600–602
Row decoder, 613, 620–622

S

Safe operating area (SOA), 532, 577
Salicide, 166
Saturation mode, 244–245
Saturation region, 186
Schmitt trigger, 603, 604
Schottky diode, 139
Schrödinger's equation, 58
Self-alignment, 166
Self-biasing circuit, 498

Self-interstitial defects, 173
Semiconductor fabrication
  bipolar junction transistors (BJT)
    basic structure, 169–170
    BiCMOS, 171–172
    modified structure, 170–171
  CMOS fabrication
    silicon-on-insulator (SOI) technology,
      168–169
    transistor formation, 164–168
    transistor isolation, 163–164
  crystal defects, 173–174
  deposition processes
    chemical vapor deposition (CVD), 158
    clean room environment, 159–160
    diffusion, 157–158
    ion implantation, 157
    metallization, 158–159
  device fabrication
    capacitors, 163
    pn junction diode, 162–163
    resistors, 161–162
  packaging, 160–161
  patterning processes
    etching, 156
    lithography, 153–156
    masks preparation, 152–153
  silicon wafer preparation
    crystal growth, 150
    high-purity silicon, 150
    oxidation, 151–152
    wafer production, 151
Semiconductor photoelectric devices, 127–128
  charged-coupled device (CCD), 128
  photocell, 126–127
  photoconductive cell, 124
  photoconductive current, 124–125
  photodiode, 125–126
Semiconductors
  carrier concentrations
    carrier generation and recombination,
      87–88
    density of states, 85–87
    energy bands, *pin* structure, 92–94
    Fermi level, extrinsic semiconductor,
      91–92
    Fermi level position, 89
    intrinsic semiconductor, 88
    n-type semiconductor, 89–90
    p-type semiconductor, 90–91
  carrier mobility
    dopant concentration variation, 94–95
    temperature variation, 95

  carrier recombination
    Auger recombination, 96
    equilibrium, 95
    imperfections, 96
    minority carrier lifetime, 97
    traps, 96
  crystal structure, 74–75
  electric conduction
    density of electron states, 71–73
    effective mass, 68
    electrons in a periodic potential, 66–67
    Fermi–Dirac distribution, 73–74
    hole conduction, 69–71
  electrochemical potential
    chemical potential, 79–81
    expression, 81
    Fermi level, 84–85
    generalized Ohm's law, 82–83
    nonuniformly doped semiconductor, 84
    state of equilibrium, 83–84
  extrinsic type, 77–79
  intrinsic type, 75–77
Sense amplifier and precharge circuit
  dynamic random access memory (DRAM),
    619
  precharge and equalization circuit, 617, 618
  static random access memory (SRAM), 617, 618
Series–series feedback amplifier, 424–428
Series–shunt feedback
  factor and feedback amplifier, 301
  high-gain amplifier, 302–303
  test current source, 303
  Thevenin's impedance, 302
Series–shunt feedback amplifier
  dc bias conditions, 422
  h-parameter equivalent circuit, 417
  loading effect of feed back circuit and factor,
    determination, 421
  modified amplifier circuit, 423
  negative feedback configuration, 418
  nonideal, non inverting op-amp circuit,
    419, 420
  signal-flow diagram, 419
  transistor amplifier, 422–424
  two-port circuit representation, 417
Short-base diode, 118
Shot noise, 286
Shunt–series feedback, 307–308
Shunt–series feedback amplifier, 428–432
Shunt–shunt feedback
  amplifier, 433–436
  input and output impedances, 310
  test current source, 309

SiGe technology, 172–173
Silicides, 166
Silicon controlled rectifier (SCR), 572
Silicon epitaxy, *see* Epitaxy
Silicon-on-insulator (SOI) technology,
      168–169, 212
Silicon unilateral switch (SUS), 574
Silicon wafer preparation
    crystal growth, 150
    high-purity silicon, 150
    oxidation, 151–152
    wafer production, 151
Single-stage transistor amplifiers
    amplifier configurations, 395–398
        CD amplifier implementation, 351
        circuit to ac circuit conversion, 348
        common-base amplifier, 363
        common-collector amplifier, 353–356
        common-drain amplifier, 350–353
        common-emitter amplifier, 348–350
        common-emitter amplifier with emitter
            resistor, 358–359
        common-gate amplifier, 361–363
        common-source amplifier, 350
        common source amplifier with source
            resistor, 356–358
        comparison, 365, 367
        conventional circuit analysis, 360
        current and voltage source,
            transformation, 343, 348
        current buffer, 364
        diode-connected BJT transistor, 368
        diode-connected enhancement
            MOSFET, 368
        effective resistance, 367
        emitter follower, 355
        equivalent transconductance, 347
        input resistance, determination,
            358–359
        n-channel MOSFET transistor, 348
        output resistance, determination,
            353, 357, 359, 362, 364
        series-series negative feedback, 357
        series-shunt feedback, 356
        small-signal analysis, 348, 353
        small-signal equivalent circuit, 349,
            351–352, 361
        small-signal parameters, 346–347
        small signal relations, 364–366
        source follower, 351, 353
        Thevenin's or Norton's equivalent circuits,
            346–347
        unbypassed emitter resistor, 358

        unbypassed source resistor, 356
        voltage buffer, 364
    cascode current sources and mirrors,
            402–403
        amplifier configurations, 392
        BiCMOS technology, 393
        cascode configuration, 391
        discrete amplifiers, 392
        wide-swing current mirror, 391
    composite transistor connections, 400–402
        BiCMOS cascode, 389–390
        BJT cascode, 386–389
        cascode amplifier, 384–389
        cascode amplifier, frequency response, 389
        CC-CE cascade
            comparison, 381–383
            frequency response, 383–384
        common collector-common emitter
            cascade, 381–384
        MOSFET cascode, 384–386
        npn and *pnp* transistor, 379
        OCSCTC method, 383, 385
        OCTC method, 382
        open-circuit voltage gain, 387
        optoisolators or optocoupler, 380–381
        photo Darlington pair, 380
    high-frequency response, 399–400
        BJT amplifiers, frequency response,
            377–378
        common-base/gate amplifier, 376–377
        common collector/drain amplifier,
            373–375
        common emitter/source amplifier,
            370–373
        feedback resistor, common-emitter/
            source amplifier, 375–376
        frequency response, determination, 371
        Miller approximation, 369, 372, 375
        Miller's theorem, 368–370
        modified equivalent circuit, 374
        poles and zeros, transfer function, 369–370
        rigorous analysis, 372
        simplified equivalent circuit, 370–371
        unbypassed emitter resistor, 375
    transistor biasing, 394–395
        basic and Widlar current mirror,
            comparison, 342–343
        basic biasing circuit, 337
        basic E-NMOS current mirror, 344–345
        BJT current mirror, 341–342
        CE amplifiers, 336
        current expression, 344
        current mirror, 341–346

current source and voltage source, transformation, 343
current-steering circuit, 344
dc blocking capacitor, 337
discrete transistors, 336–341
E-NMOS and E-PMOS transistors, 345–346
E-NMOS biasing, 341
load-line construction, output characteristics, 336
MOSFET current mirror, 344–346
negative voltage supply, 339
shunt feedback, 340
small signal equivalent circuit, 343
source resistance effects, threshold variations, 340–341
Thevenin's equivalent circuit, 337
using voltage divider, 338–339
Si–SiO$_2$ interface, 174–175
Sizing and scaling effects, CMOS, 661–663
Skin effect, 632, 633
Slew rate, 491–492, 495–497
Small-signal current gains
CE current gain, 251
common-base current gain, 250
high-level injection, 251
summary of, 253–254
variations, 252–254
Small-signal equivalent circuit, 254–255; *see also* Single-stage transistor amplifiers
Small-signal model, 6–7
Solder bumps, 160
Source-drain extensions, 165
Source resistor, 28, 341, 356, 412; *see also* Resistors
Sputter etching, *see* Ion milling
Sputtering, 159
SR latch
latching property, 605
symbols, 604
Stacking fault, 174
State of equilibrium, 83–84
Static memory cell
read operation, 614–615
write operation, 615–616
Step coverage, CVD, 158
Stepper, 155
Substitutional defect, 173
Surface-mount-technology (SMT) package, 160
Surface states, 96
Surge protectors, 36
Switched-capacitor circuits
control voltage waveforms, 517–518
inverting integrator, 518–519
noninverting integrator, 519
switched-capacitor integrator, 517

T

Taylor series expansion, 703
Telescopic cascode, 409
Temperature effects,
biased *pn* junction, 120–121
BJT, 265
MOSFET, 201
Temperature-sensitive resistors, 446; *see also* Resistors
Temperature variation, carrier mobility, 95
Tetrahedral crystal lattice, *see* Crystal lattice
Thermal CVD, *see* Epitaxy
Thermal noise, 286
Thermal resistance
thermal equivalent circuit, 534
thermal resistance, 533–534
transistor derating curve, 535
Thermionic emission, 23
Thermistors, 445
Thevenin resistance, *see* Source resistor
Thevenin's impedance, 302
Thin-film transistors, 213–214
Through-hole technology (THT) package, 160
Thyristor
breakover voltage, 572
forward blocking state, 572–573
gate turn-off (GTO) switch, 574
i–v characteristic, 572–573
light-activated silicon-controlled rectifier (LASCR), 574
pnp–npn regenerative pair, 573
silicon controlled rectifier (SCR), 572
silicon unilateral switch (SUS), 574
Total harmonic distortion (THD), 536–537
Transconductance, 194, 199; *see also* Device transconductance parameter; Process transconductance parameter
Transient voltage suppression (TVS), *see* Surge protectors
Transistor amplifiers, *see* Single-stage transistor amplifiers
Transistor biasing, 394–395
basic *vs.* Widlar current mirrors, 342–343
basic biasing circuit, 337
basic E-NMOS current mirror, 344–345
BJT current mirror, 341–342
CE amplifiers, 336
current expression, 344
current mirror, 341–346

current source and voltage,
    transformation, 343
current-steering circuit, 344
dc blocking capacitor, 337
discrete transistors, 336–341
E-NMOS and E-PMOS transistors,
    345–346
E-NMOS biasing, 341
load-line construction, output
    characteristics, 336
MOSFET current mirror, 344–346
negative voltage supply, 339
shunt feedback, 340
small signal conditions, 343
small signal equivalent circuit, 343
source resistance effects, threshold variations,
    340–341
Thevenin's equivalent circuit, 337
using voltage divider, 338–339
Transistor formation, 164–168
Transistor isolation, 163–164
Transistor–transistor logic (TTL)
    advanced low-power schottky TTL,
        685–686
    inverter, 682–685
    logic circuits, 700–702
Transition region, 115
Transmission gate
    circuit, 222
    conduction of transistors, 223
    simulation example, 224–225
    symbol, 222–223
Traps, 96
Triode region, 186
Tuned amplifiers
    CC–CB cascade, 450
    intermediate frequency, 448
    Miller effect, 450
Tuned and magnetically coupled
    circuits, 466
Tunnel diode, 131–134
Tunneling, 62–64
Twin-well construction, 163
Two-port circuits, amplifiers, and feedback
    equivalent circuits, 298
    h-parameters, 297–298
    ideal amplifiers
        equivalent circuits, 298
        methods, 299–300
        types, 299
    parameters interpretation, 296–297
    reverse and forward transmission, 298

U

Uninterruptible power supplies (UPS), 559–560
Unit cell, 74
Unity-gain bandwidth, 207, 271
Updiffusion, 170

V

Vacant lattice site, 173
Valence band, 65
Vapor-phase epitaxy, *see* Epitaxy
Varactors, 120
V_{BE} multiplier, 410
Voltage doubler, *see* Voltage multiplication
Voltage limiters
    using back-to-back zener diodes, 34
    circuit diagram, 32
    definition, 32
    negative voltage levels, 32–33
    overload protection of meter, 35
    piecewise-linear approximation, 33
    positive voltage levels, 32–33
    protection of transistor switch, 35–36
    simulation, 34
    surge protection, 36
Voltage multiplication, 41–42
Voltage multiplier, 42
Voltage proportional to absolute temperature
    (PTAT), 275
Volume defects, 173–174

W

Wafer production, 151
Wave mechanics
    behavior as particles or radiant energy, 57
    electron in infinite potential well, 60–62
    energy band structure, 64–66
    finite energy barrier and tunneling, 62–64
    free atoms, 64
    Heisenberg's uncertainty principle, 57
    quantum theory, 56
    radiant energy, 56
    Schrödinger's equation, 58
    wave packets, 59
Wave packets, 59
Wave vector, 59
Wet etching, 156
Wet oxidation, 151–152
Wide-swing current mirror, 391
Wien-Bridge oscillator, 445–446

X

X-ray tube, 23

Z

Zener breakdown, 122–123
Zener diodes, 123
Zener voltage regulator

analysis, 25–28
circuit diagram, 25
design example, 29–30
load regulation and line regulation, 28
simulation, 30–31
voltage–current characteristics,
    24–25
Zero-nulling resistor, 491; *see also*
    Resistors

Milton Keynes UK
Ingram Content Group UK Ltd.
UKHW020828141024
449569UK00008B/586